Table of Atomic Weights Based on Carbon-12

Name	Symbol	Atomic Number	Atomic Weight[a]
Actinium	Ac	89	(227)
Aluminum	Al	13	26.98
Americium	Am	95	(243)
Antimony	Sb	51	121.75
Argon	Ar	18	39.95
Arsenic	As	33	74.92
Astatine	At	85	(210)
Barium	Ba	56	137.33
Berkelium	Bk	97	(247)
Beryllium	Be	4	9.01
Bismuth	Bi	83	208.98
Boron	B	5	10.81
Bromine	Br	35	79.90
Cadmium	Cd	48	112.41
Calcium	Ca	20	40.08
Californium	Cf	98	(251)
Carbon	C	6	12.01
Cerium	Ce	58	140.12
Cesium	Cs	55	132.91
Chlorine	Cl	17	35.45
Chromium	Cr	24	52.00
Cobalt	Co	27	58.93
Copper	Cu	29	63.55
Curium	Cm	96	(247)
Dysprosium	Dy	66	162.50
Einsteinium	Es	99	(252)
Erbium	Er	68	167.26
Europium	Eu	63	151.96
Fermium	Fm	100	(257)
Fluorine	F	9	19.00
Francium	Fr	87	(223)
Gadolinium	Gd	64	157.25
Gallium	Ga	31	69.72
Germanium	Ge	32	72.59
Gold	Au	79	196.97
Hafnium	Hf	72	178.49
Helium	He	2	4.00
Holmium	Ho	67	164.93
Hydrogen	H	1	1.01
Indium	In	49	114.82
Iodine	I	53	126.90
Iridium	Ir	77	192.22
Iron	Fe	26	55.85
Krypton	Kr	36	83.80
Lanthanum	La	57	138.91
Lawrencium	Lr	103	(260)
Lead	Pb	82	207.2
Lithium	Li	3	6.94
Lutetium	Lu	71	174.97
Magnesium	Mg	12	24.31
Manganese	Mn	25	54.94
Mendelevium	Md	101	(258)
Mercury	Hg	80	200.59
Molybdenum	Mo	42	95.94
Neodymium	Nd	60	144.24
Neon	Ne	10	20.18
Neptunium	Np	93	237.05
Nickel	Ni	28	58.70
Niobium	Nb	41	92.91
Nitrogen	N	7	14.01
Nobelium	No	102	(259)
Osmium	Os	76	190.2
Oxygen	O	8	16.00
Palladium	Pd	46	106.4
Phosphorus	P	15	30.97
Platinum	Pt	78	195.09
Plutonium	Pu	94	(244)
Polonium	Po	84	(209)
Potassium	K	19	39.10
Praseodymium	Pr	59	140.91
Promethium	Pm	61	(145)
Protactinium	Pa	91	231.04
Radium	Ra	88	226.03
Radon	Rn	86	(222)
Rhenium	Re	75	186.21
Rhodium	Rh	45	102.91
Rubidium	Rb	37	85.47
Ruthenium	Ru	44	101.07
Samarium	Sm	62	150.4
Scandium	Sc	21	44.96
Selenium	Se	34	78.96
Silicon	Si	14	28.09
Silver	Ag	47	107.87
Sodium	Na	11	22.99
Strontium	Sr	38	87.62
Sulfur	S	16	32.06
Tantalum	Ta	73	180.95
Technetium	Tc	43	98.91
Tellurium	Te	52	127.60
Terbium	Tb	65	158.93
Thallium	Tl	81	204.37
Thorium	Th	90	232.04
Thulium	Tm	69	168.93
Tin	Sn	50	118.69
Titanium	Ti	22	47.90
Tungsten	W	74	183.85
Unnilennium	Une	109	(266)
Unnilhexium	Unh	106	(263)
Unniloctium	Uno	108	(265)
Unnilpentium	Unp	105	(262)
Unnilquadium	Unq	104	(261)
Unnilseptium	Uns	107	(262)
Uranium	U	92	238.03
Vanadium	V	23	50.94
Xenon	Xe	54	131.30
Ytterbium	Yb	70	173.04
Yttrium	Y	39	88.91
Zinc	Zn	30	65.38
Zirconium	Zr	40	91.22

[a] Numbers in parentheses are masses of most stable isotopes.

Chemistry for Changing Times

Chemistry for Changing Times

Seventh Edition

John W. Hill
University of Wisconsin–River Falls

Doris K. Kolb
Bradley University

With Special Contributions by
Cynthia S. Hill, R.D.

Prentice Hall, Englewood Cliffs, New Jersey 07632

Library of Congress Cataloging-in-Publication Data

Hill, John William, 1933–
Chemistry for changing times.—7th ed. / John W. Hill, Doris K. Kolb ; with special contributions by Cynthia S. Hill.
p. cm.
Includes bibliographical references and index.
ISBN 0-02-355100-3
1. Chemistry. I. Kolb, Doris K. II. Hill, Cynthia S. III. Title
QD33.H65 1995 94-16250
540—dc20 CIP

Editor: Paul F. Corey
Production Supervisor: Elisabeth H. Belfer
Production Manager: Nicholas Sklitsis
Text Designer: Laura Ierardi
Cover Design and Art: Hothouse Designs
Photo Researcher: Barbara Scott

A Simon and Schuster Company
Englewood Cliffs, New Jersey 07632

Printed in the United States of America

10 9 8 7 6 5 4 3 2 1

ISBN 0-02-355100-3

Prentice-Hall International (UK) Limited, London
Prentice-Hall of Australia Pty. Limited, Sydney
Prentice-Hall Canada, Inc., Toronto
Prentice-Hall Hispanoamericana, S.A., Mexico
Prentice-Hall of India Private Limited, New Delhi
Prentice-Hall of Japan, Inc., Tokyo
Simon & Schuster Asia Pte. Ltd., Singapore
Editora Prentice-Hall do Brasil, Ltda., Rio de Janeiro

Preface

Chemistry for Changing Times is now in its seventh edition, and times have indeed changed since the first edition appeared in 1972. The book has changed accordingly—perhaps a bit more than usual with this edition because a co-author has been added. The fact that there now are two of us shouldn't really make much difference, though, because we share the same philosophy when it comes to teaching chemistry.

We believe that a chemistry course for students who are not majoring in science should be quite different from the course we offer our science majors. It should present basic chemical concepts with intellectual honesty, but it should not focus on esoteric theories or rigorous mathematics. It should include lots of modern everyday applications. The textbook should be appealing to look at, easy to understand, and interesting to read.

Three-fourths of the legislation considered by the United States Congress involves questions having to do with science or technology, yet only rarely does a scientist or engineer enter the field of politics. Most of the people who make important decisions regarding our health and our environment are not trained in science, but it is critical that these decision makers have some measure of scientific literacy. A chemistry course for students who are not science majors should emphasize practical applications of chemistry to problems involving such things as environmental pollution, radioactivity, energy sources, and human health. The students who take our liberal arts chemistry courses include future teachers, lawyers, accountants, journalists, and judges. There are probably some future legislators, too.

Objectives

Our main objectives in a chemistry course for students who are not majoring in science are these:

- To attract as many students as possible. If they are not enrolled in the course, we cannot teach them.
- To use topics of current interest to illustrate chemical principles. We want students to appreciate the importance of chemistry in the real world.
- To relate chemical problems to the everyday lives of our students. Chemical problems seem more significant to students if they can see a personal connection.
- To instill in students an appreciation for chemistry as an open-ended learning experience. We hope that they will want to continue learning throughout their lives.

- To acquaint students with scientific methods. We want them to be able to read about science and technology with some degree of critical judgment.
- To impart to students a sense of scientific literacy. We want our students to develop such a comfortable knowledge of science that they find news articles relating to science interesting rather than intimidating.

Major Changes in This Revision

The text has been updated and many topics have been expanded. New material includes recent developments regarding energy sources, global warming, acid rain, the ozone hole, cigarette smoke, alcohol, waste disposal, water treatment, air pollution, and health and fitness. There are new diagrams, tables, and photographs.

In order to keep the size of the book as small and easy to handle as possible, we have cut the number of chapters from 25 to 20 by combining chapters. Thus, no major omissions from the sixth edition were necessary. Hydrocarbons are now discussed in Chapter 9, ''Organic Chemistry''; Chapter 14, ''Energy,'' now includes future energy sources; farm chemistry is included in Chapter 16, ''Food''; cosmetics are considered in Chapter 17, ''Household Chemicals''; and chemical therapy is included in Chapter 19, ''Drugs.''

A summary has been added at the end of every chapter, along with a list of Key Terms, in order to help students understand what they are expected to know from each chapter.

Use of Color

New color photographs and diagrams have been added. Visual material adds greatly to the general appeal of a textbook. Color diagrams can also be highly instructive, and colorful photographs relating to descriptive chemistry do much to enhance the learning process.

Readability

Over the years students have told us that they have found this textbook easy to read. The language is simple, and the style is conversational. Explanations are clear and easy to understand. We trust that the friendly tone of the book has been maintained in this seventh edition.

Units of Measurement

The United States continues to cling to the traditional English system for many kinds of measurement even though the metric system has long been used internationally. A modern version of the metric system, the Système International (SI), is now widely used, especially by scientists. So what units should be used in a text for liberal arts students? In presenting chemical principles, we use SI units for the most part. In other parts of the book we use whatever units the students are most likely to come across elsewhere in that same context.

Chemical Structures

The structures of many complicated molecules are given in the text, especially in the later chapters. They are presented mainly to emphasize that these structures are actually known and to illustrate the fact that substances with similar properties often have similar structures. Students should not feel that they must learn these structures, but they should take the time to look at them. We hope that they will come to recognize familiar features in these molecules.

Glossary

The glossary in Appendix E gives definitions for terms that appear in **boldface** throughout the text. These include all the Key Terms listed at the end of each chapter.

Questions and Problems

The end-of-chapter exercises include review questions, problems, and suggested projects. Answers to many review questions and to all the odd-numbered problems are given in Appendix F. Problems are given within some of the chapters, with worked out examples followed by similar exercises. Answers to all the in-chapter exercises are also given in Appendix F.

References and Suggested Readings

An updated list of recommended books and articles appears at the end of each chapter. A student whose interest has been sparked by a particular topic can delve more deeply into the subject in the library. Instructors might also find the lists useful.

Supplementary Materials

The most important learning aid in any course is the teacher. In order to make the instructor's job easier and to enrich the education of students, we have provided a variety of supplementary materials.

Study Guide by Diane Bunce, Catholic University of America.

Instructor's Resource Manual by Doris K. Kolb.

Chemical Investigations for Changing Times, seventh edition (laboratory manual), by Alton Hassell and Paula Marshall, Baylor University, with Instructor's Guide by Alton Hassell.

Test Item File by John S. Phillips, Wilkes University.

Prentice Hall Test Manager, Version 2.0 (Macintosh and IBM), a computerized version of Test Item File.

Color Transparencies

The New York Times Themes of the Times Program, a compilation of current articles from *The New York Times* illustrating issues pertaining to chemistry in the world around us..

ABC News/Prentice Hall Video Library, video segments selected from ABC's documentary and news coverage.

Acknowledgments

The seventh edition has benefited greatly from the critical input of these most helpful reviewers who have taught from the sixth edition.

Benjamin H. Bruckner, University of Maryland, Baltimore County
Diane Bunce, Catholic University of America
Ronald Distefano, North Hampton Community College
Barbara L. Edgar, University of Minnesota
Norman Fogel, The University of Oklahoma
Emerson E. Garver, University of Wisconsin–River Falls
Edwin Goller, Virginia Military Institute
Alton Hassell, Baylor University
Stan Johnson, Orange Coast College
Dominick A. Labianca, Brooklyn College of The City University of New York
Colin R. McArthur, York University, Ontario, Canada
Glenn McElhattan, Clarion University–Venango Campus
Daniel Meloon, State University of New York College at Buffalo
Timothy D. Rhines, Bioanalytical Systems, West Lafayette, Indiana
Ron Roth, George Mason University
Dennis L. Stevens, University of Nevada–Las Vegas
Linda M. Sweeting, Towson State University
Berton C. Weberg, Mankato State University

Cynthia S. Hill prepared much of the original material on biochemistry, food, and health and fitness. Her special contributions are acknowledged on the title page.

Four of the verses that appear in this volume were first published in the *Journal of Chemical Education* (1978, pp. 47 and 732; and 1979, pp. 53 and 469). We acknowledge with thanks the permission to reprint them here.

We also want to thank our colleagues at the University of Wisconsin–River Falls and Bradley University for all their help and support.

We have been blessed with a team of careful and considerate editors. We especially appreciate all the help we have received from Paul F. Corey and from our outstanding production supervisor, Elisabeth H. Belfer.

We owe a very special kind of thanks to our wonderful spouses, Ina and Ken. Ina has done typing, library research, and so many other things. Ken has done chapter reviews, made countless suggestions, and given invaluable help with this seventh edition. Most of all, we are grateful to both of them for their enduring love and their boundless patience.

We thank all those many students whose enthusiasm has made teaching such a joy. It is gratifying to have students learn what you are trying to teach them, but it is a supreme pleasure to find that they want to learn even more.

Finally, we want to thank all of you who have made so many helpful suggestions. We welcome and appreciate all your comments, corrections, and criticisms.

J. W. H.
D. K. K.

Contents

1 Chemistry

A Science for All Seasons 1

2 Atoms

Are They for Real? 35

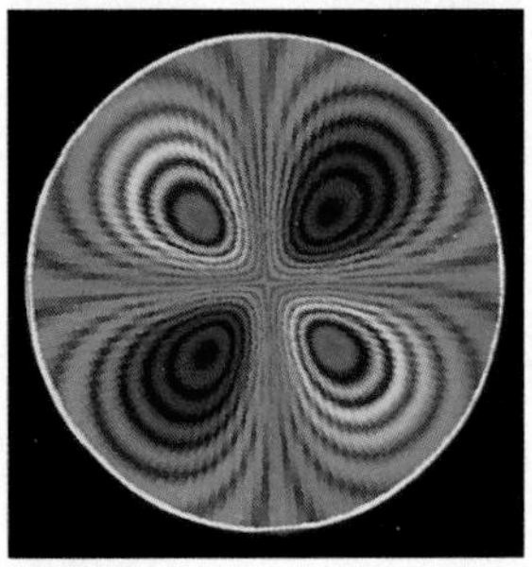

3
Atomic Structure
Images of the Invisible 51

4
Nuclear Chemistry
The Heart of Matter 85

5 Chemical Bonds

The Ties That Bind 121

6 Names, Formulas, and Equations

The Language of Chemistry 159

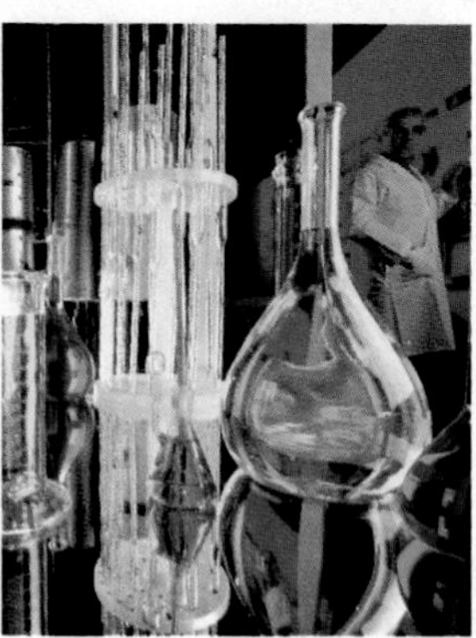

7
Acids and Bases
Please Pass the Protons 197

8
Oxidation and Reduction
Burn and Unburn 217

9
Organic Chemistry
The Infinite Variety of Carbon Compounds 243

10
Polymers
Giants Among Molecules 281

14 Energy

A Fuels Paradise 405

15
Biochemistry
A Molecular View of Life 453

16
Food
Those Incredible Edible Chemicals 493

17
Household Chemicals
Helps and Hazards 559

18
Fitness and Health
The Chemical Connection 601

20
Poisons
Chemical Toxicology 693

Appendix A

The International System of Measurement A-1

To the Student

Welcome to Our Chemical World!

Chemistry is fun. Through this book, we would like to share with you some of the excitement of chemistry and some of the joy in learning about it. We hope to convince you that chemistry does not need to be excluded from your learning experiences. Learning chemistry will enrich your life—now and long after this course is over—through a better understanding of the natural world, the technological questions now confronting us, and the choices we must face as citizens within a scientific and technological society.

Chemistry Directly Affects Our Lives

How does the human body work? How does aspirin cure our headaches? Do steroids enhance athletic ability? Is table salt poisonous? Can scientists cure genetic diseases? Why do most weight loss diets seem to work in the short run but fail in the long run? Does fasting "cleanse" the body? Why do our moods swing from happy to sad? Can a chemical test on urine predict possible suicide attempts? How does penicillin kill bacteria without harming our healthy body cells? Chemists have found answers to questions like these and continue to seek the knowledge that will unlock still other secrets of our universe. As these mysteries are resolved, the direction of our lives often changes—sometimes dramatically.

We live in a chemical world—a world of drugs, biocides, food additives, fertilizers, detergents, cosmetics, and plastics. We live in a world with toxic wastes, polluted air and water, and dwindling petroleum reserves. Knowledge of chemistry will help you to better understand the benefits and hazards of this world and enable you to make intelligent decisions in the future.

Chemical Dependency

We are all chemically dependent. Even in the womb, we depend on a constant supply of oxygen, water, glucose, and a multitude of other chemicals.

Our bodies are intricate chemical factories. They are durable but delicate systems. A myriad of chemical reactions are constantly taking place within us that allow our bodies to function properly. Thinking, learning, exercising, feeling happy or sad, putting on too much weight or not gaining enough, and virtually all life processes are made possible by these chemical reactions. Everything that we ingest is part of a complex process that determines whether our bodies work effectively or not. The consumption of some substances can initiate chemical reactions that will stop body functions altogether. Other substances, if consumed, can cause permanent handicaps, and others can make living less comfortable. A proper balance of the right foods provides the chemicals and generates the reactions we need in order to function at our best. The knowledge of chemistry that you will soon be gaining will help you to better understand how your body works so that you will be able to take proper care of it.

Changing Times

We live in a world of increasingly rapid change. It has been said that the only constant is change itself. At present, we are facing some of the greatest problems that humans have ever encountered, and the dilemmas with which we are now confronted seem to have no perfect solutions. We are sometimes forced to make a best choice among only bad alternatives, and our decisions often provide only temporary solutions to our problems. Nevertheless, if we are to choose properly, we must understand what our choices are. Mistakes can be costly, and they cannot always be rectified. It is easy to pollute, but cleaning up pollution once it is there is enormously expensive. We can best avoid mistakes by collecting as much information as possible before making critical decisions. Science is a means of gathering and evaluating information, and chemistry is central to all the sciences.

Chemistry and the Human Condition

Above all else, our hope is that you will learn that chemistry need not be dull and difficult. Rather, it can enrich your life in so many ways—through a better understanding of your body, your mind, your environment, and the world in which you live. After all, the search to understand the universe is an essential part of what it means to be human.

Chemistry for Changing Times

1 Chemistry

A Science for All Seasons

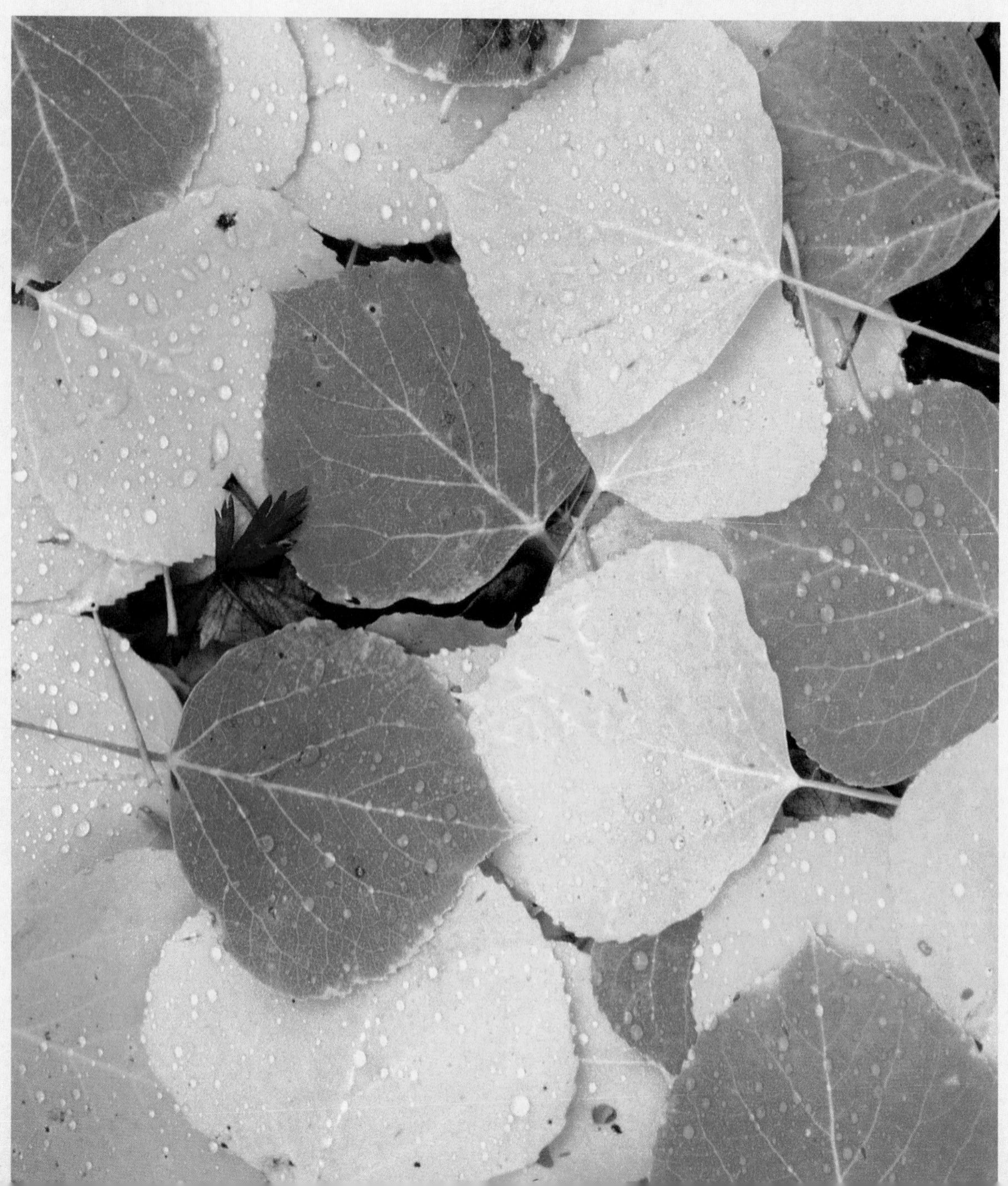

The vast and fascinating field of chemistry

Should spark your interest for many reasons,

For chemistry is everywhere *about you!*

It truly is a Science for All Seasons.

Whether you realize it or not, you are constantly making use of chemistry. Do you ever wear an acrylic sweater, or a nylon jacket, or a polyester shirt? All those materials are products of industrial chemistry. Or perhaps you are wearing a woolen sweater, or a leather jacket, or a cotton shirt. Well, those natural materials come from chemistry, too. They are made by chemical reactions that go on inside plants and animals. Your swim suit, your shoes, and your jewelry are all chemical products as well. In fact, *everything* you wear involves chemistry.

Everything you *eat* also involves chemistry. Chemical reactions are needed to produce your food, to cook it, and to digest it. The fact is that every single thing you touch has something to do with chemistry. Whenever you drive a car, take a shower, bake a cake, light a match, or simply take a walk, you are using chemistry. Even when you are sound asleep, chemical reactions are going on throughout your body. Absolutely *everything* you do involves chemistry!

Newspapers often mention chemicals that are harmful. They tell us about dangerous chemical spills, toxic chemicals that are polluting our air and water, and common chemicals that have been found to cause cancer. It is true that certain chemicals do sometimes cause problems. But chemicals can also be very helpful. Some relieve pain, and some kill bacteria that cause disease. Some increase food production, some provide strong materials for building our machines, and some provide fuel for our transportation. Many chemicals are highly important to our civilization. In fact, life itself would be impossible without chemistry.

So just what is chemistry anyhow? According to the usual definition: *chemistry* is the study of matter and the changes it undergoes. Matter is anything that has mass, which means that if you can weigh it, it is matter. Even air, which you cannot see, has weight, and therefore it is matter. Matter is the stuff of which all material things are made.

And what about those changes that matter undergoes? Sometimes matter changes on its own, as when a piece of iron gets rusty. But often we change

Chemistry is everywhere! Spectacular chemical changes occur in autumn as the chlorophyll in green leaves breaks down, revealing the brilliant colors of other pigments. [*Craig Aurness/West Light.*]

matter to make it more useful, as when we light a candle or cook an egg. Most changes in matter are accompanied by changes in energy. For example, when we burn gasoline, the reaction gives off energy that we can use to propel an automobile or a lawn mower.

More about matter and energy later. Indeed, matter and energy are what chemistry is all about. You should be interested in this subject because your own body is an incredibly marvelous chemical factory. It takes the food you eat and turns it into things like skin, and bones, and blood, and muscle, while also generating energy for all your many activities. Your body is a miraculous chemical plant that operates continuously 24 hours a day for as long as you live.

Chemistry has provided us with better and cheaper housing. It has increased our food supply and improved our nutrition. It has given us beautiful, easy-care fabrics for making our clothing. It has increased our wealth and improved our leisure time. It has enabled us to travel quickly to the far reaches of the world, and even to other worlds. Chemistry has provided us with luxuries that were not available even to the mightiest of kings in ages past.

So what is chemistry? It is the science that deals with every kind of matter, from the tiniest parts of atoms to the most complex materials of living plants and animals. Not only does it affect your own individual life every moment, but it also affects society as a whole. Chemistry in a very real way shapes our civilization.

Aristotle (384–322 B.C.), Greek philosopher and tutor of Alexander the Great, believed that we could understand nature through logic. The idea of experimental science did not triumph over Aristotelian logic until about A.D. 1500. [*Napoli Museo Nazionale/Art Resource.*]

SECTION 1.1

Science and Technology: The Roots of Knowledge

Chemistry is a **science,** but what is science? Let's examine the roots of science. Our study of the material universe has two facets: the *technological* (or *factual*) and the *philosophical* (or *theoretical*).

Technology arose long before science, having its origins in antiquity. The ancients used fire to bring about chemical changes. For example, they cooked food, baked pottery, and smelted ores to produce metals such as copper. They made beer and wine by fermentation, and obtained dyes and drugs from plant materials. These things—and many others—were accomplished without an understanding of the scientific principles involved.

The Greek philosophers, about 2500 years ago, were perhaps the first to formulate theories explaining the behavior of matter. They generally did not test their theories by experimentation, however. Nevertheless, their view of nature—attributed mainly to Aristotle—dominated natural philosophy for 2000 years.

The experimental roots of chemistry are planted in **alchemy**, a mystical chemistry that flourished in Europe during the Middle Ages, about A.D. 500 to 1500 (Figure 1.1). Modern chemists inherited from the alchemists an abiding

Figure 1.1 "The Alchemist," a painting done by the Dutch artist Cornelis Vega around 1660, depicts a laboratory of the seventeenth century. [*Palazzo Vecchio, Florence/Art Resource.*]

Modern chemists can transform matter in ways that would astound the alchemists. They can change ordinary salt into lye and laundry bleach. They can change sand into transistors and computer chips, and they can convert crude oil into plastics, pesticides, drugs, detergents, and a host of other products. Perhaps the ultimate example of this "modern alchemy" was achieved by James E. Butler of the Naval Research Laboratory in Washington, D.C. Butler made diamonds from sewer gas. Many of the products of modern chemistry are more valuable than gold.

interest in human health and the quality of life. Alchemists not only searched for a philosophers' stone that would turn cheaper metals into gold but also sought an elixir that would confer immortality on those exposed to it. Alchemists never achieved these goals, but they discovered many new chemical substances and perfected techniques such as distillation and extraction that are still used today.

Technology also developed rapidly during the Middle Ages in Europe, in spite of the generally nonproductive Aristotelian philosophy that prevailed. The beginnings of modern science were more recent, however, coming with the emergence of the experimental method. What we now call science grew out of **natural philosophy**, that is, out of philosophical speculation about nature. Science had its true beginnings in the seventeenth century, when astronomers, physicists, and physiologists began to rely on experimentation.

Technology is the practical application of knowledge.

Francis Bacon (1561–1626), English philosopher and Lord Chancellor to King James I. [*The Bettmann Archive.*]

SECTION 1.2

The Baconian Dream

It was a philosopher, Francis Bacon, who first dreamed about how science could enrich human life with new inventions and increased prosperity. By the middle of the twentieth century, science and its application in technology appeared to have made the Baconian dream come true. Many dread diseases—smallpox, polio, plague—had been virtually eliminated. Fertilizers, pesticides, and scientific animal breeding had increased and enriched our food supply. Transportation was swift, and communication nearly instantaneous. New power sources had been discovered. Nuclear energy seemed to promise an unlimited quantity of power for our every need. New materials—plastics, fibers, metals, ceramics—were developed to improve our clothing and shelter.

Much of twentieth-century technology has grown out of scientific discoveries, and technological developments are used by scientists as tools for even more discoveries. These developments in science and technology are, to a considerable extent, the base of what we mean by the "modern" world.

SECTION 1.3

The Carsonian Nightmare

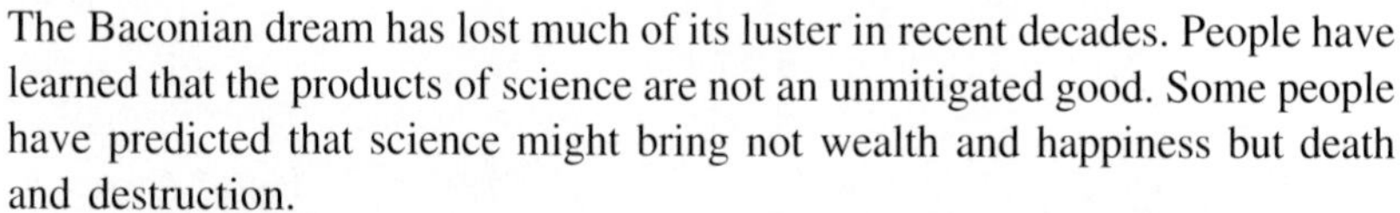

The Baconian dream has lost much of its luster in recent decades. People have learned that the products of science are not an unmitigated good. Some people have predicted that science might bring not wealth and happiness but death and destruction.

Rachel Carson (1907–1964) at Woods Hole, Massachusetts, in 1951. [*Edwin Gray Studio, copyright © 1951.*]

Perhaps most noteworthy among these critics of modern technology was Rachel Carson, a biologist. Her poetic and polemic book *Silent Spring* was published in 1962. The book's main theme is that through our use of chemicals to control insects, we are threatening the destruction of all life, including ourselves. People in the pesticide industry (and their allies) roundly denounced Carson as a "propagandist," while other scientists rallied to her support. By the late 1960s, though, we had experienced massive fish kills, the threatened extinction of several species of birds, and the disappearance of fish from rivers, lakes, and areas of the ocean that had long been productive. The majority of scientists had moved into Carson's camp. Popular support for Carson's views was overwhelming.

Carson was not the first prophet of doom. As early as 1798, Thomas Malthus, in his "Essay upon the Principles of Population," had predicted that an increase in population more rapid than the increase in food supply would lead to great famine. During the nineteenth century and for more than half of the twentieth, science and technology seemed to make a fool of Malthus. Food was abundant, at least in developed countries, and scientific discoveries and technological developments enabled us to increase food production as rapidly as the population grew.

The last few decades have brought changes, however. In spite of widespread efforts at birth control, population growth does threaten to outpace even the most optimistic projections of food production. Some scientists project a dismal future; others confidently predict that science and technology, properly applied, will save us from disaster.

SECTION 1.4

Science: Testable, Explanatory, and Tentative

What *is* science if scientists dispute what is and what will be? Is science merely a guessing game in which one guess is as good as another? We cannot *define* science precisely. Rather, we must resort to *describing* it.

Scientific Hypotheses

One essential characteristic of science is that its tenets are *testable.* Scientists make **hypotheses** (guesses) that can be tested by *experiment.* This is the main characteristic that distinguishes science from the arts and humanities. In the humanities people are still arguing about some of the same questions that were being debated thousands of years ago. (What is truth? for example.) In science it is different because experiments can be devised to answer most scientific questions. Scientists can test their ideas so they can be either verified or rejected. As a result, a firm foundation of knowledge has been built up in the sciences so that each new generation can build upon the past.

Just as students in a laboratory can get different values when making the same measurement, so scientists may not always agree about results for the same experiment. But other scientists can repeat the experiment, or devise new ones, and eventually decide whose ideas are more correct. Certain measurements have been so well verified that they have become scientific facts. The freezing point of water and the speed of light, for example, are such thoroughly tested values that they are treated as indisputable facts.

Scientists must make careful observations and accurate measurements. They record facts based upon their observations, but nothing really counts as science until those observations have been verified by others. If something is false, a scientist can't get away for long with saying that it is true.

Scientific Laws

Large amounts of scientific data can sometimes be summarized in brief statements called **scientific laws**. For example, Robert Boyle (1627–1691), an Englishman, conducted many experiments on gases. In each experiment, he found the volume of the gas to decrease when the pressure applied to the gas was increased. Many scientific laws can be stated mathematically. For example, Boyle's law can be written as

$$PV = k$$

where P is the pressure, V is the volume, and k is a constant number. If P is doubled, V will be cut in half.

Experimental observations are just the beginning of the intellectual processes of science. There are many different paths to scientific discovery, and there is no general set of rules. Science is not just a straightforward logical process for cranking out discoveries.

Scientific Theories

Science is a way of *explaining* nature, but the explanations have to be tested against a sometimes less-than-agreeable nature. The most beautiful hypothesis can be destroyed by one ugly fact. Our ideas about the universe must correspond to our observations. Scientists test ideas by predicting what they should observe if the ideas are true. Their understanding of nature is refined constantly by the interplay of ideas and observations.

Science is a body of knowledge, but the knowledge is always *tentative.* Scientists use detailed explanations, called **theories**, as a framework for the organization of scientific knowledge. These theories represent the best explanations of phenomena based on current knowledge. Sometimes a theory has to be modified or discarded in the light of new observations. Theories generally become more accurate as scientists gain more information. The body of knowledge that we call science is alive, ever changing, and rapidly growing.

Theories are useful mainly for their *predictive value.* Predictions based on theories are then tested by further experiments. Those theories that successfully make predictions that come true when new experiments are performed are generally widely accepted by the scientific community.

Science provides ways for us to cope with our environment. It involves the establishment of cause and effect. For example, scientists have learned that water vapor condenses on dust particles to form raindrops. Therefore scientists try to induce rainfall by seeding clouds with other types of particles. Unfortunately, cloud seeding is only somewhat successful in producing rain. It also raises innumerable economic, political, and ethical questions, which serve to illustrate some of the limitations of science.

Scientific Models

Scientists often use models to help explain complicated phenomena. **Scientific models** use tangible items or pictures to represent invisible processes. For example, the invisible particles of a gas can be visualized as billiard balls or marbles, or as dots or circles on paper. We know that when a glass of water stands for a period of time, the water disappears through a process called evaporation (Figure 1.2). Scientists use the kinetic-molecular theory to ex-

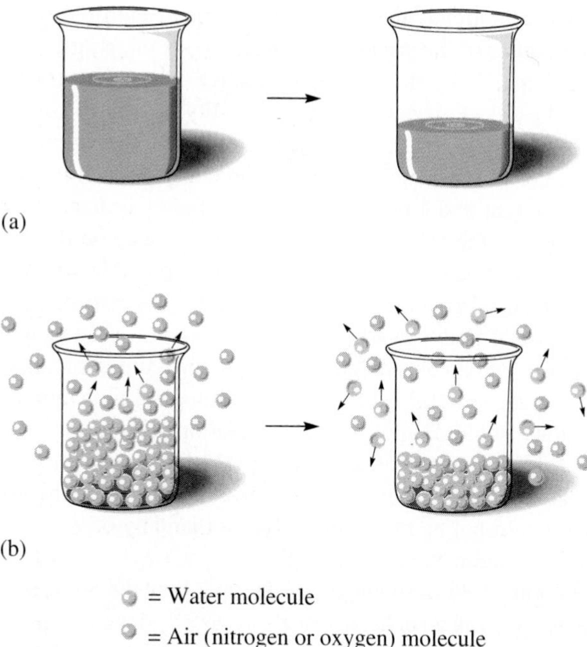

Figure 1.2 The evaporation of water. (a) When a container of water is left standing open to the air, the water slowly disappears. (b) Scientists explain evaporation in terms of the motion of molecules.

plain evaporation. According to the kinetic-molecular model, the liquid (water, in this case) is made up of tiny, invisible particles called *molecules.* The particles are in constant motion. In the bulk of the liquid, the molecules are held together by forces of attraction. The molecules collide with one another like billiard balls on a playing table. Through collisions, some of the molecules gain sufficient energy to break the attraction of their neighbors, escape from the liquid, and disperse among the widely spaced air molecules (see Figure 1.2). The water in the glass gradually disappears. It is much more satisfying to understand evaporation through the use of a model than merely to have a name for it.

What is science, then? We can only state some of its characteristics: it is *testable, explanatory,* and *tentative.* Contrary to popular notion, scientific knowledge is not absolute. Science is cumulative, but the body of knowledge is growing, changing, and never final. New facts and new concepts are always being added. Old concepts, or even old ''facts,'' are discarded when new tools, new questions, and new techniques reveal new data and generate new concepts.

Three characteristics of science are that it is testable, explanatory, and tentative.

SECTION 1.5

What Science Cannot Do

We sometimes hear scientists and nonscientists alike state that we could solve all our problems if we would only attack them using the scientific method. We have seen already that there is no single scientific method. But why can't the procedures of the scientist be applied to social, political, ethical, and economic problems? Why do scientists disagree when they try to predict the future?

The answer usually lies in the ability to control *variables.* If, for example, we wanted to study in the laboratory how the volume of gases varies with changes in pressure, we would hold constant such factors as temperature and the amount and kind of matter. If, on the other hand, an economist wished to determine the effect of increased interest rates on the rate of unemployment, he or she would find it difficult, if not impossible, to control such variables as the level of governmental expenditures, the rate of business expansion, the number of high-school and college graduates (and dropouts) entering the job market, and so on. Imagine, then, the difficulties encountered by a sociologist trying to predict the effect of a technological innovation, such as a communications satellite, upon whole populations.

We cannot control variables in social ''experiments'' (for example, public-school desegregation) as we can in laboratory experiments. Therefore, a scientist would not be in a better position than any other citizen to decide whether desegregation is good or evil.

Figure 1.3 is a rough graph showing how the number of variables increases as we go from an exact science such as physics to the complex social sciences. Notice that there is overlap between disciplines. The boundaries in the graph are crude approximations at best. A more accurate representation (but one much harder to draw) would show overlap between *all* the disciplines—even between physics and the social sciences.

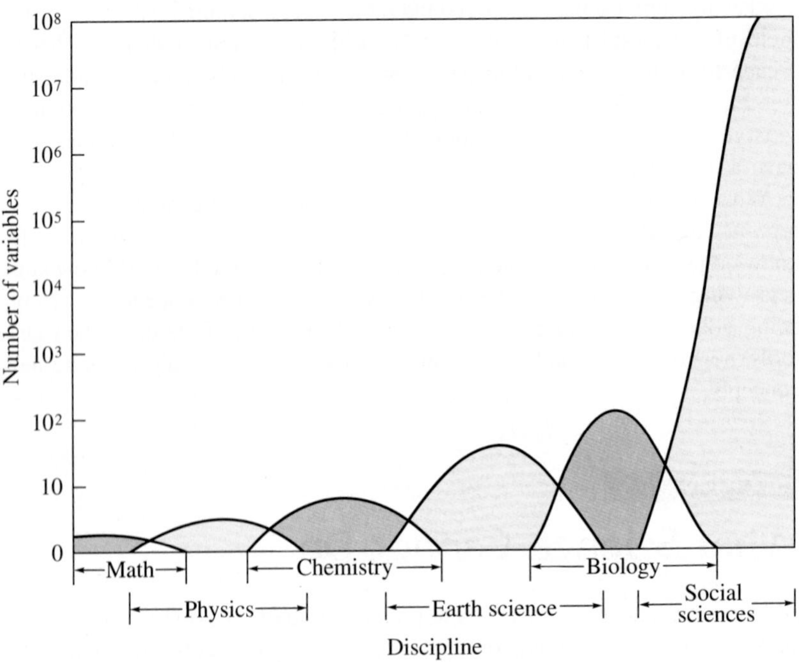

Figure 1.3 A rough estimate of the number of variables involved in scientific disciplines.

Social scientists have become more productive by using some of the methods of science. We can make observations, formulate hypotheses, and conduct experiments even if most of the variables are not subject to control. Interpretation of the results, however, is much more difficult and much more subject to disagreement. Other nonscientists use some of the methods and language of scientists. Artists experiment with new techniques and new materials. Playwrights and novelists observe life as it is before trying to express its essence in their writings. It is easy to see some of the methods of science, as well as the influence of science, in nearly all of the endeavors of modern men and women.

SECTION 1.6

Science and Technology: Risks and Benefits

Science and technology are interrelated. In everyday life, people often fail to distinguish between the two. A distinction can be useful, however. **Technology** is the sum total of the processes by which humans modify the materials of nature to better satisfy their needs and wants. These processes need not be based on scientific principles. For example, ancient peoples were able to smelt ores to produce metals, such as copper and iron, without having any understanding of the chemistry involved.

Few people question that modern Western society has benefited from science and technology, but there are risks associated with technological advances. How can we determine when the benefits outweigh the risks? One approach, called **risk–benefit analysis**, involves the calculation of a desirability quotient (DQ).

$$\text{DQ} = \frac{\text{Benefits}}{\text{Risks}}$$

A *benefit* can be described as anything that promotes well-being or has a positive effect. Benefits may be economic, social, or psychological. A *risk* can be defined as any hazard that leads to loss or injury. Some of the risks in modern technology have led to disease, death, economic loss, and environmental deterioration. Risks may involve one individual, a group, or society as a whole, and they may be local, regional, or worldwide.

Individual risk can be defined as the product of the probability that an incident will occur and the degree of severity of the incident.

$$\text{Individual risk} = \text{Probability} \times \text{Severity}$$

For example, the automobile is a technological development that benefits us by making it easy to travel from one place to another. Most of us know the

Figure 1.4 For most people in Canada, Northern Europe, and the United States, milk is a wholesome food; its benefits far outweigh its risks. In other countries, with high rates of lactose intolerance among adults, the desirability quotient for milk will be much smaller. [*Hank Morgan/ Photo Researchers, Inc.*]

risks involved in driving a car. A severe risk is the possibility of death in a traffic accident. The probability that this will happen is remote if an individual drives carefully and within the speed limit. Chances are, if we want to go to a place 50 kilometers (km) away, we will find that the benefits of driving outweigh the chance that we might be injured or killed in an accident.

Societal risk involves more people. We must therefore add another factor—the number of people affected—to our equation.

$$\text{Societal risk} = \text{Probability} \times \text{Severity} \times \text{Population affected}$$

Weighing the benefits and risks associated with a product of modern technology is decidedly more difficult when one considers a group of people. For example, for most people the benefit of pasteurized milk is that it is a safe, clean beverage that provides needed nutrition (Figure 1.4). For some people, however, milk poses a risk. Some people can't tolerate lactose, the sugar in milk. People with cardiovascular disease may be harmed by the saturated fats in whole milk. To these people, and to others who are allergic to milk protein, milk is a harmful substance. Fortunately, milk allergies and lactose intolerance are relatively rare in North America and northern Europe. In these societies, milk is generally beneficial.

In a risk–benefit analysis for milk, we find large benefits and small risks, resulting in a large DQ.

$$\frac{\text{Large benefits}}{\text{Small risks}} = \text{Large DQ}$$

Most of us readily accept the risks associated with milk. Interestingly, however, this analysis generally fits only people of northern European descent. Adults in much of the rest of the world are lactose intolerant and would find that milk has a small DQ. Thus, milk is not always suitable in programs to relieve malnutrition.

An important technological contribution is the production of natural and synthetic chemicals used as drugs. One example—among many—was the introduction of the drug thalidomide to prevent morning sickness in pregnant women. The drug was introduced in the 1950s, but was soon shown to be of little benefit. Several years later, it was shown to present a large risk: many malformed infants were born to women who took the drug during pregnancy. For thalidomide, then, we have small benefits, large risks, and a small DQ.

$$\frac{\text{Small benefits}}{\text{Large risks}} = \text{Small DQ}$$

Thalidomide was eventually judged to present unacceptable risks and was banned. (Other drugs are discussed in Chapter 19. Many have high DQs, some have low DQs, and others are more difficult to evaluate.)

It is easy to judge the desirability of milk and thalidomide, but other products present difficult choices. Consider a product whose benefits and risks both are small: the artificial sweetener aspartame. Studies have shown that artificial sweeteners generally are of little benefit to those who use them to replace sugar in an effort to lose weight. Aspartame is the most studied of all food additives. There are anecdotal reports of problems with the sweetener, but these have not been confirmed in controlled studies. To most people, the risk involved in using aspartame is small. This leads to an uncertain DQ—and, it seems, to endless debate over the safety of aspartame.

$$\frac{\text{Small benefits}}{\text{Small risks}} = \text{Uncertain DQ}$$

(Aspartame may provide some benefit to diabetics, however, because sugar consumption presents a large risk to them.)

Other technologies provide large benefits and present large risks. For these technologies, too, the DQ is uncertain.

$$\frac{\text{Large benefits}}{\text{Large risks}} = \text{Uncertain DQ}$$

An example is the conversion of coal to liquid fuels. Liquid fuels provide large benefits to society in the areas of transportation, home heating, and industry. The risks associated with coal conversion are also large, however. These include air and water pollution and the exposure of workers to toxic chemicals. The result again is an uncertain DQ and political controversy.

There are yet other problems in risk–benefit analysis. Some technologies benefit one group of people (population A) while presenting a risk to another (population B). For example, it may be economically advantageous to a community to spend as little as possible on a sewage treatment plant and to dump raw wastes into a nearby stream. These wastes might present a hazard to downstream communities, however. Difficult political decisions are needed in such a case.

Other technologies provide benefits now but present risks later. For example, although nuclear power now provides useful electricity, wastes from nuclear power plants, if improperly stored, might present hazards for centuries. Thus, the use of nuclear power is highly controversial.

It is important to remember that science and technology involve *both* risks and benefits. The determination of benefits is almost entirely a social judgment; risk assessment also involves social decisions, but scientific investigation can help considerably in risk evaluation.

EXAMPLE 1.1

Chloramphenicol is a powerful antibacterial drug that often will destroy bacteria that are unaffected by other drugs. It is highly dangerous to some people, however, causing fatal aplastic anemia in about 1 in 30,000 people.

Do a risk–benefit analysis for the use of chloramphenicol in (a) sick farm animals, from which people might consume milk or meat with residues of the drug, and (b) a person with Rocky Mountain spotted fever, with high probability of death or permanent disability.

SOLUTION

a. This use has been judged to be too risky by the U.S. Food and Drug Administration; chloramphenicol has been banned from such use.
b. The sick person might well choose to take the drug, with a risk of 1 in 30,000, to avoid a much greater risk. (Both answers involve judgments that are not clearly scientific; people can differ in their assessment of each.)

Exercise 1.1

Coal is a widely used fuel for generating electricity. It is abundant and relatively inexpensive. But burning coal produces large amounts of soot, which pollutes the atmosphere, and generates much carbon dioxide, which can contribute to global warming. Burning high sulfur coal also leads to acid rain, which can kill the fish in a lake or the trees in a forest. Do a risk–benefit analysis for using coal to produce electricity.

SECTION 1.7

Chemistry: A Study of Matter and Energy

At the beginning of this chapter, we defined chemistry as a study of matter and the changes it undergoes. Now let's look at matter and energy a little more closely.

Since the entire physical universe is made up of nothing more than matter and energy, the field of chemistry extends from atoms to stars, from rocks to living organisms. Matter and energy are such fundamental concepts that definitions are difficult. **Matter** is the stuff that makes up all material things. It has *mass*. In other words, you can weigh it. Wood, sand, water, air, and people have mass and are therefore matter. **Mass** is a measure of the quantity of matter that an object contains. The greater the mass of a thing, the more difficult it is to change its velocity. You can easily deflect a tennis ball coming toward you at 30 meters per second (m/s), but you would have difficulty stopping a cannonball of the same size moving at the same speed. A cannonball has more mass than a tennis ball of equal size.

The mass of an object does not vary with location. An astronaut has the same mass on the moon as on Earth (Figure 1.5). **Weight**, on the other hand, measures a force. On Earth, it measures the force of attraction between our planet and the mass in question. On the moon, where gravity is one-sixth that

Figure 1.5 Astronaut John W. Young leaps from the lunar surface, where gravity pulls at him with only one-sixth the force on Earth. [*NASA photo.*]

on Earth, an astronaut weighs only one-sixth as much as on Earth. Weight varies with gravity; mass does not.

▶ EXAMPLE 1.2

On Mars gravity is one-third that on Earth. (a) What would be the mass on Mars of a person who has a mass of 60 kilograms (kg) on Earth? (b) What would be the weight on Mars of a person who weighs 120 pounds (lb) on Earth?

SOLUTION

a. The person's mass would be the same (60 kg) as on Earth; the quantity of matter has not changed.
b. The person would weigh only 40 lb; the force of attraction between planet and person is only one-third that on Earth.

Exercise 1.2

On Jupiter the force of gravity is 2.4 times that on Earth. (a) How much would a standard mass of 1 kg weigh on Jupiter? (b) A man who weighs 200 lb on Earth would weigh how much on Jupiter?

Physical and Chemical Change

Using our knowledge of chemistry, we can change matter to make it more useful. We can change crude oil into gasoline, plastics, pesticides, drugs, detergents, and thousands of other products. Changes in matter are accompanied by changes in energy. Often we change matter to extract a part of its energy. For example, we burn gasoline to get energy to propel our automobiles.

Matter is characterized by its *properties.* This means that we can distinguish water from gasoline by the characteristic properties of the two liquids. They differ in odor, flammability, and many other properties. **Chemical properties** describe how one substance reacts with other substances. To demonstrate a chemical property, a substance must undergo a change in composition. Sulfur burns (combines with oxygen) to form an acrid gas called sulfur dioxide. Sulfur also combines with carbon to form a liquid called carbon disulfide and with iron to form a solid called iron sulfide. Sulfur dioxide, carbon disulfide, and iron sulfide have different properties than sulfur. Each also has a different composition. Substances that undergo a change in chemical properties are said to undergo **chemical change**.

An *acrid* gas is one with a sharp, irritating odor and taste. You can smell sulfur dioxide when you strike a match. Sulfur dioxide is an important air pollutant.

Physical properties are those that can be observed and specified without reference to any other substance. Characteristics such as color, hardness, density, melting point, and boiling point are physical properties (Figure 1.6). A physical property of sulfur, for example, is that it is a brittle yellow solid under usual conditions. Another is that sulfur is denser than water; that is, sulfur sinks in water. Sulfur melts into a liquid form at 115 °C. When sulfur melts it has undergone a physical change; it has changed from a solid to a liquid. However, its *composition* has not changed; it has *not* changed chemically. A **physical change** occurs without a change in composition, and it does not alter the chemical nature of the substance. Although liquid sulfur looks different from solid sulfur and displays different physical properties, its chemistry has not changed.

A burning candle demonstrates both physical and chemical changes. After the candle is lit, the solid wax near the burning wick melts. This melting is a physical change because the composition of the wax is the same in both the solid and the liquid form. Some of the melted wax is drawn into the burning wick where a chemical change occurs. The wax combines with oxygen in the air to form carbon dioxide gas and water vapor. As the candle burns and the wax undergoes this chemical change, the candle becomes smaller and smaller. The apparent disappearance is not necessarily a sign of chemical change.

Figure 1.6 A comparison of the physical properties of two elements. Copper, obtained as pellets, can be hammered into a thin foil or drawn into a wire. When hammered, lumps of sulfur crumble into a fine powder. [*Carey B. Van Loon.*]

When water evaporates, it seems to disappear. However, it merely changes from a liquid to a gas; it has undergone a physical change. The composition of the water does not change during evaporation. It is not always easy to distinguish a physical from a chemical change. The critical question is: Has the composition of the substance been changed? In a chemical change it has; the substance has combined with another substance or has been decomposed to simpler substances. In a physical change no new substances are formed.

EXAMPLE 1.3

Which of the following is a chemical change and which is a physical change?

a. Your hair is trimmed by a barber.
b. A female mosquito uses your blood as food from which she produces her eggs.
c. Water boils to form steam.
d. Water is broken down into hydrogen gas and oxygen gas.

SOLUTION

a. Physical change: the hair is not changed by clipping.
b. Chemical change: eggs differ in composition from blood.
c. Physical change: liquid water and water vapor (steam) have the same composition; water merely changes from a liquid to a gas.
d. Chemical change: new substances, hydrogen and oxygen, are formed.

Exercise 1.3

Which of the following is a chemical change and which is a physical change?

a. A car fender gets rusty.
b. A stick of butter is melted.
c. A pile of leaves is burned.
d. A piece of wood is ground up into sawdust.

Energy: A Matter of Moving Matter

Physical and chemical changes are always accompanied by changes in energy. **Energy** is the ability to change matter, either chemically or physically. Energy is required to make something happen that wouldn't happen by itself. The ability to modify matter is the basis for change in the material world.

Energy exists in several forms. Energy due to position or arrangement is called **potential energy**. The water at the top of a dam has potential energy due to gravitational attraction. When the water is allowed to flow through a

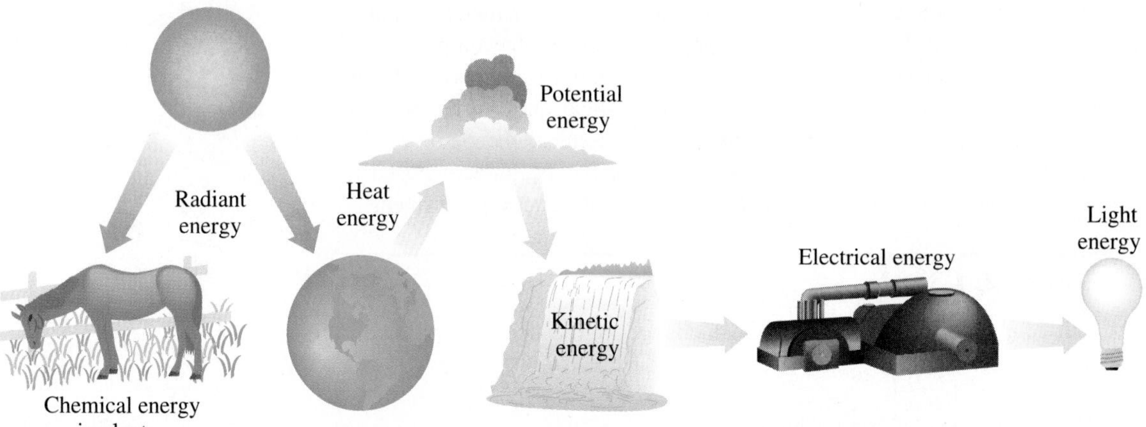

Figure 1.7 Energy can be changed from one form to another.

turbine to a lower level, the potential energy is converted to **kinetic energy** (the energy of motion). In Figure 1.7, the water at the top of the dam, as it flows through the hydroelectric plant, moves the blades of a turbine. As the water falls, it moves faster. Its kinetic energy becomes greater as its potential energy decreases. The turbine can convert a part of the kinetic energy of the water into electrical energy. The electricity thus produced can be carried by wires to homes and factories where it can be converted to light energy or to heat or to mechanical energy.

Nearly all the energy available on Earth comes ultimately from the sun. Much of what we use today was captured long ago by green plants through photosynthesis. We reap the remains of those plants of ages past as fossil fuels—coal, petroleum, and natural gas.

Chemistry is concerned not only with matter and its changes but with the energy transformations that accompany these changes.

SECTION 1.8

The States of Matter

There are three familiar *states of matter*: solid, liquid, and gas (Figure 1.8). **Solid** objects ordinarily maintain their shape and volume regardless of their location. A **liquid** occupies a definite volume, but assumes the shape of the occupied portion of its container. If you have a 12-oz soft drink, you have 12 oz whether the soft drink is in a can, a bottle, or, through some slight mishap, on the floor—which demonstrates another property of liquids. Unlike solids, liquids flow readily. A **gas** maintains neither shape nor volume. It expands to fill completely whatever container it is in. Gases can be easily

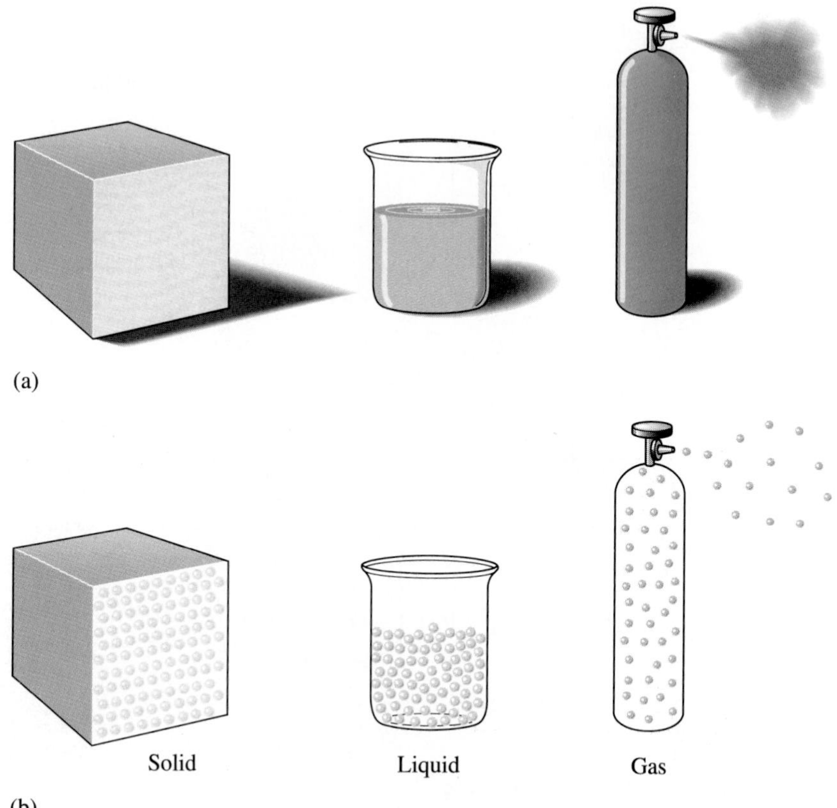

Figure 1.8 Solids, liquids, and gases. (a) Bulk properties. (b) Interpretation of bulk properties in terms of the kinetic molecular theory. In liquids, the particles are close together, but they are free to move about. In gases, the particles are far apart and in rapid random motion.

compressed. For example, enough air for many minutes of breathing can be compressed into a steel tank for underwater diving. We take up the topic of the states of matter in more detail in Chapter 5.

SECTION 1.9

Matter: Pure Substances and Mixtures

Matter can also be classified into pure substances and mixtures (Figure 1.9). **Pure substances** have a definite, or fixed, composition. The composition of **mixtures** may vary. Water is a pure substance; it always contains 11% hydrogen and 89% oxygen by weight. Similarly, pure gold is pure gold, that is, 100% gold. A milk shake, on the other hand, is a mixture. The proportions of milk, ice cream, and flavorings change depending on who is preparing the shake. Mixed nuts are a mixture; the ratio of peanuts to pecans depends on how much you are willing to pay per pound.

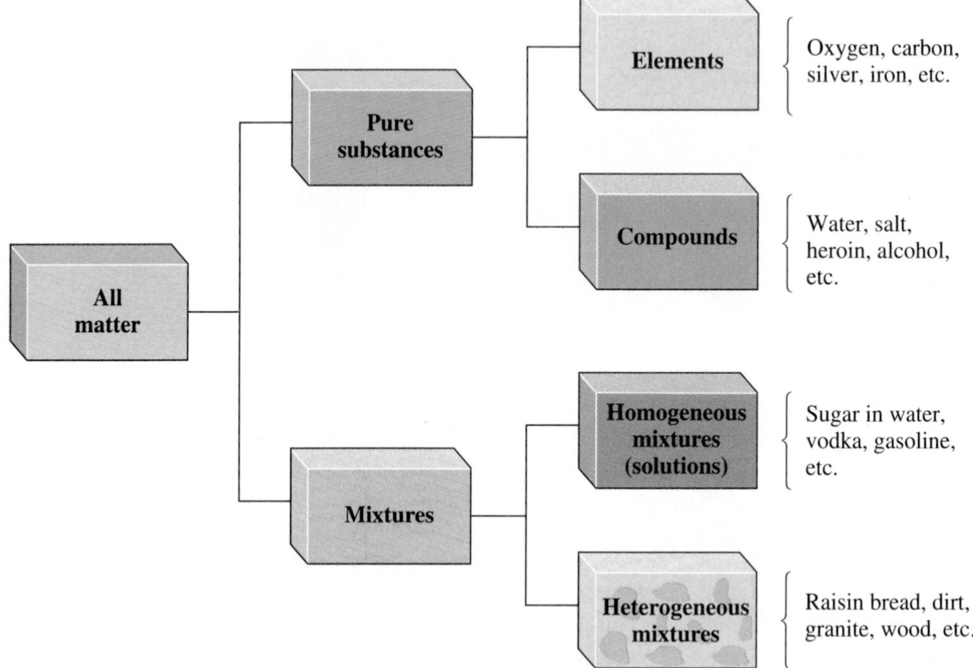

Figure 1.9 A scheme for classifying matter.

Elements and Compounds

At present 109 elements are known. We deal with only about one-third of them in any detail in this book.

Pure substances may be either elements or compounds. **Elements** are those fundamental substances from which all material things are constructed. **Compounds** are pure substances that are made up of two or more elements chemically combined in constant or fixed proportions. Our ideas about elements have changed during historical times. Fire was once considered an element, but it is now regarded as nonmaterial, that is, as a form of energy. Water was once thought to be an element, but we now know it to be a compound composed of two elements, hydrogen and oxygen. We presently regard as elements a few more than 100 pure substances that cannot be broken down by chemical means into simpler substances. Sulfur, oxygen, carbon, and iron are elements. Sulfur dioxide, carbon disulfide, and iron sulfide are compounds.

Some synthetic elements have symbols of three letters. These elements are rare and warrant little mention here.

Because elements are so fundamental to our study of chemistry, we find it useful to refer to them in a shorthand form. Each element can be represented by a **chemical symbol** made up of one or two letters derived from the name of the element (or, sometimes, from the Latin name of the element). The first letter of the symbol is always capitalized; the second is always lowercase. (It makes a difference. For example, Hf is the symbol for hafnium, an element, but HF is the formula for hydrogen fluoride, a compound. Similarly, Co is cobalt, an element; CO is carbon monoxide, a compound.)

Symbols are the alphabet of chemistry. Symbols for the more important elements studied in this book are given in Table 1.1. Memorization of the

Table 1.1 Names, Symbols, and Physical Characteristics of Some Common Elements

Name (Latin name)	Symbol	Selected Properties
Aluminum	Al	Light, silvery metal
Argon	Ar	Colorless gas
Arsenic	As	Grayish white solid
Barium	Ba	Silvery white metal
Beryllium	Be	Steel gray, hard, light solid
Boron	B	Black or brown powder; several crystal forms
Bromine	Br	Reddish brown liquid (Br_2)
Calcium	Ca	Silvery white metal
Carbon	C	Soft black solid (graphite) or hard, brilliant crystal (diamond)
Chlorine	Cl	Greenish yellow gas (Cl_2)
Copper (Cuprum)	Cu	Light reddish brown metal
Fluorine	F	Pale yellow gas (F_2)
Gold (Aurum)	Au	Yellow, malleable metal
Helium	He	Colorless gas
Hydrogen	H	Colorless gas (H_2)
Iodine	I	Lustrous black solid (I_2)
Iron (Ferrum)	Fe	Silvery white, ductile, malleable metal
Lead (Plumbum)	Pb	Bluish white, soft, heavy metal
Lithium	Li	Silvery white, soft, light metal
Magnesium	Mg	Silvery white, ductile, light metal
Mercury (Hydrargyrum)	Hg	Silvery white, liquid, heavy metal
Neon	Ne	Colorless gas
Nickel	Ni	Silvery white, ductile, malleable metal
Nitrogen	N	Colorless gas (N_2)
Oxygen	O	Colorless gas (O_2)
Phosphorus	P	Yellowish white waxy solid (P_4) or red powder
Plutonium	Pu	Silvery white, radioactive metal
Potassium (Kalium)	K	Silvery white, soft metal
Silicon	Si	Lustrous gray solid
Silver (Argentum)	Ag	Silvery white metal
Sodium (Natrium)	Na	Silvery white, soft metal
Sulfur	S	Yellow solid (S_8)
Tin (Stannum)	Sn	Silvery white, soft metal
Uranium	U	Silvery, radioactive metal
Zinc	Zn	Bluish white metal

A chemical symbol in a formula stands for one atom of the element. If more than one atom is to be indicated in a formula, a subscripted number is used after the symbol. For example, the formula Cl_2 represents two atoms of chlorine, and the formula CCl_4 stands for one atom of carbon and four atoms of chlorine. Table 1.1 gives actual formulas (Br_2, P_4, S_8, and so on) for some of the elements as they occur when in the elemental form. You need not concern yourself with such formulas now; they are included in the table for your future reference.

names and symbols of the elements shown in that table is well worth your time. The difficult symbols to learn are those based on Latin names. There are 9 of them in Table 1.1. They are copper (Cu), gold (Au), iron (Fe), lead (Pb), mercury (Hg), potassium (K), silver (Ag), sodium (Na), and tin (Sn).

▶ EXAMPLE 1.4

Which of the following represent an element and which represent a compound?

Hg   HI   BN   In

SOLUTION

Hg and In represent elements (each is a single symbol). HI and BN are composed of two symbols each and represent compounds.

Exercise 1.4

Which of the following represent elements and which represent compounds?

He   CuO   Hf   No   NO   KI

SECTION 1.10

The Measurement of Matter

Accurate measurement of such quantities as mass (or weight), volume, time, and temperature are essential to the compilation of dependable scientific "facts." Such facts may be used by a chemist interested in basic research, but similar information is of critical importance in every science-related field. Certainly we are all aware that measurements of both temperature and blood pressure are routinely made in medicine. It is also true that modern medical diagnosis depends on a whole battery of other measurements, including careful chemical analyses of blood and urine.

The system of measurement used by most scientists is an updated metric plan called the International System of Measurements, or SI (from the French Système International). Indeed, SI has been adopted worldwide for everyday use. Even the United States is committed to change to SI, but conversion is voluntary and has proceeded rather slowly so far.

The beauty of SI is that it is based on the decimal system. This makes conversion from one unit to another rather simple. The SI has only a few basic units. For example, the basic unit of length is the **meter** (m), a distance only slightly greater than a yard. The SI unit of mass is the **kilogram** (kg), a quantity slightly greater than 2 pounds. All other units for length, mass, and volume can be derived from these basic units. For example, area (length times width) can be measured in units of meters times meters ($m \times m$) or meters squared (m^2), and volume (length times width times height) can be measured in units of meters cubed (m × m × m or m^3).

A disadvantage of the basic SI units is that they are often of awkward magnitude. We seldom work with kilogram quantities in the laboratory. A cubic meter of liquid would fill a very large test tube. More convenient units can be derived by the use of (or, in the case of the kilogram, the deletion of) prefixes (Table 1.2). For example, in the laboratory we may work with grams (g) or milligrams (mg) of material. The prefix *milli-* means 1/1000 or 0.001. Thus, 1 milligram equals 1/1000 gram or 0.001 gram. The relationship between milligrams and grams is given by

$$1 \text{ mg} = 0.001 \text{ g}$$

or

$$1000 \text{ mg} = 1 \text{ g}$$

You should learn the more common prefixes (printed in color in Table 1.2) right away.

For volume we may choose cubic centimeters (cm^3) or cubic decimeters (dm^3) as more convenient units. The cubic decimeter has a special name, the **liter** (L). From that is derived the milliliter (mL), a unit that is the same as the cubic centimeter.

$$1 \text{ dm}^3 = 1 \text{ L}$$

$$1 \text{ L} = 1000 \text{ mL}$$

$$1 \text{ mL} = 1 \text{ cm}^3$$

Table 1.2 Approved Numerical Prefixes*

Exponential Expression	Decimal Equivalent	Prefix	Pronounced	Symbol
10^{12}	1,000,000,000,000.	tera-	TER-uh	T
10^{9}	1,000,000,000	giga-	GIG-uh	G
10^{6}	1,000,000	mega-	MEG-uh	M
10^{3}	1,000	kilo-	KIL-oh	k
10^{2}	100	hecto-	HEK-toe	h
10	10	deka-	DEK-uh	da
10^{-1}	0.1	deci-	DES-ee	d
10^{-2}	0.01	centi-	SEN-tee	c
10^{-3}	0.001	milli-	MIL-ee	m
10^{-6}	0.000,001	micro-	MY-kro	μ
10^{-9}	0.000,000,001	nano-	NAN-oh	n
10^{-12}	0.000,000,000,001	pico-	PEE-koh	p
10^{-15}	0.000,000,000,000,001	femto-	FEM-toe	f
10^{-18}	0.000,000,000,000,000,001	atto-	AT-toe	a
10^{-21}	0.000,000,000,000,000,000,001	zepto-	ZEP-toe	z

*The most commonly used units are shown in color.

Conversions within the SI system are much easier than those using the more familiar units of pounds, feet, and pints. This is best shown by examples.

▶ EXAMPLE 1.5

Convert 0.742 kg to grams.

SOLUTION

$$0.742\ \cancel{\text{kg}} \times \frac{1000\ \text{g}}{1\ \cancel{\text{kg}}} = 742\ \text{g}$$

Note: If you are not familiar with the use of conversion factors or would like to review the mathematics of conversions, see Appendix C.

Exercise 1.5
Convert 6.3 mg to grams.

▶ EXAMPLE 1.6

Convert 0.742 lb to ounces.

SOLUTION

$$0.742\ \cancel{\text{lb}} \times \frac{16\ \text{oz}}{1\ \cancel{\text{lb}}} = 11.9\ \text{oz}$$

Exercise 1.6
Convert 24 ounces to pounds.

▶ EXAMPLE 1.7

Convert 1247 mm to meters.

SOLUTION

$$1247\ \cancel{\text{mm}} \times \frac{1\ \text{m}}{1000\ \cancel{\text{mm}}} = 1.247\ \text{m}$$

Exercise 1.7
Convert 2.5 meters to centimeters.

▶ EXAMPLE 1.8

Convert 1247 in. to yards.

SOLUTION

$$1247\ \cancel{\text{in.}} \times \frac{1\ \text{yd}}{36\ \cancel{\text{in.}}} = 34.64\ \text{yd}$$

Exercise 1.8
Convert 6.5 ft to inches.

Figure 1.10 A comparison of metric and customary units of measure. The left beaker contains 1 kg of candy and the right beaker contains 1 pound (1 kg = 2.2 lb). The yellow ribbon is 1 in. wide and is tied around a stick 1 yd long. The green ribbon is 1 cm wide and is tied around a stick 1 m long (1 m = 1.0936 yd; 1 cm = 0.3937 in.). The flask on the right and the bottle behind it each contain 1 L of orange juice. The flask on the left and the carton behind it each contain 1 qt of milk (1 L = 1.06 qt). [*Carey B. Van Loon.*]

In conversions involving pounds, ounces, inches, and yards, you multiply and divide by numbers such as 16 or 36. In metric conversions you multiply and divide by 10 or 100 or 1000, and so on. You just shift the decimal point.

Conversions between systems are seldom necessary. (If you need to do this, however, such conversions are discussed in Appendix C.) What you *do* need is some idea of the relative sizes of comparable units. Some comparisons are shown in Figure 1.10. Other comparisons easily remembered are that the United States 10-cent coin, the dime, is about 1 mm thick and that the 5-cent piece, the nickel, weighs about 5 g.

SECTION 1.11

Measuring Energy: Temperature and Heat

The SI unit for temperature is the **kelvin** (K), but for much of their work, scientists still use the **Celsius scale**. On this scale, the freezing temperature of water is 0 °C and the boiling point is 100 °C. The scale between these two reference points is divided into 100 equal divisions, each a Celsius degree.

Degrees on the Kelvin and Celsius scales are the same size. The Kelvin scale is called an *absolute scale* because its zero point is the coldest temperature possible, or absolute zero. This fact was determined by theoretical considerations and has been confirmed by experiment. The zero point on the Kelvin scale is equal to −273 °C. Note that there are no negative temperatures on the absolute scale. To convert from Celsius to Kelvin, you merely add 273 to the Celsius temperature.

$$K = °C + 273$$

▶ EXAMPLE 1.9

What is the boiling point of water in kelvins? The boiling point of water is 100 °C.

SOLUTION

$$100\ °C + 273 = 373\ K$$

Exercise 1.9

How would normal body temperature be expressed in kelvins?

In the United States, weather reports and food recipes still use the Fahrenheit scale. This scale defines the freezing temperature of water as 32 °F and the boiling point as 212 °F. Exact conversions are seldom necessary. Figure 1.11 compares the Fahrenheit, Celsius, and Kelvin scales. Some additional temperature equivalents and formulas for converting one scale to the other are shown in Appendix A.

Scientists often need to measure amounts of heat energy. You should not confuse heat with temperature. **Heat** is a measure of quantity, that is, of *how much energy* a sample contains. **Temperature** is a measure of intensity, that is, of *how energetic* each particle of the sample is. A glass of water at 70 °C contains less heat than a bathtub of water at 60 °C. The particles of water in

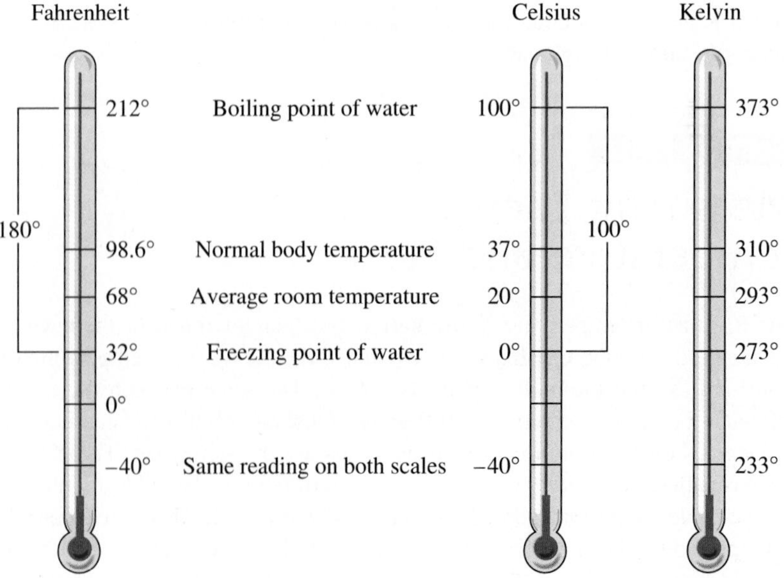

Figure 1.11 A comparison of the Fahrenheit, Celsius, and Kelvin temperature scales.

the glass are more energetic, on the average, than those in the tub, but there is far more water in the tub and its total heat content is greater.

The SI unit of heat is the **joule** (J), but we will use the more familiar **calorie** (cal).

$$1 \text{ cal} = 4.184 \text{ J}$$

$$1000 \text{ cal} = 1 \text{ kcal} = 4184 \text{ J}$$

A calorie is the amount of heat required to raise the temperature of 1 g of water 1 °C. There is a more precise definition, but this version will do for our purposes.

For measuring the energy content of foods, the large *Calorie* (note the capital *C*), or **kilocalorie** (kcal), is sometimes used. A dieter might be aware that a banana split contains 1500 Calories. If the same dieter realized that that meant 1,500,000 calories, giving up the banana split might be easier.

SECTION 1.12

Density

An important property of matter, particularly in scientific work, is **density**. When one speaks of lead as ''heavy'' or aluminum as ''light,'' one is referring to the density, *d*, of these metals. This term is defined as the amount of mass, *m* (or weight) per unit of volume, *V*.

$$d = \frac{m}{V}$$

Rearrangement of this equation gives

$$m = d \times V \qquad \text{and} \qquad V = \frac{m}{d}$$

These equations are useful for calculations. Densities are usually reported in grams per milliliter (g/mL) or grams per cubic centimeter (g/cm^3).

▶ EXAMPLE 1.10

What is the density of iron if 156 g of iron occupies a volume of 20 cm^3?

SOLUTION

$$d = \frac{m}{V} = \frac{156 \text{ g}}{20 \text{ cm}^3} = 7.8 \text{ g/cm}^3$$

Exercise 1.10

What is the density of gold if a cube that measures 2.00 cm on an edge weighs 154.4 g?

EXAMPLE 1.11

How much will a liter of gasoline weigh if its density is 0.66 g/mL?

SOLUTION

$$m = d \times V = \frac{0.66 \text{ g}}{1 \cancel{\text{mL}}} \times 1 \cancel{\text{L}} \times \frac{1000 \cancel{\text{mL}}}{1 \cancel{\text{L}}} = 660 \text{ g}$$

Exercise 1.11

What volume would be occupied by a 10-g piece of aluminum? (Density of Al is 2.7 g/mL.)

The density of water is 1 g/mL, or 1 g/cm^3 (remember that 1 mL = 1 cm^3). This nice round number for the density of water is no accident. The metric system was originally set up to ensure that this was the case.

Measurement is discussed further in Appendix A, and various conversion tables are collected there for convenient reference.

SECTION 1.13

Chemistry: Its Central Role

Chemistry is not only useful in itself but also fundamental to other scientific disciplines. Biology has been revolutionized by the application of chemical principles. Psychology, too, has been profoundly influenced by chemistry and stands to be even more radically altered as the chemistry of the nervous system is unraveled. The social goals of better health and more and better food, housing, and clothing are dependent to a large extent on the knowledge and techniques of chemists. The recycling of basic materials—paper, glass, and metals—is primarily a matter of chemical processes. Devising new, more specific pesticides that entail less risk to useful organisms will require the application of chemical principles and skills. Chemistry is indeed a central science (Figure 1.12). There is scarcely a single area of our daily lives that is not affected by chemistry.

Chemistry is also important to the *economy* of industrial nations. In the United States, the chemical industry employs over a million people in more than 10,000 plants. It is the nation's fifth largest industry with sales of about $200 billion per year. The exports of the chemical industry help to keep the United States' international trade deficit from being even worse than it now is. From 1981 to 1991 the nation's chemical industry had a trade surplus every year, with exports exceeding imports by about $126 billion. In 1992 the trade surplus was a record $20.5 billion. Incidentally, according to the National Safety Council, the chemical industry ranked either first or second throughout the 1980s in worker safety among 42 basic industries. For each year of the decade it had the fewest or next fewest incidents of occupational illness and

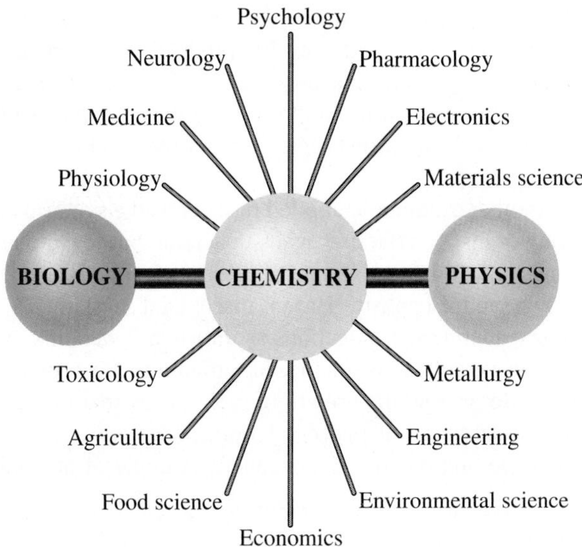

Figure 1.12 Chemistry has a central role among the sciences.

injury involving death or days away from work. The rate of such incidents in the chemical industry is only one-third of the average rate for all United States industries. This is in sharp contrast to the popular belief that chemicals are so dangerous. (Some are, of course, but even those can be used safely with proper precautions.)

An understanding of chemistry is essential to an understanding of many of the problems facing society. If you want to understand how a drug or a pesticide acts, how to clean up a pollution problem, or how to avoid consumer rip-offs in cosmetics, you must know some chemistry. Much of the remainder of this book deals with such subjects. Before going on, though, let's look briefly at some of the things chemists do.

SECTION 1.14

Solving Society's Problems: Applied Research

Most chemists work in the area of *applied research.* They analyze polluted soil, air, and water. They synthesize new chemical compounds for use as drugs or pesticides. They formulate plastics for new applications. They analyze foods, fuels, cosmetics, detergents, and drugs. These are examples of **applied research**—work oriented toward the solution of a particular problem in industry or the environment.

George Washington Carver (1860–1943), research scientist, in his laboratory at the Tuskegee Institute. [*The Bettmann Archive.*]

Among the most monumental accomplishments in the realm of applied research were those of George Washington Carver. Born in slavery, Carver attended Simpson College and later graduated from Iowa State University. A

Rachel Brown (left) and Elizabeth Hazen, codiscoverers of nystatin, a safe fungal antibiotic. [*Courtesy of Rachel Fuller Brown, Research Corporation, New York.*]

botanist and agricultural chemist, Carver taught and did research at Tuskegee Institute. He developed over 300 products from peanuts. He made other new products from sweet potatoes, pecans, and clay. Carver also taught southern farmers to rotate crops and to use legumes to replenish the nitrogen removed from the soil by cotton crops. Carver's work helped to revitalize the economy of the South.

Another example of applied research is the synthesis of an antifungal drug. Many drugs are effective against bacteria, but few are useful and safe against fungal infections in humans. Seeking an effective fungal antibiotic, Rachel Brown and Elizabeth Hazen discovered nystatin, a compound effective against such fungal infections as thrush and vaginitis. Nystatin also has been used in veterinary medicine, in agriculture (to protect fruit), and in the art world (to prevent the growth of mold on art treasures in Florence, Italy, after the objects were damaged by floods). Brown and Hazen set out to find a fungicide and did so—an excellent example of applied research.

SECTION 1.15

In Search of Knowledge: Basic Research

Many chemists are involved in basic research, the search for knowledge for its own sake. Some chemists work out the fine points of atomic and molecular structure. Others measure the intricate energy changes that accompany complex chemical reactions. Chemists synthesize new compounds and determine their properties. This type of investigation is called **basic research**. Done for the sheer joy of unraveling the secrets of nature and discovering order in our universe, basic research is characterized by the absence of any predictable, marketable product.

Joel Hildebrand (1881–1983), noted chemist and teacher, whose theoretical studies on solubility led to the use of helium in deep-sea diving. [*Courtesy of Joel Hildebrand.*]

Findings from basic research often *are* applied at some point, but that is not the primary goal of the researcher. In fact, most of our modern technology is based on results obtained in basic research. Without this base of factual information, technological innovation would be haphazard and slow. Warnings have been sounded that the United States is falling behind Germany, Japan, and other nations because it has failed to provide adequate support for basic research. How costly is research? All research done since the time of Aristotle has cost less than 2 weeks' output in the modern industrial world.

Applied research is carried on mainly by industries seeking a competitive edge with a novel, better, or more salable product. Its ultimate aim is usually profit for the stockholders. Basic research is conducted mainly at universities and research institutes. Most of its support comes from federal and state governments and foundations, although some of the larger industries also support it.

An example of basic research that was later applied to improving human welfare is the work of Joel Hildebrand of the University of California at

Berkeley. His work on solubility of gases, published in 1916, showed that helium was less soluble in liquids than other gases. He suggested that helium might replace nitrogen in air tanks used by deep sea divers. When divers breathe air, nitrogen dissolves in their blood and then bubbles out when they return to the surface, causing a painful condition known as the "bends." Today deep sea divers commonly breathe helium–oxygen mixtures, instead of air, so that they can avoid suffering the bends.

Consider also the work done by Gertrude Elion, who was studying some compounds called purines at Burroughs Wellcome Research Laboratories in North Carolina. Although she was doing basic research, her work led to the discovery of a number of valuable new drugs for carrying out successful organ transplants and for treating various diseases, such as gout, malaria, herpes, and cancer. Elion shared the 1988 Nobel prize in physiology and medicine, and in 1991 she became the first woman to be inducted into the National Inventors Hall of Fame.

Gertrude Elion, a basic research chemist, won the Nobel prize because of a chance discovery. [*AP/Wide World Photos.*]

> There are two compelling reasons why society must support basic science. One is substantial: The theoretical physics of yesterday is the nuclear defense of today; the obscure synthetic chemistry of yesterday is curing disease today. The other reason is cultural. The essence of our civilization is to explore and analyze the nature of man and his surroundings. As proclaimed in the Bible in the Book of Proverbs: "Where there is no vision, the people perish."
>
> Arthur Kornberg (1918—), American Biochemist;
> Nobel Prize in Physiology and Medicine, 1959

Summary

1. Chemistry is the study of *matter* and the changes it undergoes. Matter is anything that has *mass,* and the changes it undergoes may be chemical or physical.
2. Having its roots in natural philosophy and alchemy, chemistry finally became a *science* in the seventeenth century when chemists began to rely on experimentation. Much older than science is *technology,* which consists of all the processes we use to modify nature; these processes need not be based on science.
3. Francis Bacon suggested that science and technology might solve all the world's problems and bring about universal prosperity; but Rachel Carson pointed out that science and technology also can create some serious problems.
4. Observations about nature often lead to explanations called *hypotheses.* If the hypotheses can stand up to the tests of further experimentation, they may eventually become *theories.* Scientific theories have been tested many times, but they are always tentative. If a theory does not continue to stand up to repeated testing, it can be rejected. A valid theory can be used to predict new scientific facts.
5. Science and technology have provided many benefits for our world, but they have also introduced new

risks. A *risk–benefit analysis* can help us decide whether the risks outweigh the benefits.

6. Whenever matter undergoes a *physical* or *chemical change,* there is also a change in *energy.* Energy is either given off or absorbed during each process.

7. *Calories* of *heat* are used to measure the quantity of energy; the intensity of energy is measured in terms of *temperature.*

8. One way to classify matter is according to physical state: *solids, liquids,* and *gases.* Heat can convert a solid to a liquid or a liquid to a gas.

9. Another way to classify matter is according to purity: pure substances and mixtures. The pure substances are either *elements* or *compounds.*

10. There are only about a hundred elements. They are represented by *chemical symbols.* A compound contains two or more elements that are chemically combined.

11. Scientific measurements are made using *SI* metric units: the *meter* (m) for length, the *kilogram* (kg) for mass, the *liter* (L) for volume, and the *kelvin* (K) for temperature. Prefixes make the basic units larger or smaller by factors of ten.

12. *Density* is a measure of heaviness. It is the amount of mass per unit volume. The density of water is 1 g/mL.

13. Chemistry plays a central role in science because it is important to other sciences. Chemists are usually involved in some kind of research. *Applied research* has to do with efforts to make particular kinds of useful products. *Basic research* may happen to result in a useful product, but its purpose is simply to obtain new knowledge, or to answer some fundamental question.

Key Terms

alchemy 1.1
applied research 1.14
basic research 1.15
calorie 1.11
Celsius scale 1.11
chemical change 1.7
chemical properties 1.7
chemical symbol 1.9
compounds 1.9
density 1.12
elements 1.9
energy 1.7
heat 1.11
hypothesis 1.4
kelvin 1.11
kilocalorie 1.11
kilogram 1.10
kinetic energy 1.7
joule 1.11
liquid 1.8
liter 1.10
mass 1.7
matter 1.7
meter 1.10
mixture 1.9
natural philosophy 1.1
physical change 1.7
physical properties 1.7
potential energy 1.7
pure substance 1.9
risk–benefit analysis 1.6
science 1.1
scientific law 1.4
scientific models 1.4
solid 1.8
technology 1.6
temperature 1.11
theory 1.4
weight 1.7

Review Questions

1. Define chemistry.
2. What is matter?
3. Which of the following are examples of matter?
   a. iron b. air
   c. love d. the human body
   e. gasoline f. an idea
4. State three distinguishing characteristics of science. Which characteristic best serves to distinguish science from other disciplines?
5. Why were the ancient Greek philosophers, such as Aristotle, not successful as scientists?
6. What is alchemy?
7. What is natural philosophy?
8. What did Francis Bacon envision for us as a result of science?
9. What was the main theme of Rachel Carson's *Silent Spring*?

10. Why have Thomas Malthus's predictions not been fulfilled in developed countries?

11. What is a scientific hypothesis? How are hypotheses tested?

12. What is a scientific law?

13. What is a theory?

14. Why can't scientific methods always be used to solve social, political, ethical, and economic problems?

15. How does technology differ from science?

16. What is risk–benefit analysis?

17. What sort of judgments go into the evaluation of benefits?

18. What sort of judgments go into the evaluation of risks?

19. What is a desirability quotient?

20. Why is it often difficult to estimate desirability quotients?

21. Synthetic food colors make food more attractive and increase sales. Some such dyes are suspected carcinogens (cancer inducers). Who derives most of the benefits from the use of food colors? Who assumes most of the risk associated with use of these dyes?

22. Penicillin kills bacteria, thus saving the lives of thousands of people who otherwise might die of infectious diseases. Penicillin causes allergic reactions in some people; in extreme cases the allergic reaction can lead to death if the resulting condition is not treated. Do a risk–benefit analysis of the use of penicillin for society as a whole.

23. Do a risk–benefit analysis of the use of penicillin for a person who is allergic to it. (See Question 22.)

24. An artificial sweetener is 4000 times as sweet as table sugar, but exhibits possible toxic side effects. Do a risk–benefit analysis of the sweetener.

25. Nitrogen mustard is extremely toxic, but is an effective anticancer drug. Do a risk–benefit analysis of nitrogen mustard.

26. Do a risk–benefit analysis for the smoking of cigarettes.

27. Explain the difference between mass and weight.

28. Two samples are weighed under identical conditions in a laboratory. Sample A weighs 1 lb and Sample B weighs 2 lb. Does Sample B have twice the mass of Sample A?

29. What has changed when a formerly chubby person completes a successful diet, the person's weight or the person's mass?

30. Sample A, which is on the moon, has exactly the same mass as Sample B, which is on Earth. Do the two samples weigh the same? Explain your answer.

31. Distinguish between chemical and physical properties.

32. Which describes a physical change and which describes a chemical change?
    a. Sheep are sheared and the wool is spun into yarn.
    b. Silkworms feed on mulberry leaves and produce silk.

33. Which describes a physical change and which describes a chemical change?
    a. Because a lawn is watered and fertilized it grows thicker.
    b. An overgrown lawn is manicured by mowing it with a lawnmower.

34. Which describes a physical change and which describes a chemical change?
    a. Ice cubes form when a tray filled with water is placed in a freezer.
    b. Milk, which has been left outside a refrigerator for many hours, turns sour.

35. What is energy?

36. What is the difference between kinetic energy and potential energy?

37. How do pure substances and mixtures differ?

38. Every sample of the sugar glucose (collected anywhere on Earth) consists of 8 parts (by weight) oxygen, 6 parts carbon, and 1 part hydrogen. Is glucose a pure substance or a mixture?

39. Identify each of the following as a pure substance or a mixture.
    a. carbon dioxide b. oxygen
    c. smog

40. Identify each of the following as a pure substance or a mixture.
    a. gasoline b. mercury c. soup

41. Identify each of the following as a pure substance or a mixture.
    a. a carrot b. 24-karat gold

42. How do gases, liquids, and solids differ in their properties?

43. Which of the following represent elements and which represent compounds?
    a. H b. He c. HF d. Ca

44. Which of the following represent elements and which represent compounds?
a. C b. CO c. Cl d. $CaCl_2$

45. Without consulting Table 1.1, write symbols for each of the following.
a. carbon b. chlorine
c. phosphorus d. calcium
e. potassium f. plutonium

46. Without consulting Table 1.1, name each of the following.
a. H b. O c. Na
d. N e. Fe f. U

47. What are the names and symbols of the basic SI units for mass, length, and temperature? What derived units are more often used in the laboratory?

48. What is the SI derived unit for volume? What volume units are more often used in the laboratory?

49. What is applied research? Give an example.

50. What is basic research? Give an example.

Problems

Helpful hint: You must work a large proportion of these practice problems. You cannot learn to work problems by reading them or watching your teacher work them any more than you could become a good piano player solely by reading about piano skills or watching a performance.

The Metric System

51. How many millimeters and how many centimeters are there in 1 m?

52. For each of the following, indicate which is the larger unit.
a. mm or cm b. kg or g c. dL or μL

53. For each of the following, indicate which is the larger unit.
a. L or cm^3 b. cm^3 or mL

54. How many meters are there in each of the following?
a. 10 km b. 75 cm

55. How many millimeters are there in each of the following?
a. 7.5 m b. 46 cm

56. How many liters are there in each of the following?
a. 2056 mL b. 47 kL

57. Make the following conversions.
a. 45,000 mg to g b. 0.086 g to mg

58. Make the following conversions.
a. 0.149 L to mL b. 47 mL to L

59. How many milliliters are there in 1 cm^3? In 15 cm^3?

60. How many millimeters are there in 1 cm? In 1.83 m?

61. Convert the following to kelvins.
a. 25 °C b. 273 °C

62. Convert the following to degrees Celsius.
a. 301 K b. 473 K

63. How many calories are there in each of the following?
a. 0.82 kcal b. 3.55 Cal

64. Convert the following to calories.
a. 6.5 kcal b. 33.0 Cal

Unit Conversions

65. Carry out the following conversions.
a. 25 km to meters
b. 675 mm to meters
c. 8.3 kg to grams
d. 27 mL to liters
e. 289 μs to milliseconds
f. 118 mm to centimeters

66. Carry out the following conversions.
a. 0.063 km to meters
b. 72.8 mm to meters
c. 0.035 kg to grams
d. 375 mL to liters
e. 16 μs to milliseconds
f. 25 mm to centimeters

Density

67. What is the density of a salt solution if 50.0 mL of the solution has a mass of 67.0 g?

68. What is the density of a liquid if 25.0 mL of the liquid has a mass of 48.2 g?

69. Calculate the density of a 50.0-mL sample of cognac that has a mass of 47.0 g.

70. What is the density of lead if a 59.0-g piece of lead has a volume of 5.20 cm^3?

71. What is the mass of 50.0 mL of grenadine, which has a density of 1.32 g/mL?

72. What is the mass of 60.0 mL of Sambuca liqueur, which has a density of 1.06 g/mL?

73. What is the volume occupied by 80.0 g of a brandy that has a density of 0.950 g/mL?

74. What is the volume occupied by 25.0 g of ouzo, which has a density of 0.940 g/mL?

Projects

75. List five chemical activities that you have engaged in today.

76. Prepare a brief biographical report on the life of one of the following.

    a. Francis Bacon
    b. Rachel Carson
    c. Thomas Malthus
    d. George Washington Carver

References and Readings

1. Asimov, Isaac. *Asimov on Chemistry.* Garden City, NY: Anchor Books (Doubleday), 1975.
2. Asimov, Isaac. ''The Relativity of Wrong.'' *The Skeptical Inquirer,* Fall 1989, pp. 35–44. Discusses the nature of scientific theories.
3. Bronowski, J. *The Common Sense of Science.* New York: Vantage Books (Random House), 1960. Chapter 1, ''Science and Sensibility.'' Written for the nonscientist; shows the sciences and the arts to be complementary.
4. Callery, Michael L., and Helen G. Koritz. ''What Science Is and Is Not: Ten Myths,'' *Journal of College Science Teaching,* Dec. 1992–Jan. 1993, pp. 154–157.
5. Carson, Rachel. *Silent Spring.* Boston: Houghton Mifflin, 1962. The classic book on dangers to the environment.
6. Ihde, A. J. *The Development of Modern Chemistry.* New York: Harper and Row, 1964. Chapter 1.
7. Kauffman, George B., and H. Harry Szmant. *The Central Science: Essays on the Uses of Chemistry.* Fort Worth: Texas Christian University Press, 1984.
8. Lett, James. ''A Field Guide to Critical Thinking.'' *The Skeptical Inquirer,* Winter 1990, pp. 153–160. Six rules for evaluation of claims.
9. Marston, Charles R. ''Catalysis: The Legacy of Alchemy.'' *Today's Chemist,* October 1988, pp. 12–15.
10. MIT Commission on Productivity. ''Chemicals!'' *ChemTech,* April 1990, pp. 224–229. Contributions of the U.S. chemical industry to the economy.
11. National Science Foundation. ''How Basic Research Reaps Unexpected Rewards.'' Washington, DC: U.S. Government Printing Office, February 1980.
12. Pacey, Arnold. *Technology in World Civilization: A Thousand-Year History.* Cambridge, MA: MIT Press, 1990.
13. Pimentel, George B., and Janice A. Coonrod. *Opportunities in Chemistry: Today and Tomorrow.* Washington, DC: National Academy Press, 1987.
14. Pauling, Linus C. ''Chemistry and the World of Tomorrow.'' *Chemical and Engineering News,* 16 April 1984, pp. 54–56.
15. Rothman, Milton. ''Myths About Science . . . and Belief in the Paranormal.'' *The Skeptical Inquirer,* Fall 1989, pp. 25–34.
16. Segal, Marian. ''Determining Risk.'' *FDA Consumer,* June 1990, pp. 7–11.
17. Webb, Michael J. ''Physical and Chemical Change: What's the Difference?'' *The Science Teacher,* March 1982, pp. 39–40.
18. Wilson, Richard, and E. A. C. Crouch. ''Risk Assessment and Comparisons: An Introduction.'' *Science,* 17 April 1987, pp. 267–270. Risk assessment is the focus of several articles in this issue.
19. Zeckhauser, Richard J., and W. Kip Viscusi. ''Risk Within Reason.'' *Science,* 4 May 1990, pp. 559–564.

2 Atoms

Are They for Real?

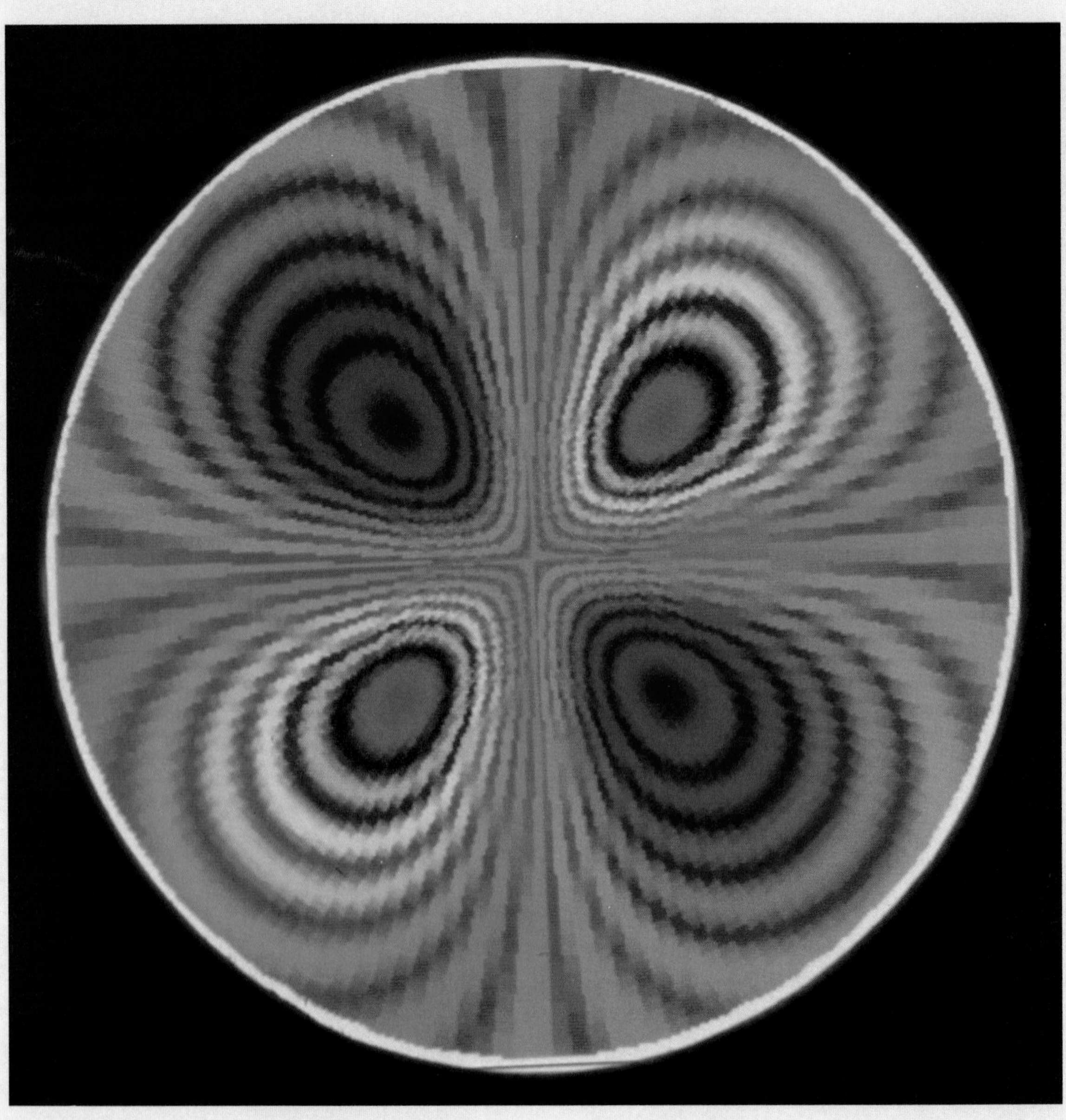

Though single atoms are too small
For you to see or feel,
There isn't any question that
These particles are real.

An atom is the smallest characteristic particle of an element. There are about 90 elements that occur in nature, and as far as we can tell, the entire universe is made up of those same elements. From these 90 elements millions of different chemical compounds are possible. More than 10 million compounds have already been identified by chemists around the world.

Each element is composed of atoms that are alike in a fundamental way. Just how atoms of a given element are alike—and how they differ from atoms of any other element—is the subject of this chapter and of Chapter 3.

Although the concept of atoms has been around for more than 2000 years, most people doubted their existence until about 200 years ago, and many refused to accept the idea until well into the twentieth century. Since atoms are so extremely small that they cannot be seen even with a microscope, it was hard for people to believe that they actually existed.

Just how small are atoms? Well, they are so tiny that we cannot really comprehend anything so minuscule. If you can imagine the atoms in a single penny being blown up until they were barely visible, like tiny grains of sand, then there would be enough "sand" to cover the whole state of Texas with a layer several inches deep! Comparing the size of an atom to a tiny grain of sand is like comparing a penny to a sandbox as big as the state of Texas!

If they are so insignificant, then who cares about atoms? You certainly should! You and your entire world are made of atoms. You may not be able to see them, but they are there just the same, and they are constantly affecting whatever you are doing. Everything is made of atoms, including you yourself.

One important thing to remember is that atoms are eternal, for all practical purposes. We can't create more atoms, and we can't get rid of the ones we already have. When we dispose of unwanted waste materials, we can hide them by burying them or throwing them into the sea, but the atoms are still there. Even when we burn waste materials, we are just combining their atoms with other kinds of atoms so that the materials are changed, but the atoms really don't disappear. Keep that in mind when you think about waste disposal.

The existence of atoms is well established, but our images of them are based on indirect evidence. [*Dr. Fred Espenak/Science Photo Library/Photo Researchers, Inc.*]

Democritus appeared on a 1983 postage stamp from Greece. [*Professor Marvin Lang and Mr. Gary Shulfer, University of Wisconsin–Stevens Point (Scott Cat. Greece, 1469).*]

SECTION 2.1

Atoms: The Greek Idea

A pool of water can be separated into drops, and then each drop can be split into smaller and smaller drops. Suppose you could keep splitting these drops into still smaller ones even after they became much too small to see. Would you ever reach the point at which the tiny drop could no longer be separated into smaller droplets of water?

Leucippus, the Greek philosopher, and his pupil Democritus might well have discussed this question as they strolled along the beach of the Aegean Sea back in the fifth century B.C. Based only on his intuition, Leucippus felt that there must be ultimate tiny particles of water that could not be subdivided. After all, from a distance the sand on the beach looked continuous, but closer inspection showed it to be made up of tiny grains.

Democritus (460–370 B.C.) expanded on this idea of Leucippus and gave the tiny particles a name. He described them as "atomos" (indivisible). We still refer to the tiny unit particles of elements as **atoms.** Democritus believed that each kind of atom was distinct in shape and size (Figure 2.1). Real substances were thought to be mixtures of various kinds of atoms. The Greeks at that time believed that there were four basic elements: earth, air, fire, and water. The relationships among these elements and the four "principles"—hot, moist, dry, and cold—are shown in Figure 2.2.

Five centuries later the Roman poet Lucretius wrote his long poem "On the Nature of Things," in which he gave strong arguments for the atomic nature of matter. But a few centuries earlier Aristotle had declared that matter was continuous, not atomistic, and the ancient Greeks and Romans had no way to determine which of these two views of matter was correct. The ancients almost never used experimentation, preferring instead to use reason and logic. The continuous view of matter seemed more logical and reasonable to most of them, and so they accepted the view of Aristotle, which prevailed for 2000 years, even though it was wrong.

Figure 2.1 Democritus imagined that "atoms" of water might be smooth round balls and that "atoms" of fire could have sharp edges.

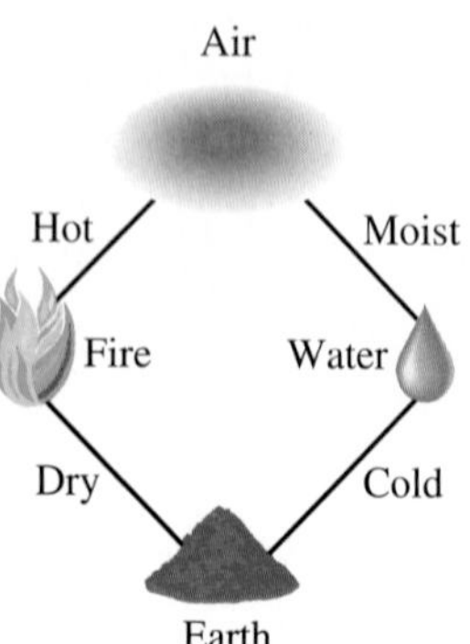

Figure 2.2 The Greek view of matter was that there were only four elements connected by four "principles."

SECTION 2.2

Lavoisier: The Law of Conservation of Mass

Antoine Lavoisier and his wife, painted by Jacques Louis David in 1788 (detail). [*The Metropolitan Museum of Art, Purchase, Mr. and Mrs. Charles Wrightsman Gift, 1977 (1977.10).*]

By about A.D. 1700, scientists were observing more carefully and measuring more accurately. Antoine Laurent Lavoisier, a Frenchman (1743–1794), perhaps did more than anyone else to establish chemistry as a quantitative science. He found that when a chemical reaction was carried out in a closed system, the total mass of the system was not changed. Perhaps the most important chemical reaction that Lavoisier performed was the decomposition of the red oxide of mercury to form metallic mercury and a gas he named oxygen. The reaction had been carried out before—by Karl Wilhelm Scheele, a Swedish apothecary (1742–1786), and by Joseph Priestley, a Unitarian minister who later fled England and settled in America—but Lavoisier was the first to weigh all the substances present before and after the reaction. He was also the first to interpret the reaction correctly.

Lavoisier carried out many quantitative experiments. He found that when coal was burned it united with oxygen to form carbon dioxide. He also experimented with animals. When a guinea pig breathed, oxygen was consumed and carbon dioxide was formed. Lavoisier therefore concluded that respiration was related to combustion. In each of these reactions he found that matter was conserved.

Lavoisier summarized his findings in a scientific law. The **law of conservation of mass** states that matter is neither created nor destroyed during a chemical change. The total mass of the reaction products is always equal to the total mass of the reactants (starting materials).

Lavoisier, a tax official for King Louis XVI, was beheaded on the guillotine by the revolutionary forces when the King was deposed and replaced by a republican government.

Scientists had by this time abandoned the Greek idea of the four elements and were almost universally using Robert Boyle's operational definition, put forth over a century before. Boyle, an Englishman, in his book *The Sceptical Chymist* (published in 1661), said that supposed *elements* must be tested to see if they really were simple. If a substance could be broken down into simpler substances, it was not an element. The simpler substances might be elements and would be so regarded until such time (if it ever came) as they in turn could be broken down into still simpler substances. On the other hand, two or more elements might combine to form a complex substance, called a *compound.*

Using Boyle's definition, Lavoisier included a table of elements in his book *Elementary Treatise on Chemistry.* The table included some substances we now know to be compounds. Lavoisier was the first to use systematic names for the chemical elements. He is often called the "father of modern chemistry," and his book is usually regarded as the first chemistry textbook.

The law of conservation of mass is the basis for many chemical calcula-

Robert Boyle. [*Courtesy of The Smithsonian Institution, Washington DC (#46834-a).*] ▶

tions. We can calculate the mass of a product that can be made from a given mass of reactant (starting material) or the mass of reactant that must be used to yield a certain mass of product (Chapter 6). The law is not just a matter of academic interest. It states that we cannot create materials from nothing; we therefore deduce that we can make new materials only by changing the way atoms are combined. Nor can we get rid of wastes by the destruction of matter. We must put the wastes somewhere. Chemistry offers alternatives, however. Through chemical reactions, we can change some kinds of potentially hazardous wastes to less harmful forms. Such transformations of matter from one form to another are what chemistry is all about.

SECTION 2.3

Proust: The Law of Definite Proportions

Proust, like Lavoisier, was a member of the French nobility. He was working in Spain, temporarily safe from the ravages of the French revolution. His laboratory was destroyed and he was reduced to poverty, however, when the French troops of Napoleon Bonaparte occupied Madrid in 1808.

By the end of the eighteenth century, Lavoisier and other scientists noted that many substances were composed of two or more elements. Each compound had the same elements in the same proportions, regardless of where the compound came from or who prepared it. The painstaking work of Joseph Louis Proust (1754–1826) convinced most chemists of the general validity of these observations. In one set of experiments, for example, Proust found that copper carbonate, whether prepared in the laboratory or obtained from natural sources, always was composed by mass of 5.3 parts of copper, 1.0 part of carbon, and 4.0 parts of oxygen (Figure 2.3). To summarize these and many other experiments, Proust in 1799 formulated a new scientific law. A compound, he said, always contains elements in certain definite proportions and in no other combinations. This generalization he called the **law of definite proportions**. (It is also sometimes called the *law of constant composition.*)

One of the earliest illustrations of the law of definite proportions is found in

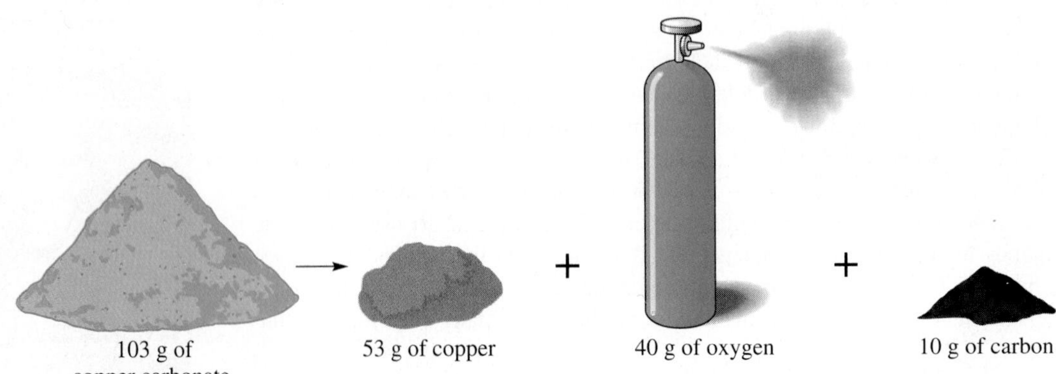

Figure 2.3 Whether synthesized in the laboratory or obtained from various natural sources, copper carbonate always has the same composition. Analysis of this compound led Proust to formulate the law of definite proportions.

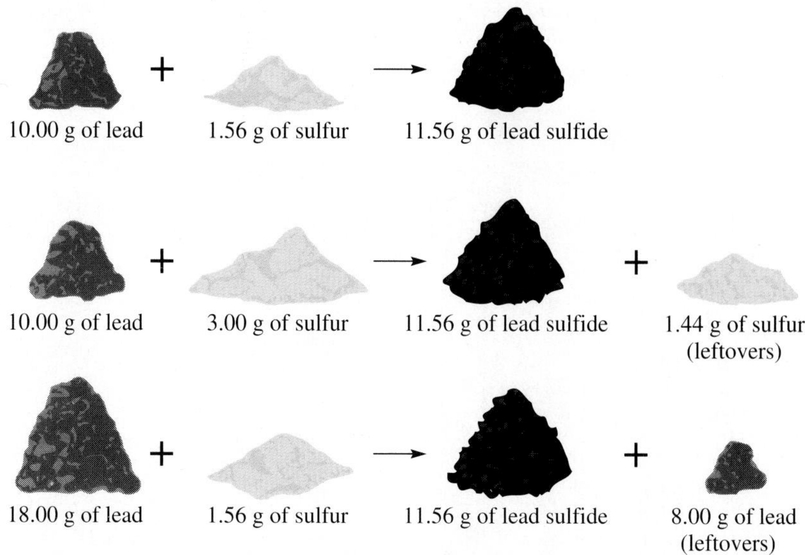

Figure 2.4 Berzelius's experiment illustrating the law of definite proportions.

the work of the noted Swedish chemist J. J. Berzelius (1779–1848). Berzelius heated 10.00 g of lead with various amounts of sulfur to form lead sulfide. Lead is a soft, grayish metal. Sulfur is a yellow solid. Lead sulfide is a shiny, black solid. It was easy, therefore, to tell when all the lead had reacted. Excess sulfur was washed away with carbon disulfide, a solvent that dissolves sulfur but not lead sulfide. As long as he used at least 1.56 g of sulfur with 10.00 g of lead, Berzelius got exactly 11.56 g of lead sulfide. Any sulfur in excess of 1.56 g was left over, unreacted. If he used more than 10.00 g of lead with 1.56 g of sulfur, he got 11.56 g of lead sulfide, with lead left over. These reactions are illustrated in Figure 2.4. (An explanation is given later in Figure 2.6.)

Jöns Jakob Berzelius was the first person to prepare a list of atomic weights. Published in 1828, it agrees remarkably well with most of our accepted values today. [*Professor Marvin Lang and Mr. Gary Shulfer, University of Wisconsin–Stevens Point (Scott Cat. Sweden 293, portrait 295).*]

The law of definite proportions is further illustrated by the electrolysis of water. In 1783, a wealthy and eccentric Englishman, Henry Cavendish (1731–1810), found that water is produced when hydrogen burns in oxygen. (It was Lavoisier, however, who correctly interpreted Cavendish's experiment and who first used the modern names for hydrogen and oxygen.) In 1800 two English chemists, William Nicholson and Anthony Carlisle, decomposed water into hydrogen gas and oxygen gas by passing an electric current through the water (Figure 2.5). The two gases are always produced in a 2:1 volume ratio. By mass, 1.00 g of water is always composed of 0.11 g of hydrogen and 0.89 g of oxygen.

Nicolson and Carlisle performed their experiment only 6 weeks after the Italian scientist Alessandro Volta (1745–1827) invented the chemical battery. This scientific breakthrough led to rapid developments in chemistry and provided the death blow to the old Greek idea of water as an element.

The law of definite proportions is the basis for chemical formulas (Chapter 6). It also has a wider meaning. Not only does a compound have a constant composition, it has constant properties. Pure water always dissolves salt and

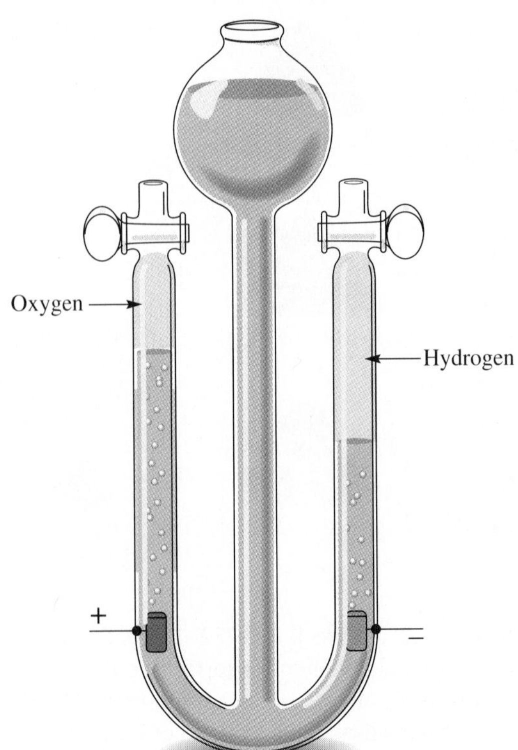

Figure 2.5 Electrolysis of water. Hydrogen and oxygen are always produced in a volume ratio of 2:1.

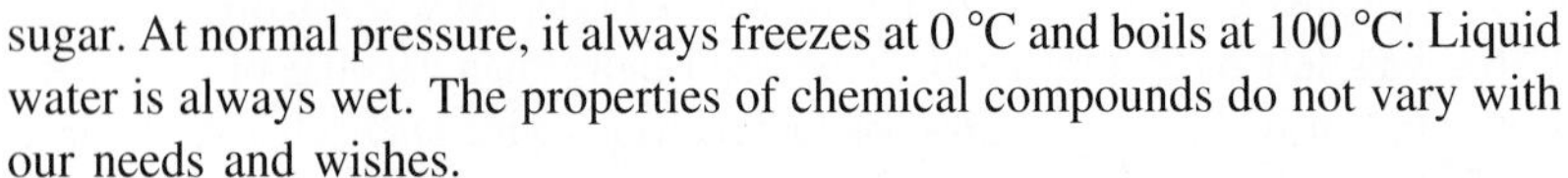
sugar. At normal pressure, it always freezes at 0 °C and boils at 100 °C. Liquid water is always wet. The properties of chemical compounds do not vary with our needs and wishes.

John Dalton (1766–1884). [*The Bettmann Archive.*]

SECTION 2.4

John Dalton and the Atomic Theory of Matter

Lavoisier's law of conservation of mass and Proust's law of definite proportions were repeatedly verified by experiment. This led to attempts to develop theories to explain these laws.

In 1803 John Dalton, an English schoolteacher, proposed a model to explain the accumulating experimental data. As he developed his model, he uncovered another law that his theory would have to explain. Proust had stated that a compound contains elements in certain proportions and only those proportions. Dalton's new law, called the **law of multiple proportions**, stated that elements might combine in *more* than one set of proportions, with each set corresponding to a different compound. For example, carbon combines with oxygen in a mass ratio of 3.0 to 8.0 to form carbon dioxide, a gas familiar as a product of respiration and of the burning of coal and wood. But Dalton found that carbon also combines with oxygen in a mass ratio of 3.0 to 4.0 to form carbon monoxide, a poisonous gas formed when a fuel is burned in limited air.

Dalton then proposed his **atomic theory**, a model that he used to explain the various laws. The most important points to Dalton's atomic theory are:

1. All matter is composed of extremely small particles called *atoms*. [Dalton assumed atoms to be indivisible. That isn't quite so, as we see in the case of radioactivity (Section 3.3).]
2. All atoms of a given element are alike, but atoms of one element differ from the atoms of any other element. (Dalton assumed that all the atoms of a given element have the same mass. That too is now known to be incorrect, as we see in Section 3.6.)
3. Compounds are formed when atoms of different elements combine in fixed proportions. (The numbers of each kind of atom in a compound usually form a simple ratio. For example, the ratio of carbon atoms to oxygen atoms is 1:1 in carbon monoxide and 1:2 in carbon dioxide.)
4. A chemical reaction involves a *rearrangement* of atoms. No atoms are created or destroyed or broken apart in a chemical reaction.

Explanations Using Atomic Theory

Dalton's theory clearly explains the difference between elements and compounds. *Elements* are composed of only one kind of atom. (Exactly what we mean by "kind" is explained in Section 3.6.) *Compounds* are made up of two or more kinds of atoms chemically combined in definite proportions.

To explain the law of definite proportions, Dalton's reasoning went something like this. Why should 1.0 g of hydrogen always combine with 19.0 g of fluorine? Why shouldn't 1.0 g of hydrogen also combine with 18.9 g of fluorine? Or 19.1 g of fluorine? Or any other mass of fluorine? If an atom of fluorine has a mass 19 times that of a hydrogen atom, then the compound formed by the union of one atom of each element would have to consist of 1 part by mass of hydrogen and 19 parts by mass of fluorine. Matter must be atomic for the law of definite proportions to be valid. Figure 2.6 shows Berzelius's experiment interpreted in terms of Dalton's atomic theory.

The atomic theory also explains the law of multiple proportions. For example, 1.000 g of carbon combines with 1.332 g of oxygen to form carbon monoxide and with 2.664 g of oxygen to form carbon dioxide. Carbon dioxide has twice the mass of oxygen per gram of carbon as does carbon monoxide. This is explained in that one atom of carbon combines with *one* atom of oxygen to form carbon monoxide and one atom of carbon combines with *two* atoms of

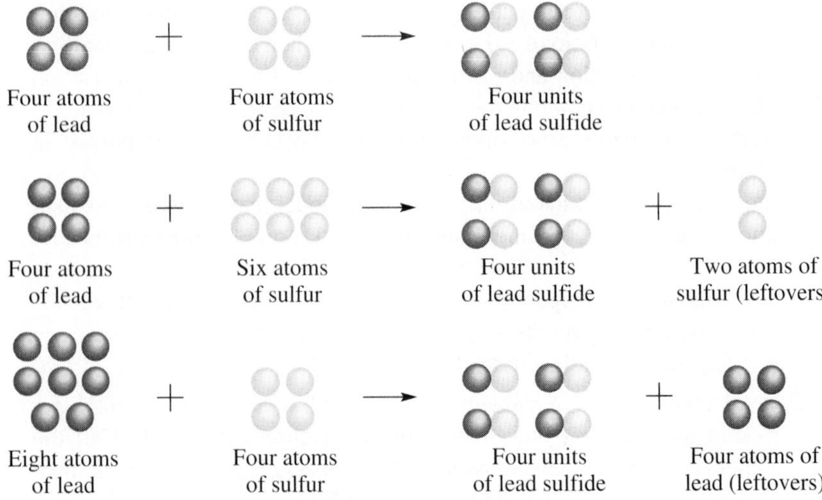

Figure 2.6 The law of definite proportions: Berzelius's experiment interpreted in terms of Dalton's atomic theory.

oxygen to form carbon dioxide. Table 2.1 shows some of the proportions in which nitrogen and oxygen combine.

Finally, the law of conservation of mass is also explained by atomic theory. When carbon atoms combine with oxygen atoms to form carbon dioxide, the atoms are merely rearranged (Figure 2.7). Matter is neither lost nor gained; the mass does not change.

Dalton set up a table of relative (atomic) masses as a part of his theory. In this table—based on hydrogen as 1—oxygen had a mass of 8, carbon 6. Thus, carbon monoxide was made up of one atom of carbon combined with one atom of oxygen to give a mass ratio of 6 parts of carbon to 8 of oxygen (or 3:4). Carbon dioxide was made up of one atom of carbon combined with two atoms of oxygen to give a mass ratio of 6 parts of carbon to 16 parts of oxygen (or 3:8). Many of Dalton's atomic masses were incorrect, as we might expect with the equipment available at that time. Dalton also invented a set of sym-

Table 2.1 The Law of Multiple Proportions

Compound	Representation*	Mass of N per 1.000 g of O	Ratio of the masses of N†
Nitrous oxide	●●●	1.750 g	(1.750 ÷ 0.4375) = 4
Nitric oxide	●●	0.8750 g	(0.8750 ÷ 0.4375) = 2
Nitrogen dioxide	●●●	0.4375 g	(0.4375 ÷ 0.4375) = 1

*● = nitrogen atom and ● = oxygen atom.

†To obtain the "ratio of masses of N," divide each quantity in the third column by the smallest (0.4375 g).

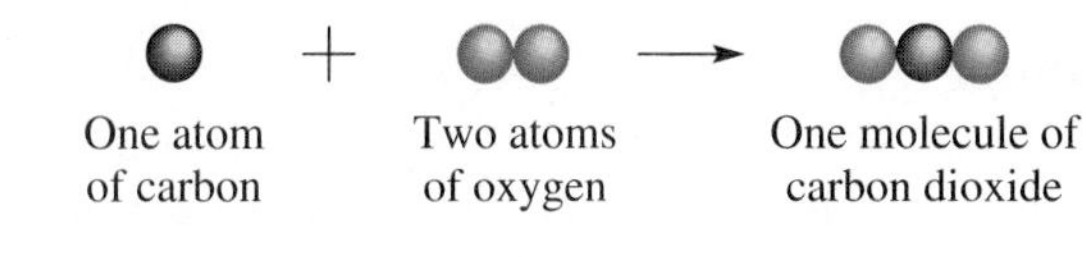

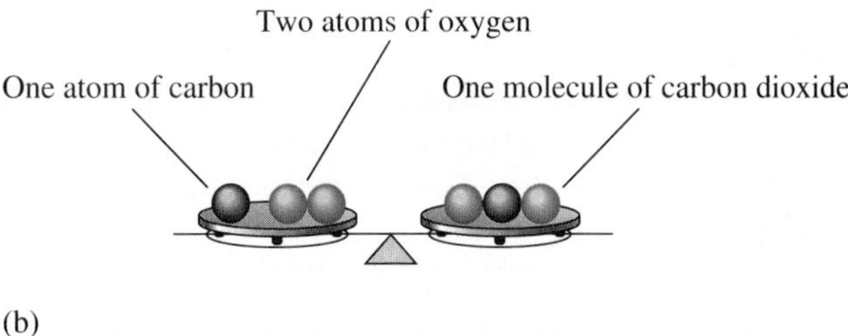

Figure 2.7 The law of conservation of mass. When one atom of carbon reacts with two atoms of oxygen to form carbon dioxide (a), the atoms are merely rearranged (not created or destroyed). Mass is therefore conserved (b).

bols to represent the different kinds of atoms. In fact, symbols similar to Dalton's are used in Table 2.1 and Figure 2.7. Those symbols have been replaced by modern symbols of one or two letters (see Table 1.1, Section 1.9).

We can use proportions, such as those determined by Dalton, to calculate the amount of one substance needed to combine with a given amount of another substance. To learn how to do this, let's look at some examples.

▶ EXAMPLE 2.1

Carbon combines with hydrogen in a ratio of 3 parts by mass of carbon to 1 part by mass of hydrogen to form a gas called methane. How much hydrogen is needed to combine with 900 g of carbon to form methane?

SOLUTION

$$900\ \cancel{\text{g carbon}} \times \frac{1\ \text{g hydrogen}}{3\ \cancel{\text{g carbon}}} = 300\ \text{g hydrogen}$$

Exercise 2.1

Nitrous oxide, sometimes called laughing gas, can be decomposed to give 7 parts by mass of nitrogen and 4 parts by mass of oxygen. What mass of nitrogen is obtained if enough nitrous oxide is decomposed to give 36 g of oxygen?

▶ EXAMPLE 2.2

Hydrogen sulfide gas can be decomposed to give sulfur and hydrogen in a mass ratio of 16:1. If the relative mass of sulfur is 32 when hydrogen is taken to be 1, how many hydrogen atoms are combined with each sulfur atom in the gas?

SOLUTION

$$\frac{32\ \cancel{\text{units of sulfur}}}{1\ \text{atom of sulfur}} \times \frac{1\ \cancel{\text{unit of hydrogen}}}{16\ \cancel{\text{units of sulfur}}} \times \frac{1\ \text{atom of hydrogen}}{1\ \cancel{\text{unit of hydrogen}}} = \frac{2\ \text{atoms of hydrogen}}{1\ \text{atom of sulfur}}$$

Exercise 2.2

Methane gas can be decomposed to give carbon and hydrogen in a mass ratio of 3 : 1. If the relative mass of carbon is 12 when hydrogen is taken to be 1, how many hydrogen atoms are combined with each carbon atom in the gas?

Despite its inaccuracies, Dalton's atomic theory was a great success. Why? Because it served—and still serves—to explain a large body of experimental data. It also successfully predicts how matter will behave under a wide variety of circumstances. Dalton arrived at his atomic theory purely on the basis of reasoning power, and with modest modification it has stood the test of time and the assault of modern, highly sophisticated instrumentation. Formulation of so successful a theory was quite a triumph for a Quaker schoolteacher in the year 1803!

SECTION 2.5

Atoms: Real and Relevant

Are atoms real? Certainly they are real as a concept, and a highly useful concept at that. Scientists can even observe computer-enhanced *images* of individual atoms. These portraits reveal little detail of the atoms, but they provide powerful (though still indirect) evidence that atoms exist.

Are atoms relevant? Much of modern science and technology—including the production of new materials and the technology of pollution control—is ultimately based on the concept of atoms. We have even built monuments to atoms. The 1958 World's Fair in Brussels, Belgium, featured a large structure called the Atomium, consisting of nine large spheres representing atoms in a cubic crystal of iron (Figure 2.8). Each "atom" is 18 meters in diameter and houses an exhibit that demonstrates some kind of progress in science and technology (except for the top sphere, which is a restaurant). The "bonds" connecting the "atoms" are giant tubes containing escalators, so that visitors can move easily from one atom to another.

We have seen that atoms are conserved in chemical reactions. Thus, material things—things made of atoms—can be recycled, for the atoms are not destroyed no matter how we use them. The one way we might lose a material from a practical standpoint is to spread the atoms so thinly that it would take too much time and energy to put them back together again.

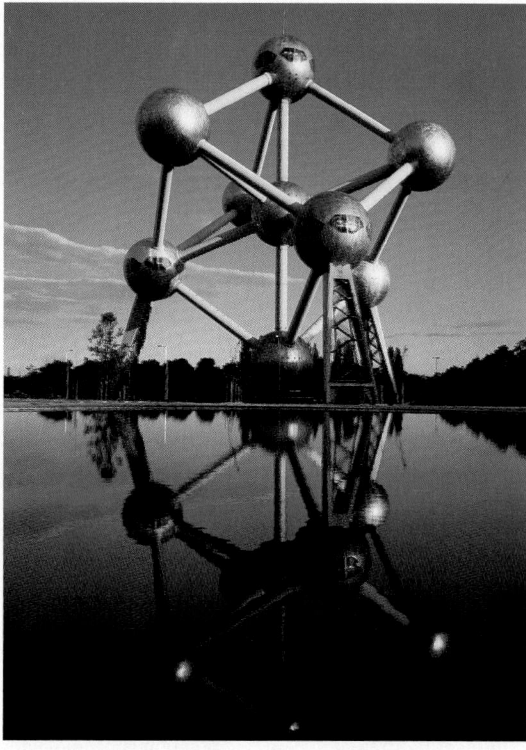

Figure 2.8 The Atomium in Brussels was built for the 1958 World's Fair. The spheres represent atoms in an iron crystal. [*Travelpix/ FPG International.*]

Consider iron atoms in a sample of hematite, an iron ore. The hematite might be converted into pig iron and then into steel and made into an automobile. After the automobile is worn out, the steel could be recovered and used again in a new automobile. Thus, the atoms could be changed from one combination to another, but in each conversion they would be conserved.

But suppose at one stage or another the iron was dissolved in sulfuric acid. The iron sulfate formed would be soluble in water. If someone poured the iron sulfate down the drain and it eventually wound up in the ocean, the iron atoms would still be there, but they would spread so thinly in the vast waters of the ocean that there would be no practical way to recover them. We may conclude, then, that atoms can be recycled—provided we do not spread them too thinly.

SECTION 2.6

Leucippus Revisited: Molecules

Now back to Leucippus's musings by the seashore. We now know that if we keep dividing those drops of water into smaller drops, we will ultimately obtain a small particle—called a **molecule**—that is still water. If we divide that particle still further, we will obtain two *atoms* of hydrogen and one *atom* of oxygen. And if we divide those . . . but that is a story for another time.

A **molecule** is a group of atoms that are chemically bonded together. Molecules are represented by chemical formulas. The symbol H represents an *atom* of hydrogen; the formula H_2 represents a *molecule* of hydrogen, which is composed of two hydrogen atoms. The formula H_2O represents a molecule of water, which is composed of two hydrogen atoms and one oxygen atom.

Dalton regarded the atom as indivisible, as did his successors up until the discovery of radioactivity in 1895. We examine the changing concept of the atom in the next chapter.

Summary

1. An *atom* is the smallest unit particle of an element. Although the concept of atoms was first suggested by Leucippus and Democritus in ancient Greece, the *atomic theory* was not proposed until 1803 by John Dalton.

2. The *law of conservation of mass* resulted from careful experiments of Lavoisier and others, who weighed all the reactants and all the products for a number of chemical reactions and found that no change in mass occurred.

3. The *law of definite proportions* (or the law of constant composition) was formulated by Proust, based on the work of Berzelius plus his own experiments. A given compound always contains the same elements in exactly the same proportions.

4. The *law of multiple proportions* states that the same group of elements might combine in several different sets of proportions, each set corresponding to a different compound. John Dalton discovered this law while working on his atomic theory.

5. The *atomic theory* states that
   (a) All matter is made up of tiny particles called atoms.
   (b) All atoms of the same element are alike.
   (c) Compounds are formed when atoms of different elements combine in certain proportions.
   (d) Atoms are only rearranged during chemical reactions, not destroyed or broken apart.

6. Since atoms are conserved in chemical reactions, matter (which is made of atoms) can be recycled. But if we hope to recycle a particular kind of matter, we should take care not to let it spread too thinly throughout nature.

7. Just as atoms are the smallest unit particles of elements, *molecules* are the smallest unit particles of compounds. A molecule is a group of atoms chemically bonded together.

Key Terms

atomic theory 2.4
atom 2.1
law of conservation of mass 2.2
law of definite proportions 2.3
law of multiple proportions 2.4
molecule 2.6

Review Questions

1. What is the distinction between the atomistic view and the continuous view of matter?

2. Why did the theory that matter was continuous (rather than atomic) prevail for so long?

3. What discoveries finally refuted the theory that matter is continuous?

4. What is Democritus's contribution to atomic theory?

5. If foods were described as atomistic or continuous, which designation would you use for each of the following?
   a. milk
   b. mashed potatoes
   c. peas
   d. olives
   e. hard-boiled eggs
   f. scrambled eggs

6. Describe Lavoisier's contribution to the development of modern chemistry.

7. State the law of conservation of mass.

8. How does the ancient Greek definition of an element differ from the modern one?

9. How did Robert Boyle define an element?

10. A balloon filled with helium floats near the ceiling. After several days, the balloon is deflated and on the floor. Have the helium atoms been destroyed? If so, how? If not, where are they?

11. A copper coin is dissolved in a solution of nitric acid. Have the copper atoms been destroyed? If so, how? If not, where are they?

12. Sugar consists of carbon, hydrogen, and oxygen atoms. When a sugar cube is burned in a crucible, nothing remains in the vessel. Have the carbon, hydrogen, and oxygen atoms of the sugar been destroyed? If so, how? If not, where are they?

13. Aspirin consists of carbon, hydrogen, and oxygen atoms. A student in the organic chemistry laboratory dissolves some aspirin in hot water in order to purify it by recrystallization. No crystals form upon cooling, so the student pours out the solution and starts the experiment over. Have the carbon, hydrogen, and oxygen atoms of the aspirin been destroyed? If so, how? If not, where are they?

14. Shakerag City has to come up with a plan to solve its solid waste problem. The solid wastes consist of many different kinds of materials, and the materials are comprised of many different kinds of atoms. The options for disposal include burying them in a landfill, incineration, and dumping at sea. Which method, if any, will get rid of the atoms that make up the waste? Which method, if any, will change the chemical form of the waste?

15. Polychlorinated biphenyls (PCBs), frequently found as environmental contaminants, are composed of carbon, hydrogen, and chlorine atoms. A news report of a new method for disposal of PCBs describes the process as converting the compounds completely to carbon dioxide (made of carbon and oxygen atoms) and water (composed of hydrogen and oxygen atoms). The oxygen atoms can come from atmospheric oxygen. Examine the report in the light of the law of conservation of mass.

16. When we burn a 10-kg piece of wood, only 0.05 kg of ash remains. Explain this apparent contradiction of the law of conservation of mass.

17. State the law of definite proportions.

18. State the law of multiple proportions.

19. When 3.00 g of carbon is burned in 8.00 g of oxygen, 11.00 g of carbon dioxide is formed. What mass of carbon dioxide is formed when 3.00 g of carbon is burned in 50.00 g of oxygen? What law does this illustrate?

20. Heptane always is composed of 84.0% carbon and 16.0% hydrogen. What law does this illustrate?

21. Outline the main points of Dalton's atomic theory.

22. Sulfur and oxygen form two compounds. One is 50.0% sulfur and 50.0% oxygen; the other is 40.0% sulfur and 60.0% oxygen. What law does this illustrate?

23. A photographic flash bulb weighing 0.750 g contains magnesium and air. The flash produces magnesium oxide. After cooling the bulb weighs 0.750 g. What law does this illustrate?

24. Consider the following set of compounds. What principle does the group illustrate?

N_2O NO NO_2 N_2O_4

25. Consider the following set of compounds. What principle does the group illustrate?

FeO Fe_2O_3 Fe_3O_4

26. An atom of calcium has a mass of 40 atomic mass units (amu) and an atom of vanadium has a mass of 50 amu. Are these findings in agreement with Dalton's atomic theory. Explain your answer.

27. An atom of calcium has a mass of 40 amu and an atom of potassium has a mass of 40 amu. Are these findings in agreement with Dalton's atomic theory? Explain your answer.

28. An atom of calcium has a mass of 40 amu and another atom of calcium has a mass of 44 amu. Are these findings in agreement with Dalton's atomic theory? Explain your answer.

29. Use Dalton's atomic theory to explain the law of conservation of mass. Give an example that illustrates the law.

30. Use Dalton's atomic theory to explain the law of definite proportions. Give an example that illustrates the law.

31. Use Dalton's atomic theory to explain the law of multiple proportions. Give an example that illustrates the law.

32. The ancient Greeks thought that water was an element. In 1800 Nicholson and Carlisle decomposed water into hydrogen and oxygen. What did their experiment prove?

33. What did each of the following contribute to the development of modern chemistry?
a. J. J. Berzelius b. Henry Cavendish
c. Joseph Proust

34. Jan Baptista van Helmont, a Flemish alchemist (1579–1644), performed an experiment in which he planted a young willow tree in a weighed bucket of soil. After five years he found that the tree had gained 75 kg, yet the soil had lost only 0.057 kg. He had added only water to the system, so he concluded that the substance of the tree had come from water. Criticize his conclusion.

Problems

Chemical Compounds

35. A colorless liquid is thought to be a pure compound. Analyses of three samples of the material gives the following results.

	Mass of sample	Mass of carbon	Mass of hydrogen
Sample 1	1.000 g	0.862 g	0.164 g
Sample 2	1.549 g	1.335 g	0.254 g
Sample 3	0.988 g	0.852 g	0.162 g

Could the material be a pure compound?

36. A blue solid is thought to be a pure compound. Analysis of three samples of the material gives the following results.

	Mass of sample	Mass of carbon	Mass of hydrogen
Sample 1	1.000 g	0.937 g	0.0629 g
Sample 2	0.244 g	0.229 g	0.0153 g
Sample 3	0.100 g	0.094 g	0.0063 g

Could the material be a pure compound?

37. Roger heats 1.0000 g of zinc powder with 0.2000 g of sulfur. He reports that he obtains 0.0608 g of zinc sulfide and recovers 0.0592 g of unreacted zinc. Show by calculation whether or not his results obey the law of conservation of mass.

38. Karl heats 0.5585 g of iron with 0.3550 g of sulfur. He reports that he obtains 0.8792 g of iron(II) sulfide and recovers 0.0433 g of unreacted sulfur. Show by calculation whether or not his results obey the law of conservation of mass.

Definite Proportions

39. When 18.0 g of water is decomposed by electrolysis, 16.0 g of oxygen and 2.0 g of hydrogen are formed. According to the law of definite proportions, how much hydrogen is formed by the electrolysis of 360 g of water?

40. Hydrogen, from the decomposition of water, has been promoted as the fuel of the future (Chapter 14). How much water would have to be electrolyzed to produce 100 kg of hydrogen? (See Problem 39.)

41. In plentiful air, 3.0 parts of carbon react with 8.0 part of oxygen to produce carbon dioxide. Use this mass ratio to calculate how much carbon is required to produce 990 g of carbon dioxide.

42. In limited air, 3.0 parts of carbon react with 4.0 parts of oxygen to produce carbon monoxide. Use this mass ratio to calculate how much carbon monoxide can be formed from 42 g of carbon.

43. The gas silane can be decomposed to yield silicon and hydrogen in a ratio of 7 parts by mass of silicon to 1 part by mass of hydrogen. If the relative mass of silicon atoms is 28 when the mass of hydrogen atoms is taken to be 1, how many hydrogen atoms are combined with each silicon atom?

References and Readings

1. Asimov, Isaac. *A Short History of Chemistry*. Garden City, NY: Doubleday, 1965. Chapters 1, 5, 12, and 13.

2. Brock, William H. *The Norton History of Chemistry*. New York: Norton, 1993.

3. Cole K. C. ''On Imagining the Unseeable.'' *Discover,* December 1982, pp. 70–72.

4. Jaffe, Bernard. *Crucibles: The Story of Chemistry*. New York: Fawcett World Library, 1957.

5. Kolb, Doris. ''Chemical Principles Revisited: But If

Atoms Are So Tiny. . . .'' *Journal of Chemical Education,* September 1977, pp. 543–547.

6. Kolb, Doris. ''Chemical Principles Revisited: What Is an Element?'' *Journal of Chemical Education,* November 1977, pp. 696–700.

7. Lloyd, G. E. R. *Methods and Problems in Greek Science.* New York: Cambridge University Press, 1991. Troubles with the approach of ancient Greek philosophers to questions of science.

8. Patterson, Elizabeth C. *John Dalton and the Atomic Theory.* New York: Doubleday, 1970. The story of Dalton and his formulation of the atomic theory.

9. Ritchie-Calder, Lord. ''The Lunar Society of Birmingham.'' *Scientific American,* June 1982, pp. 136–145. John Watt, Ben Franklin, Erasmus Darwin, Josiah Wedgwood, and Joseph Priestley were members of the Lunar Society.

10. Salzberg, Hugh W. *From Caveman to Chemist: Circumstances and Achievements.* Washington, DC: American Chemical Society, 1991. A history of chemistry from ancient times up to 1900.

11. Stillman, John Maxon. *The Story of Alchemy and Early Chemistry.* New York: Dover Publications, 1960. Chapter 14, ''Lavoisier and the Chemical Revolution.''

12. Waterman, Edward L. ''An Annotated Asimov: A Resource Bibliography for Chemistry Teachers.'' *Journal of Chemical Education,* October 1981, pp. 826–827. Contains references to Asimov's writings relevant to an introductory chemistry course.

13. Young, Louise B. (Ed.). *The Mystery of Matter.* New York: Oxford University Press, 1965. Contains excerpts from the original works of Lucretius, Dalton, and others.

3 Atomic Structure

Images of the Invisible

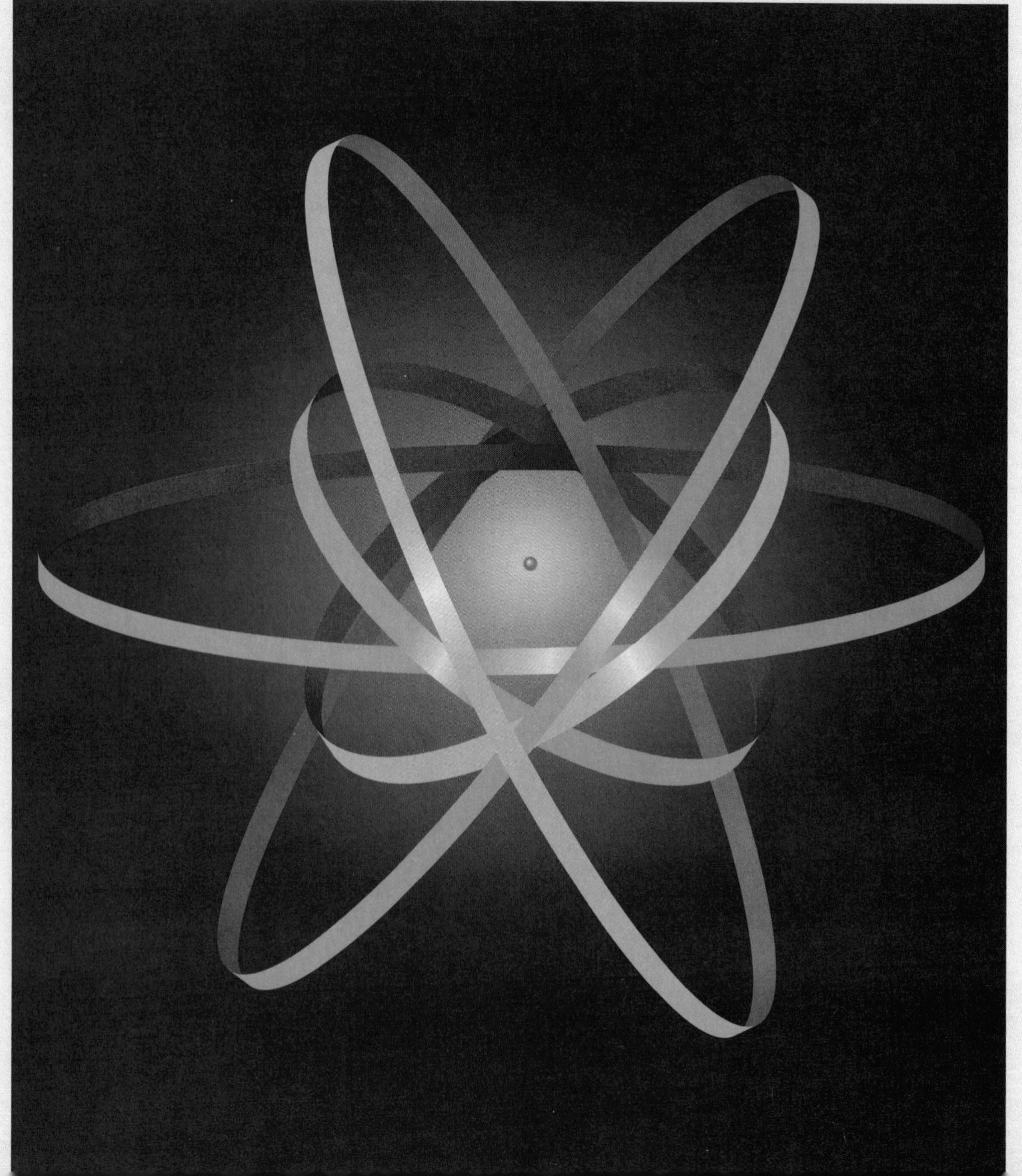

Electrons whirling swiftly

About a tiny core—

Thus do we view the atom

Since Rutherford and Bohr.

If atoms are much too small to see, even with the most powerful optical microscope, then how is it possible for us to study their structure? It turns out that we have learned quite a lot about the structure of the atom by gathering information indirectly. Using a number of clever experiments and exercising their powers of deduction, scientists have managed to construct a picture of the atom in amazing detail, even though they have never actually looked at an atom.

It is not quite true to say that no one has ever seen an atom. In 1970 Albert Crewe at the University of Chicago used an electron microscope to make the photograph shown in Figure 3.1. The seven spots are images of individual uranium atoms. Similar photographs have been taken by others. More recently scientists have used scanning tunneling microscopy (STM) to get pictures

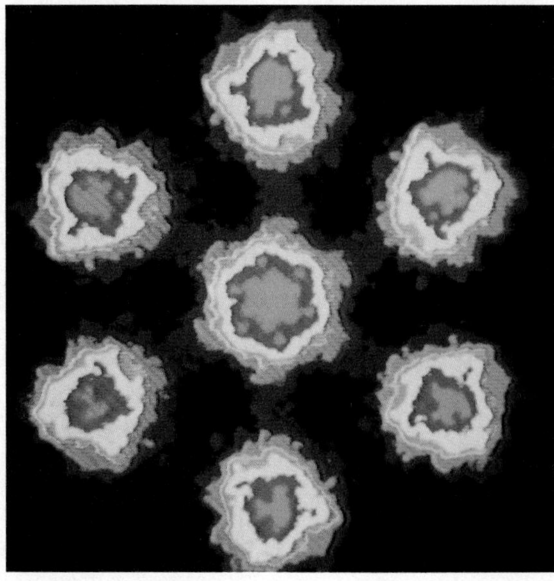

Figure 3.1 Shown here are images of seven individual uranium atoms (the colored spots with red-orange centers). The images are made with an electron microscope. The atoms are in the form of a compound called uranyl acetate and are pictured on an extremely thin carbon layer that appears black in this photograph. The atoms are 0.34 nm apart. [*Courtesy of M. Isaacson, Cornell University, and M. Ohtsuki, The University of Chicago.*]

The simple Bohr–Rutherford model of the atom is just a tiny, dense nucleus with orbiting electrons.

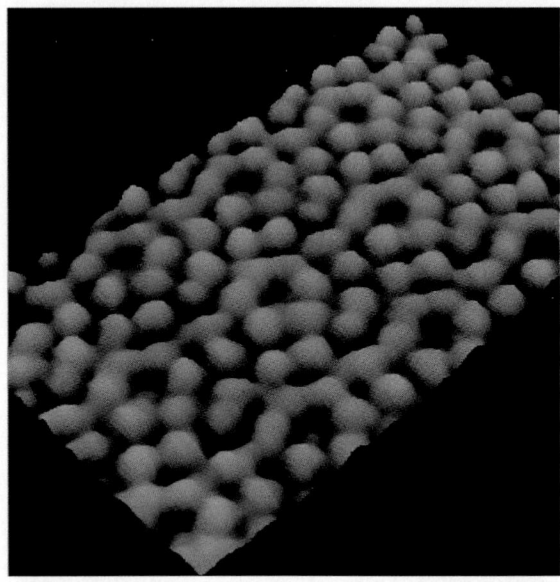

Figure 3.2 Image of the surface of silicon, produced by a scanning tunneling microscope. The blue spots are individual silicon atoms, which are arranged in a regular pattern that repeats itself across the surface. Images such as this aid our understanding of the surface structure of many different materials; silicon, in particular, is a material of vital significance to the semiconductor and computer industries. [*Courtesy of IBM Thomas J. Watson Research Center, NY.*]

such as the one in Figure 3.2. This shows the pattern of atoms lying along the surface of a silicon crystal. We can see outlines of atoms in such photographs, and we can tell quite a bit about how the atoms are arranged, but these pictures tell us nothing about the inner structure of the atom. No one has ever seen *inside* an atom.

Why do we care about the structure of atoms? Because it is the arrangement of the parts of atoms that determines the properties of different kinds of matter. Only by understanding atomic structure can we know how atoms combine to make the many different substances in nature and, even more important, how we can modify materials to meet our needs more precisely. A knowledge of atomic structure is even essential to your health. Many medical diagnoses are based on the analysis of body fluids such as blood and urine. Many such analyses depend on knowledge of how the structure of atoms is changed when energy is absorbed.

Perhaps of greater immediate interest to you is the fact that your understanding of chemistry (as well as much of biology and other sciences) will depend at least in part on your knowledge of atomic structure. Let's start our study of atomic structure by going back to the time of John Dalton.

SECTION 3.1

Electricity and the Atom

Dalton, who set forth his atomic theory in 1803, regarded the atom as hard and indivisible. It wasn't long, however, before evidence accumulated to show that matter is electrical in nature. Indeed, the electrolytic decomposition of water by Nicholson and Carlisle in 1800 (Chapter 2) had already indicated as much. Electricity played an important role in the unraveling of the structure of the atom.

Static electricity has been known since ancient times, but continuous electric current was born with the nineteenth century. Alessandro Volta invented what is now known as the voltaic pile in 1800. This device is an electrochemical cell, much like a modern battery. If the poles of a pile are connected by a wire, current flows through the wire. The current is sustained by chemical reactions inside the pile. Volta's invention soon was applied in many areas of science and everyday life.

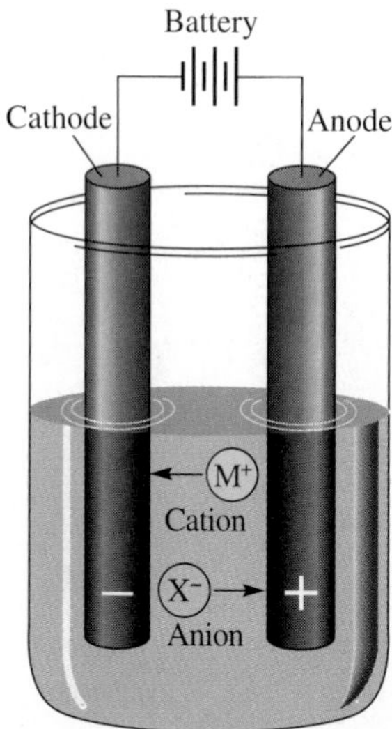

Figure 3.3 Electrolysis apparatus.

Electrolysis

Soon after Volta's invention, Humphry Davy (1778–1829), a British chemist, built a powerful battery that he used to pass electricity through molten salts. Davy quickly discovered several new elements. In 1807 he liberated highly reactive potassium metal from molten (melted) potassium hydroxide. Shortly thereafter he produced sodium metal by passing electricity through molten sodium hydroxide. Within a year Davy had also produced magnesium, strontium, barium, and calcium metals for the first time. The science of *electrochemistry* was born.

Davy's protégé Michael Faraday (1791–1867) greatly extended this new science and defined many of the terms we still use today. **Electrolysis** is the splitting of compounds by electricity (Figure 3.3). An **electrolyte** is a compound that conducts electricity when melted or dissolved in water. **Electrodes** are carbon rods or metal strips that are inserted into a molten compound or solution to carry the electric current. In electrolysis (see Figure 3.3), the *anode* is the electrode that bears a positive charge, and the *cathode* is the electrode that is negatively charged. Faraday made the hypothesis that the electric current is carried through the melted compound or solution by charged atoms—later named **ions**. An **anion** is an ion with a negative charge; anions travel toward the anode. A **cation** is a positively charged ion; cations move toward the cathode.

Some **ions** are simply electrically charged atoms. Other ions consist of a group of atoms that act as a unit (see Section 6.3).

Faraday's electrochemical work established that atoms are electrical in nature, but further detail of the structure of atoms had to wait several decades for the development of gas discharge tubes and for more powerful sources of electrical voltage. Actually, Faraday himself tried and failed to pass electricity through a tube that had part of the air pumped out. His vacuum was not good enough for the voltage that he had available.

Electrochemistry, which includes the study of electrochemical cells and electrolysis, is discussed in detail in Chapter 8.

Cathode Ray Tubes

By 1875 tubes with a better vacuum than Faraday could achieve were available. William Crookes (1832–1919), an English chemist, passed an electric current through such a tube containing air at low pressure (Figure 3.4). His experiment can be repeated today. The tube has metal electrodes sealed in it. It is connected to a vacuum pump and most of the air is removed. A beam of

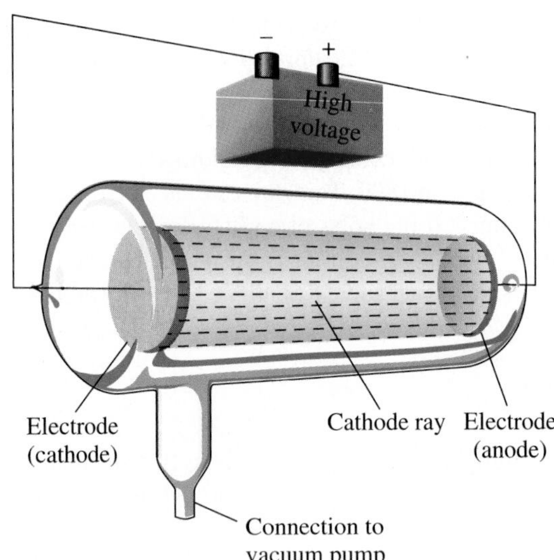

Figure 3.4 A simple gas discharge tube.

Cathode rays travel in straight lines in the absence of an external applied field.

current is seen as a green fluorescence, which is observed when the beam strikes a screen coated with zinc sulfide. This beam seems to leave the cathode and travel to the anode (Figure 3.5). The beam is called a **cathode ray**.

Thomson's Experiment: Mass-to-Charge Ratio

Considerable speculation arose as to the nature of cathode rays. In general, British scientists believed the rays to be beams of particles. Most German scientists held that the rays were more likely to consist of a form of energy much like visible light. The answer came, as scientific answers should, from an experiment performed by the English physicist Joseph John Thomson (1856–1940) in 1897. Thomson showed that the cathode rays were deflected in an electric field (see Figure 3.5). The beam was attracted to the positive plate and repelled by the negative plate. Thomson therefore concluded that cathode rays consist of negatively charged particles. His experiments also showed that the particles were the same regardless of the materials from which the electrodes were made or the type of gas in the tube. He concluded that these negative particles are constituents of all kinds of atoms. The name **electron** was given to these units of negative charge. Cathode rays, then, are beams of electrons.

The greater the charge on a particle, the more it is deflected in an electric (or magnetic) field. (The attraction or repulsion is greater the greater the charge.) The greater the mass of a particle, the less it is deflected by a force. (Consider two spheres coming at you; it is much easier to deflect a Ping-Pong ball than a cannonball.)

Cathode rays are deflected in magnetic fields as well as in electric fields. By measuring the amount of deflection in fields of known strength, Thomson was able to calculate the ratio of the mass of the electron to its charge. He could not measure either the mass or the charge separately. That is like knowing that you weigh 25 lb for each foot of height, but not knowing either your weight or height. Once either your weight or height is determined, however, it is easy to calculate the other from the known value and the 25 lb/ft ratio. Thomson was awarded the Nobel prize in physics in 1906.

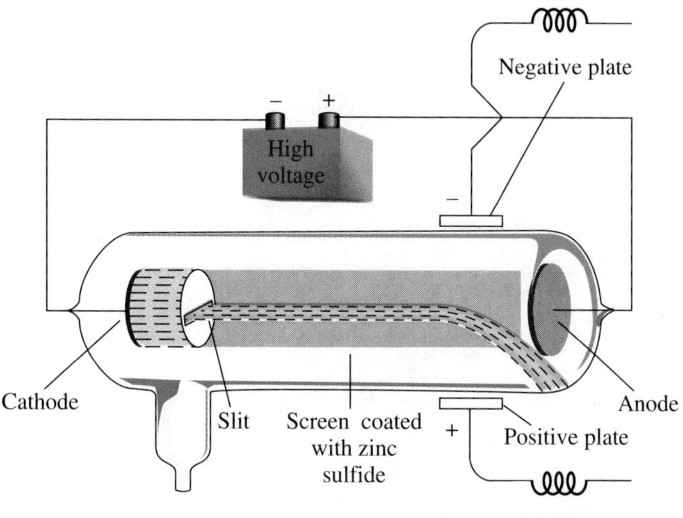

Figure 3.5 Thomson's apparatus showing deflection of cathode rays (a beam of electrons). Cathode rays are themselves invisible but are observed through the green fluorescence they produce when they strike a zinc sulfide–coated screen. The diagram (a) shows deflection of the beam in an electric field. The photograph (b) shows the deflection in a magnetic field. The magnetic field is created by the magnet to the right of and slightly behind the screen. [*(b) Carey B. Van Loon.*]

Goldstein's Experiment: Positive Particles

In 1886 a German scientist named Eugen Goldstein performed experiments with gas discharge tubes that had perforated cathodes (Figure 3.6). He found that although electrons were formed and sped off toward the anode as usual, positive ions were also formed, which shot in the opposite direction toward the cathode. Some of these positive particles went through the holes in the cathode. In 1907 a study of the deflection of these particles in a magnetic field indicated that they were of varying mass. The lightest particles, formed when there was a little hydrogen gas in the tube, were later shown to have a mass 1837 times that of an electron.

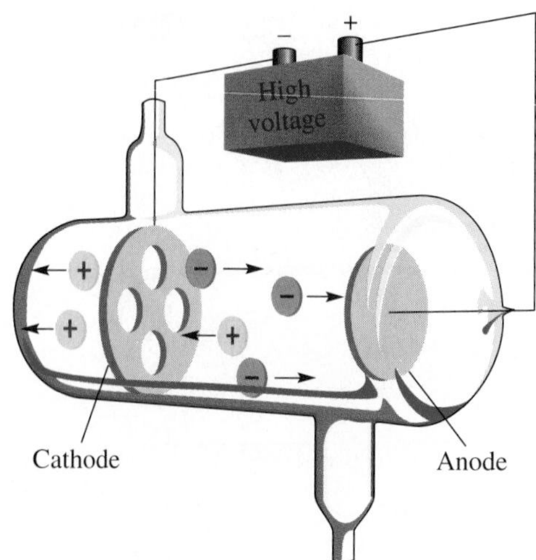

Figure 3.6 Goldstein's apparatus for the study of positive particles.

Millikan's Oil-Drop Experiment: Electron Charge

For many purposes the electron is considered to be the unit of negative charge. That charge is shown as a superior minus (which means one minus). In indicating charges on ions, for example, we use $^{-}$ to indicate a net charge of one electron, $^{2-}$ to indicate a charge of two electrons, and so on.

Thomson measured the mass-to-charge ratio for an electron, but he could not measure the mass or the charge. The charge on the electron was not determined until 1909. An American physicist, Robert A. Millikan, experimented with electrically charged oil drops in an electric field. A diagram of Millikan's apparatus is shown in Figure 3.7. A spray bottle is used to inject tiny droplets of oil. Some of the oil droplets pass into a chamber where they can be viewed through a microscope. Some of the droplets acquire negative charges by picking up electrons from the friction involved as the particles rub against the opening of the spray nozzle and against each other. (The charge is static electricity, just like the charge you get by walking across a nylon carpet.) The negative droplets could acquire one or more extra electrons by this process. Charges can also be produced by irradiation with X-rays.

The mass of the droplets can be determined by the rate at which they fall in air under the influence of gravity. As they fall between the charged plates,

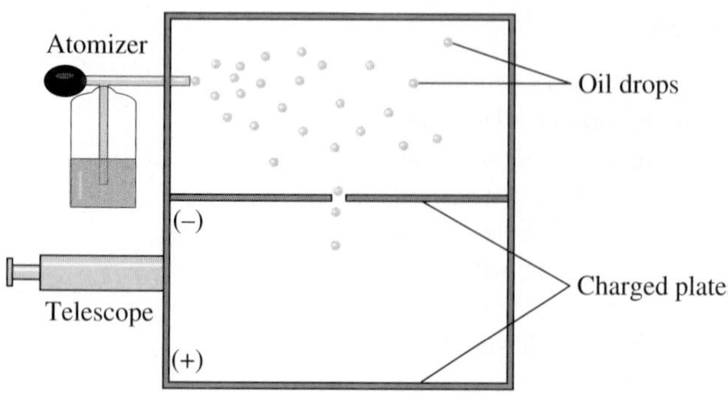

Figure 3.7 The Millikan oil drop experiment. Oil drops irradiated with X-rays pick up electrons and become negatively charged. Their fall due to gravity can be balanced by adjusting the voltage of the electric field. From the applied voltage and the mass of the oil drop, the charge on the oil drop can be determined. Each drop carries a charge corresponding to some whole number of electrons.

those that bear a negative charge are attracted upward toward the positive plate. The voltage differential between the plates can be adjusted to cause the particle to remain in place, with the upward attraction of the positive plate just balanced by the downward pull of gravity. From the voltage necessary to establish this stalemate, the mass-to-charge ratio of the particle can be calculated. Since the mass has already been determined from the free fall of the droplet, its charge can be calculated. The smallest possible difference in charge between two droplets was taken to be the charge of an individual electron. Millikan was awarded the Nobel prize in physics in 1923.

From Millikan's value for the charge and Thomson's value for the mass-to-charge ratio, the mass of the electron was readily calculated. Electrons are extremely light particles, having a mass of only 9.1×10^{-28} g. (If you do not understand numbers like 9.1×10^{-28}, you can find out about them in Appendix B.)

SECTION 3.2

Out of Chaos: The Periodic Table

Before moving on, let's take a look at a remarkable parallel development. New elements were being discovered with surprising frequency. By 1830, there were 55 known elements, all with different properties and with no apparent order in these properties. John Dalton had set up a table of relative atomic masses in his book *A New System of Chemical Philosophy* in 1808. Many of Dalton's values were wrong. They were improved in subsequent years, notably by Berzelius, who published a table of atomic weights in 1828 that contained 54 elements. Except for three cases, Berzelius's values are in quite good agreement with modern values.

Relative Atomic Weights

Please note that in the 1800s it was impossible to determine actual masses of atoms. Chemists were able, however, to determine relative atomic weights by the amounts of the various elements that combined with a given mass of another element. Dalton's atomic weights were based on an atomic mass of 1 for hydrogen. As more accurate atomic weights were determined, this standard was replaced by one in which oxygen was assigned a value of 16.0000. The oxygen standard survived until 1961 when it was replaced by a slightly more logical one based on an isotope of carbon. Adoption of this new standard caused little change in the atomic weights. These relative atomic weights are usually expressed in **atomic mass units** (amu).

Relative atomic masses were determined by comparison to a standard mass, a technique called *weighing*. For that historical reason, many chemists refer to the relative masses as *atomic weights*. The term *atomic mass* is then used to refer to the mass of an individual atom.

Mendeleev's Periodic Table

Several attempts were made to arrange the elements in some sort of systematic fashion. The first successful arrangement, and one that soon became widely

◀ Dmitri Mendeleev, the Russian chemist who invented the periodic table of the elements, continues to be honored in his native land. [*Sovfoto and Professor Marvin Lang and Mr. Gary Shulfer, University of Wisconsin–Stevens Point (Scott Cat. Russia, #3607).*]

accepted by chemists, was published in 1869 by Dmitri Ivanovich Mendeleev (1834–1907), a Russian chemist. Mendeleev's **periodic table** arranged the elements primarily in order of increasing atomic weight, although in a few cases he placed a slightly heavier element before a lighter one in order to get elements with similar chemical properties in the same column. For example, he placed tellurium, with an atomic weight of 127.6 amu, ahead of iodine, which has an atomic weight of 126.9 amu. He did this in order to get tellurium into the same column as sulfur and selenium, which it resembles in chemical properties. His rearrangement also put iodine into the same column as chlorine and bromine, which it resembles.

Mendeleev left gaps in his table. This was also necessary in order to get elements into groups with similar properties. Instead of looking upon these blank spaces as defects, he boldly predicted the existence of elements yet undiscovered. Further, he even predicted the properties of some of those missing elements.

In subsequent years, many of the gaps were filled in by the discovery of new elements. The properties often were quite close to those predicted by Mendeleev. It was the predictive value of this great innovation that led to wide acceptance of Mendeleev's table.

The modern periodic table (inside front cover) contains over 100 elements.

SECTION 3.3

Serendipity in Science: X-rays and Radioactivity

Serendipity is an aptitude for making fortunate discoveries by accident. It is so called because of Horace Walpole's *The Three Princes of Serendip,* who made many such discoveries.

Let's return now to the structure of the atom and look at a little scientific serendipity. Often scientific discoveries are described as happy accidents. Have you ever wondered why these accidents always seem to happen to scientists? It is probably because scientists are trained observers. The same accident could happen right before the eyes of an untrained person and go unnoticed. Or, if it were noticed, its significance might not be grasped.

Roentgen: The Discovery of X-rays

Two such happy accidents occurred in the last years of the nineteenth century. In 1895, a German scientist, Wilhelm Conrad Roentgen, was working in a dark room, studying the glow produced in certain substances by cathode rays. To his surprise, he noted this glow on a chemically treated piece of paper some distance from the cathode ray tube. The paper even glowed when taken

Figure 3.8 An early example of the use of X-rays in medicine. Professor Michael Purpin of Columbia University made this X-ray in 1896 to aid in the removal by surgery of gunshot pellets (the dark spots) from the hand of a patient. [*Courtesy of Burndy Library, Norwalk CT.*]

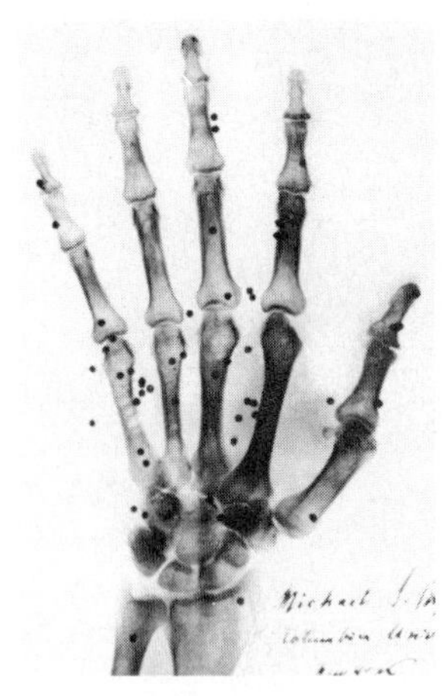

to the next room. Roentgen had discovered a new type of ray that could travel through walls! The rays were given off from the anode whenever the cathode ray tube was operating. With seeming lack of imagination, Roentgen called the mysterious, penetrating rays **X-rays**.

The Discovery of Radioactivity

Certain chemicals exhibit *fluorescence*; when exposed to strong sunlight, they continue to glow even when taken into a dark room. In 1895, Antoine Henri Becquerel (1852–1908), a French physicist, was studying fluorescence by wrapping photographic film in black paper, placing a few crystals of the fluorescing chemical on top of the paper, and then placing the paper in strong sunlight. If the glow was like ordinary light, it would not pass through the paper. On the other hand, if it was similar to X-rays, it would pass through the black paper and fog the film.

Before Becquerel learned much more about fluorescence, he made an important accidental discovery. He was working with a uranium compound. When placed in sunlight (which caused the compound to fluoresce), the film was fogged. On several cloudy days when exposures to sunlight were not possible, he prepared samples and placed them in a drawer. To his great surprise, the photographic film was fogged even though the uranium compound had not been exposed to sunlight. Further experiments showed that the radiation coming from the uranium compound had nothing to do with fluorescence, but was a characteristic of the element uranium.

Other scientists immediately began to study this new radiation. Marie Sklodowska Curie gave it a name: radioactivity. **Radioactivity** is the spontaneous emission of radiation from certain unstable elements. Marie Sklodowska was born in Poland in 1867. As a young woman, she went to Paris to work for her doctor's degree in mathematics and physics. There she met and married Pierre Curie (b. 1859), a French physicist of some note. Marie and Pierre discovered the radioactive elements radium and polonium. Pierre Curie was killed in a traffic accident in 1906, three years after he, Marie, and Becquerel shared the Nobel prize in physics. Marie Curie continued to work with radioactive substances, winning the Nobel prize for chemistry in 1911. For more than 50 years she was the only person ever to have received two Nobel prizes. She died in 1934 of leukemia, probably brought on by long exposure to the radiation from the compounds with which she worked.

Marie Sklodowska Curie in her laboratory and Marie and Pierre Curie on a French postage stamp. [*Culver Pictures and Professor Marvin Lang and Mr. Gary Shulfer, University of Wisconsin–Stevens Point (Scott Cat. France #376).*]

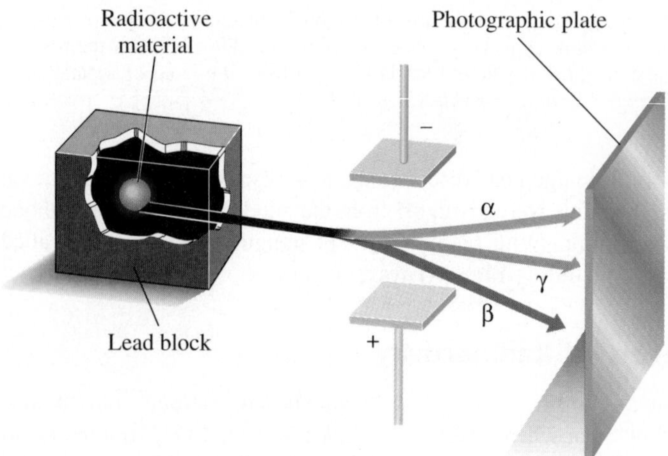

Figure 3.9 Behavior of radioactive rays in an electric field.

SECTION 3.4

Three Types of Radioactivity

Scientists soon showed that three types of radiation emanated from various radioactive elements. When passed through a strong magnetic or electric field, the first type, called *alpha rays* were deflected in a manner that indicated that the rays were beams of positive particles (Figure 3.9). Later experiments showed that the **alpha particles** have a mass four times that of a hydrogen atom and a charge twice the size of but opposite in sign to that of an electron.

Alpha particles are identical to helium ions, that is, helium atoms with both electrons removed (He^{2+}).

The second type of radiation, called *beta rays,* was shown to be composed of negatively charged particles identical with those of cathode rays. These **beta particles** are therefore electrons.

The third type of radiation, called *gamma rays,* is not deflected by a magnetic field. **Gamma rays** are a form of energy, much like the X-rays used in medical work, but even more penetrating. The three types of radioactivity are summarized in Table 3.1.

The discoveries of the late nineteenth century paved the way for an entirely new picture of the atom. The new concept developed rapidly during the early years of the twentieth century.

Table 3.1 Types of Radioactivity

Name	Greek Letter	Mass (atomic mass units)	Charge
Alpha	α	4	2+
Beta	β	$\frac{1}{1837}$	1−
Gamma	γ	0	0

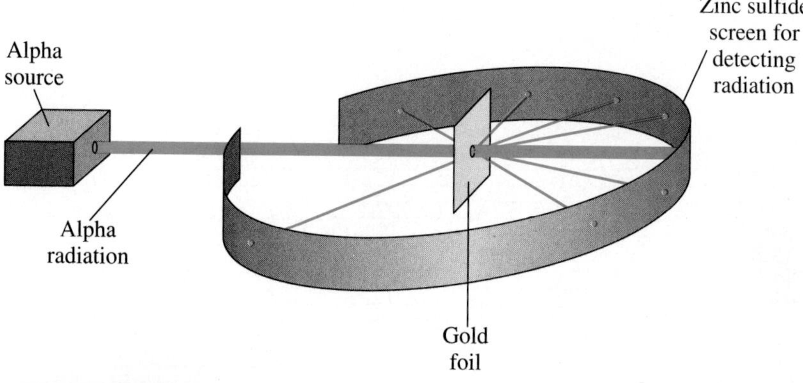

Figure 3.10 Rutherford's gold-foil experiment. Most alpha rays passed right through the gold foil, but now and then a ray was deflected.

SECTION 3.5

Rutherford's Experiment: The Nuclear Model of the Atom

Ernest Rutherford (1871–1937), a New Zealander who spent his career in Canada and Great Britain, chose the names alpha, beta, and gamma for the three types of radiation. At Rutherford's suggestion, two of his coworkers used alpha particles in an experiment that indicated a surprising structure for atoms. Hans Geiger (1882–1945), a German physicist, and an English undergraduate student, Ernest Marsden (1889–1970), carried out experiments in which very thin foils of metal were bombarded with alpha particles from a radioactive source (Figure 3.10). One target was a thin piece of gold foil. Most of the particles went right through the foil undeflected or deflected only slightly. A few particles were deflected sharply, however, and once in a while one would be sent right back in the direction from which it came. Rutherford had expected the alpha bullets to go right through the foil with little or no scattering. He assumed the positive charge to be spread smoothly over all the space occupied by the atom. Obviously it was not. To explain the experiment, Rutherford concluded that all the positive charge and nearly all the mass of an atom are concentrated at the center of the atom in a tiny core called the **nucleus**.

When an alpha particle, which is positively charged, approached the positively charged nucleus, it was strongly repelled and therefore sharply deflected (Figure 3.11). Since only a few alpha particles were deflected, Rutherford concluded that the nucleus must occupy only a tiny fraction of the volume of an atom. Most of the particles passed right through because most of an atom is empty space. The space outside the nucleus isn't completely empty, however. It is here that Rutherford placed the negatively charged electrons. He concluded that the electrons had so little mass that they were no match for the alpha bullets. It would be analogous to a mouse trying to stop the charge of a bull elephant.

Rutherford's nuclear theory of the atom, set forth in 1911, was a revolutionary idea. He postulated that all the positive charge and nearly all the mass

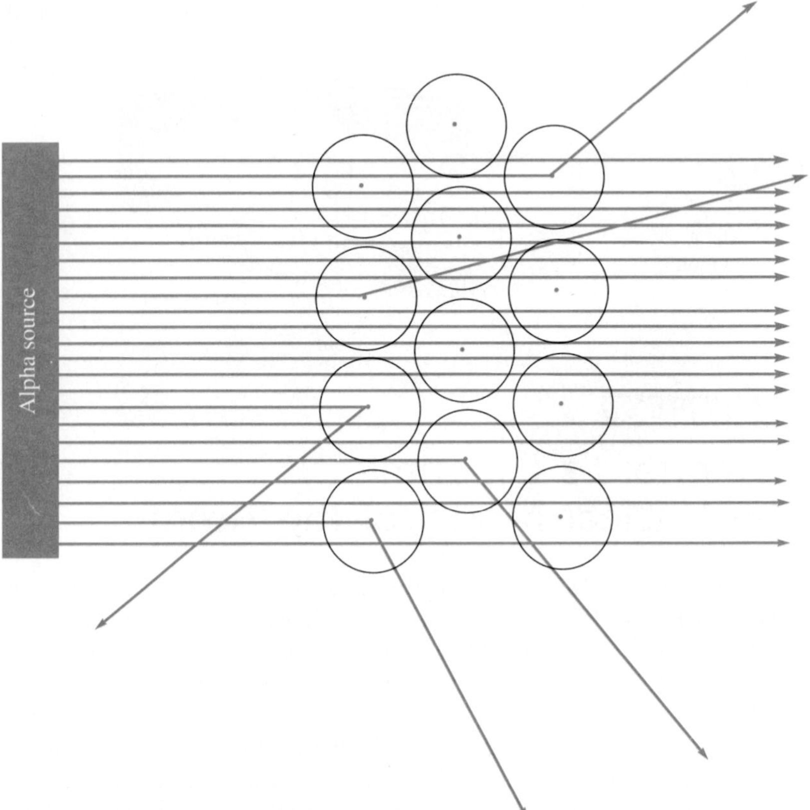

Figure 3.11 Model explaining the results of Rutherford's gold-foil experiment.

of an atom are concentrated in a tiny, tiny nucleus. The negatively charged electrons have almost no mass, yet they occupy nearly all the volume of an atom. To picture Rutherford's model, visualize a sphere as big as a giant indoor football stadium. The nucleus at the middle of the sphere is as small as a pea but weighs several million tons. A few flies flitting here and there throughout the sphere represent the electrons.

SECTION 3.6

The Structure of the Nucleus

In 1914, Rutherford suggested that the smallest positive-ray particle (i.e., that which is formed when there is hydrogen gas in the Goldstein apparatus—see Section 3.1) is the unit of positive charge in the nucleus. This particle, called a **proton**, has a charge equal in magnitude to that of the electron and has nearly

Table 3.2 Subatomic Particles				
Particle	Symbol	Mass (amu)	Charge	Location in Atom
Proton	p^+	1	1+	Nucleus
Neutron	n	1	0	Nucleus
Electron	e^-	$\frac{1}{1837}$	1−	Outside nucleus

the same mass as the hydrogen atom. Rutherford's suggestion was that protons constitute the positively charged matter in all atoms. The hydrogen atom's nucleus consisted of one proton, and the nuclei of larger atoms contained a number of protons. Except for hydrogen atoms, though, atomic nuclei were found to be heavier than would be indicated by the number of positive charges (number of protons). For example, the helium nucleus was found to have a charge of 2+ (and, therefore, it contained two protons, according to Rutherford's theory), but its mass was four times that of hydrogen. This excess mass puzzled scientists until 1932, when the English physicist James Chadwick discovered a particle with about the same mass as a proton, but with *no* electrical charge. This particle was called a **neutron**, and its existence made possible an explanation of the unexpectedly high mass of the helium nucleus. Whereas the hydrogen nucleus contained only one proton of 1 amu, the helium nucleus contained not only two protons (2 amu) but also two neutrons (2 amu), giving a total mass to the nucleus of 4 amu.

With the discovery of the neutron, the list of "building blocks" we will need for "constructing" atoms is complete. The properties of these particles are summarized in Table 3.2.

An *element* is a substance in which all the atoms have the same atomic number. All the atoms of a given element have the same number of protons. For neutral atoms (those without an electric charge), the atomic number also gives the number of electrons.

The number of protons in the nucleus of an atom of any element is equal to the **atomic number** of that element. This number determines the kind of atom, that is, the identity of the element. Dalton had said that the mass of an atom determines the element. We now say it is not the mass but the number of protons that determines the element. For example, an atom with 26 protons (one whose atomic number is 26) is an atom of iron (Fe). An atom with 92 protons is an atom of uranium (U).

Isotopes

The number of neutrons in the nuclei of atoms of a given element may vary. For example, most hydrogen (H) atoms have a nucleus consisting of a single proton and no neutrons (and therefore a mass of 1 amu.) About 1 hydrogen atom in 5000, however, does have a neutron as well as a proton in the nucleus. This heavier hydrogen is called **deuterium** and has a mass of 2 amu. Both kinds are hydrogen atoms (any atom with atomic number 1, that is, with one proton, is a hydrogen atom). Atoms that have this sort of relationship—the same number of protons but differing numbers of neutrons—are called **isotopes** (Figure 3.12). A third, very rare isotope of hydrogen is **tritium**, which

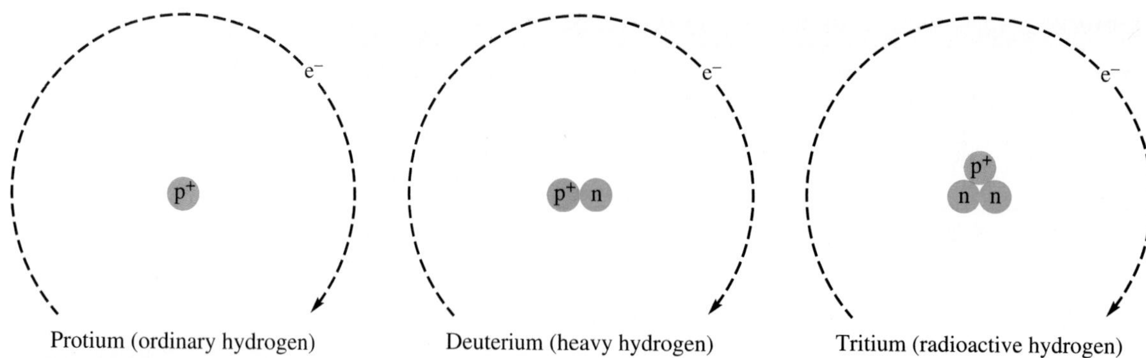

Figure 3.12 The three isotopes of hydrogen. Each has one proton and one electron, but they differ in the number of neutrons in the nucleus.

has two neutrons and one proton in the nucleus (and thus a mass of 3 amu). Most, but not all, elements exist in nature in isotopic forms. This fact also requires a major modification of Dalton's original theory. He said that all atoms of the same element have the same mass. We now say that most elements have several isotopes, that is, atoms with different numbers of neutrons and, therefore, different masses.

SECTION 3.7

Electron Arrangement: The Bohr Model

Let us now turn our attention once more to the electrons. Rutherford demonstrated that atoms have a tiny positively charged nucleus with electrons outside the nucleus. Evidence soon accumulated that the electrons were not randomly distributed, but rather they were arranged in a structured fashion. We will examine the evidence soon. But first let's take a side trip into some fun, some practical chemistry, and some physics that provide a background for our study of the electron structure of atoms.

Figure 3.13 Certain chemical elements can be identified by the characteristic colors that their compounds impart to flames. Five examples are shown here. [*Carey B. Van Loon.*]

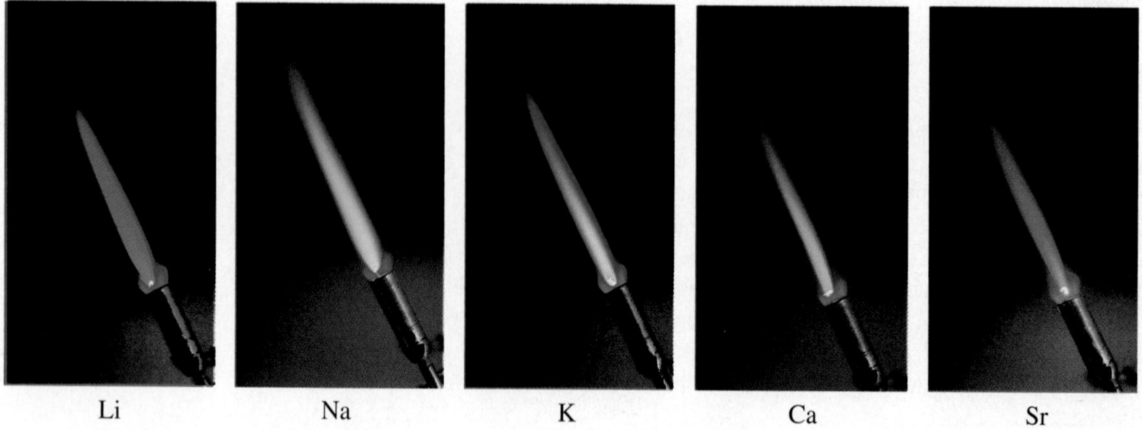

Fireworks and Flame Tests

Chemists of the eighteenth and nineteenth centuries developed *flame tests* that used the color of flames to identify several elements (Figure 3.13). Sodium salts give a persistent yellow flame; potassium salts, a fleeting lavender; lithium salts, a brilliant red. Like those of fireworks, these flame colors are due to the electron structure of the atoms of the specific elements.

Fireworks originated earlier than flame tests, in ancient China. The brilliant colors of aerial displays still mark our celebrations of patriotic holidays (Figure 3.14). The colors of the fires are due to specific elements. Brilliant reds are

Figure 3.14 Flame colors (Figure 3.13) also are the basis for the brilliant colors of fireworks displays. Strontium compounds produce red, copper compounds produce blue, and sodium compounds give yellow. [*Richard Megna/Fundamental Photographs.*]

produced by strontium compounds, while barium compounds give bright green, sodium compounds yield yellow, and copper salts produce a greenish blue.

The colors of fireworks and flame tests are not what they seem to the unaided eye. If the light from the flame is passed through a prism, it is separated into lines of light.

Figure 3.15 The glass prism separates white light into a continuous spectrum or rainbow of colors. [*David Parker/Science Source/Photo Researchers, Inc.*]

Continuous and Line Spectra

When white light from an incandescent lamp is passed through a prism, it produces a *continuous spectrum,* or rainbow of colors (Figure 3.15). When sunlight passes through a raindrop, the same thing occurs (Figure 3.16). The different colors of light represent different wavelengths. Blue light has shorter wavelengths than red light, but there is no sharp transition as one moves from one color to the next. All wavelengths are present in a continuous spectrum. White light is simply a combination of all the various colors.

If the light from a gas discharge tube that contains a particular element is passed through a prism, only narrow colored lines are observed (Figure 3.17). Each line corresponds to light of a particular wavelength. The pattern of lines emitted by an element is called its *line spectrum.* The line spectrum of an element is characteristic of that element and can be used to identify it. Not all the lines in the spectrum of an atom are visible; some lines appear as infrared or ultraviolet radiation.

The line spectrum of hydrogen is fairly simple; it consists of four lines in the visible portion of the electromagnetic spectrum. It was to explain this

Figure 3.16 The rainbow is an example of a continuous spectrum. [*Dick Canby/DRK Photo.*]

Spectroscopy

Light given off by an element when it is heated is called its *emission spectrum.* Simple devices called spectroscopes are used to analyze emission spectra. Atoms also can absorb energy, and we may choose to observe an element's *absorption spectrum.* Emission spectroscopy and atomic absorption spectroscopy are both useful in the laboratory when we wish to determine how much of a certain element is in a particular sample.

Consider the fact that the only thing reaching Earth from the stars or the planets is light. Until recently, everything we knew about these heavenly bodies had to be deduced from our examination of this light. Scientists have used absorption spectra to determine the chemical makeup of the stars and the atmospheres of the planets. Elements present in the atmosphere of a heavenly body absorb a part of the light given off by these bodies. The absorption spectra have dark lines that correspond to the line spectra of the elements present in the atmosphere of the planet or star.

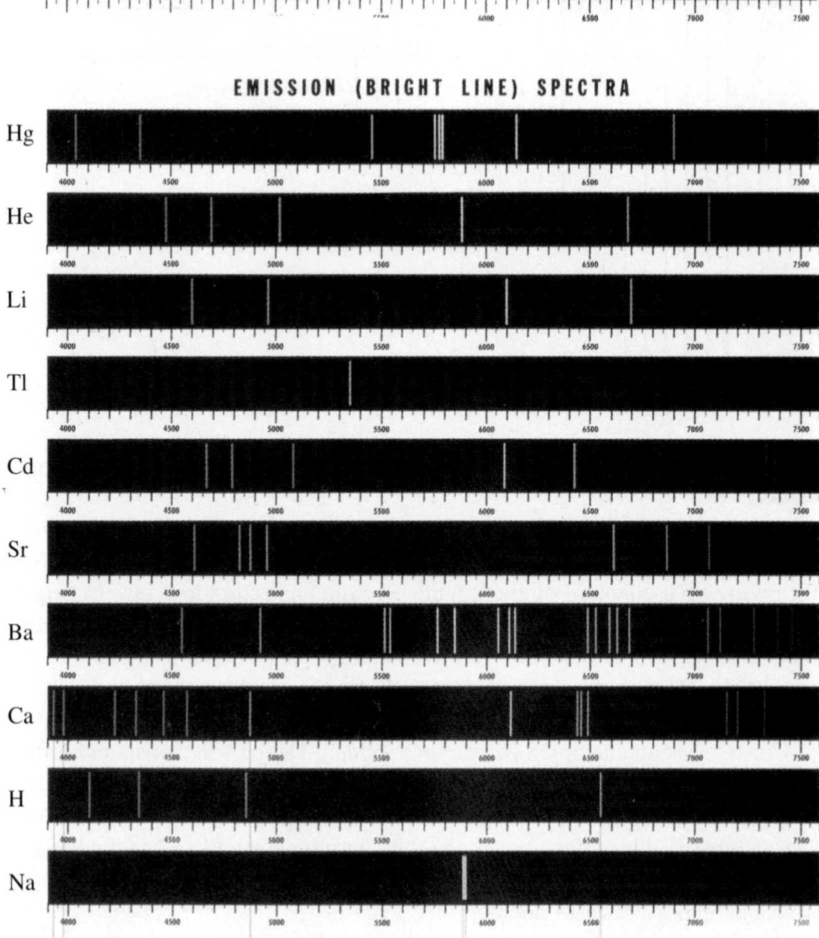

Figure 3.17 Line spectra of selected elements are shown here. Some of the components of the light emitted by excited atoms appear as colored lines. A continuous spectrum is shown at the top for comparison. The numbers are wavelengths of light given in Angstrom units (1 Å$=10^{-10}$ m). Each element has its own characteristic line spectrum that is different from all others. In addition to their practical use in analyzing matter, atomic spectra are the basis for many of the concepts of atomic structure. [*Courtesy of Wabash Instrument Co.*]

spectrum that Niels Bohr worked out his model for the electron structure of the hydrogen atom.

Bohr's Explanation of Line Spectra

Niels Bohr put forth his explanation of line spectra in 1913. He made the suggestion that electrons cannot have just any amount of energy but can have only certain specified amounts; that is to say, the energy of an electron is

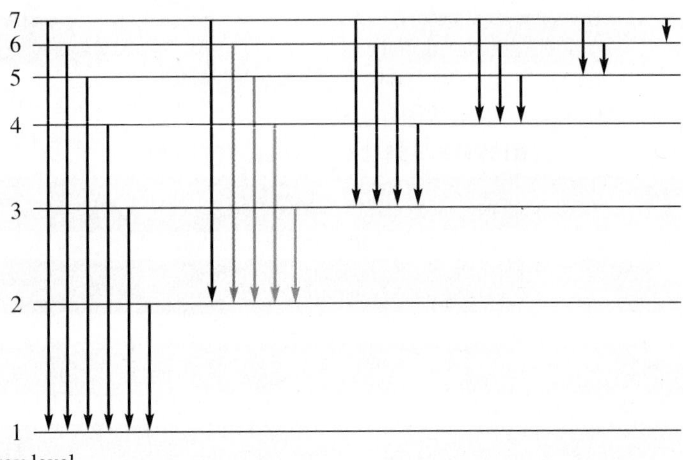

Figure 3.18 Possible electron shifts between energy levels in atoms to produce the lines found in spectra. Not all the lines are in the visible portion of the spectrum. The four colored lines correspond to the colored lines in the hydrogen spectrum.

quantized. The specified energy values for an atom are called its **energy levels**.

An electron, by absorbing a **quantum** of energy (for example, when atoms of the element are heated), is elevated to a higher energy level (Figure 3.18). By giving up a quantum of energy, the electron can return to a lower energy level. The energy released or absorbed in these transitions shows up as a line spectrum. Each line has a specific wavelength, corresponding to a quantum of energy. An electron moves instantaneously from one energy level to another; there are no intermediate stages.

Consider as an analogy a person on a ladder. The person can stand on the first rung, the second rung, the third rung, and so on, but is unable to stand between rungs. As the person goes from one rung to another, the potential energy (energy due to position) changes by definite amounts or quanta. For an electron, its total energy (both potential and kinetic) is changed as it moves from one energy level to another.

Bohr's model of the atom was based on the laws of planetary motion that had been set down by the German astronomer Johannes Kepler (1571–1630) three centuries before. Bohr imagined the electrons to be orbiting about the nucleus much as planets orbit the sun (Figure 3.19). Different energy levels were pictured as different orbits. The modern picture is different, as we shall see in subsequent sections.

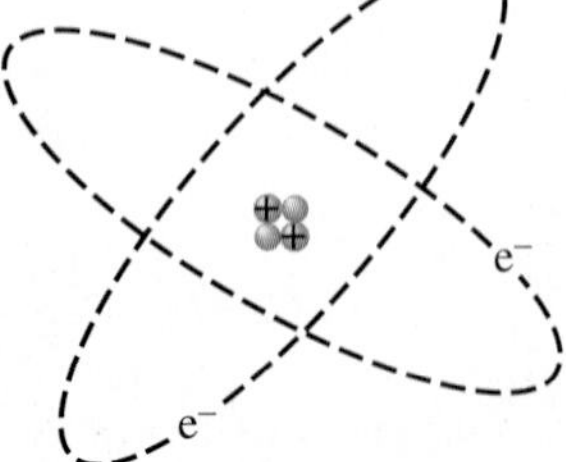

Figure 3.19 The nuclear atom, as envisioned by Bohr, has most of its mass in an extremely small nucleus. Electrons orbit about the nucleus, occupying most of the volume of the atom but contributing little to its mass.

Ground States and Excited States

The electron in the hydrogen atom is usually in the first energy level. Given the choice, electrons usually remain in their lowest possible energy levels (those nearest the nucleus); atoms whose electrons are in this situation are said to be in their **ground state**. When a flame or other source supplies energy to an atom (hydrogen, for example) and an electron jumps from the lowest possible level to a higher level, the atom is said to be in an **excited state**. An atom in an excited state eventually emits a photon of energy as the electron jumps back down to one of the lower levels and ultimately reaches the ground state.

Bohr's theory was a spectacular success in explaining the line spectrum of hydrogen. It established the important idea of energy levels in atoms. Bohr was awarded the Nobel prize in physics in 1922 for this work.

A **photon** is a quantum or "particle" of light.

Atoms larger than hydrogen have more than one electron, and Bohr was also able to deduce that the various energy levels of an atom could only handle a certain number of electrons at one time. We shall simply state Bohr's findings in this regard. The maximum number of electrons that can be in a given level is indicated by the formula $2n^2$, where n is equal to the energy level being considered. For the first energy level ($n = 1$), the maximum population is 2×1^2 or 2. For the second energy level ($n = 2$), the maximum number of electrons is 2×2^2 or 8. For the third level, the maximum is 2×3^2 or 18. (However, the outermost level can have no more than 8 electrons.)

EXAMPLE 3.1

What is the maximum number of electrons in the fifth energy level?

SOLUTION

For the fifth level, $n = 5$, so we have

$$2 \times 5^2 = 2 \times 25 = 50$$

Exercise 3.1

What is the maximum number of electrons in the fourth energy level?

Building Atoms

Imagine building up atoms by adding one electron to the proper energy level as protons are added to the nucleus, keeping in mind that electrons will go to the lowest energy level available. For hydrogen (H), with a nucleus of only one proton, the one electron goes into the first energy level. For helium (He), with a nucleus of two protons (and two neutrons), both electrons go into the first energy level. According to Bohr, two electrons is the maximum population of the first energy level; that level is filled in the helium atom.

With lithium (Li), which has three electrons, two electrons go into the first level; the other must go into the second energy level. This process of adding electrons is continued until the second energy level is filled with eight electrons, as in the neon (Ne) atom, which has two of its ten electrons in the first energy level and the remaining eight in the second energy level.

Niels Bohr in 1922 received the Nobel prize in physics for his planetary model of the atom with its quantized electron energy levels. [*Professor Marvin Lang and Mr. Gary Shulfer, University of Wisconsin–Stevens Point (Scott Cat. Denmark #410.*]

The sodium (Na) atom has 11 electrons. Of these, two are in the first energy level, the second level is filled with eight electrons, and the remaining electron is in the third energy level. We can use a modified Bohr diagram to indicate this **electron configuration** (or arrangement).

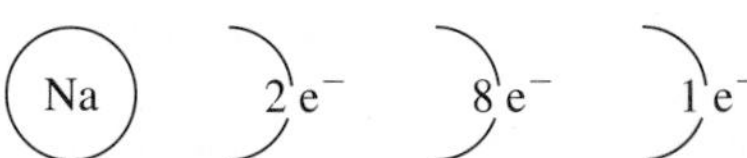

The circle with the symbol indicates the sodium nucleus. The arcs represent the energy levels, with the one closest to the nucleus being the first energy level, the next the second, and so on.

We could now continue to add electrons to the third energy level until we get to argon (Ar).

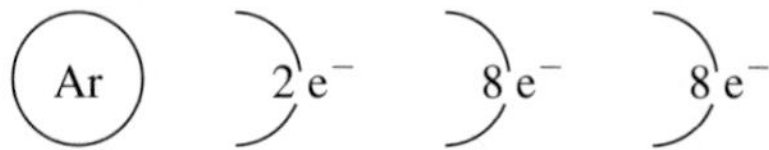

Even though the third energy level can have up to 18 electrons, it is temporarily filled with only eight. The next element, potassium (K), has the electron configuration.

Similarly, calcium (Ca) has two electrons in the fourth energy level.

Ca) 2 e^-) 8 e^-) 8 e^-) 2 e^-

After calcium, things get a bit more complicated. The next ten electrons resume filling the third energy level.

The buildup of atoms to calcium is diagrammed in Figure 3.20. The entire topic of atomic structure is discussed in more detail in Section 3.8. Using rules outlined there, we could (in our imagination, at least) draw Bohr diagrams for all the known elements. For now, though, let us practice with some simple examples.

Sometimes the Bohr configuration is written without the arcs to indicate energy levels. The following illustrate the method.

Na	2, 8, 1
Ar	2, 8, 8
K	2, 8, 8, 1
F	2, 7

▶ EXAMPLE 3.2

Draw a Bohr diagram for fluorine (F).

SOLUTION

Fluorine is element number 9; it has nine electrons. Two of these go into the first energy level, the remaining seven into the second.

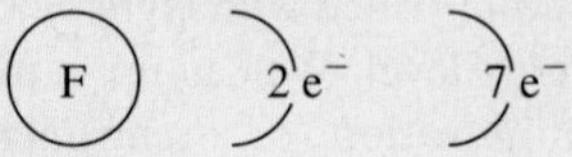

F) 2 e^-) 7 e^-

Exercise 3.2

Draw Bohr diagrams for magnesium (Mg) and phosphorus (P).

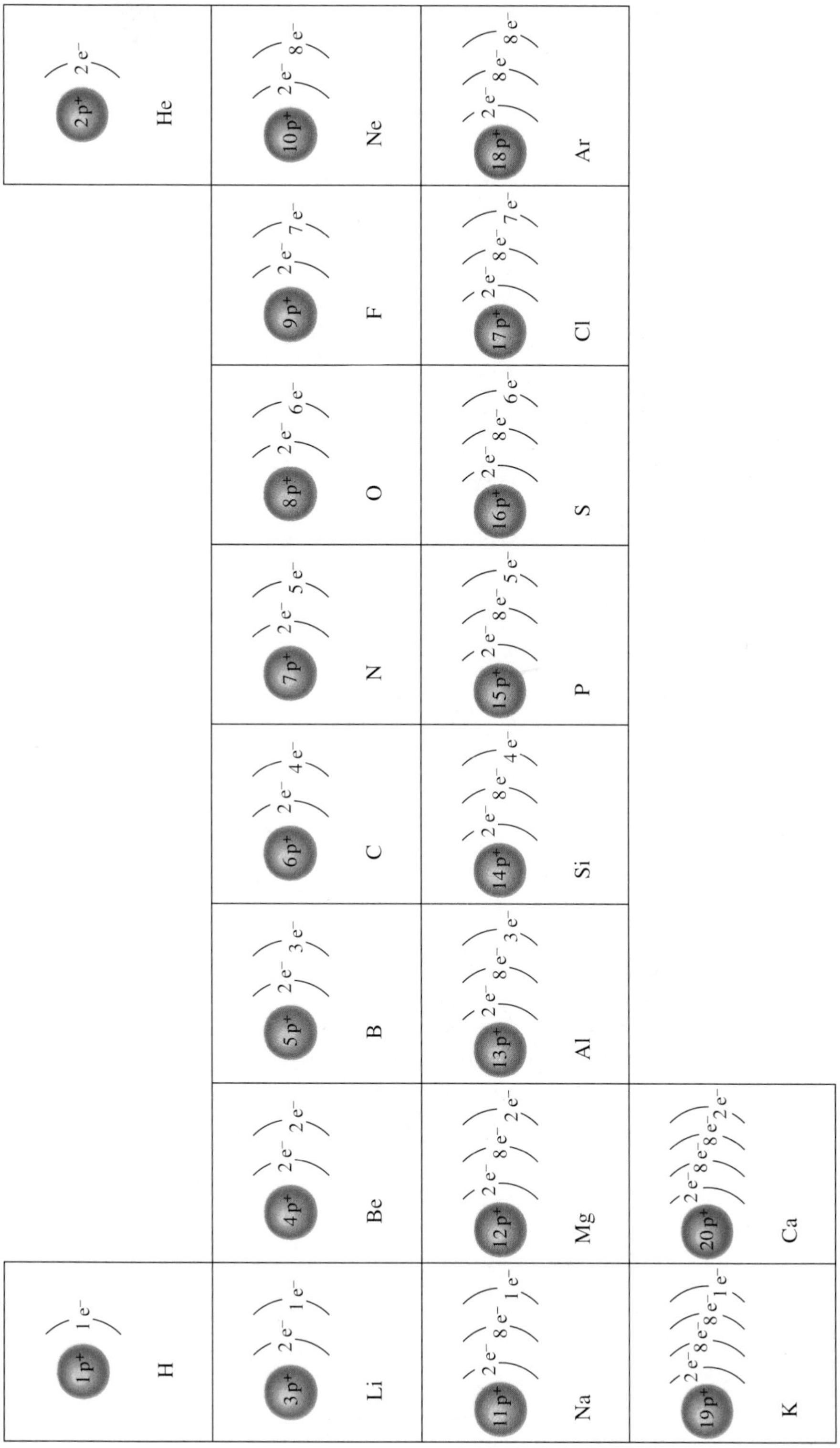

$1p^+$ $1e^-$ H							$2p^+$ $2e^-$ He
$3p^+$ $2e^-$ $1e^-$ Li	$4p^+$ $2e^-$ $2e^-$ Be	$5p^+$ $2e^-$ $3e^-$ B	$6p^+$ $2e^-$ $4e^-$ C	$7p^+$ $2e^-$ $5e^-$ N	$8p^+$ $2e^-$ $6e^-$ O	$9p^+$ $2e^-$ $7e^-$ F	$10p^+$ $2e^-$ $8e^-$ Ne
$11p^+$ $2e^-$ $8e^-$ $1e^-$ Na	$12p^+$ $2e^-$ $8e^-$ $2e^-$ Mg	$13p^+$ $2e^-$ $8e^-$ $3e^-$ Al	$14p^+$ $2e^-$ $8e^-$ $4e^-$ Si	$15p^+$ $2e^-$ $8e^-$ $5e^-$ P	$16p^+$ $2e^-$ $8e^-$ $6e^-$ S	$17p^+$ $2e^-$ $8e^-$ $7e^-$ Cl	$18p^+$ $2e^-$ $8e^-$ $8e^-$ Ar
$19p^+$ $2e^-$ $8e^-$ $8e^-$ $1e^-$ K	$20p^+$ $2e^-$ $8e^-$ $8e^-$ $2e^-$ Ca						

Figure 3.20 Bohr diagrams of the first 20 elements.

SECTION 3.8

The Quantum Mechanical Atom

It is interesting to note that the distance given by Bohr for the radius of the first orbit, 52.9 picometers (pm), is the same as the "most probable" distance from the nucleus to an electron in the first energy level—as revealed by wave mechanical calculations. Here is a point worth commenting on. A new theory replaces an old one when the new theory explains something the old one can't. However, the new theory must also answer all those questions the old theory was able to handle, and the new answers must be at least as good as the old.

The simple planetary Bohr model of the atom has been replaced for many purposes by more sophisticated models that are more difficult to understand. Electrons are treated as waves, and their locations are indicated only in terms of probabilities.

A young French physicist, Louis de Broglie, first suggested (in 1924) that the electron should have wavelike properties. In other words, de Broglie said that a beam of electrons should behave very much like a beam of light. Since Thomson had proved (in 1897) that electrons were particles, this suggestion that they be treated as waves was hard to accept. Nevertheless, de Broglie's theory was experimentally verified within a few years. Electron microscopes, which make use of the wave nature of electrons, now are found in many scientific laboratories.

Erwin Schrödinger, an Austrian physicist, used highly mathematical quantum mechanics in the 1920s to develop equations that describe the properties of electrons in atoms. (Fortunately, we need not understand the elaborate mathematics to make use of some of the results.) The solutions to these equations can provide a measure of the probability of finding an electron in a given volume of space. The planetary orbits of the Bohr model are replaced by shaped volumes of space in which electrons move. The term **orbital** (replacing Bohr's *orbits*) is used for this new description of the location of electrons.

Suppose you had a camera that could photograph electrons (there is no such thing, but we are just supposing) and you left the shutter open while an electron zipped about the nucleus. When you developed the picture, you would have a record of where the electron had been. Doing the same thing with an electric fan that was turned on would give you a picture in which the blades of the fan look like a disk of material. The blades move so rapidly that their photographic image is blurred. Similarly, electrons in the first energy level of an atom would appear in our imaginary photograph as a fuzzy ball (Figure 3.21). The fuzzy ball (frequently referred to as a *charge cloud* or *electron cloud*) is the rough equivalent of an orbital.

Like Bohr, Schrödinger concluded that only two electrons could occupy the first energy level in an atom. In the quantum mechanical atom, this level is referred to as the 1*s* orbital (which is spherical). Also like Bohr, Schrödinger

Figure 3.21 Charge-cloud representations of atomic orbitals.

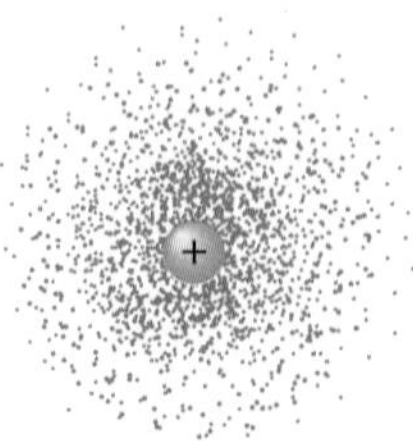

An *s* orbital

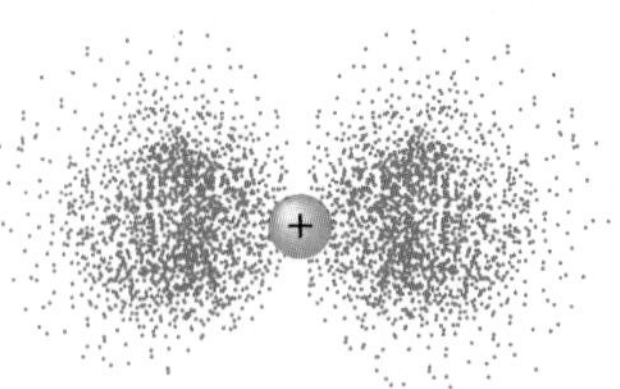

A *p* orbital

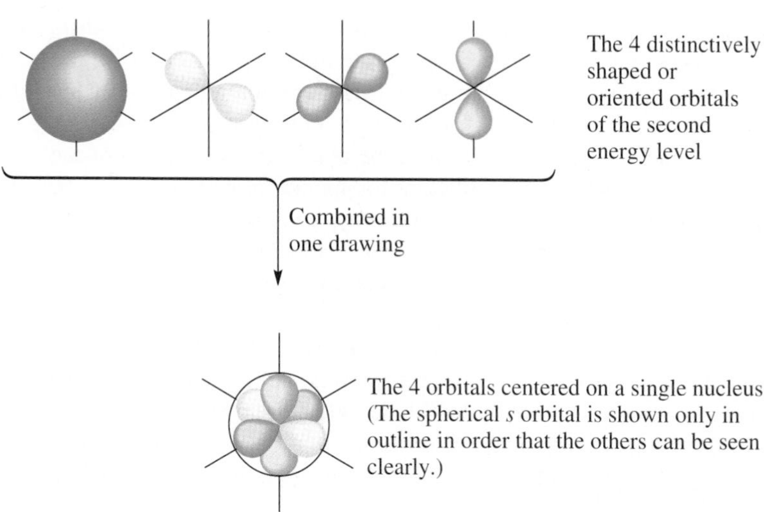
The 4 distinctively shaped or oriented orbitals of the second energy level
Combined in one drawing
The 4 orbitals centered on a single nucleus. (The spherical s orbital is shown only in outline in order that the others can be seen clearly.)

Figure 3.22 Electron orbitals. In these drawings the nucleus of the atom is located at the intersection of the axes. The eight electrons that would be placed in the second energy level of Bohr's model are distributed among these four orbitals in the current model of the atom, two electrons per orbital.

stated that the second energy level of an atom could contain a maximum of eight electrons. However, Schrödinger also concluded that these eight electrons must be located in four different orbitals. Each individual orbital could contain a maximum of two electrons. One of the orbitals of the second energy level is spherical in shape and is referred to as the 2*s* orbital. The remaining three orbitals of the second level have identical dumbbell shapes (Figure 3.21). They differ in their orientation in space (i.e., the direction in which they point) (Figure 3.22). These are designated the 2*p* orbitals. Again, the 2 indicates that these orbitals are in the second energy level, and the *p* means that they have the dumbbell shape. To distinguish the different *p* orbitals, they are sometimes referred to as the $2p_x$, $2p_y$, and $2p_z$ orbitals.

Building Atoms by Orbital Filling

An orbital can be occupied by one electron or by two paired electrons, but no more. Thus, the two electrons of the lowest energy level can be accommodated by the 1*s* orbital. The second level requires four different orbitals to accommodate its eight electrons. The third level (with 18 electrons) requires nine orbitals; a 3*s* orbital, three 3*p* orbitals, and an additional set of orbitals, the 3*d*, of which there are five. The shape of a *d* orbital resembles a four-leaf clover. We shall not bother to draw all of these orbitals. The point is that higher energy levels have more orbitals to accommodate larger numbers of electrons.

In the second energy level, the *p* electrons are in a slightly higher energy **sublevel** than the *s* electrons. In building up the electron configuration of atoms of the various elements, the lower sublevels are filled first. Let's illustrate by "building" a few atoms.

A hydrogen atom (H), with atomic number 1, has only one electron. That single electron goes into the first energy level, which has only one kind of orbital, an *s*. We write the electron configuration of hydrogen, then, as $1s^1$. The superscript (1 in this case) gives the number of electrons in the orbital. Similarly, the two electrons of helium (He) are in the first energy level; helium's electron structure is $1s^2$. Lithium (Li) has three electrons. Two go into the first energy level; the third must be placed in the *s* orbital of the second energy level. The electron structure for lithium is written $1s^22s^1$. Beryllium (Be) has four electrons; its electron configuration is $1s^22s^2$. Boron (B) has five electrons. Two are placed in the 1*s* orbital, two in the 2*s*, and the one remaining is placed in one of the three 2*p* orbitals. The electron structure of boron, then, is $1s^22s^22p^1$.

The three *p* orbitals (that is, the *p* sublevel) hold a maximum of six electrons (two per orbital). The elements boron, carbon, nitrogen, oxygen, fluorine, and neon correspond to the filling of the 2*p* sublevel with one through six electrons, respectively. The electron configuration of neon, then, is $1s^22s^22p^6$. This corresponds to a filled *p* sublevel—and, of course, to a completely filled second energy level.

▶ EXAMPLE 3.3

Write out the electron configuration, using sublevel notation, for nitrogen (N).

SOLUTION

Nitrogen has seven electrons. Place them in the lowest unfilled energy sublevels. Two go into the 1*s* orbital and two into the 2*s* orbital. That leaves three electrons to be placed in the *p* sublevel. The electron configuration is $1s^22s^22p^3$.

Exercise 3.3

Write out the electron configuration for fluorine (F).

We could continue to build up electron configurations by following a simple procedure. First, find the atomic number for the element (use the periodic table), then place that number of electrons in the lowest possible energy sublevels. Keep in mind that the maximum number of electrons in an *s* sublevel is two and in a *p* sublevel is six.

▶ EXAMPLE 3.4

Write out the electron configuration for sulfur (S).

SOLUTION

Sulfur atoms have 16 electrons each. The electron configuration is $1s^22s^22p^63s^23p^4$. Note that the total of the superscripts is 16, and that we have not exceeded the maximum capacity for any sublevel.

Exercise 3.4

Write out the electron configuration for chlorine (Cl).

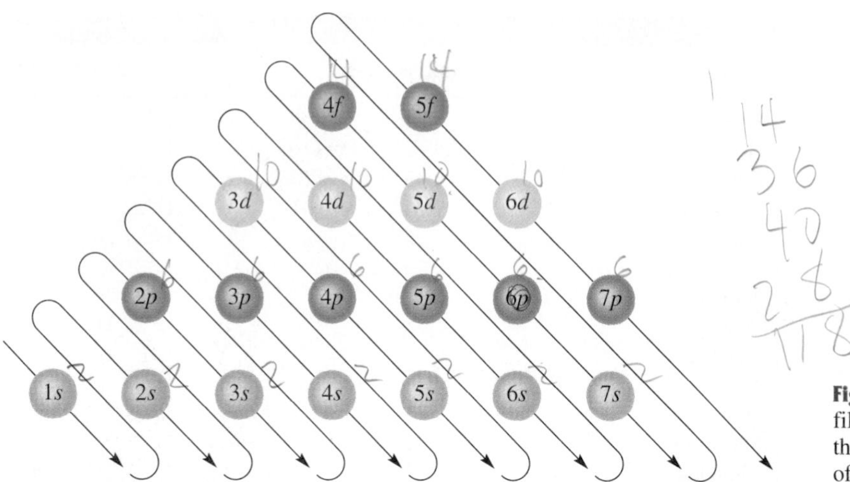

Figure 3.23 An order-of-filling chart for determining the electron configurations of atoms.

Next, let's write the electron configuration for argon (Ar). Argon has 18 electrons; its configuration is $1s^22s^22p^63s^23p^6$. Note that the highest occupied energy sublevel, the $3p$, is filled. Recall, however, that the third energy level has a maximum of 18 electrons. We might reasonably expect that potassium (K), with 19 electrons, would have the argon configuration plus the nineteenth electron in a $3d$ sublevel. It doesn't. Rather, the potassium atom has the electron structure $1s^22s^22p^63s^23p^64s^1$. The $4s$ sublevel fills before the $3d$. Calcium, with 20 electrons, has the structure $1s^22s^22p^63s^23p^64s^2$. With scandium (Sc), the $3d$ sublevel begins to fill.

We will not further pursue here the topic of electron configurations. For those who wish to go further, an "order-of-filling" chart is given in Figure 3.23. Using this chart and the additional information that the maximum capacity for a *d* sublevel is 10 electrons and that of an *f* sublevel is 14 electrons, you can write out the electron structure of nearly any element. (There are, as you might expect, a few exceptions that do not follow these rules exactly.) We will deal in this book, for the most part, with the first 20 elements. The electron structure of atoms of these elements is given, for convenience, in Table 3.3.

SECTION 3.9

Electron Configurations and the Periodic Table

The modern periodic table (inside front cover) has vertical columns called **groups** or (sometimes) *families*. The horizontal rows of the periodic table are called **periods**. Elements in a group have similar chemical properties. The properties of elements vary periodically across a period.

Table 3.3 Electron Structures for Atoms of the First 20 Elements

Name	Atomic Number	Electron Structure
Hydrogen	1	$1s^1$
Helium	2	$1s^2$
Lithium	3	$1s^22s^1$
Beryllium	4	$1s^22s^2$
Boron	5	$1s^22s^22p^1$
Carbon	6	$1s^22s^22p^2$
Nitrogen	7	$1s^22s^22p^3$
Oxygen	8	$1s^22s^22p^4$
Fluorine	9	$1s^22s^22p^5$
Neon	10	$1s^22s^22p^6$
Sodium	11	$1s^22s^22p^63s^1$
Magnesium	12	$1s^22s^22p^63s^2$
Aluminum	13	$1s^22s^22p^63s^23p^1$
Silicon	14	$1s^22s^22p^63s^23p^2$
Phosphorus	15	$1s^22s^22p^63s^23p^3$
Sulfur	16	$1s^22s^22p^63s^23p^4$
Chlorine	17	$1s^22s^22p^63s^23p^5$
Argon	18	$1s^22s^22p^63s^23p^6$
Potassium	19	$1s^22s^22p^63s^23p^64s^1$
Calcium	20	$1s^22s^22p^63s^23p^64s^2$

In the United States, the groups are usually indicated by a Roman numeral followed by the letter A or B. The International Union of Pure and Applied Chemistry (IUPAC) recommends numbering the groups from 1 to 18. Both systems are indicated on the periodic table on the inside front cover, but we follow the traditional U.S. method in this book.

In the United States, the letter A identifies the **main group elements** and B indicates the **transition elements**. (Nearly opposite usage prevailed in Europe, leading IUPAC to develop the new system.)

Metals, Nonmetals, and Metalloids

Elements also are divided into two classes by a heavy, stepped diagonal line. Those to the left of the line are **metals**, elements that have a characteristic luster and generally are good conductors of heat and electricity. Except for mercury, which is a liquid, all the metals are solids at room temperature. Metals generally are *malleable;* that is, they can be hammered into thin sheets. Most also are *ductile;* they can be drawn into wires.

Elements to the right of the stepped line are **nonmetals**. These elements lack metallic properties. Several are gases (oxygen, nitrogen, fluorine, chlorine). Others are solids (carbon, sulfur, phosphorus, iodine). Bromine is the only nonmetal that is a liquid at room temperature.

Some of the elements bordering the stepped line are **metalloids**, elements that have intermediate properties. Metalloids have properties that resemble those of both the metals and the nonmetals.

Family Groups

Elements within a group have similar properties. For example, the elements in Group IA are all (except hydrogen) soft, highly reactive metals. These elements, often called the **alkali metals**, react vigorously with water to evolve hydrogen gas. There are important *trends* within a family. For example, lithium is the hardest metal of the group. Sodium is softer than lithium; potassium is softer still, and so on down the group. Lithium is also the least reactive toward water. Sodium, potassium, rubidium, and cesium are progressively more reactive. (Francium is highly radioactive and extremely rare; few of its properties have been measured.) Hydrogen is the odd one of Group IA. It is *not* an alkali metal. Indeed, it is a rather characteristic nonmetal. As far as its properties are concerned, hydrogen probably should be put in a group of its own.

Group IIA elements are sometimes known as the **alkaline earth metals**. The metals in this group are fairly soft and moderately reactive with water. Beryllium is an odd member of the group in that it is rather hard and does not react with water. There are trends in properties within the group—as there are in the other families. For example, magnesium, calcium, strontium, barium, and radium are progressively more reactive toward water.

Group VIIA elements, often called **halogens**, also consist of reactive elements. For example, the halogens react vigorously with the alkali metals to form crystalline solids (more about this in Chapter 5). There are trends in the halogen family. Fluorine is the most reactive of the halogens toward the alkali metals, chlorine next, and so on. Fluorine and chlorine are gases at room temperature, bromine is a liquid, and iodine is a solid. (Astatine, like francium, is highly radioactive and extremely rare; few of its properties have been determined.)

The group to the far right of the periodic table (sometimes called Group VIIIA) is known as the **noble gases**. The noble gases are known for their lack of chemical reactivity; their main chemical property is that they undergo few chemical reactions.

Family Features: Outer Electron Configurations

The period in which an element appears in the periodic table tells us how many main electron energy levels there are in that atom. Sulfur, for example, is in the third period, so the sulfur atom has three main energy levels. The group number (for main group elements) tells us how many electrons are in the outermost energy level. These are called **valence electrons**. Since sulfur is in Group VIA, it has six valence electrons. Two of these are in an s orbital, and the other four are in *p* orbitals. We can indicate the outer electron configuration of the sulfur atom as follows: $3s^2 3p^4$.

It is these outermost electrons that mainly determine the chemistry of an atom. Since all the elements in the same group of the periodic table have the same number of valence electrons, they should have very similar chemistry; and they do. All the elements in Group IA have only one valence electron, and all are very active metals (except for hydrogen, which is a unique nonmetal). All have the outer electron structure ns^1, where n denotes the number of the outermost main energy level. All the elements in Group VIIA (halogens) have seven valence electrons in the configuration ns^2np^5, and all are very active nonmetals.

In general the properties of all the elements can be correlated with their electron configurations. We explore bond formation using electron configurations in Chapter 5.

▶ EXAMPLE 3.5

Write out the sublevel notation for the electrons in the highest main energy level for strontium (Sr) and arsenic (As).

SOLUTION

Strontium is in Group IIA and the fifth period of the periodic chart. Its outer electron configuration is $5s^2$. (Each period of the periodic table corresponds to the filling, or at least partial filling, of a main energy level.) Arsenic is in Group VA and the fourth period. Its five outer (valence) electrons have the configuration denoted by $4s^2 4p^3$.

Exercise 3.5

Write out the sublevel notation for the electrons in the highest main energy level for rubidium (Rb) and selenium (Se).

Figure 3.24 relates the sublevel configurations to the various groups of the periodic table.

SECTION 3.10

Which Model to Choose?

For some purposes in this text, Bohr diagrams of atoms are used to picture the distribution of electrons among the main energy levels. At other times, the electron clouds of the quantum mechanical model are more useful. Even Dalton's model will sometimes prove to be the best way to describe certain phenomena (the behavior of gases, for example). The choice of model will always be based on which one is most helpful for understanding a particular concept. That, after all, is the whole purpose of scientific models.

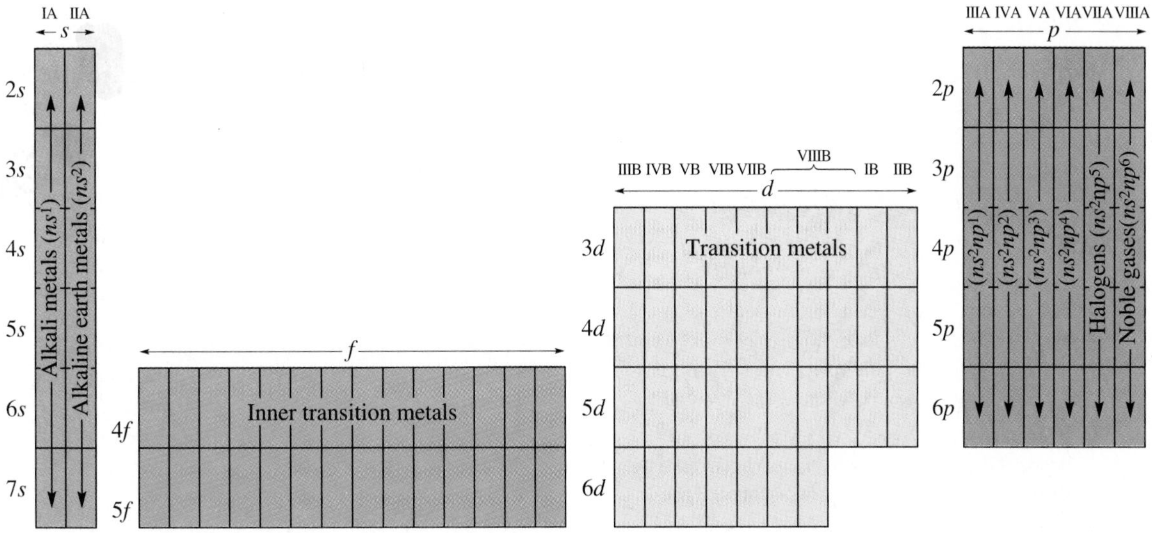

Figure 3.24 Valence shell electron configurations and the periodic table.

Summary

1. The fact that matter is electrical in nature was established during the eighteenth century by Davy, Faraday, and others when they were able to split compounds into their component elements by *electrolysis.*

2. Experiments with *cathode ray* tubes showed that matter contained negatively charged particles, which were called *electrons.* By deflecting cathode rays with a magnet, Thomson was able to determine the charge/mass ratio for the electron. Later Millikan's oil drop experiment measured the charge on the electron, so the mass of the electron could then be calculated.

3. Goldstein's experiment using gas discharge tubes with perforated cathodes showed that matter also contained positively charged particles, but different gases produced positive particles of different mass. Those of lightest mass were formed when the gas in the tube was hydrogen. These smallest positive particles were later called *protons.*

4. Other important chemical developments of the nineteenth century included: the establishment of a scale of relative atomic weights by Berzelius; introduction of the *periodic table* by Mendeleev; the discovery of *X-rays* by Roentgen; and the discovery of *radioactivity* by Becquerel.

5. Rutherford's gold foil experiment indicated that the atom must have a tiny and very dense *nucleus* with a positive charge. It appeared that the protons in the atom must all be located in the nucleus, with electrons occupying most of the atom's space. It was not until 1932 that Chadwick discovered the *neutron,* a particle in the nucleus as massive as a proton but bearing no charge.

6. The number of protons determines the amount of positive charge on a particular nucleus, and this is known as the *atomic number* for that element. When atoms have the same atomic number but differ in mass, they are called *isotopes.* Isotopes of the same element differ only in their number of neutrons.

7. A study of the spectral lines of hydrogen led Bohr to propose an atom with concentric shells of electrons surrounding the positively charged nucleus. Each shell represents an electron *energy level.* The farther the level is from the nucleus, the higher the electron energy and the greater the electron capacity. The first four energy levels have capacities of 2, 8, 18, and 32 electrons, respectively.

8. Although the simple Bohr model is very useful, our modern view of the atom is much more complicated. Electrons act as waves as well as particles, and they can be treated by the mathematical methods of quantum mechanics. The energy levels of Bohr can be further split into sublevels that differ slightly in energy because of the differing shapes of their electron *orbitals.* (An orbital is an electron cloud that can hold one pair of electrons.) For example, *s* orbitals are spherical, *p* orbitals are shaped like dumbbells, and *d* orbitals are shaped like four-leaf clovers. The first main energy level can hold only one *s* orbital; the second level can hold one *s* and three *p* orbitals; and the third level can hold one *s*, three *p*, and five *d* orbitals.
9. The periodic table is made up of horizontal rows called *periods* and vertical rows called *groups* (or families). A slanting diagonal line near the right end of the table (from B to At) divides the *metals* on the left from the *nonmetals* on the right. Most of the elements are metals. Many elements along the dividing line are *metalloids.* The period number tells you how many electron energy levels there are in an atom. The group number (for main group elements) tells you how many valence electrons the atom has.
10. The first group in the periodic table is made up of *alkali metals* (the most active of all metals), and the second group is the *alkaline earth metals.* The last group is made up of the *noble gases* (the least reactive of all the elements), and the next to last group is the *halogens* (highly reactive nonmetals). The ten short groups across the middle portion of the table are the *transition metals.*

Key Terms

alkali metals 3.9
alkaline earth metals 3.9
alpha particle 3.4
anion 3.1
atomic mass unit 3.2
atomic number 3.6
beta particle 3.4
cathode ray 3.1
cation 3.1
deuterium 3.6
electrodes 3.1
electrolysis 3.1
electrolyte 3.1
electron 3.1
electron configuration 3.7
energy levels 3.7
excited state 3.7
gamma rays 3.4
ground state 3.7
groups 3.9
halogens 3.9
ions 3.1
isotopes 3.6
main group elements 3.9
metalloids 3.9
metals 3.9
neutron 3.6
noble gases 3.9
nonmetals 3.9
nucleus 3.5
orbital 3.8
periodic table 3.2
periods 3.9
photon 3.7
proton 3.6
quantum 3.7
radioactivity 3.3
sublevel 3.8
transition elements 3.9
tritium 3.6
valence electrons 3.9
X-rays 3.3

Review Questions

1. What did each of the following scientists contribute to our knowledge of the atom?
   a. Crookes b. J. J. Thomson
   c. Goldstein d. Millikan
   e. Roentgen f. Becquerel
2. What evidence is there that electrons are particles?
3. What is radioactivity? How did the discovery of radioactivity contradict Dalton's atomic theory?
4. Define or identify each of the following.
   a. alpha particle b. beta particle
   c. gamma ray d. deuterium
   e. tritium
5. What are isotopes?
6. How are X-rays and gamma rays similar? How are they different?

7. The table below describes four atoms.

	Atom A	Atom B	Atom C	Atom D
Number of protons	10	11	11	10
Number of neutrons	11	10	11	10
Number of electrons	10	11	11	10

Are atoms A and B isotopes? A and C? A and D? B and C?

8. What are the masses of the atoms in Question 7?

9. Discuss Rutherford's gold-foil experiment. What did it tell us about the structure of the atom?

10. What is the atomic nucleus?

11. Give the distinguishing characteristics of the proton, the neutron, and the electron.

12. Should the proton and electron attract or repel one another?

13. Should the neutron and proton attract or repel one another?

14. Which subatomic particles are found in the nuclei of atoms?

15. In an atom, what are the extranuclear (outside the nucleus) subatomic particles?

16. Compare Dalton's model of the atom with the nuclear model of the atom.

17. If the nucleus of an atom contains ten protons, how many electrons are there in the neutral atom?

18. What is the symbol of the element with atomic number 98?

19. What is the name of the element that has 18 protons in the nucleus of its atoms?

20. Explain what is meant by the term atomic weight.

21. How did Bohr refine the nuclear model of the atom?

22. What particles travel in the "orbits" of the Bohr model of the atom?

23. According to Bohr, what is the maximum number of electrons in the fourth energy level ($n = 4$)?

24. If the third energy level of an atom contains two electrons, what is the total number of electrons in the atom?

25. Define the following terms.
a. ground state b. excited state

26. When light is emitted by an atom, what change has occurred within the atom?

27. Which atom absorbed more energy: one in which an electron moved from the second energy level to the third level or an otherwise identical atom in which an electron moved from the first to the third energy level?

28. Use the periodic table to determine the number of protons in atoms of the following elements.
a. helium (He) b. sodium (Na)
c. chlorine (Cl) d. oxygen (O)
e. magnesium (Mg) f. sulfur (S)

29. How many electrons are there in the neutral atoms of the elements listed in Question 28?

30. Draw Bohr diagrams for the elements listed in Question 28.

31. The following Bohr diagram is supposed to represent the neutral atoms of an element. The diagram is incorrectly drawn. Identify the error.

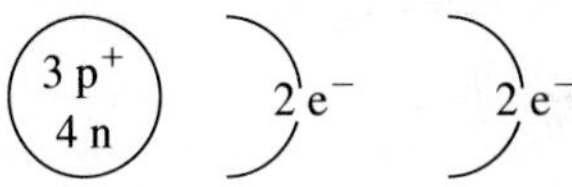

32. Elements are defined on a theoretical basis as being composed of atoms that share the same atomic number. On the basis of this theory, would you think it possible for someone to discover a new element that would fit between magnesium (atomic number 12) and aluminum (atomic number 13)?

33. What is the difference between a Bohr orbit and the orbital of wave mechanics?

34. Where are the metals and nonmetals located on the periodic table?

35. List some characteristic properties of metals.

36. List some characteristic properties of nonmetals.

37. Identify the following elements as metals or nonmetals. You may refer to the periodic table. The numbers in parentheses are the atomic numbers of the elements.
a. sulfur (16) b. chromium (24)
c. iodine (53)

38. Indicate the group number of the following families.
a. alkali metals
b. halogens
c. alkaline earth metals

39. Identify the period of the following elements. You may refer to the periodic table. The numbers in parentheses are the atomic numbers of the elements.
a. chlorine (17) b. osmium (76)
c. hydrogen (1)

40. List some of the properties of alkali metals.

41. What is the most distinguishing property of the noble gases?

42. Which of the following elements are halogens?
a. Ag b. At c. As

43. Which of the following elements are alkali metals?
a. K b. Y c. W

44. Which of the following elements are noble gases?
a. Fe b. Ne c. Ge
d. He e. Xe

45. Which of the following elements are transition metals?
a. Ti (22) b. Tc (43) c. Te (52)

46. Which of the following elements are alkaline earth metals?
a. Bi b. Ba c. Be d. Br

47. How many electrons are in the outermost energy level of Group IIA elements?

48. In what period of elements are electrons first introduced into the fourth energy level?

Problems

Quantum Mechanical Notation for Electron Structure

49. In the quantum mechanical notation $2s^2$ how many electrons are described? What is the general shape of the orbitals described in the notation? How many orbitals are included in the notation?

50. In the quantum mechanical notation $2p^6$ how many electrons are described? What is the general shape of the orbitals described in the notation? How many orbitals are included in the notation?

51. Use quantum mechanical notation to describe the electron configuration of the atom represented in the following Bohr diagram.

52. Give the electron configurations (using quantum mechanical notation) for the elements in Question 28. You may refer to the periodic table.

53. Identify the elements from their electron configurations. You may refer to the periodic table.
a. $1s^22s^2$ b. $1s^22s^22p^3$
c. $1s^22s^22p^63s^23p^1$

54. Without referring to the periodic table, give the atomic numbers of the elements described in Problem 53.

55. None of the following electron configurations is reasonable. In each case explain why.
a. $1s^22s^23s^2$ b. $1s^22s^22p^23s^1$
c. $1s^22s^22p^62d^5$

56. None of the following electron configurations is reasonable. In each case explain why.
a. $1s^12s^1$ b. $1s^22s^22p^7$
c. $1s^22p^2$

57. Referring only to the periodic table, indicate what similarity in electron structure is shared by fluorine (F) and chlorine (Cl). What is the difference in their electron structures?

58. Referring only to the periodic table, indicate what similarity in electron structure is shared by oxygen (O) and sulfur (S). What is the difference in their electron structure?

59. What is the difference in the electron configurations of sulfur (S) and chlorine (Cl)?

60. What is the difference in the electron configurations of fluorine (F) and sulfur (S)?

61. If three electrons were added to the outermost energy level of a phosphorus atom, the new electron configuration would resemble that of what element?

62. If two electrons were removed from the outermost energy level of a magnesium atom, the new electron configuration would resemble that of what element?

References and Readings

1. Ainslie, Donald S. ''Spectrum on a Screen.'' *The Science Teacher,* April 1979, p. 44.
2. Andrade, E. N. da C. *Rutherford and the Nature of the Atom.* Garden City, NY: Doubleday, 1964.
3. Asimov, Isaac. *Atom: Journey Across the Subatomic Cosmos.* New York: Dutton, 1991. A history of atomic and subatomic research.
4. Conkling, John A. ''Pyrotechnics.'' *Scientific American,* July 1990, pp. 96–102. Discusses the science of fireworks.
5. Freedman, David H. ''Weird Science,'' *Discover,* November 1990, pp. 62–68. Reveals some of the latest experiments in quantum mechanics.
6. Houwink, R. *Data: Mirrors of Science.* New York: American Elsevier, 1970. Chapter 3 includes marvelous illustrations showing the minute size of atoms.
7. Jaffe, Bernard. *Moseley and the Numbering of the Elements.* London: Heinemann Educational Books, 1971. Tells how atomic numbers were determined.
8. Knight, David M. *Humphry Davy: Science and Power.* Cambridge, MA: Blackwell, 1992. A fascinating biography of a highly creative scientist.
9. Radtke, Neil. ''Atomic 'Cities.''' *The Science Teacher,* January 1978, p. 35. Relates the filling of electron shells to the placement of people in houses and neighborhoods.
10. Wolff, Peter. *Breakthroughs in Chemistry.* New York: Signet Science Library, 1967. Chapters 6–9 relate the research of Faraday, Mendeleev, Marie Curie, and Bohr.
11. Walker, Jearl. ''The Amateur Scientist: The Spectra of Street Lights Illuminate Basic Principles of Quantum Mechanics.'' *Scientific American,* January 1984, pp. 138–143.

4 Nuclear Chemistry

The Heart of Matter

▼

Nuclei of atoms are

As tiny as can be;

But they have enormous mass

And awesome energy!

In Chapter 3 we discussed the structure of the atom. We briefly mentioned the atomic nucleus, but our attention was mainly focused on the electrons in the atom. It is the electrons, after all, especially the outermost electrons, that determine the chemistry of an atom. However, let us pause for a moment to take a closer look at that tiny speck of matter in the center of the atom—the atomic *nucleus.*

If the size of an atom is incomprehensibly small, then the atomic nucleus is completely beyond our imagination.The diameter of an atom is 100,000 times as great as the diameter of its nucleus. If an atom could be blown up in size until it was as large as your classroom, the size of the nucleus would be about as large as the period at the end of this sentence. The atomic nucleus is exceedingly tiny.

Yet the nucleus contains almost all of the atom's mass. How enormously dense the atomic nucleus must be! A cubic centimeter of water weighs 1 gram, a cubic centimeter of lead about 11 grams, and a cubic centimeter of gold about 19 grams. A cubic centimeter of pure atomic nuclei would weigh about 120 billion kilograms!

But even more amazing than its density is the incredible energy that is contained within the atomic nucleus. Some atomic nuclei give off radiation that can be used to treat cancer; some nuclei undergo reactions that can provide electricity for cities with millions of people; and some atomic nuclei have been used to fuel nuclear bombs.

In this chapter we discuss radioactivity and its applications in medicine, agriculture, and archaeology. We also talk about nuclear fallout and the effect of radiation on living things. The important subject of nuclear power plants will be covered later in Chapter 14 along with other sources of energy.

It seems that much of what we hear today about nuclear processes is negative. We hear about bombs and about problems at nuclear power plants. But there is a positive side, too. Nuclear medicine saves lives. Diseases once regarded as incurable can be diagnosed and treated effectively with radioactive isotopes. Applications of nuclear chemistry to biology, industry, and agri-

◀ Enormous amounts of energy are constantly being generated in the sun by reactions of incredibly small atomic nuclei. [*NASA photo.*]

culture have improved the human condition significantly. The use of radioisotopes in biological and agricultural research has led to increased crop production, which provides more food for a hungry world.

SECTION 4.1

A Partial Parts List for the Atomic Nucleus

We saw in Chapter 3 that atomic nuclei are made up of protons and neutrons. Although nuclear physicists have extended this parts list to include at least a hundred particles, most of these particles have a transitory existence and are of little interest to a chemist. In this discussion, we take the oversimplified, but useful, view that atomic nuclei are made up of protons and neutrons.

A proton and a neutron have virtually the same mass, 1.0073 amu and 1.0087 amu, respectively. That's equivalent to saying that two different people weigh 100.7 kg and 100.9 kg. The difference is so small that it usually can be ignored. Thus, for many purposes, we assume the masses of the proton and the neutron to be the same, 1 amu. The proton has a charge equal in magnitude but opposite in sign to that of an electron. This charge on a proton is written as 1+. The electron has a charge of 1− and a mass of 0.0005 amu. The electrons in an atom contribute so little of its total mass that they are usually disregarded and are treated as if their mass were 0. The subatomic particles of greatest interest to us are summarized in Table 4.1.

Recall that the number of protons in the nucleus of an atom of any element is the atomic number of that element. Atoms of an element have differing numbers of neutrons, that is, they exist as isotopes (Chapter 3). Most, but not all, elements exist in nature in isotopic forms. An interesting and easy-to-remember example is the element tin (Sn), which exists in 10 isotopic forms. (Tin . . . ten isotopes.) Tin also has 15 radioactive isotopes that do not occur in nature.

Water in which both hydrogen atoms are deuterium is called *heavy water,* often written D_2O. Heavy water boils at 101.4 °C and freezes at 3.8 °C. Its density is 1.108 g/cm^3. (The density of ordinary water is 1.000 g/cm^3.)

Isotopes usually are of little importance in ordinary chemical reactions. All three hydrogen isotopes react with oxygen to form water. Since the isotopes differ in mass, compounds formed with different hydrogen isotopes have different physical properties, but such differences are usually slight. In nuclear reactions, however, isotopes are of utmost importance.

Table 4.1 Subatomic Particles

Particle	Symbol	Approximate Mass (amu)	Charge
Proton	p^+	1	1+
Neutron	n	1	0
Electron	e^-	0	1−

SECTION 4.2

Nuclear Arithmetic: Symbols for Isotopes

The two principal nuclear particles, protons and neutrons, collectively are called **nucleons.** Isotopes are represented by symbols with subscripts and superscripts. In the general symbol

$$^{A}_{Z}X$$

Z is the nuclear charge, or the atomic number (the number of protons), and A is the **mass number**, or the **nucleon number** (the number of protons plus the number of neutrons). As an example, in the symbol

$$^{35}_{17}Cl$$

the number of protons is 17 and the number of nucleons is 35. The number of neutrons is therefore $35 - 17 = 18$.

We will refer to A as the *nucleon number* in order to stress the fact that the mass number of an atom is equal to the total number of nucleons.

Isotopes also are identified by placing the nucleon number as a suffix to the name of the element. The three hydrogen isotopes are therefore represented either as

$$^{1}_{1}H \quad ^{2}_{1}H \quad ^{3}_{1}H$$

or as hydrogen-1, hydrogen-2, and hydrogen-3. The latter are easier to type; the former are useful for balancing nuclear equations.

EXAMPLE 4.1

How many neutrons are there in the $^{235}_{92}U$ nucleus?

SOLUTION

Simply subtract the atomic number (number of protons) from the nucleon number (number of protons plus neutrons).

$$\text{Nucleon number} - \text{atomic number} = \text{neutron number}$$

$$235 - 92 = 143$$

There are 143 neutrons in the nucleus.

Exercise 4.1

How many neutrons are there in the $^{90}_{38}Sr$ nucleus?

▶ EXAMPLE 4.2

How many neutrons are there in the chlorine-37 nucleus?

SOLUTION

The atomic number of chlorine is 17. The nucleon number is given as 37. The number of neutrons is therefore

$$37 - 17 = 20$$

Exercise 4.2

How many neutrons are there in the molybdenum-90 nucleus?

▶ EXAMPLE 4.3

(a) Which of the following are isotopes of the same element? (We are using the letter X as the symbol for all so that the symbol will not identify the elements.) (b) Which of the five isotopes have the same number of neutrons?

$$^{16}_{8}\text{X} \qquad ^{16}_{7}\text{X} \qquad ^{14}_{7}\text{X} \qquad ^{14}_{6}\text{X} \qquad ^{12}_{6}\text{X}$$

SOLUTION

a. $^{16}_{7}\text{X}$ and $^{14}_{7}\text{X}$ are isotopes of nitrogen (N). $^{14}_{6}\text{X}$ and $^{12}_{6}\text{X}$ are isotopes of carbon (C). $^{16}_{8}\text{X}$ and $^{16}_{7}\text{X}$ have the same nucleon number. The first is an isotope of oxygen, the second an isotope of nitrogen. $^{14}_{7}\text{X}$ and $^{14}_{6}\text{X}$ have the same nucleon number. The first is an isotope of nitrogen, the second an isotope of carbon.

b. $^{16}_{8}\text{X}$ and $^{14}_{6}\text{X}$ each have eight neutrons ($16 - 8 = 8$ and $14 - 6 = 8$, respectively).

Exercise 4.3

Which of the following are isotopes of the same element?

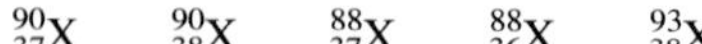

$$^{90}_{37}\text{X} \qquad ^{90}_{38}\text{X} \qquad ^{88}_{37}\text{X} \qquad ^{88}_{36}\text{X} \qquad ^{93}_{38}\text{X}$$

SECTION 4.3

Natural Radioactivity: Nuclear Equations

Some nuclei are unstable as they occur in nature; they undergo **radioactive decay**. Radium atoms with a nucleon number of 226, for example, break down spontaneously, giving off alpha particles. Since alpha particles are identical with helium nuclei, this process can be summarized by the equation

$$^{226}_{88}\text{Ra} \longrightarrow {}^{4}_{2}\text{He} + {}^{222}_{86}\text{Rn}$$

The new element, with two fewer protons, is identified by its atomic number (86) as radon (Rn). A second example of alpha decay is shown in Figure 4.1(a).

Tritium nuclei are also unstable. Tritium is one of the heavy isotopes of hydrogen. Like all hydrogen nuclei, the tritium nucleus contains one proton. Unlike the most common isotope of hydrogen, however, the tritium nucleus contains two neutrons, and its mass is therefore 3 amu ($^{3}_{1}\text{H}$). Tritium decomposes by beta decay. Since a beta particle is identical to an electron, this process may be written

$$^{3}_{1}\text{H} \longrightarrow {}^{0}_{-1}\text{e} + {}^{3}_{2}\text{He} \qquad \text{or} \qquad {}^{3}_{1}\text{H} \longrightarrow \beta + {}^{3}_{2}\text{He}$$

The product isotope is identified by its atomic number as helium (He).

How can the original nucleus, which contains only a proton and two neutrons, emit an *electron?* We can envision one of the neutrons in the original tritium nucleus changing into a proton and an electron.

$$^{1}_{0}\text{n} \longrightarrow {}^{1}_{1}\text{p} + {}^{0}_{-1}\text{e}$$

The new proton is retained by the nucleus (therefore, the atomic number of the product increased by one), and the almost massless electron or beta particle is kicked out (the product nucleus has the same mass as the original). A second example of beta decay is pictured in Figure 4.1(b).

The third type of radioactivity is called gamma decay. Since this type of

Figure 4.1 Nuclear emission of (a) an alpha particle and (b) a beta particle.

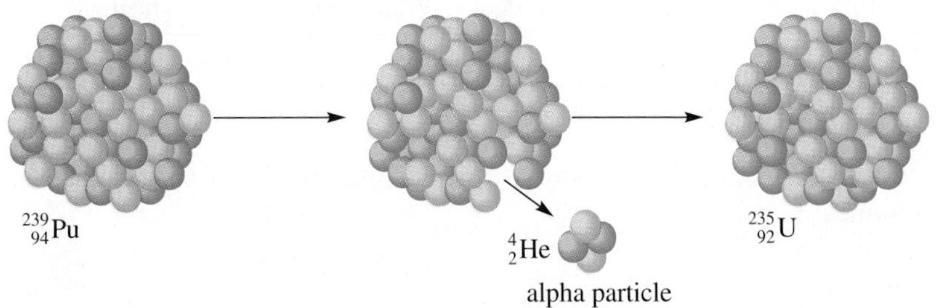

(a) Nuclear changes accompanying alpha decay.

$^{14}_{6}$C $\longrightarrow$ $^{14}_{7}$N + $^{0}_{-1}$e beta particle

(b) Nuclear changes accompanying beta decay.

Table 4.2 Radioactive Decay and Nuclear Change

Type of Radiation	Greek Letter	Nucleon Number	Charge	Change in Nucleon Number	Change in Atomic Number
Alpha	α	4	2+	Decreases by 4	Decreases by 2
Beta	β	0	1−	No change	Increases by 1
Gamma	γ	0	0	No change	No change

emission involves no particle, no equation is needed. Gamma emission often accompanies the emission of alpha or beta particles.

The major types of radioactive decay and the ensuing nuclear changes are summarized in Table 4.2.

▶ EXAMPLE 4.4

Plutonium-239 emits an alpha particle when it decays. What new isotope is formed?

$$^{239}_{94}\text{Pu} \longrightarrow \ ? + {}^{4}_{2}\text{He}$$

SOLUTION

Nucleon number and nuclear charge are conserved. The new element must have a nucleon number of 239 − 4 = 235 and a charge of 94 − 2 = 92. The atomic number (nuclear charge), $Z = 92$, identifies the element as uranium.

$$^{239}_{94}\text{Pu} \longrightarrow {}^{235}_{92}\text{U} + {}^{4}_{2}\text{He}$$

Exercise 4.4

Fermium-250 undergoes alpha decay. What new isotope is formed?

▶ EXAMPLE 4.5

Cobalt-61 emits a beta particle when it decays. What new isotope is formed?

$$^{61}_{27}\text{Co} \longrightarrow \ ? + {}^{\ 0}_{-1}\text{e}$$

SOLUTION

Nucleon number and nuclear charge are conserved. The new element must have a mass number of 61 − 0 = 61 and a charge of 27 − (−1) = 28. The atomic number (nuclear charge), $Z = 28$, identifies the element as nickel.

$$^{61}_{27}\text{Co} \longrightarrow {}^{61}_{28}\text{Ni} + {}^{\ 0}_{-1}\text{e}$$

Exercise 4.5

Selenium-85 undergoes beta decay. What new isotope is formed?

SECTION 4.4

Half-life

Thus far we have discussed radioactivity as applied to single atoms. In the laboratory, we generally deal with great numbers of atoms—numbers far larger than the number of people on all the Earth. If we could see the nucleus of an individual atom, we could tell *whether or not* it would undergo radioactive decay by noting its composition. Certain combinations of protons and neutrons are unstable. We could not, however, determine *when* the atom would undergo a change. Radioactivity is a random process, generally independent of outside influences.

With large numbers of atoms, the process of radioactive decay becomes more predictable We can measure the *half-life,* a property characteristic of each radioisotope. The **half-life** of a radioactive isotope is the period in which one-half the original number of atoms undergo radioactive decay to form a new element. Suppose, for example, we had 16 billion atoms of tritium, the radioactive isotope of hydrogen. The half-life of tritium is 12.3 years. This means that in 12.3 years, 8 billion atoms of the tritium will have decayed, and there will be 8 billion atoms left. In another 12.3 years, half the remaining 8 billion atoms will have decayed. After two half-lives, then, one-quarter of the original tritium atoms will remain. Two half-lives, then, do not make a whole! The concept of half-life is illustrated by the graph in Figure 4.2.

The half-life of an element can be very long (millions of years) or extremely short (tiny fractions of a second). The half-life of uranium-238 is 4.5 billion years; that of boron-9 is 8×10^{-19} s.

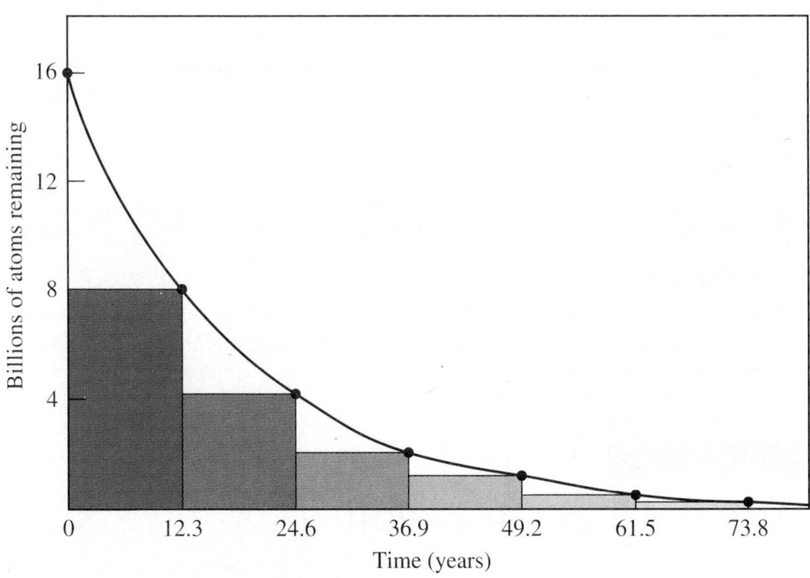

Figure 4.2 The radioactive decay of tritium (hydrogen-3). Successive colored blocks each represent one half-life.

We can calculate the fraction of the original isotope that remains after a given number of half-lives from the relationship

$$\text{Fraction remaining} = \frac{1}{2^n}$$

where n is the number of half-lives.

EXAMPLE 4.6

You obtain a new sample of cobalt-60, half-life 5.25 years, with a mass of 400 mg. How much cobalt-60 remains after 15.75 years (three half-lives)?

SOLUTION

The fraction remaining after three half-lives is

$$1/2^n = 1/2^3 = 1/(2 \times 2 \times 2) = 1/8$$

The amount of cobalt-60 remaining is (1/8)(400 mg) = 50 mg.

Exercise 4.6

You have 1.224 mg of freshly prepared gold-189, half-life 30 min. How much of the gold-189 sample remains after five half-lives?

It is impossible to say when *all* the atoms of a radioactive isotope will have decayed. For most samples, we can assume that the activity is essentially gone after about 10 half-lives. (After 10 half-lives, the activity is $1/2^{10} = 1/1024$ of the original value.) Generally, we can say that one-thousandth of the original activity remains.

EXAMPLE 4.7

You obtain a 20.0-mg sample of mercury-190, half life 20 min. How much of the mercury-190 sample remains after 2 hr?

SOLUTION

There are 120 min in 2 hr. There are (120/20) = 6 half-lives in 2 hr. The fraction remaining after six half-lives is

$$1/2^n = 1/2^6 = 1/(2 \times 2 \times 2 \times 2 \times 2 \times 2) = 1/64$$

The amount of mercury-190 remaining is (1/64)(20.0) mg = 0.313 mg.

Exercise 4.7

A sample of 16.0 mg of nickel-57, half-life 36.0 hr, is produced in a nuclear reactor. How much of the nickel-57 sample remains after 7.5 days?

SECTION 4.5

Artificial Transmutation

The forms of radioactivity encountered thus far occur in nature. Other nuclear reactions may be brought about by bombardment of stable nuclei with alpha particles, neutrons, or other subatomic particles. These particles, given sufficient energy, penetrate the formerly stable nucleus and bring about some form of radioactive emission. Like natural radioactivity, this sort of nuclear change brings about a **transmutation**: one element is changed into another. Because the change would not have occurred naturally, the process is called *artificial transmutation.*

Ernest Rutherford carried out the first nuclear bombardment experiment. [*Professor Marvin Lang and Mr. Gary Shulfer, University of Wisconsin–Stevens Point (Scott Cat. New Zealand #487).*]

Ernest Rutherford, a few years after his famous gold-foil experiment (Chapter 3), studied the bombardment of a variety of light elements with alpha particles. One such experiment, in which he bombarded nitrogen, resulted in the production of protons.

$$^{14}_{7}\text{N} + ^{4}_{2}\text{He} \longrightarrow ^{17}_{8}\text{O} + ^{1}_{1}\text{H}$$

(Recall that the hydrogen nucleus is a proton; hence the alternative symbol $^{1}_{1}\text{H}$ for the proton.) Notice that the sum of the nucleon numbers on the left equals the sum of the nucleon numbers on the right. The atomic numbers are also balanced.

Rutherford had postulated the existence of protons in nuclei in 1914. An experiment published in 1919 gave the first empirical vertification of the existence of these fundamental particles. Eugen Goldstein had earlier produced protons in his discharge tube experiments (Chapter 3). He obtained these particles from hydrogen gas in the tube by knocking an electron away from the hydrogen atom. The significance of Rutherford's experiment lay in the fact that he obtained protons from the nucleus of an atom other than hydrogen, thus establishing their nature as fundamental particles. By **fundamental particles** we mean basic units from which more complicated structures (such as the nitrogen nucleus) can be fashioned. Rutherford's experiment was the first induced nuclear reaction.

A great many transmutations were carried out during the 1920s. In the 1930s one such reaction led to the discovery of another fundamental particle. James Chadwick, in 1932, bombarded beryllium with alpha particles.

$$^{9}_{4}\text{Be} + ^{4}_{2}\text{He} \longrightarrow ^{12}_{6}\text{C} + ^{1}_{0}\text{n}$$

Among the products was the neutron.

▶ EXAMPLE 4.8

When potassium-39 is bombarded with neutrons, chlorine-36 is produced. What other particle is emitted?

$$^{39}_{19}\text{K} + ^{1}_{0}\text{n} \longrightarrow ^{36}_{17}\text{Cl} + ?$$

SOLUTION

Write a balanced nuclear equation. To balance the equation, we need four mass units and two charge units (that is, a particle with a nucleon number of 4 and an atomic number of 2). That's an alpha particle.

$$^{39}_{19}\text{K} + ^{1}_{0}\text{n} \longrightarrow ^{36}_{17}\text{Cl} + ^{4}_{2}\text{He}$$

Exercise 4.8

Technetium-97 is produced by bombarding molybdenum-96 with a deuteron (hydrogen-2 nucleus). What other particle is emitted?

$$^{96}_{42}\text{Mo} + ^{2}_{1}\text{H} \longrightarrow ^{97}_{43}\text{Tc} + ?$$

SECTION 4.6

Induced Radioactivity

The first artificial nuclear reactions produced isotopes already known to occur in nature. This was perhaps fortuitous, for it was inevitable that an unstable nucleus would be produced sooner or later. Irène Curie (daughter of the 1903 Nobel Prize winners) and her husband, Frédéric Joliot, were studying the bombardment of aluminum with alpha particles. Neutrons were produced, leaving behind an isotope of phosphorus.

$$^{27}_{13}\text{Al} + ^{4}_{2}\text{He} \longrightarrow ^{30}_{15}\text{P} + ^{1}_{0}\text{n}$$

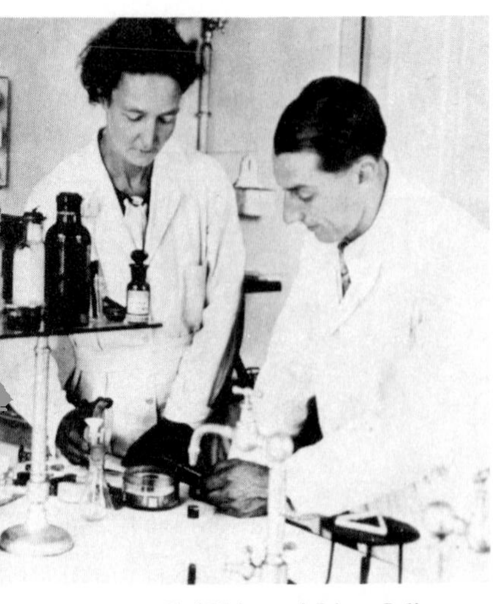

Frédéric and Irène Joliot-Curie discovered artificially induced radioactivity in 1934. They are shown here working in their laboratory. [*The Bettmann Archive.*]

Much to their surprise the target continued to emit particles after the bombardment was halted. The isotope of phosphorus was radioactive, emitting particles equal in mass to the electron but opposite in charge. These particles are called **positrons**. The reaction they were observing is written

$$^{30}_{15}\text{P} \longrightarrow {}_{1+}^{0}\text{e} + ^{30}_{14}\text{Si}$$

Once again the question arises: Where does this particle come from if the nucleus contains only protons and neutrons? Previously we accounted for a beta particle (an electron) popping out of a nucleus by saying a neutron changed into a proton and an electron. Perhaps a similar happening can account for the appearance of a positron. Imagine a proton changing to a neutron and a positron (a proton is the same as a hydrogen nucleus).

$$^{1}_{1}\text{H} \longrightarrow {}_{1+}^{0}\text{e} + ^{1}_{0}\text{n}$$

Everything balances rather nicely in this equation. When the positron is emitted, the original radioactive nucleus suddenly has one less proton and one

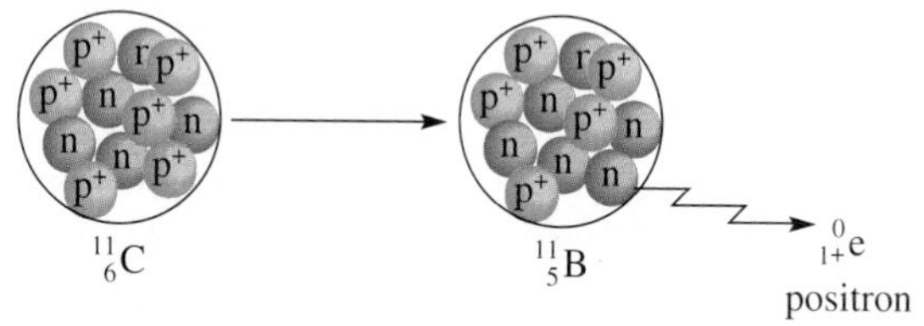

Figure 4.3 Nuclear change accompanying positron emission.

more neutron than before. Therefore, the mass of the product nucleus is the same, but its atomic number is one less than that of the original nucleus.

This work gave the Joliot-Curies a Nobel prize of their own in 1935. (The Joliot-Curies adopted the combined surname to perpetuate the Curie name. Marie and Pierre Curie had two daughters, but no son.)

Figure 4.3 presents another example of nuclear decay involving positron emission.

▶ EXAMPLE 4.9

Carbon-10 decays by positron emission. Write a balanced nuclear equation for the process.

$$^{10}_{6}C \longrightarrow \, ^{0}_{1+}e + ?$$

SOLUTION

To balance the equation, an isotope with a nucleon number of 10 and an atomic number of 5 is required. The element with atomic number 5 is boron (B).

$$^{10}_{6}C \longrightarrow \, ^{0}_{1+}e + \, ^{10}_{5}B$$

Exercise 4.9

Gold-188 decays by positron emission. Write a balanced nuclear equation for the process.

SECTION 4.7

Penetrating Power of Radiation

Radioactive materials can be dangerous because the radiation emitted can damage living tissue. The ability to inflict injury depends in part on the power of the radiation to penetrate the tissue.

All other things being equal, the more massive the particle, the less its penetrating power. Alpha particles, which are helium nuclei with a mass of 4 amu, are the least penetrating of the three main types of radioactivity. Beta particles, which are identical to the almost massless electrons, are somewhat more penetrating. Gamma rays, like X-rays, truly have no mass; they are considerably more penetrating than the other two types.

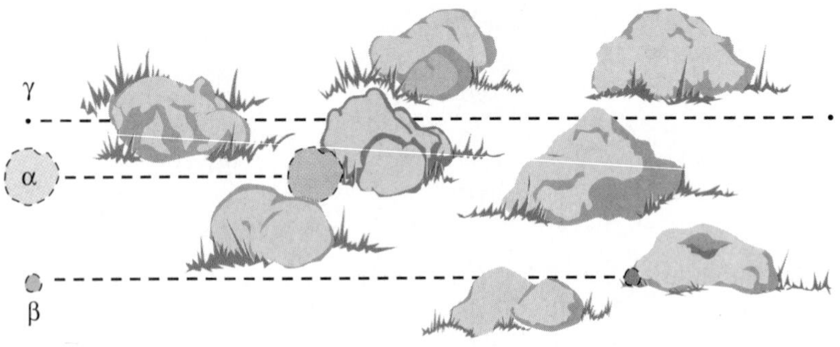

Figure 4.4 Shooting radioactive particles through matter is like rolling rocks through a field of boulders—the larger rocks are more quickly stopped.

It may seem contrary to common sense that the biggest particles make the least headway. Consider that penetrating power reflects the ability of the radiation to make its way through a sample of matter. It is as if you were trying to roll some rocks through a field of boulders. The alpha particle acts as if it were a boulder itself. Because of its size, it cannot get very far before it bumps into and is stopped by other boulders. The beta particle acts like a small stone. It can sneak between boulders and perhaps ricochet off one or another until it has made its way farther into the field (Figure 4.4). The gamma ray can be compared to a grain of sand that can get through the smallest openings.

At the beginning of this discussion we said that, all other things being equal, this is how things worked. But all other things are not always equal.

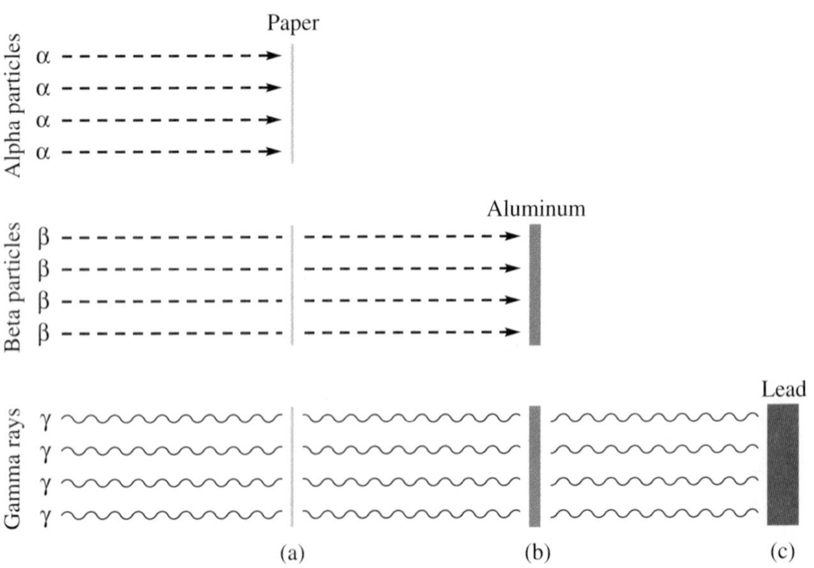

Figure 4.5 The relative penetrating power of alpha, beta, and gamma radiation. Alpha particles are stopped by a sheet of paper (a). Beta particles will not penetrate a sheet of aluminum foil (b). It takes several centimeters of lead to block gamma rays (c).

The faster a particle moves or the more energetic the radiation is, the more penetrating power it has.

If a radioactive substance is *outside* the body, alpha particles of low penetrating power are least dangerous; they are stopped by the outer layer of skin. Beta rays also usually are stopped before reaching vital organs. Gamma rays readily pass through tissues; an external gamma source can be quite dangerous. When the radioactive source is *inside* the body, the situation is reversed. The nonpenetrating alpha particles can do great damage. All such particles are trapped within the body, which must then absorb all the energy released by the particle. Alpha rays inflict all their damage in a very small area because they do not travel far. Beta rays distribute the damage over a somewhat larger area because they travel farther. Tissue may recover from limited damage spread over a large area; it is less likely to survive concentrated damage.

People working with radioactive materials can do several things to protect themselves. The simplest is to move away from the source, for intensity of radiation decreases with distance from the source. Workers can also be protected by shielding. A sheet of paper will stop most alpha particles. A block of wood or a thin sheet of aluminum will stop beta particles. But it takes several meters of concrete or several centimeters of lead to stop gamma rays (Figure 4.5).

SECTION 4.8

Uses of Radioisotopes

Scientists in a wide variety of fields make use of radioactive isotopes (**radioisotopes**) as **tracers** in physical, chemical, and biological systems. Isotopes of a given element, whether radioactive or not, behave nearly identically in chemical and physical processes. Since radioactive isotopes are easily detected, it is relatively easy to trace their movements, even through a complicated system.

As a simple example, let's consider the flow of a liquid through a pipe. Suppose there is a leak in a pipe buried beneath a concrete floor. We could locate the leak by digging up extensive areas of the floor, or we could add a small amount of radioactive material to liquid going into the drain and trace the flow of liquid with a Geiger counter (an instrument that detects radioactivity). Once the leak was located, only a small area of the floor would have to be dug up to repair the leak. Short-lived isotopes—which disappear soon after doing their job—usually are employed for such purposes.

In a similar manner, we could trace the uptake of phosphorus by a green plant. The plant is fed some fertilizer containing radioactive phosphorus. A simple method of detection involves placing the plant on a photographic film. Radiation from the phosphorus isotopes exposes the film, much as light does. This type of exposure, called a *radiograph*, shows the distribution of phospho-

(a)

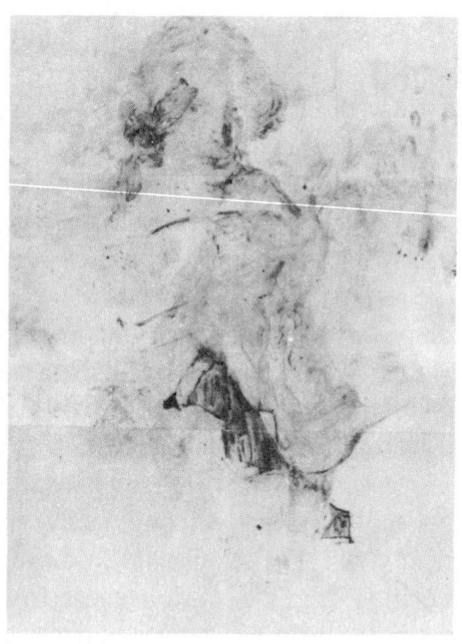

(b)

Figure 4.6 (a) Saint Rosalie Interceding for the Plague-Stricken of Palermo by Anthony van Dyck (1599–1641). (b) Radiograph reveals the presence of an earlier painting on the same canvas. [*(a) The Metropolitan Museum of Art, Purchase, 1871 (71.41); (b) courtesy of Brookhaven National Laboratory with permission of the Paintings Conservation Department, The Metropolitan Museum of Art.*]

rus in the plant. Figure 4.6 shows another application of radiography—in art.

Radioactive tracers also are put to good use in agricultural research. They are used to study the effectiveness of fertilizers and weed killers, to compare the nutritional value of various feeds, and to determine the best methods for controlling insects. The purposeful mutation of plants by irradiation has produced new and improved strains of commercially valuable crop plants ranging from tobacco to peanuts.

Radioisotopes are also used as sources for the irradiation of foodstuffs as a method of preservation (Figure 4.7). The radiation destroys microorganisms that cause food spoilage. Irradiated food shows little change in taste or appearance. Some people are concerned about possible harmful effects of chemical substances produced by the radiation, but there is no good evidence of harm to

Figure 4.7 Gamma radiation delays the decay of mushrooms. Those on the right were irradiated; the ones on the left were not. [*Richard Megna/Fundamental Photographs.*]

laboratory animals fed irradiated food, nor are there any known adverse effects in humans in countries where irradiation has been used for several years. There is no residual radiation in the food after the sterilization process.

Radioisotopes have been used extensively in basic scientific research. The mechanism of photosynthesis was worked out in large part by using carbon-14 as a tracer. For example, to determine how plants make the sugar glucose from carbon dioxide and water, the plant is exposed to radioactive carbon dioxide. The compounds formed from these starting materials and their order of formation then is followed by determining which new compounds become radioactive, and in what sequence. Using data from radioactive tracer experiments, scientists determine metabolic pathways in plants, animals, and humans.

SECTION 4.9

Nuclear Medicine

Nuclear medicine involves two distinct uses of radioisotopes: therapeutic and diagnostic. In radiation therapy, an attempt is made to treat or cure disease with radiation. The diagnostic use of radioisotopes is aimed at obtaining information about the state of an individual's health.

Radiation Therapy

Cancer is not one disease but many. Some forms are particularly susceptible to radiation therapy. Radiation is carefully aimed at the cancerous tissue, and exposure of normal cells is minimized. If the cancer cells are killed by the destructive effects of the radiation, the malignancy is halted. But persons undergoing radiation therapy often get sick from the treatment. Nausea and vomiting are the usual symptoms of radiation sickness. Thus, the aim of radiation therapy is to destroy the cancerous cells before too much damage is done to healthy tissue. Radiation is most lethal to rapidly reproducing cells, and this is precisely the characteristic of cancer cells that allows the therapy to be successfully applied.

Diagnostic Uses of Radioisotopes

Radioisotopes are used for diagnostic purposes to provide information about the type or extent of illness. Radioactive iodine-131 is used to determine the size, shape, and activity of the thyroid gland, as well as to treat cancer located in this gland and to control a hyperactive thyroid. First, one drinks a solution of potassium iodide incorporating iodine-131. The body concentrates iodide in the thyroid. Large doses are used for treatment of thyroid cancer; the radiation from the isotope concentrates in the thyroid cancer cells even if the cancer has spread to other parts of the body. For diagnostic purposes, however, only a small amount of the radioisotope is needed. Again the material is concentrated in the thyroid. A detector is set up so that readings are translated into a permanent visual record showing the differential uptake of the isotope. The "picture" that results is referred to as a **photoscan**, and it can pinpoint the location of a tumor in that area of the body.

The radioisotope most widely used in medicine is gadolinium-153. This isotope is used to determine bone mineralization. Its popularity is an indication of the large number of people, mostly women, who suffer from *osteoporosis* (reduction in the quantity of bone) as they grow older. Gadolinium-153 gives off two characteristic radiations, a gamma ray and an X-ray. A scanning device compares these radiations after they pass through bone. Bone densities are then determined by differences in absorption of the rays.

Technetium-99^{m} is used in a variety of diagnostic tests (Figure 4.8). The *m* stands for *metastable,* which means that this isotope will give up some energy to become a more stable version of the same isotope (same atomic number, same atomic weight). The energy it gives up is the gamma ray needed to detect the isotope.

$$ {}^{99m}_{43}\mathrm{Tc} \longrightarrow {}^{99}_{43}\mathrm{Tc} + \gamma $$

Notice that the decay of technetium-99^{m} produces no alpha or beta particles, which could cause unnecessary damage to the body. Technetium-99^{m} also has a short half-life (about 6 hr), which means that the radioactivity does not linger in the body long after the scan has been completed. With this short a half-life, use of the isotope must be carefully planned. In fact, the isotope itself is not what is purchased. Technetium-99^{m} is formed by the decay of molybdenum-99.

$$ {}^{99}_{42}\mathrm{Mo} \longrightarrow {}^{99m}_{43}\mathrm{Tc} + {}^{0}_{-1}\mathrm{e} + \gamma $$

A container of this molybdenum isotope is obtained, and the decay product, technetium-99^{m}, is "milked" from the container as needed.

Using modern computer technology, **positron emission tomography**

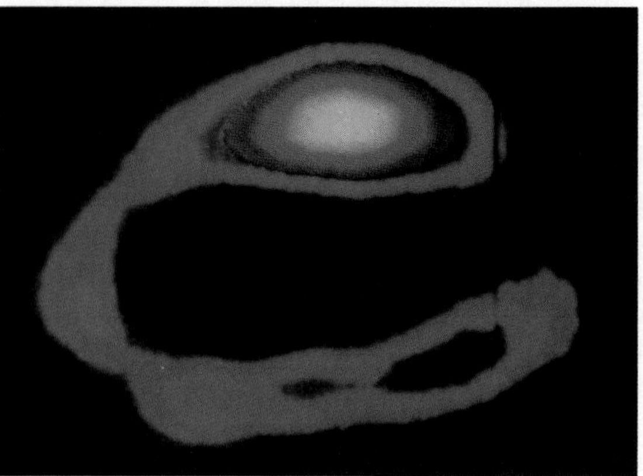

Figure 4.8 Blood flow patterns in a healthy heart (left) and in a damaged heart (right). The highlighted images from a technetium-99^{m} compound indicate regions receiving adequate blood flow. [*Courtesy of Du Pont Merck Pharmaceutical Company.*]

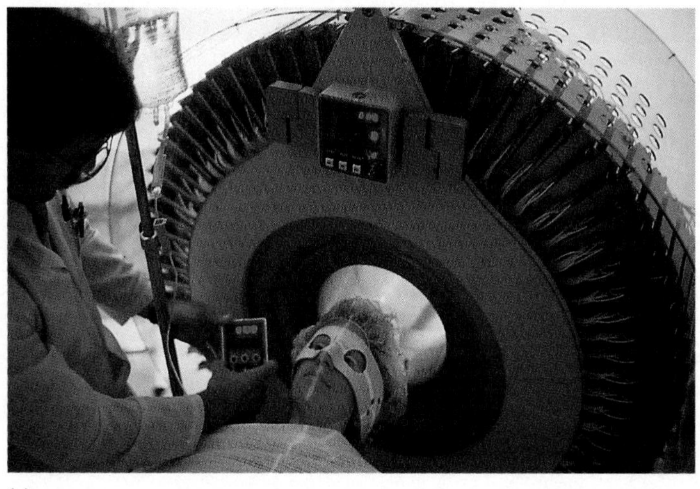

(a)

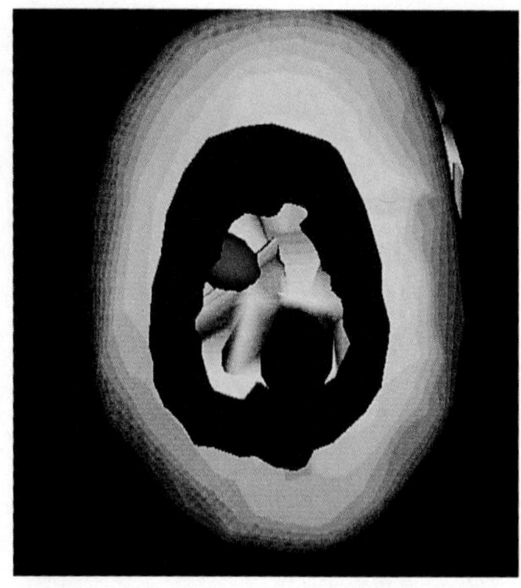

(b)

Figure 4.9 Modern computer technology used for medical diagnosis. (a) Patient in position for positron emission tomography (PET), a technique that uses radioisotopes to scan internal organs. (b) Image created by computer tomography (CT), a scanning technique that uses X-rays rather than radioisotopes, looking through the skull at a pituitary brain tumor. [*(a) Courtesy of The University of Chicago Medical Center; (b) Dan McCoy/Rainbow Stock.*]

(PET), can measure dynamic processes occurring in the body, such as blood flow or the rate at which oxygen or glucose is being metabolized. PET scans are being used to pinpoint the area of brain damage that triggers severe epileptic seizures. Compounds incorporating positron-emitting isotopes, such as carbon-11, are inhaled or injected prior to the scan. Before the emitted positron can travel very far in the body, it encounters an electron (in any ordinary matter there are numerous electrons) and two gamma rays are produced.

$$^{11}_{6}\text{C} \longrightarrow {}^{11}_{5}\text{B} + {}^{0}_{1+}\text{e}$$

$$^{0}_{1+}\text{e} + {}^{0}_{1-}\text{e} \longrightarrow 2\,\gamma$$

These exit from the body in exactly opposite directions. Detectors positioned on opposite sides of the patient record the gamma rays. By setting the recorders so that two simultaneous gamma rays must be ''seen,'' gamma rays resulting from natural background radiation are ignored. A computer is then used to calculate the point within the body at which the annihilation of the positron and electron occurred, and an image of that area is produced (Figure 4.9).

Table 4.3 is a list of some radioisotopes in common use in medicine. The list is necessarily incomplete. Even this abbreviated discussion should give you an idea of the importance of radioisotopes in medicine. The claim that nuclear science has saved many more lives than nuclear bombs have destroyed is not an idle one.

Positron emission tomography can reveal metabolic changes that occur in the brain during tactile learning (learning by the sense of touch).

▶ EXAMPLE 4.10

One of the isotopes used for PET scans is oxygen-15, a positron emitter. What new element is formed when the oxygen-15 decays?

Table 4.3 Some Radioisotopes and Their Application in Medicine

Isotope	Name	Radiation	Uses
^{51}Cr	Chromium-51	γ	Determination of volume of red blood cells and total blood volume
^{57}Co	Cobalt-57	γ	Determination of uptake of vitamin B_{12}
^{60}Co	Cobalt-60	β, γ	Radiation treatment of cancer
^{153}Gd	Gadolinium-153	γ	Determination of bone density
^{131}I	Iodine-131	β, γ	Detection of thyroid malfunction; measurement of liver activity and fat metabolism; treatment of thyroid cancer
^{59}Fe	Iron-59	β, γ	Measurement of rate of formation and lifetime of red blood cells
^{32}P	Phosphorus-32	β	Detection of skin cancer or cancer of tissue exposed by surgery
^{226}Ra	Radium-226	α, γ	Radiation therapy for cancer
^{24}Na	Sodium-24	β, γ	Detection of constrictions and obstructions in the circulatory system
^{99m}Tc	Technetium-99^{m}	γ	Imaging of brain, thyroid, liver, kidney, lung, and cardiovascular system
^{3}H	Tritium	β	Determination of total body water

SOLUTION

First write the nuclear equation.

$$^{15}_{8}O \longrightarrow {}^{0}_{1+}e + ?$$

The nucleon number does not change, but the atomic number becomes 8 − 1, or 7; so the new product is nitrogen-15.

$$^{15}_{8}O \longrightarrow {}^{0}_{1+}e + {}^{15}_{7}N$$

Exercise 4.10

Phosphorus-30 is a positron-emitting radioisotope suitable for use in PET scans. What new element is formed when phosphorus-30 decays?

SECTION 4.10

Radioisotopic Dating

The half-lives of certain isotopes can be used to estimate the age of rocks and archaeological artifacts. Uranium-238 decays with a half-life of 4.5 billion years. The initial products of this decay are also radioactive, and breakdown continues until an isotope of lead (lead-206) is formed. By measuring the relative amounts of uranium-238 and lead-206, chemists can estimate the age of a rock. Some of the older rocks on the Earth have been found to be from 3.0

to 4.5 billion years old. Moon rocks and meteorites have been dated at a maximum age of about 4.5 billion years. Thus, the age of the Earth is generally estimated to be about 4.5 billion years.

Carbon-14 Dating

The dating of artifacts usually involves a radioactive isotope of carbon. Carbon-14 is formed in the upper atmosphere by the bombardment of ordinary nitrogen by neutrons from cosmic rays.

$$^{14}_{7}N + ^{1}_{0}n \longrightarrow ^{14}_{6}C + ^{1}_{1}H$$

This process leads to a steady-state concentration of carbon-14 in the earth's CO_2. Living plants and animals incorporate this isotope into their own cells. When they die, however, the incorporation of carbon-14 ceases, and the carbon-14 in the plants and organisms decays—with a half-life of 5730 years—to nitrogen-14. Thus, we merely need to measure the carbon-14 activity remaining in an artifact of plant or animal origin to determine its age. For example, a sample that has half the carbon-14 activity of new plant material is 5730 years old; it has been dead for one half-life. Similarly, an artifact with 25% of the carbon-14 activity of new plant material is 11,460 years old; it has been dead for two half-lives.

Carbon-14 dating, as outlined here, assumes that the formation of the isotope was constant over the years. This is not quite the case. However, for the most recent 7000 years or so, carbon-14 dates have been correlated with those obtained from the annual growth rings of trees. Calibration curves have been constructed from which accurate dates can be determined. Generally, carbon-14 is reasonably accurate for dating objects up to about 50,000 years old. Objects older than that have too little of the isotope left for accurate measurement.

Charcoal from the fires of an ancient people, dated by determining the carbon-14 activity, is used to estimate the age of other artifacts found at the same archaeological site. Carbon-14 dating also has been used to detect forgeries of supposedly ancient artifacts.

The Shroud of Turin

Perhaps you have heard of the Shroud of Turin. It is a very old piece of linen cloth, about 4 meters long, bearing a faint human likeness. Since about A.D. 1350 it had been alleged to be part of the burial shroud of Christ. However, carbon-14 dating studies in 1988 by three different nuclear laboratories indicated that the flax used in making the cloth was not grown until sometime between A.D. 1260 and 1390. Therefore, the cloth could not possibly have existed at the time of Christ. Unlike the Dead Sea Scrolls, which were shown by carbon-14 dating to be authentic records from a civilization that existed about 2000 years ago, the Shroud of Turin was exposed as a hoax.

Table 4.4 Several Isotopes Useful in Radioactive Dating

Isotope	Half-life (years)	Useful Range	Dating Applications
Carbon-14	5730	500 to 50,000 years	Charcoal, organic material
Tritium (^3_1H)	12.3	1 to 100 years	Aged wines
Potassium-40	1.3×10^9	10,000 years to the oldest Earth samples	Rocks, the Earth's crust, the moon's crust
Rhenium-187	4.3×10^{10}	4×10^7 years to oldest samples in the universe	Meteorites
Uranium-238	4.5×10^9	10^7 years to the oldest Earth samples	Rocks, the Earth's crust

Tritium Dating

Tritium, the radioactive isotope of hydrogen, also is useful for dating. Its half-life of 12.3 years makes it useful for dating items up to about 100 years old. An interesting application is the dating of brandies. These alcoholic beverages are quite expensive when aged from 10 to 50 years. Tritium dating can be used to check the veracity of advertising claims about the aging process of the most expensive kinds.

Many other isotopes are useful for estimating the ages of objects and materials. Several of the more important ones are listed in Table 4.4.

EXAMPLE 4.11

A piece of fossilized wood has a carbon-14 activity that is one-eighth that of new wood. How old is the artifact? The half-life of carbon-14 is 5730 years.

SOLUTION

The carbon-14 has gone through three half-lives.

$$\frac{1}{8} = \left(\frac{1}{2}\right)^3 = \frac{1}{2} \times \frac{1}{2} \times \frac{1}{2}$$

It is therefore about $3 \times 5730 = 17{,}190$ years old.

Exercise 4.11

How old is a piece of cloth that has a carbon-14 activity one-sixteenth that of new cloth fibers? The half-life of carbon-14 is 5730 years.

Albert Einstein on a Chinese postage stamp. [*Professor Marvin Lang and Mr. Gary Shulfer, University of Wisconsin–Stevens Point (Scott Cat. China #1468).*]

SECTION 4.11

Einstein and the Equivalence of Mass and Energy

One of the most famous and most unusual scientists was Albert Einstein. Whereas most scientists work with glassware and instruments in laboratories,

Einstein worked with a pencil and a note pad. By 1905, at the age of 26, he had already worked out his special theory of relativity and had developed his famous **mass–energy equation**

$$E = mc^2$$

where E represents energy, m is mass, and c is the speed of light. The equation suggests that mass and energy are just two different aspects of the same thing.

According to Einstein's equation, a chemical reaction that gives off heat must lose mass in the process, although the change in mass is much too small to measure. Reaction energy must be enormous in order for the mass loss to be measurable—for example, the energy given off by nuclear explosions. (Converting a single gram of matter to energy would provide enough heat to warm an average home for 1000 years, if conversion were complete; but even when a hydrogen bomb explodes, less than 1% of the matter is converted to energy.)

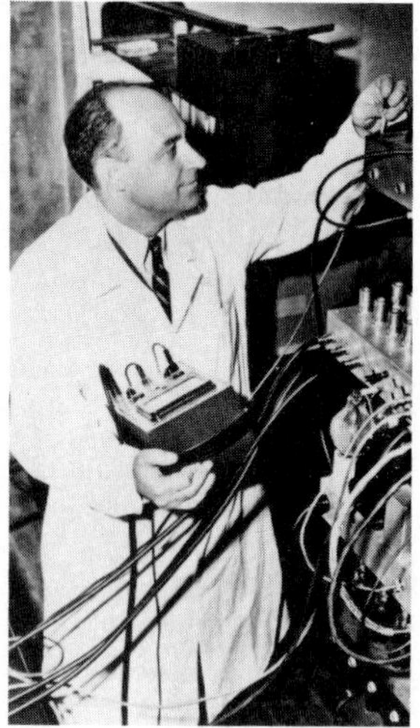

Enrico Fermi. [*Courtesy of Mrs. Laura Fermi.*]

SECTION 4.12

The Building of the Bomb

In 1934 the Italian scientists Enrico Fermi and Emilio Segrè bombarded uranium atoms with neutrons. They were trying to make elements higher in atomic number than uranium, which then had the highest known number. To their surprise, they found *four* radioactive species among the products. One presumably was element 93, formed by the initial conversion of uranium-238 to uranium-239.

$$^{238}_{92}\mathrm{U} + ^{1}_{0}\mathrm{n} \longrightarrow ^{239}_{92}\mathrm{U}$$

The latter then underwent beta decay.

$$^{239}_{92}\mathrm{U} \longrightarrow {}_{-1}^{0}\mathrm{e} + ^{239}_{93}\mathrm{Np}$$

The two scientists were unable to explain the remaining radioactivity.

Lise Meitner is portrayed on an Austrian postage stamp. [*Professor Marvin Lang and Mr. Gary Shulfer, University of Wisconsin–Stevens Point (Scott Cat. Austria #31093).*]

Nuclear Fission

In repeating the Fermi–Segrè experiment in 1938, German chemists Otto Hahn and Fritz Strassman were perplexed to find isotopes of barium (Ba) among the reaction products. Hahn wrote to Lise Meitner, his long-time former colleague, to ask what she thought about these strange results.

Lise Meitner was an Austrian physicist who had worked with Hahn in Berlin. Because she was Jewish she had recently escaped to Sweden when the Nazis took over Austria in 1938. Upon hearing about Hahn's work, she noted that barium atoms were only about half the size of uranium atoms. Was it possible that the uranium nucleus might be splitting into fragments? She made

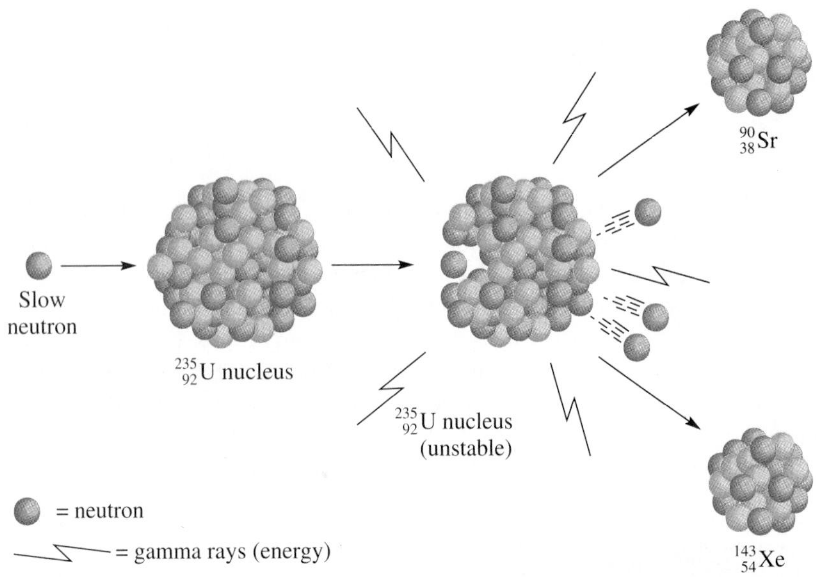

Figure 4.10 The splitting of a uranium atom. The neutrons produced in the fission can split other uranium atoms, thus sustaining a chain reaction. The splitting of one uranium-235 atom yields 8.9×10^{-18} kWh of energy. Fission of a mole of uranium-235 (6.02×10^{23} atoms) produces 5.3 million kWh of energy.

some calculations that convinced her that uranium nuclei had indeed been split apart! Her nephew, Otto Frisch, was visiting for the Christmas holidays, and they discussed this new discovery with great excitement. It was Frisch who later coined the term **nuclear fission**. (See Figure 4.10.)

Frisch was working with Niels Bohr at the University of Copenhagen, and when he returned to Denmark he took the news about the fission reaction to Bohr, who happened to be going to the United States to attend a physics conference. The discussions in the corridors about this new reaction would be the most important talks to take place at that meeting.

Meanwhile, Enrico Fermi had just received the 1938 Nobel prize in physics. Since Fermi's wife Laura was Jewish, and the fascist Italian dictator Mussolini was an ally of Hitler, Fermi accepted the award in Stockholm and then immediately fled with his wife and child to the United States. Thus it was that by 1939 the United States had received news about the German discovery of nuclear fission and had also acquired from Italy one of the world's foremost nuclear scientists.

Leo Szilard persuaded Einstein to send the letter to President Franklin Roosevelt that resulted in the Manhattan Project. But later he was one of the scientists who begged that the bomb not be used. [*UPI/Bettmann.*]

Nuclear Chain Reaction

Leo Szilard was one of the first scientists to realize that nuclear fission could be a practical chain reaction. Szilard had been born in Hungary and educated in Germany, but he came to the United States in 1937, another Jewish refugee. He saw that neutrons released in the fission of one atom could trigger the fission of other uranium atoms, thus setting off a **chain reaction** (Figure 4.11). Since massive amounts of energy could be obtained from the fission of uranium, he saw that the fission process might produce a bomb with tremendous explosive force.

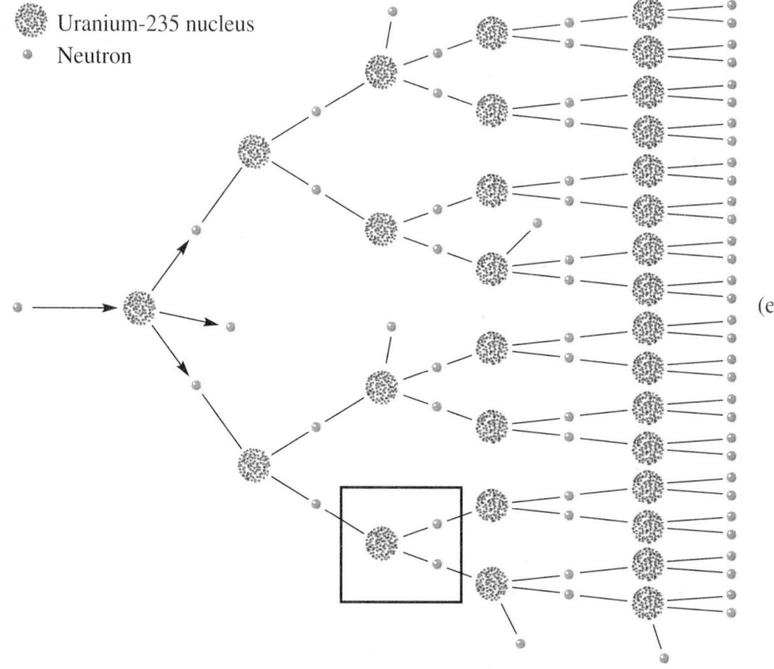

Figure 4.11 Schematic representation of a nuclear chain reaction. Neutrons released in the fission of one uranium-235 nucleus can strike another nucleus, causing it to split and release more neutrons. For simplicity, fission fragments are not shown. An area equivalent to that inside the box is shown in detail in Figure 4.10.

Aware of the destructive forces that could be produced, and concerned that Germany might develop such a bomb, Szilard prevailed on Einstein to sign a letter to President Franklin D. Roosevelt indicating the importance of the discovery. The U.S. government launched a massive research project—called the Manhattan Project—for the study of atomic energy. Uranium had to be collected and the isotopes separated, for only the relatively rare uranium-235 isotope is fissionable. The neutrons, it was found, had to be slowed by graphite to increase the probability that they would hit a uranium nucleus. Fermi and his group achieved the first sustained nuclear reaction on 2 December 1942 under the bleachers at Stagg Field at The University of Chicago.

Isotopic Enrichment

Uranium-235 makes up only 0.7% of natural uranium; the rest is nonfissile uranium-238. To make a bomb, the uranium has to be enriched to about 90% uranium-235. This isotope *enrichment* proceeded slowly at a top-secret installation in Oak Ridge, Tennessee. Separation could not be accomplished by chemical reactions because the isotopes are nearly identical chemically. Rather, separation was accomplished by conversion of all the uranium to volatile uranium hexafluoride. Molecules of the latter containing the uranium-235 isotope are slightly lighter and move slightly more rapidly than molecules containing the uranium-238 isotope. The vapors of uranium hexafluoride were allowed to pass through thousands of consecutive pinholes, a process in which

A *fissile* isotope is one that undergoes fission (splitting). A *nonfissile* isotope does not undergo fission.

the molecules that contained uranium-235 gradually outdistanced the others. The scientists eventually obtained 15 kg of the enriched uranium with enough uranium-235 to make a small explosive device.

The Synthesis of Plutonium

Seaborg was involved in the discovery of several other transuranium elements. In 1994 he was honored by having element 106 named for him: Seaborgium (Sg). Seaborg is the first living person to be so honored.

While the tedious work of separating uranium isotopes was under way at Oak Ridge, other workers, led by Glenn T. Seaborg, approached the problem of obtaining fissionable material by another route. It was known that uranium-238 would not fission when bombarded by neutrons. However, it had been determined that when uranium-238 was bombarded by neutrons a new element named neptunium (Np) was formed, and this product quickly decayed to another new element, plutonium (Pu).

$$^{238}_{92}U + ^{1}_{0}n \longrightarrow ^{239}_{92}U$$

$$^{239}_{92}U \longrightarrow ^{239}_{93}Np + ^{0}_{-1}e$$

$$^{239}_{93}Np \longrightarrow ^{239}_{94}Pu + ^{0}_{-1}e$$

The isotope plutonium-239 was found to be fissile and, thus, was suitable material for the making of a bomb. A series of large reactors was built near Hanford, Washington, to produce plutonium.

Critical Mass

Before a fissile material can sustain a chain reaction, a certain minimum amount, called the **critical mass**, must be brought together. There must be enough fissionable nuclei that the neutrons released in one fission process will have a good chance of finding another fissile nucleus before escaping from the mass. For uranium-235, this critical mass is about 4 kg, a mass about the size of a baseball. To construct a bomb, separate smaller masses are used. These subcritical masses are then brought together forcefully to trigger the runaway chain reaction of a nuclear explosion.

By July 1945, enough plutonium had been made for a bomb to be assembled. The first atomic bomb was tested in the desert near Alamogordo, New Mexico, on 16 July 1945. The heat from the explosion vaporized the 30-m steel tower on which it was placed and melted the sand for several hectares around the site. The light released was the brightest anyone had ever seen.

Some of the scientists were so awed by the force of the blast that they argued against its use on Japan. A few, led by Leo Szilard, argued for a demonstration of its power at an uninhabited site. But fear of a well-publicized "dud" and the desire to avoid millions of casualties in an invasion of Japan led President Harry S Truman to order the dropping of bombs on Japanese cities. A uranium bomb called "Little Boy" was dropped on Hiroshima on 6 August 1945, causing over 100,000 casualties (Figures 4.12 and 4.13). Three days later, a plutonium bomb called "Fat Man" was dropped on Nagasaki

Figure 4.12 A nuclear bomb of the type exploded over Hiroshima. The bomb is 71 cm in diameter and 305 cm long; it weighs 4000 kg and has an explosive power equivalent to about 18 million kg of high explosive. [*Courtesy of National Air and Space Museum, The Smithsonian Institution, Washington DC.*]

Figure 4.13 (a) The now familiar mushroom cloud that follows a nuclear explosion. (b) A cloud and fireball produced by explosion of a hydrogen bomb. [*Courtesy of the Naval Photographic Center, Washington DC (a) and U.S. Department of Defense (b).*]

with comparable results. World War II ended with the surrender of Japan on 14 August 1945.

SECTION 4.13

Radioactive Fallout

When a nuclear explosion occurs in the open atmosphere, radioactive materials can rain upon parts of the Earth thousands of miles away, days and weeks later. This is called *radioactive fallout.*

The uranium atom can split many different ways. Some examples are

$$^{235}_{92}\text{U} + ^{1}_{0}\text{n} \longrightarrow ^{90}_{38}\text{Sr} + ^{143}_{54}\text{Xe} + 3\,^{1}_{0}\text{n}$$

$$^{235}_{92}\text{U} + ^{1}_{0}\text{n} \longrightarrow ^{94}_{36}\text{Kr} + ^{139}_{56}\text{Ba} + 3\,^{1}_{0}\text{n}$$

$$^{235}_{92}\text{U} + ^{1}_{0}\text{n} \longrightarrow ^{90}_{37}\text{Rb} + ^{144}_{55}\text{Cs} + 2\,^{1}_{0}\text{n}$$

The primary fission products are radioactive. They decay to *daughter isotopes,* many of which are also radioactive. In all, over 200 different fission products are produced, with half-lives varying from less than a second to more than a billion years. In addition, the neutrons produced in the explosion act on molecules in the atmosphere to produce carbon-14, tritium, and other radioisotopes. Fallout is therefore exceedingly complex. We consider only three of the more worrisome isotopes here.

Of all the isotopes, strontium-90 presents the greatest hazard to people. The isotope has a half-life of 28 years. Strontium-90 reaches us primarily through

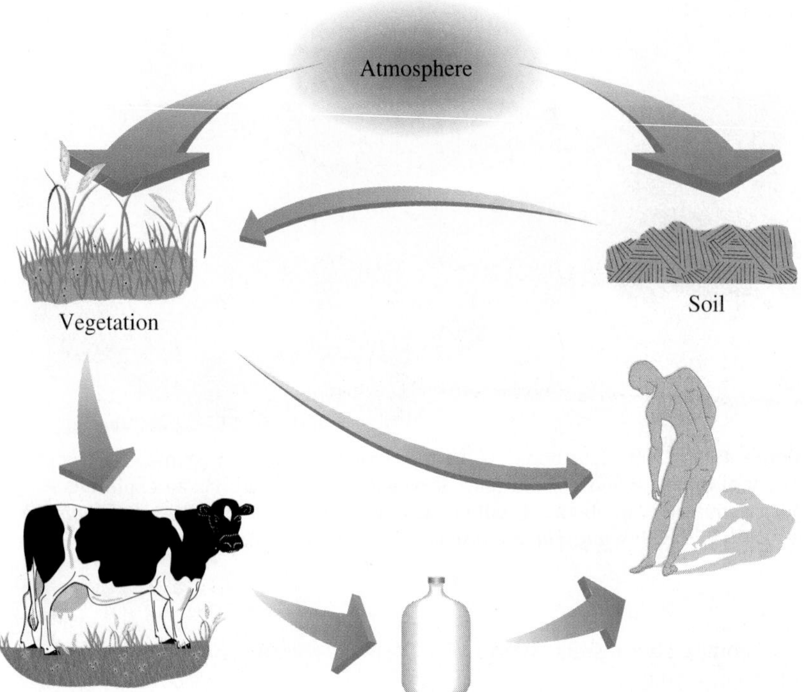

Figure 4.14 Pathways of strontium-90 from fallout. [*Adapted from C. L. Comer,* Fallout *(Oak Ridge, TN: U.S. Department of Energy, 1966).*]

milk and vegetables (Figure 4.14). Because of its similarity to calcium (both are Group IIA elements), strontium-90 is incorporated into bone. There it remains a source of internal radiation for many years.

Iodine-131 may present a greater threat immediately after a nuclear explosion. Its half-life is only 8 days, but it is produced in relatively large amounts. Iodine-131 is efficiently carried through the food chain. In the body it is concentrated in a small area, the thyroid gland. It is precisely this characteristic that makes this isotope so useful for diagnostic scanning. However, for a healthy individual, the incorporation of radioactive iodine in the thyroid gland offers no useful information, only damaging side effects.

Another important isotope in fallout is cesium-137. Cesium is similar to potassium (both are Group IA elements), and it mimics potassium in the body. Cesium-137 is a gamma emitter and has a half-life of 30 years. It is less of a threat than strontium-90, though, because it is removed from the body more readily. We get cesium-137 through vegetables, milk, and meat.

By the late 1950s, radioactive isotopes from atmospheric testing of nuclear weapons had been detected in the environment. Concern over radiation damage from nuclear fallout led to a movement to ban atmospheric testing. Many scientists were leaders in the movement. Linus Pauling, who won the Nobel prize in chemistry in 1954 for his bonding theories and for his work in determining the structure of proteins, was a particularly articulate advocate of banning atmospheric nuclear testing. In 1963 a nuclear test ban treaty was signed by the major powers—with the exception of France and the People's

Republic of China, who continued above-ground tests. Since the signing of the treaty, other countries have joined the nuclear club. Pauling, who had endured being called a communist and a traitor for his outspoken position, was awarded the Nobel prize for peace in 1962.

SECTION 4.14

Nuclear Winter

Radioactive fallout isn't the only concern in the aftermath of nuclear explosions. The nations of planet Earth have acquired nuclear weapons with an explosive power equal to more than a million Hiroshima bombs. Studies suggest that explosion of only half these weapons would produce enough soot, smoke, and dust to blanket the Earth, block out the sun, and bring on a ''**nuclear winter**'' that would threaten the survival of the human race. Computer simulations suggest that as much as 90% of the sunlight could be prevented from reaching the surface of the Earth, dropping the temperature to −25°C for several months. Crops would freeze, farm animals would die of thirst and hunger, and many species would be extinguished, including perhaps our own.

SECTION 4.15

Radiation and Us

Radiation with enough energy to knock electrons from atoms and molecules, converting them into ions (electrically charged atoms or groups of atoms), is called **ionizing radiation**. Nuclear radiation and X-rays are examples. Ionizing radiation can devastate living cells by interfering with the normal chemical processes of the cell. Because radiation is invisible and because it has such great potential for harm, we are much concerned with exposure to it.

Background Radiation

Humans have always been exposed to radiation. Some of it, called **cosmic rays**, comes from the sun and outer space. Other radiation reaches us from natural radioactive isotopes in air, water, soil, and rocks. This ever-present radiation is called **background radiation**.

Human activities have added to our exposure to radiation. Figure 4.15 shows that about two-thirds of the average radiation exposure comes from background radiation. Most of the remaining third comes from medical irradiation such as X-rays. Other sources, such as fallout, releases from the nuclear industry, and occupational exposure, account for only a minute fraction of our total exposure. Nevertheless, accidents such as those at the Three Mile Island

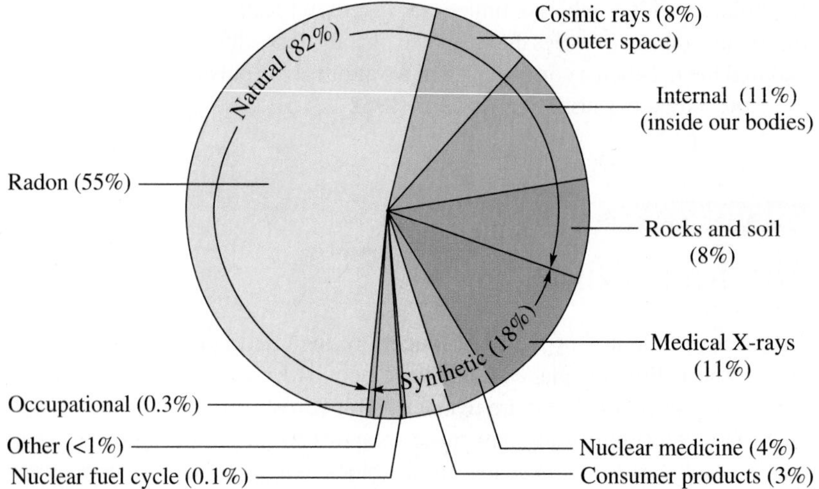

Figure 4.15 Most of our exposure to ionizing radiation comes from natural sources, including radon that seeps into our homes from the underlying rocks. [*Source: National Council on Radiation Protection and Measurement.*]

In 1993 a Ukrainian Academy of Science study group led by Vladimir Chernousenko investigated the aftereffects of the Chernobyl accident. Chernousenko claims that 15,000 of those who helped clean up the accident site have died and another 250,000 have been left as invalids. He also says that 200,000 children have experienced radiation-induced illnesses and that half the children in Ukraine and Belarus have symptoms. The incident at Chernobyl greatly influenced the desire of Ukrainians for independence upon the breakup of the Soviet Union in 1991.

plant near Harrisburg, Pennsylvania, in 1979 and at Chernobyl in the Ukraine, U.S.S.R., in 1986 did much to increase public apprehension about nuclear power. Even though no one was hurt at Three Mile Island, some people are still concerned about the long-term effect of exposure to the small amounts of radioactivity released. The accident at Chernobyl was more severe. Some people were killed outright; others died later of severe radiation poisoning. Increased levels of radioactivity throughout Europe brought great concern about possible harmful effects.

We discuss nuclear power, with its promise and peril, in Chapter 14.

Radiation Damage to Cells

Although radiation has always been with us, we didn't know it until the discovery of X-rays and radioactivity in the 1890s. Shortly after that, it was noted that such radiation could be both beneficial and harmful. We are all familiar with the use of X-rays in medical diagnosis. We have seen how radioisotopes are used in beneficial ways. But radiation also presents a hazard to living things. High-energy particles and rays knock electrons from atoms, forming ions. Such chemical changes, when they occur in living cells, can be highly disruptive. Water in the cells can be transformed to highly reactive hydrogen peroxide (H_2O_2), which can disrupt the delicate chemical balance in the cells. Particularly vulnerable are the white blood cells, the body's first line of defense against bacterial infection. Radiation also affects bone marrow, causing a drop in the production of red blood cells, which results in anemia. Radiation also has been shown to induce leukemia, a cancerlike disease of the blood-forming organs.

Radiation also causes changes in the molecules of heredity (DNA) in reproductive cells. Such changes show up as *mutations* in the offspring of exposed parents. Little is known of the effects of such exposure on humans. However,

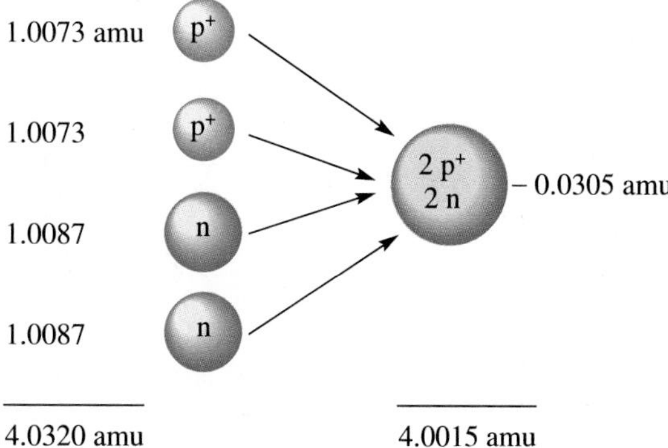

Figure 4.16 Nuclear binding energy in ^4_2He. The mass of a helium-4 nucleus is 4.0015 amu, which is 0.0305 amu less than the masses of two protons and two neutrons. The missing mass is equivalent to the binding energy of the helium-4 nucleus.

many of the mutations during the evolution of present species may have been caused by background radiation.

SECTION 4.16

Binding Energy

Nuclear fission involves a tremendous discharge of energy. Where does all this energy come from? It is locked inside the atomic nucleus. When protons and neutrons are combined to form atomic nuclei, a small amount of mass is converted to energy. This is the **binding energy** that holds the nucleons together in the nucleus. For example, the helium nucleus contains two protons and two neutrons. The mass of these four particles is 4.0320 amu (Figure 4.16). However, the actual mass of the helium nucleus is only 4.0015 amu. The missing mass amounts to 0.0305 amu. Using Einstein's equation $E = mc^2$ (in which c is the speed of light), we can calculate a value of 28.3 MeV (million electron volts) for the binding energy of the helium nucleus. This is the amount of energy it would take to separate one helium nucleus into two protons and two neutrons. If we divide by 4 (the number of nucleons), the binding energy per nucleon is 7.1 MeV.

The word *nucleons* includes both protons and neutrons.

When binding energy per nucleon is calculated for all the elements and plotted against mass number, a graph such as that in Figure 4.17 is obtained. Those elements with the highest binding energies per nucleon have the most stable nuclei. They include iron and nearby elements. When uranium atoms undergo nuclear fission, they split into fragments with higher binding energies; in other words, the fission reaction converts large atoms into smaller ones with greater nuclear stability.

We can also see from the graph that even more energy can be obtained by combining small atoms, such as hydrogen or deuterium, to form larger atoms with more stable nuclei. This kind of reaction is called **nuclear fusion**. It is what happens when a hydrogen bomb explodes, and it is also the source of the sun's energy.

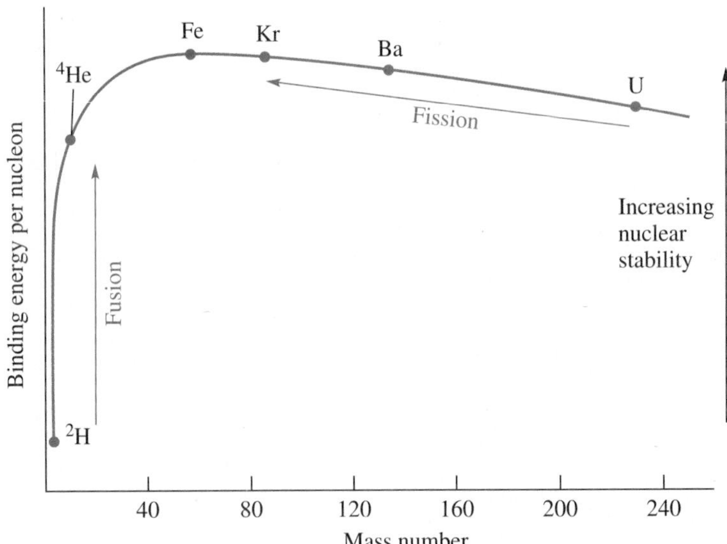

Figure 4.17 Nuclear stability is greatest near iron in the periodic table. Fission of very large atoms or fusion of very small ones results in greater nuclear stability.

SECTION 4.17

Thermonuclear Reactions

Almost all of our energy here on Earth comes from the thermonuclear reactions taking place in the sun. They are called **thermonuclear reactions** because very high temperatures (millions of degrees) are required in order to initiate them. The intense temperatures of the sun cause nuclei to fuse and release enormous amounts of energy. The principal reaction is believed to be the fusion of four hydrogen nuclei to produce one helium nucleus.

$$4\ {}^{1}_{1}\mathrm{H} \longrightarrow {}^{4}_{2}\mathrm{He} + 2\ {}^{0}_{+1}\mathrm{e}$$

The fusion of one single gram of hydrogen releases an amount of energy equivalent to burning nearly 20 tons of coal.

Although controlled nuclear fusion is not yet a reality on Earth, we can carry out uncontrolled fusion in the form of the **hydrogen bomb**. In this case a small fission bomb using uranium or plutonium must be used as the core in order to provide enough heat to achieve the million degree temperatures needed to start the nuclear fusion reaction. The outer layer of a hydrogen bomb is mainly lithium deuteride (${}^{6}_{3}\mathrm{Li}\ {}^{2}_{1}\mathrm{H}$). The lithium produces tritium by neutron bombardment from the fission reaction,

$${}^{6}_{3}\mathrm{Li} + {}^{1}_{0}\mathrm{n} \longrightarrow {}^{4}_{2}\mathrm{He} + {}^{3}_{1}\mathrm{H}$$

and then the deuterium and tritium undergo nuclear fusion.

$${}^{2}_{1}\mathrm{H} + {}^{3}_{1}\mathrm{H} \longrightarrow {}^{4}_{2}\mathrm{He} + {}^{1}_{0}\mathrm{n}$$

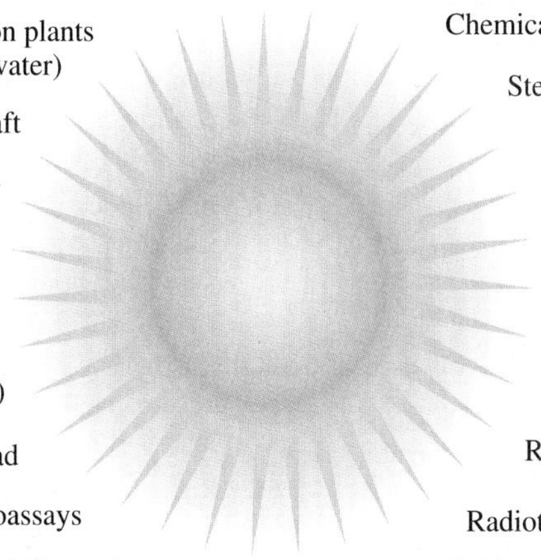

Generation of electricity
Power for desalinization plants (removing salt from seawater)
Power for spacecraft
Power for ocean vessels
Wear testing (auto engines, tires)
Flow rate indicators (pipelines)
Thickness gauges (metal sheet, plastic film)
Sterilization of suture thread
Radioimmunoassays
Medical diagnosis
Synthesis of new elements
Chemical analysis (by neutron activation)
Sterilization of insects
Preservation of foods
Smoke detectors
Bomb detectors at airports
Manufacture of semiconductors
Manufacture of radioisotopes
Radiodating
Radiotracer research
Crosslinking of polymers

Figure 4.18 Some peacetime uses of nuclear energy.

SECTION 4.18

The Nuclear Age

We live in exciting times. The goal of the alchemists, to change one element into another, has been achieved through application of scientific principles. New elements have been formed, and the periodic chart has been extended beyond uranium ($Z = 92$) to element 109. This modern alchemy produces plutonium by the ton; neptunium ($Z = 93$), americium ($Z = 95$), and curium ($Z = 96$) by the kilogram; and berkelium ($Z = 97$) and einsteinium ($Z = 99$) by the milligram. These new elements have been used in medicine and to power spacecraft and build bombs.

We live in an age in which fantastic forces have been unleashed. The threat of nuclear war has been a constant specter for the last five decades. Still, it is hard to believe that the world would be a better place if we had not discovered the secrets of the atomic nucleus (Figure 4.18).

Summary

1. The *atomic nucleus* is a tiny speck in the middle of an atom containing all of its protons and neutrons and therefore almost all of its mass.
2. A *nuclear symbol* includes the atomic number as a subscript and the *mass number* (or *nucleon number*) as a superscript, both placed in front of the symbol.
3. *Nuclear equations* are written using nuclear symbols, with nucleon numbers and atomic numbers each in balance on both sides of the arrow.
4. *Radioactive decay* may involve alpha emission, beta emission, or gamma radiation. There are also other modes of decay, such as positron emission.
5. The *half-life* of a radioactive isotope is the time it takes for half of a sample to decay.

6. Artificial transmutation can be achieved by nuclear bombardment of stable atomic nuclei with protons, neutrons, alpha particles, or other subatomic particles.
7. Alpha particles (helium nuclei) are relatively slow and low in penetrating power. Beta particles (electrons) are much faster and more penetrating. Gamma rays (photons) travel at the speed of light and have great penetrating power.
8. Scientists often use radioactive isotopes as "tracers." A radioactive atom in a molecule labels it so that it can be followed by a radiation detector.
9. In medicine *radioisotopes* are used both for diagnosis and for therapy. The diagnostic procedures (such as *PET scans*) use radioisotopes as tracers. Radiation therapy uses radiation to kill cancer cells.
10. Half-lives of radioisotopes can be used to measure the age of certain objects. The age of rocks that contain uranium can be estimated by measuring their uranium/lead ratios. *Carbon-14 dating* can be used to estimate the age of artifacts containing carbon.
11. Einstein's *mass-energy equation,* $E = mc^2$, was derived in 1905, but it was not fully appreciated until the explosion of the atom bomb 40 years later.
12. The *nuclear fission* reaction, although discovered in Germany, became the center of a massive but secret U.S. Manhattan Project. Its goals were (a) to achieve sustained nuclear fission and determine the critical mass required; (b) to enrich the amount of fissile uranium-235 in ordinary uranium-238; (c) to synthesize plutonium-239, which is also fissile; and (d) to construct a nuclear fission bomb before the Germans were able to do it.
13. In addition to the devastation at the site of a nuclear explosion, there is much radioactive debris produced, resulting in radioactive fallout all over the planet.
14. It is believed that full-scale nuclear war would produce enough smoke and soot to block out the sunlight and bring on a "*nuclear winter.*"
15. Radiation is hazardous because it can knock electrons from molecules, turning them into ions. When this happens to molecules in human cells, the cells often die.
16. The total mass of the individual nucleons in an atom is greater than the actual mass of the nucleus. The missing mass is present as *binding energy* that is holding the nucleons together.
17. Whereas fission reactions break up large atomic nuclei into smaller ones, *nuclear fusion* reactions combine small nuclei to form larger ones. Nuclear fusion produces even more energy than nuclear fission and is the basis for the *hydrogen bomb.* It is also the source of the sun's energy.

Key Terms

background radiation 4.15
binding energy 4.16
carbon-14 dating 4.10
chain reaction 4.12
cosmic rays 4.15
critical mass 4.12
fundamental particles 4.5
half-life 4.4
hydrogen bomb 4.17
ionizing radiation 4.15
mass–energy equation 4.11
mass number 4.2
nuclear fission 4.12
nuclear fusion 4.16
nuclear winter 4.14
nucleon number 4.2
nucleons 4.2
PET scans 4.9
photoscan 4.9
positron 4.6
radioactive decay 4.3
radioisotopes 4.8
thermonuclear reaction 4.17
tracers 4.8
transmutation 4.5

Review Questions

1. Define or identify each of the following.
   a. half-life
   b. positron
   c. background radiation
   d. radioisotope
   e. fission
   f. fusion
   g. artificial transmutation
   h. binding energy.

2. How does the size of the nucleus compare to that of the atom as a whole?

3. Why are isotopes important in nuclear reactions?

4. What is meant by the nucleon number of an isotope?

5. Give the nuclear symbols for protium, deuterium, and tritium (which are hydrogen-1, hydrogen-2, and hydrogen-3, respectively).

6. Give the nuclear symbol for an isotope with a nucleon number of 8 and an atomic number of 5.

7. Give the nuclear symbol for an isotope with $Z = 35$ and $A = 83$.

8. Give the nuclear symbol for an isotope with 53 protons and 72 neutrons.

9. Give the nuclear symbols for the following isotopes. You may refer to the periodic table.
a. gallium-69 b. molybdenum-98
c. molybdenum-99 d. technetium-98

10. Indicate the number of protons and the number of neutrons in atoms of the following isotopes.
a. $^{62}_{30}Zn$ b. $^{241}_{94}Pu$ c. $^{99m}_{43}Tc$ d. $^{81m}_{36}Kr$

11. Which of the following pairs represent isotopes?
a. $^{70}_{34}X$ and $^{70}_{33}X$ b. $^{57}_{28}X$ and $^{66}_{28}X$
c. $^{186}_{74}X$ and $^{186}_{74}X$ d. $^{8}_{2}X$ and $^{6}_{4}X$
e. $^{22}_{11}X$ and $^{44}_{22}X$

12. The longest lived isotope of fermium (Fm) has a nucleon number of 257. How many neutrons are there in the nucleus of this isotope?

13. The longest lived isotope of technetium (Tc) has 54 neutrons. What is the nucleon number of this isotope?

14. The two principal isotopes of lithium are lithium-6 and lithium-7. The atomic mass of lithium is 6.9 amu. Which is the predominant isotope of lithium?

15. Out of every five atoms of boron, one has a mass of 10 amu and four have a mass of 11 amu. What is the atomic mass of boron? Use the periodic table only to check your answer.

16. Give the nuclear symbols for the following subatomic particles.
a. the alpha particle b. the beta particle
c. the neutron d. the positron

17. When a nucleus emits a beta particle, what changes occur in the nucleon number and atomic number of the nucleus?

18. When a nucleus emits a neutron, what changes occur in the nucleon number and the atomic number of the nucleus?

19. When a nucleus emits a proton, what changes occur in the nucleon number and the atomic number of the nucleus?

20. When a nucleus emits an alpha particle, what changes occur in the nucleon number and atomic number of the nucleus?

21. When a nucleus emits a gamma ray, what changes occur in the nucleon number and the atomic number of the nucleus?

22. When a nucleus emits a positron, what changes occur in the nucleon number and the atomic number of the nucleus?

23. Explain how radioisotopes can be used for therapeutic purposes.

24. Which radioisotope has been used extensively for treatment of overactive or cancerous thyroid glands?

25. Describe the use of a radioisotope as a diagnostic tool in medicine.

26. What are some of the characteristics that make technetium-99^{m} such a useful radioisotope for diagnostic purposes?

27. A pair of gloves would be sufficient to shield the hands from which type of radiation: the heavy alpha particles or the massless gamma rays?

28. Heavy lead shielding is necessary as protection from which type of radiation: alpha, beta, or gamma?

29. Plutonium is especially hazardous when inhaled or ingested because it emits alpha particles. Why would alpha particles cause more damage to tissue than beta particles?

30. What form of radiation is detected in PET scans?

31. List two ways in which workers can protect themselves from the radioactive materials with which they work.

32. Which subatomic particles are responsible for carrying on the chain of reactions that are characteristic of nuclear fission?

33. Compare nuclear fission and nuclear fusion. Why is energy liberated in each case?

34. Discuss nuclear winter.

35. What is the source of the greatest proportion of our exposure to artificial radiation?

36. What is the source of the energy that triggers the thermonuclear reactions of the hydrogen bomb?

37. Did President Harry S Truman make the right decision when he decided to drop the nuclear bombs on Japanese cities? Would your answer be the same if you were living in 1945 and had relatives among the troops preparing for the invasion of Japan? If you were an inhabitant of one of the cities bombed?

38. Discuss the impact of nuclear science on the following topics. (See the list of references and readings for this chapter for resource materials.)
   a. war and peace
   b. industrial progress
   c. medicine
   d. agriculture
   e. human, animal, and plant genetics

Problems

Nuclear Equations

39. Write a balanced equation for the emission of a positron by sulfur-31.

40. Write a balanced equation for the emission of a neutron by bromine-87.

41. Write a balanced equation for the emission of a proton by magnesium-21.

42. Lead-209 undergoes beta decay. Write a balanced equation for this reaction.

43. Thorium-225 undergoes alpha decay. Write a balanced equation for this reaction.

44. Write an equation representing the emission of a gamma ray by gold-186.

45. Complete the following equations.
   a. ${}^{179}_{79}\text{Au} \rightarrow {}^{175}_{77}\text{Ir} + ?$
   b. ${}^{23}_{10}\text{Ne} \rightarrow {}^{23}_{11}\text{Na} + ?$

46. Complete the following equations.
   a. ${}^{10}_{5}\text{B} + {}^{1}_{0}\text{n} \rightarrow {}^{4}_{2}\text{He} + ?$
   b. ${}^{12}_{6}\text{C} + {}^{2}_{1}\text{H} \rightarrow {}^{13}_{6}\text{C} + ?$
   c. ${}^{121}_{51}\text{Sb} + ? \rightarrow {}^{121}_{52}\text{Te} + {}^{1}_{0}\text{n}$
   d. ${}^{154}_{62}\text{Sm} + {}^{1}_{0}\text{n} \rightarrow 2\,{}^{1}_{0}\text{n} + ?$

47. When magnesium-24 is bombarded with a neutron, a proton is ejected. What new element is formed? (*Hint:* Write a balanced nuclear equation.)

48. When chlorine-37 is bombarded with a neutron, a proton is ejected. What new element is formed?

49. A radioactive isotope decays to give an alpha particle and bismuth-211. What was the original element?

50. A radioisotope decays to give an alpha particle and protactinium-233. What was the original element?

Half-life

51. C. E. Bemis and colleagues at Oak Ridge National Laboratory confirmed the synthesis of Unq (element 104), the half-life of which was only 4.5 s. Only 3000 atoms of the element were created in the tests. How many atoms were left after 4.5 s? After a total of 9.0 s?

52. Krypton-81^{m} is used for lung ventilation studies. Its half-life is 13 s. How long will it take the activity of this isotope to reach one-quarter of its original value?

53. A 100-mg technetium-99^{m} sample is used in a medical study. How much of the technetium-99^{m} sample remains after 24 h? The half-life of technetium-99^{m} is 6.0 hr.

54. The half-life of molybdenum-99 is 67 hr. How much time passes before a sample with an activity of 160 counts per minute has decreased to 5.0 counts per minute?

Radioisotopic Dating

55. Living matter has a carbon-14 activity of 16 counts per minute per gram of carbon. What is the age of an artifact for which the carbon-14 activity is 8 counts per minute per gram of carbon?

56. A piece of wood from an Egyptian tomb has a carbon-14 activity of 980 counts per hour. A piece of new wood of the same size gave 3920 counts per hour. What is the age of the wood from the tomb?

57. The ratio of carbon-14 to carbon-12 in a piece of charcoal from an archaeological excavation is found to be one-half the ratio in a sample of modern wood.

Approximately how old is the site? How old would it be if the ratio was 25% of the ratio in a sample of modern wood?

58. How old is a bottle of wine if the tritium activity is 25% that of new wine? The half-life of tritium is 12.3 years.

Additional Problems

59. To make Unh (element 106), a 0.25-mg sample of ^{249}Cf was used as the target. Four neutrons were emitted to yield a nucleus with 106 protons and a mass of 263 amu. What was the bombarding particle?

60. One atom of Une (element 109) with a nucleon number of 266 was produced in 1982 by bombarding a target of bismuth-209 with iron-58 nuclei for 1 week. How many neutrons are released in the process?

61. Unnilennium (element 109) undergoes alpha emission to form Uns (element 107), which in turn also emits an alpha particle. What are the atomic number and nucleon number of the isotope formed by these two steps? Write balanced nuclear equations for the two reactions.

62. Radium-223 nuclei usually decay by alpha emission. For every billion alpha decays, one atom emits a carbon-14 nucleus. Write a balanced nuclear equation for each type of emission.

63. Uranium has a density of 19 g/cm^3. What volume is occupied by a critical mass of 8 kg of uranium?

64. Neptunium-237 undergoes a series of seven alpha and four beta decays. What stable isotope results from this radioactive decay series?

References and Readings

1. Budavari, Susan (Ed.). *The Merck Index,* 11th ed. Rahway, NJ: Merck and Co., 1989. Contains extensive tables of radioisotopes (pp. MISC-31–MISC-45) and radioisotopes used in medical diagnosis and therapy (pp. MISC-46–MISC-52).
2. Chivian, E., et al. (Eds.). *Last Aid: The Medical Dimensions of Nuclear War.* San Francisco: Freeman, 1981.
3. Ehrlich, Paul R., et al. "Long-Term Biological Consequences of Nuclear War." *Science,* 23 December 1983, pp. 1293–1300.
4. Keen, Judy. "Chernobyl: Coping With Nuclear Disaster," *USA Today,* 17 September 1991, p. 6A.
5. Marshall, Eliot. "Academy Panel Raises Radiation Risk Estimate." *Science,* 5 January 1990, pp. 22–23.
6. Marshall, Eliot. "Recalculating the Cost of Chernobyl." *Science,* 8 May 1987, pp. 658–659. The accident may cause 39,000 extra cancer deaths.
7. Rawls, Rebecca. "Element 109 Made by West German Researchers." *Chemical and Engineering News,* 11 October 1982, pp. 27–28.
8. Rhodes, Richard. *The Making of the Atomic Bomb.* New York: Simon & Schuster, 1986. Part One, "Profound and Necessary Truth."
9. Sagan, Leonard. "Radiation and Human Health." *EPRI Journal,* September 1979, pp. 6–13.
10. Schecter, Bruce. "The Short, Bright Life of Element 109." *Discover,* December 1982, pp. 98–106.
11. Seaborg, Glenn T., and Walter D. Loveland. *The Elements Beyond Uranium.* New York: Wiley, 1990.
12. Sun, Marjorie. "Renewed Interest in Food Irradiation." *Science,* 17 February 1984, pp. 667–668.
13. Taubes, Gary. "The Case of the Cosmic Rays." *Discover,* September 1989, pp. 52–60.
14. Turco, R. P., et al. "Nuclear Winter: Global Consequences of Multiple Nuclear Explosions." *Science,* 23 December 1983, pp. 1283–1292.
15. Wetherill, George W. "Dating Very Old Objects." *Natural History,* September 1982, pp. 14–20.
16. Yalow, Rosalyn S. "Radioactivity in the Service of Man." *Journal of Chemical Education,* September 1982, pp. 735–738.

5 Chemical Bonds

The Ties That Bind

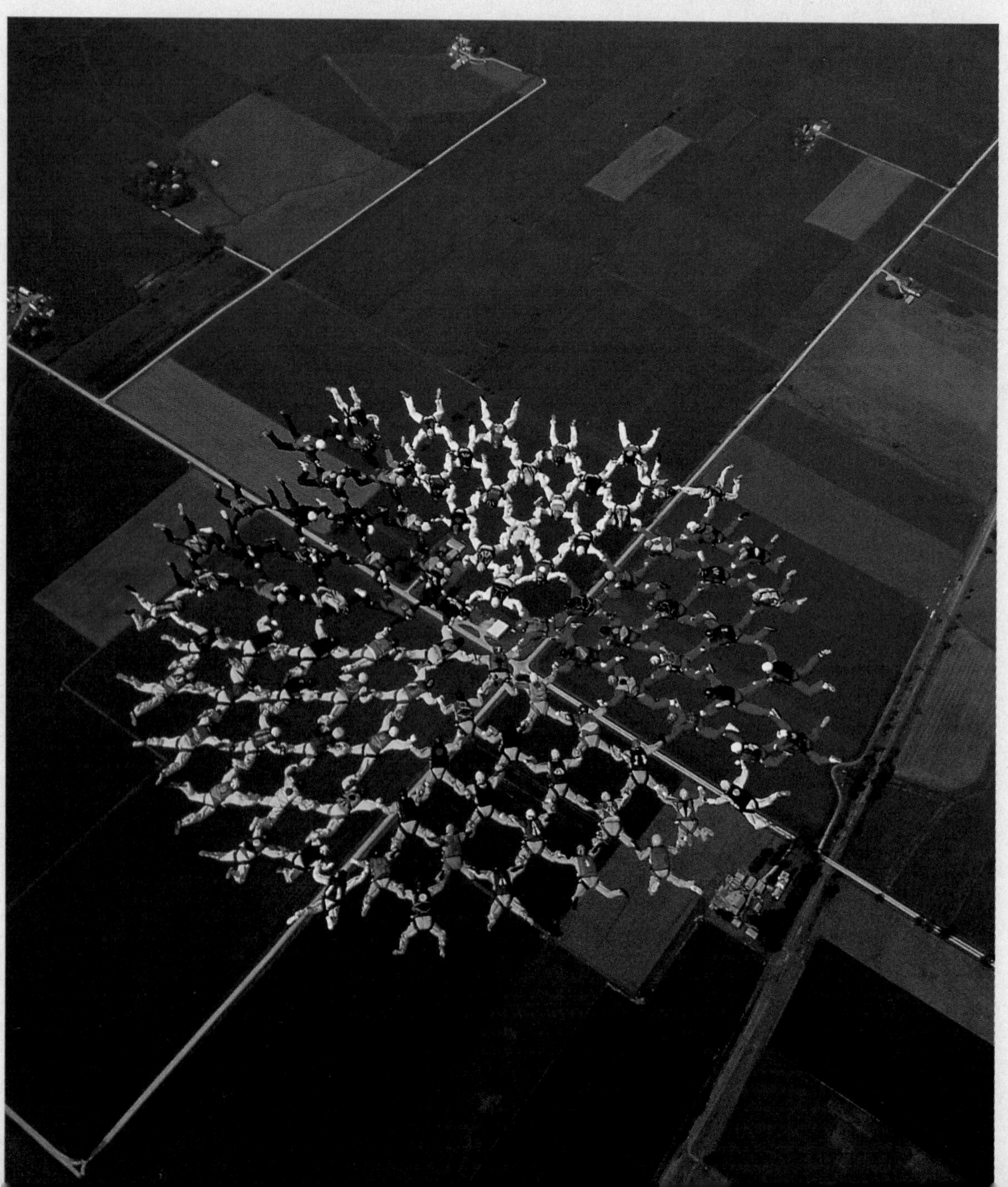

▼

The nature of materials

In part depends on whether

Strong bonds or weak attractions

Are holding them together.

Have you ever wondered why salt crystals are so hard and high melting while butter is so soft and easy to melt? The reason has to do with the very different kinds of chemical bonds in these materials.

In Chapter 3 we learned a bit about atomic structure, and in Chapter 4 we took a brief look at the atomic nucleus. Now we are ready to consider chemical bonds, the ties that bind atoms together.

Chemical bonds are those forces that hold atoms together in molecules or ions together in crystals. Whether a substance is a solid, a liquid, or a gas depends mainly on its chemical bonding. A solid material can be hard and strong or soft and waxy depending on what kind of bonds are holding it together.

Bonds are of great importance to chemists. Most of the properties of materials can be explained in terms of the forces holding their atoms and molecules together. Chemical bonds are also of concern to other scientists, and even to those who are not scientists, if they want to understand the world around them. Let us consider a few examples of why chemical bonds are so important.

Carbon atoms have a unique ability to bond to each other and to other kinds of atoms to form millions of different molecules (Chapter 9). These carbon compounds include carbohydrates, fats, proteins, and nucleic acids, compounds that make up all living things, including us.

Carbon monoxide can kill us because it forms stronger bonds with hemoglobin than oxygen does.

The shape of drug molecules is determined by the bonds within the molecules. The action of most drugs depends on the shape of the molecule. The drugs act by bonding to receptor molecules in our cell membranes or to receptors on bacteria or viruses.

Knowledge of the structure and bonding of molecules has enabled chemists to design not only drugs but also synthetic fibers and plastics, insecticides, and thousands of other chemical compounds with specific properties.

◀ Just as these skydivers bond together to form a pattern, chemical bonds hold atoms in place in molecules and crystals, giving structure to matter. [*Tom Sanders/The Stock Market.*]

There are several kinds of bonds, each of which results in unique properties in the materials characterized by that type of bonding. Each type is examined in some detail in this chapter.

Let us begin our consideration of chemical bonding so that we, too, can understand the forces that control the structure of matter, living and nonliving.

SECTION 5.1

The Art of Deduction: Stable Electron Configurations

In our discussion of the atom and its structure (Chapters 2 and 3), we followed the historical development of some of the more important atomic concepts. Some of the nuclear concepts (Chapter 4) were approached in the same manner. We could continue to look at chemistry in this manner, but that would require several volumes of print—if we got very far—and perhaps more of your time than you care to spend. We won't abandon the historical approach entirely, but we will emphasize that other aspect of scientific enterprise: deduction.

The art of deduction works something like this.

Fact	*Theory*	*Deduction*
The noble gases, such as helium, neon, and argon, are inert (i.e., they undergo few, if any, chemical reactions).	The inertness of the noble gases is due to their electron structures (each has a filled outermost energy level).	If other elements could alter their electron structures to become more like noble gases, they would become less reactive.

To illustrate, let's look at an atom of the element sodium (Na). It has 11 electrons, 2 in the first energy level, 8 in the second, and 1 in the third. If the atom could get rid of an electron, it would have the same electron structure as an atom of the noble gas neon (Ne).

$$\underset{Na}{(11\,p^+)\ 2e^-\ 8e^-\ 1e^-} \longrightarrow \underset{Na^+}{(11\,p^+)\ 2e^-\ 8e^-} + 1\,e^-$$

Recall that neon has the structure

$$\underset{Ne}{(10\,p^+)\ 2e^-\ 8e^-}$$

If a chlorine atom (Cl) could gain an electron, it would have the same structure as argon (Ar).

$$\underset{\text{Cl}}{(17\,p^+)\ 2\,e^-\ 8\,e^-\ 7\,e^-} + 1\,e^- \longrightarrow \underset{\text{Cl}^-}{(17\,p^+)\ 2\,e^-\ 8\,e^-\ 8\,e^-}$$

The structure of the argon atom is

$$\underset{\text{Ar}}{(18\,p^+)\ 2\,e^-\ 8\,e^-\ 8\,e^-}$$

The sodium, having lost an electron, becomes positively charged. It has 11 protons (11+) and only 10 electrons (10−). It is written Na^+ and is called a *sodium ion.* The chlorine atom, having gained an electron, becomes negatively charged. It has 17 protons (17+) and 18 electrons (18−). It is written Cl^- and is called a *chloride ion.* Note that a positive charge, as in Na^+, indicates that one electron has been lost. Similarly, a negative charge, as in Cl^-, indicates that one electron has been gained.

Let us emphasize that chlorine does not *become* argon. The chloride ion and argon atom have the same electron configuration, but the chloride ion has 17 protons in its nucleus and a net charge of 1−. The argon atom has 18 protons in its nucleus and no net charge; it is electrically neutral. Neither does sodium become neon by giving up an electron. The sodium ion has the same electron configuration as neon, but the nuclei differ in the number of protons.

SECTION 5.2

Electron-Dot Structures

In forming ions, the cores of sodium atoms and chlorine atoms do not change. It is convenient therefore to let the symbol represent the *core* of the atom (nucleus plus inner electrons). The **valence** (outer) **electrons** are then represented by dots. The equations of the preceding section then can be written as follows.

$$\text{Na}\cdot \longrightarrow \text{Na}^+ + 1\,e^-$$

and

$$:\ddot{\underset{\cdot\cdot}{\text{Cl}}}\cdot + 1\,e^- \longrightarrow :\ddot{\underset{\cdot\cdot}{\text{Cl}}}:^-$$

Such forms, in which the symbol of the element represents the core and dots stand for valence electrons, are called **electron-dot symbols.**

Electron-Dot Symbols and the Periodic Table

It is especially easy to write electron-dot symbols for most of the main group elements. The number of valence electrons for most of these elements is equal to the group number (Table 5.1).

Table 5.1 Electron-Dot Symbols for Selected Main Group Elements

IA	IIA	IIIA	IVA	VA	VIA	VIIA	Noble Gases
H·							He:
Li·	·Be·	·B·	·C·	:N·	:O·	:F·	:Ne:
Na·	·Mg·	·Al·	·Si·	:P·	:S·	:Cl·	:Ar:
K·	·Ca·				:Se·	:Br·	:Kr:
Rb·	·Sr·				:Te·	:I·	:Xe:
Cs·	·Ba·						

EXAMPLE 5.1

Without referring to Table 5.1, give electron-dot symbols for magnesium, oxygen, and phosphorus.

SOLUTION

Magnesium is in Group IIA, oxygen is in Group VIA, and phosphorus is in Group VA. The electron-dot symbols are therefore

·Mg· :O: :P·

Exercise 5.1

Without referring to Table 5.1, give electron-dot symbols for each of the following elements.

a. Ar b. Ca c. F d. N e. K f. S

SECTION 5.3

Chemical Symbolism

The mystery of chemistry to the nonchemist is probably due in large part to chemists' use of symbolism. Chemists find it *convenient* to represent the sodium atom as Na· rather than the more complex

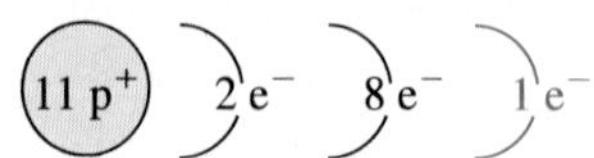

And it is easier to write :Cl· than

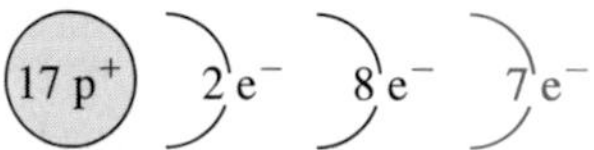

Thus, chemists often use the shorter form.

Symbolism is a convenient, shorthand way to convey a lot of information in a compact form. It is a chemist's most efficient and economical form of communication. Learning this symbolism is a good deal like learning a foreign language. Once you have learned a basic "vocabulary," the rest is a lot easier.

SECTION 5.4

Sodium Reacts with Chlorine: The Facts

Sodium is a highly reactive metal. It is soft enough to be cut with a knife. When freshly cut, it is bright and silvery, but it dulls rapidly because it reacts with oxygen in the air. In fact, it reacts so readily in air that it is usually stored under oil or kerosene. Sodium reacts violently with water also, getting so hot that it melts. A small piece will form a spherical bead after melting and race around on the surface of the water as it reacts.

Chlorine is a greenish yellow gas. It is familiar as a disinfectant for swimming pools and city water supplies. (The actual substance added may be a compound that reacts with water to form chlorine.) Who hasn't been swimming in a pool that had "so much chlorine in it that you could taste it"? Chlorine is extremely irritating to the respiratory tract. In fact, chlorine was used as a poison gas in World War I.

If a piece of sodium is dropped into a flask containing chlorine gas, a violent reaction ensues. A white solid that is quite unreactive is formed. It is a familiar compound—sodium chloride (table salt) (Figure 5.1).

Actually, chlorine gas is composed of Cl_2 molecules. Each atom of the molecule takes an electron from a sodium atom. Two sodium ions and two chloride ions are formed.

$$Cl_2 + 2\,Na \rightarrow 2\,Cl^- + 2\,Na^+$$

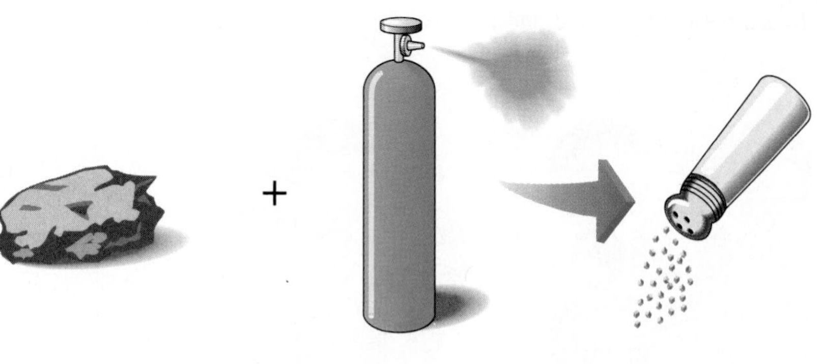

Figure 5.1 Sodium, a soft silvery metal, reacts with chlorine, a greenish gas, to form sodium chloride (ordinary table salt), a white crystalline solid.

SECTION 5.5

Sodium Reacts with Chlorine: The Theory

A sodium atom becomes less reactive by *losing* an electron. A chlorine atom becomes less reactive by *gaining* an electron. What happens when sodium atoms come into contact with chlorine atoms? The obvious: chlorine extracts an electron from a sodium atom.

$$\underset{\mathrm{Na}}{(11\,p^+)\ 2e^-\ 8e^-\ 1e^-} + \underset{\mathrm{Cl}}{(17\,p^+)\ 2e^-\ 8e^-\ 7e^-} \longrightarrow$$

$$\underset{\mathrm{Na^+}}{(11\,p^+)\ 2e^-\ 8e^-} + \underset{\mathrm{Cl^-}}{(17\,p^+)\ 2e^-\ 8e^-\ 8e^-}$$

In the abbreviated electron-dot form, this reaction is written

$$\mathrm{Na}\cdot + :\underset{\cdot\cdot}{\overset{\cdot\cdot}{\mathrm{Cl}}}\cdot \longrightarrow \mathrm{Na}^+ + :\underset{\cdot\cdot}{\overset{\cdot\cdot}{\mathrm{Cl}}}:^-$$

Ionic Bonds

The two ions formed from sodium and chlorine atoms have opposite charges. They are strongly attracted to one another. Remember, however, that for even a tiny grain of salt, there are billions and billions of each kind of ion. These ions arrange themselves in an orderly fashion (Figure 5.2). The arrangements are repeated many times in all directions—above and below, left and right, top and bottom—to make a crystal of sodium chloride. Each sodium ion attracts (and is attracted by) six chloride ions (the ones to its front and back, its top and bottom, and its two sides). Each chloride ion attracts (and is attracted by) six sodium ions. The forces holding the crystal together (the attractive forces between positive and negative ions) are called **ionic bonds**.

A **crystal** is a solid substance with a regular arrangement of its constituent particles. The solid as a whole has a well-defined, regular shape.

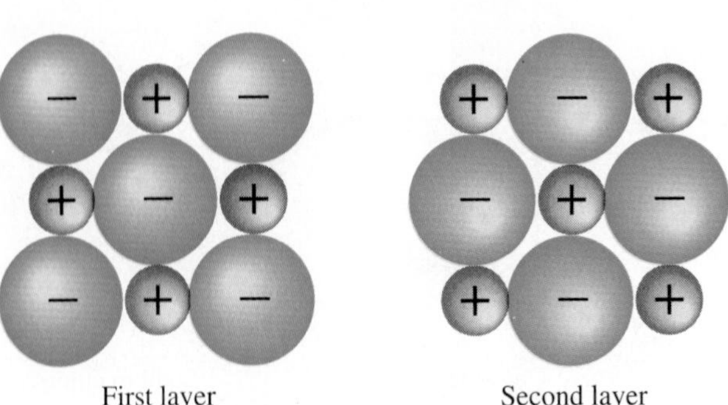

Figure 5.2 The arrangement of ions in a sodium chloride crystal. This arrangement is repeated for billions of particles in all directions.

Scientists sometimes use different models to represent the same system. The model employed in Figure 5.2 is a space-filling model showing the relative sizes of the sodium and chloride ions. Sometimes a ball-and-stick model is employed to better show the geometry of the crystal (Figure 5.3). From this model it is easy to see the cubic arrangement of the ions. In the crystal as a whole, for each sodium ion there is one chloride ion; thus, the ratio of ions is 1:1, and the simplest formula for the compound is NaCl. The symbols Na and Cl, written together, stand for the compound sodium chloride. The formula is also used to represent one sodium ion and one chloride ion.

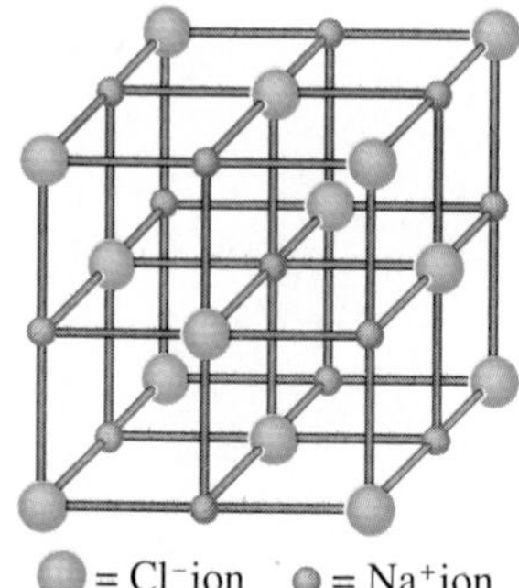

Figure 5.3 Ball-and-stick model of a sodium chloride crystal.

SECTION 5.6

Using Electron-Dot Symbols: More Ionic Compounds

As we might expect, potassium, a metal in the same family as sodium, also reacts with chlorine. The reaction yields potassium chloride (KCl).

$$K\cdot + \cdot\underset{\cdot\cdot}{\overset{\cdot\cdot}{Cl}}: \longrightarrow K^+ + :\underset{\cdot\cdot}{\overset{\cdot\cdot}{Cl}}:^-$$

And potassium reacts with bromine, a reddish brown liquid in the same family as chlorine, to form a stable white crystalline solid called potassium bromide (KBr).

$$K\cdot + \cdot\underset{\cdot\cdot}{\overset{\cdot\cdot}{Br}}: \longrightarrow K^+ + :\underset{\cdot\cdot}{\overset{\cdot\cdot}{Br}}:^-$$

Sodium also reacts with bromine to form sodium bromide.

EXAMPLE 5.2

Use electron-dot symbols to show the transfer of electrons from sodium atoms to bromine atoms to form ions with noble gas configurations.

SOLUTION

Sodium has one valence electron, and bromine has seven. Transfer of the single electron from sodium to bromine leaves each with a noble gas configuration.

$$Na\cdot + \cdot\underset{\cdot\cdot}{\overset{\cdot\cdot}{Br}}: \longrightarrow Na^+ + :\underset{\cdot\cdot}{\overset{\cdot\cdot}{Br}}:^-$$

Exercise 5.2

Use electron-dot symbols to show the transfer of electrons from lithium atoms to fluorine atoms to form ions with noble gas configurations.

Magnesium, a Group IIA metal, is harder and less reactive than sodium. Magnesium reacts with oxygen, a Group VIA element (a colorless gas), to form another stable white crystalline solid called magnesium oxide (MgO).

$$\cdot\mathrm{Mg}\cdot + \cdot\ddot{\underset{\cdot}{\mathrm{O}}}: \longrightarrow \mathrm{Mg}^{2+} + :\ddot{\underset{\cdot\cdot}{\mathrm{O}}}:^{2-}$$

Magnesium must give up two electrons and oxygen must gain two electrons for each to have the same configuration as the noble gas neon.

An atom such as oxygen, which needs two electrons to complete a noble gas configuration, may react with potassium atoms, which have only one electron each to give. In this case, two atoms of potassium are needed for each oxygen atom. The product is potassium oxide (K_2O).

$$\begin{matrix}\mathrm{K}\cdot \\ \mathrm{K}\cdot\end{matrix} + \cdot\ddot{\underset{\cdot}{\mathrm{O}}}: \longrightarrow \begin{matrix}\mathrm{K}^+ \\ \mathrm{K}^+\end{matrix} + :\ddot{\underset{\cdot\cdot}{\mathrm{O}}}:^{2-}$$

By this process, each potassium atom achieves the argon configuration. Oxygen again assumes the neon configuration.

EXAMPLE 5.3

Use electron-dot symbols to show the transfer of electrons from magnesium atoms to nitrogen atoms to form ions with noble gas configurations.

SOLUTION

$$\begin{matrix}\cdot\mathrm{Mg}\cdot \\ \cdot\mathrm{Mg}\cdot \\ \cdot\mathrm{Mg}\cdot\end{matrix} + \begin{matrix}\cdot\ddot{\underset{\cdot}{\mathrm{N}}}\cdot \\ \cdot\ddot{\underset{\cdot}{\mathrm{N}}}\cdot\end{matrix} \longrightarrow \begin{matrix}\mathrm{Mg}^{2+} \\ \mathrm{Mg}^{2+} \\ \mathrm{Mg}^{2+}\end{matrix} + \begin{matrix}:\ddot{\underset{\cdot\cdot}{\mathrm{N}}}:^{3-} \\ :\ddot{\underset{\cdot\cdot}{\mathrm{N}}}:^{3-}\end{matrix}$$

Each of three magnesium atoms gives up two electrons (a total of six), and each of the two nitrogen atoms acquires three (a total of six). Notice that the total positive and negative charges on the products are equal (6+ and 6−). Magnesium reacts with nitrogen to give magnesium nitride (Mg_3N_2).

Exercise 5.3

Use electron-dot symbols to show the transfer of electrons from aluminum atoms to oxygen atoms to form ions with noble gas configurations.

Generally, speaking, metallic elements in Groups IA, IIA, and IIIA of the periodic table react with nonmetallic elements in Groups VA, VIA, and VIIA to form stable crystalline solids. The metals tend to give up electrons to the nonmetals. The ions so formed have noble gas configurations. Atoms of the Group IA metals give up one electron to form 1+ ions, and those of the Group

IA	IIA	IIIB	IVB	VB	VIB	VIIB	VIIIB			IB	IIB	IIIA	IVA	VA	VIA	VIIA	Noble gases
Li^+														N^{3-}	O^{2-}	F^-	
Na^+	Mg^{2+}											Al^{3+}		P^{3-}	S^{2-}	Cl^-	
K^+	Ca^{2+}						Fe^{2+} Fe^{3+}			Cu^+ Cu^{2+}	Zn^{2+}					Br^-	
Rb^+	Sr^{2+}									Ag^+						I^-	
Cs^+	Ba^{2+}																

Figure 5.4 The periodic relationship of some simple ions. The B group elements often have ions of more than one kind.

IIA metals give up two electrons to form 2+ ions. The Group VIIA nonmetal atoms take on one electron to form 1− ions, and Group VIA atoms tend to pick up two electrons to form 2− ions. Atoms of the B group metals can give up varying numbers of electrons to form positive ions with varying charges. These periodic relationships are summarized in Figure 5.4. Some of the ions are discussed further in Chapter 6.

EXAMPLE 5.4

What is the formula of the compound formed by the reaction of sodium and sulfur?

SOLUTION

Sodium is in Group IA; the sodium atom has one valence electron. Sulfur is in Group VIA; the sulfur atom has six valence electrons

$$Na\cdot \quad \cdot\overset{\cdot\cdot}{\underset{\cdot\cdot}{S}}\cdot$$

Sulfur needs two electrons to gain an argon configuration, but sodium has only one to give. The sulfur atom must react with two sodium atoms.

$$\begin{matrix} Na\cdot \\ \quad + \cdot\overset{\cdot\cdot}{\underset{\cdot\cdot}{S}}\cdot \\ Na\cdot \end{matrix} \longrightarrow \begin{matrix} Na^+ \\ \quad + :\overset{\cdot\cdot}{\underset{\cdot\cdot}{S}}:^{2-} \\ Na^+ \end{matrix}$$

The formula of the compound, called sodium sulfide, is Na_2S.

Exercise 5.4

What is the formula of the compound formed by the reaction of calcium with fluorine?

SECTION 5.7

Covalent Bonds: Shared Electron Pairs

One might expect a hydrogen atom, with its one electron, to acquire another electron and assume the helium configuration. Indeed, hydrogen atoms do just that in the presence of atoms of a reactive metal such as lithium—that is, a metal that finds it easy to give up an electron.

$$\mathrm{Li}\cdot + \mathrm{H}\cdot \longrightarrow \mathrm{Li}^{+} + \mathrm{H}\!:^{-}$$

But what if there are no other kinds of atoms around? What if there are only hydrogen atoms? One atom can't gain an electron from another, for among hydrogen atoms all have an equal attraction for electrons. They can compromise, however, by *sharing a pair* of electrons.

$$\mathrm{H}\cdot + \cdot\mathrm{H} \longrightarrow \mathrm{H}\!:\!\mathrm{H}$$

By sharing electrons, the two hydrogen atoms form a hydrogen molecule. The bond formed by a shared pair of electrons is called a **covalent bond**.

$$\mathrm{H}\!:\!\mathrm{H}$$

covalent bond (shared pair of electrons)

Consider next the case of chlorine. A chlorine atom will pick up an extra electron from anything willing to give one up. But, again, what if the only thing around is another chlorine atom? Chlorine atoms also can attain a more stable arrangement by sharing a pair of electrons.

$$:\underset{..}{\overset{..}{\mathrm{Cl}}}\cdot + \cdot\underset{..}{\overset{..}{\mathrm{Cl}}}: \longrightarrow\ :\underset{..}{\overset{..}{\mathrm{Cl}}}:\underset{..}{\overset{..}{\mathrm{Cl}}}:$$

The shared pair of electrons in the chlorine molecule is another example of a covalent bond; they are called a **bonding pair**. The other electrons that stay on one atom and are not shared are called **nonbonding pairs**.

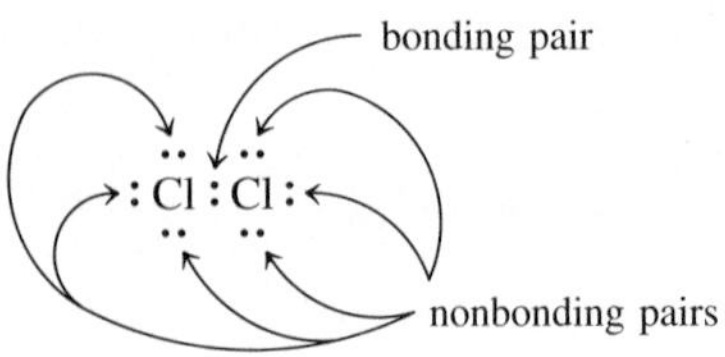

For simplicity, the hydrogen molecule is often represented as H_2 and the chlorine molecule as Cl_2. In each case, the covalent bond between the atoms is understood. Sometimes the covalent bond is indicated by a dash, H—H and Cl—Cl. Nonbonding pairs of electrons often are not shown.

The Octet Rule

Each chlorine atom in the chlorine molecule has eight electrons around it, an arrangement like that of the noble gas argon. This stable *octet* of electrons is the arrangement characteristic of all the noble gases except helium. When atoms react with each other, they often appear to be trying to attain the same stable electron configuration that noble gases have. They are said to be following the **octet rule**, or the rule of eight. In the case of helium, only two electrons can exist in its single energy level, so very small atoms, such as hydrogen, follow the rule of two.

SECTION 5.8

Unequal Sharing: Polar Covalent Bonds

So far, we have seen that atoms combine in two different ways. Some that are quite different in electron structure (from opposite ends of the periodic table) react by the complete transfer of an electron from one atom to another to form an ionic bond. Atoms that are identical combine by sharing one or more pairs of electrons to form covalent bonds. Now let's consider bond formation between atoms that are different, but not different enough to form ionic bonds.

Both hydrogen and chlorine actually consist of diatomic molecules; the reaction is more accurately represented by the scheme

$$\text{H}:\text{H} + :\ddot{\underset{\cdot\cdot}{\text{Cl}}}:\ddot{\underset{\cdot\cdot}{\text{Cl}}}: \rightarrow 2\,\text{H}:\ddot{\underset{\cdot\cdot}{\text{Cl}}}:$$

We use the individual atoms in order to focus on the sharing of electrons to form a covalent bond.

Hydrogen Chloride

Hydrogen and chlorine react to form a colorless gas called hydrogen chloride. This reaction may be represented as

$$\text{H}\cdot + \cdot\ddot{\underset{\cdot\cdot}{\text{Cl}}}: \longrightarrow \text{H}:\ddot{\underset{\cdot\cdot}{\text{Cl}}}: \quad (\text{or H—Cl})$$

Both hydrogen and chlorine need an electron to achieve a noble gas configuration, so they share a pair and form a covalent bond.

EXAMPLE 5.5

Use electron-dot structures to show the formation of a covalent bond between

a. two fluorine atoms.

b. a fluorine atom and a hydrogen atom

SOLUTION

a. $:\ddot{\underset{\cdot\cdot}{\text{F}}}\cdot + \cdot\ddot{\underset{\cdot\cdot}{\text{F}}}: \longrightarrow :\ddot{\underset{\cdot\cdot}{\text{F}}}:\ddot{\underset{\cdot\cdot}{\text{F}}}:$ ← bonding pair

b. $\text{H}\cdot + \cdot\ddot{\underset{\cdot\cdot}{\text{F}}}: \longrightarrow \text{H}:\ddot{\underset{\cdot\cdot}{\text{F}}}:$ ← bonding pair

Exercise 5.5

Use electron-dot structures to show the formation of a covalent bond between

a. two bromine atoms

b. a hydrogen atom and a bromine atom

c. an iodine atom and a chlorine atom

One might reasonably ask why the hydrogen molecule and the chlorine molecule react at all. Have we not just explained that they themselves were formed to provide a more stable arrangement of electrons? Yes, indeed, we did say that. But there is stable, and there is more stable. The chlorine molecule represents a more stable arrangement than two separate chlorine atoms. But, given the opportunity, a chlorine atom would rather form a bond with a hydrogen atom than with another chlorine atom.

For the sake of convenience and simplicity, the reaction of hydrogen (molecule) and chlorine (molecule) to form hydrogen chloride often is reduced to

$$H_2 + Cl_2 \longrightarrow 2\,HCl$$

Molecules of hydrogen chloride consist of one atom of hydrogen and one atom of chlorine. These unlike atoms share a pair of electrons. Sharing does not mean sharing equally, though. Chlorine atoms have a greater attraction for a shared pair of electrons than do hydrogen atoms; chlorine is said to be more *electronegative* than hydrogen.

Electronegativity

The **electronegativity** of an element is a measure of the attraction of an atom in a molecule for a pair of shared electrons. The atoms to the right in the periodic table are, in general, more electronegative than those to the left. The ones on the right are precisely those atoms that, in forming ions, tend to gain electrons and form negative ions. The ones on the left, the metals, tend to give up electrons to become positive ions. The more electronegative an atom, the greater its tendency to pull the electrons in the bond toward its end of the bond when it is involved in covalent bonding.

Figure 5.5 Chlorine hogs the electron blanket, leaving hydrogen partially, but positively, exposed.

Chlorine is more electronegative than hydrogen. In the hydrogen chloride molecule, the shared electrons are held more tightly by the chlorine atom, and this results in the chlorine end of the molecule being more negative than the hydrogen end (Figure 5.5). When the electrons in a covalent bond are not equally shared, the bond is said to be *polar*. Thus, the bonding in hydrogen chloride is described as **polar covalent**, whereas the bonding in the hydrogen molecule or in the chlorine molecule is **nonpolar covalent**. The polar covalent bond is not an ionic bond. In an ionic bond, one atom completely loses an electron. In a polar covalent bond, the atom at the positive end of the bond (hydrogen in HCl) still has some share in the bonding pair of electrons (Figure

5.6). To distinguish this arrangement from that in an ionic bond, the following notation is used.

$$\overset{\delta+}{\text{H}}\text{—}\overset{\delta-}{\text{Cl}}$$

The line between the atoms represents the covalent bond, a pair of shared electrons. The $\delta+$ and $\delta-$ (read ''delta plus'' and ''delta minus'') signify which end is partially positive and which is partially negative. (The word *partially* is used to distinguish this charge from the full charge on an ion).

δ+ δ−
H :Cl:
δ+ δ−

(a) (b)

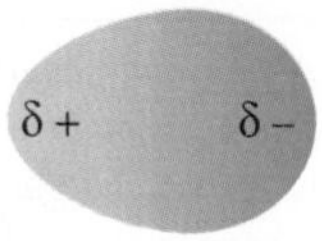

Figure 5.6 Representation of the polar hydrogen chloride molecule. (a) The electron-dot formula, with the shared electron pair shown nearer the chlorine atom. The symbols $\delta+$ and $\delta-$ indicate partial positive and partial negative charges, respectively. (b) A diagram depicting the unequal distribution of electron density in the hydrogen chloride molecule.

SECTION 5.9

Polyatomic Molecules: Water, Ammonia, and Methane

To obtain an octet of electrons, an oxygen atom must share electrons with two hydrogen atoms, a nitrogen atom must share with three hydrogen atoms, and a carbon atom must share electrons with four hydrogen atoms. In general, many nonmetals often form a number of covalent bonds equal to eight minus the group number. Oxygen, which is in Group VIA, forms $8 - 6 = 2$ covalent bonds in most compounds. Nitrogen, in Group VA, forms $8 - 5 = 3$ covalent bonds in most of its compounds. Carbon, in Group IVA, forms $8 - 4 = 4$ covalent bonds in most carbon compounds, including the great host of organic compounds (Chapter 9). These simple rules will enable you to write formulas for a large number of molecules.

Carbon forms 4 bonds.

Nitrogen forms 3 bonds.

Oxygen forms 2 bonds.

Hydrogen forms 1 bond.

Water

Water is one of the most familiar chemical substances. The electrolysis experiment of Nicholson and Carlisle (Section 2.3) and ample evidence since their time indicate that the molecular formula for water is H_2O. In order to be surrounded by an octet, oxygen shares two pairs of electrons. But a hydrogen atom shares only one pair of electrons. An oxygen atom must therefore bond with two hydrogen atoms.

Water is discussed in considerable detail in Chapter 13.

$$\cdot\ddot{\text{O}}\text{:} + 2\,\text{H}\cdot \longrightarrow \text{H:}\ddot{\text{O}}\text{:}$$
$$\qquad\qquad\qquad\qquad\ddot{\ }\text{H}$$

A *polar molecule* has separate centers of positive and negative charge, just as a magnet has north and south poles.

This arrangement completes the octet in the valence energy level of oxygen, giving it the neon structure. It also completes the outer energy level of the hydrogen atoms, each of which now has the helium structure. Since the water molecule is bent, it is *polar* (Section 5.14).

Ammonia

Ammonia (NH_3) is a gas at room temperature. Vast quantities of it are compressed into tanks and used as fertilizer.

An atom of the element nitrogen has five electrons in its valence energy level. It can assume the neon configuration by sharing three pairs of electrons with *three* hydrogen atoms. The result is the compound ammonia.

$$\cdot\ddot{\text{N}}\cdot + 3\,\text{H}\cdot \longrightarrow \text{H}:\ddot{\underset{\cdot\cdot}{\text{N}}}:\text{H} \quad (\text{H below N})$$

In ammonia, the bond arrangement is that of a tripod with a hydrogen atom at the end of each ''leg'' and the nitrogen atom with its unshared pair of electrons sitting at the top. This pyramidal shape causes the ammonia molecule to be polar (Section 5.14).

Methane

Methane (CH_4) is the simplest of the hydrocarbons, a group of organic compounds discussed in detail in Chapter 9. It is the principal component of natural gas, used as a fuel.

An atom of carbon has four electrons in its valence energy level. It can assume the neon configuration by sharing pairs of electrons with four hydrogen atoms, forming the compound methane.

$$\cdot\dot{\underset{\cdot}{\text{C}}}\cdot + 4\,\text{H}\cdot \longrightarrow \begin{matrix} & \text{H} & \\ & \cdot\cdot & \\ \text{H}: & \text{C} & :\text{H} \\ & \cdot\cdot & \\ & \text{H} & \end{matrix}$$

The methane molecule has a tetrahedral shape (Section 5.14).

SECTION 5.10

Multiple Bonds

Atoms can share more than one pair of electrons. In carbon dioxide, for example, the carbon atom shares two pairs of electrons with each of the two oxygen atoms.

$$:\ddot{\text{O}}::\text{C}::\ddot{\text{O}}:$$

A shared single pair of electrons is called a **single bond**.

Note that each atom has an octet of electrons about it as a result of this sharing. We say that the atoms are joined by a **double bond,** a covalent linkage in which the two atoms share two pairs of electrons.

Atoms also can share three pairs of electrons. In the nitrogen (N_2) molecule, for example, each nitrogen atom shares three pairs of electrons with the other.

$$:N:::N:$$

The atoms are joined by a **triple bond**, a covalent linkage in which two atoms share three pairs of electrons. Note that each of the nitrogen atoms has an octet of electrons about it.

Covalent bonds often are represented as dashes. We can therefore show the three kinds of bonds as follows.

H—Cl   O═C═O   N≡N

SECTION 5.11

Rules for Writing Electron-Dot Formulas

Recall that electrons are transferred (Section 5.5) or shared (Section 5.7) in ways that leave most atoms with an octet of electrons in their outermost energy level. In this section we describe the procedure we can follow in writing **electron-dot formulas** for molecules. First we must put the atoms of the molecules in their proper places. The only way to do this for sure is from experiment, but there are some generalizations that will help us to get many structures right.

The *skeletal structure* of a molecule tells us the order in which the atoms are attached to one another. In the absence of experimental evidence, the following rules help us to devise likely skeletal structures.

1. Hydrogen atoms form only one bond; they are always at the end of a sequence of atoms. Hydrogen often is bonded to carbon, nitrogen, or oxygen.

```
    H                                       H
    |                                       |
H—C—H     H—N—H     H—O     H—C—O—H
    |          |          |         |
    H          H          H         H
```

2. Polyatomic molecules and ions often consist of a central atom surrounded by more electronegative atoms. (Hydrogen is an exception; it is always on the outside, even when bonded to a more electronegative element.)

```
              O           Cl
              |           |
O—C—O     O—S—O     Cl—C—Cl
                          |
                          Cl
```

After a skeletal formula for a polyatomic molecule or ion has been chosen, we can use the following steps to write an electron-dot formula.

1. Calculate the total number of valence electrons. The total for a molecule is the sum of the valence electrons for all the atoms. For a polyatomic anion, add the number of negative charges. For a polyatomic cation, subtract the number of positive charges.
Examples:

Polyatomic ions are discussed in Chapter 6.

N_2O_4 has $(2 \times 5) + (4 \times 6) = 34$ valence electrons.

NO_3^- has $5 + (3 \times 6) + 1 = 24$ valence electrons.

NH_4^+ has $5 + (4 \times 1) - 1 = 8$ valence electrons.

2. Write the skeletal structure, and connect bonded pairs of atoms by a dash (one electron pair).
3. Place electrons about outer atoms so that each (except hydrogen) has an octet.
4. Subtract the number of electrons assigned so far from the total calculated in Step 1. Any electrons that remain are assigned in pairs to the central atom(s).
5. If a central atom has fewer than eight electrons after Step 4, a multiple bond is likely. Move one or more nonbonding pairs from an outer atom to the space between the atoms to form a double or triple bond. A deficiency of two electrons suggests a double bond, and a shortage of four electrons is indicative of a triple bond or two double bonds to the central atom.

▶ EXAMPLE 5.6

Give the electron-dot formula for methanol, CH_4O.

SOLUTION

1. The total number of valence electrons is $4 + (4 \times 1) + 6 = 14$.
2. The skeletal structure in which all the hydrogen atoms are on the outside and the least electronegative carbon atom is most central is

```
     H
     |
  H—C—O—H
     |
     H
```

3. Now, counting each bond as two electrons gives 10 electrons. The 4 remaining electrons are placed (as two nonbonding pairs) on the oxygen atom.

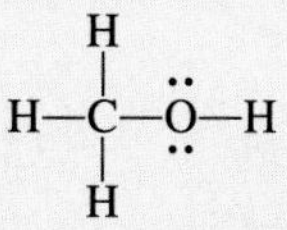

```
      H
      |  ..
   H—C—O—H
      |  ..
      H
```

(The remaining steps are not necessary; both carbon and oxygen have an octet of electrons.)

Exercise 5.6
Give the electron-dot formula for ethyl chloride, C_2H_5Cl.

▶ EXAMPLE 5.7

Give the electron-dot formula for nitrogen trifluoride, NF_3.

SOLUTION

1. There are $5 + (3 \times 7) = 26$ valence electrons.
2. The skeletal structure is

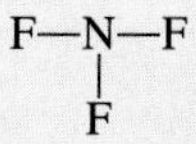

```
 F—N—F
   |
   F
```

3. Place three nonbonding pairs on each fluorine atom.

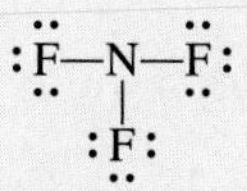

```
 ..       ..
:F—N—F:
 ..  |   ..
    :F:
     ..
```

4. We have assigned 24 electrons. Place the remaining two as a nonbonding pair on the nitrogen atom.

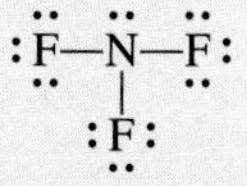

```
 ..  ..  ..
:F—N—F:
 ..  |   ..
    :F:
     ..
```

(Each atom has an octet; Step 5 is not needed.)

Exercise 5.7
Give the electron-dot structure for oxygen difluoride, OF_2.

▶ EXAMPLE 5.8

Give the electron-dot structure for the BF_4^- ion.

SOLUTION

1. There are 3 + (4 × 7) + 1 = 32 electrons.
2. The skeletal structure is

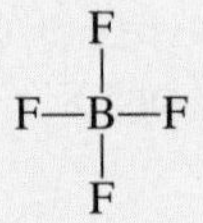

```
     F
     |
  F—B—F
     |
     F
```

3. Place three nonbonding pairs on each fluorine atom.

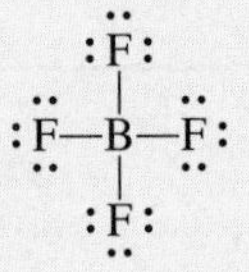

```
      ..
     :F:
  ..  |  ..
 :F—B—F:
  ..  |  ..
     :F:
      ..
```

4. We have assigned 32 electrons. None remains to be assigned.

Exercise 5.8
Give the electron-dot structure for PH_4^+ ion.

▶ EXAMPLE 5.9

Give the electron-dot structure for carbon dioxide, CO_2.

SOLUTION

1. There are 4 + (2 × 6) = 16 valence electrons.
2. The skeletal structure is

O—C—O

3. Place three nonbonding pairs on each oxygen atom.

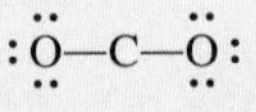

```
 ..     ..
:O—C—O:
 ..     ..
```

4. We have assigned 16 electrons. None remains to be placed.
5. The carbon atom has only four electrons. It needs to form two double bonds in order to have an octet. (It is not reasonable to expect that carbon would form a triple bond to one of the oxygen atoms. Why?) Move a nonbonding pair from each oxygen atom to the space between the atoms to form a double bond on each side of the carbon atom.

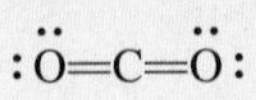

```
 ..     ..
:O=C=O:
```

Exercise 5.9
Give the electron-dot formula for nitryl fluoride, NO_2F.

Table 5.2 Number of Bonds Formed by Selected Elements

Electron-Dot Picture	Bond Picture	Number of Bonds	Representative Molecules	
H·	H—	1	H—H H—Cl	HCl
He:		0		
·C· (4 dots)	—C— (4 bonds)	4	CH_4 (H—C—H with H above and below); H—C(=O)—F	CH_4
·N· (lone pair + 3 dots)	—N— (3 bonds)	3	H—N—H with H below; H—C(=N—O—H)—H	NH_3
·O:	—O (2 bonds)	2	H—O—H; H—C(=O)—H	H_2O
·F:	—F	1	H—F F—F	F_2
·Cl:	—Cl	1	Cl—Cl; H—C(H)(H)—Cl	CH_3Cl

The rules we have used here lead to results that are summarized and illustrated for selected elements in Table 5.2. Figure 5.7 relates the number of covalent bonds to the periodic table.

SECTION 5.12

Exceptions to the Octet Rule

Many molecules made of atoms of the main group elements have electron structures that follow the octet rule. There are numerous exceptions, however. The exceptions fall into three main groups. Each type is readily identified by some structural characteristic.

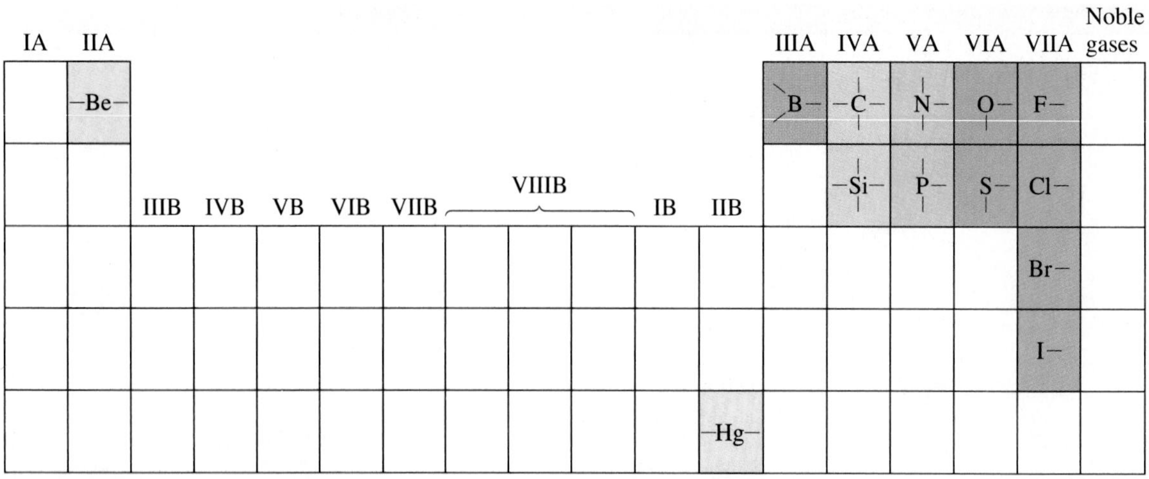

IA IIA IIIA IVA VA VIA VIIA Noble gases

—Be— B— —C— N— O— F—

IIIB IVB VB VIB VIIB VIIIB IB IIB —Si— P— S— Cl—

Br—

I—

—Hg—

Figure 5.7 Covalent bonding of representative elements of the periodic table.

Odd-Electron Molecules: Free Radicals

Molecules with an odd number of valence electrons obviously cannot satisfy the octet rule. Examples of such molecules are nitrogen monoxide (NO, also called nitric oxide), with $5 + 6 = 11$ valence electrons; nitrogen dioxide (NO_2), with 17 valence electrons; and chlorine dioxide (ClO_2), which has 19 outer electrons. Obviously one of the atoms in each of these molecules will have an odd number of electrons and therefore cannot have an octet.

Atoms and molecules with an unpaired electron are called **free radicals**. Most free radicals are highly reactive and have only transitory existence as intermediates in chemical reactions. An example is the chlorine atom that is formed from the breakdown of chlorofluorocarbons in the stratosphere and that leads to the depletion of the ozone shield (Chapter 12). Some free radicals are quite stable, however. The nitrogen oxides are major components of smog. Chlorine dioxide is made in multiton quantities and is used for bleaching flour.

All atoms and molecules with an odd number of electrons must have one unpaired electron. Filled energy levels and sublevels have all their electrons paired, with two electrons in each orbital (Section 3.8). We need only consider valence electrons to determine whether or not an atom or molecule is a free radical. Electron-dot structures of NO, NO_2, and ClO_2 show that one atom of each has an unpaired electron; that atom obviously does not have an octet of electrons in its outer energy level.

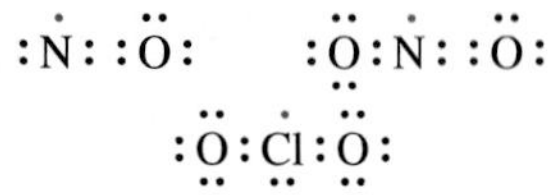

:N::O: :O:N::O:

:O:Cl:O:

Molecules with Incomplete Octets

Boron atoms have three valence electrons; fluorine atoms have seven. When boron reacts with fluorine, it shares those electrons with three fluorine atoms to form boron trifluoride.

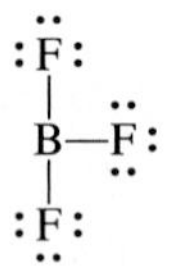

:F:
|
B—F:
|
:F:

This structure uses all 24 of the valence electrons; there are none left to put on the central boron atom. Experimental evidence indicates that this structure is consistent with the reactivity of BF_3 toward molecules with unshared pairs. For instance, BF_3 reacts readily with ammonia to form BF_3NH_3, a compound in which all atoms (except hydrogen) have an octet of electrons.

$$BF_3 + :NH_3 \longrightarrow F_3B\text{—}NH_3$$

Beryllium has two valence electrons. It reacts with the halogens to form compounds with the simplest formula BeX_2, where X = F, Cl, Br, or I. In the vapor state these compounds exist as discrete BeX_2 molecules. Beryllium bromide, for example, sublimes at about 490 °C, to form separate $BeBr_2$ molecules. In these molecules beryllium shares a pair of electrons with each of two bromine atoms.

$$:\ddot{\underset{\cdot\cdot}{Br}}:Be:\ddot{\underset{\cdot\cdot}{Br}}:$$

It has only four electrons in its valence shell.

Molecules with Expanded Octets

The second period elements—carbon, nitrogen, oxygen, and fluorine—nearly always obey the octet rule. (The odd-electron compounds are obvious exceptions.) The valence electron level of the second period elements holds a maximum of eight electrons ($2s^22p^6$). The third main energy level can hold up to 18 electrons ($3s^23p^63d^{10}$). Third period elements therefore can violate the octet rule by having more than eight electrons in their valence level. These so-called *expanded octets* are evident in the following compounds.

$$PCl_5 \qquad SF_6$$

Phosphorus pentachloride   Sulfur hexafluoride

Elements in the third period and beyond, then, are not limited to an octet. Yet many of their compounds still follow the octet rule.

It is a good idea to use the octet rule except in cases where it obviously doesn't apply: when there is an odd number of electrons, when there are too few electrons to make an octet, and (third period and beyond) where there are obviously more than eight electrons that must be in the valence level.

SECTION 5.13

Molecular Shapes: The VSEPR Theory

We have represented molecules in two dimensions on paper, but molecules have three-dimensional shapes. We can use electron-dot structures as part of

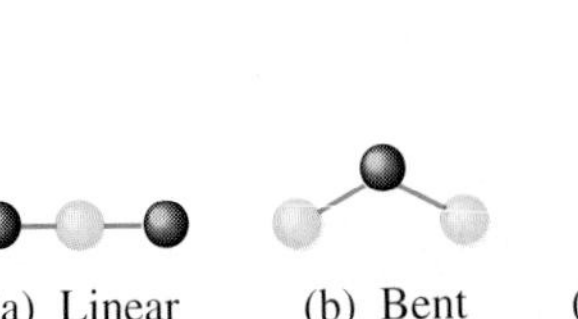 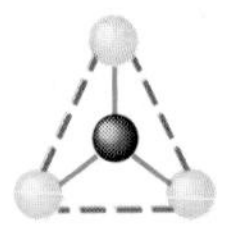 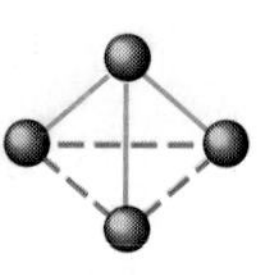 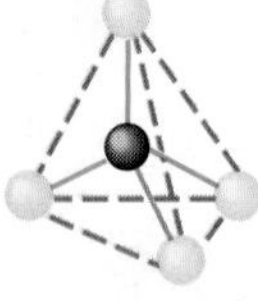

(a) Linear (b) Bent (c) Triangular (d) Pyramidal (e) Tetrahedral

Figure 5.8 Shapes of molecules. In a *linear* molecule (a), all the atoms are along a line; the bond angle is 180°. The *bent* molecule (b) has an angle less than 180°. Connecting the three outer atoms of the *triangular* molecule (c) with imaginary lines produces a triangle with an atom at the center. Imaginary lines connecting all four atoms of a *pyramidal* molecule (d) form a three-sided pyramid. Connecting the four outer atoms of the *tetrahedral* molecule (e) with imaginary lines produces a tetrahedron (a four-sided figure in which each side is a triangle) with an atom at the center.

Table 5.3 Bonding and the Shape of Molecules

Number of Bonded Atoms	Number of NBP*	Number of Sets	Molecular Shape	Examples				
2	0	2	Linear	$BeCl_2$	$HgCl_2$	CO_2	HCN	$BeCl_2$
3	0	3	Triangular	BF_3	$AlBr_3$	CH_2O		BF_3
4	0	4	Tetrahedral	CH_4	CBr_4	$SiCl_4$		CH_4
3	1	4	Pyramidal	NH_3	PCl_3			NH_3
2	2	4	Bent	H_2O	H_2S	SCl_2		H_2O
2	1	3	Bent	SO_2	O_3			SO_2

*NBP = nonbonding pair of electrons.

the process of predicting molecular shapes. The shapes that we consider in this book are shown in Figure 5.8.

The **valence shell electron pair repulsion theory** (**VSEPR**) often is used to predict the arrangement of atoms about a central atom. The basis of the VSEPR theory is that electron pairs will arrange themselves about a central atom in a way that minimizes repulsion between the like-charged particles. This means that they will get as far apart as possible. Table 5.3 gives the geometric shapes associated with arrangement of two, three, or four entities about a central atom.

The farthest apart two substituent atoms can get are the opposite sides of the central atom at an angle of 180°.

Three groups assume a triangular arrangement about the central atom, forming angles of separation of 120°. Four groups form a tetrahedral array around the central atom, giving a separation of about 109.5°.

You can determine the shapes of many molecules (and polyatomic ions, Chapter 6) by following these simple rules.

The shapes of molecules are of considerable importance. We are interested in the shapes of biologically active molecules (Chapters 15, 16, 18, and 19). These molecules must have the right groups in the right places in order to function properly. For now, we limit our discussion to a few simple molecules for which we can use the VSEPR theory to predict shapes.

1. Draw an electron-dot structure. In the structure, indicate a shared electron pair (bonding pair, BP) by a line. Use dots to indicate any nonbonding pairs (NBP) of electrons.

2. To determine shape, count the number of atoms *and* NBP attached to the central atom. Note that a multiple bond counts only as 1 *set*. Examples

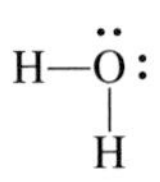

Four sets
(2 atoms, 2 NBP)

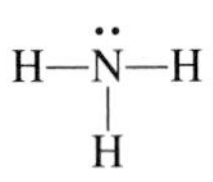

Four sets
(3 atoms, 1 NBP)

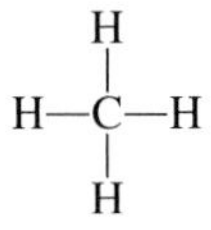

Four sets
(4 atoms)

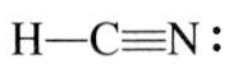

Two sets
(2 atoms)

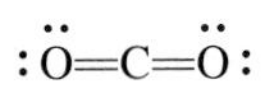

Two sets
(2 atoms)

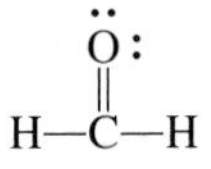

Three sets
(3 atoms)

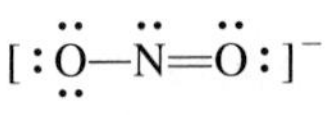

Three sets
(2 atoms, 1 NBP)

3. Determine the number of electron sets and draw a shape *as if* all were bonding pairs.

4. Sketch that shape, placing the electron pairs as far apart as possible (see Table 5.3). If there is *no* NBP, that is the shape of the molecule. If there *are* NBPs, remove them, leaving the BPs exactly as they were. (This may seem strange, but it stems from the fact that *all* the sets determine the geometry, but only the arrangement of bonded atoms is considered in the shape of the molecule.)

▶ EXAMPLE 5.10

What is the shape of the BH_3 molecule?

SOLUTION

1. The electron-dot structure is

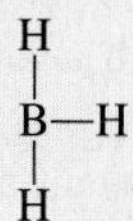

2. There are three sets to consider

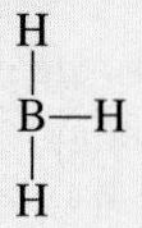

```
H
|
B—H
|
H
```

3. The three sets get as far apart as possible, giving a triangular arrangement of the sets.

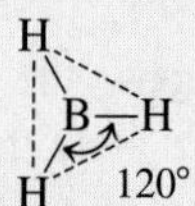

4. All the sets are bonding pairs; the molecular shape is the same as the arrangement of the electrons.

Exercise 5.10
What is the shape of the BeH_2 molecule?

▶ EXAMPLE 5.11

What is the shape of the SCl_2 molecule?

SOLUTION

1. The electron-dot structure is

```
 ..  ..
:S—Cl:
 |  ..
:Cl:
 ..
```

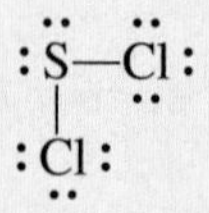

2. There are four sets on the sulfur atom to consider.

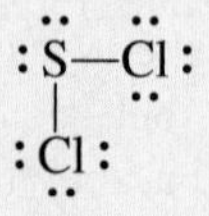

```
 ..  ..
:S—Cl:
 |  ..
:Cl:
 ..
```

3. The four sets get as far apart as possible, giving a tetrahedral arrangement of the sets about the central atom.

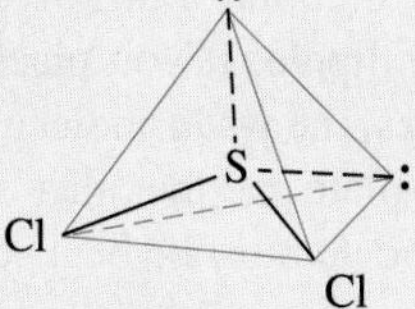

4. Two of the sets are bonding pairs and two are NBP. Remove (ignore) the NBP; the molecular shape is bent with a bond angle of 109.5 °.

Exercise 5.11
What is the shape of the PH_3 molecule?

SECTION 5.14

Shapes and Properties: Polar and Nonpolar Molecules

In Section 5.8 we discussed polar and nonpolar bonds. Diatomic molecules are polar if their bonds are polar and nonpolar if their bonds are nonpolar.

$$\overset{\delta+}{\text{H}}\text{—}\overset{\delta-}{\text{Cl}} \qquad \text{Cl—Cl}$$

Polar Nonpolar

For molecules with three or more atoms, we also must consider the orientation of the bonds to determine whether or not the molecule as a whole is polar.

Water: A Bent Molecule

We should expect the bonds in water to be polar because oxygen is more electronegative than hydrogen. (Like chlorine, oxygen is to the right in the periodic table, and electronegativity increases as one moves to the right in the table.) Just because a molecule contains polar bonds, however, does not mean that the molecule as a whole is polar. If the atoms in the water molecule were in a straight row (that is, in a linear arrangement), the two polar bonds would cancel one another out.

$$\overset{\delta+}{\text{H}}\text{—}\overset{\delta-}{\text{O}}\text{—}\overset{\delta+}{\text{H}}$$ *(incorrect structure)*

Dipoles often are represented by an arrow with a plus at the tail end (+→).

$\overset{+\!\!\longrightarrow}{\text{H—Cl}}$

The positive end of the arrow is obvious; the head of the arrow indicates the negative end of the dipole. Using these arrows, we see that carbon dioxide is nonpolar despite its polar bonds.

$\overset{\longleftarrow\!+\;\;+\!\longrightarrow}{\text{O=C=O}}$

The two cancel one another out. In water, both dipoles point toward the oxygen, giving a net dipole toward that end of the molecule.

H—O—H (dipole arrows toward O) or H—O—H (net dipole arrow)

Instead of having one end of the molecule positive and the other end negative, the electrons would be pulled toward the right in one bond and toward the left in the other. Overall there would be no net dipole. By a **dipole**, we mean a molecule that has a positive end and a negative end.

But water *does* act like a dipole. If you place a sample of water between two electrically charged plates, the water molecules align themselves, one end attracted toward the positive plate and the other end toward the negative plate. To act like a dipole, the molecule must be bent so that the bonds do not cancel one another out.

$^{\delta+}\text{H}:\ddot{\underset{\cdot\cdot}{\text{O}}}:^{\delta-}$ with $\text{H}^{\delta+}$ below, or $^{\delta+}\text{H}-\text{O}^{\delta-}$ with $|$ $\text{H}^{\delta+}$ below

Such molecules would align themselves between charged plates as shown in Figure 5.9.

The shape of the water molecule can be accounted for by a modification of the VSEPR theory. According to VSEPR, the two bonds and two NBP should form a tetrahedral arrangement.

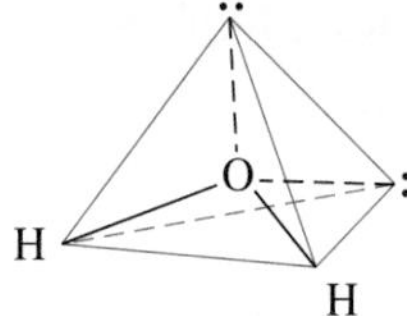

Ignoring the NBP, the molecular shape has the atoms in a bent arrangement, with a bond angle of 109.5° (the tetrahedral angle).

The predicted bond angle of 109.5° for water is a bit larger than the measured angle of 104.5°. The difference is explained by the fact that the NBP occupy a greater volume than do the bonding pairs. These larger orbitals push the smaller BP closer together.

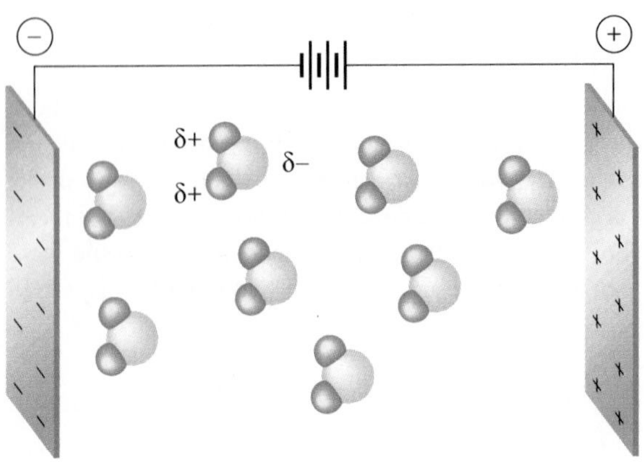

Figure 5.9 Polar molecules are aligned in an electric field, with the positive end of the molecule preferentially pointing toward the negative plate and the negative end of the molecule directed toward the positive plate.

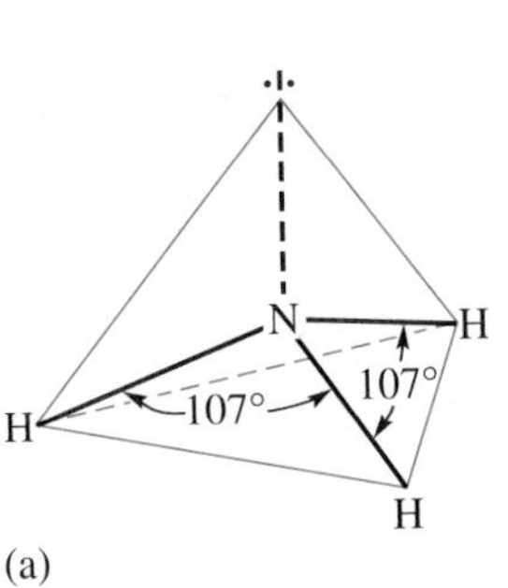

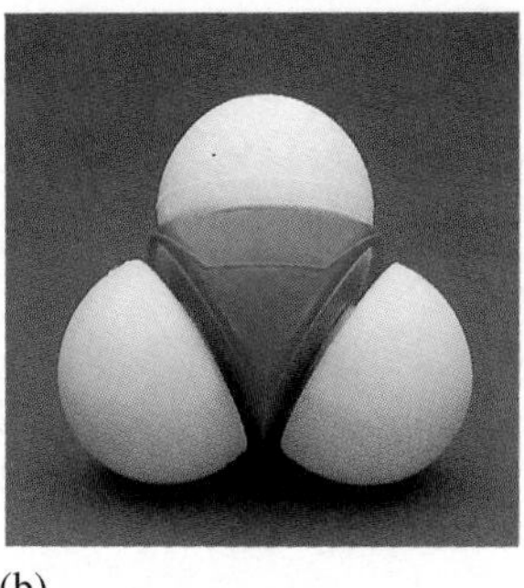

Figure 5.10 The ammonia molecule. In the drawing (a) solid lines indicate covalent bonds; the color lines outline a tetrahedron. The photograph (b) shows a space-filling model of an ammonia molecule. [*(b) Fundamental Photographs/Macmillan Science Files.*]

Ammonia: A Pyramidal Molecule

There are three BP and one NBP about the nitrogen atom of ammonia. The VSEPR theory predicts a tetrahedral arrangement with bond angles of 109.5°. The actual bond angles are 107°, close to the theoretical value. Presumably, the unshared pair of electrons occupies a greater volume than does a shared pair, pushing the latter slightly closer together. The arrangement is, therefore, that of a tripod with a hydrogen atom at the end of each leg and the nitrogen atom with its unshared pair sitting at the top (Figure 5.10). Each nitrogen-to-hydrogen bond is somewhat polar, making the ammonia molecule polar.

Methane: A Tetrahedral Molecule

There are four pairs of electrons on the central carbon atom in methane. Using the VSEPR theory, we would expect a tetrahedral arrangement and bond angles of 109.5°. The actual bond angles are 109.5°, in perfect agreement with theory (Figure 5.11). All four electron pairs are shared with hydrogen atoms; thus, all four pairs occupy identical volumes.

Each carbon-to-hydrogen bond is slightly polar, but the methane molecule as a whole is symmetrical. The slight bond polarities cancel out, leaving the methane molecule, as a whole, nonpolar.

Many of the properties of compounds—such as melting point, boiling point, and solubility—depend on the polarity of the molecules of the compounds.

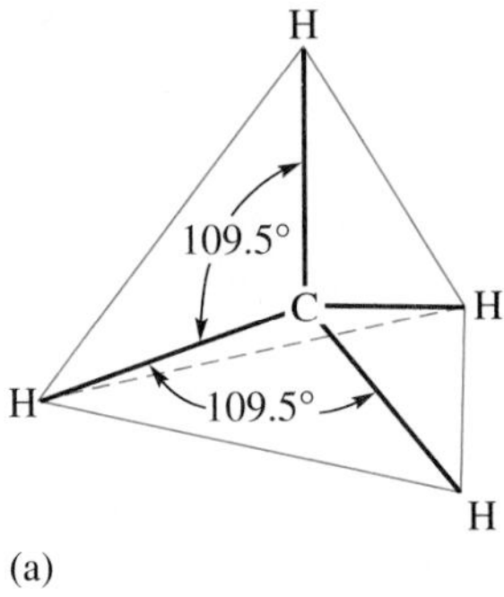

Figure 5.11 The methane molecule. In the drawing (a) solid lines indicate covalent bonds; the color lines outline the tetrahedron. All bond angles are 109.5°. The photograph (b) shows a space-filling model of the methane molecule. [*(b)Fundamental Photographs/Macmillan Science Files.*]

SECTION 5.15

Bonding Forces and the States of Matter

The states of matter—solid, liquid, and gas—are obviously different from one another (Chapter 1). Chemists offer a model to explain these differences. The model is referred to as the **kinetic-molecular theory**. The basic postulates of this theory are

1. All matter is composed of tiny, discrete particles called molecules.
2. Molecules are in rapid constant motion and move in straight lines.
3. The molecules of a gas are very small compared to the distances between them.
4. There is very little attraction between molecules of a gas.
5. Molecules collide with one another, and energy is conserved in these collisions—although one molecule can gain energy at the expense of another.
6. Temperature is a measure of the *average* kinetic energy of the gas molecules.

Right now we need only consider how the model pictures solids, liquids, and gases. *Solids* are viewed as highly ordered assemblies of particles in close contact with one another. *Liquids* are pictured as much more loosely organized collections of particles. In a liquid, the particles are still in close contact with one another, but they are much more free to move about. Finally, in *gases* the particles are no longer in close contact with one another, but are separated by relatively great distances and are moving about at random.

Table salt (sodium chloride) is a typical crystalline solid. In this solid, ionic bonds hold the ions in position and maintain the orderly arrangement. Not all solids are held together by ionic bonds, but some attractive force is necessary to maintain the characteristic orderly array of particles in a solid.

To get a better image of a liquid at the molecular level, think of a box of marbles that is being shaken continuously. The marbles move back and forth, rolling over one another. The particles of a liquid (like the marbles) are not so rigidly held in place as are particles in a solid. The particles in a liquid are being held close together, however, and that means there must be some force attracting them.

Gas particles do not experience significant attractive forces. They move about at great distances from one another and interact only during occasional collisions.

Solids can be changed to liquids; that is, they can be *melted*. The solid is heated, and the heat energy is absorbed by the particles of the solid. The energy causes the particles to vibrate in place with more and more vigor until, finally, the forces holding the particles in a particular arrangement are over-

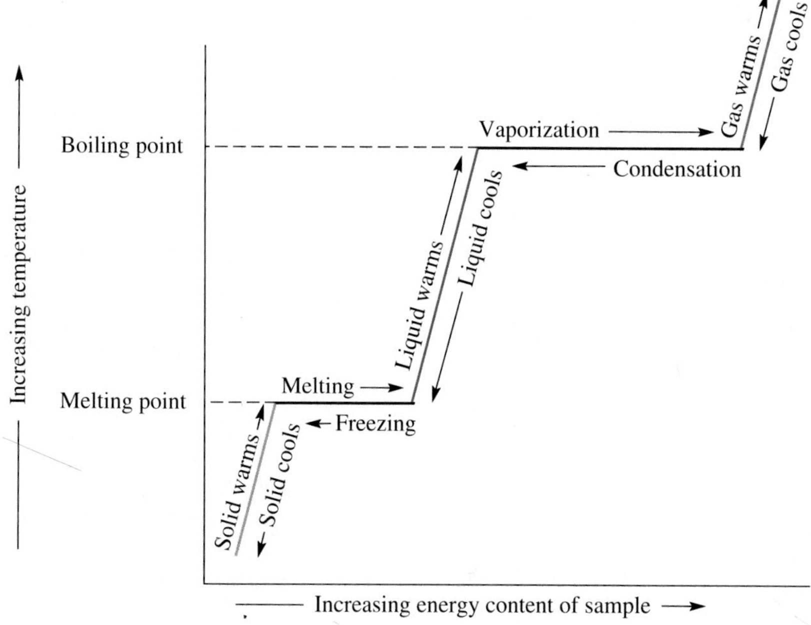

Figure 5.12 Diagram of change in state of matter on heating or cooling.

come. The solid has become a liquid. The temperature at which this happens is called the **melting point** of that solid. A high melting point is one indication that the forces holding a solid together are very strong.

A liquid can change to a gas or vapor in a process called **vaporization**. Again, one need only supply sufficient heat to achieve this change. Energy is absorbed by the liquid particles, which move faster and faster as a result. Finally the attractive forces holding the liquid particles in contact are overcome by this increasingly violent motion and the particles fly away from one another. The liquid has become a gas.

The entire sequence of changes can be reversed by removing energy from the sample and slowing down the particles. Vapor changes to liquid in a process referred to as **condensation**; liquid changes to solid in a process called **freezing**. Figure 5.12 presents a diagram of the changes in state that occur as energy is added to or removed from a sample.

The amount of energy required to accomplish these changes depends on the type of forces responsible for maintaining the solid or the liquid state. The ionic bonds found in salt crystals are very strong. Sodium chloride must be heated to about 800 °C before it melts. Generally, interionic forces are the strongest of all the forces that hold solids and liquids together. We will now consider some other interactions that hold the particles of solids and liquids together.

Dipole Forces

Hydrogen chloride melts at −112 °C and boils at −85 °C (it is a gas at room temperature). The attractive forces between molecules are not nearly as strong

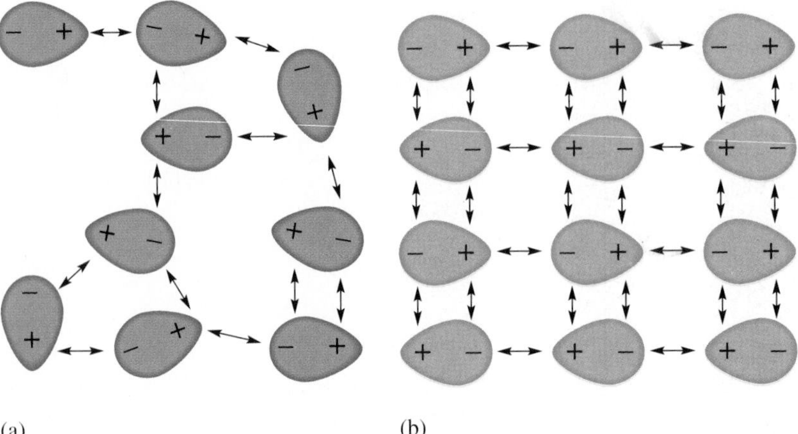

Figure 5.13 An idealized representation of dipole forces in (a) a liquid and (b) a solid. In a real liquid or solid, interactions are more complex.

as the interionic forces in salt crystals. We know that covalent bonds hold the hydrogen and chlorine *atoms* together to form the hydrogen chloride *molecule,* but what makes one molecule interact with another in the solid or liquid states?

Remember that the hydrogen chloride molecule is a dipole. It has a positive and a negative end. Two dipoles brought close enough will attract one another. The positive end of one molecule attracts the negative end of another. Such forces may exist throughout the structure of a liquid or solid (Figure 5.13). In general, attractive forces between dipoles are much weaker than the attractive forces between ions. **Dipole interactions** are, however, stronger than the forces between nonpolar molecules of comparable size.

Hydrogen Bonds

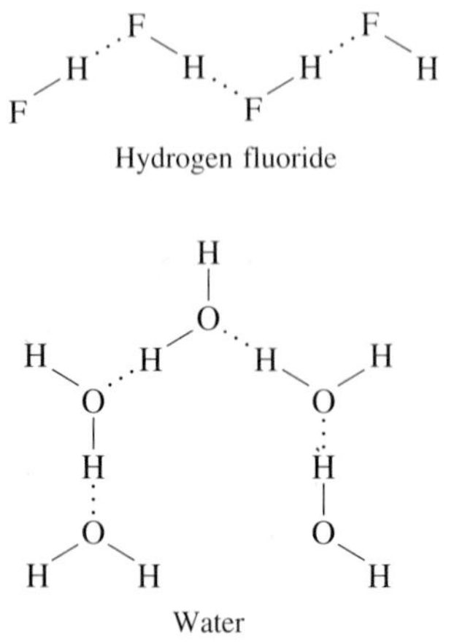

Certain polar molecules exhibit stronger attractive forces than expected on the basis of ordinary dipolar interactions. These forces are strong enough to be given a special name, *hydrogen bonds.* Note that "*hydrogen* bond" is a somewhat misleading name, since it emphasizes only one component of the interaction. Not all compounds containing hydrogen exhibit this strong attractive force; the hydrogen *must* be attached to fluorine, oxygen, or nitrogen. It is the presence of these atoms that permits us to offer an explanation for the extra strength of hydrogen bonds as compared with other dipolar forces. Fluorine, oxygen, and nitrogen all have a high electron-attracting power (they are very electronegative), and they are small (they are at the top of the periodic table). A hydrogen–fluorine bond, for example, is very strongly polarized with a negative fluorine end and a positive hydrogen end. Both hydrogen and fluorine are small atoms, so the negative end of one dipole can approach very closely the positive end of a second dipole. This results in an unusually strong interaction between the hydrogen atom of one molecule and an unshared pair of electrons on another. This special force between two molecules is called a **hydrogen bond**. Hydrogen bonds are often explicitly represented by *dotted*

Hydrogen fluoride

Water

Figure 5.14 Hydrogen bonding in hydrogen fluoride and in water.

lines to emphasize their exceptional strength compared to ordinary dipolar interactions. A *dotted* line is used to distinguish a hydrogen bond from the much stronger covalent bond, which is represented by a *solid* line (Figure 5.14).

Hydrogen bonds may be as much as 5 to 10% as strong as covalent bonds.

Water has both an unusually high melting point and an unusually high boiling point for a compound with molecules of such small size. These abnormal values are attributed to water's ability to form hydrogen bonds. Water is discussed in detail in Chapter 13.

Dispersion Forces

If one understands that positive attracts negative, then it is easy enough to understand how ions or polar molecules maintain contact with one another. But how can we explain the fact that nonpolar compounds can exist in the liquid and solid states? Even hydrogen can exist as a liquid or a solid if the temperature is low enough (its melting point is −259 °C). Some force must be holding these molecules in contact with one another in the liquid and solid states.

Dispersion forces are weak for small, nonpolar molecules such as H_2, N_2, and CH_4. These forces can be substantial between large molecules. The properties of polymers such as polyethylene (Chapter 10) are determined to a large degree by dispersion forces between long chains of repeating $\leftarrow\!\!CH_2CH_2\!\!\rightarrow_n$ units. (In this formula, n is several hundred or even several thousand.)

Up to this point we have pictured the electrons in a covalent bond as being held in place between the two atoms sharing the bond. But the electrons are *not* really static; they actually move about in the bonds. On the average, the two electrons in the hydrogen molecule (or any nonpolar bond) are between and equidistant from the two nuclei. At any given instant, however, the electrons may be at one end of the molecule. At some other time, a moment later, the electrons may be at the other end of the molecule. At the instant the electrons of one molecule are at one end, the electrons in the next molecule will move away from its adjacent end. Thus, at this instant there will be an attractive force between the electron-rich end of one molecule and the electron-poor end of the next. These momentary, usually weak, attractive forces between molecules are called **dispersion forces**. To a large extent, dispersion forces determine the physical properties of nonpolar compounds.

Forces in Solutions

To complete our look at chemical bonding, we shall briefly examine the interactions that occur in solutions. A **solution** is an intimate, homogeneous mixture of two or more substances. By *intimate* we mean that the mixing occurs down to the level of individual ions and molecules. In a salt–water solution, for example, there are not clumps of ions floating around, but single, randomly distributed ions among the water molecules. **Homogeneous** means that the mixing is thorough. All parts of the solution have the same distribution of components. For the salt solution, the saltiness is the same at the top, bottom, and middle of the solution. The substance being dissolved and usually present in lesser amount (salt in a salt solution) is called the **solute**. The substance doing the dissolving and usually present in greater amount is the **solvent** (water in the salt–water solution) (Figure 5.15).

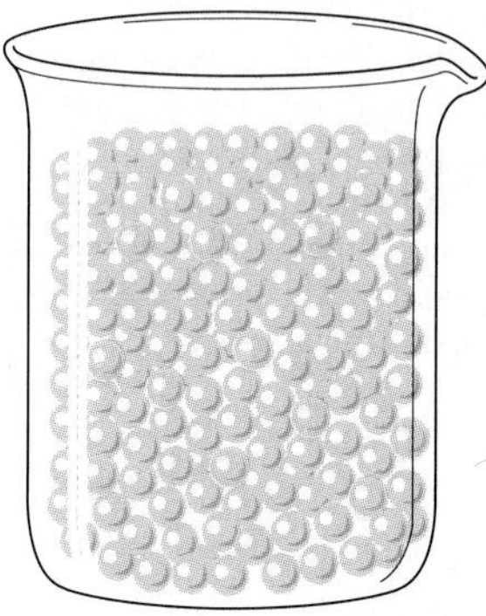

Figure 5.15 In a solution the solute molecules are randomly distributed among the solvent molecules.

(a)

(b) (c)

Figure 5.16 (a) Lawn mowers with two-cycle engines are fueled and lubricated with a solution of nonpolar lubricating oil in nonpolar gasoline. (b) In Italian dressing, polar vinegar and nonpolar olive oil are mixed. However, the two liquids do not form a solution and will separate on standing. (c) Wine is a solution of polar alcohol in polar water.

Ordinarily, solutions form most readily when the substances involved have *similar* bonding characteristics. An old chemical rule is ''Like dissolves like.'' Nonpolar solutes dissolve best in nonpolar solvents. For example, oil and gasoline, both nonpolar, mix (Figure 5.16a), but oil and water do not (Figure 5.16b). Alcohol is a liquid held together by hydrogen bonds. So is water. Alcohol readily dissolves in water because the two substances can hydrogen bond with one another (Figure 5.16c). In general, a solute dissolves when attractive forces between it and the solvent overcome the attractive forces operating in the pure solute and in the pure solvent.

Why, then, does salt dissolve in water? Ionic solids are held together by strong ionic bonds. We have already indicated that very high temperatures are required to melt ionic solids and break these bonds. Yet, simply by placing sodium chloride in water at room temperature we can dissolve the salt (or, rather, the water can). And when such a solid dissolves, its bonds *are* broken. The difference between the two processes is the difference between brute force and persuasion. In the melting process, we are simply pouring in enough energy (as heat) to pull the crystal apart. In the dissolving process, we offer the ions an attractive alternative to the ionic interactions in the crystal.

It works this way. Water molecules surround the crystal. Those that approach a negative ion align themselves so that the positive ends of their dipoles point toward the ion. With a positive ion the process is reversed, and the negative end of the water dipole points toward the ion. Still, the attraction between a dipole and an ion is not as strong as that between two ions. To compensate for their weaker attractive power, several molecules surround each ion, and in this way the many *ion–dipole* interactions overcome the *ion–ion* interactions (Figure 5.17).

In an ionic solid, the positive and negative ions are strongly bonded together in an orderly crystalline arrangement. In solution, cations and anions move about more or less independently, each surrounded by a cage of solvent molecules. Water—including the water in our bodies—is an excellent solvent for many ionic compounds. Water also dissolves molecules that are polar

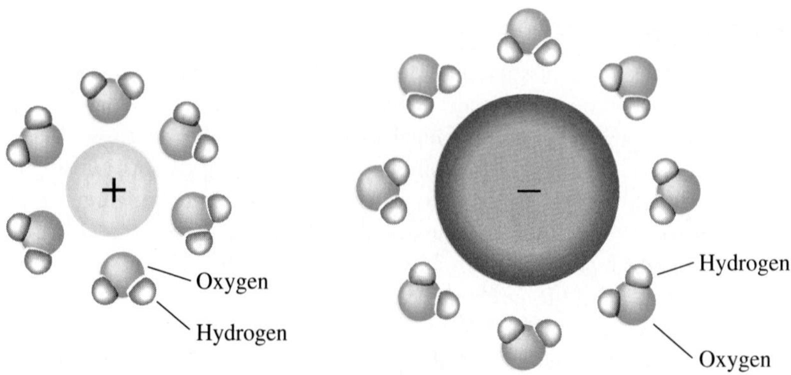

Figure 5.17 The interaction of polar water molecules with ions.

covalent like itself. These solubility principles explain how nutrients get to the cells of our bodies (dissolved in blood, which is mostly water) and how pollutants get into our water supplies.

Summary

1. The most unreactive of all the elements are the noble gases. They all have filled outer electron shells, so this electron configuration must confer stability.
2. *Electron-dot symbols* are chemical symbols surrounded by dots representing valence electrons.
3. When a very active metal reacts with a very active nonmetal, the metal gives up electrons and the nonmetal takes them. The metal atoms become cations and the nonmetal atoms anions, all being held together by *ionic bonds*.
4. When two similar atoms react, they combine by sharing electrons. A pair of shared electrons is called a *covalent bond*.
5. The valence shell of a noble gas atom contains eight electrons (except for helium), and other atoms seem to be striving for that same stable octet of electrons when they react. They are said to be following the *octet rule* or the "rule of eight." In the case of helium only two electrons can exist in its only electron shell, so small atoms such as hydrogen follow the "rule of two."
6. Two similar but different atoms will combine by sharing electrons, but one atom will have a stronger attraction for electrons than the other, so that the electrons will be unequally shared. The attraction of an atom for a pair of shared electrons is called *electronegativity*, and an unequally shared pair of electrons is a *polar covalent bond*. An equally shared pair of electrons between two identical atoms is a *nonpolar covalent bond*.
7. When two atoms share two pairs of electrons, the covalent bond is a *double bond*. Three shared electron pairs is a *triple bond*.
8. Using electron dot symbols for each atom, you can write *electron dot formulas* for molecules.
9. There are exceptions to the octet rule. In some molecules there are incomplete octets of electrons and in others expanded octets. Some molecules have unpaired electrons, which make them *free radicals*.
10. According to the *VSEPR theory* if you assume that sets of electrons around a central atom will get as far away from each other as possible, you can predict the shapes of simple molecules.
11. If you know the shape of a molecule, you can decide whether or not it is polar. The water molecule (H_2O) is bent, and so it is polar.
12. Small nonpolar molecules are held together only by weak *dispersion forces*, which makes for low melting points and boiling points. Polar molecules are held together by *dipole forces* as well as dispersion forces. Molecules that contain H—O, H—N, or H—F bonds have special attractive forces known as *hydrogen bonds*.

Key Terms

ammonia 5.9
bonding pair 5.7
condensation 5.15
covalent bond 5.7
crystal 5.5
dipole 5.14
dipole forces 5.15
dispersion forces 5.15
double bond 5.10
electron-dot formulas 5.11
electron-dot symbols 5.2
electronegativity 5.8
free radicals 5.12
freezing 5.15
homogeneous 5.15
hydrogen bonds 5.15
ionic bond 5.5
kinetic-molecular theory 5.15
melting point 5.15
methane 5.9
nonbonding pair 5.7
nonpolar covalent bond 5.8
octet rule 5.7
polar covalent bond 5.8
polar molecule 5.9
single bond 5.10
solute 5.15
solution 5.15
solvent 5.15
triple bond 5.10
valence electrons 5.2
vaporization 5.15
VSEPR theory 5.13
water 5.9

Review Questions

1. Which group of elements in the periodic table is characterized by stable electron arrangements?
2. What is the structural difference between a sodium atom and a sodium ion?
3. How does sodium metal differ from sodium ions (in sodium chloride, for example) in properties?
4. What is the structural difference between a sodium ion and a neon atom? What is similar?
5. What is the structural difference among chlorine atoms, chlorine molecules, and chloride ions? How do their properties differ?
6. Give electron-dot symbols for each of the following elements. You may use the periodic table.
   a. sodium b. oxygen
   c. fluorine d. aluminum
7. Give electron-dot symbols for each of the following elements. You may use the periodic table.
   a. carbon b. potassium
   c. magnesium d. chlorine
   e. nitrogen
8. Give Bohr diagrams for each of the following.
   a. K^+ b. S^{2-}
   c. F^- d. Al^{3+}
9. Give Bohr diagrams for each of the following.
   a. Mg^{2+} b. Cl^-
   c. Li^+ d. N^{3-}
10. Using electron-dot formulas, show the formation of ion from atom for each of the following.
    a. barium b. bromine
    c. aluminum d. sulfur
11. Use electron-dot symbols to show the transfer of electrons from calcium atoms to bromine atoms to form ions with noble gas configurations.
12. Use electron-dot symbols to show the transfer of electrons from magnesium atoms to sulfur atoms to form ions with noble gas configurations.
13. Use electron-dot symbols to show the transfer of electrons from aluminum atoms to sulfur atoms to form ions with noble gas configurations.
14. Use electron-dot symbols to show the transfer of electrons from magnesium atoms to phosphorus atoms to form ions with noble gas configurations.
15. Use electron-dot symbols to show the sharing of electrons between two iodine atoms to form an iodine (I_2) molecule. Label all electron pairs as bonding or nonbonding.
16. Use electron-dot symbols to show the sharing of electrons between a hydrogen atom and a fluorine atom. Label ends of the molecule with symbols that indicate polarity.
17. Classify the following bonds as ionic or covalent. For those bonds that are covalent, indicate whether they are polar or nonpolar.
    a. KF b. IBr c. MgO
18. Classify the following bonds as ionic or covalent. For those bonds that are covalent, indicate whether they are polar or nonpolar.
    a. NO b. CaO c. NaBr
19. Classify the following bonds as ionic or covalent. For those bonds that are covalent, indicate whether they are polar or nonpolar.
    a. Br_2 b. F_2 c. HCl
20. Classify the following covalent bonds as polar or nonpolar.
    a. H—O b. N—Cl c. B—F
21. Classify the following covalent bonds as polar or nonpolar.
    a. H—N b. Be—F c. P—Cl
22. Use the symbol $\leftarrow\!\!\!+$ to indicate the direction of the dipole in each of the bonds in Question 20.
23. Use the symbol $\leftarrow\!\!\!+$ to indicate the direction of the dipole in each of the bonds in Question 21.
24. The molecule BeF_2 is linear. Is it polar or nonpolar? Use the symbol $\leftarrow\!\!\!+$ to explain your answer.
25. The molecule SF_2 is bent. Is it polar or nonpolar? Use the symbol $\leftarrow\!\!\!+$ to explain your answer.
26. How many covalent bonds do each of the following usually form? You may refer to the periodic table.
    a. H b. C c. O
    d. F e. N f. Br
27. List three elements that readily form double bonds.
28. List two elements that readily form triple bonds.
29. List three main types of compounds that are exceptions to the octet rule. Give an example of each.
30. In what ways are liquids and solids similar? In what ways are they different?
31. List four types of interactions between particles in

the liquid and in the solid states. Give an example of each type.

32. Define
a. melting b. vaporization
c. condensation d. freezing

33. In which process is energy absorbed by the material undergoing the change of state?
a. melting or freezing
b. condensation or vaporization

34. Label the arrows with the term listed in Question 32 that correctly identifies the process presented.

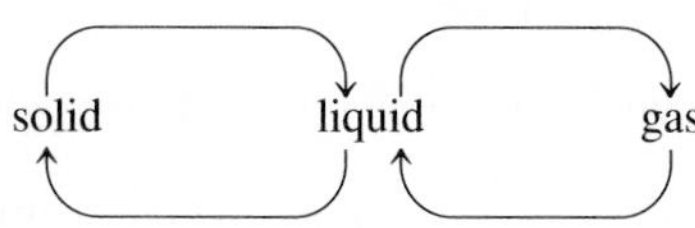

35. Use the kinetic-molecular theory to explain the properties (Chapter 1) of solids, liquids, and gases.

36. Use the kinetic-molecular theory to explain the process of melting.

37. Use the kinetic-molecular theory to explain the process of vaporization.

38. With which of the following would hydrogen bonding be an important intermolecular force?

```
a. H—S
      \
       H

        H
        |
b. H—C—N—H
        |  |
        H  H

        H
        |
c. H—C—F
        |
        H

        H
        |
d. H—C—O
        |   \
        H    H

        H  H
        |  |
e. H—C—C—H
        |  |
        H  H
```

39. Define
a. solution b. solute c. solvent

40. Alcohol that is used as a disinfectant to clean the skin prior to an injection is actually a solution of 3 parts of water and 7 parts of alcohol. Which component is the solvent and which is the solute?

41. Explain why a salt dissolves in water.

42. Benzene (C_6H_6) is a nonpolar solvent. Would you expect NaCl to dissolve in benzene? Explain.

43. Motor oil is nonpolar. Would you expect it to dissolve in water? In benzene? Explain.

44. A 250-mL can of motor oil is poured into a can that contains 2.0 L of gasoline. Which component is the solute and which is the solvent?

▸ Problems

Ions and Ionic Bonding

45. Give electron-dot symbols for the following.
a. Al and Al^{3+} b. Br and Br^-
c. Mg and Mg^{2+} d. O^{2-} and Ne

46. Give electron-dot symbols for the following.
a. Ca and Ca^{2+} b. S and S^{2-}
c. Rb and Rb^+ d. P and P^{3-}

47. Draw Bohr diagrams for the elements and ions in Problem 45.

48. Draw Bohr diagrams for the elements and ions in Problem 46.

49. Give electron-dot formulas for each of the following ionic compounds.
a. magnesium fluoride
b. calcium chloride
c. sodium oxide
d. potassium sulfide

50. Give electron-dot formulas for each of the following ionic compounds.
a. sodium fluoride
b. potassium chloride
c. potassium fluoride

51. Give electron-dot formulas for each of the following ionic compounds.
a. magnesium oxide b. aluminum nitride
c. aluminum sulfide

52. Give electron-dot formulas for each of the following ionic compounds.
a. sodium nitride b. aluminum chloride
c. magnesium nitride

Molecules: Covalent Bonds

53. Use electron-dot symbols to show the sharing of electrons between a phosphorus atom and hydrogen atoms to form a molecule in which phosphorus has an octet of electrons.

54. Use electron-dot symbols to show the sharing of electrons between a silicon atom and hydrogen atoms to form a molecule in which silicon has an octet of electrons.

55. Use electron-dot symbols to show the sharing of electrons between a carbon atom and fluorine atoms to form a molecule in which each atom has an octet of electrons.

56. Use electron-dot symbols to show the sharing of electrons between a nitrogen atom and chlorine atoms to form a molecule in which each atom has an octet of electrons.

Electronegativity: Polar Covalent Bonds

57. Use the symbols $\delta+$ and $\delta-$ to indicate partial charges, if any, on the following bonds.
a. Si—Cl b. Cl—Cl c. O—F

58. Use the symbols $\delta+$ and $\delta-$ to indicate partial charges, if any, on the following bonds.
a. N—H b. C—F c. C—C

Electron-Dot Formulas

59. Give electron-dot formulas that follow the octet rule, showing all valence electrons as dots, for the following covalent molecules.
a. CH_4O b. NOH_3
c. CH_5N d. N_2H_4

60. Give electron-dot formulas that follow the octet rule, showing all valence electrons as dots, for the following covalent molecules.
a. NF_3 b. C_2H_4
c. C_2H_2 d. CH_2O

61. Shared pairs of electrons can be represented by dashes. Give dash line formulas for the molecules in Problem 59.

62. Shared pairs of electrons can be represented by dashes. Give dash line formulas for the molecules in Problem 60.

63. Give electron-dot formulas that follow the octet rule for the following covalent molecules.
a. COF_2 b. PCl_3
c. H_3PO_3 d. HCN

64. Give electron-dot formulas that follow the octet rule for the following covalent molecules.
a. SCl_2 b. H_2SO_4
c. XeO_3 d. $HClO_4$

65. Give electron-dot formulas that follow the octet rule for the following ions.
a. ClO^- b. HPO_4^{2-}
c. ClO_2^- d. BrO_3^-

66. Give electron-dot formulas that follow the octet rule for the following ions.
a. CN^- b. IO_4^-
c. HSO_4^- d. PO_4^{3-}

Molecules That Are Exceptions to the Octet Rule

67. Give an electron-dot formula for each of the following.
a. NO b. BeI_2 c. BCl_3

68. Give an electron-dot formula for each of the following.
a. PF_5 b. $AlBr_3$

VSEPR Theory: The Shapes of Molecules

69. Use the valence shell electron pair repulsion (VSEPR) theory to predict the shape of each of the following molecules.
a. silane (SiH_4)
b. hydrogen selenide (H_2Se)

70. Use the VSEPR theory to predict the shape of each of the following molecules.
a. beryllium chloride ($BeCl_2$)
b. boron chloride (BCl_3)

71. Use the VSEPR theory to predict the shape of each of the following molecules.
a. arsine (AsH_3)
b. carbon tetrafluoride (CF_4)

72. Use the VSEPR theory to predict the shape of each of the following molecules.
a. oxygen difluoride (OF_2)
b. silicon tetrachloride ($SiCl_4$)

73. Use the VSEPR theory to predict the shape of each of the following molecules.
a. nitrogen trichloride (NCl_3)
b. sulfur dichloride (SCl_2)

74. Use the VSEPR theory to predict the shape of each of the following molecules.
a. phosphorus trifluoride (PF_3)
b. dichlorodifluoromethane (CCl_2F_2)

Additional Problems

75. Why does neon tend not to form chemical bonds?

76. Draw a charge cloud picture for the HF molecule. Use the symbols $\delta+$ and $\delta-$ to indicate the polarity of the molecule.

77. There are two different covalent molecules with the formula C_2H_6O. Give electron-dot formulas for the two molecules.

78. Fill this table assuming that elements X, Y, and Z are all in A subgroups in the periodic table.

	Element X	Element Y	Element Z
Group number	IA	___	___
Electron-dot formula	___	$\cdot\dot{\mathrm{Y}}\cdot$	___
Charge on ion	___	___	2−

79. Consider the hypothetical elements X, Y, and Z with electron-dot formulas:

$$:\ddot{\underset{\cdot\cdot}{\mathrm{X}}}\cdot \qquad :\ddot{\underset{\cdot}{\mathrm{Y}}}\cdot \qquad :\dot{\underset{\cdot}{\mathrm{Z}}}\cdot$$

a. To which group in the periodic table would each belong?
b. Give the electron-dot formula for the simplest compound of each with hydrogen.
c. Give electron-dot formulas for the ions formed when X and Y react with sodium.

80. Solutions of iodine chloride (ICl) are used as disinfectants. Are the molecules of ICl ionic, polar covalent, or nonpolar covalent?

81. Potassium is a soft silvery metal that reacts violently with water and ignites spontaneously in air. Your doctor recommends you take a potassium supplement. Would you take potassium metal? What would you take?

82. Is there any such thing as a sodium chloride molecule? Explain.

83. Use sublevel notations to give electron configurations for the most stable simple ion formed by each of the following elements.
a. Ba b. K c. Se

84. Use sublevel notations to give electron configurations for the most stable simple ion formed by each of the following elements.
a. I b. N c. Te

References and Readings

1. Asimov, Isaac. *A Short History of Chemistry.* Garden City, NY: Doubleday, 1965. Chapter 7, ''Molecular Structure.''
2. Benfey, O. T. *Classics in the Theory of Chemical Combination.* New York: Dover Publications, 1963. A history of bonding.
3. Brady, James E., and John R. Holum. *Chemistry: The Study of Matter and Its Changes.* New York: Wiley, 1993. Chapters 6 and 7 discuss chemical bonding.
4. Companion, Audrey. *Chemical Bonding,* 2d ed. New York: McGraw-Hill, 1979.
5. Dahl, Peter. ''The Valence-Shell Electron-Pair Repulsion Theory.'' *Chemistry,* March 1973, pp 17–19.
6. Ebbing, Darrell D. *General Chemistry,* 4th ed. Boston: Houghton Mifflin, 1993. Chapter 9 discusses bonding.
7. Gillespie, Ronald J. *Molecular Geometry,* London: Van Nostrand Reinhold, 1972.
8. Pauling, L., and R. Hayward. *The Architecture of Molecules.* San Francisco: W. H. Freeman, 1964. Parts 1–14. Easy reading, great artwork.
9. Zumdahl, Steven S. *Chemistry,* 3rd ed. Lexington, MA: D. C. Heath, 1993. Chapter 8 discusses bonding.

6 Names, Formulas, and Equations

The Language of Chemistry

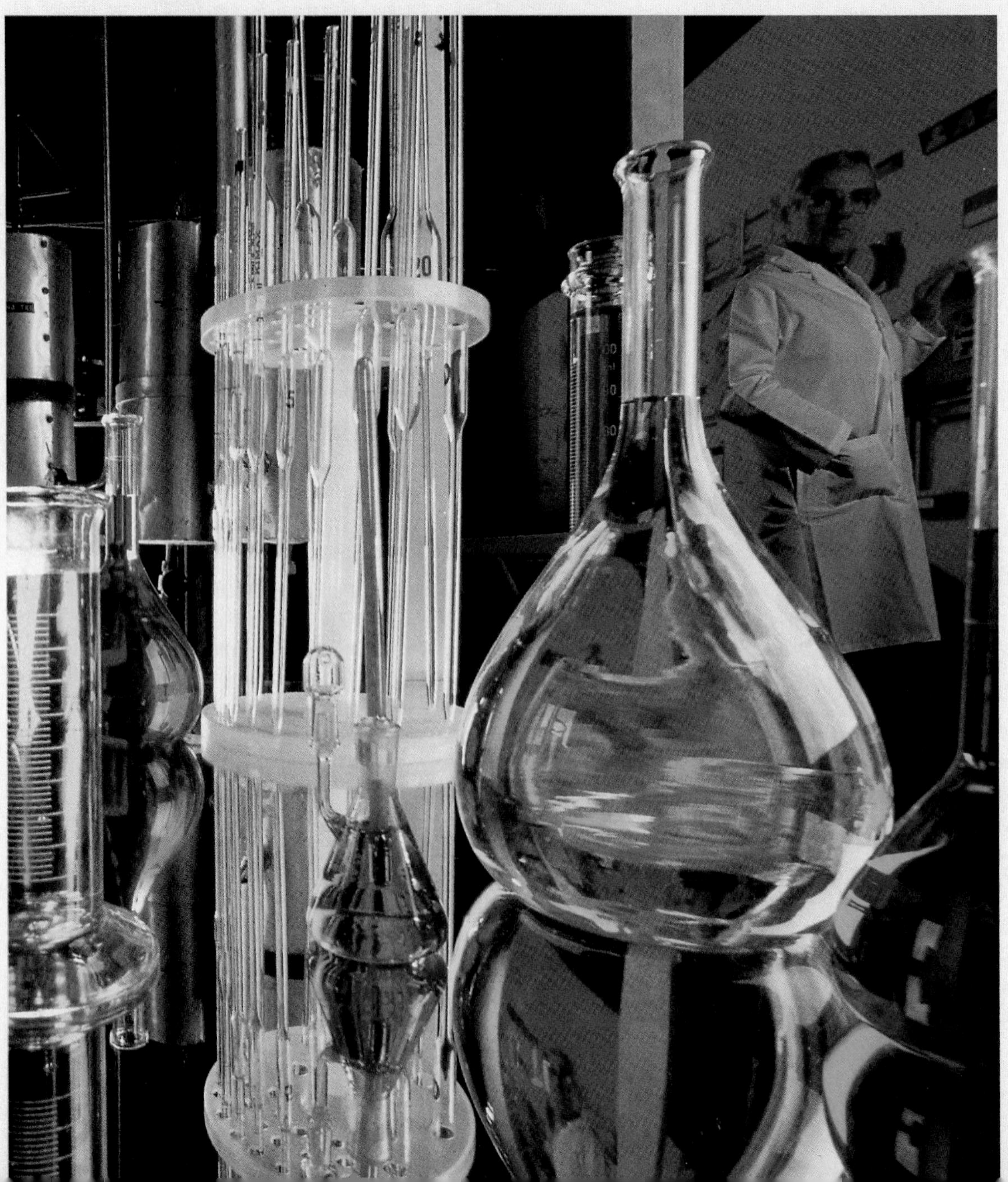

Some names of compounds end in -ide,

And some in -ate or -ite.

It takes a little practice

If you want to get them right.

Have you ever listened to people talking in a language you didn't understand? To you it probably seemed like a stream of meaningless sounds, but the people speaking understood each other perfectly. The language that chemists use is much like a foreign language. An equation such as

$$4\,Fe + 3\,O_2 \longrightarrow 2\,Fe_2O_3$$

does not mean much to someone who has no knowledge of chemistry. But if you know the "language," you can readily translate the chemical equation into English:

Iron combines with oxygen (in the air) to form rust.

There are two ways to find out what is being said in a foreign language: you can use a translator, or you can learn the language yourself. It can be very awkward to have to depend on a translator all the time. If you are taking a course in chemistry, you need to be able to speak the language. In this chapter we offer a short, intensive course.

The language of chemistry includes symbols, formulas, and equations, as well as names for specific compounds. Learning a few simple rules will enable you to name thousands of chemical compounds. Sometimes the language chemists use is closely related to that of mathematics. But don't worry! The only math that we will be dealing with is much like the arithmetic you use in your everyday life. If you are willing to give this chapter a little serious effort, you should have no trouble learning to speak the language of chemistry.

SECTION 6.1

Names and Symbols for Simple Ions

You are familiar with many kinds of symbols (Figure 6.1). In Chapter 1, we introduced chemical symbols. A symbol of one or two letters is used to repre-

To describe the great variety of chemical substances and their properties, chemists use a special language of symbols, formulas, and equations. [*Barry Bomzer/Tony Stone Worldwide.*]

♂ ♀ $ ¢

Male Female Money Music

Figure 6.1 The use of symbols is not unique to chemistry. Symbols can be quite helpful—when you know what they mean.

sent each of the chemical elements. If you have not yet memorized the list in Table 1.1, you should do so now. You will save time in the long run.

In the last chapter, we saw how certain metals (those from the left side of the periodic chart) react with nonmetals (those from the right side) to form ionic compounds. Recall that in forming compounds each atom of metal tends to give up the electrons in its outer shell, and each atom of nonmetal takes on enough electrons to complete its valence shell. For example, aluminum atoms with three electrons in their outermost energy levels give up three electrons to form triply charged ions. In electron-dot symbols this reaction may be written

$$\cdot\dot{\mathrm{Al}}\cdot \longrightarrow \mathrm{Al}^{3+} + 3\,\mathrm{e}^-$$

Oxygen, with six electrons in its outermost energy level, tends to acquire two more.

$$:\ddot{\underset{\cdot}{\mathrm{O}}}\cdot + 2\,\mathrm{e}^- \longrightarrow :\ddot{\underset{\cdot\cdot}{\mathrm{O}}}:^{2-}$$

The charged atoms formed by the gain or loss of electrons are called **ions**. Table 6.1 lists symbols and names for some simple ions formed in this manner. Note that the charge on an ion of a Group IA element is 1+ (usually written simply as +). The charge on an ion of a Group IIA element is 2+, and the charge on an ion of a Group IIIA element is 3+. You can calculate the charge on the negative ions in the table by subtracting 8 from the group number. For example, the charge on the oxide ion (oxygen is in Group VIA) is $6 - 8 = -2$. The charge on a nitride ion (nitrogen is in Group VA) is $5 - 8 = -3$.

There is no simple way to determine the most likely charge on ions formed from Group VIII elements and from those in B subgroups. Indeed, you may have noticed that these can form ions with different charges. In such cases, chemists use Roman numerals with the names to indicate the charge. Thus,

Table 6.1 Symbols and Names for Some Simple Ions

Group	Element	Name of Ion	Symbol for Ion
IA	Hydrogen	Hydrogen ion*	H^+
	Lithium	Lithium ion	Li^+
	Sodium	Sodium ion	Na^+
	Potassium	Potassium ion	K^+
IIA	Magnesium	Magnesium ion	Mg^{2+}
	Calcium	Calcium ion	Ca^{2+}
IIIA	Aluminum	Aluminum ion	Al^{3+}
VA	Nitrogen	Nitride ion	N^{3-}
VIA	Oxygen	Oxide ion	O^{2-}
	Sulfur	Sulfide ion	S^{2-}
VIIA	Fluorine	Fluoride ion	F^-
	Chlorine	Chloride ion	Cl^-
	Bromine	Bromide ion	Br^-
	Iodine	Iodide ion	I^-
IB	Copper	Copper(I) ion (cuprous ion)	Cu^+
		Copper(II) ion (cupric ion)	Cu^{2+}
	Silver	Silver ion	Ag^+
IIB	Zinc	Zinc ion	Zn^{2+}
VIIIB	Iron	Iron(II) ion (ferrous ion)	Fe^{2+}
		Iron(III) ion (ferric ion)	Fe^{3+}

*Does not exist independently in aqueous solution.

iron(II) ion means Fe^{2+} and *iron(III) ion* means Fe^{3+}. In an older system of terminology, Fe^{2+} was called a *ferrous ion* and Fe^{3+} was called a *ferric ion.* See similar terms for the two copper ions in Table 6.1.

Names of simple positive ions (*cations*) are derived from those of their parent elements by the addition of the word *ion.* A sodium atom (Na), upon losing an electron, becomes a *sodium ion* (Na^+). A magnesium atom (Mg), upon losing two electrons, becomes a *magnesium ion* (Mg^{2+}). Names of simple negative ions (*anions*) are derived from those of their parent elements by changing the usual ending to *-ide* and adding the word *ion.* A chlor*ine* atom (Cl), upon gaining an electron, becomes a chlor*ide ion* (Cl^-). A sulf*ur* atom (S), upon gaining two electrons, becomes a sulf*ide ion* (S^{2-}).

We cannot emphasize too strongly the difference between ions and the atoms from which they are made. They are as different as a whole peach (an atom) and a peach pit (a positive ion). The names and symbols may look a lot alike, but the substances themselves are quite different (Figure 6.2). Unfortunately, the situation is confused because people talk about needing ''iron'' to perk up ''tired blood'' and ''calcium'' for healthy teeth

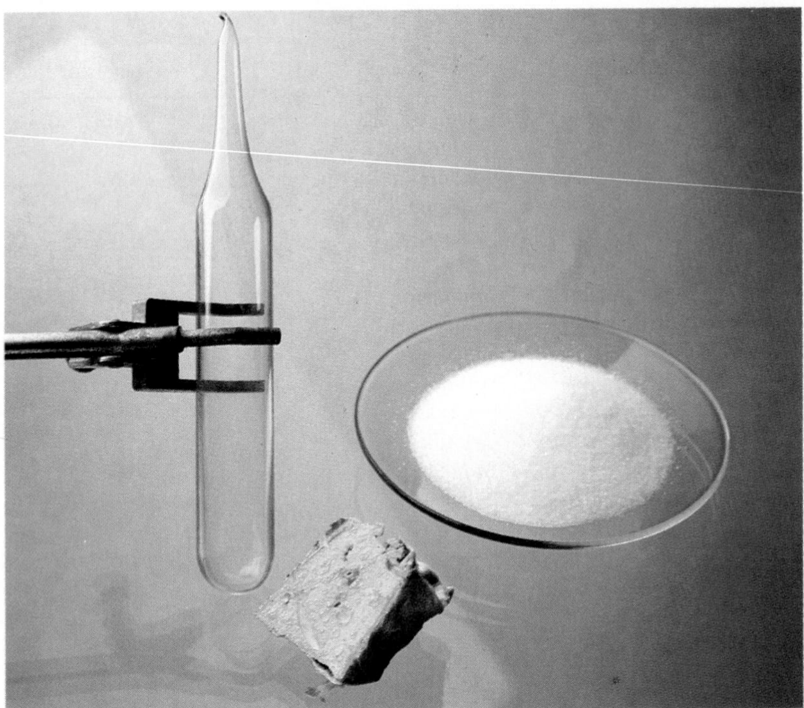

Figure 6.2 Ions differ greatly from the atoms from which they are made. Sodium atoms are the constituents of a soft, highly reactive metal. Chlorine atoms—paired in chlorine molecules—make up a corrosive greenish yellow gas. Sodium ions and chloride ions make up ordinary table salt. [*Richard Megna/Fundamental Photographs.*]

and bones. What they really mean is iron(II) *ions* (Fe^{2+}) and calcium *ions* (Ca^{2+}). You wouldn't think of eating iron nails to get "iron." Nor would you eat highly reactive calcium metal. Although careful distinction is not always made by persons who are not chemists, we try to use precise terminology here.

SECTION 6.2

Formulas and Names for Binary Ionic Compounds

Simple ions of opposite charge can be combined to form **binary** (two-component) **compounds**. To get the correct **formula** for a binary compound, simply write each ion with its charge (positive ion to the left), then cross over the numbers (but not the plus or minus signs) and write them as subscripts. The process is best learned by practice. Work through Examples 6.1–6.5 and Exercises 6.1–6.5. There are also problems at the end of the chapter for further practice.

EXAMPLE 6.1

Give the formula for calcium chloride.

SOLUTION
First, write the symbols for the ions.

$$Ca^{2+} \quad Cl^{1-}$$

Then cross over the numbers as subscripts.

$$Ca^{(2)+} \quad Cl^{(1)-} \quad \text{gives} \quad Ca_1^{2+}Cl_2^{-}$$

Then rewrite the formula, dropping the charges. The formula for calcium chloride is

$$Ca_1Cl_2 \quad \text{or} \quad CaCl_2$$

Exercise 6.1
Give the formula for potassium oxide.

Note that the cross-over method works because it is based on the transfer of electrons and the conservation of charge. Two aluminum atoms lose three electrons each (that's six electrons lost), and three oxygen atoms gain two electrons each (that's six electrons gained). Electrons lost equal electrons gained, and all is well. Similarly, two aluminum ions have six positive charges (three each) and three oxide ions have six negative charges (two each). The net charge on aluminum oxide is 0, just as it should be.

EXAMPLE 6.2

Give the formula for aluminum oxide.

SOLUTION
Write the symbols for the ions.

$$Al^{3+} \quad O^{2-}$$

Cross over the numbers as subscripts.

$$Al^{(3)+} \quad O^{(2)-} \quad \text{gives} \quad Al_2^{3+}O_3^{2-}$$

Then rewrite the formula, dropping the charges. The formula for aluminum oxide is

$$Al_2O_3$$

Exercise 6.2
Give the formula for calcium nitride.

EXAMPLE 6.3

Give the formula for magnesium oxide.

SOLUTION

The ions are

$$Mg^{2+} \quad O^{2-}$$

Crossing over,

$$Mg^{2+} \times O^{2-}$$

we get

$$Mg_2^{2+}O_2^{2-}$$

Dropping the charges, we get

$$Mg_2O_2$$

Such formulas usually are reduced to the lowest ratio, and we write magnesium oxide as MgO.

Exercise 6.3

Give the formula for calcium sulfide.

Now you are able to translate the "English," such as aluminum oxide, into the "chemistry," Al_2O_3. You also can translate in the other direction.

EXAMPLE 6.4

What is the name for MgS?

SOLUTION

Table 6.1 tells us that MgS is made up of Mg^{2+} (magnesium ions) and S^{2-} (sulfide ions). The name is simply magnesium sulfide.

Exercise 6.4

What is the name for CaF_2?

EXAMPLE 6.5

What is the name for $FeCl_3$?

SOLUTION

The ions are

$$Fe^{3+} \quad Cl^-$$

(How do we know the iron is Fe^{3+} and not Fe^{2+}? Since there are three Cl^- ions, each 1−, the one Fe ion must be 3+ because the compound $FeCl_3$ is neutral.) The names of these ions are iron(III) ion (or ferric ion) and chloride ion. Therefore, the compound is iron(III) chloride (or, by the older system, ferric chloride).

Exercise 6.5

What is the name for $CuBr_2$?

SECTION 6.3

Polyatomic Ions

Many compounds contain both ionic and covalent bonds. Sodium hydroxide, commonly known as lye, consists of sodium ions (Na^+) and hydroxide ions (OH^-). The hydroxide ion contains an oxygen atom covalently bonded to a hydrogen atom, plus an "extra" electron. Whereas the sodium atom becomes a cation by giving up an electron, the hydroxide group becomes an anion by gaining an electron.

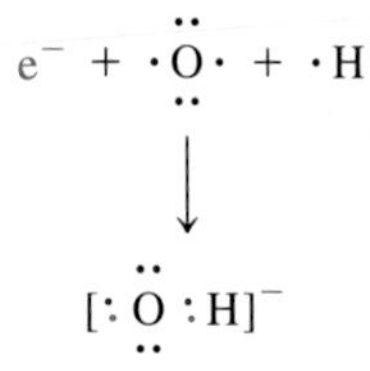

$$e^- + \cdot\ddot{\underset{\cdot\cdot}{O}}\cdot + \cdot H \longrightarrow [:\ddot{\underset{\cdot\cdot}{O}}:H]^-$$

The formula for sodium hydroxide is NaOH; for each sodium ion there is one hydroxide ion.

There are many groups of atoms that (like hydroxide ion) remain together through most chemical reactions. **Polyatomic ions** are charged particles containing two or more covalently bonded atoms (Figure 6.3). A list of common polyatomic ions is given in Table 6.2. You can use these ions, in combination with the simple ions in Table 6.1, to determine formulas for compounds that contain polyatomic ions.

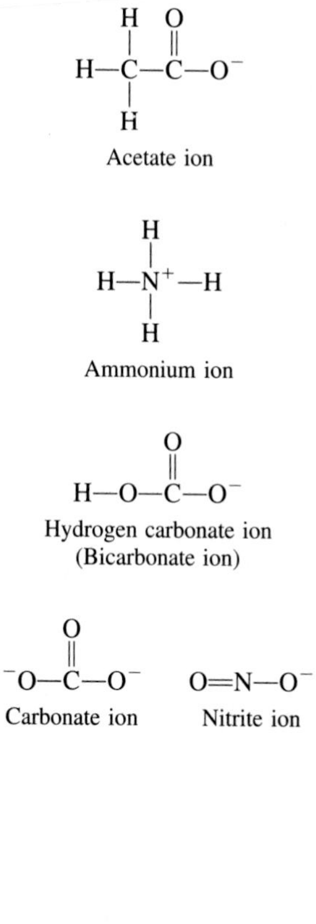

```
   H  O
   |  ||
H—C—C—O⁻
   |
   H
```

Acetate ion

```
    H
    |
H—N⁺—H
    |
    H
```

Ammonium ion

```
         O
         ||
H—O—C—O⁻
```

Hydrogen carbonate ion (Bicarbonate ion)

```
      O
      ||
⁻O—C—O⁻        O=N—O⁻
```

Carbonate ion   Nitrite ion

Figure 6.3 Polyatomic ions have both covalent bonds (dashes) and ionic charges (superior + or −).

Table 6.2 Some Common Polyatomic Ions

Charge	Name	Formula
1+	Ammonium ion	NH_4^+
	Hydronium ion	H_3O^+
1−	Hydrogen carbonate (bicarbonate) ion	HCO_3^-
	Hydrogen sulfate (bisulfate) ion	HSO_4^-
	Acetate ion	$CH_3CO_2^-$ (or $C_2H_3O_2^-$)
	Nitrite ion	NO_2^-
	Nitrate ion	NO_3^-
	Cyanide ion	CN^-
	Hydroxide ion	OH^-
	Dihydrogen phosphate ion	$H_2PO_4^-$
	Permanganate ion	MnO_4^-
2−	Carbonate ion	CO_3^{2-}
	Sulfate ion	SO_4^{2-}
	Monohydrogen phosphate ion	HPO_4^{2-}
	Oxalate ion	$C_2O_4^{2-}$
	Dichromate ion	$Cr_2O_7^{2-}$
3−	Phosphate ion	PO_4^{3-}

EXAMPLE 6.6

What is the formula for sodium sulfate?

SOLUTION

First, write the formulas for the ions.

$$Na^+ \quad SO_4^{2-}$$

Crossing over,

$$Na^{+} \quad SO_4^{2-}$$

we get

$$Na_2^{+}(SO_4)_1^{2-}$$

Then, dropping the charges, we have

$$Na_2(SO_4)_1 \quad \text{or simply} \quad Na_2SO_4$$

Exercise 6.6

What is the formula for potassium phosphate?

▶ EXAMPLE 6.7

What is the formula for ammonium sulfide?

SOLUTION

The ions are

$$NH_4^+ \qquad S^{2-}$$

Crossing over, we get

$$NH_4^{+} \quad S^{2-}$$

we get

$$(NH_4)_2^+S_1^{2-}$$

Dropping the charges gives

$$(NH_4)_2S$$

The parentheses with a subscript 2 indicate that the entire ammonium unit is taken twice; there are two nitrogen atoms and eight ($4 \times 2 = 8$) hydrogen atoms.

Exercise 6.7

What is the formula for calcium acetate?

▶ EXAMPLE 6.8

What is the formula for ammonium nitrate?

SOLUTION

The ions are

$$NH_4^+ \qquad NO_3^-$$

Both superscripts are 1 (understood). The formula for ammonium nitrate is NH_4NO_3.

Exercise 6.8

What is the formula for calcium monohydrogen phosphate?

▶ EXAMPLE 6.9

What is the name for the compound NaCN?

SOLUTION

The ions are

$$Na^+ \qquad CN^-$$

The name is sodium cyanide.

Exercise 6.9

What is the name for the compound $CaCO_3$?

▶ EXAMPLE 6.10

What is the name for KH_2PO_4?

SOLUTION

The ions are

$$K^+ \qquad H_2PO_4^-$$

The name is potassium dihydrogen phosphate.

Exercise 6.10

What is the name for $K_2Cr_2O_7$?

▶ EXAMPLE 6.11

What is the name of $Cu_3(PO_4)_2$?

SOLUTION

The ions are

$$Cu^{2+} \qquad PO_4^{3-}$$

The name is copper(II) phosphate.

Exercise 6.11

What is the name of $Fe_2(CO_3)_3$?

SECTION 6.4

Names for Covalent Compounds

Many molecular compounds have common and widely used names. Examples are water (H_2O), methane (CH_4), and ammonia (NH_3). For other compounds, the prefixes *mono, di-, tri-,* and so on, are used to indicate the number of atoms of each element in the molecule. A list of these prefixes for up to ten atoms is given in Table 6.3.

Simply use the prefixes to indicate the number of each kind of atom. For example, the compound N_2O_4 is called *dinitrogen tetroxide.* (The *a* often is dropped from tetra- and other prefixes when it precedes another vowel.) We often leave off the mono- prefix (NO_2 is nitrogen dioxide), but do include it to distinguish between two compounds of the same pair of elements (CO is carbon monoxide; CO_2 is carbon dioxide).

Table 6.3
Prefixes That Indicate the Number of Atoms of an Element in a Compound

Prefix	Number of Atoms
Mono-	1
Di-	2
Tri-	3
Tetra-	4
Penta-	5
Hexa-	6
Hepta-	7
Octa-	8
Nona-	9
Deca-	10

EXAMPLE 6.12

What is the name for SCl_2? For SF_6?

SOLUTION

With one sulfur atom and two chlorine atoms, SCl_2 is sulfur dichloride. With one sulfur atom and six fluorine atoms, SF_6 is sulfur hexafluoride.

Exercise 6.12

What is the name for BrF_3? For BrF_5?

EXAMPLE 6.13

Give the formula for carbon tetrachloride.

SOLUTION

The name indicates one carbon atom and four chlorine atoms. The formula is CCl_4.

Exercise 6.13

Give the formula for dinitrogen pentoxide.

EXAMPLE 6.14

Give the formula for tetraphosphorus hexoxide.

SOLUTION

The name indicates four phosphorus atoms and six oxygen atoms. The formula is P_4O_6.

Exercise 6.14

Give the formula for tetraphosphorus triselenide. (The symbol for selenium is Se.)

SECTION 6.5

Chemical Sentences: Equations

Chemistry is a study of matter and the changes it undergoes. More than that, it is a study of the energy that brings about those changes, or that is released when those changes occur. So far, we have discussed the symbols and formulas that have been invented to represent elements and compounds. Now that we have learned the letters (symbols) and words (formulas) of our chemical language, we are ready to write sentences (chemical equations). A **chemical equation** is a shorthand way of describing chemical change, using symbols and formulas to represent the elements and compounds that are involved in the change.

Writing Chemical Equations

Carbon reacts with oxygen to form carbon dioxide. In the chemical shorthand, this reaction is written

$$C + O_2 \longrightarrow CO_2$$

The plus sign (+) indicates addition of carbon to oxygen (or vice versa) or a mixing of the two in some manner. The arrow (→) is often read "yields." Substances on the left of the arrow are **reactants**, or *starting materials*. Those on the right are the **products** of the reaction. The conventions here are like those we used in writing nuclear equations (Section 4.3). Now, however, the nucleus will remain untouched. The chemical reactions we are going to look at involve only electrons.

Sometimes the physical states of the reactants and products are indicated. The initial letter of the state is written immediately following the formula. Thus (g) indicates a gaseous substance, (l) a liquid, and (s) a solid. The label (aq) indicates an aqueous solution, that is, a water solution. Using these labels, our equation becomes

$$C(s) + O_2(g) \rightarrow CO_2(g)$$

Chemical equations have meaning on the atomic and molecular levels. The equation

$$C + O_2 \longrightarrow CO_2$$

means that one atom of carbon (C) reacts with one molecule of oxygen (O_2) to produce one molecule of carbon dioxide (CO_2).

Balancing Chemical Equations

Not all chemical reactions are as simply represented as that between carbon and oxygen to form carbon dioxide. Hydrogen reacts with oxygen to form water. We can write this reaction as

$$H_2 + O_2 \longrightarrow H_2O \quad (\textit{not balanced})$$

Note that we could not balance the equation by changing the subscript for oxygen in water.

$$H_2 + O_2 \rightarrow H_2O_2 \quad (\textit{not correct})$$

The equation would be balanced, but it would not mean "hydrogen reacts with oxygen to form water." The formula H_2O_2 represents hydrogen peroxide, *not* water.

This representation, however, is not consistent with the law of conservation of matter. There are two oxygen atoms shown among the reactants (as O_2), and only one among the products (in H_2O). For the equation to represent correctly the chemical happening, it must be balanced. To balance the oxygen atoms, we need only place the coefficient 2 in front of the formula for water.

$$H_2 + O_2 \longrightarrow 2\,H_2O \qquad (\textit{not balanced})$$

This coefficient means that there are two molecules of water involved. As is the case with subscripts, a coefficient of 1 is understood when no other number appears. A coefficient preceding a formula multiplies everything in the formula. In the above equation, the coefficient 2 not only increases the number of oxygen atoms to two, but also increases the number of hydrogen atoms to four.

But the equation is still not balanced. We took care of oxygen at the expense of messing up hydrogen. To balance hydrogen, we place a coefficient 2 in front of the H_2.

$$2\,H_2 + O_2 \longrightarrow 2\,H_2O \qquad (\textit{balanced})$$

Now there are enough hydrogen atoms on the left. In fact, there are four hydrogen atoms and two oxygen atoms on each side of the equation. Atoms are conserved: the equation is balanced (Figure 6.4).

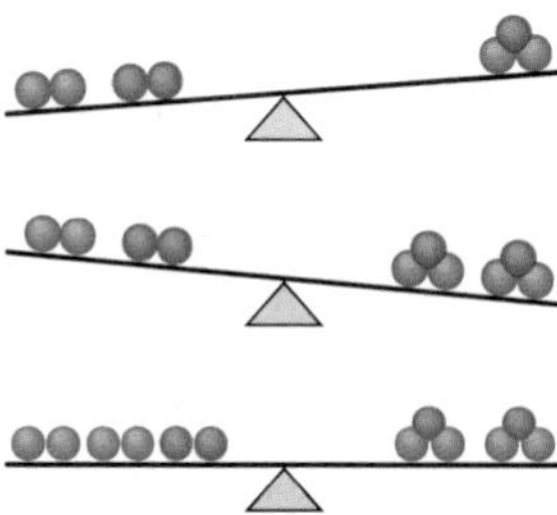

● Hydrogen
● Oxygen

Figure 6.4 To balance the equation for the reaction of hydrogen with oxygen that forms water, the same number of each kind of atom must appear on each side (atoms are conserved). When the equation is balanced, there are four hydrogen atoms and two oxygen atoms on each side.

▶ EXAMPLE 6.15

Balance the following equation.

$$N_2 + H_2 \longrightarrow NH_3 \qquad (\textit{not balanced})$$

SOLUTION

For this sort of problem, we will use the concept of the least common multiple. We will balance the hydrogen first. There are two hydrogen atoms on the left and three on the right. The least common multiple of 3 and 2 is 6. Six is the smallest number of hydrogen atoms that can be evenly converted from reactants to products. We therefore need three molecules of H_2 and two of NH_3.

$$N_2 + 3\,H_2 \longrightarrow 2\,NH_3 \qquad (\textit{balanced})$$

We have balanced the hydrogen atoms and, in the process, the nitrogen atoms. There are two nitrogen atoms on the left and two on the right. The entire equation is balanced (Figure 6.5).

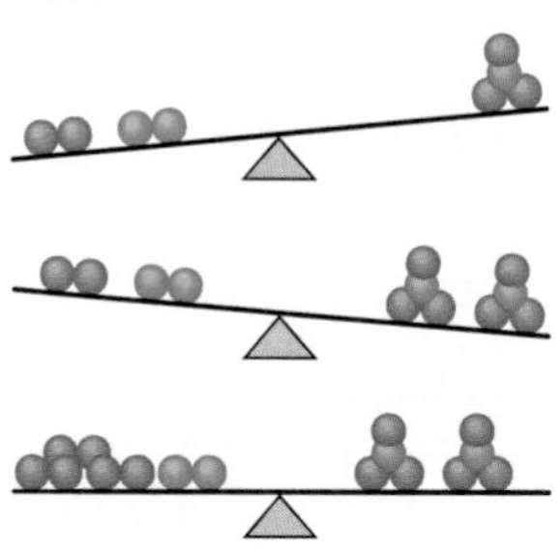

● Hydrogen
● Nitrogen

Figure 6.5 To balance the equation for the reaction of nitrogen with hydrogen that forms ammonia, the same number of each kind of atom must appear on each side of the equation. When the equation is balanced, there are two nitrogen atoms and six hydrogen atoms on each side.

Exercise 6.15

Balance the following equation.

$$P_4 + H_2 \longrightarrow PH_3$$

▶ EXAMPLE 6.16

Balance the following equation.

$$Fe + O_2 \longrightarrow Fe_2O_3 \quad \textit{(not balanced)}$$

SOLUTION

Begin by balancing the oxygen. The least common multiple is 6. We need three molecules of O_2 and two of Fe_2O_3

$$Fe + 3\,O_2 \longrightarrow 2Fe_2O_3 \quad \textit{(not balanced)}$$

We now have four atoms of iron on the right side. We can get four on the left by placing the coefficient 4 in front of Fe.

$$4\,Fe + 3\,O_2 \longrightarrow 2\,Fe_2O_3 \quad \textit{(balanced)}$$

The equation is now balanced.

Exercise 6.16

Balance the following equation

$$V + O_2 \longrightarrow V_2O_5$$

▶ EXAMPLE 6.17

Balance the following equation.

$$CH_4 + O_2 \longrightarrow CO_2 + H_2O \quad \textit{(not balanced)}$$

SOLUTION

In this equation, oxygen appears in two different products; we will leave the oxygen for last and balance the other two elements first. Carbon is already balanced, with one atom on either side of the equation. The least common multiple of 2 and 4 is 4, so to balance hydrogen we place the coefficient 2 in front of H_2O. Now we have four hydrogen atoms on each side.

$$CH_4 + O_2 \longrightarrow CO_2 + 2\,H_2O \quad \textit{(not balanced)}$$

Now for the oxygen. There are four oxygen atoms on the right. If we place a 2 in front of O_2, the oxygen atoms balance.

$$CH_4 + 2\,O_2 \longrightarrow CO_2 + 2\,H_2O \quad \textit{(balanced)}$$

The equation is balanced.

Exercise 6.17
Balance the following equation.

$$C_3H_8 + O_2 \longrightarrow CO_2 + H_2O$$

▶ **EXAMPLE 6.18**

Balance the following equation.

$$H_2SO_4 + NaCN \longrightarrow HCN + Na_2SO_4 \quad \textit{(not balanced)}$$

SOLUTION
Here we have an equation that involves compounds with polyatomic ions. The SO_4 group (actually SO_4^{2-}) should be treated as a unit and balanced as a whole. The same is true of the CN group (actually CN^-). As the equation is presently written, the SO_4 groups and the CN groups are balanced, but the hydrogen atoms and the sodium atoms are not. The least common multiple for sodium is 2, so to get two sodium atoms on the left we place a 2 before NaCN. The least common multiple for hydrogen is 2, so to get two hydrogen atoms on the right we place a 2 before HCN.

$$H_2SO_4 + 2\,NaCN \longrightarrow 2\,HCN + Na_2SO_4 \quad \textit{(balanced)}$$

The same coefficients balance the CN groups and the SO_4 groups, and the entire equation is balanced.

Exercise 6.18
Balance the following equation.

$$H_3PO_4 + Ca(OH)_2 \longrightarrow Ca_3(PO_4)_2 + H_2O$$

We have made the task of balancing equations deceptively easy by considering simple reactions. It is more important at this point for you to understand the principle than to be able to balance complicated equations. You should know what is meant by a balanced equation and be able to handle simple systems.

SECTION 6.6

Volume Relationships in Chemical Equations

Chemists generally cannot work with individual atoms and molecules. Even the tiniest speck of matter that we can see contains billions of billions of atoms. John Dalton postulated that atoms of different elements had different

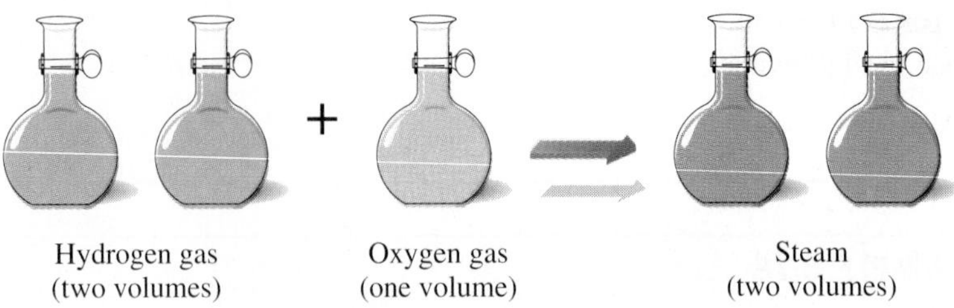

Figure 6.6 Two volumes of hydrogen gas react with one volume of oxygen gas to give two volumes of steam.

masses. Therefore, equal masses of different elements would contain different numbers of atoms. Consider the analogous situation of golf balls and Ping-Pong balls. A kilogram of golf balls contains a smaller number of balls than a kilogram of Ping-Pong balls. One could determine the number of balls in each case simply by counting them. For atoms, however, such a straightforward method is not available. It was in the experiments of a French chemist and the mind of an Italian scientist that approaches to the problem of numbering atoms were found.

In 1809 the French scientist Joseph Louis Gay-Lussac (1778–1850) announced the results of some chemical reactions that he had carried out with gases. These experiments were summarized in his **law of combining volumes**, which states that when all measurements are made at the same temperature and pressure, the volumes of gaseous reactants and products are in a small whole-number ratio. For example, when he allowed hydrogen to react with oxygen to form steam at 100 °C, two volumes of hydrogen united with one volume of oxygen to give two volumes of steam (Figure 6.6). The small whole-number ratio was 2:1:2.

In another experiment (Figure 6.7), Gay-Lussac found that if hydrogen is permitted to react with nitrogen to form ammonia, the combining volumes are three of hydrogen with one of nitrogen to give two of ammonia (3:1:2).

Gay-Lussac thought there must be some relationship between the *numbers* of molecules and the *volumes* of gaseous reactants and products. But it was the Italian chemist Amedeo Avogadro who first explained the law of combining volumes. **Avogadro's hypothesis**, based on shrewd interpretation of experimental facts, was that equal volumes of all gases (at the same temperature and pressure) contain the same number of molecules (Figure 6.8).

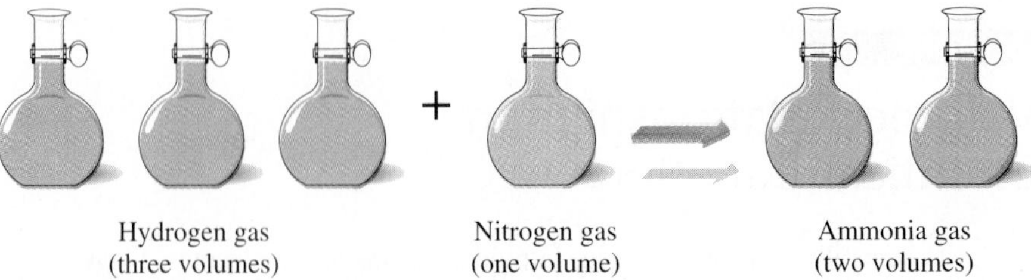

Figure 6.7 Three volumes of hydrogen gas react with one volume of nitrogen gas to give two volumes of ammonia gas.

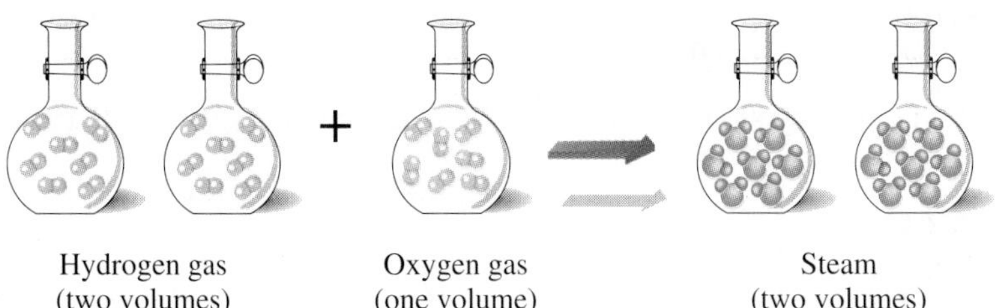

Figure 6.8 Avogadro's explanation of Gay-Lussac's law of combining volumes. Equal volumes of each of the gases contain the same number of molecules.

The equation for the combination of hydrogen and oxygen to form water (steam) is

$$2\,H_2(g) + O_2(g) \longrightarrow 2\,H_2O(g)$$

The coefficients of the molecules are the same as the combining ratio of the gas volumes, 2:1:2 (see Figure 6.6). Similarly, the formation of ammonia is described in the equation

$$3\,H_2(g) + N_2(g) \longrightarrow 2\,NH_3(g)$$

The coefficients are identical to the factors of the combining ratio (see Figure 6.7). The equation says that a nitrogen molecule reacts with three hydrogen molecules to produce two ammonia molecules. It also indicates that if you had 1 million nitrogen molecules, you would need 3 million hydrogen molecules to produce 2 million ammonia molecules. The equation provides the combining ratios. If identical volumes of gases contain identical numbers of molecules, then, according to the equation, one volume of nitrogen reacts with three volumes of hydrogen to produce two volumes of ammonia.

It was also Avogadro who first suggested that certain elements such as hydrogen, oxygen, and nitrogen were made up of diatomic molecules. If these substances were monatomic, the equations would be

$$2\,H(g) + O(g) \rightarrow H_2O(g)\ (\textit{incorrect})$$

$$3\,H(g) + N(g) \rightarrow NH_3(g)\ (\textit{incorrect})$$

Note that these equations would give the wrong ratios of combining volumes. The first, for example, shows a ratio of 2:1:1, rather than the observed volume ratio of 2:1:2. On the other hand, if hydrogen, oxygen, and nitrogen molecules are diatomic, we get the observed ratios.

EXAMPLE 6.19

According to the equation

$$CH_4(g) \longrightarrow C(s) + 2\,H_2(g)$$

what volume of hydrogen would be obtained from 1.00 L of methane (CH_4)?

SOLUTION

The equation indicates that two volumes of hydrogen are obtained from the reaction of one volume of methane. Therefore, 1.00 L of methane would yield 2.00 L of hydrogen.

Exercise 6.19

Using the equation in Example 6.19, calculate what volume of hydrogen is obtained from the reaction of 25.0 L of methane.

Note that no calculations such as these can be made that involve carbon, a solid [C(s)]. The law of combining volumes applies only to gases.

EXAMPLE 6.20

Using the equation in Example 6.19, calculate how much methane must react to produce 10.0 L of H_2.

SOLUTION

$$10.0\ \text{L}\ H_2 \times \frac{1\ \text{L}\ CH_4}{2\ \text{L}\ H_2} = 5.00\ \text{L}\ CH_4$$

Exercise 6.20

How many liters of oxygen must react with hydrogen to form 10.0 L of steam?

SECTION 6.7

Avogadro's Number: 6.02×10^{23}

Avogadro's hypothesis, which has been verified repeatedly over the years, states that equal volumes of gas at the same temperature and pressure contain equal numbers of molecules. That means that if we weigh equal volumes of several gases, the ratio of their weights should be the same as the weight ratio of the molecules themselves.

It is not possible to weigh out individual atoms and molecules, since they are much too small. However, by knowing the relative weights of several atoms, it is possible to weigh out equal numbers of their atoms by using their weight ratios. Since we know that carbon atoms and oxygen atoms have a weight ratio of 12 to 16, then 12 grams of carbon should contain the same number of atoms as 16 grams of oxygen. Therefore, if we react 12 g of carbon with 16 g of oxygen, we should get 28 g of carbon monoxide (CO), with no reactants left over. We can do this without having any idea as to how many carbon atoms there are in 12 g of carbon. All we need to know is that it is the same as the number of oxygen atoms in 16 g of oxygen (*one gram atomic weight*), or the number of CO molecules in 28 g of carbon monoxide (*one gram molecular weight*).

Scientists knew that the number of carbon atoms in 12 g of carbon (or the number of atoms in any gram atomic weight, or the number of molecules in any gram molecular weight) must be very large. But they had no idea how large until Josef Loschmidt in 1865 tried to measure the size of air molecules. He found the molecules to be about a "millionth of a millimeter" in diameter, which meant that a gram molecular weight must contain 4×10^{22} molecules. This estimate was not bad for a first attempt. Later measurements, using various approaches, have shown the actual diameter of air molecules to be a bit smaller than Loschmidt had determined, and the number of molecules in a gram molecular weight to be 6×10^{23}. In German speaking countries 6×10^{23} is usually called the "Loschmidt number," but in most places it is known as "Avogadro's number," in spite of the fact that Avogadro never knew how big the number was. It was not measured until after his death.

There are 6×10^{23} carbon atoms in 12 g of carbon, and 6×10^{23} water molecules in 18 g of H_2O. No matter what the substance is, its formula weight expressed in grams contains 6×10^{23} formula units (atoms, molecules, or whatever the units happen to be) of that substance. Sulfur has an atomic weight of 32, so 32 g of sulfur contains Avogadro's number of sulfur atoms. Methane (CH_4) has a molecular weight of 16, so 16 g of methane contains Avogadro's number of methane molecules.

The most careful and reliable value that has been measured for Avogadro's number is 6.0221367×10^{23}. But that value is much more precise than it needs to be for most purposes. Ordinarily the number is rounded off to three significant figures, 6.02×10^{23}, since that corresponds with the precision of most laboratory data.

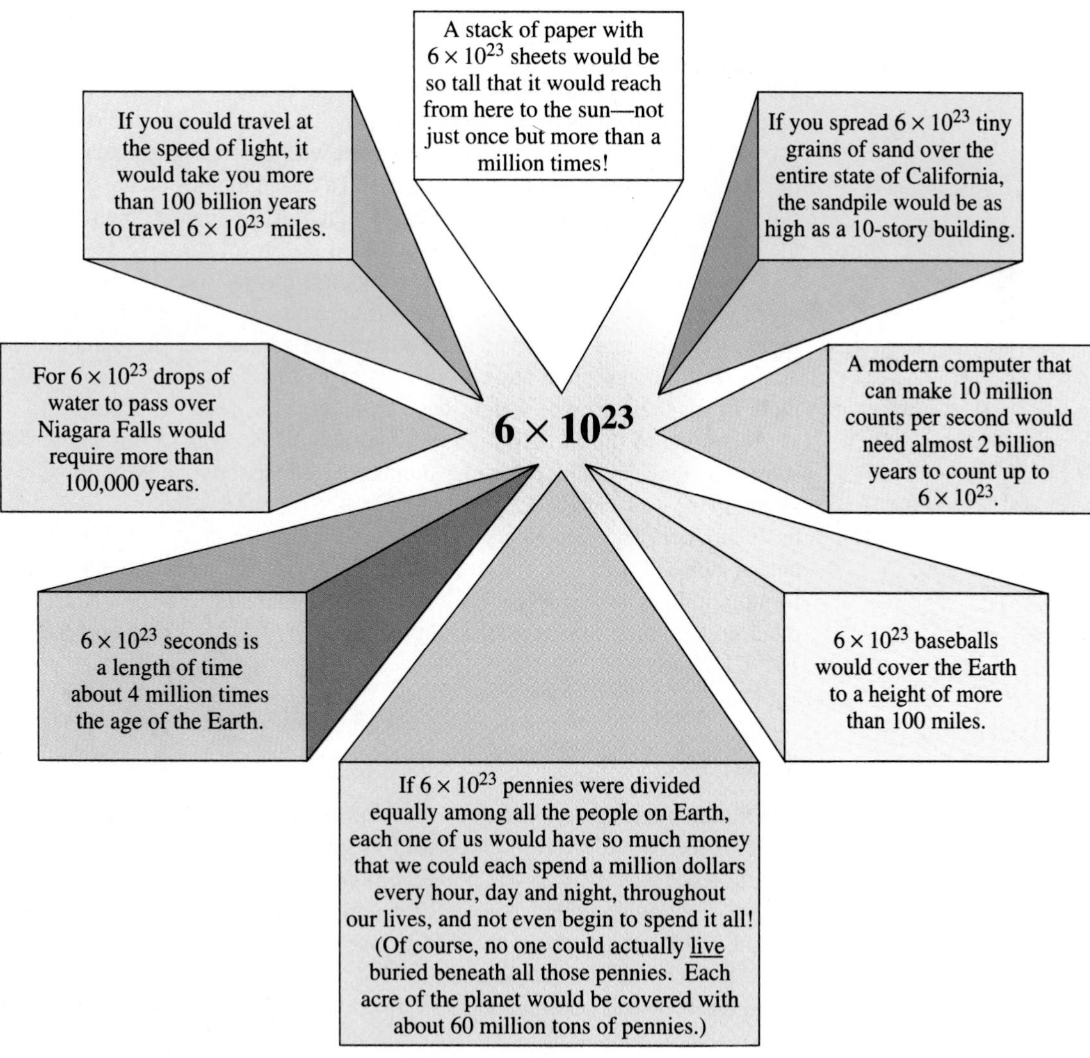

Figure 6.9 How big is Avogadro's number?

Avogadro's number is an enormous number, so large that it staggers the imagination. If you had 6×10^{23} dollars, for example, you could spend a billion dollars every second for as long as you lived, and you would still have more than 99.999% of your money left to pass on to your children. Or if 6×10^{23} snowflakes fell evenly all across the United States, the blanket of snow would completely cover up every building in the country, including the tallest skyscrapers. See Figure 6.9 for further illustrations as to how gigantic Avogadro's number is.

SECTION 6.8

The Mole: "A Dozen Eggs and a Mole of Sugar, Please"

We buy socks by the pair (2 socks), eggs by the dozen (12 eggs), soda by the case (24 cans), pencils by the gross (144 pencils), and paper by the ream (500 sheets). A dozen is the same *number* whether we are counting a dozen melons or a dozen oranges. But a dozen oranges and a dozen melons do not *weigh* the same. If a melon weighs three times as much as an orange, then a dozen melons will weigh three times as much as a dozen oranges.

Chemists count atoms and molecules by the *mole.* (A single carbon atom is much too small to see, but a *mole* of carbon atoms will fill a tablespoon.) A mole of carbon and a mole of magnesium both contain the same number of atoms. But a magnesium atom weighs twice as much as a carbon atom, so a mole of magnesium will weigh twice as much as a mole of carbon.

According to the SI definition, a **mole** (abbreviated *mol*) is an amount of substance that contains the same number of elementary units as there are atoms in 12 g of carbon-12. That number is 6.02×10^{23}, known as Avogadro's number. The elementary units may be atoms (such as S or Ca), molecules (such as O_2 or CO_2), ions (such as K^+ or SO_4^{2-}), or any other kind of formula unit. A mole of NaCl, for example, contains 6.02×10^{23} NaCl formula units, which means that it contains 6.02×10^{23} Na^+ ions and 6.02×10^{23} Cl^- ions.

Molar Mass

The **molar mass** of a substance is the mass of 1 mole of that substance. It is equal to the formula weight expressed in grams. In fact, the molar mass is sometimes referred to as the *gram formula weight.* Carbon has a formula weight of 12 and a molar mass of 12 g; carbon dioxide (CO_2) has a formula weight of 44 and a molar mass of 44 g; and sucrose (table sugar, $C_{12}H_{22}O_{11}$) has a formula weight of 342 and a molar mass of 342 g.

Figure 6.10 is a photograph showing 1-mole samples of several different

Figure 6.10 One mole of each of several familiar substances: salt (left), sugar (top), copper (right), and carbon (center). [*Richard Megna/Fundamental Photographs.*]

chemical substances. Each dish contains Avogadro's number of formula units of that substance. There are just as many carbon atoms in the smallest dish as there are sugar molecules in the largest one.

Molar Volume: 22.4 L at STP

A mole of gas contains Avogadro's number of molecules, no matter what kind of gas it is. Furthermore, Avogadro's number of molecules will occupy the same amount of volume (at a given temperature and pressure) regardless of how big, how small, or how heavy the individual molecules are. The volume occupied by 1 mole of gas is the **molar volume** of a gas.

Since the volume of a gas is altered by changes in temperature or pressure, a particular set of conditions has been chosen for reference purposes as "standard." Standard pressure is 1 atmosphere (atm), which is the normal pressure of the air at sea level; and standard temperature is 0 °C, which is the freezing point of water. A mole of any gas at standard temperature and pressure **(STP)** occupies a volume of about 22.4 L. This is known as the standard molar volume of a gas.

A pressure of 1 atmosphere (standard atmospheric pressure) is equal to the pressure exerted by a column of mercury 760 mm high.

$$1 \text{ atm} = 760 \text{ mm Hg}$$

The SI unit of pressure is the pascal (Pa).

$$1 \text{ atm} = 101{,}325 \text{ Pa}$$
$$= 101 \text{ kPa}$$

What Is a Mole?

A mole is a particular amount
Of substance—just its formulary weight
Expressed in grams, with Avogadro's count
Of units making up the aggregate.

A mole is a specific quantity:
Its volume measures twenty-two point four
In liters, for a gas at STP.
A mole's a counting unit, nothing more.

A mole is but a single molecule
By Avogadro's number multiplied;
One entity, extremely minuscule,
A trillion trillion times intensified.

A mole is a convenient amount,
For molecules are just too small to count.

A cube that measures 28.2 cm along each edge has a volume of 22.4 L (Figure 6.11). It is just a bit smaller than a cubic foot. At a temperature of 0 °C and 1 atm pressure, a 22.4 L container will hold 28 g of N_2, 32 g of O_2, 44 g of CO_2, or 1 mole of any gas.

It is easy to calculate the density of a gas under standard conditions, since the mass of 1 mole of gas is its molecular weight and the volume is 22.4 L. Just divide the molar mass by 22.4 L. Conversely, if you know the density of a gas at STP, it is a simple matter to calculate its molar weight. Merely multiply the mass per liter by 22.4 L.

EXAMPLE 6.21

Calculate the density of helium gas at STP.

SOLUTION

At STP 1 mol (4.00 g) of He occupies (approximately) 22.4 L.

$$\frac{4.00\text{ g}}{1\text{ mol}} \times \frac{1\text{ mol}}{22.4\text{ L}} = 0.179\text{ g/L}$$

Exercise 6.21

Calculate the density of argon gas at STP.

Figure 6.11 The molar volume of a gas compared to a basketball, football, and soccer ball. At STP the cube holds 16.0 g of CH_4, 28.0 g of N_2, 32.0 g of O_2, 39.9 g of Ar, and so on. It holds 1 mol of any gas at STP. [*Carey B. Van Loon.*]

EXAMPLE 6.22

The density of oxygen gas is 1.43 g/L at STP. Calculate the molecular weight of oxygen.

SOLUTION

$$\frac{1.43 \text{ g}}{1 \not{L}} \times \frac{22.4 \not{L}}{1 \text{ mol}} = 32 \text{ g/mol}$$

Exercise 6.22
The density of diethyl ether vapor at STP is 3.30 g/L. Calculate its molecular weight.

Calculations: Grams to Moles and Moles to Grams

Chemists and other scientists and engineers often are confronted with questions such as: "How many grams of phosphoric acid can I make from 660 g of phosphorus?" and "How many grams of hydrogen peroxide do I need to convert 5.82 g of lead sulfide to lead sulfate?" It is not possible to calculate grams of one substance directly from grams of another. To answer such questions, it is necessary to convert grams of a substance to moles of that substance and to write balanced equations to determine molar ratios. Let's take it one step at a time. First, let's convert grams to moles and moles to grams. Like most other things, such calculations are best learned by practice.

▶ EXAMPLE 6.23

Calculate the molar mass of sulfuric acid, H_2SO_4.

SOLUTION
First, calculate the formula weight of H_2SO_4.

$$\begin{aligned} 2 \times \text{atomic weight H} &= 2 \times 1.0 \text{ amu} = 2.0 \text{ amu} \\ 1 \times \text{atomic weight S} &= 1 \times 32.1 \text{ amu} = 32.1 \text{ amu} \\ 4 \times \text{atomic weight O} &= 4 \times 16.0 \text{ amu} = 64.0 \text{ amu} \\ \text{Formula weight } H_2SO_4 &= 98.1 \text{ amu} \end{aligned}$$

The molar mass of H_2SO_4 is therefore 98.1 g/mol.

Exercise 6.23
Calculate the molar mass of butyric acid, $C_4H_8O_2$.

▶ EXAMPLE 6.24

Calculate the mass of 0.260 mol of sulfuric acid, H_2SO_4.

SOLUTION
The molar mass of sulfuric acid is 98.1 g/mol (Example 6.23). The mass of 0.260 mol of sulfuric acid is therefore

$$0.260 \cancel{\text{mol } H_2SO_4} \times \frac{98.1 \text{ g } H_2SO_4}{1 \cancel{\text{mol } H_2SO_4}} = 25.5 \text{ g } H_2SO_4$$

Exercise 6.24
Calculate the mass of 1.37 mol of butyric acid, $C_4H_8O_2$.

▶ EXAMPLE 6.25

How many moles of sulfuric acid are there in 19.6 g of pure H_2SO_4?

SOLUTION
Use the molar mass from Example 6.23 and invert the conversion factor (to give an answer with the unit *moles*).

$$19.6 \cancel{\text{g } H_2SO_4} \times \frac{1 \text{ mol } H_2SO_4}{98.1 \cancel{\text{g } H_2SO_4}} = 0.200 \text{ mol } H_2SO_4$$

Exercise 6.25
Calculate the number of moles in each of the following.

a. 3.71 g of C_4H_{10}
b. 76.0 g of H_3PO_4
c. 47.8 g of C_2H_6O
d. 1.99 g of K_2CrO_4

SECTION 6.9

Mole and Mass Relationships in Chemical Equations

Chemical equations (Section 6.5) not only represent ratios of atoms and molecules but also give us mole ratios. The equation

$$C + O_2 \longrightarrow CO_2$$

tells us that one atom of carbon reacts with one molecule (two atoms) of oxygen to form one molecule of carbon dioxide. The equation also indicates that 1 mol (6.02×10^{23} atoms) of carbon reacts with 1 mol (6.02×10^{23} molecules) of oxygen to give 1 mol (6.02×10^{23} molecules) of carbon dioxide. Since the molar mass in grams of a substance is numerically equal to the formula weight of the substance in atomic mass units, the equation also tells us (indirectly) that 12.0 g (1 mol) of carbon reacts with 32.0 g (1 mol) of oxygen to give 44.0 g (1 mol) of CO_2 (Figure 6.12).

We need not use exactly 1 mol of each reactant. The important thing is to keep the ratio constant. For example, in the reaction above, the mass ratio of oxygen to carbon is 32.0 to 12.0 or 8.0 to 3.0. We could use 8.0 g of oxygen and 3.0 g of carbon to produce 11.0 g of CO_2. In fact, to calculate the amount of oxygen needed to react with a given amount of carbon, we need only multiply the amount of carbon by the factor 32.0/12.0.

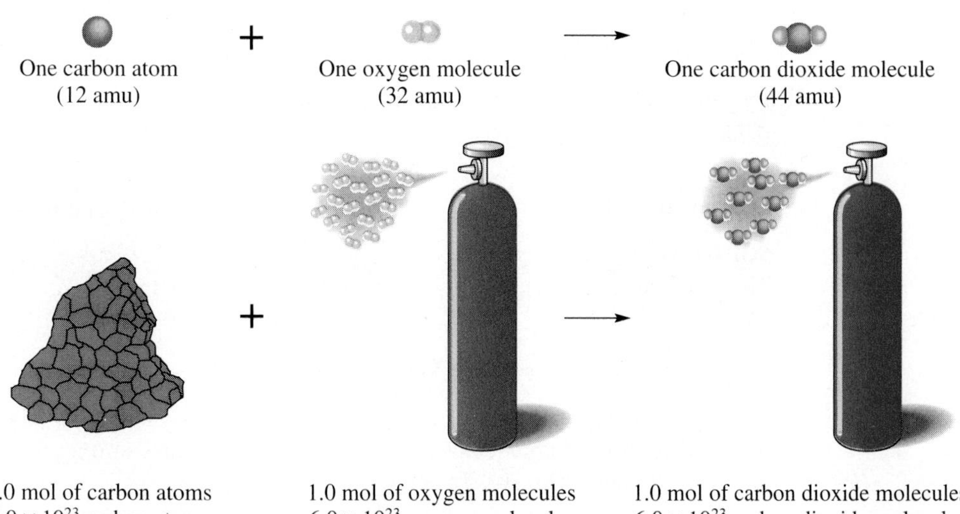

Figure 6.12 We cannot weigh single atoms or molecules, but we can weigh equal numbers of these fundamental particles.

▶ EXAMPLE 6.26

Nitrogen monoxide (nitric oxide) combines with oxygen to form nitrogen dioxide according to the equation

$$2\,NO + O_2 \longrightarrow 2\,NO_2$$

State the molecular, molar, and mass relationships indicated by the equation.

SOLUTION

Molecular: 2 molecules of NO react with 1 molecule of O_2 to form 2 molecules of NO_2. Molar: 2 mol of NO react with 1 mol of O_2 to form 2 mol of NO_2. Mass: 60.0 g of NO react with 32.0 g of O_2 to form 92.0 g of NO_2.

Exercise 6.26

Hydrogen sulfide burns in air to produce sulfur dioxide and water according to the equation

$$2\,H_2S + 3\,O_2 \longrightarrow 2\,SO_2 + 2\,H_2O$$

State the molecular, molar, and mass relationships indicated by the equation.

Molar Relationships in Chemical Equations

The ratio of moles of reactants and products is given by the coefficients in a balanced chemical equation.

▶ EXAMPLE 6.27

Propane burns in air to form carbon dioxide and water.

$$C_3H_8 + 5\,O_2 \longrightarrow 3\,CO_2 + 4\,H_2O$$

How many moles of oxygen are required to burn 0.105 mol of propane?

SOLUTION

The equation tells us that 5 mol O_2 is required to burn 1 mol C_3H_8. We can write

$$1 \text{ mol } C_3H_8 \approx 5 \text{ mol } O_2$$

where the symbol $\approx$ means "is chemically equivalent to." From this relationship we can construct conversion factors to relate moles of oxygen to moles of propane. The possible conversion factors are

$$\frac{5 \text{ mol } O_2}{1 \text{ mol } C_3H_8} \quad \text{and} \quad \frac{1 \text{ mol } C_3H_8}{5 \text{ mol } O_2}$$

Multiply the given quantity (0.105 mol C_3H_8) by the factor on the left to get an answer with the asked-for units (moles of oxygen).

$$0.105 \text{ mol } C_3H_8 \times \frac{5 \text{ mol } O_2}{1 \text{ mol } C_3H_8} = 0.525 \text{ mol } O_2$$

Exercise 6.27

Use the equation in Example 6.27 to answer the following questions. (a) How many moles of carbon dioxide are produced when 0.529 mol of propane is burned? (b) How many moles of water are produced when 76.2 mol of propane is burned? (c) How many moles of carbon dioxide are produced when 1.020 mol of oxygen is consumed?

Stoichiometry: Mass Relationships in Chemical Equations

Problems are seldom formulated in moles. Typically, you are given an amount of one substance in grams and are asked to calculate how many grams of another substance can be made from it. Such calculations involve several steps.

1. Write a balanced chemical equation for the reaction.
2. Determine the molar masses of the substances involved in the calculation.
3. Write down the quantity that is given and use the molar mass to convert that quantity to moles.
4. Use the balanced chemical equation to convert moles of given substance to moles of desired substance.
5. Use the molar mass to convert moles of desired substance to grams of desired substance.

The conversion process is diagrammed in Figure 6.13. It is best learned from examples and by working out exercises.

▶ EXAMPLE 6.28

Calculate the mass of oxygen needed to react with 10.0 g of carbon in the reaction that forms carbon dioxide.

SOLUTION

1. The balanced equation is

$$C + O_2 \longrightarrow CO_2$$

2. The molar mass of oxygen is 32.0 g and that of carbon is 12.0 g.

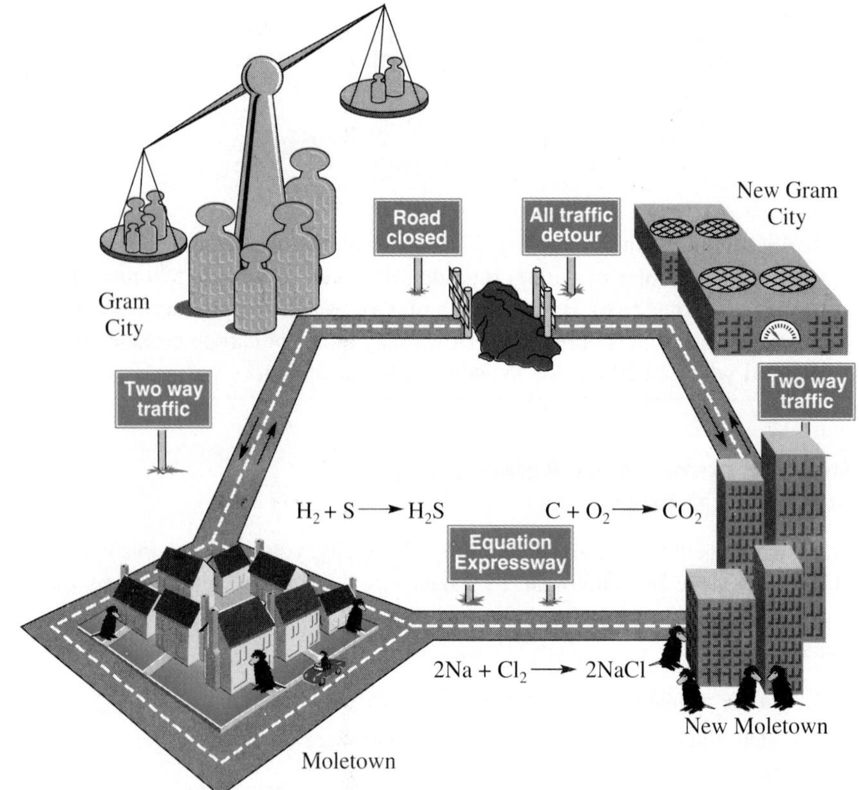

Figure 6.13 The chemical equation relates moles of reactants to moles of products. For mass relationship between reactants and products, we must convert masses to moles, then moles back to masses.

3. $$10.0\,\cancel{\text{g C}} \times \frac{1 \text{ mol C}}{12.0\,\cancel{\text{g C}}} = 0.833 \text{ mol C}$$

4. From the equation,

$$1 \text{ mol C} \approx 1 \text{ mol O}_2$$

$$0.833\,\cancel{\text{mol C}} \times \frac{1 \text{ mol O}_2}{1\,\cancel{\text{mol C}}} = 0.833 \text{ mol O}_2$$

5. $$0.833\,\cancel{\text{mol O}_2} \times \frac{32.0 \text{ g O}_2}{1\,\cancel{\text{mol O}_2}} = 27.0 \text{ g O}_2$$

The problem also can be set up and solved in one step.

$$10.0 \text{ g C} \times \frac{1\,\cancel{\text{mol C}}}{12.0\,\cancel{\text{g C}}} \times \frac{1\,\cancel{\text{mol O}_2}}{1\,\cancel{\text{mol C}}} \times \frac{32.0 \text{ g O}_2}{1\,\cancel{\text{mol O}_2}} = 27.0 \text{ g O}_2$$

Note that the units in the denominators of the conversion factors are chosen so that each cancels the unit in the numerator of the preceding term.

Exercise 6.28
Calculate the mass of oxygen needed to react with 0.334 g of sulfur in the reaction that forms sulfur dioxide. The equation is

$$S_8 + 8\,O_2 \longrightarrow 8\,SO_2$$

Note that we can set up conversion factors for any two compounds involved in a reaction. (Conversion factors are explained in Appendix C.) Typical problems will ask you to calculate how much of one compound is equivalent to a given amount of one of the other compounds. All you need to do to solve such a problem is put together a conversion factor relating the two compounds. The following involve additional conversion factors derived from the equation

$$C + O_2 \longrightarrow CO_2$$

EXAMPLE 6.29

Calculate the mass of carbon that can be burned to carbon dioxide in the presence of 400 g of oxygen. [*Hint:* You are given the amount of oxygen and are asked for the amount of carbon. You need a conversion factor that relates carbon to oxygen.]

SOLUTION

$$400\,\cancel{g\,O_2} \times \frac{1\,\cancel{mol\,O_2}}{32.0\,\cancel{g\,O_2}} \times \frac{1\,\cancel{mol\,C}}{1\,\cancel{mol\,O_2}} \times \frac{12.0\text{ g C}}{1\,\cancel{mol\,C}} = 150\text{ g C}$$

Exercise 6.29
What mass of carbon dioxide can be obtained from 48.0 g of carbon, assuming sufficient oxygen? [*Hint:* You are given the amount of carbon and are asked for the amount of carbon dioxide. You need a conversion factor that relates carbon dioxide to carbon.]

The examples and exercises so far are based on an exceptionally simple reaction. In the balanced equation, there are no coefficients other than 1. That is, 1 mol of oxygen reacts with 1 mol of carbon to produce 1 mol of carbon dioxide. The following examples and exercises involve balanced equations that have coefficients other than 1. We shall work out the sort of problem industrial chemists face every day. Whether chemists are making drugs, obtaining metals from their ores, synthesizing plastics, or whatever, they must use materials economically.

EXAMPLE 6.30

Ammonia (for fertilizer and other uses) is made by causing hydrogen and nitrogen to react at high temperature and pressure. The equation is

$$N_2 + H_2 \longrightarrow NH_3 \quad \textit{(not balanced)}$$

How many grams of ammonia can be made from 60.0 g of hydrogen?

SOLUTION

First, we must balance the equation.

$$N_2 + 3\,H_2 \longrightarrow 2\,NH_3 \quad \textit{(balanced)}$$

Next, we calculate the molar masses of the compounds involved in the problem, that is, of hydrogen and ammonia.

H_2 has a molar mass of 2.02 g.

NH_3 has a molar mass of 17.0 g.

We are given an amount of hydrogen and are asked for a mass of ammonia. We must convert the mass given to moles.

$$60.0\ \cancel{g\ H_2} \times \frac{1\ mol\ H_2}{2.02\ \cancel{g\ H_2}} = 29.7\ mol\ H_2$$

Next, we must use the coefficients given in the equation to determine the number of moles of ammonia that can be obtained from the calculated moles of hydrogen gas. The equation tells us that 3 mol of H_2 produces 2 mol of NH_3, thus

$$29.7\ \cancel{mol\ H_2} \times \frac{2\ mol\ NH_3}{3\ \cancel{mol\ H_2}} = 19.8\ mol\ NH_3$$

Finally, we must convert moles of ammonia to grams of ammonia.

$$19.8\ \cancel{mol\ NH_3} \times \frac{17.0\ g\ NH_3}{1\ \cancel{mol\ NH_3}} = 337\ g\ NH_3$$

Or we could do it all in one step.

$$60.0\ \cancel{g\ H_2} \times \frac{1\ \cancel{mol\ H_2}}{2.02\ \cancel{g\ H_2}} \times \frac{2\ \cancel{mol\ NH_3}}{3\ \cancel{mol\ H_2}} \times \frac{17.0\ g\ NH_3}{1\ \cancel{mol\ NH_3}} = 337 g\ NH_3$$

Exercise 6.30

Phosphorus reacts with oxygen to form tetraphosphorus decoxide. The equation is

$$P_4 + O_2 \longrightarrow P_4O_{10} \quad \textit{(not balanced)}$$

How many grams of tetraphosphorus decoxide can be made from 3.50 g of phosphorus?

EXAMPLE 6.31

How many grams of oxygen must be consumed if the following reaction is to produce 9.00 g of water?

$$H_2 + O_2 \longrightarrow H_2O \quad \textit{(not balanced)}$$

SOLUTION

The balanced equation is

$$2\,H_2 + O_2 \longrightarrow 2\,H_2O$$

The molar mass for oxygen gas is 32.0 g, and the molar mass for water is 18.0 g.

$$9.00\ \cancel{g\ H_2O} \times \frac{1\ \cancel{mol\ H_2O}}{18.0\ \cancel{g\ H_2O}} \times \frac{1\ \cancel{mol\ O_2}}{2\ \cancel{mol\ H_2O}} \times \frac{32.0\ g\ O_2}{1\ \cancel{mol\ O_2}} = 8.00\ g\ O_2$$

Exercise 6.31

a. What mass of oxygen can be made from 2.47 g of potassium chlorate? The equation (not balanced) is

$$KClO_3 \longrightarrow KCl + O_2$$

b. What mass of carbon dioxide is formed when 73.9 g of propane is burned? The equation (not balanced) is

$$C_3H_8 + O_2 \longrightarrow CO_2 + H_2O$$

c. What mass of magnesium metal is required to reduce 83.6 g of titanium(IV) chloride to titanium metal? The equation (not balanced) is

$$TiCl_4 + Mg \longrightarrow Ti + MgCl_2$$

Summary

1. Simple *binary compounds* are named by naming the more positive element first and then putting an -ide ending on the name of the more negative element.
2. Names of *polyatomic ions* often end in -ate, but if there are several anions containing the same elements, the name of the one with less oxygen may end in -ite.
3. When a metal forms several different compounds with the same nonmetal, each compound is named by mentioning the charge on the metal ion as a parenthetical Roman numeral.
4. When two nonmetals form several different compounds, prefixes such as mono, di, tri, etc. are used to indicate numbers of atoms of the same element.

5. Chemical equations are statements of chemical reactions written in chemical shorthand. They are balanced by making the number of atoms of each element equal on both sides of the arrow.

6. Gay-Lussac's *law of combining volumes* states that gases at the same pressure and temperature combine in small whole number ratios by volume.

7. A *mole* is an amount of substance that contains 6×10^{23} molecules (or formula units) of that substance. (The number of formula units in a mole is known as *Avogadro's number.*) A mole is equal to the formula weight of the substance expressed in grams. This is called its *molar mass.* If the substance is a gas at standard temperature and pressure (*STP*), its *molar volume* is 22.4 L, regardless of what gas it is.

8. In calculating how much product should be formed from a given amount of reactant it is much more convenient to use moles than to use molecules (because molecules are too small to see). One mole of a substance contains the same number of molecules (or formula units) as 1 mole of any other substance. The mass relationship between reactants and products in chemical reactions is called *stoichiometry.*

Key Terms

Avogadro's hypothesis 6.6
Avogadro's number 6.7
binary compounds 6.2
chemical equation 6.5
formula 6.2
ions 6.1
law of combining volumes 6.6
molar mass 6.8
molar volume 6.8
mole 6.8
polyatomic ions 6.3
products 6.5
reactants 6.5
stoichiometry 6.9
STP 6.8

Review Questions

1. Define or illustrate each of the following.
   a. binary compound
   b. polyatomic ion
   c. mole
   d. Avogadro's number
   e. molar mass
   f. molar volume

2. Explain the difference between "the atomic weight of oxygen" and "the formula weight of oxygen (gas)."

3. What is Avogadro's hypothesis? How did it explain Gay-Lussac's law of combining volumes?

4. How do the law of combining volumes and Avogadro's hypothesis indicate that hydrogen gas is composed of diatomic molecules rather than individual atoms?

5. How many oxygen molecules are there in 1.00 mol of O_2? How many oxygen atoms are there in 1.00 mol of O_2?

6. How many calcium ions and how many chloride ions are there in 1.00 mol of $CaCl_2$?

7. What is the formula weight of CO_2? What is the molar mass of CO_2? State in words how each is determined from the formula.

8. Consider the law of conservation of mass in explaining why we must work with *balanced* chemical equations.

9. What is the molar volume of each of the following gases at STP?
   a. He   b. H_2 c. C_2H_6

10. What is the mass of a molar volume of each of the gases in Question 9?

11. Indicate charges on simple ions formed from the following elements.
    a. Group IIIA   b. Group VIA
    c. Group IA   d. Group VIIA

12. In what group of the periodic table would elements that form ions with the following charges likely be found?
    a. 2+   b. 3−   c. 1−

13. Chlorine dioxide is used to bleach flour. Give the formula for chlorine dioxide.

14. Tetraphosphorus trisulfide is used in the tops of "strike anywhere" matches. Give the formula for tetraphosphorus trisulfide.

15. The gas phosphine (PH_3) is used as a fumigant to protect stored grain and other durable produce from pests. Phosphine is generated *in situ* by adding water to aluminum phosphide or magnesium phosphide. Give formulas for the two phosphides.

16. How many hydrogen atoms are indicated in one formula unit of each of the following?
a. NH_4NO_3 b. CH_3OH
c. $CH_3CH_2CH_3$ d. C_6H_5COOH

17. How many hydrogen atoms are indicated in one formula unit of each of the following?
a. $(NH_4)_2HPO_4$ b. $Al(C_2H_3O_2)_3$

18. How many oxygen atoms are indicated in one formula unit of each of the following?
a. $Al_2(C_2O_4)_3$ b. $Ca_3(PO_4)_2$

19. How many atoms of each kind (Al, C, H, and O) are indicated by the notation $2\ Al(C_2H_3O_2)_3$?

20. How many atoms of each kind (N, P, H, and O) are indicated by the notation $6\ (NH_4)_2HPO_4$?

21. Indicate whether the equations are balanced. (You need not balance the equation; just determine whether it is balanced as written.)
a. $Mg + H_2O \rightarrow MgO + H_2$
b. $FeCl_2 + Cl_2 \rightarrow FeCl_3$
c. $F_2 + H_2O \rightarrow 2\ HF + O_2$
d. $Ca + 2\ H_2O \rightarrow Ca(OH)_2 + H_2$
e. $2\ LiOH + CO_2 \rightarrow Li_2CO_3 + H_2O$

22. Indicate whether the equations are balanced as written.
a. $2\ KNO_3 + 10\ K \rightarrow 6\ K_2O + N_2$
b. $2\ NH_3 + O_2 \rightarrow N_2 + 3\ H_2O$
c. $4\ LiH + AlCl_3 \rightarrow 2\ LiAlH_4 + 2\ LiCl$
d. $SF_4 + 3\ H_2O \rightarrow H_2SO_3 + 4\ HF$
e. $4\ BF_3 + 3\ H_2O \rightarrow H_3BO_3 + 3\ HBF_4$

23. Indicate whether the equations are balanced as written.
a. $2\ Sn + 2\ H_2SO_4 \rightarrow 2\ SnSO_4 + SO_2 + 2\ H_2O$
b. $3\ Cl_2 + 6\ NaOH \rightarrow 5\ NaCl + NaClO_3 + 3\ H_2O$

24. The following equation appeared in an article on the incineration of toxic gases.

$$(CH_3)_3As + 15\ O_2 \rightarrow As_4O_6 + 12\ CO_2 + 6\ H_2O$$

Count the number of atoms of each kind on each side of the equation. Is the equation balanced?

25. Propane burns in air to form carbon dioxide and water. The equation is

$$C_3H_8 + 5\ O_2 \rightarrow 3\ CO_2 + 4\ H_2O$$

State the molecular, molar, and mass relationships indicated by the equation.

26. The poison gas phosgene reacts with water in the lungs to form hydrogen chloride and carbon dioxide. The equation is

$$COCl_2 + H_2O \rightarrow 2\ HCl + CO_2$$

State the molecular, molar, and mass relationships indicated by the equation.

Problems

Names and Symbols for Simple Ions

27. Name the following ions.
a. K^+ b. Ca^{2+} c. Zn^{2+}
d. Br^- e. Li^+ f. S^{2-}

28. Name the following ions.
a. Na^+ b. Mg^{2+} c. Al^{3+}
d. Cl^- e. O^{2-} f. N^{3-}

29. Name the following ions.
a. Fe^{2+} b. Cu^+ c. I^-

30. Name the following ions.
a. Fe^{3+} b. Cu^{2+} c. Ag^+

31. Give symbols for the following ions.
a. sodium ion b. aluminum ion
c. oxide ion d. copper(II) ion

32. Give symbols for the following ions.
a. bromide ion b. calcium ion
c. potassium ion d. iron(II) ion

Names and Formulas for Polyatomic Ions

33. Name the following ions.
a. CO_3^{2-} b. HPO_4^{2-} c. MnO_4^- d. OH^-

34. Name the following ions.
a. NO_3^- b. SO_4^{2-} c. $H_2PO_4^-$ d. HCO_3^-

35. Give formulas for the following ions.
a. ammonium ion
b. hydrogen sulfate ion
c. cyanide ion
d. nitrite ion

36. Give formulas for the following ions.
a. phosphate ion
b. hydrogen carbonate ion
c. dichromate ion
d. oxalate ion

Names and Formulas for Ionic Compounds

37. Name the following ionic compounds.
a. NaBr b. $CaCl_2$ c. $FeCl_3$
d. LiI e. K_2S f. CuBr

38. Name the following ionic compounds.
a. KCl b. $MgBr_2$ c. CuI_2
d. CaS e. $FeCl_2$ f. Al_2O_3

39. Give formulas for the following ionic compounds.
a. magnesium sulfate
b. sodium hydrogen carbonate
c. potassium nitrate
d. calcium monohydrogen phosphate

40. Give formulas for the following ionic compounds.
a. calcium carbonate
b. potassium dihydrogen phosphate
c. magnesium cyanide
d. lithium hydrogen sulfate

41. Give formulas for the following ionic compounds.
a. iron(II) phosphate
b. potassium dichromate
c. copper(I) iodide
d. ammonium nitrite

42. Give formulas for the following ionic compounds.
a. iron(III) oxalate
b. sodium permanganate
c. copper(II) bromide
d. zinc monohydrogen phosphate

43. Name the following ionic compounds.
a. KNO_2 b. LiCN
c. NH_4I d. $NaNO_3$
e. $KMnO_4$ f. $CaSO_4$

44. Name the following ionic compounds.
a. $NaHSO_4$ b. $Al(OH)_3$
c. Na_2CO_3 d. $KHCO_3$
e. NH_4NO_2 f. $Ca(HSO_4)_2$

45. Name the following ionic compounds.
a. Na_2HPO_4 b. $(NH_4)_3PO_4$
c. $Al(NO_3)_3$ d. NH_4NO_3

46. Name the following ionic compounds.
a. Li_2CO_3 b. $Na_2Cr_2O_7$
c. $Ca(H_2PO_4)_2$ d. $(NH_4)_2C_2O_4$

Names and Formulas for Covalent Compounds

47. Give formulas for the following covalent compounds.
a. dinitrogen monoxide
b. tetraphosphorus trisulfide
c. phosphorus pentachloride
d. sulfur hexafluoride

48. Give formulas for the following covalent compounds.
a. oxygen difluoride
b. dinitrogen pentoxide
c. phosphorus tribromide
d. tetrasulfur tetranitride

49. Name the following covalent compounds.
a. CS_2 b. N_2S_4 c. PF_5 d. S_2F_{10}

50. Name the following covalent compounds.
a. CBr_4 b. Cl_2O_7 c. P_4S_{10} d. I_2O_5

Balancing Chemical Equations

51. Balance the following equations.
a. $Cl_2O_5 + H_2O \rightarrow HClO_3$
b. $V_2O_5 + H_2 \rightarrow V_2O_3 + H_2O$
c. $Al + O_2 \rightarrow Al_2O_3$
d. $Sn + NaOH \rightarrow Na_2SnO_2 + H_2$
e. $PCl_5 + H_2O \rightarrow H_3PO_4 + HCl$

52. Balance the following equations.
a. $TiCl_4 + H_2O \rightarrow TiO_2 + HCl$
b. $C_4H_{10} + O_2 \rightarrow CO_2 + H_2O$
c. $WO_3 + H_2 \rightarrow W + H_2O$
d. $Al_4C_3 + H_2O \rightarrow Al(OH)_3 + CH_4$
e. $Al_2(SO_4)_3 + NaOH \rightarrow Al(OH)_3 + Na_2SO_4$

53. Balance the following equations.
a. $Na_3P + H_2O \rightarrow NaOH + PH_3$
b. $Cl_2O + H_2O \rightarrow HClO$
c. $CH_3OH + O_2 \rightarrow CO_2 + H_2O$
d. $Zn(OH)_2 + H_3PO_4 \rightarrow Zn_3(PO_4)_2 + H_2O$
e. $C_3H_8 + O_2 \rightarrow CO_2 + H_2O$

54. Balance the following equations.
a. $Ca_3P_2 + H_2O \rightarrow Ca(OH)_2 + PH_3$
b. $Cl_2O_7 + H_2O \rightarrow HClO_4$
c. $MnO_2 + HCl \rightarrow MnCl_2 + Cl_2 + H_2O$
d. $Fe + O_2 \rightarrow Fe_3O_4$
e. $C_5H_{12} + O_2 \rightarrow CO_2 + H_2O$

Formula Weights

You may round all atomic weights to one decimal place.

55. Calculate the formula weights of each of the following compounds.
a. CH_4 b. AlF_3 c. UF_6

56. Calculate the formula weights of each of the following compounds.
a. SO_3 b. $KBrO_3$ c. $CaSO_4$

57. Calculate the formula weights of each of the following compounds.
a. C_6H_5Br b. $H_4P_2O_7$
c. $K_2Cr_2O_7$ d. $Al_2(SO_4)_3$

58. Calculate the formula weights of each of the following compounds.
a. $(NH_4)_3PO_4$ b. $Fe(NO_3)_3$
c. $C_2H_5NO_2$ d. MgS_2O_3

Gay-Lussac's Law of Combining Volumes

59. According to the equation

$$C_3H_8 + 5\,O_2 \rightarrow 3\,CO_2 + 4\,H_2O$$

a. How many liters of CO_2 are formed when 0.529 L of C_3H_8 is burned?
b. How many liters of O_2 are required to burn 16.1 L of C_3H_8?

60. According to the equation

$$C_3H_8 + 5\,O_2 \rightarrow 3\,CO_2 + 4\,H_2O$$

a. How many liters of H_2O are formed when 2.93 L of C_3H_8 is burned?
b. How many liters of O_2 are required to form 0.370 L of H_2O?

Molar Volume, Gas Densities, and Molecular Weights

61. Calculate the density of argon (Ar) gas, in grams per liter, at STP.

62. Calculate the density of nitrogen (N_2) gas, in grams per liter, at STP.

63. Calculate the molecular weight of a gas which has a density of 2.35 g/L at STP.

64. Calculate the molecular weight of a gas which has a density of 1.98 g/L at STP.

65. Calculate the molecular weight of a liquid the vapor of which has a density of 3.02 g/L at STP.

66. Calculate the molecular weight of a liquid the vapor of which has a density of 2.81 g/L at STP.

Molar Masses

67. Calculate the mass of 0.377 mol of sodium nitrite, $NaNO_2$.

68. Calculate the mass of 1.67 mol of potassium hydrogen oxalate, KHC_2O_4.

69. Calculate the number of moles of copper(I) sulfate, Cu_2SO_4, in 16.3 g of the compound.

70. Calculate the number of moles of octane, C_8H_{18}, in 102 g of octane.

71. Calculate the number of grams in each of the following.
a. 0.00500 mol MnO_2
b. 1.12 mol CaH_2
c. 0.250 mol $C_6H_{12}O_6$
d. 4.61 mol $AlCl_3$

72. Calculate the number of moles in each of the following.
a. 98.6 g HNO_3 b. 9.45 g CBr_4
c. 9.11 g $FeSO_4$ d. 11.8 g $Pb(NO_3)_2$

Mole and Mass Relationships in Chemical Equations

73. Consider the reaction for the combustion of octane.

$$2\,C_8H_{18} + 25\,O_2 \rightarrow 16\,CO_2 + 18\,H_2O$$

a. How many moles of CO_2 are produced when 2.09 mol of octane is burned?
b. How many moles of oxygen are required to burn 4.47 mol of octane?

74. Consider the reaction for the combustion of octane.

$$2\,C_8H_{18} + 25\,O_2 \rightarrow 16\,CO_2 + 18\,H_2O$$

a. How many moles of H_2O are produced when 2.81 mol of octane is burned?
b. How many moles of CO_2 are produced when 4.06 mol of oxygen is consumed?

75. How many grams of ammonia can be made from 440 g of H_2? The equation (not balanced) is

$$N_2 + H_2 \rightarrow NH_3$$

76. How many grams of hydrogen are needed to react completely with 892 g of N_2? The equation (not balanced) is

$$N_2 + H_2 \rightarrow NH_3$$

77. How many grams of oxygen can be prepared from 24.0 g of H_2O_2? The equation (not balanced) is

$$H_2O_2 \rightarrow H_2O + O_2$$

78. Nitric acid is used in the production of trinitrotoluene (TNT), an explosive. The equation (not balanced) is

$$\underset{\text{Toluene}}{C_7H_8} + HNO_3 \rightarrow \underset{\text{TNT}}{C_7H_5N_3O_6} + H_2O$$

How much nitric acid is required to react with 454 g of toluene?

79. Use the equation (not balanced) in Problem 78 to calculate the amount of TNT that can be made from 829 g of toluene.

80. What mass of quicklime (calcium oxide) can be made when 4.72 g of limestone (calcium carbonate) is decomposed by heating?

$$CaCO_3 \rightarrow CaO + CO_2$$

81. What mass of nitric acid can be made from 971 g of ammonia? The equation (not balanced) is

$$NH_3 + O_2 \rightarrow HNO_3 + H_2O$$

82. Acetylene (C_2H_2) burns in pure oxygen with a very hot flame. The equation (not balanced) is

$$C_2H_2 + O_2 \rightarrow CO_2 + H_2O$$

How much oxygen is required to react with 52.0 g of acetylene?

Additional Problems

83. Joseph Priestley discovered oxygen in 1774 by heating "red calx of mercury," mercury(II) oxide. The calx decomposed to the elements. The equation (not balanced) is

$$HgO \rightarrow Hg + O_2$$

How much oxygen is produced by the decomposition of 10.8 g of HgO?

84. How much iron can be converted to the magnetic oxide of iron (Fe_3O_4) by 8.80 g of pure oxygen? The equation (not balanced) is

$$Fe + O_2 \rightarrow Fe_3O_4$$

85. Laughing gas (dinitrogen monoxide, N_2O, also called nitrous oxide) can be made by heating ammonium nitrate with great care. The equation (not balanced) is

$$NH_4NO_3 \rightarrow N_2O + H_2O$$

How much N_2O can be made from 4.00 g of ammonium nitrate?

86. Small amounts of hydrogen gas often are made by the reaction of calcium metal with water. The equation (not balanced) is

$$Ca + H_2O \rightarrow Ca(OH)_2 + H_2$$

How many grams of hydrogen are formed by the action of water on 0.413 g of calcium?

87. For many years the noble gases were called the "inert gases," and it was thought that they formed no chemical compounds. Neil Bartlett made the first noble gas compound in 1962. Xenon hexafluoride is made according to the equation (not balanced)

$$Xe + F_2 \rightarrow XeF_6$$

How many grams of fluorine are required to make 0.112 g of XeF_6?

88. Phosphine gas (used as a fumigant to protect stored grain) is generated by the action of water on magnesium phosphide. The equation (not balanced) is

$$Mg_3P_2 + H_2O \rightarrow PH_3 + Mg(OH)_2$$

How much magnesium phosphide is needed to produce 134 g of PH_3?

▸ References and Readings

1. Chang, Raymond, *Chemistry,* 4th ed. New York: Random House, 1991. Chapter 3 considers the arithmetic of chemistry.
2. DeLorenzo, Ronald A. *Problem Solving in General Chemistry.* Lexington, MA: D. C. Heath, 1981. Chapters 4 and 6 discuss moles and equations.
3. Ebbing, Darrell D. *General Chemistry,* 4th ed. Boston: Houghton Mifflin, 1993. Chapters 2 and 4 relate to the material in this chapter.
4. Gizara, Jeanne M. ''Bridging the Stoichiometry Gap.'' *The Science Teacher,* April 1981. pp. 36–37. Presents a road-map method for solving problems based on equations.
5. Kolb, Doris. ''The Chemical Formula, Part I: Development.'' *Journal of Chemical Education,* January 1978, pp. 44–47. A brief historical account of how chemical formulas were developed.
6. Kolb, Doris. ''The Mole.'' *Journal of Chemical Education,* November 1978, pp. 728–732. A discussion of the mole concept and Avogadro's number.
7. Petrucci, Ralph H., and William S. Harwood. *General Chemistry: Principles and Modern Applications,* 6th ed. New York: Macmillan, 1993. Chapter 3 discusses formulas of chemical compounds and the mole concept.
8. Priesner, Claus. ''How the Language of Chemistry Developed.'' *Chemistry International,* June 1989, pp. 216–238. The story of how chemical nomenclature evolved.
9. Zumdahl, Steven S. *Chemistry,* 3rd ed. Lexington, MA: D. C. Heath, 1993. Chapter 3 considers mass and mole relationships in chemical equations.

7 Acids and Bases

Please Pass the Protons

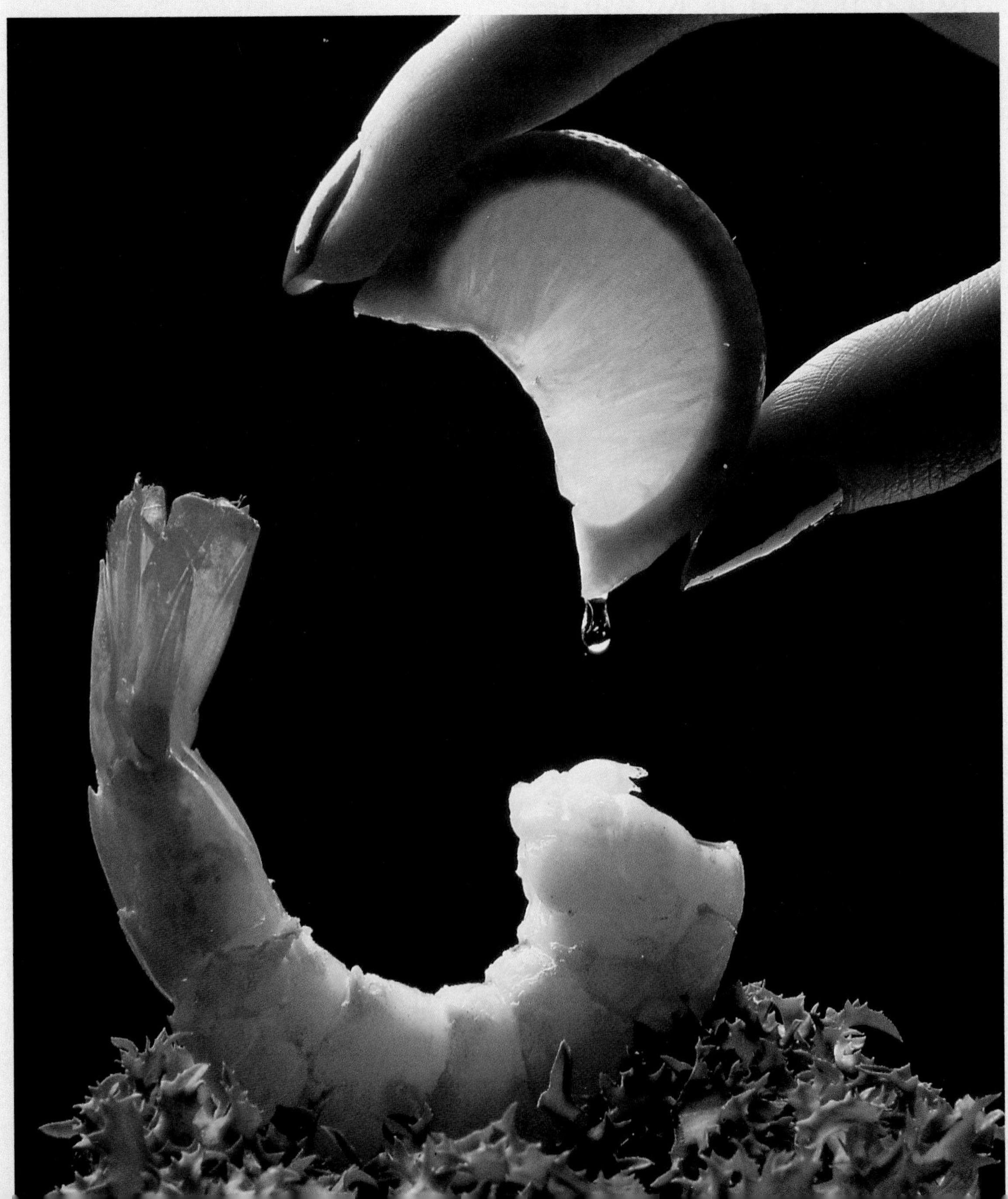

Bases have a bitter taste
And potent cleaning power;
But acids are in many foods
Like lemons, they taste sour.

There are two classes of chemical compounds that are related to each other and are quite important in our daily lives. They are *acids* and *bases*. Some familiar acids are vinegar (acetic acid), vitamin C (ascorbic acid), and battery acid (sulfuric acid). Some familiar bases are lye (sodium hydroxide), baking soda (sodium bicarbonate), and ammonia.

Air and water pollution often involve acids and bases. Acid rain, for example, is a serious environmental problem, and alkaline (basic) water is sometimes undrinkable.

Did you know that the four tastes are related to acid–base chemistry? Acids taste *sour,* bases taste *bitter,* and the compounds formed when acids react with bases (salts) taste *salty*. The sweet taste is more complicated. To taste sweet a compound must have both an acidic-type group and a basic-type group, plus just the right geometry to fit the sweet-taste receptor.

In this chapter we discuss some of the chemistry of acids and bases. You use them every day, and you will probably be hearing and reading about them as long as you live. We hope that what you learn here will help you gain a better understanding of these important classes of compounds.

SECTION 7.1

Acids and Bases: Experimental Definitions

How can you tell if a compound is an acid or a base? Let us begin by listing a few of their properties.

Acids are compounds that

1. cause litmus indicator dye to turn red.
2. taste sour.

The chemical substances we call acids and bases are all around us, even in the food we eat. [*Gerald Zanetti/The Stock Market.*]

Figure 7.1 Some common acids (left) and bases (right). Acids and bases are components of many familiar consumer products. [*Robert Mathena/Fundamental Photographs.*]

3. dissolve active metals (such as zinc or iron) producing hydrogen gas.
4. react with bases to form water and ionic compounds called *salts*.

Acids and bases are chemical opposites, so their properties are quite different. **Bases** are compounds that

1. cause litmus indicator dye to turn blue.
2. taste bitter.
3. feel slippery on the skin.
4. react with acids to form water and salts.

Foods that are acidic can be identified by their sour taste. Vinegar and lemon juice are good examples. Vinegar is a solution of acetic acid (around 5%) in water. Lemons, limes, and other citrus fruits contain citric acid. Lactic acid gives yogurt its tart taste, and phosphoric acid is often added to carbonated drinks to impart tartness. The bitter taste of tonic water, on the other hand, is due to the presence of quinine, which is a base.

Although all acids have a sour taste and all bases are bitter, a taste test is hardly the best general purpose test for determining whether a substance is an acid or a base. Some acids and bases are highly poisonous, and nearly all of them are quite corrosive unless they have been greatly diluted. Some common acids and bases are pictured in Figure 7.1.

The most common way to identify a substance as an acid or a base is the litmus test. If you dip a strip of neutral (violet-colored) litmus paper into an

Figure 7.2 Some properties of acids and bases. The paper strip is impregnated with methyl red, an acid–base indicator. Lemon juice is acidic, as demonstrated by the red color of the indicator. The soap is basic, as shown by the fact that the red dye is changed to yellow. [*Carey B. Van Loon.*]

unknown solution and it turns pink, the solution is acidic. If it turns blue, the solution is basic. If the strip does not turn pink or blue, the solution is neither acidic or basic.

Litmus is only one of several hundred known compounds that are acid–base **indicators** (Figure 7.2). Many natural food colors, such as those in grape juice, red cabbage, and blueberries, are acid–base indicators. So are the colors in most flower petals.

SECTION 7.2

Acids Explained: Hydronium Ions

We know that acids have certain characteristic properties, but *why* do they have these properties? Experimental evidence indicates that what acids have in common are *hydrogen ions* (H^+). (Since these are hydrogen atoms from which the electrons have been removed, they are also called *protons*.)

Table 7.1 lists some common acids. Notice that each formula contains one or more hydrogen atoms. In water solution the hydrogen ions attach themselves to molecules of water, thus forming **hydronium ions** (H_3O^+).

$$\underset{\text{Water}}{\mathrm{H}:\ddot{\underset{\cdot\cdot}{\mathrm{O}}}:\atop\mathrm{H}} + \mathrm{H}^+ \longrightarrow \underset{\text{Hydronium ion}}{\left[\mathrm{H}:\ddot{\underset{\cdot\cdot}{\mathrm{O}}}:\mathrm{H}\atop\mathrm{H}\right]^+}$$

Table 7.1 Some Familiar Acids

Name	Formula	Classification
Sulfuric acid	H_2SO_4	Strong
Nitric acid	HNO_3	Strong
Hydrochloric acid	HCl	Strong
Phosphoric acid	H_3PO_4	Moderate
Hydrogen sulfate ion	HSO_4^-	Moderate
Lactic acid	$CH_3CHOHCOOH$	Weak
Acetic acid	CH_3COOH	Weak
Carbonic acid	H_2CO_3	Weak
Boric acid	H_3BO_3	Very Weak
Hydrocyanic acid	HCN	Very Weak

Acids are *proton* (H^+) *donors.*

The hydrogen ions, or protons, have been donated by the acid molecules to the water molecules, so the acids are acting as **proton donors**. In the case of HCl, hydrogen chloride gas dissolves in water to form hydrochloric acid.

$$\text{H}:\ddot{\underset{\cdot\cdot}{\text{Cl}}}: + \text{H}:\ddot{\underset{\cdot\cdot}{\text{O}}}:\text{H} \longrightarrow \left[\text{H}:\ddot{\underset{\cdot\cdot}{\text{O}}}:\text{H}_2\right]^+ + :\ddot{\underset{\cdot\cdot}{\text{Cl}}}:^-$$

Acid (proton donor) — Hydrochloric acid

Notice that the HCl molecule donates a proton to the water molecule producing a hydronium ion. Other acids react in a similar way, donating hydrogen ions to water to produce hydronium ions. Even when the solvent is something other than water, the acid acts as a proton donor, transferring H^+ ions to the solvent molecules.

Nonmetal Oxides: Acid Anhydrides

Many acids are made by the reaction of nonmetal oxides with water. For example, sulfur trioxide reacts with water to form sulfuric acid.

$$SO_3 + H_2O \longrightarrow H_2SO_4$$

Similarly, carbon dioxide reacts with water to form carbonic acid.

$$CO_2 + H_2O \longrightarrow H_2CO_3$$

This is a general reaction of nonmetal oxides.

$$\text{Nonmetal oxide} + H_2O \longrightarrow \text{Acid}$$

Nonmetal oxides are called *acid anhydrides.* **Anhydride** means ''without water.'' These reactions explain why rainwater is acidic, particularly that which forms in air polluted with sulfur oxides (Chapter 12).

EXAMPLE 7.1

Give the formula for the acid formed when sulfur dioxide reacts with water.

SOLUTION

Simply write the equation for the reaction.

$$SO_2 + H_2O \longrightarrow H_2SO_3$$

Exercise 7.1

Give the formula for the acid formed when dinitrogen pentoxide (N_2O_5) reacts with water. [*Hint: Two* molecules of acid are formed.]

SECTION 7.3

Bases Explained: Hydroxide Ions

Experimental evidence indicates that the properties of bases in water are due to OH^-, the **hydroxide ion**. Table 7.2 lists some common bases. Most of these are ionic compounds containing positive metal ions, such as Na^+ or Ca^{2+}, and negative hydroxide ions (OH^-). When the compounds dissolve in water, they all provide OH^- ions, and thus they are all bases.

Ammonia seems out of place in Table 7.2, because it contains no hydroxide ions. How can it be a base? The only way to get hydroxide ions from ammonia

Table 7.2 Common Bases

Name	Formula	Classification
Sodium hydroxide	$NaOH$	Strong
Potassium hydroxide	KOH	Strong
Lithium hydroxide	$LiOH$	Strong
Calcium hydroxide	$Ca(OH)_2$	(see text)
Magnesium hydroxide	$Mg(OH)_2$	(see text)
Ammonia	NH_3	Weak

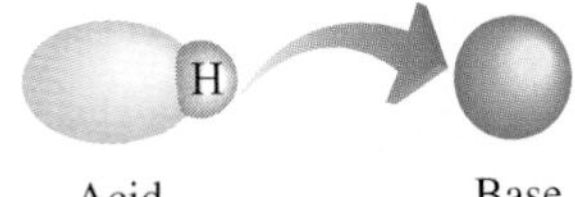

Figure 7.3 An acid is a proton donor. A base is a proton acceptor.

in water is for the ammonia molecule to accept a proton from water, leaving a hydroxide ion.

$$NH_3 + H_2O \longrightarrow NH_4^+ + OH^-$$

The NH_4^+ is the ammonium ion.

How can ammonia accept a proton from water? Recall that the nitrogen atom has an unshared pair of electrons. This pair can be used to attach a proton. Removal of a proton from water leaves a hydroxide ion, which is negatively charged because it still has the electrons that bound the ''lost'' hydrogen.

```
    H                    ..          [    H    ]+   [  ..    ]-
    ..                               [    ..   ]    [        ]
H : N :             (H) : O :   ---> [H : N : H] +  [: O : H ]
    ..                    ..         [    ..   ]    [  ..    ]
    H                     H          [    H    ]    [        ]

     Base                Acid          Ammonium     Hydroxide
(proton acceptor)   (proton donor)       ion           ion
```

In water the properties of bases are those of hydroxide ions, just as the properties of acids are those of hydronium ions. When acids react with bases, then, it is really the reaction of hydronium ions with hydroxide ions. In fact, the reaction is simply a proton transfer from H_3O^+ to OH^-.

$$\underset{\text{Proton donor}}{H_3O^+} + \underset{\text{Proton acceptor}}{OH^-} \longrightarrow 2\,H_2O$$

Bases are *proton* (H^+) *acceptors.*

We can now generalize the definition of bases. Just as an acid is a proton donor, a *base* is a **proton acceptor** (Figure 7.3). This definition includes not only hydroxide ions but also neutral molecules such as ammonia. It also includes other negative ions such as carbonate (CO_3^{2-}) or bicarbonate (HCO_3^-).

$$\underset{\text{Proton donor}}{H_3O^+} + \underset{\text{Proton acceptor}}{CO_3^{2-}} \longrightarrow H_2O + HCO_3^-$$

$$\underset{\text{Proton donor}}{H_3O^+} + \underset{\text{Proton acceptor}}{HCO_3^-} \longrightarrow H_2O + H_2CO_3$$

The idea of an acid as a proton donor and a base as a proton acceptor greatly expands our concept of acids and bases.

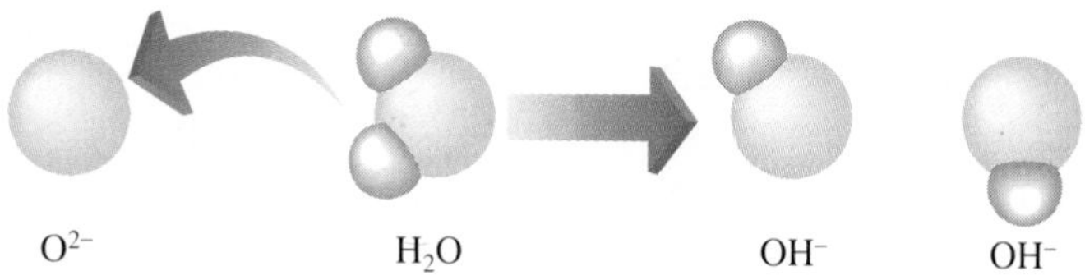

Figure 7.4 Metal oxides are basic because the oxide ion reacts with water to form two hydroxide ions.

Metal Oxides: Basic Anhydrides

Just as acids can be made from nonmetal oxides, many common bases can be made from metal oxides. For example, calcium oxide (lime) reacts with water to form calcium hydroxide (slaked lime).

$$CaO + H_2O \longrightarrow Ca(OH)_2$$

Another example is the reaction of lithium oxide with water to form lithium hydroxide.

$$Li_2O + H_2O \longrightarrow 2\,LiOH$$

In general, metal oxides react with water to form bases (Figure 7.4). These metal oxides are called *basic anhydrides.*

$$\text{Metal oxide} + H_2O \longrightarrow \text{Base}$$

▶ EXAMPLE 7.2

What base is formed by the addition of water to barium oxide (BaO)?

SOLUTION

Simply write the equation for the reaction.

$$BaO + H_2O \longrightarrow Ba(OH)_2$$

Exercise 7.2

What base is formed by the addition of water to potassium oxide (K_2O)? [*Hint:* Two moles of base are formed for each mole of potassium oxide.]

SECTION 7.4

Strong and Weak Acids and Bases

When gaseous hydrogen chloride (HCl) reacts with water, it reacts completely to give hydronium ions and chloride ions. Essentially no HCl molecules remain.

$$HCl + H_2O \longrightarrow H_3O^+ + Cl^-$$

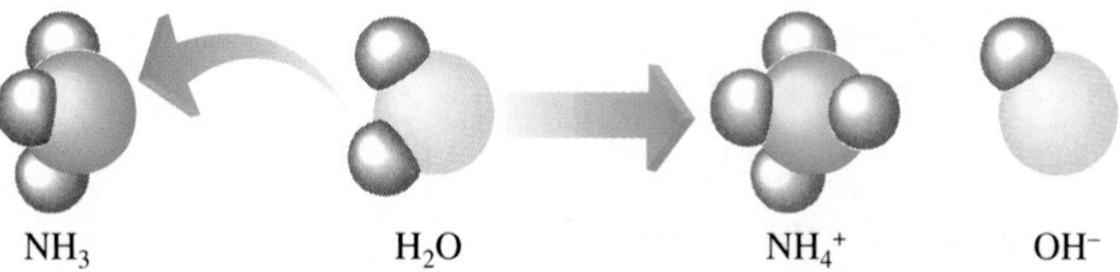

Figure 7.5 Ammonia is a base because it accepts a proton from water. A solution of ammonia in water contains ammonium ions and hydroxide ions. Only a small fraction of the ammonia molecules react, however; most remain unchanged. Ammonia is therefore a weak base.

The poisonous gas hydrogen cyanide (HCN) also reacts with water to produce hydronium ions and cyanide ions.

$$HCN + H_2O \longrightarrow H_3O^+ + CN^-$$

A strong acid (or base) is one that exists almost completely as ions in solution. The word *strong* does not refer to the amount of acid (or base) in the solution. A solution that contains a relatively large amount of acid or base in a given volume of solution is called a *concentrated* solution. (The acid or base may be strong or weak.) A solution with only a little solute in that same volume of solution is said to be *dilute.*

But in the latter case the reaction takes place only to a slight extent. Less than 1% of the HCN molecules react to produce hydronium ions. Most of them remain intact as HCN molecules. Acids like HCl that react completely with water are called **strong acids**. Those that react only slightly with water are **weak acids**. There are not many strong acids. The first three acids listed in Table 7.1 (sulfuric, nitric, and hydrochloric) are the only common ones. Most acids are weak.

Bases are also classified as strong or weak, depending on whether they react completely or only slightly with water to form hydroxide ions. Perhaps the most familiar **strong base** is sodium hydroxide (NaOH). It exists as sodium ions and hydroxide ions even in the solid state. Other strong bases include potassium hydroxide (KOH) and the hydroxides of all the other Group IA metals. Group IIA hydroxides are also strong bases [except for $Be(OH)_2$], but $Ca(OH)_2$ is only slightly soluble in water and $Mg(OH)_2$ is nearly insoluble, so these are not very practical strong bases. The most familiar **weak base** is ammonia (NH_3). It reacts with water to a slight extent to produce ammonium ions (NH_4^+) and hydroxide ions (Figure 7.5).

$$NH_3 + H_2O \longrightarrow NH_4^+ + OH^-$$

Neutralization

When an acid reacts with a base, the products are water and a **salt**. If a solution containing hydronium ions (an acid) is mixed with another solution containing exactly the same amount of hydroxide ions (a base), the resulting solution no longer affects litmus, and it no longer tastes sour or bitter (it tastes salty). It is no longer either acidic or basic; it is neutral. The reaction of an acid with a base is called **neutralization** (Figure 7.6). In water it is simply the reaction of hydronium ions with hydroxide ions to form water molecules.

(a) (b) (c)

Figure 7.6 The amount of acid (or base) in a solution is determined by careful neutralization. In the experiment shown here, a 5.00-mL sample of vinegar, a small amount of water, and a few drops of phenolphthalein (an acid–base indicator) are added to a flask (a). A solution of 0.1000 M NaOH is added slowly from a buret (a device for precise measurement of volumes of solutions) (b). As long as the acid is in excess, the solution is colorless. When the acid has been neutralized and a tiny excess of base is present, the phenolphthalein indicator turns pink (c). [*Carey B. Van Loon.*]

$$H_3O^+ + OH^- \longrightarrow 2\,H_2O$$

If sodium hydroxide is being neutralized by hydrochloric acid, the products are water and sodium chloride (ordinary table salt).

$$\underset{\text{Base}}{NaOH} + \underset{\text{Acid}}{HCl} \longrightarrow \underset{\text{Water}}{H_2O} + \underset{\text{Salt}}{NaCl}$$

EXAMPLE 7.3

Write the equation for the neutralization reaction between potassium hydroxide and nitric acid.

SOLUTION

$$KOH + HNO_3 \longrightarrow KNO_3 + H_2O$$

Exercise 7.3

Write the equation for the neutralization reaction between calcium hydroxide and hydrochloric acid.

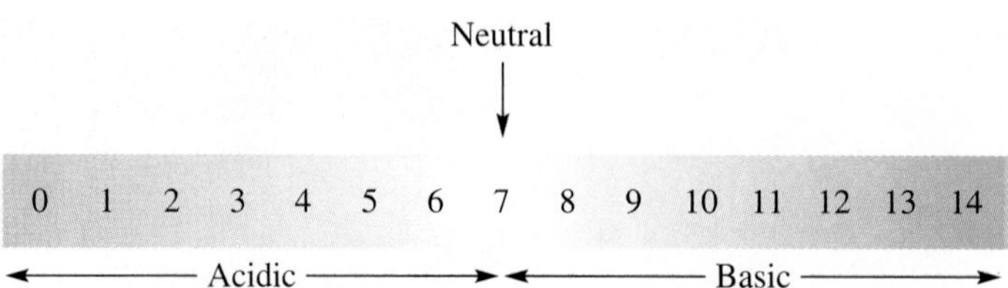

Figure 7.7 The pH scale. A change in pH of one unit means a tenfold change in the hydronium ion concentration.

SECTION 7.5

The pH Scale

Because they are so very tiny, atoms, molecules, and ions are usually counted in *moles* (see Section 6.8). In solutions the concentrations of ions are measured in moles per liter. A solution of 1 molar hydrochloric acid (1 M HCl), for example, contains 1 mole of H_3O^+ ions per liter of solution. In other words, 1 L of 1 M HCl contains 6×10^{23} hydronium ions; 0.5 L of 0.001 M HCl would contain 3×10^{20} hydronium ions.

In describing the acidity of a particular solution one might say: "The hydronium ion concentration of this solution is 1×10^{-3} mole per liter." But rarely will you hear anyone make such a statement. Instead the solution will simply be identified as "pH 3." When people talk about degree of acidity or basicity, they usually do it in terms of pH.

The pH scale is a convenient acidity scale proposed in 1909 by the Danish biochemist S. P. L. Sorensen. It mainly extends from 0 to 14. The neutral point on the scale is 7, with values below 7 becoming increasingly acidic and those above 7 increasingly basic. Thus pH 6 is only very slightly acidic, whereas pH 12 is strongly basic (Figure 7.7).

The numbers on the scale were not chosen arbitrarily. They were directly related to the concentration of hydronium ions. We might expect that pure water would be completely in the form of H_2O molecules; but it turns out that one out of every 500 million molecules is split into ions. That causes the concentration of hydronium ions in pure water to be 0.0000001 mole per liter, or 1×10^{-7} M. Can you see why 7 is the pH of pure water? It is simply the power of 10 for the molar concentration of H_3O^+, with the negative sign removed. (The H in pH stands for *hydrogen* and the p for *power*.) We can define **pH** as the negative logarithm of the molar concentration of hydronium ion (Table 7.3).

Table 7.3 Relationship Between pH and Concentration of Hydronium Ions

Concentration of H_3O^+ (mol/L)	pH
1×10^{0}	0
1×10^{-1}	1
1×10^{-2}	2
1×10^{-3}	3
1×10^{-4}	4
1×10^{-5}	5
1×10^{-6}	6
1×10^{-7}	7
1×10^{-8}	8
1×10^{-9}	9
1×10^{-10}	10
1×10^{-11}	11
1×10^{-12}	12
1×10^{-13}	13
1×10^{-14}	14

Notice that although pH is an acidity scale, the pH goes down when acidity goes up. Not only is the relationship an inverse one, but it is also logarithmic. A decrease of one pH unit represents a tenfold increase in acidity, and when pH goes down by two units, acidity increases by a factor of 100. The relationship may seem strange at first, but once you understand the pH scale, you appreciate its convenience.

Table 7.3 summarizes the relationship between hydronium ion concentration and pH. A pH of 4 means a hydronium ion concentration of 1 ×

Table 7.4 The Approximate pH of Some Common Solutions

Solution	pH
Hydrochloric acid (4%)	0
Gastric juice	1.6–1.8
Lemon juice	2.1
Vinegar (4%)	2.5
Soft drinks	2.0–4.0
Rainwater (thunderstorm)	3.5–4.2
Milk	6.3–6.6
Urine	5.5–7.0
Rainwater (unpolluted)	5.6
Saliva	6.2–7.4
Pure water	7.0
Blood	7.4
Fresh egg white	7.6–8.0
Bile	7.8–8.6
Milk of magnesia	10.5
Washing soda	12.0
Sodium hydroxide (4%)	13.0

10^{-4} mol/L, or 0.0001 M. If the concentration of hydronium ion is 0.01 M, or 1×10^{-2} M, then the pH is 2. The pH values for various common solutions are listed in Table 7.4.

The pH of a solution is defined as the negative logarithm of the hydronium ion concentration,

$$\text{pH} = -\log [H_3O^+]$$

where the brackets indicate molar concentration. Perhaps the relationship is easier to see when the equation is written:

$$[H_3O^+] = 10^{-\text{pH}}$$

EXAMPLE 7.4

What is the pH of a solution that has a hydronium ion concentration of 1.0×10^{-5} M?

SOLUTION

The exponent is -5; the pH is therefore 5.

Exercise 7.4

What is the pH of a solution that has a hydronium ion concentration of 1.0×10^{-11} M?

EXAMPLE 7.5

What is the hydronium ion concentration in a solution that has a pH of 4?

SOLUTION

The pH value is the negative exponent of 10, so the hydronium ion concentration is 1.0×10^{-4} M.

Exercise 7.5
What is the hydronium ion concentration of a solution that has a pH of 2?

Normally pH is used to describe the acidity of a solution (rather than pOH), whether the solution is acidic or basic.

In the case of basic solutions, pOH is related to $[OH^-]$ just as pH is related to $[H_3O^+]$. In 0.01 M NaOH solution the $[OH^-] = 0.01 = 1 \times 10^{-2}$, and therefore the pOH = 2. To convert pOH to pH, simply subtract the pOH from 14.

$$pH + pOH = 14$$

$$pH = 14 - pOH$$

If the pOH = 2, then the pH = 14 − 2 = 12.

EXAMPLE 7.6

What is the pH of a KOH solution that has a hydroxide ion concentration of 1×10^{-4}?

SOLUTION
The exponent is −4; therefore, the pOH is 4. Since pH = 14 − pOH, the pH = 14 − 4 = 10.

Exercise 7.6
What is the pH of a solution of ammonia that has a hydroxide ion concentration of 1×10^{-3}?

SECTION 7.6

Antacids: A Basic Remedy

Drugs such as Zantac and Tagamet are used to treat people with ulcers. These drugs inhibit the release of hydrochloric acid by the stomach.

The stomach secretes hydrochloric acid (HCl) as an aid in the digestion of food. Sometimes overindulgence or emotional stress leads to a condition of **hyperacidity** (too much acid). There are thousands of brands of **antacids** (Figure 7.8) sold in the United States to treat this condition. Despite the many brand names, there are only a few different antacid ingredients, primarily sodium bicarbonate, calcium carbonate, aluminum hydroxide, and magnesium hydroxide (Table 7.5.)

All antacids are basic compounds. They act by neutralizing hydronium ions in stomach acid.

Sodium bicarbonate ($NaHCO_3$), commonly called baking soda, is probably safe and effective for most people, but overuse can make the blood too alkaline, a condition called **alkalosis**.

Calcium carbonate ($CaCO_3$) is safe in small amounts, but regular use can cause constipation. It also appears that calcium carbonate can actually cause *increased* acid secretion after a few hours.

pH

For coffee it's 5; for tomatoes it's 4;
While household ammonia's 11 or more.
It's 7 for water, if in a pure state,
But rainwater's 6, and seawater is 8.
It's basic at 10, quite acidic at 2,
And well above 7 when litmus turns blue.
Some find it a puzzlement. Doubtless their fog
Has something to do with that negative log!

Table 7.5 Some Common Antacids

Commercial Product	Antacid Ingredient(s)
Alka-Seltzer	$NaHCO_3$, citric acid, aspirin
Amphojel	$Al(OH)_3$
Baking soda	$NaHCO_3$
DiGel	$CaCO_3$
Maalox	$Al(OH)_3$, $Mg(OH)_2$
Milk of magnesia	$Mg(OH)_2$
Mylanta	$Al(OH)_3$, $Mg(OH)_2$
Rolaids	$AlNa(OH)_2CO_3$
Rolaids Sodium-Free*	$CaCO_3$, $Mg(OH)_2$
Tums	$CaCO_3$

*Sodium-containing antacids are not recommended for people with hypertension (high blood pressure).

You can make your own aspirin-free "Alka-Seltzer." Simply place half a teaspoon of baking soda in a glass of orange juice. (What is the acid and what is the base in this reaction?)

Figure 7.8 A great variety of antacids are available to consumers. All antacids are basic compounds. [*Robert Mathena/Fundamental Photographs.*]

Claims of "fast action" are almost meaningless. All acid–base reactions are almost instantaneous. Some tablets may dissolve a little slower than others. You can speed their action by chewing them.

Aluminum hydroxide [$Al(OH)_3$], like calcium carbonate, can cause constipation. There is also some concern that antacids containing aluminum ions can deplete the body of essential phosphate ions.

A suspension of magnesium hydroxide [$Mg(OH)_2$] in water is sold as "milk of magnesia." Magnesium carbonate ($MgCO_3$) is also used as an antacid. In small doses magnesium compounds act as antacids, but in large doses they act as laxatives.

Although antacids are generally safe for occasional use, they can interact with other medications; and anyone who has severe or repeated attacks of indigestion should consult a physician. Self-medication can sometimes be dangerous.

SECTION 7.7

Acids, Bases, and Human Health

Concentrated strong acids and bases are corrosive poisons that can cause serious chemical burns. Once the chemical agents are removed, the injuries are similar to burns from heat, and they are often treated the same way.

Sulfuric acid (H_2SO_4) is by far the leading chemical product in the United States. About 40 billion kg is produced each year, most of it for making other industrial chemicals, especially fertilizers. Around the home we use sulfuric acid in automobile batteries and in some drain cleaners. Besides being a strong acid, sulfuric acid is also a powerful dehydrating agent that can react with water in the cells. Acid rain mainly results from sulfuric acid mists produced by the burning of high sulfur coal.

Hydrochloric acid (also called muriatic acid) is used in industry to remove rust from metal, in construction to remove excess mortar from bricks, and in the home to remove lime deposits from toilet bowls. Concentrated solutions (about 38% HCl) cause severe burns, but dilute solutions can be used safely in the home if handled carefully. The gastric juice in your stomach is a solution containing around 0.5% hydrochloric acid.

Lime (CaO) is the cheapest and most widely used commercial base. It is made by heating limestone ($CaCO_3$) to drive off CO_2. It is the fifth most produced industrial chemical in the United States with an annual production of about 17 billion kg. It is used to make mortar and cement and also to "sweeten" acidic soil (Figure 7.9).

Sodium hydroxide (commonly known as lye) is the strong base most often used in the home. It is used as an oven cleaner in products such as Easy Off, and it is used to open clogged drains with products such as Drano.

Both acids and bases, even in dilute solutions, break down the protein molecules in living cells. Generally, the fragments are not able to carry out the functions of the original proteins. In cases of severe exposure, the fragmentation continues until the tissue has been completely destroyed. And, within

Figure 7.9 Treating the soil with lime makes it "sweeter" (less acid). [*Michael P. Gadomski/Photo Researchers, Inc.*]

living cells, proteins function properly only at an optimum pH. If the pH changes much in either direction, the proteins can't carry out their usual functions.

When they are misused, acids and bases can be damaging to human health. But acids and bases affect human health in more subtle—and ultimately more important—ways. A delicate balance must be maintained between acids and bases in the blood and body fluids. If the acidity of the blood changes very much, the blood loses its capacity to carry oxygen. Fortunately, the body has a complex but efficient mechanism for maintaining proper acid–base balance. (Consult reference 3 for an explanation of this mechanism.)

Summary

1. *Acids* are *proton donors.* They taste sour, turn litmus red, react with active metals to form hydrogen, and react with bases to form salts and water.
2. *Bases* are *proton acceptors.* They taste bitter, turn litmus blue, feel slippery to the skin, and react with acids to form salts and water.
3. When an acid dissolves in water, water molecules pick up hydrogen ions (H^+), or protons, from the acid molecules to produce *hydronium ions* (H_3O^+).
4. When a base dissolves in water, *hydroxide ions* (OH^-) are formed.
5. Nonmetal oxides (acid *anhydrides*) react with water to form acids; metal oxides (basic *anhydrides*) react with water to form bases.

6. A *strong acid* or *strong base* when dissolved in water is almost completely in the form of ions.
7. A *weak acid* or *weak base* reacts only slightly with water to produce ions.
9. Sulfuric, hydrochloric, and nitric acids are strong. Acetic and most other acids are weak.
9. Group IA hydroxides are strong bases, and so are most of the Group IIA hydroxides. Ammonia is a weak base.
10. When an acid reacts with a base, the products are a *salt* and water, and the process is called *neutralization*.
11. The *pH* scale is an acidity scale. A pH of 7 is neutral; pH values lower than 7 are increasingly acidic; and pH values greater than 7 are increasingly basic.
12. *Antacids* are products used to neutralize excess stomach acid; but taking too much antacid can produce a condition of *alkalosis*.

Key Terms

acid 7.1	hydronium ion 7.2	pH 7.5	strong base 7.4
alkalosis 7.6	hydroxide ion 7.3	proton acceptor 7.3	weak acid 7.4
anhydride 7.2	hyperacidity 7.6	proton donor 7.2	weak base 7.4
antacid 7.6	indicator 7.1	salt 7.4	
base 7.1	neutralization 7.4	strong acid 7.4	

Review Questions

1. Define and illustrate the following terms.
   a. acid b. base c. salt
2. List four general properties of acidic solutions.
3. List four general properties of basic solutions.
4. What ion is responsible for the properties of acidic solutions (in water)?
5. What ion is responsible for the properties of basic solutions (in water)?
6. Give the formulas and the names of two strong acids and two weak acids.
7. Give the formulas and the names for two strong bases and one weak base.
8. Strong acids and weak acids both have properties characteristic of hydronium ions. How do strong and weak acids differ?
9. Give a general definition for an acid. Write an equation that illustrates your definition.
10. What is meant by the proton as used in acid–base chemistry? How does it differ from the proton of nuclear chemistry (Chapter 4)?
11. What is an acid anhydride?
12. What is a basic anhydride?
13. How does one brand of household ammonia differ from another?
14. What is meant by the neutralization of an acid or base?
15. Describe the taste and effect on litmus of a solution that has been neutralized.
16. Magnesium hydroxide is completely ionic, even in the solid state; yet it can be taken internally as an antacid. Explain why it does not cause injury as sodium hydroxide would.
17. Why do people take antacids?
18. Name some of the active ingredients in antacids.
19. What is alkalosis?
20. Indicate whether each of the following pH values represents an acidic, basic, or a neutral solution.
    a. 4 b. 7 c. 3.5 d. 9
21. Lime juice has a pH of about 2. Is lime juice acidic or basic?

22. What is aqueous ammonia? Why is it sometimes called ammonium hydroxide?

23. What is the leading chemical product of United States industry?

24. What is the effect of strong acids on clothing?

25. What is the effect of strong acids and strong bases on the skin?

26. How would you define pOH?

Problems

Acids and Bases

27. Use the definitions to identify the first compound in each equation as an acid or base. [*Hint:* What is *produced* by the reaction?]
   a. $C_5H_5N + H_2O \rightarrow C_5H_5NH^+ + OH^-$
   b. $C_6H_5OH + H_2O \rightarrow C_6H_5O^- + H_3O^+$
   c. $CH_3COCOOH + H_2O \rightarrow CH_3COCOO^- + H_3O^+$

28. Use the definitions to identify the first compound in each equation as an acid or base.
   a. $C_6H_5SH + H_2O \rightarrow C_6H_5S^- + H_3O^+$
   b. $CH_3NH_2 + H_2O \rightarrow CH_3NH_3^+ + OH^-$
   c. $C_6H_5SO_2NH_2 + H_2O \rightarrow C_6H_5SO_2NH^- + H_3O^+$

29. Give formulas for the following acids.
   a. hydrochloric acid
   b. sulfuric acid
   c. carbonic acid
   d. hydrocyanic acid

30. Give formulas for the following acids.
   a. nitric acid
   b. sulfurous acid
   c. phosphoric acid
   d. hydrosulfuric acid

31. Give formulas for the following bases.
   a. lithium hydroxide
   b. magnesium hydroxide
   c. sodium hydroxide

32. Give formulas for the following bases.
   a. calcium hydroxide
   b. potassium hydroxide
   c. ammonia

33. Write the equation that shows hydrogen chloride gas reacting with water to form ions. What is the name of the acid formed?

34. Write the equation that shows how ammonia acts as a base in water.

Acidic and Basic Anhydrides

35. Give the formula for the compound formed when sulfur trioxide reacts with water. Is the product an acid or a base?

36. Give the formula for the compound formed when magnesium oxide reacts with water. Is the product an acid or a base?

37. Give the formula for the compound formed when potassium oxide reacts with water. Is the product an acid or a base?

38. Give the formula for the compound formed when carbon dioxide reacts with water. Is the product an acid or a base?

Strong and Weak Acids and Bases

39. Cesium hydroxide (CsOH) is ionic in the solid state and is quite soluble in water. Classify the compound as a strong acid, weak acid, weak base, or strong base.

40. Hydrogen iodide (HI) gas reacts completely with water to form hydronium ions and iodide ions. Classify the compound as a strong acid, weak acid, weak base, or strong base.

41. Hydrogen sulfide (H_2S) gas reacts slightly with water to form relatively few hydronium ions and hydrogen sulfide ions (HS^-). Classify the compound as a strong acid, weak acid, weak base, or strong base.

42. Methylamine (CH_3NH_2) gas reacts slightly with water to form relatively few hydroxide ions and methylammonium ions ($CH_3NH_3^+$). Classify the compound as a strong acid, weak acid, weak base, or strong base.

Neutralization

43. Write the equation for the reaction of sodium hydroxide with hydrochloric acid.

44. Write the equation for the reaction of lithium hydroxide with nitric acid.

45. Write the equation for the reaction of 1 mol calcium hydroxide with 2 mol hydrochloric acid.

46. Write the equation for the reaction of 1 mol sulfuric acid with 2 mol potassium hydroxide.

47. Write the equation for the reaction of 1 mol phosphoric acid with 3 mol sodium hydroxide.

48. Write the equation for the reaction of 1 mol sulfuric acid with 1 mol calcium hydroxide.

The pH Scale

49. What is the pH of a solution that has a hydronium ion concentration of 1.0×10^{-3} M?

50. What is the pH of a solution that has a hydronium ion concentration of 1.0×10^{-10} M?

51. What is the pH of a solution of 0.001 M KOH that has a pOH of 3?

52. What is the pH of a solution of 0.01 M NaOH with a hydroxide ion concentration of 1×10^{-2} M?

53. What is the hydronium ion concentration of a solution that has a pH of 5?

54. What is the hydronium ion concentration of a solution that has a pH of 11?

55. What is the hydroxide ion concentration of a solution that has a pOH of 6?

56. What is the hydroxide ion concentration of a solution that has a pH of 13?

Additional Problems

57. How much sulfur trioxide can be made from 40 g of sulfur? The equation (not balanced) is

$$S + O_2 \rightarrow SO_3$$

58. How much magnesium hydroxide can be made from 282 g of magnesium oxide?

59. Lime has been used to neutralize water in lakes that have become too acidic from acid precipitation. Assume that the acid is sulfuric acid and write the equation for the reaction. How many metric tons of lime does it take to neutralize 10 t of sulfuric acid?

Projects

60. Examine the labels of at least five antacid preparations. Make a list of the ingredients in each. Look up the properties (medical use, side effects, toxicity, etc.) of each ingredient in a reference book such as *The Merck Index* (Reference 1).

61. Examine the labels of at least five toilet-bowl cleaners and five drain cleaners. Make a list of the ingredients in each. Look up the formulas and properties of each ingredient in a reference book such as *The Merck Index* (Reference 1). Which ingredients are acids? Which are bases?

References and Readings

1. Budavari, Susan (Ed.). *The Merck Index,* 11th ed. Rahway, NJ: Merck and Co., 1989.
2. Ebbing, Darrell D. *General Chemistry,* 4th ed. Boston: Houghton Mifflin, 1993. Chapter 16 discusses acid–base chemistry.
3. Hill, John W., Dorothy M. Feigl, and Stuart J. Baum. *Chemistry and Life,* 4th ed. New York: Macmillan, 1993. See especially Chapters 10 and 11.
4. "How to Choose an Antacid." *Consumer Reports,* August 1983, pp. 412–418.

5. Jensen, William B. ''Acids and Bases: Ancient Concepts in Modern Science.'' *ChemMatters,* April 1983, pp. 14–15. Part of a special issue on acid–base chemistry.

6. Kohn, Harold W. ''The Rolaids Caper or Titration in the Classroom.'' *Chemistry,* March 1973, pp. 27–28.

7. Kolb, Doris, ''Chemical Principles Revisited: Acids and Bases.'' *Journal of Chemical Education,* July 1978, pp. 459–464.

8. Kolb, Doris. ''Chemical Principles Revisited: The pH Concept.'' *Journal of Chemical Education,* January 1979, pp. 49–53.

9. Kotz, John C., and Keith F. Purcell. *Chemistry & Chemical Reactivity,* 2nd ed. Philadelphia: Saunders, 1991. Chapters 17 and 18 deal with acids and bases.

10. Lowenstein, Jerome. *Acids and Basics.* New York: Oxford University Press, 1993. A discussion of acid–base physiology.

11. Morris, D. L. ''Brønsted–Lowry Acid–Base Theory—A Brief Survey.'' *Chemistry,* March 1970, pp. 18–19.

12. Zumdahl, Steven S. *Chemistry,* 3rd ed. Lexington, MA: D. C. Heath, 1993. Chapter 14 considers acids and bases.

8 Oxidation and Reduction

Burn and Unburn

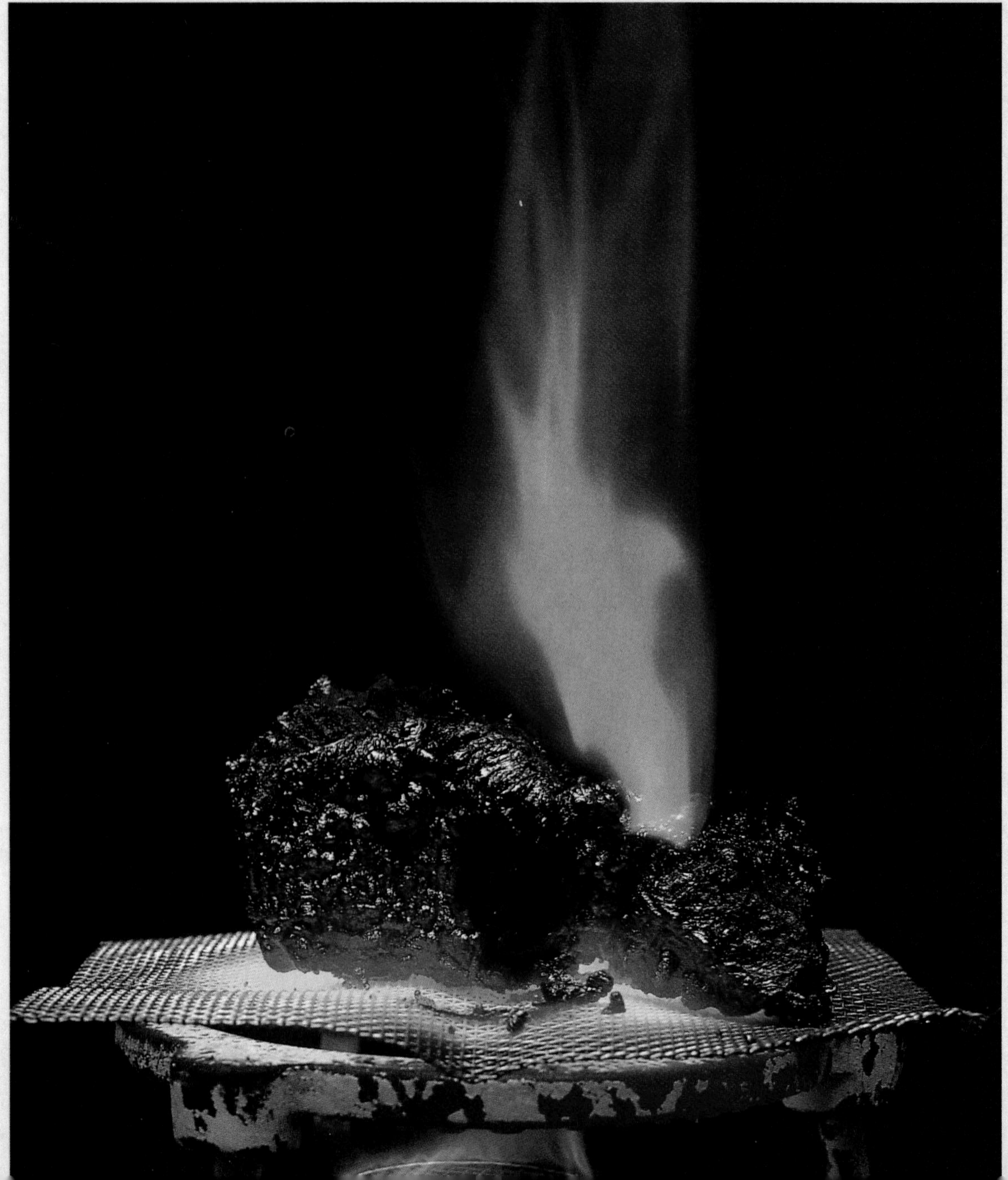

The burning of a lump of coal . . .

Your body's respiration . . .

The rusting of an iron pole . . .

They all are OXIDATION.

When coal burns, releasing its chemical energy as heat, the carbon in the coal is oxidized to carbon dioxide as oxygen from the air is reduced to water. [*Craig Hammell/The Stock Market.*]

Chemical reactions can be classified in several ways. In this chapter, we consider an important group of reactions called *reduction–oxidation* (or *redox*) *reactions.* The two processes—oxidation and reduction—always occur together. You can't have one without the other. When one substance is oxidized, another is reduced (Figure 8.1). For convenience, however, we may choose to talk about only a part of the process—the oxidation part or the reduction part.

Our cells obtain energy to maintain themselves by oxidizing foods. Green plants, using energy from sunlight, produce food by the reduction of carbon dioxide. We win metals from their ores by reduction, then lose them again to

Figure 8.1 Oxidation and reduction always occur together. Pictured here on the left is ammonium dichromate. In the reaction (center), the ammonium ion (NH_4^+) is oxidized, and the dichromate ion ($Cr_2O_7^{2-}$) is reduced. Considerable heat and light are evolved. The equation for the reaction is

$$(NH_4)_2Cr_2O_7 \rightarrow Cr_2O_3 + N_2 + 4\,H_2O$$

The water is driven off as vapor, and the nitrogen gas escapes, leaving pure Cr_2O_3 as the visible product (right). [*Carey B. Van Loon.*]

corrosion as they are oxidized. We maintain our technological civilization by oxidizing fossil fuels (coal, natural gas, and petroleum) to obtain the chemical energy that was stored in these materials eons ago by green plants.

Reduced forms of matter—food, coal, gasoline—are high in energy (Figure 8.2). Oxidized forms—carbon dioxide and water—are low in energy. Let's examine the processes of oxidation and reduction in some detail, in order that we might better understand the chemical reactions that keep us alive and enable us to maintain our civilization.

(a)

(b)

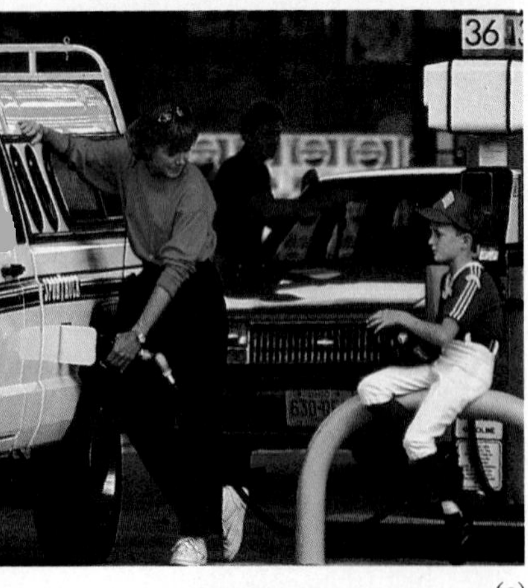

(c)

Figure 8.2 Some reduced forms of matter. The energy in food and fossil fuels is released when these materials are oxidized.
[*(a) Romilly Lockyer and (b) Derek Redfearn/The Image Bank; (c) Chris Jones/The Stock Market.*]

SECTION 8.1

Oxygen: Abundant and Essential

Oxygen is surely one of the most important elements on Earth. It is one of the 20 or so elements that are essential to life, and it is extremely abundant, making up about half of Earth's crust by weight. Oxygen occurs in all three subdivisions of the crust—the *atmosphere,* the *hydrosphere,* and the *lithosphere.* It occurs in the atmosphere (the gaseous mass surrounding Earth) as molecular oxygen (O_2). In the hydrosphere (the oceans, seas, rivers, and lakes) it occurs in combination with hydrogen in the remarkable compound water. In the lithosphere (the solid portion of Earth) it occurs in combination with silicon (sand is SiO_2) and various other elements.

Oxygen is found in most of the compounds that are important to living organisms. Foodstuffs—carbohydrates, fats, and proteins—all contain oxygen. The human body is approximately 65% water (by weight). Since water is 89% oxygen by weight, and many other compounds in your body also contain oxygen, about two-thirds of your body weight is oxygen.

The atmosphere is about 21% elemental oxygen (O_2) by volume. [The rest is mainly nitrogen (N_2), which is rather unreactive.] The free, uncombined oxygen in the air is taken into our lungs, passes into our bloodstreams, is carried to our body tissues, and combines with the food we eat. This is the process that provides us with all our energy. Fuels such as natural gas, gasoline, and coal also need oxygen to burn and release their stored energy (Figure 8.3). Combustion of such ''fossil fuels'' currently supplies about 86% of the energy that turns the wheels of civilization.

Pure oxygen is obtained by liquefying air and then letting the nitrogen and argon boil off. (Nitrogen boils at −196 °C, argon at −186 °C, and oxygen at −183 °C.) Over 20 billion kg of oxygen is produced annually in the United States, but most of it is used directly by industry, much of it by steel plants. About 1% is compressed into tanks for use in welding, hospital respirators, and other purposes (Figure 8.4).

Not everything that oxygen does is desirable. It causes iron to rust and copper to corrode. It causes food to spoil, and it aids in the decay of wood.

When iron rusts, it combines with oxygen from the air to form a reddish brown powder.

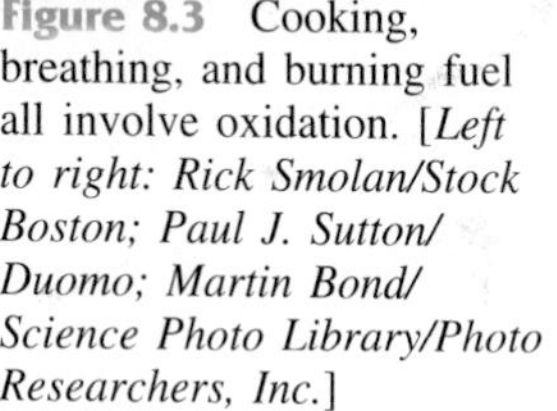
Figure 8.3 Cooking, breathing, and burning fuel all involve oxidation. [*Left to right: Rick Smolan/Stock Boston; Paul J. Sutton/Duomo; Martin Bond/Science Photo Library/Photo Researchers, Inc.*]

$$4\,Fe + 3\,O_2 \longrightarrow 2\,Fe_2O_3$$

The chemical name for rust is iron(III) oxide. (Sometimes it is called by an older name, ferric oxide.) Many metals react with oxygen to form metal oxides. Most nonmetals also react with oxygen to form oxides. For example, sulfur and carbon burn in oxygen to form carbon dioxide and sulfur dioxide, respectively.

$$C + O_2 \longrightarrow CO_2$$

$$S + O_2 \longrightarrow SO_2$$

Figure 8.4 Fuels burn more rapidly in pure oxygen than in air. Huge quantities of liquid oxygen are used to burn the fuels that blast rockets into orbit. [*NASA photo*]

EXAMPLE 8.1

Magnesium combines readily with oxygen when ignited in air. Write the equation for the reaction.

SOLUTION

Magnesium is an element with the symbol Mg. Oxygen occurs as diatomic molecules (O_2). The two react to form MgO (Section 6.2). The reaction is

$$Mg + O_2 \longrightarrow MgO \quad \textit{(not balanced)}$$

To balance the equation, we need 2 MgO and then 2 Mg.

$$2\,Mg + O_2 \longrightarrow 2\,MgO$$

Exercise 8.1

Zinc burns in air to form zinc oxide (ZnO). Write the equation for the reaction.

Oxygen also reacts with many compounds. Methane, the principal ingredient in natural gas, burns in air to produce carbon dioxide and water

$$CH_4 + 2\,O_2 \longrightarrow CO_2 + 2\,H_2O$$

Hydrogen sulfide, a gaseous compound with a rotten-egg odor, burns, producing water and sulfur dioxide.

$$2\,H_2S + 3\,O_2 \longrightarrow 2\,H_2O + 2\,SO_2$$

In each example, oxygen combines with each of the elements in the compound to form their oxides.

EXAMPLE 8.2

Carbon disulfide is highly flammable; it combines readily with oxygen, burning with a blue flame. What products are formed? Write the equation.

SOLUTION

Carbon disulfide is CS_2. The products are carbon dioxide and sulfur dioxide. The balanced equation is

$$CS_2 + 3\,O_2 \longrightarrow CO_2 + 2\,SO_2$$

Exercise 8.2

When heated in air, lead sulfide (PbS) combines with oxygen to form lead(II) oxide (PbO) and sulfur dioxide. Write a balanced equation for the reaction.

SECTION 8.2

Hydrogen: Lightweight and Reactive

Another important element is **hydrogen**. Oxygen and hydrogen are especially vital elements because they are the components that make up water. By weight, hydrogen makes up only about 0.9% of the Earth's crust. However, hydrogen is a very light element, so on the basis of number of atoms it ranks rather high in abundance. If we had a random sample of 10,000 atoms from the Earth's crust, 5330 would be oxygen, 1590 would be silicon, and 1510 would be hydrogen. Unlike oxygen, hydrogen is seldom found in a free uncombined state on Earth. Most of it is combined with oxygen in water. Some is combined with carbon in petroleum and natural gas, which are mixtures of *hydrocarbons*. Nearly all compounds derived from plants and animals contain combined hydrogen.

Hydrogen ranks low in abundance on Earth. If we look beyond our home planet, however, hydrogen becomes much more significant. The sun, for example, is made up largely of hydrogen. In fact, in the universe, hydrogen is by far the most abundant element.

Small amounts of elemental hydrogen can be made for laboratory use by reacting zinc with hydrochloric acid.

$$Zn + 2\,HCl \longrightarrow ZnCl_2 + H_2$$

Since hydrogen does not dissolve in water, it can be collected by water displacement (Figure 8.5). Commercial quantities of hydrogen are obtained as by-products of petroleum refining or by reaction of natural gas with steam. About 200 million kg of hydrogen is produced each year in the United States, at least two-thirds of it being used to make ammonia.

Hydrogen is a colorless, odorless gas and the lightest of all substances. Its density is only one-fourteenth that of air, and for this reason it was once used

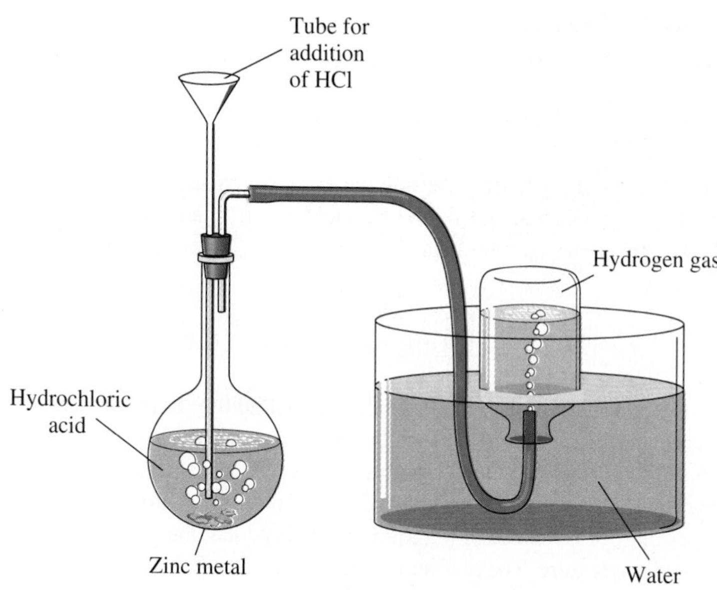

Figure 8.5 Hydrogen gas in the laboratory is prepared by the reaction of zinc with hydrochloric acid. The gas bubbles from the reaction flask and is trapped in the inverted bottle. Initially, the bottle was filled with water, but the hydrogen pushes the water out as it collects in the bottle.

Figure 8.6 Hydrogen is the most buoyant gas, but it is highly flammable. The disastrous fire in the hydrogen-filled German zeppelin *Hindenburg* led to the replacement of hydrogen by nonflammable helium, which buoys the Fuji blimp. [*The Bettmann Archive and courtesy of Fuji Photo Film USA, Inc.*]

Hot air balloons, as their name implies, are buoyed by hot air, which is less dense than the ambient air.

to fill lighter-than-air craft (Figure 8.6). Unfortunately, hydrogen can be ignited by a spark, which is what occurred in 1937 when the German airship *Hindenburg* was destroyed in a disastrous fire and explosion as it was landing in New Jersey. The use of hydrogen in airships was discontinued after that, and the dirigible industry never recovered. Today the few airships that are still in service are filled with nonflammable helium, but they are used mainly in advertising (Figure 8.6).

A stream of pure hydrogen will burn quietly in air with an almost colorless flame; but when a mixture of hydrogen and oxygen is ignited by a spark or a flame, an explosion results. The product in both cases is water.

$$2\,H_2 + O_2 \longrightarrow 2\,H_2O$$

A **catalyst** is a substance that increases the *rate* of a chemical reaction without itself being used up. For example, unsaturated fats ordinarily react very slowly with hydrogen, but in the presence of nickel metal, the reaction proceeds readily. The nickel increases the *rate* at which the reaction takes place, but it does not increase the *amount* of product formed.

Hydrogen has such strong attraction for oxygen that it can remove oxygen atoms from many metal oxides to yield the free metal. For example, when hydrogen is passed over heated copper oxide, metallic copper and water are formed.

$$CuO + H_2 \longrightarrow Cu + H_2O$$

Certain metals, such as platinum and palladium, have unusual affinity for hydrogen, being able to *absorb* large volumes of the gas. Palladium can absorb up to 900 times its own volume of hydrogen! It is interesting to note that hydrogen and oxygen can be mixed at room temperature with no perceptible reaction. But if a piece of platinum gauze is added, the gases react violently at room temperature. The platinum is acting as a *catalyst.* It lowers the *activation energy* for the reaction. The heat from the initial reaction heats up the platinum, making it glow; it then ignites the hydrogen–oxygen mixture, causing an explosion.

Platinum and palladium, as well as nickel, are often used as catalysts for reactions involving hydrogen. The metals have greatest catalytic activity when they are finely divided and have lots of active surface area. Hydrogen adsorbed on the surface of these metals is more reactive than ordinary hydrogen gas.

The *activation energy* for a chemical reaction is the minimum energy needed to get the reaction started. A catalyst acts by lowering the energy required to start a reaction.

SECTION 8.3

Oxidation and Reduction: Three Definitions

When oxygen combines with other elements or compounds, the process is called **oxidation**. The substances that combine with oxygen are said to have been *oxidized.* Originally the term *oxidation* was limited to reactions involving combination with oxygen. Then chemists came to realize that combination with chlorine (or bromine, or other active nonmetals) was not all that different from reaction with oxygen. So they broadened the definition of oxidation.

Reduction is exactly the opposite of oxidation. When hydrogen burns, the hydrogen combines with oxygen to form water.

$$2\,H_2 + O_2 \longrightarrow 2\,H_2O$$

The hydrogen is being oxidized in this reaction, but at the same time the oxygen is being *reduced.* Whenever oxidation occurs, reduction must occur also. Oxidation and reduction always happen at the same time and in exactly equivalent amounts.

Since oxidation and reduction are chemical opposites, and constant companions, it is convenient to link their definitions together. We can define oxidation and reduction at least three different ways (Figure 8.7).

1. Oxidation is a gain of oxygen atoms.
   Reduction is a loss of oxygen atoms.

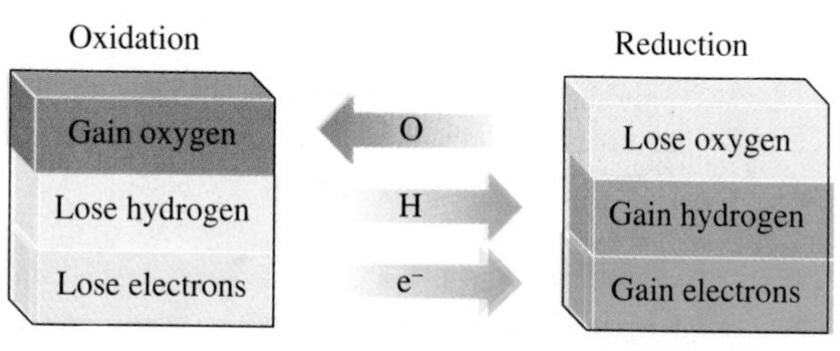

Figure 8.7 Three different definitions of oxidation and reduction.

At high temperatures (such as those in automobile engines) nitrogen, which is normally quite unreactive, combines with oxygen to form nitric oxide.

$$N_2 + O_2 \longrightarrow 2\,NO$$

Since nitrogen is gaining oxygen atoms, it is being oxidized. When the compound methane is burned to carbon dioxide and water,

$$CH_4 + 2\,O_2 \longrightarrow CO_2 + 2\,H_2O$$

both carbon and hydrogen gain oxygen atoms, so both elements are being oxidized.

When lead dioxide is heated at high temperatures, it decomposes as follows.

$$2\,PbO_2 \longrightarrow 2\,PbO + O_2$$

The lead dioxide is losing oxygen, so it is being reduced.

EXAMPLE 8.3

In each of the following, is the reactant undergoing oxidation or reduction? (These are not complete chemical equations.)

a. $Pb \longrightarrow PbO_2$

b. $SnO_2 \longrightarrow SnO$

c. $KClO_3 \longrightarrow KCl$

d. $Cu_2O \longrightarrow 2\,CuO$

SOLUTION

a. Pb *gains* oxygen atoms (it has none on the left and two on the right); it is oxidized.

b. Sn *loses* an oxygen atom (it has two on the left and only one on the right); it is reduced.

c. There are three oxygen atoms on the left and none on the right. The compound *loses* oxygen; it is reduced.

d. The two copper atoms on the left share a single oxygen atom; they have half an oxygen each. On the right, each copper atom has an oxygen atom all its own. Cu has *gained* oxygen; it is oxidized.

Exercise 8.3

In each of the following, is the reactant undergoing oxidation or reduction? (These are not complete chemical equations.)

a. $3\,Fe \longrightarrow Fe_3O_4$

b. $NO \longrightarrow NO_2$

c. $Cr_2O_3 \longrightarrow CrO_3$

d. $C_3H_6O \longrightarrow C_3H_6O_2$

2. Oxidation is a loss of hydrogen atoms.
   Reduction is a gain of hydrogen atoms.

Methyl alcohol, when passed over hot copper gauze, forms formaldehyde and hydrogen gas.

$$CH_4O \longrightarrow CH_2O + H_2$$

Since the methyl alcohol loses hydrogen, it is oxidized in this reaction.

Methyl alcohol can be made by reaction of carbon monoxide with hydrogen.

$$CO + 2\,H_2 \longrightarrow CH_4O$$

Since the carbon monoxide has gained hydrogen atoms, it has been reduced.

EXAMPLE 8.4

In each of the following, is the reactant undergoing oxidation or reduction? (These are not complete chemical equations.)

a. $C_2H_6O \longrightarrow C_2H_4O$ b. $C_2H_2 \longrightarrow C_2H_6$

SOLUTION

a. There are six hydrogen atoms in the compound on the left and only four in the one on the right. The compound *loses* hydrogen atoms; it is oxidized.
b. There are two hydrogen atoms in the compound on the left and six in the one on the right. The compound *gains* hydrogen atoms; it is reduced.

Exercise 8.4

In each of the following, is the reactant undergoing oxidation or reduction? (These are not complete chemical equations.)

a. $C_6H_6 \longrightarrow C_6H_{12}$ b. $C_3H_6O \longrightarrow C_3H_4O$

3. Oxidation is loss of electrons.
   Reduction is gain of electrons.

Just remember Leo the lion.

LEO says *GER*

Loss of
Electrons is
Oxidation.

Gain of
Electrons is
Reduction.

When magnesium metal reacts with chlorine, magnesium ions and chloride ions are formed.

$$Mg + Cl_2 \longrightarrow Mg^{2+} + 2\,Cl^-$$

Since the magnesium atom is losing electrons, it is being oxidized; and since the chlorine atoms are gaining electrons, they are being reduced.

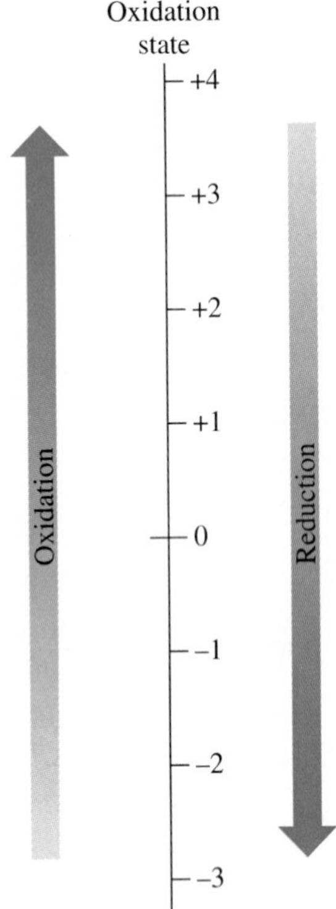

Figure 8.8 An increase in oxidation state means a loss of electrons and is therefore oxidation. A decrease in oxidation state means a gain of electrons and is therefore reduction.

It is easy enough to figure out that when Mg becomes Mg^{2+} it must be losing electrons, or that when chlorine becomes Cl^- ions it must be gaining electrons. However, many a student has become confused as to which is oxidation and which is reduction. (Is reduction a gain or a loss of electrons? Surely reduction would not be an *increase!*) The student who decides that reduction must be *loss* of electrons is doomed to get things backwards.

Perhaps Figure 8.8 can help. The charge on an ion is often referred to as its ''oxidation state.'' Increase in oxidation state is oxidation; decrease in oxidation state is reduction. For Mg the charge is zero, so conversion to Mg^{2+} is an increase in oxidation state, which is oxidation. For Cl_2 the charge is also zero, so a change to Cl^- is a decrease in oxidation state, which is reduction.

▶ EXAMPLE 8.5

In each of the following, is the reactant undergoing oxidation or reduction? (These are not complete chemical equations.)

a. $Zn \longrightarrow Zn^{2+}$ b. $Fe^{3+} \longrightarrow Fe^{2+}$

c. $S^{2-} \longrightarrow S$ d. $AgNO_3 \longrightarrow Ag$

SOLUTION

a. To form a 2+ ion, zinc *loses* two electrons. (Note the *increase* in oxidation state.) Zinc is oxidized.

b. To go from a 3+ ion to a 2+ ion, iron must *gain* an electron. (Note the *decrease* in oxidation state.) It is reduced.

c. To go from a 2− ion to an atom with no charge, sulfur *loses* two electrons. (Note the *increase* in oxidation state.) It is oxidized.

d. To answer this one, you must recognize that $AgNO_3$ is an ionic compound with Ag^+ and NO_3^- ions. In going from Ag^+ to Ag, silver *gains* an electron. (Note the *decrease* in oxidation state.) Silver is reduced.

Exercise 8.5

In each of the following, is the reactant undergoing oxidation or reduction? (These are not complete chemical equations.)

a. $Cu^{2+} \longrightarrow Cu$ b. $MnO_4^{2-} \longrightarrow MnO_4^-$

c. $Sn^{2+} \longrightarrow Sn^{4+}$ d. $Cu \longrightarrow CuSO_4$

Why do we have so many different ways to define oxidation and reduction? Simply for convenience. Which one should we use? Whichever is most convenient. For the reaction

$$C + O_2 \longrightarrow CO_2$$

it is most convenient to see that carbon is being oxidized because it is gaining oxygen atoms. Similarly, for the reaction

$$CH_2O + H_2 \longrightarrow CH_4O$$

it is convenient to see that the reactant is being reduced because it is gaining hydrogen atoms. Finally, in the case of the reaction

$$3\,Sn^{2+} + 2\,Bi^{3+} \longrightarrow 3\,Sn^{4+} + 2\,Bi$$

it is convenient to see that tin is being oxidized because it is losing electrons [or going up in oxidation state (+2 to +4)], while bismuth is being reduced because it is gaining electrons [or going down in oxidation state (+3 to 0)].

SECTION 8.4

Oxidizing and Reduction Agents

Oxidation and reduction go hand in hand. You can't have one without the other. When one substance is oxidized, another is reduced. For example, in the reaction

$$CuO + H_2 \longrightarrow Cu + H_2O$$

copper oxide is reduced and hydrogen is oxidized. Further, if one substance is being oxidized, the other must be causing it to be oxidized. In the example, CuO is causing H_2 to be oxidized. Therefore, CuO is called the **oxidizing agent**. Conversely, H_2 is causing CuO to be reduced, so H_2 is the **reducing agent**. Each oxidation–reduction reaction has an oxidizing agent and a reducing agent among the reactants. The reducing agent is the substance being oxidized; the oxidizing agent is the substance being reduced.

Reduction:
copper oxide is being reduced;
CuO is the oxidizing agent.

$$CuO + H_2 \longrightarrow Cu + H_2O$$

Oxidation:
hydrogen is being oxidized;
H_2 is the reducing agent.

OXIDATION is electron DRAIN,
While REDUCTION is electron GAIN;
Forever linked, they must have one another,
For one cannot occur without the other.

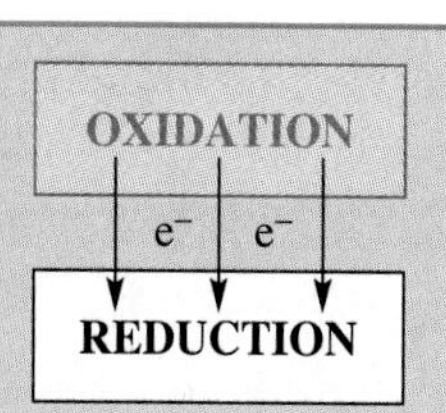

▶ EXAMPLE 8.6

Circle the oxidizing agents and underline the reducing agents in the following reactions.

a. $2\,C + O_2 \longrightarrow 2\,CO$

b. $N_2 + 3\,H_2 \longrightarrow 2\,NH_3$

c. $SnO + H_2 \longrightarrow Sn + H_2O$

d. $Mg + Cl_2 \longrightarrow Mg^{2+} + 2\,Cl^-$

SOLUTION

The answers are determined using the definitions previously given.

a. C gains oxygen and is oxidized, so it must be the reducing agent. Therefore, O_2 is the oxidizing agent.

$$\underline{C} + \textcircled{O_2}$$

b. N_2 gains hydrogen and is reduced, so it is the oxidizing agent. Therefore, H_2 is the reducing agent.

$$\textcircled{N_2} + \underline{H_2}$$

c. SnO loses oxygen and is reduced, so it is the oxidizing agent. H_2 is the reducing agent.

$$\textcircled{SnO} + \underline{H_2}$$

d. Mg loses electrons and is oxidized, so it is the reducing agent. Cl_2 is the oxidizing agent.

$$\underline{Mg} + \textcircled{Cl_2}$$

Exercise 8.6

Circle the oxidizing agents and underline the reducing agents in the following reactions.

a. $Se + O_2 \longrightarrow SeO_2$

b. $CH_3C{\equiv}N + 2\,H_2 \longrightarrow CH_3CH_2NH_2$

c. $V_2O_5 + 2\,H_2 \longrightarrow V_2O_3 + 2\,H_2O$

d. $2\,K + Br_2 \longrightarrow 2\,K^+ + 2\,Br^-$

SECTION 8.5

Electrochemistry: Cells and Batteries

We have seen that electricity can produce chemical change by the process of electrolysis (Chapter 3). Molten salt (sodium chloride), for example, can be changed to sodium metal and chlorine gas by passing an electric current through it. Conversely, chemical change can produce electricity. This is what happens in dry cell and storage batteries.

If a strip of zinc metal is placed in a solution of copper(II) sulfate, the following reaction takes place.

$$Zn\colon + Cu^{2+} \longrightarrow Zn^{2+} + Cu\colon$$

The zinc atoms give up their outer electrons to the copper ions. (Since the sulfate ions do not change, we can omit them from the equation.) The zinc metal dissolves, going into solution as zinc ions, while the copper ions come out of solution as copper metal (Figure 8.9). The zinc is oxidized; the copper ions are reduced. Zinc atoms are giving up their outer electrons directly to the copper ions.

Figure 8.9 The left photograph shows a blue solution of Cu^{2+} ions and a sample of zinc metal. When the zinc is added to the Cu^{2+} solution, the more active zinc displaces the less active copper from solution. The products of the displacement reaction (right) are a red-brown precipitate of copper metal and a colorless solution of Zn^{2+} ions. The equation for the reaction is

$$Cu^{2+} + Zn \rightarrow Cu + Zn^{2+}$$

[*Carey B. Van Loon.*]

If we separate the copper ions from the zinc, placing them in two separate compartments but connecting them with a wire, the electrons must flow through the wire in order to get from the zinc to the copper ions. This flow of electrons through the wire constitutes an electric current, and it can be used to run a motor or light a lamp.

In the cell pictured in Figure 8.10, there are two separate compartments. One contains zinc metal in a solution of zinc sulfate, and the other contains copper metal in a solution of copper(II) sulfate (which is blue). Zinc atoms give up electrons much more readily than copper atoms, so electrons flow away from the zinc and toward the copper. The zinc metal slowly dissolves as zinc atoms give up electrons to form zinc ions. The electrons flow through the wire to the copper, where copper ions pick up the electrons to become copper atoms. As time goes by, the zinc bar will slowly disappear, while the copper bar will get bigger, and the blue solution will gradually lose its color. Meanwhile something else also happens. Sulfate ions move from the blue copper sulfate solution to the zinc sulfate solution. Since more and more positively charged zinc ions are being added to the compartment on the left, while fewer and fewer copper ions are left in the compartment at the right, some of the negative sulfate ions must move from the right to the left in order to keep the two solutions electrically neutral. Notice that the porous partition in the cell allows the sulfate ions to move through it. (If the sulfate ions were unable to move through the barrier, the cell would not work.) Each time a zinc atom

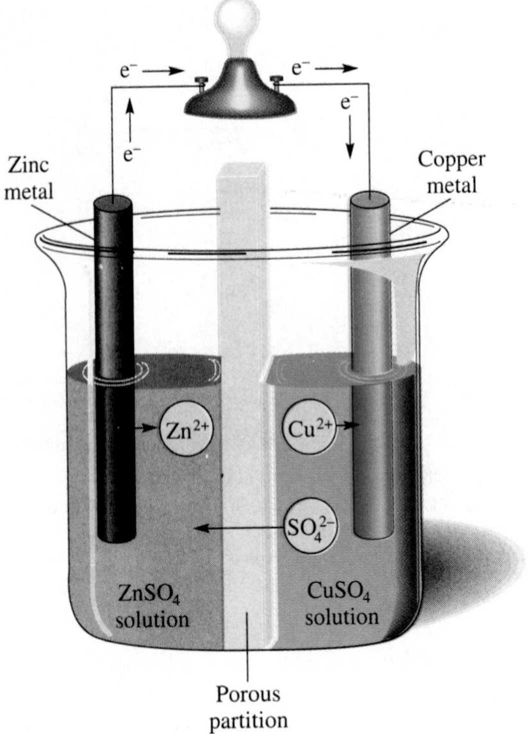

Figure 8.10 A simple electrochemical cell. A battery may consist of one or more such cells.

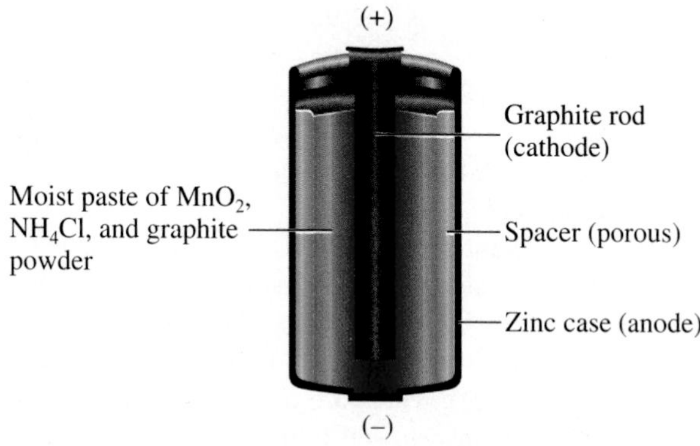

Figure 8.11 Cross section of a zinc-carbon cell.

gives up two electrons, a copper ion picks up two electrons, and one sulfate ion moves from the right compartment to the left.

The two pieces of metal are acting as electrodes. The electrode where *oxidation* occurs is called the **anode**. The one where *reduction* occurs is the **cathode**. Since zinc is giving up electrons, it is being oxidized, and it is therefore the *anode*. Copper ions are gaining electrons, so they are being reduced, and the copper electrode is the *cathode*.

Note that *anode* and *oxidation* both begin with vowels, while *cathode* and *reduction* begin with consonants.

The familiar dry cell (Figure 8.11) used in flashlights and many other small portable devices also has a zinc anode. In this case it is the container itself. A carbon rod in the center of the cell serves as the cathode. It is surrounded by a moist paste of graphite powder (carbon), manganese dioxide (MnO_2), and ammonium chloride (NH_4Cl). The anode reaction is the oxidation of the zinc cylinder to zinc ions. The cathode reaction involves reduction of manganese dioxide. A simplified version of the overall reaction is

$$Zn + 2\,MnO_2 + H_2O \longrightarrow Zn^{2+} + Mn_2O_3 + 2\,OH^-$$

Alkaline cells are similar, but the zinc case is porous and the paste around the carbon cathode is moist manganese dioxide and potassium hydroxide. They are more expensive than ordinary zinc–carbon cells, but they maintain a high voltage longer.

Although we often refer to dry cells as "batteries," a **battery** is actually a series of **electrochemical cells**. The 12-volt storage battery used in automobiles, for example, is a series of six 2-volt cells. Each cell contains a pair of electrodes, one lead and the other lead dioxide, in a chamber filled with sulfuric acid. A distinctive feature of the lead storage battery is its capacity for being recharged. It will supply electricity (i.e., it will discharge) when you turn on the ignition to start the car or when the motor is off and the lights are on. But it can be recharged when the car is moving and an electric current is supplied *to* the battery by the mechanical action of the car. The net reaction during discharge is

$$Pb + PbO_2 + 2\,H_2SO_4 \longrightarrow 2\,PbSO_4 + 2\,H_2O$$

That during recharge is just the reverse:

$$2\,PbSO_4 + 2\,H_2O \longrightarrow Pb + PbO_2 + 2\,H_2SO_4$$

These lead storage batteries are durable, but they are heavy and contain corrosive sulfuric acid.

Other important commercial batteries include: mercury cells (zinc anode/HgO cathode), which are used in small electronic devices such as calculators and hearing aids: silver oxide cells (Zn anode/Ag_2O cathode), which are used in many watches and cameras; nickel–cadmium or ''Ni-Cad'' cells (Cd anode/NiO cathode), which are rechargeable and popular for portable radios and cordless appliances; and lithium cells (Li anode/MnO_2 cathode), which are very light in weight and used in devices such as pacemakers. All of these cells have a potassium hydroxide paste between the electrodes.

An interesting kind of battery is the **fuel cell**. When fossil fuels, our major energy source, are burned to generate electricity, only about 35–40% of their energy of combustion is actually harnessed. In a fuel cell, the fuel is oxidized at the anode and oxygen is reduced at the cathode with 70–75% efficiency. Some fuel cells have been developed using fuels such as propane gas, but most fuel cells use hydrogen as fuel with platinum, nickel, or rhodium electrodes, in a solution of potassium hydroxide. Thus far fuel cells have been used mainly in spacecraft.

Somewhat like a fuel cell is the aluminum–air battery. Aluminum metal acts as a solid fuel that reacts with oxygen.

$$4\,Al + 3\,O_2 + 6\,H_2O \longrightarrow 4\,Al(OH)_3$$

Recharging the battery means adding more water and aluminum. Because of its light weight and high voltage, it is being considered for use in electric cars.

SECTION 8.6

Corrosion

An oxidation reaction of particular economic importance is the corrosion of metals. It is estimated that in the United States alone corrosion costs $100 billion a year. Perhaps 20% of all the iron and steel production in the United States each year goes to replace corroded items. Let's look first at the corrosion of iron.

The Rusting of Iron

In moist air, iron is oxidized, particularly at a nick or scratch.

$$Fe \longrightarrow Fe^{2+} + 2e^-$$

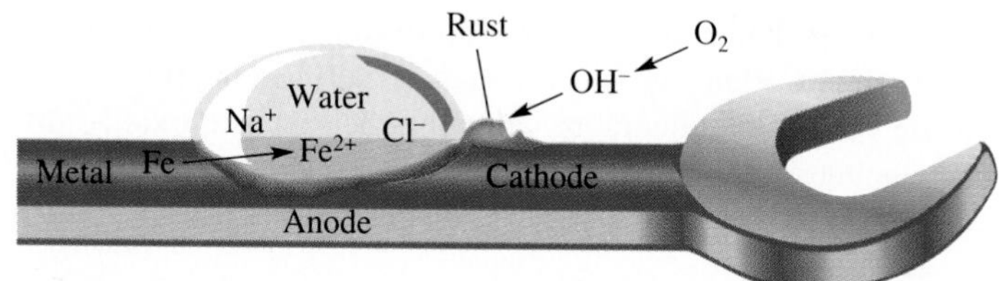

Figure 8.12 The corrosion of iron requires water, oxygen, and an electrolyte.

In order for iron to be oxidized, oxygen must be reduced.

$$O_2 + 2\,H_2O + 4\,e^- \longrightarrow 4\,OH^-$$

The net result, initially, is the formation of insoluble iron(II) hydroxide.

$$2\,Fe + O_2 + 2\,H_2O \longrightarrow 2\,Fe(OH)_2$$

This product is usually further oxidized to iron(III) hydroxide.

$$4\,Fe(OH)_2 + O_2 + 2\,H_2O \longrightarrow 4\,Fe(OH)_3$$

The $Fe(OH)_3$ is sometimes written as $Fe_2O_3{\cdot}3H_2O$, it is the familiar iron rust.

The formula $Fe_2O_3{\cdot}3H_2O$ is the same as $2\,Fe(OH)_3$.

Oxidation and reduction often occur at separate points on the metal surface. Electrons are transferred through the iron metal. The circuit is completed by an electrolyte in aqueous solution. In the snowbelt, this solution is often the slush from road salt and melting snow. The metal is pitted in an *anodic* area, where iron is oxidized to Fe^{2+}. These ions migrate to the *cathodic* area, where they react with the hydroxide ions formed by reduction of oxygen.

$$Fe^{2+} + 2\,OH^- \longrightarrow Fe(OH)_2$$

As indicated above, this iron(II) hydroxide is then oxidized to $Fe(OH)_3$, or rust. This process is diagrammed in Figure 8.12. Notice that the anodic area is protected from oxygen by the water film, while the cathodic area is exposed to air.

Protection of Aluminum

Aluminum is more reactive than iron, and yet corrosion is not a serious problem with aluminum. We use aluminum foil, we cook with aluminum pots, and we buy beverages in aluminum cans. Even after many years, they have not corroded. How is that possible?

The aluminum surface reacts with oxygen in the air to form a thin layer of oxide. However, instead of being porous and flaky like iron oxide, aluminum oxide is hard and tough, providing protection to the metal from further oxidation.

We should add, however, that corrosion can sometimes be a problem with aluminum. Certain substances, such as salt, can interfere with the protective oxide coating on aluminum, allowing the metal to oxidize. Mag wheels on automobiles have cracked, and planes with aluminum landing gear have had their wheels sheared off because of this problem.

Silver Tarnish

The tarnish on silver is due to the oxidation of the silver surface by hydrogen sulfide (H_2S) in the air. It produces a film of black silver sulfide (Ag_2S) on the metal surface. You can use a silver polish to remove the tarnish, but in doing so, you also lose part of the silver. An alternative method involves the use of aluminum metal to reduce the silver ions back to silver metal.

$$3\,Ag^{+} + Al \longrightarrow 3\,Ag + Al^{3+}$$

This reaction also requires an electrolyte. Sodium bicarbonate ($NaHCO_3$) is usually used. The tarnished silver is placed in contact with aluminum foil and covered with a solution of sodium bicarbonate. A precious metal is conserved at the expense of a cheaper one.

SECTION 8.7

Some Common Oxidizing Agents

Oxygen itself is the most common oxidizing agent. Making up one-fifth of the air, it oxidizes the wood in our campfires and the gasoline in our automobiles. It even ''burns'' the food we eat to give us the energy to move and to think. Large amounts of purified oxygen are used in hospital respirators and welding torches, but most of it goes into steel furnaces.

Another common oxidizing agent is hydrogen peroxide (H_2O_2). Pure hydrogen peroxide is a syrupy liquid. It is available (in the laboratory) as a dangerous 30% solution that has powerful oxidizing power or as a 3% solution sold in stores for various uses around the home. It has the advantage of being converted to water in most reactions.

An oxidizing agent often used in the laboratory is potassium dichromate ($K_2Cr_2O_7$). It is an orange material that turns green when it is reduced to Cr^{3+} ions. One of the compounds that it oxidizes is alcohol. The Breathalyzer test for intoxication makes use of a dichromate solution. The exhaled breath of the person being tested mixes with the acidic dichromate and the degree of color change indicates the level of alcohol.

The Breathalyzer test for intoxication (used to test suspected drunken drivers, for instance) makes use of the color change associated with the reduction of dichromate ions. If present in the breath, alcohol reduces the orange-red dichromate to green chromium(III) ions.

Many antiseptics are mild oxidizing agents. (Antiseptics are compounds applied to living tissue to kill microorganisms or prevent their growth.) For example, a 3% solution of hydrogen peroxide is often used to treat minor cuts, and tincture of iodine has long been a household antiseptic. Ointments for treating acne often contain 5 to 10% of benzoyl peroxide, a powerful antiseptic and also a skin irritant. It causes old skin to slough off and be replaced by new, fresher-looking skin. When used on areas exposed to sunlight, however, benzoyl peroxide may promote skin cancer.

A tincture is a solution made up in alcohol.

Oxidizing agents are also used as disinfectants. A good example is chlorine, which is used to kill disease-causing microorganisms in drinking water. Swimming pools are often chlorinated with calcium hypochlorite [$Ca(OCl)_2$]. Since calcium hypochlorite is alkaline, it also raises the pH of the water. (When a pool becomes too alkaline, the pH is lowered by adding hydrochloric acid. Swimming pools are usually maintained at pH 7.2 to 7.8.)

Bleaches are oxidizing agents, too. Bleaches remove unwanted color from fabrics or other material. Nearly any oxidizing agent would do the job, but some might be unsafe, or harmful to fabrics, or perhaps too expensive. Laundry bleaches are usually sodium hypochlorite (NaOCl) as an aqueous solution (in products such as Purex or Clorox) or calcium hypochlorite [$Ca(OCl)_2$], known as bleaching powder. The powder is usually preferred for large industrial operations, such as the whitening of paper or fabrics. Nonchlorine bleaches often contain sodium perborate (a loose combination of $NaBO_2$ and H_2O_2).

For lightening hair color, the bleaches are usually 6% or 12% solutions of hydrogen peroxide, which oxidizes the dark pigment (melanin) in the hair to colorless products. Hydrogen peroxide can also be used to lighten certain paints by oxidizing sulfides (S^{2-}) to sulfates (SO_4^{2-}). When lead-based paints are exposed to air containing hydrogen sulfide (H_2S), they turn black because of the formation of lead sulfide (PbS). Hydrogen peroxide oxidizes the black sulfide to white sulfate.

$$\underset{\text{black}}{PbS} + 4\,H_2O_2 \longrightarrow \underset{\text{white}}{PbSO_4} + 4\,H_2O$$

Hydrogen peroxide can be used to restore the once-white areas of old paintings that have darkened by the reaction of white lead compounds (paint pigments) with sulfur compounds. The darkened pigments (black PbS) are converted to white $PbSO_4$ by the hydrogen peroxide.

Stain removal is more complicated than bleaching. A few stain removers are oxidizing agents, but some are reducing agents, some are solvents or detergents, and some have quite different action. Some stains require rather specific stain removers.

SECTION 8.8

Some Reducing Agents of Interest

In every reaction involving oxidation, the oxidizing agent gets reduced, and so the substance being oxidized is acting as a reducing agent. But let us consider here reactions in which the purpose of the reaction is reduction.

Most metals occur in nature as compounds. In order to prepare the free metals, the compounds must be reduced. Metals are often freed from their ores with coal or coke (elemental carbon obtained by heating coal to drive off volatile matter). Tin oxide is one of the many ores that can be reduced with coal or coke.

$$SnO_2 + C \longrightarrow Sn + CO_2$$

Sometimes a metal can be obtained by heating its ore with a more active metal. Chromium oxide, for example, can be reduced by heating it with aluminum.

$$Cr_2O_3 + 2\,Al \longrightarrow Al_2O_3 + 2\,Cr$$

Hydrogen is an excellent reducing agent that can free many metals from their ores, but it is generally used to produce more expensive metals, such as tungsten.

$$WO_3 + 3\,H_2 \longrightarrow W + 3\,H_2O$$

Hydrogen can be used to reduce many kinds of chemical compounds. Ethylene, for example, can be reduced to ethane.

$$C_2H_4 + H_2 \longrightarrow C_2H_6$$

(Nickel is used as a catalyst in this reaction.) Hydrogen also reduces nitrogen, from the air, in the industrial production of ammonia.

$$N_2 + 3\,H_2 \longrightarrow 2\,NH_3$$

(An iron catalyst is used in this case.)

Perhaps a more familiar reducing agent, by use if not by name, is the developer used in black and white photography. Photographic film is coated with a silver *salt* (Ag^+Br^-). Silver ions that have been exposed to light react with the developer, a reducing agent (such as the organic compound hydroquinone), to form metallic silver.

$$\underset{\text{Hydroquinone}}{C_6H_4(OH)_2} + 2\,Ag^+ \longrightarrow C_6H_4O_2 + \underset{\text{Silver metal}}{2\,Ag} + 2\,H^+$$

Silver ions not exposed to light are not reduced by the developer. The film is then treated with "hypo," a solution of sodium thiosulfate ($Na_2S_2O_3$), which washes out unexposed silver bromide to form the negative. That leaves the negative dark where the metallic silver has been deposited (where it was originally exposed to light) and transparent where light did not strike it. Light is then shined through the negative onto light-sensitive paper to make the positive print. Figure 8.13 shows positive and negative prints.

Figure 8.13 A photographic negative (left) and a positive print. [*Tomas Sennett for WORLD BANK.*]

In food chemistry reducing agents are sometimes referred to as **antioxidants**. Ascorbic acid (vitamin C) can prevent the browning of fruit (such as sliced apples or pears) by inhibiting air oxidation. Whereas vitamin C is water soluble, tocopherol (vitamin E) is a fat soluble antioxidant. Both of these vitamins are believed to retard various oxidation reactions that are potentially damaging to vital components of living cells.

SECTION 8.9

Oxidation, Reduction, and Living Things

Perhaps the most important oxidation–reduction processes are the ones that maintain life on this planet. We obtain energy for all our physical and mental activities by eating and metabolizing food through the process of respiration. The process has many steps, but eventually the food we eat is converted mainly into carbon dioxide, water, and energy.

Bread and many of the other foods we eat are largely made up of carbohydrates (Chapter 15). If we represent carbohydrates with the simple example glucose ($C_6H_{12}O_6$), we can write the overall equation for their metabolism as

Figure 8.14 Photosynthesis is occurring in these green plants. The chlorophyll pigments that catalyze the photosynthesis process give the green color to much of the land area of the Earth. [*Thomas R. Fletcher/Stock Boston.*]

follows.

$$C_6H_{12}O_6 + 6\,O_2 \longrightarrow 6\,CO_2 + 6\,H_2O + \text{energy}$$

This process is constantly occurring in animals, including humans. The carbohydrate is oxidized in the process.

Meanwhile, plants need carbon dioxide and water, from which they produce carbohydrates. The energy needed comes from the sun, and the process is called **photosynthesis** (Figure 8.14). The chemical equation is

$$6\,CO_2 + 6\,H_2O + \text{energy} \longrightarrow C_6H_{12}O_6 + 6\,O_2$$

Notice that this process going on inside plants is exactly the reverse of the process going on inside animals. Whereas food metabolism in animals is an oxidation process, photosynthesis in plants is a reduction process.

The carbohydrates produced by photosynthesis are the ultimate source of all our food, since fish, fowl, and other animals either eat plants or eat other animals that eat plants. Note that the photosynthesis process not only makes carbohydrates but it also yields free elementary oxygen, O_2. In other words, photosynthesis does not just provide all the food we eat; it also provides all the oxygen we breathe.

There are many oxidation reactions that occur in nature (with oxygen being reduced in the process). Photosynthesis is unique in that it is a natural *reduction* process (with oxygen being the element oxidized). Many reactions in nature *use* oxygen. Photosynthesis is the only natural process that *produces* it.

▸ Summary

1. When a substance gains oxygen atoms or loses hydrogen atoms, it is being oxidized.
2. When a substance gains hydrogen atoms or loses oxygen atoms, it is being reduced.
3. *Oxidation* is a loss of electrons; *reduction* is a gain of electrons.
4. Increase in oxidation state is oxidation; decrease in oxidation state is reduction.
5. Whenever oxidation occurs, an equal amount of reduction must occur simultaneously.
6. The substance being oxidized is the *reducing agent;* the substance being reduced is the *oxidizing agent.*
7. In an *electrochemical cell* oxidation occurs at the *anode* and reduction at the *cathode.*
8. An electrochemical cell generates an electric current by having oxidation and reduction occur in two different locations, so that electrons must flow through a wire to get from one to the other.
9. Metal corrosion is an oxidation process, the most familiar example being the rusting of iron.
10. Laundry *bleaches* and household antiseptics are common oxidizing agents; photographic developer and *antioxidants* in food are common reducing agents.
11. *Photosynthesis* is the most important reduction process on Earth. We could not exist without it.

Key Terms

anode 8.5
antioxidant 8.8
battery 8.5
bleach 8.7
cathode 8.5
electrochemical cell 8.5
hydrogen 8.2
oxidation 8.3
oxidizing agent 8.4
oxygen 8.1
photosynthesis 8.9
reducing agent 8.4
reduction 8.3

Review Questions

1. Define oxidation and reduction in terms of the following.
   a. oxygen atoms gained or lost
   b. hydrogen atoms gained or lost
   c. electrons gained or lost
2. What is an electrochemical cell?
3. What is the purpose of a porous plate between the two electrode compartments in an electrochemical cell?
4. From what material is the case of a carbon–zinc dry cell made? What purpose does the material serve? What happens to it as the cell discharges?
5. How does the alkaline cell differ from a regular carbon–zinc dry cell?
6. What is an electrochemical battery?
7. Describe what happens when the lead storage battery discharges.
8. What happens when the lead storage battery is charged?
9. Describe what happens when iron corrodes. How does road salt speed this process?
10. Why does aluminum corrode more slowly than iron, even though aluminum is more reactive than iron?
11. How does silver tarnish? How can tarnish be removed without the loss of silver?
12. List four common oxidizing agents.
13. List three common reducing agents.
14. Name some oxidizing agents used as antiseptics and disinfectants.
15. Relate the chemistry of photosynthesis to the chemistry that provides energy for your heartbeat.
16. Describe how a bleaching agent, such as hypochlorite (ClO^-), works.

Problems

Oxidation and Reduction: Partial Reactions

17. The following "equations" show only part of a chemical reaction. Indicate whether the reactant shown is being oxidized or reduced.
   a. $C_2H_4O \rightarrow C_2H_4O_2$
   b. $H_2O_2 \rightarrow H_2O$
18. In which of the following partial reactions is the reactant undergoing oxidation?
   a. $C_2H_4O \rightarrow C_2H_6O$ b. $WO_3 \rightarrow W$
19. In which of the following partial reactions is the reactant undergoing oxidation?
   a. $Cl_2 \rightarrow 2\,Cl^-$ b. $Fe^{3+} \rightarrow Fe^{2+}$
20. In which of the following partial reactions is the reactant undergoing oxidation?
   a. $2\,H^+ \rightarrow H_2$ b. $CO \rightarrow CO_2$

Oxidizing Agents and Reducing Agents

21. Circle the oxidizing agent and underline the reducing agent in these reactions.
   a. $4\,Al + 3\,O_2 \rightarrow 2\,Al_2O_3$
   b. $2\,SO_2 + O_2 \rightarrow 2\,SO_3$

22. Circle the oxidizing agent and underline the reducing agent in these reactions.
a. $Cl_2 + 2\,KBr \rightarrow 2\,KCl + Br_2$
b. $C_2H_4 + H_2 \rightarrow C_2H_6$

23. Circle the oxidizing agent and underline the reducing agent in these reactions.
a. $Fe + 2\,HCl \rightarrow FeCl_2 + H_2$
b. $CS_2 + 3\,O_2 \rightarrow CO_2 + 2\,SO_2$

24. Circle the oxidizing agent and underline the reducing agent in these reactions.
a. $2\,AgNO_3 + Cu \rightarrow Cu(NO_3)_2 + 2\,Ag$
b. $CuCl_2 + Zn \rightarrow ZnCl_2 + Cu$

Oxidation and Reduction: Chemical Reactions

25. In the following reactions, which element is oxidized and which is reduced?
a. $2\,HNO_3 + SO_2 \rightarrow H_2SO_4 + 2\,NO_2$
b. $2\,CrO_3 + 6\,HI \rightarrow Cr_2O_3 + 3\,I_2 + 3\,H_2O$

26. In the following reactions, which substance is oxidized? Which is the oxidizing agent?
a. $H_2CO + H_2O_2 \rightarrow H_2CO_2 + H_2O$
b. $5\,C_2H_6O + 4\,MnO_4^- + 12\,H^+ \rightarrow 5\,C_2H_4O_2 + 4\,Mn^{2+} + 11\,H_2O$

27. Acetylene (C_2H_2) reacts with hydrogen to form ethane (C_2H_6). Is the acetylene oxidized or reduced? Explain your answer.

28. Unsaturated vegetable oils react with hydrogen to form saturated fats. A typical reaction is

$$C_{57}H_{104}O_6 + 3\,H_2 \rightarrow C_{57}H_{110}O_6$$

Is the unsaturated oil oxidized or reduced? Explain.

29. To test for an iodide ion (for example, in iodized salt), a solution is treated with chlorine to liberate iodine. The reaction is

$$2\,I^- + Cl_2 \rightarrow I_2 + 2\,Cl^-$$

Which substance is oxidized? Which is reduced?

30. Molybdenum metal, used in special kinds of steel, can be manufactured by the reaction of its oxide with hydrogen. The reaction is

$$MoO_3 + 3\,H_2 \rightarrow Mo + 3\,H_2O$$

Which substance is reduced? Which is the reducing agent?

31. Green grapes are exceptionally sour because of a high concentration of tartaric acid. As the grapes ripen, this compound is converted to glucose.

$$\underset{\text{Tartaric acid}}{C_4H_6O_6} \longrightarrow \underset{\text{Glucose}}{C_6H_{12}O_6}$$

Is the tartaric acid being oxidized or reduced?

32. The dye indigo (used to color blue jeans) is formed from indoxyl by exposure of the latter to air.

$$\underset{\text{Indoxyl}}{2\,C_8H_7ON} + O_2 \longrightarrow \underset{\text{Indigo}}{C_{16}H_{10}N_2O_2} + 2\,H_2O$$

What substance is oxidized? What is the oxidizing agent?

33. When the water pump failed in the nuclear reactor at Three Mile Island in 1979, zirconium metal reacted with the very hot water to produce hydrogen gas.

$$Zr + 2\,H_2O \rightarrow ZrO_2 + 2\,H_2$$

What substance was oxidized in the reaction? What was the oxidizing agent?

34. Vitamin C (ascorbic acid) is thought to protect our stomachs from the carcinogenic effect of nitrite ions by converting the ions to NO gas.

$$NO_2^- \rightarrow NO$$

Is the nitrite ion oxidized or reduced? Is ascorbic acid an oxidizing agent or a reducing agent?

35. In the preceding reaction (Problem 34), ascorbic acid is converted to dehydroascorbic acid.

$$C_6H_8O_6 \rightarrow C_6H_6O_6$$

Is ascorbic acid oxidized or reduced in the reaction?

Combination with Oxygen

36. Give formulas for the products formed in each of the following reactions.
a. $C + O_2 \rightarrow$ b. $N_2 + O_2 \rightarrow$

37. Give formulas for the products formed in each of the following reactions.
a. $S + O_2 \rightarrow$ b. $H_2 + O_2 \rightarrow$

38. Give formulas for the products formed in each of the following.
a. $CH_4 + O_2 \rightarrow$ b. $C_3H_8 + O_2 \rightarrow$

39. Give formulas for the products formed in each of the following.
a. $CS_2 + O_2 \rightarrow$ b. $C_6H_{12}O_6 + O_2 \rightarrow$

Additional Problems

40. When a strip of copper is placed in a solution of silver nitrate ($AgNO_3$), the copper is plated with silver. Write an equation for the reaction that occurs. (Nitrate ion is not involved in the reaction.)

41. To oxidize 1 kg of fat, our bodies require about 2000 L of oxygen. A good diet contains about 80 g of fat per day. What volume of oxygen (at STP) is required to oxidize that fat?

42. The oxidizing agent we use to obtain energy from food is oxygen (from the air). If you breathe 15 times a minute (at rest), taking in and exhaling 0.5 L of air with each breath, what volume of air do you breathe each day? Air is 21% oxygen by volume. What volume of oxygen do you breathe each day?

43. In the photosynthesis reaction

$$6\,CO_2 + 6\,H_2O \rightarrow C_6H_{12}O_6 + 6\,O_2$$

which substance is oxidized? Which is the oxidizing agent? Which substance is reduced? Which is the reducing agent?

References and Readings

1. ''Auto Batteries.'' *Consumer Reports,* February 1985, pp. 94–97.
2. ''Chem I Supplement: Everyday Examples of Oxidation–Reduction Processes.'' *Journal of Chemical Education,* May 1978, pp. 332–333.
3. DeLorenzo, Ronald. ''Electrochemical Errors.'' *Journal of Chemical Education,* May 1985, p. 424.
4. Ennis, John L. ''Photography at Its Genesis.'' *Chemical and Engineering News,* 18 December 1989, pp. 26–42.
5. ''Fade Creams.'' *Consumer Reports,* January 1985, p. 12. Hydroquinone is used to get rid of unwanted ''age spots.''
6. Hill, John W., and James E. Laib. ''Acne.'' *SciQuest,* July–August 1980, pp. 7–10.
7. Kolb, Doris. ''The Chemical Equation: Part II: Oxidation–Reduction Reactions.'' *Journal of Chemical Education,* May 1978, pp. 326–331.
8. Petrucci, Ralph H. and William S. Harwood. *General Chemistry,* 6th ed. New York: Macmillan, 1993 pp. 146–155.
9. U.S. Department of Agriculture. ''Stain Removal for Washable Fabrics.'' Madison WI: Cooperative Extension Programs, 1979. North Central Regional Publication 64.
10. Worley, John D. ''Hydrogen Peroxide in Cleansing Antiseptics.'' *Journal of Chemical Education,* August 1983, p. 678.

9 Organic Chemistry

The Infinite Variety of Carbon Compounds

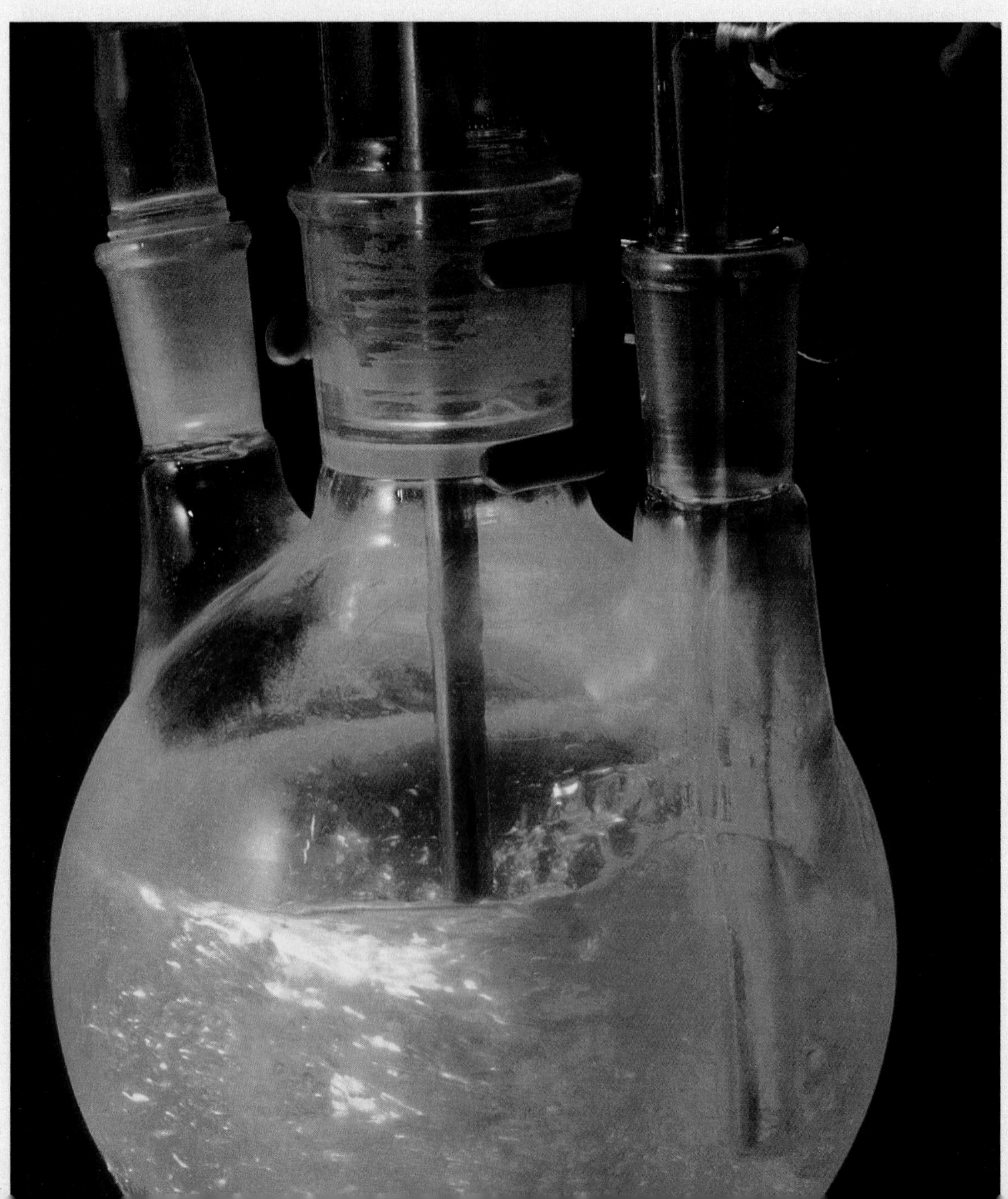

So many carbon compounds with their infinite variety!
To have to learn their names is surely reason for anxiety.
It might be hopeless if our naming methods were erratic,
But nomenclature rules are simple—and quite systematic.

The only chemical element that has a major field of chemistry devoted to the study of its compounds is carbon. The field is called "organic" chemistry because it was once believed that these compounds had to be derived from living things (Figure 9.1). Most organic compounds do come from living things, or things that were once living, but that is not necessarily the case. Perhaps the most remarkable thing about organic compounds is that there are so many of them. Of all the millions of known chemical compounds, more than 95 percent are compounds of carbon.

Figure 9.1 The word *organic* has several different meanings. Organic fertilizer is organic in the original sense—derived from a living organism. There is no legal definition of organic foods, but the term generally means foods grown without pesticides or synthetic fertilizers. Organic chemistry is the chemistry of carbon compounds.

SECTION 9.1

The Unique Carbon Atom

Carbon atoms are unique in their ability to bond to each other so strongly that they can form very long chains. This process of chain formation is called **catenation**. Silicon and a few other elements can form chains, but only short

Many simple organic chemicals are liquids. They are often separated from mixtures by distillation. [*Charles West/The Stock Market.*]

In January 1990 the Chemical Abstracts Service recorded the 10 millionth known chemical compound. More than 600,000 new compounds are added to the list each year. Most of them are compounds of carbon.

ones; carbon chains often contain thousands of carbon atoms. No other element can form chains nearly so well as carbon.

Carbon chains can also have branches, or they can form rings of various sizes. Add to this the fact that carbon atoms also bond strongly to other elements, such as hydrogen, oxygen, and nitrogen, and that these atoms can be arranged in many different ways, and it soon becomes obvious why there are so many carbon compounds.

In addition to the millions of carbon compounds that are already known, there are new ones being discovered every day. Carbon can form an almost infinite number of molecules of various shapes, sizes, and compositions.

We use thousands of carbon compounds every day without even realizing it because they are silently carrying out important chemical reactions within our bodies. Many of these carbon compounds are so vital that we literally could not live without them.

SECTION 9.2

Simple Hydrocarbons: Alkanes

The simplest organic compounds are the **hydrocarbons**, which contain only hydrogen and carbon. Each carbon atom forms four bonds, and each hydrogen atom forms only one bond, so the simplest hydrocarbon molecule that is possible is CH_4, which is called *methane.* (It is the main component of natural gas.) It has the structure

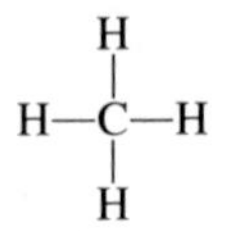

```
    H
    |
H—C—H
    |
    H
```

Alkanes often are called **saturated hydrocarbons**. They are hydrocarbons in which there are only single bonds.

Methane is the first member of a group of related compounds called **alkanes**. These are hydrocarbons that contain only single bonds. The next member of the series is *ethane* (C_2H_6).

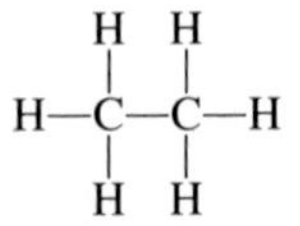

```
    H  H
    |  |
H—C—C—H
    |  |
    H  H
```

The methane molecule is tetrahedral, and the ethane molecule is also three-dimensional. Figure 9.2 shows ball-and-stick and space-filling models of these molecules. The ball-and-stick models show the bond angles best, but the space-filling models more accurately reflect the shapes of the molecules. Ordinarily we shall use simple **structural formulas** like the one shown above for ethane, since they are so much easier to draw. These formulas show you

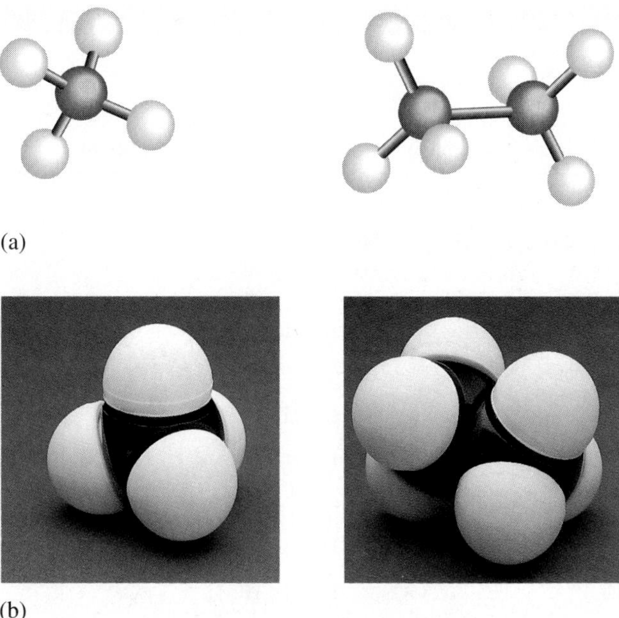

Figure 9.2 Ball-and-stick (a) and space-filling (b) models of methane and ethane. [*(b) Richard Megna/Macmillan Science Files.*]

which atoms are bonded to each other, but they do not attempt to show the actual shapes of the molecules.

The three-carbon alkane is propane. Models of propane are shown in Figure 9.3, and its structural formula is

```
    H  H  H
    |  |  |
H—C—C—C—H
    |  |  |
    H  H  H
```

A pattern is now becoming apparent. We can build alkanes of any length simply by tacking carbon atoms together in long chains and adding sufficient hydrogen atoms to give each of the carbon atoms a total of four bonds. Even the naming of these compounds follows a pattern. For compounds of five carbon atoms or more, each stem or root is derived from the Greek or Latin name for the number of carbon atoms in the molecule (Table 9.1). The compound names end in *-ane*, signifying that the compounds are alk*anes*. Table 9.2 gives structural formulas and names for continuous-chain alkanes up to 10 carbon atoms in length. We do not need to stop at 10 carbon atoms. One could hook together 100 or 1000 or 1 million. We can make an infinite number of alkanes simply by lengthening the chain. But lengthening the chain is not the only option. With four carbon atoms, chain branching is also possible.

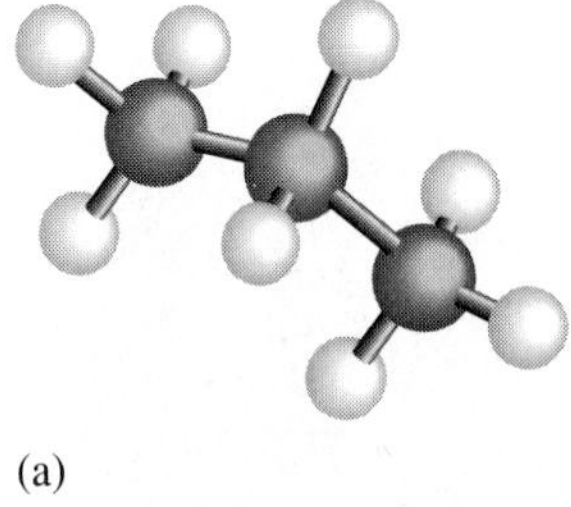

(a)

(b)

Figure 9.3 Ball-and-stick (a) and space-filling (b) models of propane. [*(b) Richard Megna/Macmillan Science Files.*]

Table 9.1
Word Stems That Indicate the Number of Carbon Atoms in Organic Molecules

Stem	Number
Meth-	1
Eth-	2
Prop-	3
But-	4
Pent-	5
Hex-	6
Hept-	7
Oct-	8
Non-	9
Dec-	10

Table 9.2 The First Ten Continuous-Chain Alkanes

Name	Molecular Formula	Structural Formula	Number of Possible Isomers
Methane	CH_4	$\begin{array}{ccccc} & & H & & \\ & & \vert & & \\ H & - & C & - & H \\ & & \vert & & \\ & & H & & \end{array}$	1
Ethane	C_2H_6	$\begin{array}{ccccccc} & & H & & H & & \\ & & \vert & & \vert & & \\ H & - & C & - & C & - & H \\ & & \vert & & \vert & & \\ & & H & & H & & \end{array}$	1
Propane	C_3H_8	$\begin{array}{ccccccccc} & & H & & H & & H & & \\ & & \vert & & \vert & & \vert & & \\ H & - & C & - & C & - & C & - & H \\ & & \vert & & \vert & & \vert & & \\ & & H & & H & & H & & \end{array}$	1
Butane	C_4H_{10}	$\begin{array}{ccccccccccc} & & H & & H & & H & & H & & \\ & & \vert & & \vert & & \vert & & \vert & & \\ H & - & C & - & C & - & C & - & C & - & H \\ & & \vert & & \vert & & \vert & & \vert & & \\ & & H & & H & & H & & H & & \end{array}$	2
Pentane	C_5H_{12}	$\begin{array}{ccccccccccccc} & & H & & H & & H & & H & & H & & \\ & & \vert & & \vert & & \vert & & \vert & & \vert & & \\ H & - & C & - & C & - & C & - & C & - & C & - & H \\ & & \vert & & \vert & & \vert & & \vert & & \vert & & \\ & & H & & H & & H & & H & & H & & \end{array}$	3
Hexane	C_6H_{14}	$\begin{array}{ccccccccccccccc} & & H & & H & & H & & H & & H & & H & & \\ & & \vert & & \vert & & \vert & & \vert & & \vert & & \vert & & \\ H & - & C & - & C & - & C & - & C & - & C & - & C & - & H \\ & & \vert & & \vert & & \vert & & \vert & & \vert & & \vert & & \\ & & H & & H & & H & & H & & H & & H & & \end{array}$	5
Heptane	C_7H_{16}	$\begin{array}{ccccccccccccccccc} & & H & & H & & H & & H & & H & & H & & H & & \\ & & \vert & & \vert & & \vert & & \vert & & \vert & & \vert & & \vert & & \\ H & - & C & - & C & - & C & - & C & - & C & - & C & - & C & - & H \\ & & \vert & & \vert & & \vert & & \vert & & \vert & & \vert & & \vert & & \\ & & H & & H & & H & & H & & H & & H & & H & & \end{array}$	9
Octane	C_8H_{18}	$\begin{array}{ccccccccccccccccccc} & & H & & H & & H & & H & & H & & H & & H & & H & & \\ & & \vert & & \vert & & \vert & & \vert & & \vert & & \vert & & \vert & & \vert & & \\ H & - & C & - & C & - & C & - & C & - & C & - & C & - & C & - & C & - & H \\ & & \vert & & \vert & & \vert & & \vert & & \vert & & \vert & & \vert & & \vert & & \\ & & H & & H & & H & & H & & H & & H & & H & & H & & \end{array}$	18
Nonane	C_9H_{20}	$\begin{array}{ccccccccccccccccccccc} & & H & & H & & H & & H & & H & & H & & H & & H & & H & & \\ & & \vert & & \vert & & \vert & & \vert & & \vert & & \vert & & \vert & & \vert & & \vert & & \\ H & - & C & - & C & - & C & - & C & - & C & - & C & - & C & - & C & - & C & - & H \\ & & \vert & & \vert & & \vert & & \vert & & \vert & & \vert & & \vert & & \vert & & \vert & & \\ & & H & & H & & H & & H & & H & & H & & H & & H & & H & & \end{array}$	35
Decane	$C_{10}H_{22}$	$\begin{array}{ccccccccccccccccccccccc} & & H & & H & & H & & H & & H & & H & & H & & H & & H & & H & & \\ & & \vert & & \vert & & \vert & & \vert & & \vert & & \vert & & \vert & & \vert & & \vert & & \vert & & \\ H & - & C & - & C & - & C & - & C & - & C & - & C & - & C & - & C & - & C & - & C & - & H \\ & & \vert & & \vert & & \vert & & \vert & & \vert & & \vert & & \vert & & \vert & & \vert & & \vert & & \\ & & H & & H & & H & & H & & H & & H & & H & & H & & H & & H & & \end{array}$	75

The alkanes can be represented by a general formula C_nH_{2n+2} in which n is the number of carbon atoms.

Isomerism

If we extend the carbon chain to four atoms and add enough hydrogen atoms to give each carbon atom four bonds, we get

$$\begin{array}{ccccccccccc} & & H & & H & & H & & H & & \\ & & | & & | & & | & & | & & \\ H & - & C & - & C & - & C & - & C & - & H \\ & & | & & | & & | & & | & & \\ & & H & & H & & H & & H & & \end{array}$$

This formula represents butane (C_4H_{10}), a compound that boils at 0 °C. A second compound, whose boiling point is −12 °C, has the same molecular formula: C_4H_{10}. The *structural formula* of the second compound is *not* the same as butane's. Instead of having four carbon atoms connected in a continuous chain, this new compound has a continuous chain of only three carbon atoms. The fourth carbon is branched off the middle carbon of this three-carbon chain.

$$\begin{array}{ccccccccc} & & H & & H & & H & & \\ & & | & & | & & | & & \\ H & - & C & - & C & - & C & - & H \\ & & | & & | & & | & & \\ & & H & & | & & H & & \\ & & & & | & & & & \\ & & H & - & C & - & H & & \\ & & & & | & & & & \\ & & & & H & & & & \end{array}$$

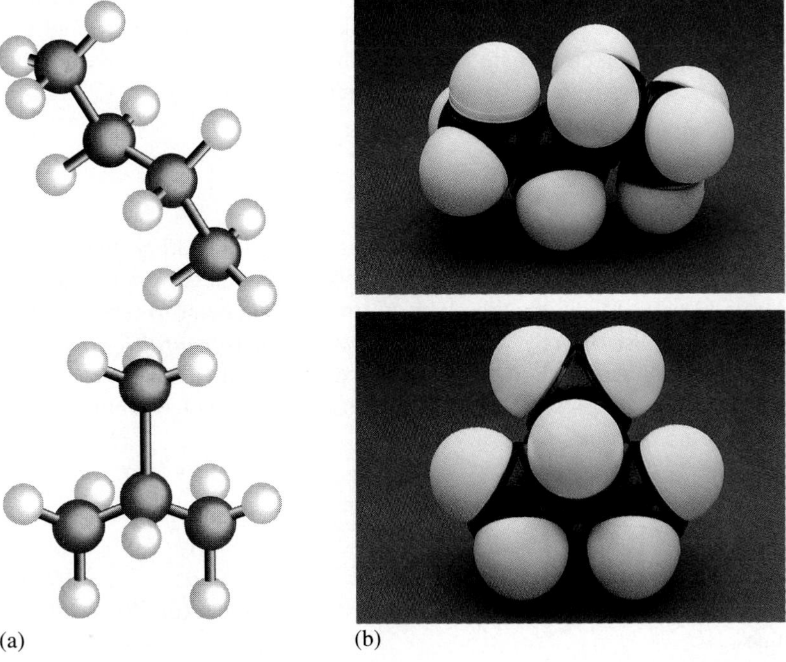

Figure 9.4 Ball-and-stick (a) and space-filling (b) models of butane and isobutane. [*(b) Richard Megna/ Macmillan Science Files.*]

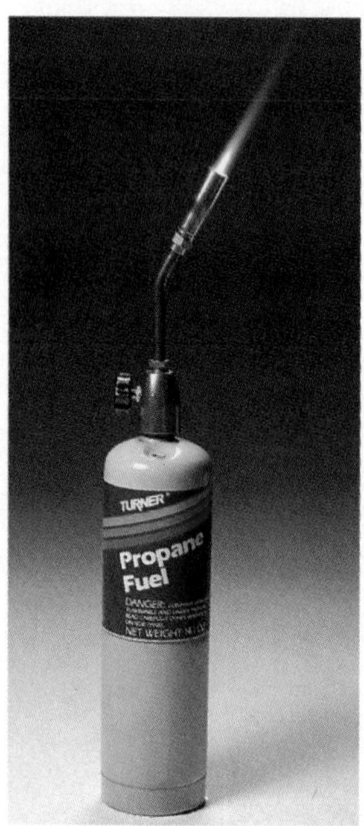

Figure 9.5 A propane torch. Propane burns in air with a hot flame. [*Carey B. Van Loon.*]

Compounds that have the same molecular formula but different structural formulas are called **isomers**. Because it is an isomer of butane, the branched four-carbon alkane is called isobutane. Figure 9.4 shows ball-and-stick and space-filling models of the two isomeric butanes.

The number of isomers increases rapidly with the number of carbon atoms (see Table 9.2). There are three pentanes, five hexanes, nine heptanes, and so on. Isomerism is common in the compounds of carbon and provides another reason for the existence of millions of organic compounds.

Propane and the butanes are familiar fuels (Figure 9.5). They usually are supplied under pressure in tanks. Although they are gases at ordinary temperatures and under normal atmospheric pressure, they are liquefied under pressure and sold as liquefied petroleum gas (LP gas). Gasoline is a mixture of hydrocarbons, mostly alkanes, with 5 to 12 carbon atoms.

Homologous Series

Notice that the molecular formula of each alkane in Table 9.2 differs from the one preceding it by precisely one carbon atom and two hydrogen atoms, that is, by a CH_2 unit. Such a series of compounds has properties that vary in a regular and predictable manner. This principle, called *homology,* gives order to organic chemistry in much the same way that the periodic table gives organization to the chemistry of the elements. Instead of studying the chemistry of a bewildering array of individual carbon compounds, organic chemists study a few members of a **homologous series** from which they can *deduce* the properties of other compounds in the series.

We should point out that not all the possible isomers of the larger molecules have been isolated. Indeed, the task rapidly becomes more and more prohibitive as you proceed up the series. There are, for example, over 4 billion possible isomers with the molecular formula $C_{30}H_{62}$.

Condensed Structural Formulas

The structural formulas we have been using show all the carbon and hydrogen atoms and how they are attached to one another. But these formulas take up a lot of space and they are quite a bit of trouble to draw or to type. For those reasons chemists usually prefer to use *condensed structural formulas.* The condensed structures show how many hydrogen atoms are attached to each carbon atom without showing each hydrogen atom. For example, the two butanes are written

$$CH_3{-}CH_2{-}CH_2{-}CH_3 \qquad \text{and} \qquad \begin{array}{c} CH_3{-}CH{-}CH_3 \\ | \\ CH_3 \end{array}$$

These can be simplified even further by omitting some (or all) of the bond lines.

$CH_3CH_2CH_2CH_3$ and CH_3CHCH_3 (with CH_3 attached to the middle carbon) or $CH_3CH(CH_3)CH_3$

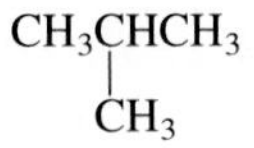

or $CH_3CH(CH_3)_2$ or $CH(CH_3)_3$

▶ EXAMPLE 9.1

Give the molecular formula, the structural formula, and the condensed structural formula for heptane.

SOLUTION

The name heptane means an alkane with seven carbon atoms, so the molecular formula is C_7H_{16}. For the structural formula, just write out a string of seven carbon atoms.

C—C—C—C—C—C—C

Then you attach enough hydrogens to the carbons to give each carbon a valence of 4. This requires three hydrogens on each end carbon and two each on the others.

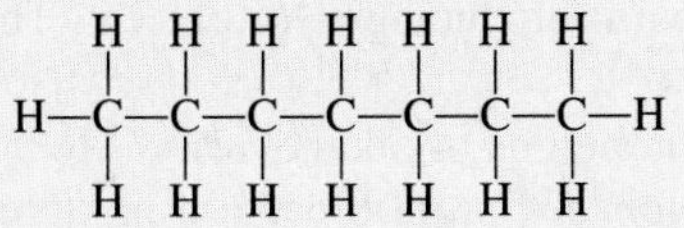

```
    H  H  H  H  H  H  H
    |  |  |  |  |  |  |
 H—C—C—C—C—C—C—C—H
    |  |  |  |  |  |  |
    H  H  H  H  H  H  H
```

For the condensed form, just write each carbon atom's set of hydrogen atoms next to the carbon. Note that each end carbon has three hydrogen atoms, while the carbons in the middle have only two each.

$CH_3CH_2CH_2CH_2CH_2CH_2CH_3$

Exercise 9.1

Give the molecular, structural, and condensed structural formulas for octane.

Properties of Alkanes

Note from Table 9.3 that after the first few members of the alkane series, the rest show about a 20 to 30 °C increase in boiling point with each added CH_2 group. You should also note from the table that at room temperature alkanes with 1 to 4 carbon atoms per molecule are gases, those with 5 to about 16 carbon atoms per molecule are liquids, and those with more than 16 carbon atoms per molecule are solids. Note, too, that the densities of the liquid alkanes are less than that of water. The alkanes are nonpolar molecules and are essentially insoluble in water; hence they float on top of water. Alkanes dissolve many organic substances of low polarity—such as fats, oils, and waxes.

Water has a density of 1.00 g/mL at room temperature.

Table 9.3 Physical Properties of Selected Alkanes

Name	Molecular Formula	Melting Point (°C)	Boiling Point (°C)	Density at 20 °C (g/mL)
Methane	CH_4	−183	−162	(Gas)
Ethane	C_2H_6	−172	−89	(Gas)
Propane	C_3H_8	−188	−42	(Gas)
Butane	C_4H_{10}	−138	0	(Gas)
Pentane	C_5H_{12}	−130	36	0.626
Hexane	C_6H_{14}	−95	69	0.659
Heptane	C_7H_{16}	−91	98	0.684
Octane	C_8H_{18}	−57	126	0.703
Decane	$C_{10}H_{22}$	−30	174	0.730
Dodecane	$C_{12}H_{26}$	−10	216	0.749
Tetradecane	$C_{14}H_{30}$	6	254	0.763
Hexadecane	$C_{16}H_{34}$	18	280	0.775
Octadecane	$C_{18}H_{38}$	28	316	(Solid)
Eicosane	$C_{20}H_{42}$	37	343	(Solid)

Alkanes undergo few chemical reactions. Their most important chemical property is that they burn, producing a lot of heat. They are mainly used as fuels.

The physiological properties of alkanes vary. Methane appears to be inert physiologically. We probably could breathe a mixture of 80% methane and 20% oxygen without ill effect. This mixture would be flammable, however, and no fire or spark of any kind could be permitted in such an atmosphere. Breathing an atmosphere of pure methane (the ''gas'' of a gas-operated stove) can lead to death—not because of the presence of methane but because of the absence of oxygen (asphyxia). Light liquid alkanes, such as those in gasoline, when spilled on the skin, dissolve body oils. Repeated contact may cause dermatitis. Swallowed, alkanes do little harm in the stomach; however, in the lungs, alkanes cause **chemical pneumonia** by dissolving fatlike molecules from the cell membranes in the alveoli. Heavier liquid alkanes, when applied to the skin, act as **emollients** (skin softeners), and petroleum jelly (Vaseline is one brand) is a semisolid mixture of hydrocarbons that can be applied as an emollient or simply as a protective film.

SECTION 9.3

Cyclic Hydrocarbons: Rings and Things

The hydrocarbons we have encountered so far (alkanes) have been composed of open-ended chains of carbon atoms. Carbon and hydrogen atoms also can hook up in other arrangements in which closed rings are formed. The simplest possible ring-containing, or **cyclic**, hydrocarbon has the molecular formula C_3H_6.

```
     H   H
      \ /
       C
      / \
H—C———C—H     or      CH2
   |     |            /  \
   H     H         CH2——CH2
```

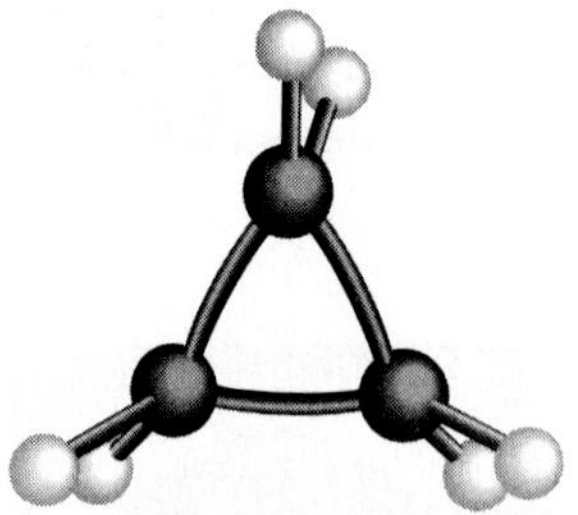

Figure 9.6 Ball-and-stick model of cyclopropane.

The compound is called cyclopropane (Figure 9.6).

Names of cycloalkanes (cyclic compounds containing only single bonds) are formed by addition of the prefix *cyclo-* to the name of the open-chain compound with the same number of carbon atoms as are in the ring.

Cyclopropane is a potent, quick-acting anesthetic with few undesirable side effects. It is seldom used in surgery, however, because it forms explosive mixtures with air at nearly all concentrations.

▶ EXAMPLE 9.2

Give the structure for cyclobutane.

SOLUTION

Cyclobutane has four carbon atoms arranged in cyclic fashion.

```
C—C
|  |
C—C
```

Each carbon atoms needs two hydrogen atoms to complete its set of four bonds.

```
CH2—CH2
 |    |
CH2—CH2
```

Exercise 9.2

Give the structure for cyclopentane.

Sometimes chemists use symbols for these rings (Figure 9.7) instead of drawing out the entire structures. For example, a triangle is used to represent the cyclopropane ring and a square to represent cyclobutane.

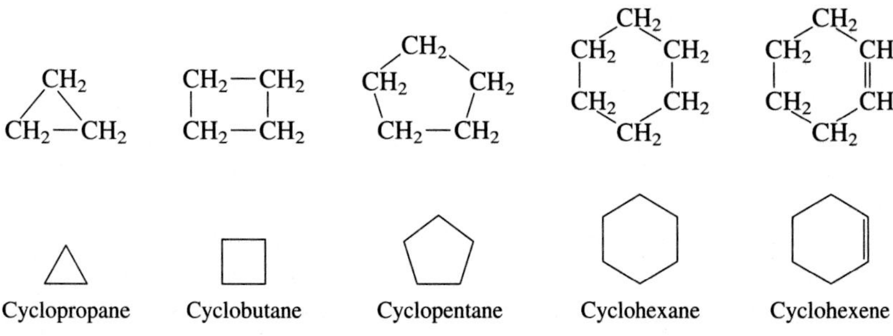

Figure 9.7 Some cyclic hydrocarbons.

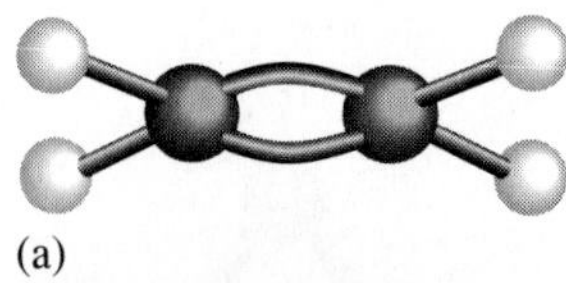
(a)

(b)

Figure 9.8 Ball-and-stick (a) and space-filling (b) models of ethylene. [*(b) Richard Megna/Macmillan Science Files.*]

SECTION 9.4

Unsaturated Hydrocarbons: Alkenes and Alkynes

Two carbon atoms can share more than one pair of electrons. In ethylene (C_2H_4), the two carbon atoms share *two* pairs of electrons and the carbon atoms are joined by a *double bond* (Figure 9.8).

$$\mathrm{H_2C::CH_2} \quad \text{or} \quad \mathrm{H_2C{=}CH_2} \quad \text{or} \quad \mathrm{CH_2{=}CH_2}$$

Ethylene is the most important commercial organic chemical. Annual United States production is about 18 billion kg. More than half of this ethylene goes into the manufacture of polyethylene, one of the most familiar plastics. Another 15% or so is converted to ethylene glycol, the major component of most brands of antifreeze for automobile radiators.

In acetylene (C_2H_2), the two carbon atoms share *three* pairs of electrons; the carbon atoms are joined by a *triple bond* (Figure 9.9).

$$\mathrm{H{:}C{:::}C{:}H} \quad \text{or} \quad \mathrm{H{-}C{\equiv}C{-}H}$$

Acetylene is used in oxyacetylene torches for cutting and welding metals. Such torches can produce very high temperatures. Acetylene is also converted to a variety of other chemical products.

Alkenes can be represented by a general formula C_nH_{2n}. The general formula for alkynes is C_nH_{2n-2}. In both formulas, n is the number of carbon atoms.

Ethylene is the first member of a family of hydrocarbons called **alkenes**. Each alkene contains a carbon-to-carbon double bond. Similarly, acetylene is the first member of the **alkyne** family. Each alkyne contains a triple bond. Collectively, alkenes and alkynes are called **unsaturated hydrocarbons** because they can add more hydrogen atoms to form saturated hydrocarbons (alkanes).

$$\mathrm{H_2C{=}CH_2} + \mathrm{H{-}H} \longrightarrow \mathrm{H{-}CH_2{-}CH_2{-}H}$$

Ethylene (unsaturated) — Ethane (saturated)

$$\mathrm{H{-}C{\equiv}C{-}H} + 2\,\mathrm{H{-}H} \longrightarrow \mathrm{H{-}CH_2{-}CH_2{-}H}$$

Acetylene (unsaturated) — Ethane (saturated)

Alkenes and alkynes typically undergo **addition reactions** in which all the atoms of the reactants are incorporated into a single product. In the reactions here ethylene (C_2H_4) adds hydrogen (H_2) to form ethane (C_2H_6) and acetylene (C_2H_2) adds 2 H_2 to form ethane (C_2H_6).

Unsaturated fats, like the alkenes, contain carbon-to-carbon double bonds.

Properties of Alkenes and Alkynes

The physical properties of alkenes and alkynes are quite similar to those of corresponding alkanes. Those with 2 to 4 carbon atoms per molecule are gases at room temperature, those with 5 to 18 carbon atoms are liquids, and those with more than 18 carbon atoms are solids. Like alkanes, alkenes and alkynes are insoluble in water, and they float on water.

Alkenes and alkynes undergo many more chemical reactions than alkanes. They mainly undergo a wide variety of addition reactions. The addition of hydrogen to double and triple bonds is shown above. Chlorine, bromine, water, and many other kinds of molecules can also add to double and triple bonds.

One of the most unusual features of alkene (and alkyne) molecules is that they can add to each other to form very large molecules called polymers. These interesting molecules are discussed in Chapter 10.

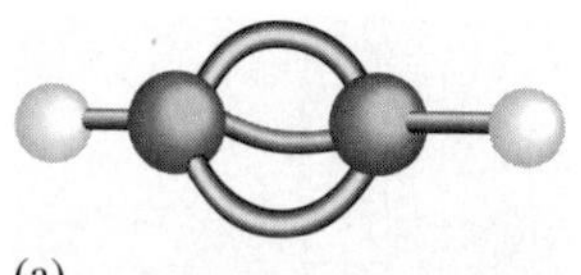

(a)

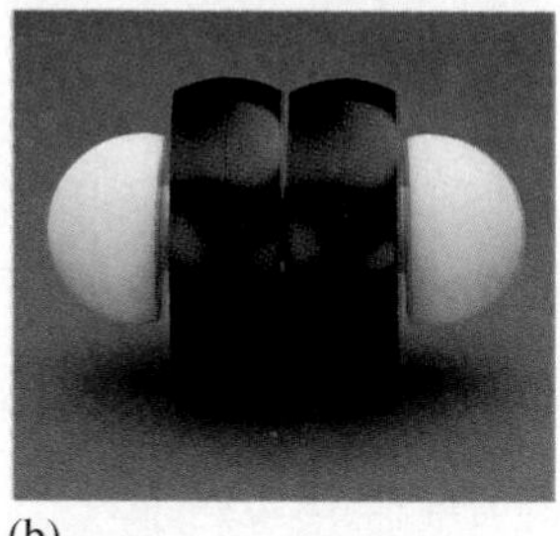

(b)

Figure 9.9 Ball-and-stick (a) and space-filling (b) models of acetylene. [*(b) Richard Megna/Macmillan Science Files.*]

Alkenes occur widely in nature. Ripening fruits and vegetables give off ethylene, which triggers further ripening. Food processors artificially introduce ethylene to hasten the normal ripening process: 1 kg of tomatoes can be ripened by exposure to as little as 0.1 mg of ethylene for 24 hr. Unfortunately, the tomatoes don't taste much like those that ripen on the vine.

Aromatic Hydrocarbons: Benzene and Relatives

Still another type of hydrocarbon is **benzene**. Discovered by Michael Faraday in 1825, it has a molecular formula of C_6H_6. Many different structures can be drawn for this formula. Some examples are:

```
  H H       H H                     H H
  | |       | |                     | |
H—C=C—C≡C—C=C—H     and     H—C≡C—C—C—C≡C—H
                                    | |
                                    H H
```

All of the possible structures you can draw for C_6H_6 have several double or triple bonds. But the remarkable thing about benzene is that it does not seem to contain any double or triple bonds. It does not readily undergo addition reactions the way unsaturated compounds usually do.

For 40 years the structure of benzene remained a mystery. Then in 1865 August Kekulé solved the puzzle by suggesting a ring of six carbon atoms, each attached to one hydrogen atom.

```
        H                      H
        |                      |
  H     C     H          H     C     H
   \  //  \  /            \  /  \\  /
    C       C               C       C
    |       ||      or      ||      |
    C       C               C       C
   /  \\  /  \            /  \  //  \
  H     C     H          H     C     H
        |                      |
        H                      H
```

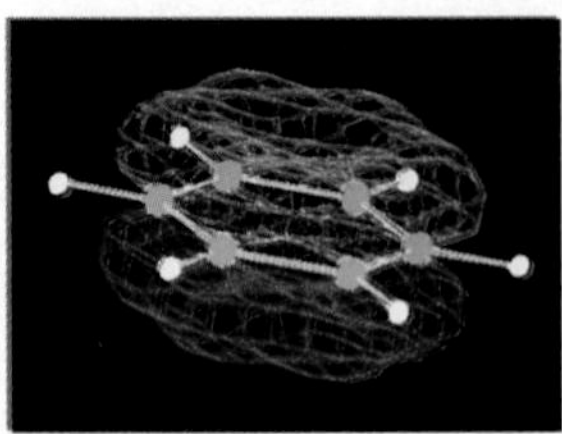

Figure 9.10 A computer-generated model of the benzene molecule. The six unassigned electrons occupy the blue and red areas above and below the plane occupied by the carbon and hydrogen atoms. [*Chemical Design Ltd., Oxford, England.*]

The two structures shown above appear to contain double bonds, but in fact they do not. Both structures represent the very same molecule, and the actual structure is a hybrid of the two. The benzene molecule actually has six identical carbon–carbon bonds that are neither single bonds nor double bonds but something in between. In other words, the three pairs of electrons that would form the three double bonds are not tied down in one location but are smeared around the ring (Figure 9.10). A popular modern way to represent the benzene ring is with a circle inside a hexagon.

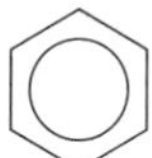

The hexagon represents the ring of six carbon atoms and the inscribed circle the ring of six unassigned electrons. Since its ring of electrons resists being disrupted, the benzene molecule is a very stable structure.

Benzene and similar compounds are often referred to as **aromatic hydrocarbons**. This is because quite a few of the first benzene-like substances to be discovered had strong aromas. Even though many benzene derivatives have turned out to be odorless, the name stuck. Today the term aromatic in organic chemistry is applied to any compound that contains a benzene ring or has certain properties similar to those of benzene.

A circle in a cyclic structure simply indicates an aromatic compound. It does not always mean six electrons. Naphthalene (Figure 9.11), for example, has a circle in each ring to indicate that naphthalene is an aromatic hydrocarbon. The total of unassigned electrons, however, is only ten. Some chemists still prefer to represent naphthalene (and other aromatic compounds) with alternate double and single bonds.

Properties of Aromatic Hydrocarbons

Structures of some common aromatic hydrocarbons are shown in Figure 9.11. Benzene, toluene, and the xylenes are all liquids that float on water. They are used mainly as solvents and fuels, but they are also used to make other benzene derivatives. Their vapors can act as narcotics when inhaled. Since benzene may cause leukemia after long exposure, its use has been restricted. Naphthalene is a volatile white crystalline solid used as an insecticide, especially as a moth repellant.

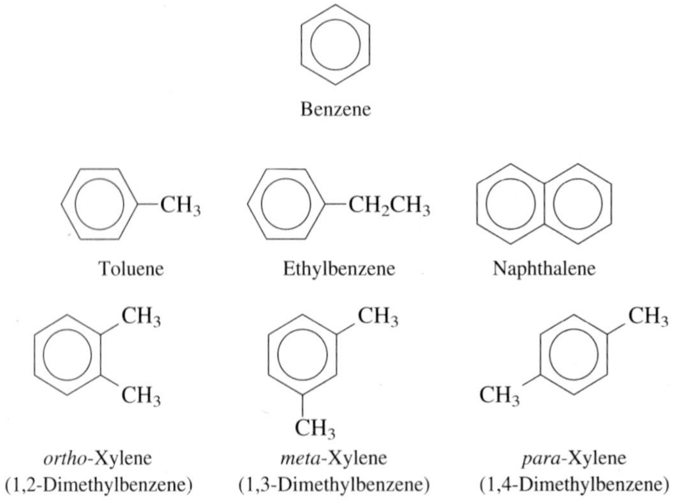

Benzene

CH_3 Toluene

CH_2CH_3 Ethylbenzene

Naphthalene

CH_3 CH_3 *ortho*-Xylene (1,2-Dimethylbenzene)

CH_3 CH_3 *meta*-Xylene (1,3-Dimethylbenzene)

CH_3 CH_3 *para*-Xylene (1,4-Dimethylbenzene)

Figure 9.11 Some aromatic hydrocarbons. Benzene, toluene, and the three xylenes are components of gasoline. They also serve as solvents. All these compounds are important intermediates for the synthesis of polymers (Chapter 10) and other organic chemicals.

SECTION 9.6

Chlorinated Hydrocarbons: Many Uses, Some Hazards

Now that you know what hydrocarbons are, let's look at some chlorinated hydrocarbons. When chlorine gas (Cl_2) is mixed with methane (CH_4) in the presence of ultraviolet light, a reaction takes place at a very rapid (even explosive!) rate. The result is a mixture of products (Figure 9.12), some of which are probably quite familiar to you.

Methyl chloride (CH_3Cl) is mainly used in making silicones (Chapter 10). Methylene chloride (CH_2Cl_2) is a solvent used, for example, as a paint remover. Chloroform was used as an anesthetic in earlier times, but such use is now considered dangerous. The dosage required for effective anesthesia is too close to a lethal dose. Carbon tetrachloride has been used as a dry-cleaning solvent and in fire extinguishers, but it is no longer recommended for either use. Exposure to carbon tetrachloride (or most other chlorinated hydrocarbons) can cause severe damage to the liver. Use of a carbon tetrachloride fire extinguisher in conjunction with water to put out a fire can be deadly. Carbon tetrachloride reacts with water to form phosgene ($COCl_2$), an extremely poisonous gas. In fact, phosgene was used in poison-gas warfare during World War I.

A variety of more complicated chlorinated hydrocarbons are of considerable interest. DDT (dichlorodiphenyltrichloroethane) and other chlorinated hydrocarbons used as insecticides are discussed in Chapter 16. In Chapter 10, we study the PCBs (polychlorinated biphenyls). For now, let's just say that all chlorinated hydrocarbons have similar properties. Most are only slightly polar, and they do not dissolve in water, which is highly polar. Instead, they dissolve (and *dissolve in*) fats, oils, greases, and other substances of low polarity. That is why certain chlorinated hydrocarbons make good dry-cleaning solvents; they remove grease and oily stains from fabrics. That is also why DDT and PCBs cause problems for fish and birds and perhaps for people; the toxic substances are concentrated in fatty animal tissues.

```
     H              H
     |              |
  H—C—Cl         H—C—Cl
     |              |
     H              Cl
  Methyl        Methylene
  chloride       chloride

     H              Cl
     |              |
 Cl—C—Cl        Cl—C—Cl
     |              |
     Cl             Cl
 Chloroform      Carbon
              tetrachloride
```

Figure 9.12 Formulas for the four chlorine derivatives of methane. Methyl chloride is a gas at room temperature. The other three compounds are liquids that are frequently used as solvents.

Chlorofluorocarbons

Carbon compounds containing fluorine as well as chlorine have been used as the dispersing gases in aerosol cans and as refrigerants. Properly called *chlorofluorocarbons (CFCs),* they are perhaps best known as *Freons,* from their Du Pont trade name. Structures of three of the Freons are illustrated in Figure 9.13. At room temperature, the chlorofluorocarbons are gases or liquids with low boiling temperatures. They are essentially insoluble in water and inert toward most other substances. These properties make them ideal propellants for use in aerosol cans of deodorants, hair sprays, and food products. Unfortunately, the inertness of these compounds allows them to persist in the environment. They diffuse into the stratosphere, where they undergo chemical reac-

```
          Cl
          |
      Cl—C—F
          |
          Cl
       Freon 11

     Cl            F  F
     |             |  |
 Cl—C—F        F—C—C—F
     |             |  |
     F             Cl Cl
  Freon 12       Freon 114
```

Figure 9.13 Three chlorofluorocarbons.

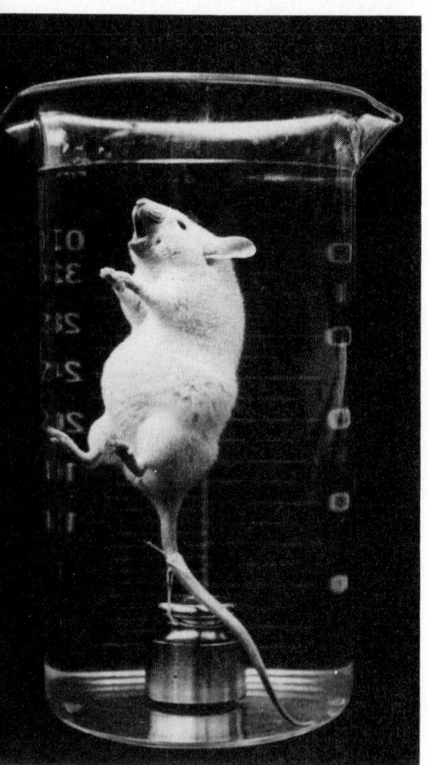

Figure 9.14 This submerged mouse survives by breathing a liquid perfluoro compound saturated with oxygen. [*Courtesy of L. C. Clark, Jr.*]

tions that may lead to the destruction of the ozone layer, which protects the Earth from harmful ultraviolet radiation. A decrease in the amount of ozone in the stratosphere might lead to an increase in the incidence of skin cancer. We look at this problem in more detail in Chapter 12.

Chlorofluorocarbons have been banned in the United States from most aerosol preparations, but they are still used as refrigerants. CFCs still escape into the atmosphere from refrigerators and air conditioners. As propellants in aerosol products, chlorofluorocarbons were replaced by methylene chloride and flammable hydrocarbons such as isobutane. Methylene chloride won't burn, but it in turn has been banned as a possible carcinogen. Other replacements for CFCs are under development.

Perfluorocarbons

Fluorinated compounds have found some interesting uses. Some have been used as blood extenders. Oxygen is quite soluble in *perfluorocarbons.* (The prefix *per-* means that all hydrogen atoms have been replaced, in this case by fluorine atoms.) These compounds can therefore serve as temporary substitutes for hemoglobin, the oxygen-carrying protein in blood (Figure 9.14). Fluosol-SA, a mixture of perfluoro compounds, has been tested in hundreds of patients in Japan. In the United States, it has been used mainly for those who reject transfusions of normal blood for religious reasons.

Teflon, which is a perfluorinated polymer, has many interesting applications because of its resistance to water and to high temperatures. It is especially noted for its unusual nonstick properties. Teflon is discussed in Chapter 10.

SECTION 9.7

The Functional Group

Methyl, ethyl, and propyl alcohols are the first three members of the homologous series of alcohols.

$CH_3—OH$	$CH_3CH_2—OH$	$CH_3CH_2CH_2—OH$
Methyl alcohol	Ethyl alcohol	Propyl alcohol

All of these alcohols have similar properties because they all have a hydroxyl (OH) group. The hydroxyl group is the **functional group** of alcohols, just as the double bond (C$=$C) is the functional group for alkenes, and the triple bond (C$\equiv$C) is the functional group for alkynes. Table 9.4 lists a number of common functional groups found in organic molecules. Each of these groups is the basis for its own homologous series.

In many simple molecules, a functional group is simply attached to a hydrocarbon stem called an **alkyl group**. An alkyl group is just an alkane from

Table 9.4 Selected Organic Functional Groups

Name of Class	Functional Group	General Formula of Class
Alkane	None	$R-H$
Alkene	$-\overset{\vert}{C}=\overset{\vert}{C}-$	$R-\overset{R'}{\overset{\vert}{C}}=\overset{R''}{\overset{\vert}{C}}-R'''$
Alkyne	$-C\equiv C-$	$R-C\equiv C-R'$
Alcohol	$-\overset{\vert}{\underset{\vert}{C}}-O-H$	$R-O-H$
Ether	$-\overset{\vert}{\underset{\vert}{C}}-O-\overset{\vert}{\underset{\vert}{C}}-$	$R-O-R'$
Aldehyde	$-\overset{O}{\overset{\Vert}{C}}-H$	$R-\overset{O}{\overset{\Vert}{C}}-H$
Ketone	$-\overset{O}{\overset{\Vert}{C}}-$	$R-\overset{O}{\overset{\Vert}{C}}-R'$
Amine	$-\overset{\vert}{\underset{\vert}{C}}-\underset{\vert}{N}-$	$R-\overset{H}{\overset{\vert}{N}}-H$
		$R-\overset{H}{\overset{\vert}{N}}-R'$
		$R-\overset{R'}{\overset{\vert}{N}}-R''$
Carboxylic acid	$-\overset{O}{\overset{\Vert}{C}}-O-H$	$R-\overset{O}{\overset{\Vert}{C}}-O-H$
Ester	$-\overset{O}{\overset{\Vert}{C}}-O-\overset{\vert}{\underset{\vert}{C}}-$	$R-\overset{O}{\overset{\Vert}{C}}-O-R'$
Amide	$-\overset{O}{\overset{\Vert}{C}}-\underset{\vert}{N}-$	$R-\overset{O}{\overset{\Vert}{C}}-\underset{\underset{H}{\vert}}{N}-H$
		$R-\overset{O}{\overset{\Vert}{C}}-\underset{\underset{H}{\vert}}{N}-R'$
		$R-\overset{O}{\overset{\Vert}{C}}-\underset{\underset{R''}{\vert}}{N}-R'$

Table 9.5 Common Alkyl Groups

Name	Structural Formula	Condensed Structural Formula
Derived from Methane		
Methyl	H—C(H)(H)—	CH_3—
Derived from Ethane		
Ethyl	H—C(H)(H)—C(H)(H)—	CH_3CH_2—
Derived from Propane		
Propyl	H—C(H)(H)—C(H)(H)—C(H)(H)—	$CH_3CH_2CH_2$—
Isopropyl	H—C(H)(H)—C(H)(—)—C(H)(H)—H	CH_3CHCH_3 (bond from central C)
Derived from Butane		
Butyl	H—C(H)(H)—C(H)(H)—C(H)(H)—C(H)(H)—	$CH_3CH_2CH_2CH_2$—
Secondary butyl (*sec*-butyl)	H—C(H)(H)—C(H)(—)—C(H)(H)—C(H)(H)—H	$CH_3CHCH_2CH_3$ (bond from second C)
Derived from Isobutane		
Isobutyl	H—C(H)(H)—C(H)(CH₃ as H—C(H)—H)—C(H)(H)—	CH_3CHCH_2— with CH_3 on CH
Tertiary butyl (*tert*-butyl)	H—C(H)(H)—C(CH₃ as H—C(H)—H)(—)—C(H)(H)—H	CH_3—C(CH_3)—CH_3 (bond from central C)

which a hydrogen atom has been removed. The methyl group (CH_3—), for example, is derived from methane (CH_4) and the ethyl group (CH_3CH_2—) from ethane (CH_3CH_3). Propane yields two different alkyl groups, depending on whether the hydrogen atom is missing from the end or the middle carbon atom.

$CH_3CH_2CH_2$—	$CH_3\underset{\vert}{C}HCH_3$
Propyl group	Isopropyl group

Table 9.5 lists the four alkyl groups mentioned here plus four others derived from the two butanes.

Often the letter R is used to stand for alkyl groups in general. Thus, ROH is a general formula for alcohols, and RCl represents any alkyl chloride.

SECTION 9.8

The Alcohol Family

When a hydroxyl group (OH) is substituted for any hydrogen atom in an alkane, the molecule becomes an **alcohol** (ROH). Like many organic compounds, alcohols can be called by their common names or by the systematic names of the International Union of Pure and Applied Chemistry (IUPAC). The IUPAC names for alcohols are based on those of the alkanes, but the names end in *-ol.* Methanol and ethanol are two simple examples. Common names for these alcohols, established by everyday usage, are methyl alcohol and ethyl alcohol. The IUPAC names are the ones used in most scientific literature.

Methyl Alcohol (Methanol)

The simplest alcohol is methanol, or methyl alcohol (CH_3OH) (Figure 9.15). Methanol is sometimes called wood alcohol because it can be made by destructive distillation of wood (Figure 9.16). The modern industrial process,

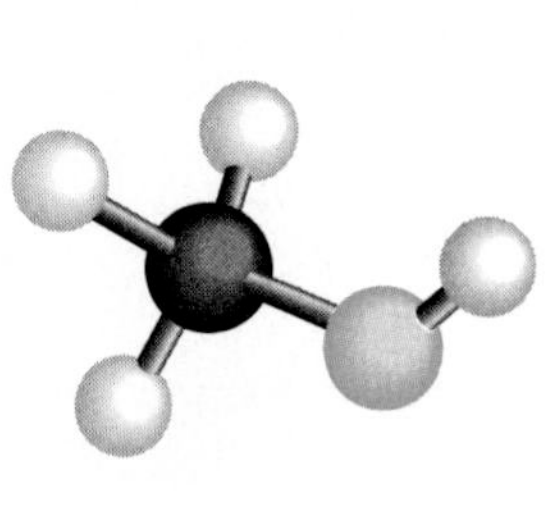

(a)

(b)

Figure 9.15 Ball-and-stick (a) and space-filling (b) models of the methanol molecule. [*(b) Richard Megna/Macmillan Science Files.*]

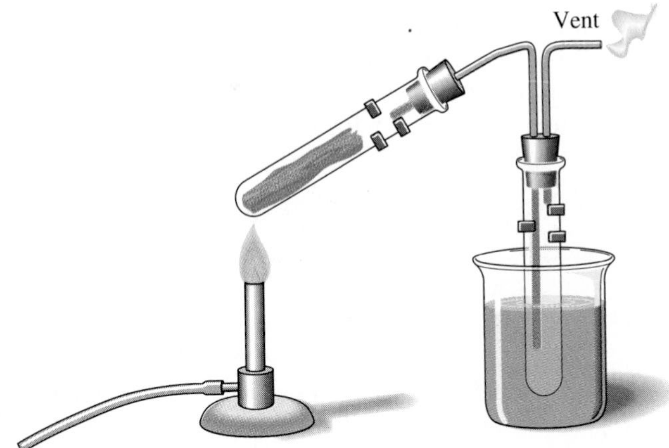

Figure 9.16 An apparatus for the destructive distillation of wood. The wood is heated in an enclosed tube, and alcohol is condensed in the second tube by the cold water in the beaker. Gases formed in the process can be burned as they emerge through the vent tube.

however, makes methyl alcohol from two gases, carbon monoxide and hydrogen, at high temperature and pressure in the presence of a catalyst.

$$CO + 2\,H_2 \xrightarrow{\text{cat.}} CH_3OH$$

Methyl alcohol is important as a solvent and as a chemical intermediate. It may even be a possible replacement for gasoline in automobiles.

Ethyl Alcohol (Ethanol)

The next member of the homologous series of alcohols is ethyl alcohol (CH_3CH_2OH), also called ethanol or grain alcohol (Figure 9.17). Most ethyl alcohol is made by fermentation of grain (or other starchy or sugary materials). If the sugar is glucose, the reaction is

$$C_6H_{12}O_6 \xrightarrow{\text{yeast}} 2\,CH_3CH_2OH + 2\,CO_2$$

All ethyl alcohol for beverages and for automotive fuel is made this way. A much smaller amount of ethanol, for industrial use only, is made by reacting ethylene with water.

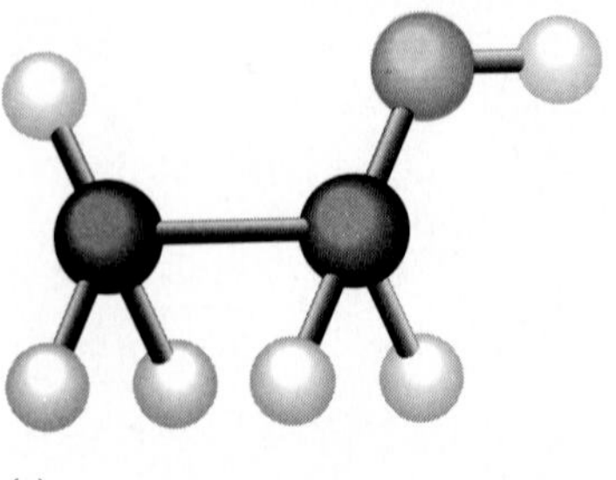

(a)

(b)

Figure 9.17 Ball-and-stick (a) and space-filling (b) models of the ethanol molecule. [*(b) Richard Megna/ Macmillan Science Files.*]

$$CH_2{=}CH_2 + H_2O \xrightarrow{\text{acid}} CH_3CH_2OH$$

This industrial alcohol is exactly the same as that made by fermentation and it is generally cheaper, but by law it cannot be used in alcoholic beverages. Since it carries no excise tax, the law requires that noxious substances be added to the alcohol in order to prevent people from drinking it. The resulting *denatured alcohol* is not fit to drink. This is the kind of alcohol commonly found on the shelves in chemical laboratories.

There is about 10 percent ethyl alcohol in the motor fuel known as "gasohol." It might be assumed that this is industrial alcohol, made from ethylene, but the fact is that it is made by fermentation. Because the United States has a large corn surplus, there is a generous government subsidy for gasohol producers who make their alcohol from corn. This has resulted in an enormous increase in the number of plants making ethyl alcohol by fermentation of corn.

Toxicity of Alcohols

Although ethyl alcohol is an ingredient in wine, beer, and other alcoholic beverages, alcohols are a rather toxic family of substances. Methyl alcohol, for example, is oxidized in the body to formaldehyde (HCHO). Drinking as little as one ounce (about 30 mL, or 2 tablespoonfuls) can cause blindness, and even death. Many poisonings each year result from mistaking methanol for its less toxic relative, ethanol.

Ethyl alcohol may not be as poisonous as methyl alcohol, but it is still toxic. One pint (about 500 mL) of pure ethyl alcohol, rapidly ingested, will kill most people. Of course, even the strongest alcoholic beverages are seldom more than 45% alcohol (90 proof).

Rubbing alcohol is a 70% solution of isopropyl alcohol (equivalent to 140 proof). Since isopropyl is also more toxic than ethyl alcohol, it is not surprising that people get very ill, and sometimes die, from drinking rubbing alcohol.

More than two-thirds of the adult population in the United States drinks alcoholic beverages at least occasionally. The majority do so responsibly. Unfortunately, the nature of people and of alcohol is such that misuse—and abuse—is all too common. It is estimated that there are about 10 million alcoholics in the United States, people so severely addicted that they are unable to hold a steady job or maintain stable family relationships.

Excessive ingestion of ethanol over a long period of time can lead to deterioration of the liver and loss of memory (Figure 9.18). In addition, over half the fatal automobile accidents involve at least one drinking driver. Babies born to alcoholic mothers often are small, deformed, and mentally retarded. Some investigators believe that this *fetal alcohol syndrome* can occur even if the mothers drink only moderately. Alcohol is by far the most abused drug in the United States.

Generally, ethanol acts as a mild *depressant;* it slows down both physical and mental activity. Table 9.6 lists the effects of various doses. Although

People often want to know whether or not something is a *poison.* The question is difficult to answer. Toxicity depends on the nature of the substance, the amount, and the route by which it is taken into the body. Toxicity is discussed in detail in Chapter 20.

The *proof* of an alcoholic beverage is merely twice the percentage of alcohol by volume. The term has its origin in an old seventeenth-century English method for testing whiskey. Dealers were perhaps too often tempted to increase profits by adding water to the booze. A qualitative method for testing the whiskey was to pour some of it on gunpowder and ignite it. If the gunpowder ignited after the alcohol had burned away, that was considered "proof" that the whiskey did not contain too much water.

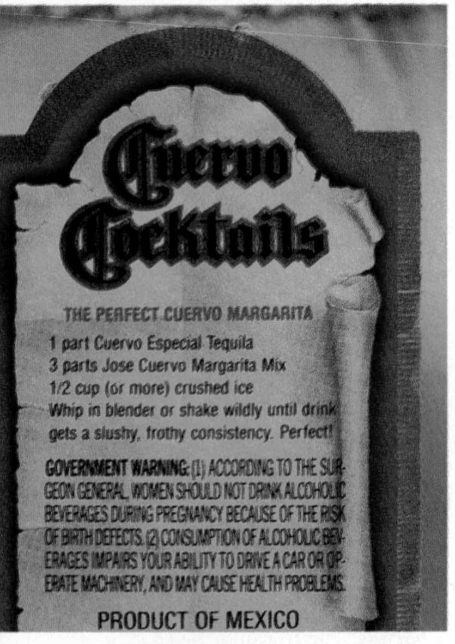

Figure 9.18 Warning labels on alcoholic beverages alert consumers to the hazards of excessive consumption. [*Richard Megna/ Fundamental Photographs.*]

Table 9.6 Approximate Relationship Between Drinks Consumed, Blood-Alcohol Level, and Behavior for a 70-kg (154-lb) Moderate Drinker

Number of Drinks*	Blood-Alcohol Level (percent by volume)	Behavior†
2	0.05	Mild sedation; tranquility
4	0.10	Lack of coordination
6	0.15	Obvious intoxication
10	0.30	Unconsciousness
20	0.50	Possible death

*Rapidly consumed 30-mL (1-oz) "shots" of 90-proof whiskey, 360-mL (12-oz) bottles of beer, or 150-mL (5-oz) glasses of wine.

†An inexperienced drinker would be affected more strongly, or more quickly, than one who is ordinarily a moderate drinker. Conversely, an experienced *heavy* drinker would be affected less.

ethanol generally is a depressant, small amounts seem to act as a stimulant, perhaps by relaxing tensions and relieving inhibitions. Heavy drinking alters brain cell function, causes nerve damage, and shortens the life span by contributing to diseases of the liver, cardiovascular system, and virtually every other organ of the body.

Multifunctional Alcohols

Several alcohols have more than one hydroxyl group. Examples are ethylene glycol, propylene glycol, and glycerol.

```
    H  H             H  H  H             H  H  H
    |  |             |  |  |             |  |  |
 H—C——C—H        H—C——C——C—H        H—C——C——C—H
    |  |             |  |  |             |  |  |
   OH OH             H OH OH            OH OH OH
Ethylene glycol   Propylene glycol       Glycerol
```

Ethylene glycol is the main ingredient in most permanent antifreeze mixtures. Its high boiling point keeps it from boiling away in the automobile radiator. Ethylene glycol is a syrupy liquid with a sweet taste, but it is quite toxic. It is oxidized in the liver to oxalic acid.

```
 OH   OH                          O  O
 |    |                           ‖  ‖
 CH2—CH2   —liver enzymes→   HO—C——C—OH
Ethylene glycol                Oxalic acid
```

The oxalic acid forms crystals of its calcium salt, calcium oxalate (CaC_2O_4), which can damage the kidneys, leading to kidney failure and death. Some manufacturers are beginning to market propylene glycol as a safer permanent antifreeze. It is a high boiling antifreeze, but it is not poisonous.

Glycerol (or glycerin) is a sweet, syrupy liquid made as a by-product from fats during soap manufacture (Chapter 17). It is used in lotions to keep the skin soft and as a food additive to keep cakes moist. It is also reacted with nitric acid to make nitroglycerin, the explosive material in dynamite. Incidentally, nitroglycerin is also important as a vasodilator, a medication taken by heart patients to relieve angina pain.

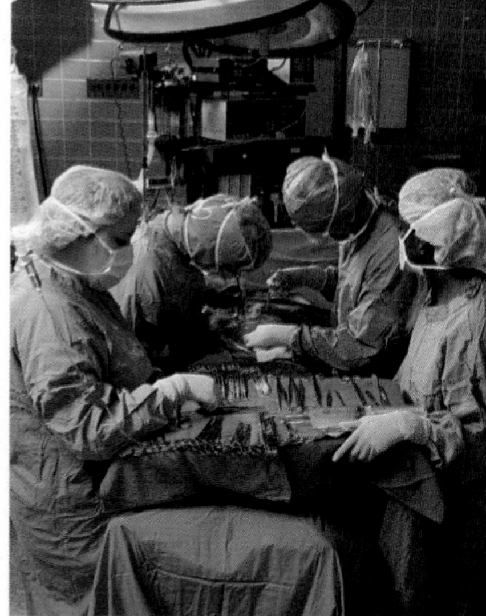

Figure 9.19 Phenolic compounds such as hexylresorcinol assure antiseptic conditions in operating rooms. [*SIU/Photo Researchers, Inc.*]

SECTION 9.9

Phenols

When a hydroxyl group is attached to a benzene ring, the compound is called **phenol**.

$$C_6H_5\text{—OH}$$

Although it may appear to be an alcohol, it actually is not. The benzene ring greatly alters the nature of the hydroxyl group. In fact, phenol is a weak acid (sometimes called *carbolic acid*), and it is highly poisonous.

Phenol was the first antiseptic used in an operating room. It was used by Joseph Lister in 1867. Up until that time surgery was not antiseptic, and many patients died from infections following surgical operations. Although phenol has strong germicidal action, it is not an ideal antiseptic because it causes severe skin burns, and it kills healthy cells along with harmful microorganisms.

Phenol is still sometimes used as a disinfectant (for floors and furniture), but other phenolic compounds are now used as antiseptics. Hexylresorcinol, for example, is a more powerful germicide than phenol, but it is less damaging to the skin and has fewer other side effects (Figure 9.19).

OH

OH

$CH_2CH_2CH_2CH_2CH_2CH_3$

Hexylresorcinol

A benzene ring with hydroxyl groups at both ends of the ring is called hydroquinone.

$$\text{HO—}C_6H_4\text{—OH}$$

Hydroquinone

It is commonly used as a photographic developer.

SECTION 9.10

Ethers

Compounds with two alkyl groups attached to the same oxygen atom are called **ethers**. The general formula is ROR or ROR′, since the alkyl groups need not be alike. Most well-known is diethyl ether ($CH_3CH_2—O—CH_2CH_3$), often called simply "ether."

Diethyl ether is mainly used as a solvent, since it dissolves many organic substances that are insoluble in water. It boils at 36 °C, so it evaporates readily, making it easy to recover dissolved materials. Although diethyl ether has little chemical reactivity, it is highly flammable, and great care must be used to avoid sparks or flames when ether is in use. People who have used ether to extract cocaine from various mixtures, a practice known as "free-basing," have often found that out the hard way.

Another problem with ethers is that over time they can react slowly with oxygen to form unstable peroxides, which may decompose explosively! Beware of previously opened containers of ether, especially old ones.

Diethyl ether was once the most widely used general anesthetic for surgery (since 1842), but nowadays it is rarely used as an anesthetic for humans, since it has some undesirable side effects, such as postanesthetic nausea and vomiting. There are many other general anesthetics available these days, including several ethers. Methyl propyl ether ($CH_3—O—CH_2CH_2CH_3$) and methoxyflurane ($CHCl_2CF_2—O—CH_3$), for example, are anesthetics known as Neothyl and Penthrane, respectively.

An ether that has been produced in very large volume over the past decade is methyl *tert*-butyl ether (MTBE).

$$CH_3—O—\overset{\displaystyle CH_3}{\underset{\displaystyle CH_3}{\overset{|}{\underset{|}{C}}}}—CH_3$$

It is being added to gasoline both to reduce the emission of carbon monoxide (CO) in automotive exhaust gas and to replace tetraethyllead as an antiknock agent (Chapter 14).

The ether produced in largest amount commercially is a compound called ethylene oxide.

$$\begin{array}{c} H_2C—CH_2 \\ \diagdown\ \diagup \\ O \end{array}$$

It is a cyclic ether with two carbon atoms and an oxygen atom forming a three-membered ring. Ethylene oxide is a toxic gas, more than 60% of which is used to make ethylene glycol.

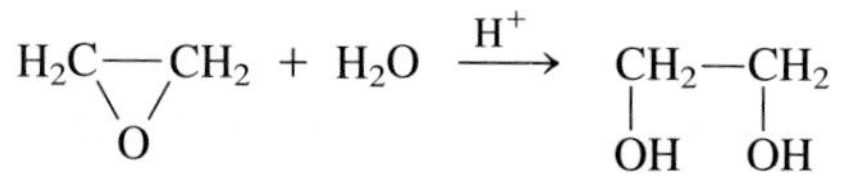

$$H_2C\text{—}CH_2 \text{ (ring with O)} + H_2O \xrightarrow{H^+} CH_2(OH)\text{—}CH_2(OH)$$

Ethylene oxide Water Ethylene glycol

Most of the glycol is used in making polyester fibers (Chapter 10) and antifreeze.

▶ EXAMPLE 9.3

Classify each of the following as an alcohol, ether, or phenol.

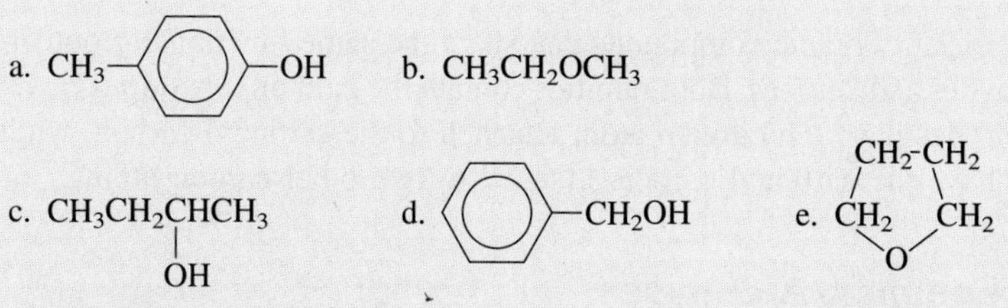

a. $CH_3\text{—}C_6H_4\text{—}OH$ b. $CH_3CH_2OCH_3$

c. $CH_3CH_2CH(OH)CH_3$ d. $C_6H_5\text{—}CH_2OH$ e. $CH_2\text{-}CH_2$, CH_2 CH_2, O (five-membered ring)

SOLUTION

a. The OH is attached directly to the benzene ring; the compound is a phenol.

b. The O atom is between two C atoms; the compound is an ether.

c. The OH group attached to an alkyl group makes the compound an alcohol.

d. An alcohol; the OH is *not* attached *directly* to the benzene ring.

e. An ether; the O atom is between two C atoms.

Exercise 9.3

Classify the following as alcohols, ethers, or phenols.

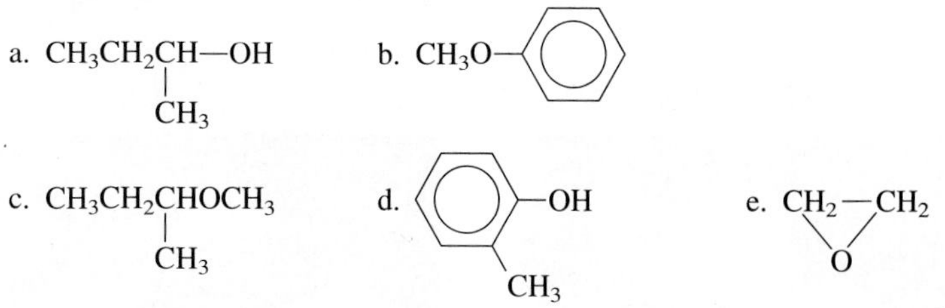

a. $CH_3CH_2CH(CH_3)\text{—}OH$ b. $CH_3O\text{—}C_6H_5$

c. $CH_3CH_2CH(CH_3)OCH_3$ d. $C_6H_4(CH_3)\text{—}OH$ e. $CH_2\text{—}CH_2$ (ring with O)

▶ EXAMPLE 9.4

Give the formula for dimethyl ether.

SOLUTION
Dimethyl ether has an oxygen atom joined to two methyl groups; the formula is CH_3OCH_3.

Exercise 9.4
Give the formula for ethyl methyl ether.

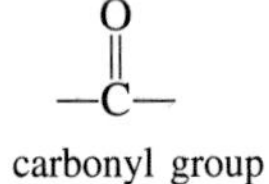

A carbonyl group

R—C(=O)—H
An aldehyde

R—C(=O)—R′
A ketone

SECTION 9.11

Aldehydes and Ketones

Two families of organic compounds that share the same functional group are the aldehydes and ketones. Both families contain the **carbonyl group** (C=O), but **aldehydes** have a hydrogen atom attached to the carbonyl carbon, while **ketones** have the carbonyl carbon attached to two other carbon atoms.

Some Common Aldehydes

The simplest aldehyde is formaldehyde (Figure 9.20). It is a gas at room temperature but is readily soluble in water. As a 40% solution called *formalin,* it is used as a preservative for biological specimens and in embalming fluid.

Figure 9.20 Ball-and-stick (a) and space-filling (b) models of formaldehyde, acetaldehyde, and acetone. [*(b) Richard Megna/Macmillan Science Files.*]

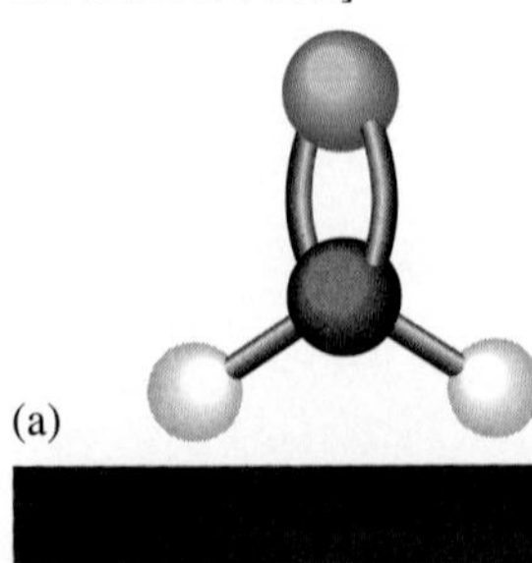
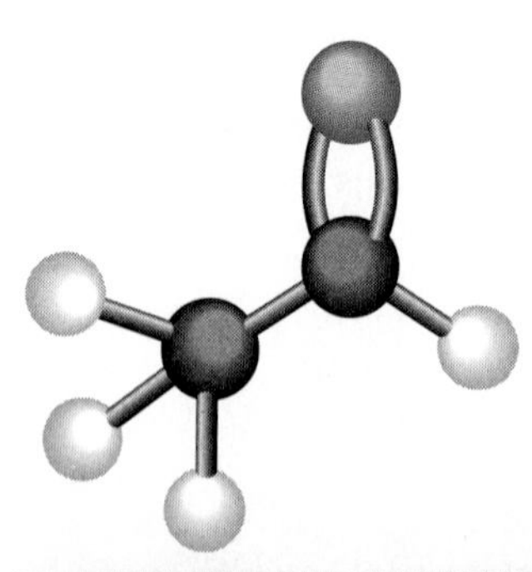
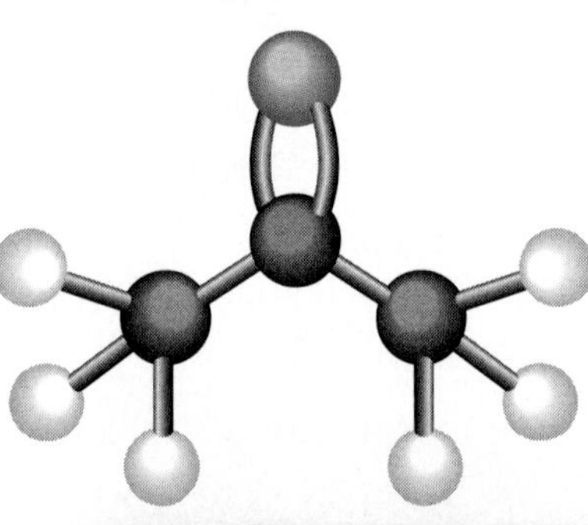

(a)

(b)

Formaldehyde is used to make certain plastics (Chapter 10). It also is used to disinfect homes, ships, and warehouses. Commercially, formaldehyde is made by the oxidation of methanol. This is the same reaction that occurs in the human body when methanol is ingested.

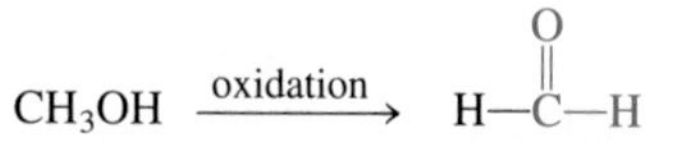

$$\underset{\text{Methanol}}{CH_3OH} \xrightarrow{\text{oxidation}} \underset{\text{Formaldehyde}}{H{-}\overset{\overset{\displaystyle O}{\|}}{C}{-}H}$$

The next member of the homologous series of aldehydes is acetaldehyde (see Figure 9.20), formed by the oxidation of ethanol.

$$\underset{\text{Ethanol}}{CH_3CH_2OH} \xrightarrow{\text{oxidation}} \underset{\text{Acetaldehyde}}{CH_3{-}\overset{\overset{\displaystyle O}{\|}}{C}{-}H}$$

The next two members of the aldehyde series are propionaldehyde and butyraldehyde. Both have strong unpleasant odors.

$$\underset{\text{Propionaldehyde}}{CH_3CH_2{-}\overset{\overset{\displaystyle O}{\|}}{C}{-}H} \qquad \underset{\text{Butyraldehyde}}{CH_3CH_2CH_2{-}\overset{\overset{\displaystyle O}{\|}}{C}{-}H}$$

Benzaldehyde has an aldehyde group attached to a benzene ring.

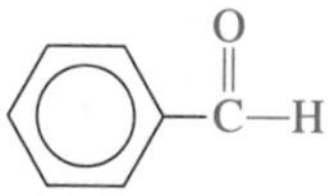

$$\underset{\text{Benzaldehyde}}{C_6H_5{-}\overset{\overset{\displaystyle O}{\|}}{C}{-}H}$$

Also called (synthetic) oil of almond, benzaldehyde is used in perfumery and flavoring. It is the flavor ingredient in maraschino cherries.

Some Common Ketones

The simplest ketone is acetone (see Figure 9.20), made by the oxidation of isopropyl alcohol.

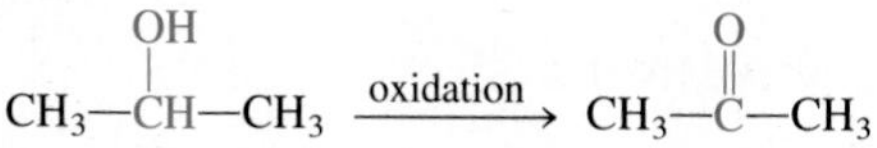

Organic chemists often write equations that show only the organic reactants and products. Inorganic substances are omitted for the sake of simplicity.

$$\underset{\text{Isopropyl alcohol}}{CH_3{-}\overset{\overset{\displaystyle OH}{|}}{C}H{-}CH_3} \xrightarrow{\text{oxidation}} \underset{\text{Acetone}}{CH_3{-}\overset{\overset{\displaystyle O}{\|}}{C}{-}CH_3}$$

Acetone is a common solvent for such organic materials as fats, rubbers, plastics, and varnishes. It also finds use in paint and varnish removers. It is the major (sometimes the only) ingredient in fingernail polish removers.

Two other familiar ketones are ethyl methyl ketone and isobutyl methyl ketone, which, like acetone, frequently are used as solvents.

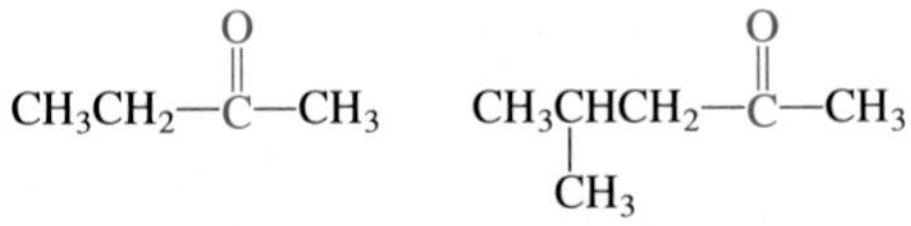

$$CH_3CH_2-\overset{\overset{\Large O}{\|}}{C}-CH_3 \qquad CH_3\underset{\underset{\Large CH_3}{|}}{C}HCH_2-\overset{\overset{\Large O}{\|}}{C}-CH_3$$

Ethyl methyl ketone   Isobutyl methyl ketone

EXAMPLE 9.5

Identify each of the following compounds as an aldehyde or ketone.

a. 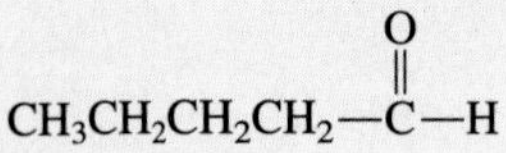 $CH_3CH_2CH_2CH_2-\overset{\overset{\Large O}{\|}}{C}-H$

b. $\begin{array}{l} CH_2-C{=}O \\ \;|\qquad\;\; | \\ CH_2-CH_2 \end{array}$

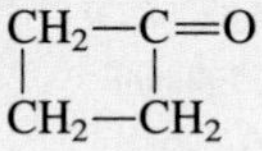

c. $C_6H_5-\overset{\overset{\Large O}{\|}}{C}-CH_3$ (benzene ring)

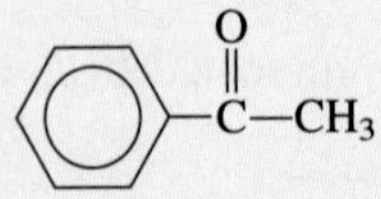

SOLUTION

a. There is an H attached to the carbonyl carbon atom; the compound is an aldehyde.

b. The carbonyl group is between two other carbon atoms; the compound is a ketone.

c. A ketone. (Remember that the corner of the hexagon stands for a carbon atom.)

Exercise 9.5

Identify each of the following compounds as an aldehyde or ketone.

a. 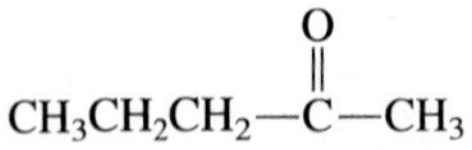 $CH_3CH_2CH_2-\overset{\overset{\Large O}{\|}}{C}-CH_3$

b. $\begin{array}{l} CH_2-CH-\overset{\overset{\Large O}{\|}}{C}-H \\ \;|\qquad\;\;\; | \\ CH_2-CH_2 \end{array}$

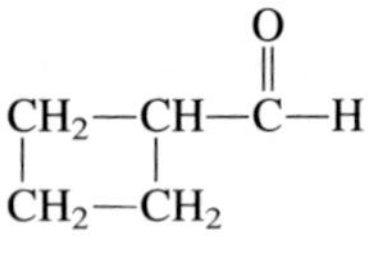

c. $CH_3-C_6H_4-\overset{\overset{\Large O}{\|}}{C}-H$ (benzene ring)

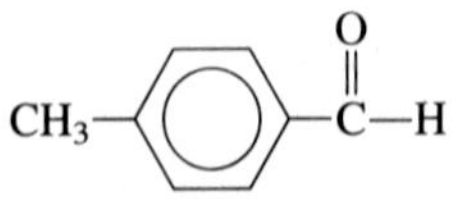

Aldehydes and Ketones

A *ketone* is a carbon chain
With carbonyl *inside.*
If carbonyl is on the *end,*
Then it's an *aldehyde.*

SECTION 9.12

Organic Acids

The functional group of the organic acids is called the **carboxyl group**, and the acids are called **carboxylic acids**. The simplest carboxylic acid is formic acid. It was first obtained by the destructive distillation of ants (Latin *formica*: ''ant''). The bite of an ant smarts because the ant injects formic acid as it bites. The stings of wasps and bees also contain formic acid (as well as other poisonous materials).

Acetic acid can be made by the aerobic fermentation of a mixture of cider and honey. This produces a solution (vinegar) that contains about 4 to 10% acetic acid, plus a number of other compounds that give vinegar its flavor. Acetic acid is probably the most familiar *weak* acid used in educational and industrial chemistry laboratories (Figure 9.21).

The third member of the homologous series of acids, propionic acid, is seldom encountered in everyday life. The fourth member is more familiar, at least by its odor. If you've ever smelled rancid butter, you know what butyric acid smells like. It is one of the most foul-smelling substances imaginable. Butyric acid can be isolated from butterfat or synthesized in the laboratory. It is one of the ingredients of body odor. Extremely small amounts of this and other chemicals enable bloodhounds to track fugitives.

The acid with a carboxyl group attached directly to a benzene ring is called benzoic acid. In general, carboxylic acids can be represented by the formula RCOOH.

$$-\overset{\displaystyle O}{\overset{\|}{C}}-OH$$
or ($-COOH$)
The carboxyl group

$$H-\overset{\displaystyle O}{\overset{\|}{C}}-OH$$
Formic acid

$$CH_3-\overset{\displaystyle O}{\overset{\|}{C}}-OH$$
Acetic acid

$$CH_3CH_2-\overset{\displaystyle O}{\overset{\|}{C}}-OH$$
Propionic acid

$$CH_3CH_2CH_2-\overset{\displaystyle O}{\overset{\|}{C}}-OH$$
Butyric acid

$$C_6H_5-\overset{\displaystyle O}{\overset{\|}{C}}-OH$$
Benzoic acid

Figure 9.21 Acetic acid is a familiar laboratory weak acid. It is also the active ingredient in vinegar. [*Richard Megna/Fundamental Photographs.*]

Carboxylic acid salts, calcium propionate, sodium benzoate, and others, are widely used as food additives to prevent molds (Chapter 16).

▶ EXAMPLE 9.6

Give the structural formula for each of the following compounds.

a. propionaldehyde
b. butyl alcohol
c. acetic acid
d. ethyl methyl ketone

SOLUTION

a. Propionaldehyde has three carbon atoms with an aldehyde function.

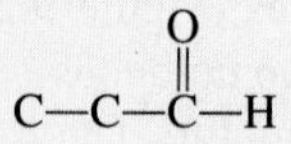

```
        O
        ‖
  C—C—C—H
```

Adding the proper number of hydrogen atoms to the other two carbon atoms gives the structure

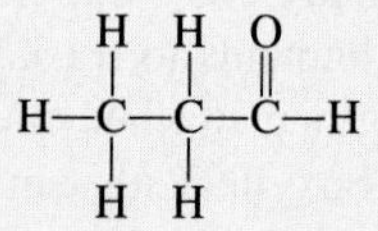

```
    H  H  O
    |  |  ‖
  H—C—C—C—H
    |  |
    H  H
```

b. Butyl alcohol has four carbon atoms with an alcohol function at one end.

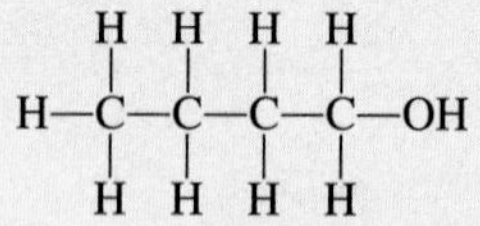

```
    H  H  H  H
    |  |  |  |
  H—C—C—C—C—OH
    |  |  |  |
    H  H  H  H
```

c. Acetic acid has two carbon atoms with a carboxylic acid function.

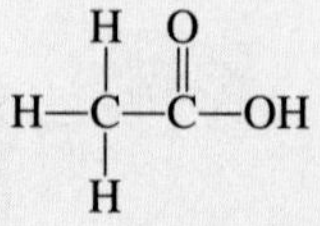

```
    H  O
    |  ‖
  H—C—C—OH
    |
    H
```

d. Ethyl methyl ketone has a ketone function between an ethyl and a methyl group.

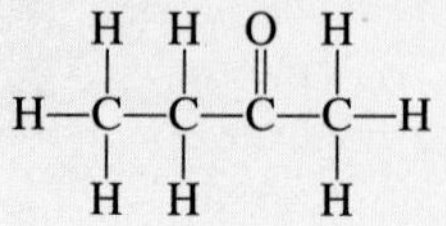

```
    H  H  O  H
    |  |  ‖  |
  H—C—C—C—C—H
    |  |     |
    H  H     H
```

Exercise 9.6

Give the structural formula for each of the following compounds.

a. butyric acid
b. propyl alcohol
c. acetaldehyde
d. diethyl ketone

Acids and Esters

Unless the hydrocarbon chain is reasonably long,
A carboxylic acid's apt to smell both foul and strong.
An ester, on the other hand, is so extremely sweet
It can be used in perfumes and delicious things to eat.

SECTION 9.13

Esters: The Sweet Smell of RCOOR′

Esters are derived from carboxylic acids and alcohols. The general reaction involves splitting out a molecule of water.

$$\underset{\text{An acid}}{\text{R}-\overset{\overset{\displaystyle\text{O}}{\|}}{\text{C}}-\text{OH}} + \underset{\text{An alcohol}}{\text{R}'\text{OH}} \xrightarrow{\text{H}^+} \underset{\text{An ester}}{\text{R}-\overset{\overset{\displaystyle\text{O}}{\|}}{\text{C}}-\text{OR}'} + \text{H}_2\text{O}$$

Although carboxylic acids often have powerfully unpleasant odors, the esters derived from them are usually quite fragrant, especially when dilute. Many esters have fruity odors and tastes. Some examples are given in Table 9.7. They are widely used as flavorings in cakes, candies, and other foods and as ingredients in perfumes.

Table 9.7 Ester Flavors and Fragrances

Ester	Flavor/Fragrance
Methyl butyrate	Apple
Ethyl butyrate	Pineapple
Propyl acetate	Pear
Pentyl acetate	Banana
Pentyl butyrate	Apricot
Octyl acetate	Orange
Methyl benzoate	Ripe kiwi
Ethyl formate	Rum
Methyl salicylate	Wintergreen
Benzyl acetate	Jasmine

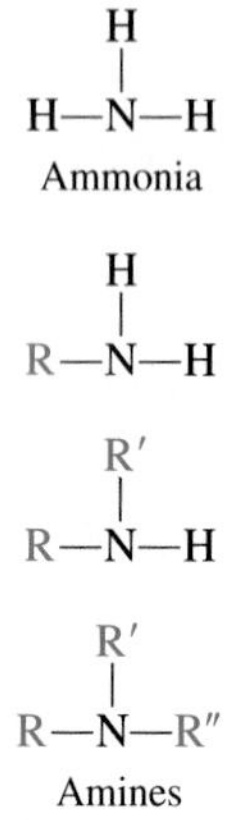

$$\begin{array}{c} H \\ | \\ H{-}N{-}H \end{array}$$

Ammonia

$$\begin{array}{c} H \\ | \\ R{-}N{-}H \end{array} \qquad \begin{array}{c} R' \\ | \\ R{-}N{-}H \end{array} \qquad \begin{array}{c} R' \\ | \\ R{-}N{-}R'' \end{array}$$

Amines

Figure 9.22 Ammonia and three kinds of amines derived from it by replacement of one, two, or all three of the hydrogen atoms by alkyl groups.

In naming an ester, the part from the alcohol is always named first, and the part from the carboxylic acid is named last, with the name ending in *-ate*.

SECTION 9.14

Amines and Amides: Nitrogen-Containing Organics

Many organic compounds contain nitrogen. We encounter a variety of such compounds in the chapters that follow. In this chapter, we introduce two families that will provide you with a vital background for the material ahead—the amines and the amides.

The **amines** contain the elements carbon, hydrogen, and nitrogen. They are derived from ammonia by replacing one, two, or three of the hydrogen atoms by an alkyl group (Figure 9.22).

The simplest amine is methylamine (CH_3NH_2). The next higher homolog is ethylamine ($CH_3CH_2NH_2$). With two carbon atoms, however, we can also have dimethylamine (CH_3NHCH_3). Note that ethylamine and dimethylamine are isomers; both have the molecular formula C_2H_7N. With three carbon atoms, there are several possibilities, including trimethylamine ($\begin{array}{c} CH_3NCH_3 \\ | \\ CH_3 \end{array}$).

▶ EXAMPLE 9.7

Give structures and names for the other three-carbon amines.

SOLUTION

Two are derived from the two propyl groups.

$$CH_3CH_2CH_2NH_2 \qquad \begin{array}{c} CH_3CHCH_3 \\ | \\ NH_2 \end{array}$$

Propylamine Isopropylamine

The other has one methyl group and one ethyl group.

$$CH_3CH_2NHCH_3$$

Ethylmethylamine

Exercise 9.7

Give structures for the following amines.

a. butylamine
b. diethylamine
c. methylpropylamine
d. isopropylmethylamine

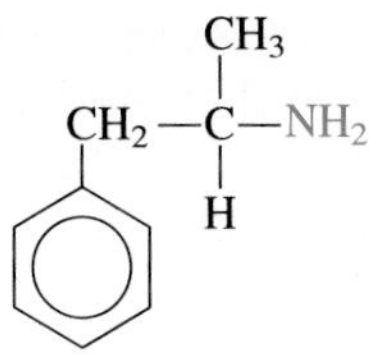

$$C_6H_5-CH_2-CH(CH_3)-NH_2$$

Amphetamine
(a stimulant drug)

$$H_2N-CH_2CH_2CH_2CH_2CH_2CH_2-NH_2$$

1,6-Hexanediamine
(an intermediate in the synthesis of nylon)

$$H_2N-CH_2CH_2CH_2CH_2CH_2-NH_2$$

Cadaverine
(odor of decaying flesh)

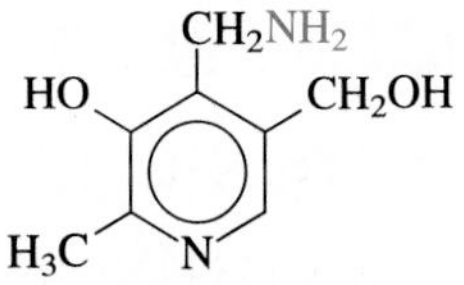

Pyridoxamine
(a B-complex vitamin)

Figure 9.23 Some amines of interest

The amine with an NH_2 group attached directly to a benzene ring has the special name *aniline*.

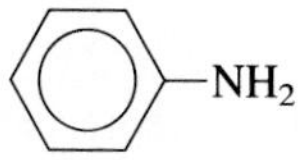

$$C_6H_5-NH_2$$

Aniline

Aniline, like many other aromatic amines, is mainly used in making dyes. Aromatic amines tend to be rather toxic, and some are strongly carcinogenic.

The simple amines are similar to ammonia in odor, basicity, and other properties. It is the higher amines that are the most interesting, though. Figure 9.23 includes a variety of these. Notice that each structure contains an $-NH_2$ unit. This is called the *amino group.*

Among the most important kinds of organic molecules are the **amino acids**. As the name implies, these compounds have carboxylic acid and amine functional groups. The amino acids are the building blocks from which proteins are constructed. The simplest amino acid is called glycine.

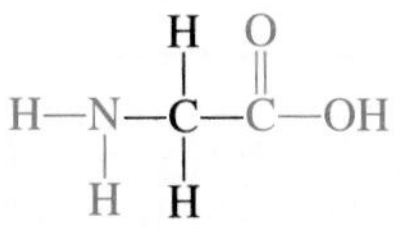

$$H-N(H)-CH_2-C(=O)-OH$$

Glycine

Amino acids and proteins are considered in detail in Chapter 15.

Another important group of nitrogen-containing compounds are the **amides**. (What a difference that one letter makes!) Amides contain oxygen as well as carbon, hydrogen, and nitrogen. In an amide the nitrogen atom is attached directly to a carbonyl group.

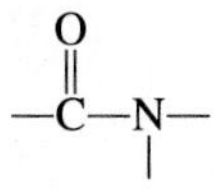

$$-C(=O)-N<$$

Amide functional group

The simple amides are of little interest to us here, but the complex amides are of tremendous importance. Nylon, silk, and wool molecules all contain hundreds of amide functional groups. Your body contains many kinds of proteins, and they all are held together by amide linkages (Chapter 15).

SECTION 9.15

Heterocyclic Compounds: Alkaloids and Others

Cyclic hydrocarbons feature rings of carbon atoms. Now let's look at some ring compounds that have atoms other than carbon within the ring. These **heterocyclic compounds** usually have one or more nitrogen, oxygen, or sulfur atoms.

EXAMPLE 9.8

Which of the following structures represent heterocyclic compounds?

a. O over $CH_2—CH_2$ b. S over $CH_2—CH_2$ c. CH_2 over $CH_2—CH_2$ d. NH over $CH_2—CH_2$

SOLUTION

Compounds a, b, and d have oxygen, sulfur, and nitrogen atoms, respectively, in a ring structure; these represent heterocyclic compounds.

Exercise 9.8

Which of the following structures represent heterocyclic compounds?

a. N, N (six-membered ring) b. (benzene ring)—NH_2 c. (six-membered ring) d. N (six-membered ring)—CH_3

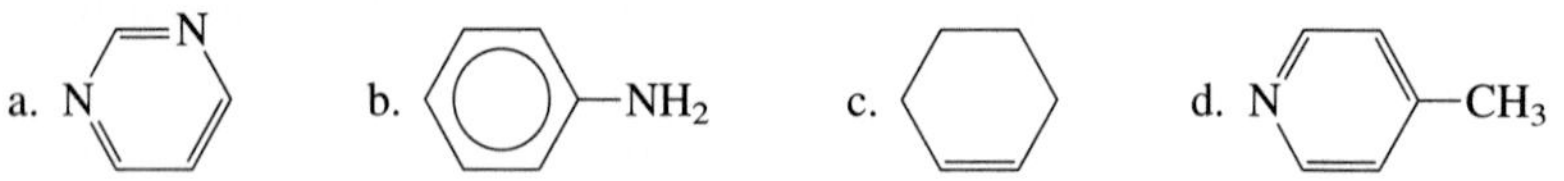

Many amines, particularly heterocyclic ones, occur naturally in plants. Like other amines, these compounds are basic. They are called **alkaloids**, which means ''like alkalis.'' Among the familiar alkaloids are morphine, caffeine, nicotine, and cocaine. The action of these compounds as drugs is considered in Chapter 19. Of more immediate interest are pyrimidine, which has two nitrogen atoms in a six-membered ring, and purine, which has four nitrogen atoms in two rings that share a common side.

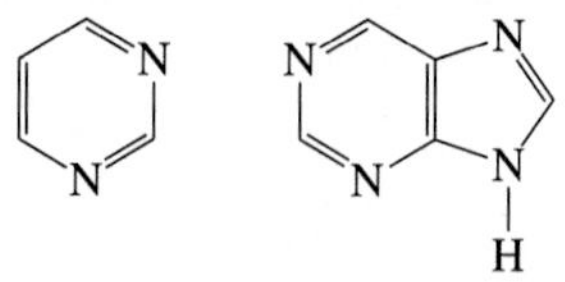

N N N N N N H

Pyrimidine Purine

Compounds related to these are constituents of the nucleic acids (Chapter 15).

Summary

1. Organic chemistry is the study of carbon compounds—hydrocarbons and their derivatives. Over 95% of all known compounds contain carbon.
2. *Hydrocarbons* contain only carbon and hydrogen. They include alkanes, alkenes, alkynes, and aromatic hydrocarbons.
3. *Alkanes,* which contain only single bonds, have names ending in -ane. They may be straight-chained, branched-chained, or cyclic.
4. *Alkenes* contain at least one double bond, and their names end in -ene. *Alkynes* contain at least one triple bond, and their names end in -yne. Both of these classes of compounds are *unsaturated hydrocarbons.*
5. The members of a *homologous series* of molecules differ one from another only by one or more CH_2 groups.
6. *Isomers* are molecules that have the same molecular formula but different structures. Because there is so much isomerism in organic chemistry, *structural formulas* are generally used rather than molecular formulas.
7. *Aromatic hydrocarbons* mainly include *benzene* and its derivatives. The symbol for benzene is a hexagon with an inscribed circle, representing a ring of six electrons.
8. Hydrocarbons form many kinds of derivatives, which behave according to the *functional groups* they contain.
9. *Alcohols* are hydrocarbon derivatives containing —OH groups.
10. An *ether* contains an oxygen atom between two alkyl groups.
11. *Aldehydes* and *ketones* both contain the *carbonyl group.* In aldehydes the carbonyl is on the end of the molecule.
12. *Carboxylic acids* contain the *carboxyl group,* an —OH group attached to the carbon atom of a carbonyl group.
13. An *ester* is a compound made by reacting a carboxylic acid with an alcohol. Most esters are very fragrant.
14. An *amine* is a hydrocarbon derivative of ammonia.
15. An *amide* contains a carbonyl group, the carbon atom of which is attached to an amino nitrogen.
16. *Alkaloids* are amines, especially *heterocyclic* amines, that occur naturally in plants.
17. *Amino acids,* which have both carboxylic acid and amine functional groups, are the building blocks of proteins.

Key Terms

alcohol 9.8
aldehyde 9.11
alkaloid 9.15
alkane 9.2
alkene 9.4
alkyl group 9.7
alkyne 9.4
amide 9.14
amine 9.14
amino acid 9.14
aromatic hydrocarbon 9.5
benzene 9.5
carbonyl group 9.11
carboxyl group 9.12
carboxylic acid 9.12
catenation 9.1
cyclic hydrocarbon 9.3
ester 9.13
ether 9.10
functional group 9.7
heterocyclic compound 9.15
homologous series 9.2
hydrocarbon 9.2
isomer 9.2
ketone 9.11
phenol 9.9
structural formula 9.2
unsaturated hydrocarbon 9.4

Review Questions

1. What is organic chemistry?
2. What is catenation?
3. List three characteristics of the carbon atom that make possible the existence of millions of organic compounds.

4. Define, illustrate, or give an example of each of these terms.
   a. hydrocarbon b. alkyne
   c. alkane d. alkene
5. What is a homologous series?
6. What are isomers? How can you tell whether or not two compounds are isomers?
7. What is an unsaturated hydrocarbon? Give examples of two types.
8. What is the meaning of the circle inside the hexagon in the modern structure of benzene?
9. What is an aromatic hydrocarbon? How can you recognize an aromatic compound from its structure?
10. Which alkanes are gases at room temperature? Which are liquids? Which are solids? State your answers by the number of carbon atoms per molecule.
11. Compare the densities of liquid alkanes to that of water.
12. You add some hexane to water in a beaker. What would you expect to observe?
13. What are the chemical names for the alcohols known by the following familiar names?
   a. grain alcohol b. rubbing alcohol
   c. wood alcohol
14. What are some of the long-term effects of excessive ethanol consumption?
15. Give an important historical use for diethyl ether. What is its main use today?
16. Give an important use for phenols.
17. What structural feature distinguishes aldehydes from ketones?
18. What is an alkaloid?
19. How do carboxylic acids and esters differ in odor?
20. For what family is $R\overset{O}{\overset{\|}{C}}OR'$ the general formula?

Problems

Organic and Inorganic

21. Classify the following compounds as organic or inorganic.
   a. C_6H_{10} b. CH_3NH_2
   c. $CoCl_2$ d. $NaNH_2$
22. Classify the following compounds as organic or inorganic.
   a. CsCl b. CH_3Cl
   c. $C_{12}H_{22}O_{11}$ d. $Cu(NH_3)_6Cl_2$

Names and Formulas

23. How many carbon atoms are there in each of the following?
   a. ethane b. cyclobutane
   c. heptane d. pentane
24. How many carbon atoms are there in each of the following?
   a. propane b. cyclopentane
   c. ethylene d. nonane
25. Name these hydrocarbons.
   a. CH_3CH_3 b. $HC{\equiv}CH$
   c. $CH_2{=}CH_2$
26. Name these hydrocarbons.
   a. 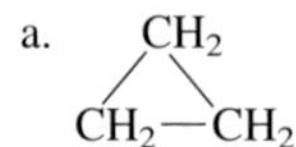

   b. 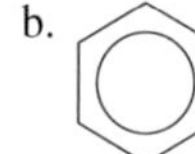
   c. □
27. Give the structural formulas of the four-carbon alkanes (C_4H_{10}). Identify butane and isobutane.
28. Give the molecular formulas and structures for these hydrocarbons.
   a. propane b. pentane
29. Name the following compounds.
   a. CH_3OH b. $CH_3\underset{\underset{OH}{|}}{C}HCH_3$
30. Give the structural formula for each of the following alcohols.
   a. ethyl alcohol b. propyl alcohol
31. Give structures for the following alkyl groups.
   a. ethyl b. isopropyl
32. Name the following alkyl groups.
   a. $CH_3CH_2CH_2—$ b. $CH_3—$

33. Give the structure for phenol.

34. Give the structure for diethyl ether.

35. Give the structure of each of the following compounds.
a. acetaldehyde b. formaldehyde

36. Give the structure of each of the following compounds.
a. butyraldehyde b. propionaldehyde

37. Name these compounds.
a. $CH_3CH_2CH_2OH$ b. $CH_3CH_2C(=O)—OH$

38. Name these compounds.
a. $CH_3CH_2C(=O)—H$ b. $C_6H_5—C(=O)—H$

39. Give structural formulas for each of the following.
a. acetic acid b. acetone

40. Give structural formulas for each of the following.
a. butyric acid
b. ethyl methyl ketone
c. formic acid

41. Give structural formulas for each of the following.
a. ethyl acetate b. methyl butyrate

42. Give structural formulas for each of the following.
a. ethyl butyrate b. methyl acetate

43. Give structural formulas for the following.
a. ethylamine b. dimethylamine

44. Give structural formulas for the following.
a. methylamine b. isopropylamine

45. Name the following compounds.
a. $CH_3CH_2CH_2NH_2$
b. $CH_3CH_2NHCH_2CH_3$

46. Name the following compounds.
a. $C_6H_5—C(=O)—OH$ b. $C_6H_5—NH_2$

Isomers and Homologs

47. Indicate whether the structures in each set represent the same compound or isomers.
a. CH_3CH_3 $CH_3—CH_3$ (vertical)
b. $CH_3CH_2—CH_3$ (CH₃ above CH₂) $CH_3CH_2CH_3$
c. $CH_3CH_2CH(CH_3)CH_2CH_3$ $CH_3CH(CH_3)CH_2CH_2CH_3$

48. Indicate whether the structures in each set represent the same compound or isomers.
a. $CH_3CH(CH_3)CH_2CH_3$ $CH_3CH_2CH(CH_3)CH_3$
b. $CH_3CH_2CH(CH_3)—CH_2(CH_3)$ $CH_2(CH_3)CH_2CH(CH_3)CH_3$

49. Classify the following pairs as homologs, identical, isomers, or none of these.
a. $CH_3CH_2CH_3$ and $CH_3CH_2CH_2CH_3$
b. cyclopropane (CH_2 ring: CH_2, $CH_2—CH_2$) and $CH_3CH_2CH_3$

50. Classify the following pairs as homologs, identical, isomers, or none of these.
a. $CH_3CH(CH_3)CH_3$ and $CH_3CH(CH_3)CH_3$
b. $CH_2{=}CHCH_3$ and cyclopropane (CH_2 ring: CH_2, $CH_2—CH_2$)

Classification of Hydrocarbons

51. Indicate whether each compound is saturated or unsaturated.
a. $CH_3C(CH_3){=}CH_2$ b. $CH_3—C(CH_3)_2—CH_3$

52. Indicate whether each compound is saturated or unsaturated.
a. $CH_3C{\equiv}CCH_3$ b. cyclohexane (hexagon)

53. Classify the compounds in Problem 51 as alkanes, alkenes, or alkynes.

54. Classify the compounds in Problem 52 as alkanes, alkenes, or alkynes.

Functional Groups

55. Give the structure of the carbonyl functional group.

56. Give the structure of the carboxyl group.

57. What is the name of the $—NH_2$ group?

58. Give the structure of the amide functional group.

Additional Problems

59. Which of the following represent heterocyclic compounds?

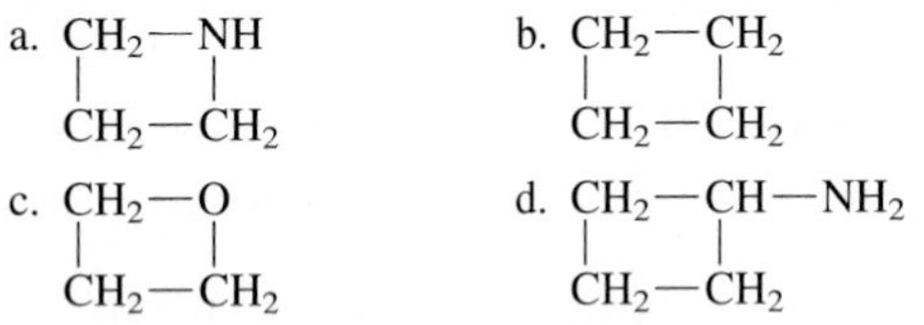

a. $CH_2—NH$ / $CH_2—CH_2$ (ring: CH_2–NH bonded above CH_2–CH_2)

b. $CH_2—CH_2$ / $CH_2—CH_2$ (ring)

c. $CH_2—O$ / $CH_2—CH_2$ (ring)

d. $CH_2—CH—NH_2$ / $CH_2—CH_2$ (ring of CH_2–CH above CH_2–CH_2)

60. What is the percent by volume of alcohol in 80 proof vodka?

61. Give the structure of the alcohol that is oxidized to each of the following.

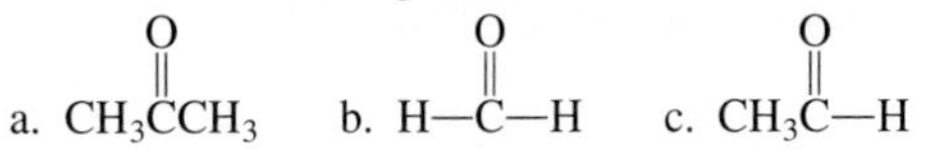

a. $CH_3\overset{O}{\overset{\|}{C}}CH_3$ b. $H—\overset{O}{\overset{\|}{C}}—H$ c. $CH_3\overset{O}{\overset{\|}{C}}—H$

62. Thymol is used as an antimold agent in preserving books.

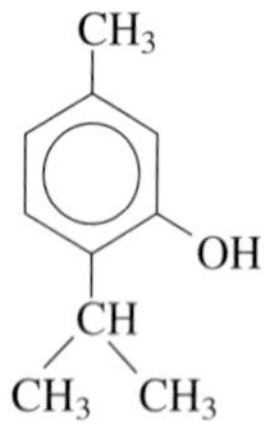

CH_3 on benzene ring; OH on ring; CH(CH_3)(CH_3) on ring

Thymol

To what family does thymol belong? Name the alkyl groups on the benzene ring.

63. Consider the following set of compounds. What principle does the series illustrate?

CH_3CH_2OH

$CH_3CH_2CH_2OH$

$CH_3CH_2CH_2CH_2OH$

$CH_3CH_2CH_2CH_2CH_2OH$

64. Consider the following set of compounds. What principle does the series illustrate?

$CH_3CH_2CH_2OCH_3$

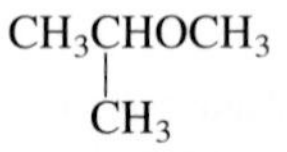

$CH_3\underset{CH_3}{\underset{|}{C}}HOCH_3$

$CH_3CH_2OCH_2CH_3$

$CH_3\overset{CH_3}{\overset{|}{\underset{OH}{\underset{|}{C}}}}—CH_3$

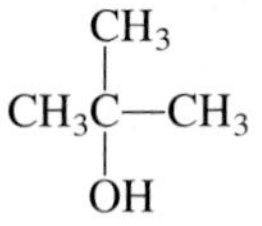

65. The complete combustion of benzene forms carbon dioxide and water.

$$C_6H_6 + O_2 \rightarrow CO_2 + H_2O$$

Balance the equation. What mass of carbon dioxide is formed by the complete combustion of 39.0 g of benzene?

66. Water adds to ethylene to form ethyl alcohol.

$$CH_2{=}CH_2 + H_2O \rightarrow CH_3CH_2OH$$

Balance the equation. What mass of ethyl alcohol is formed by the addition of water to 366 g of ethylene?

References and Readings

1. Benfey, O. T. ''August Kekulé and the Birth of the Structural Theory of Organic Chemistry in 1858.'' *Journal of Chemical Education,* January 1958, pp. 21–25.

2. Capindale, J. B. ''Organic Chlorine Compounds: A Boon or a Curse?'' *Chem 13 News,* January 1984, pp. 6–7.

3. Capindale, J. B. ''Organic Chlorine Compounds:

Insecticides and Bacteriocides.'' *Chem 13 News,* February 1984, pp. 4–5.

4. Chafetz, Morris E. ''Alcohol and Alcoholism.'' *American Scientist,* May–June 1979, pp. 293–299.

5. Eisenstein, Albert B. ''Nutritional and Metabolic Effects of Alcohol.'' *Journal of the American Dietetic Association,* September 1982, pp. 247–257.

6. Fessenden, Ralph J., and Joan S. Fessenden. *Fundamentals of Organic Chemistry.* New York: Harper & Row, 1990.

7. Hart, Harold. *Organic Chemistry: A Short Course,* 8th ed. Boston: Houghton Mifflin, 1991.

8. Hill, John W., Dorothy M. Feigl, and Stuart J. Baum. *Chemistry and Life,* 4th ed. New York: Macmillan, 1993.

9. Julian, Maureen M. ''What Compound Was Discovered as a Result of an Insurance Claim?'' *Journal of Chemical Education,* October 1981, p. 793.

10. McMurry, John. *Fundamentals of Organic Chemistry,* 3rd ed. Pacific Grove, CA: Brooks Cole, 1994.

11. Morrison, Robert T., and Robert N. Boyd. *Organic Chemistry,* 5th ed. Boston: Allyn and Bacon, 1987.

12. Ouellette, Eileen M. ''The Fetal Alcohol Syndrome.'' *Contemporary Nutrition,* March 1984, pp. 1–2.

13. Wotiz, John H., ed. *The Kekulé Riddle.* Vienna, IL: Cache River Press, 1993.

14. Zurer, Pamela. ''CFC Substitutes: Candidates Pass Early Toxicity Tests.'' *Chemical and Engineering News,* 9 October 1989, p. 4.

10 Polymers

Giants Among Molecules

▼

Polystyrene toys and dishes.
Teflon-coated cooking ware,
Cotton towels, leather jackets,
Dacron shirts with easy care,

Nylon carpets, woolen sweaters,
Wigs made of acrylic hair,
Rubber tires and vinyl cushions. . .
Polymers are everywhere!

Look around you. Almost everywhere you look you see polymers. Most of your clothes are made of polymers. In your car the dashboard, the seats, the tires, the steering wheel, the floor mats, the ceiling, and many parts that you cannot see are made of polymers. Around your home the carpets, the curtains, the upholstery, the towels, the sheets, the floor tile, the books, the furniture, and most of the toys and containers (not to mention such things as telephones, tooth brushes, and piano keys) are made of polymers.

Some of these polymers come from nature, but many are synthetic polymers, made in chemical plants in order to improve upon nature in some way. Polymers are an extremely important class of chemical compounds. Much of the food you eat contains polymers, and many important molecules in your body are polymers. It is no exaggeration to say that you could not live without polymers.

SECTION 10.1

Polymerization: Making Big Ones Out of Little Ones

Chemists often refer to polymers as **macromolecules** (from the Greek *makros* meaning ''large'' or ''long''). Macromolecules may not seem large to the human eye (in fact, many of these giant molecules are invisible), but when they are compared to other molecules, they are enormous.

Polymers (from the Greek *poly* meaning ''many,'' and *meros* meaning ''parts'') are made from much smaller molecules called **monomers** (from the Greek *monos* meaning ''one''). Sometimes there are thousands of monomer units making up a single polymer molecule. Monomers are the building blocks of polymers. They are the tiny links that make up long polymer chains. **Polymerization** is the process by which monomers are converted to polymers. A polymer is as different from its monomer as a long strand of spaghetti is from

◀ Modern plastics are made in many different colors, shapes, and textures. [*Arthur Meyerson/The Image Bank.*]

a tiny speck of flour. For example, polyethylene, the familiar waxy material used to make plastic bags, is made from the monomer ethylene, which is a gas.

SECTION 10.2

Natural Polymers

Polymers have served humanity for centuries in the form of starch and protein, which have been used for food; in the form of wood, which has been used for shelter; and in the form of wool, cotton, and silk, which have been used for clothing.

Starch is a polymer made up of glucose units. (Glucose, $C_6H_{12}O_6$, is a simple sugar molecule.) Cotton is made of cellulose, which is also a glucose polymer; and wood is largely cellulose as well. Proteins are polymers made up of amino acid monomers. Wool and silk are two of the thousands of different kinds of proteins in nature.

Living things could not exist without polymers. Each plant and animal requires many different specific types of polymers. Probably the most amazing natural polymers are the nucleic acids, which carry the coded genetic information that makes each individual unique. The polymers found in nature are discussed in Chapter 15. In this chapter we focus mainly on macromolecules that have been made in the laboratory. Let us take a brief look at the fascinating field of synthetic polymers.

SECTION 10.3

Celluloid: Billiard Balls and Collars

The oldest attempts to improve on nature simply involved chemical modification of natural macromolecules. The synthetic material **celluloid,** as its name implies, was derived from natural cellulose (from cotton and wood, for example). When cellulose is treated with nitric acid, a derivative called cellulose nitrate is formed. In response to a contest to find a substitute for ivory for use in billiard balls, John Wesley Hyatt, an American inventor, found a way to soften cellulose nitrate by treating it with ethyl alcohol and camphor. The softened material could be molded into smooth, hard balls. How many of us could afford to have a pool table in our basement if we had to buy ivory billiard balls? Hyatt brought the game of billiards within the economic reach of working people.

Celluloid was also used in movie film and for stiff collars (so they didn't require laundering and repeated starching) (Figure 10.1). Because of its dangerous flammability (cellulose nitrate also is used as smokeless gunpowder), celluloid was removed from the market when safer substitutes became available. Today, movie film is made from cellulose acetate, another semisynthetic

DU PONT AMERICAN INDUSTRIES
CHALLENGE CLEANABLE COLLARS
Reduce high cost of living—save labor and starch
THE ARLINGTON WORKS
Owned and Operated by E. I. du Pont de Nemours & Co.
725 Broadway, New York
Boston, St. Louis, Chicago, San Francisco, Toronto
DU PONT

Figure 10.1 Celluloid, a nitrocellulose product, was the first synthetic plastic. In the early twentieth century, many commercial uses were found for celluloid. [*Courtesy of E. I. du Pont de Nemours & Company, Inc., Wilmington, DE.*]

modification of cellulose. And high, stiff collars on men's shirts are out of fashion.

It didn't take long for the chemical industry to recognize the potential of synthetics. Scientists both in and out of industry began making macromolecules from small molecules rather than modifying large ones. The first such truly synthetic polymers were the phenol–formaldehyde resins, first made in 1909. These complex polymers are discussed later in this chapter. Let's look at some simpler ones first.

SECTION 10.4

Polyethylene: From the Battle of Britain to Bread Bags

Polyethylene is the simplest of the synthetic polymers, and it's the cheapest, too. It is familiar today in the plastic bags used for packaging fruit and vegetables, in garment bags for dry-cleaned clothing, in garbage-can liners, and in many other items. **Polyethylene** is made from ethylene.

$$\underset{\text{Ethylene}}{CH_2{=}CH_2}$$

This unsaturated hydrocarbon is produced in large quantities from the cracking of petroleum.

With pressure and heat and in the presence of a catalyst, ethylene molecules are made to join together in long chains. The polymerization can be represented by the reaction of a few monomer units.

$$\cdots + CH_2{=}CH_2 + CH_2{=}CH_2 + CH_2{=}CH_2 + CH_2{=}CH_2 + \cdots \longrightarrow$$
$$\sim CH_2CH_2{-}CH_2CH_2{-}CH_2CH_2{-}CH_2CH_2\sim$$

The dotted line and tildes (~) are like et ceteras; they indicate that the number of monomers and the polymer structure are extended for many units in each direction. A truer picture of the polyethylene molecule is given in Figure 10.2a, which shows a segment of a framework model; still better is the space-filling representation in Figure 10.2b. Both models are deficient in that they are much too short. Real polyethylene molecules have varying numbers of carbon atoms, from a few hundred to several thousand.

Polyethylene was invented shortly before the start of World War II. Before long, it was used for insulating cables in a top-secret invention—radar—which helped British pilots spot enemy aircraft before the aircraft became visible to the naked eye. Polyethylene proved to be tough and flexible and an excellent electrical insulator. It could withstand both high and low temperatures. Without polyethylene, the British could not have had effective radar,

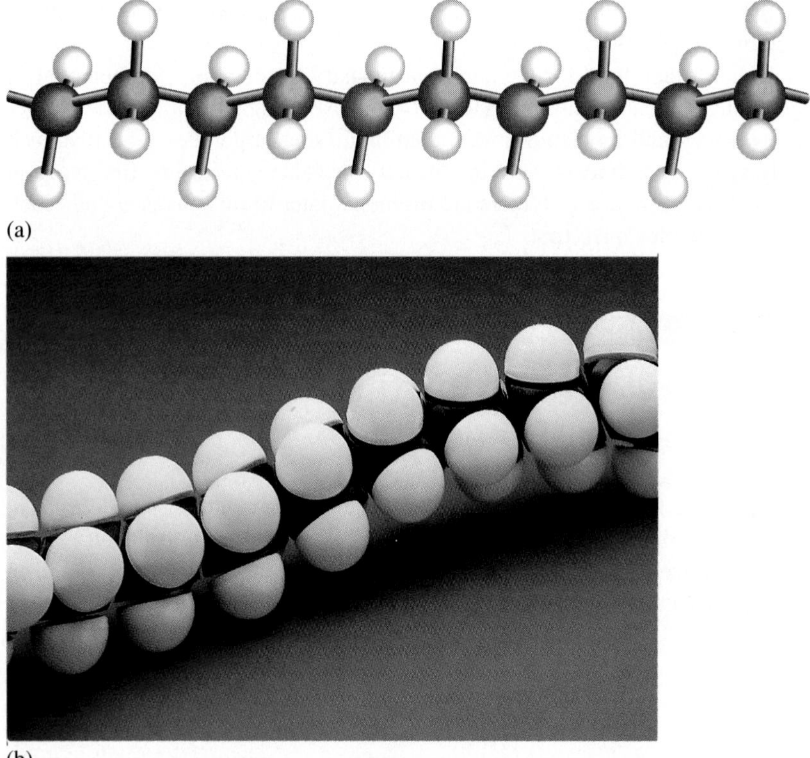

Figure 10.2 Ball-and-stick (a) and space-filling (b) models of a short segment of a polyethylene molecule. [*(b) Richard Megna/ Macmillan Science Files.*]

and without radar, the Battle of Britain might have been lost. The invention of this simple plastic may have changed the course of history.

Today, there are two principal kinds of polyethylene produced by the use of different catalysts and different reaction conditions. *High-density polyethylenes* (*HDPE*) have largely linear molecules that pack closely together. These linear molecules can assume a fairly ordered, crystalline structure; this gives high-density polyethylenes greater rigidity and higher tensile strength than other ethylene plastics. Linear polyethylenes are used for such things as threaded bottle caps, toys, bottles, and gallon milk jugs.

Low-density polyethylenes (LDPE), on the other hand, have a lot of side chains branching off the polymer molecules. The branches prevent the molecules from packing closely together and assuming a crystalline structure. Low-density polyethylenes are waxy, bendable plastics that are lower melting than high-density polyethylenes. Objects made of HDPE hold their shape in boiling water, while those made of LDPE are severely deformed (Figure 10.3). Low-density branched polyethylenes are used to make plastic bags, plastic film, squeeze bottles, electric wire insulation, and many common household products.

A third type of polyethylene, linear low-density polyethylene (LLDPE), is becoming increasingly important. LLDPE is actually a copolymer of ethylene with a higher alkene, such as 1-hexene. The structure of a segment of LLDPE would be something like

$$\sim CH_2CH_2{-}CH_2\underset{\substack{|\\ CH_2\\ |\\ CH_2\\ |\\ CH_2\\ |\\ CH_3}}{CH}{-}CH_2CH_2\sim$$

where the horizontal portion shows an ethylene residue at each end, and the middle portion depicts a segment from 1-hexene. More than 70% of the LLDPE produced is used to make film.

Figure 10.3 These bottles made of polyethylene were heated in the same oven for the same length of time. The one that melted has branched polyethylene molecules; the other is composed of unbranched chains. [*Richard Megna/Fundamental Photographs.*]

Thermoplastic and Thermosetting Polymers

Polyethylene is a **thermoplastic** material—that is, it can be softened by heat and then reformed. Many common plastics are thermoplastic. They can be melted down and remolded, and this can be done repeatedly. However, not all plastics can be readily melted. Some are *thermosetting,* which means that they harden permanently when they are formed. They cannot be softened by heat and remolded. We look at some **thermoset** polymers later in this chapter.

SECTION 10.5

Addition Polymerization: One + One + One . . . Gives One!

There are two general types of polymerization reactions: addition polymerization and condensation polymerization. In *addition polymerization,* the building blocks (or monomers) add to one another in such a way that the polymeric product contains all the atoms of the starting monomers. The polymerization of ethylene to form polyethylene is an example. Notice that in the structure of polyethylene (Section 10.4), the two carbons and the four hydrogens of each monomer molecule are incorporated into the polymer structure. In *condensation polymerization,* a portion of the monomer molecule is *not* incorporated in the final polymer but is split out as the polymer is formed. This process is discussed later in this chapter.

Polypropylene

There are many familiar **addition polymers**. Most are made from monomers that can be considered to be derivatives of ethylene in which one or more of the hydrogen atoms are replaced by another atom or group. Replacing one of the hydrogen atoms with a methyl group gives the monomer *propene,* or *propylene.*

$$CH_2{=}CH{-}CH_3$$

The polymer looks like polyethylene, except that there are methyl groups attached to every other carbon atom.

$$\sim CH_2{-}\underset{\displaystyle CH_3}{\underset{|}{CH}}{-}CH_2{-}\underset{\displaystyle CH_3}{\underset{|}{CH}}{-}CH_2{-}\underset{\displaystyle CH_3}{\underset{|}{CH}}{-}CH_2{-}\underset{\displaystyle CH_3}{\underset{|}{CH}}\sim$$

Polypropylene is a tough plastic material that is molded into hard shell luggage, battery cases, and various kinds of appliance parts. It is also made into fibers for upholstery fabrics and carpets, especially indoor-outdoor carpets.

Polystyrene

Replacing one of the hydrogen atoms in ethylene with a benzene ring gives a monomer called *styrene.*

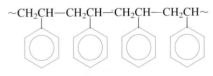

$$CH_2{=}\underset{\displaystyle C_6H_5}{\underset{|}{CH}}$$

Styrene

Polymerization of styrene produces **polystyrene**. A segment of a polystyrene molecule is illustrated below.

$$\sim CH_2\underset{\displaystyle C_6H_5}{\underset{|}{CH}}{-}CH_2\underset{\displaystyle C_6H_5}{\underset{|}{CH}}{-}CH_2\underset{\displaystyle C_6H_5}{\underset{|}{CH}}{-}CH_2\underset{\displaystyle C_6H_5}{\underset{|}{CH}}\sim$$

Polystyrene

Polystyrene is the plastic material used to make transparent ''throw-away'' drinking cups, and, with color and filler added, it is used to make thousands of inexpensive toys and household items. When gas is blown into polystyrene liquid, it foams and hardens into the familiar solid foam that is used to make disposable coffee cups. The foam is also used as packing material for shipping instruments and appliances, and it is widely used for home insulation (Figure 10.4).

Figure 10.4 Styrofoam insulation saves energy by reducing the transmission of heat from a warm house to the outside in winter or from the hot outside to the cooled inside during summer. [*Stacy Rick/Stock Boston.*]

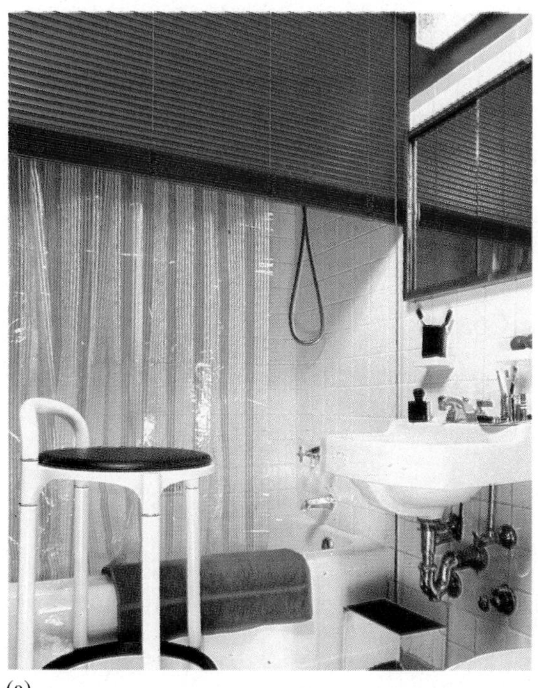
(a)

(b)

Figure 10.5 Polyvinyl chloride polymers can be formed into molded objects, films, such as shower curtains (a), and fibers, such as webbing for beach chairs (b). [*(a) David Frazier and (b) W. Eastep/The Stock Market.*]

Vinyl Polymers

Would you like a tough synthetic material that looks like leather yet costs only a fraction as much? Would you like a clear, rigid material from which unbreakable bottles could be made? Would you like an attractive, long-lasting floor covering? How about light-weight, rust-proof plumbing? Polyvinyl chloride (PVC) has all these properties—and more.

Replacing one of the hydrogen atoms of ethylene with a chlorine atom gives a compound called *vinyl chloride.* This compound is a gas at room temperature.

$$CH_2{=}CHCl$$
Vinyl chloride

Polymerization of vinyl chloride yields the tough thermoplastic material *polyvinyl chloride* (PVC). A segment of the PVC molecule is illustrated below.

$$\sim CH_2\underset{\displaystyle Cl}{\underset{|}{C}}H{-}CH_2\underset{\displaystyle Cl}{\underset{|}{C}}H{-}CH_2\underset{\displaystyle Cl}{\underset{|}{C}}H{-}CH_2\underset{\displaystyle Cl}{\underset{|}{C}}H\sim$$
Polyvinyl chloride

PVC is readily formed into various shapes (Figure 10.5). The clear, transparent polymer (Figure 10.6) is used in plastic wrap and clear plastic bottles.

Figure 10.6 A giant bubble of tough, transparent plastic film emerges from the die of an extruding machine. The film is used in packaging, consumer products, and food service operations. [*Courtesy of Reynolds Metals Company.*]

Artificial leather is made by adding color and other ingredients to vinyl plastics. Most floor tile and shower curtains are also made from vinyl plastics, and they are widely used to simulate wood in home siding panels and window frames. About 40% of the PVC produced is molded into pipes.

Unfortunately, the monomer from which vinyl plastics are made is known to be a carcinogen. Many people who have worked closely with vinyl chloride gas have later developed a kind of cancer known as angiosarcoma. Carcinogens are discussed in Chapter 20.

Teflon: The Nonstick Coating

When all the hydrogen atoms in ethylene are replaced by fluorine, the compound is called tetrafluoroethylene.

$$F_2C{=}CF_2$$

When tetrafluoroethylene is polymerized, the product is polytetrafluoroethylene (PTFE), or Teflon.

$$\sim CF_2{-}CF_2{-}CF_2{-}CF_2{-}CF_2{-}CF_2{-}CF_2{-}CF_2\sim$$

Teflon

Because the C—F bonds are very strong and resistant to heat and chemicals, Teflon is a very tough, unreactive, and non-flammable material. It is used to make electrical insulation, bearings, and gaskets. It is also widely used for coating surfaces on items such as skillets to give them nonsticking properties (Figure 10.7).

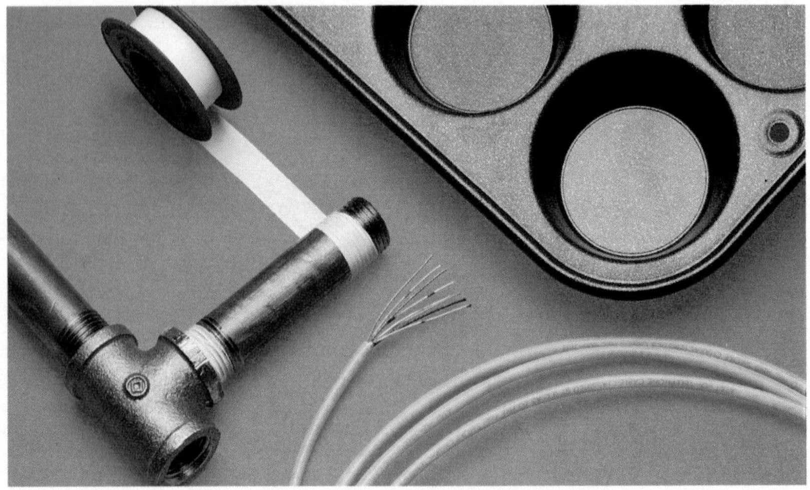

Figure 10.7 The muffin pan, plumber's tape, wire casing, and pen point are all coated with Teflon. [*Kip and Pat Peticolas/Fundamental Photographs.*]

Polymerization Equations

The polymer structures and polymerization equations we have written are quite cumbersome. There are other ways of writing formulas for polymers and equations for polymerization reactions—ways that do not involve writing out long segments of the polymer chain. We can represent the polymerization of ethylene by the equation

$$n\,CH_2{=}CH_2 \longrightarrow \left[\!-CH_2CH_2-\!\right]_n$$

In the formula for the polymer product, the repeating polymer unit (sometimes called the **segmer**, meaning "repeating segment") is placed within brackets with bonds extending to both sides. The subscript *n* indicates that this molecular fragment is repeated an unspecified number of times in the full polymer structure. It is certainly easier to write the bracketed segmer than to draw the extended chain—although the latter is probably better at conveying the concept of a polymer as a giant molecule (even though this formula also represents only a small portion of the whole molecule).

▶ EXAMPLE 10.1

What is the segmer in polyvinyl chloride? The polymer is

$$\sim CH_2\underset{\underset{Cl}{|}}{C}HCH_2\underset{\underset{Cl}{|}}{C}HCH_2\underset{\underset{Cl}{|}}{C}HCH_2\underset{\underset{Cl}{|}}{C}HCH_2\underset{\underset{Cl}{|}}{C}HCH_2\underset{\underset{Cl}{|}}{C}H\sim$$

SOLUTION

The segmer (repeating unit) is

$$CH_2\underset{\underset{Cl}{|}}{C}H$$

Exercise 10.1

What is the segmer in polyacrylonitrile? The polymer is

$$\sim CH_2\underset{\underset{CN}{|}}{C}HCH_2\underset{\underset{CN}{|}}{C}HCH_2\underset{\underset{CN}{|}}{C}HCH_2\underset{\underset{CN}{|}}{C}HCH_2\underset{\underset{CN}{|}}{C}HCH_2\underset{\underset{CN}{|}}{C}H\sim$$

The simplicity of the abbreviated formula does facilitate certain comparisons between the monomer and the polymer. Note that the monomer ethylene contains a double bond and polyethylene does not. The double bond of the reactant contains two pairs of electrons. One of these pairs is used to connect one monomer unit to the next in the polymer (indicated by the lines sticking out to the sides in the segmer). That leaves only a single pair of electrons between the two carbons of the polymer segmer—in other words, a single bond. Note that each segmer unit in the polymer has the same composition (C_2H_4) as the monomer.

▶ EXAMPLE 10.2

Give the structure of the polymer made from vinyl fluoride, $CH_2{=}CHF$. Show at least four segmer units.

SOLUTION

The carbon atoms become bonded in a chain. The fluorine atom is a substituent on the chain. There are only single bonds in the polymer. (Two of the electrons in the double bond of the monomer are used to join the segmers together.) The polymer is

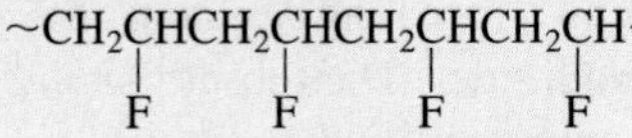

$$\sim\!CH_2\underset{\displaystyle F}{\underset{|}{C}}HCH_2\underset{\displaystyle F}{\underset{|}{C}}HCH_2\underset{\displaystyle F}{\underset{|}{C}}HCH_2\underset{\displaystyle F}{\underset{|}{C}}H\!\sim$$

Exercise 10.2

Give the structure of the polymer made from vinylidene chloride, $CH_2{=}CCl_2$. Show at least four segmer units.

Addition polymers are the highest volume products of the plastics industry, with the polyethylenes leading the list (see Table 10.1). We cannot mention all of the many kinds of addition polymers here, but Table 10.2 lists the more important ones, along with a few of their uses.

Polyacetylene

Although acetylene, $H{-}C{\equiv}C{-}H$, has a triple bond instead of a double bond, it can still undergo addition polymerization. The polymer is called polyacetylene.

$$\sim CH{=}CH{-}CH{=}CH{-}CH{=}CH{-}CH{=}CH{-}CH{=}CH{-}CH{=}CH\sim$$

Unlike polyethylene, which has a carbon chain containing only single bonds, notice that every other bond in polyacetylene is a double bond.

Table 10.1 Production of Top Plastics in 1992

Polymer	Billions of Kilograms
Polyethylenes (PE)	10.5
High density (HDPE)	4.6
Low density (LDPE)	3.4
Linear low density (LLDPE)	2.5
Polyvinyl chloride (PVC)	4.5
Polypropylene (PP)	3.9
Polystyrene (PS)	2.4

Table 10.2 Some Addition Polymers

Monomer	Polymer	Polymer Name	Some Uses
$H_2C{=}CH_2$	$\left[CH_2{-}CH_2 \right]_n$	Polyethylene	Plastic bags, bottles toys, electrical insulation
$H_2C{=}CH{-}CH_3$	$\left[CH_2{-}CH(CH_3) \right]_n$	Polypropylene	Indoor-outdoor carpeting, bottles, luggage
$H_2C{=}CH{-}C_6H_5$	$\left[CH_2{-}CH(C_6H_5) \right]_n$	Polystyrene	Simulated wood furniture, Styrofoam insulation, cups, toys, packing materials
$H_2C{=}CH{-}Cl$	$\left[CH_2{-}CHCl \right]_n$	Polyvinyl chloride (PVC)	Plastic wrap, simulated leather, plumbing, garden hoses, floor tile
$H_2C{=}CCl_2$	$\left[CH_2{-}CCl_2 \right]_n$	Polyvinylidene chloride (Saran)	Food wrap
$F_2C{=}CF_2$	$\left[CF_2{-}CF_2 \right]_n$	Polytetrafluoroethylene (Teflon)	Nonstick coating for cooking utensils, electrical insulation
$H_2C{=}CH{-}CN$	$\left[CH_2{-}CH(CN) \right]_n$	Polyacrylonitrile (Orlon, Acrilan, Creslan, Dynel)	Yarns, wigs, paints
$H_2C{=}CH{-}O{-}C({=}O){-}CH_3$	$\left[CH_2{-}CH(O{-}C({=}O){-}CH_3) \right]_n$	Polyvinyl acetate	Adhesives, textile coatings, chewing gum resin, paints
$H_2C{=}C(CH_3){-}C({=}O){-}O{-}CH_3$	$\left[CH_2{-}C(CH_3)(C({=}O){-}O{-}CH_3) \right]_n$	Polymethyl methacrylate (Lucite, Plexiglas)	Glass substitute, bowling balls

The alternating double and single bonds make it easy for electrons to travel along the chain, so the polymer is able to conduct electricity. (This is rather amazing, since plastics are normally electrical insulators!) Polyacetylene can actually be used as a light-weight substitute for metal. In fact, the plastic even *looks* like metal, having a silvery luster. When formed into film, it looks like metal foil.

Polyacetylene was discovered in 1970, and since then a number of other polymers with electrical conductivity have been made, but polyacetylene was the first one.

SECTION 10.6

Rubber

Although rubber is a polymer that comes from nature, it is discussed here among the synthetic polymers because it had a lot to do with the development of the synthetic polymer industry. During World War II the Allies' supply of natural rubber from Malaysia and Indonesia was cut off by Japan. The search for synthetic substitutes resulted in much more than just a replacement for natural rubber. The plastics industry, to a large extent, developed out of the search for synthetic rubber.

Natural **rubber** can be broken down into a simple hydrocarbon called *isoprene.*

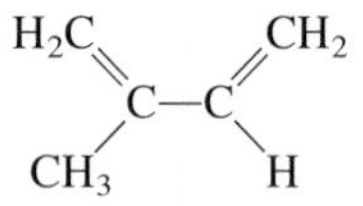

$$H_2C{=}C(CH_3){-}C(H){=}CH_2$$

Isoprene

The macromolecules from which rubber is made are now known to have the structure

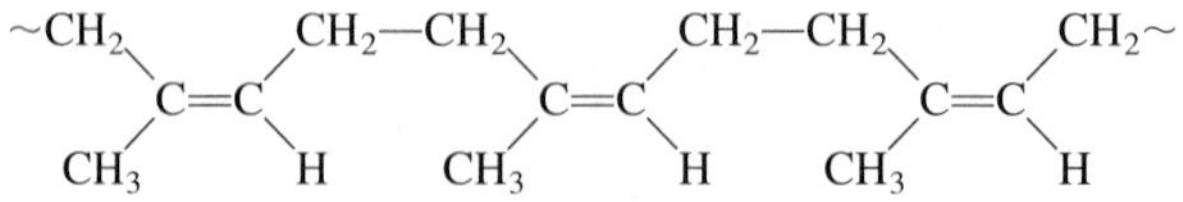

$$\sim CH_2{-}C(CH_3){=}C(H){-}CH_2{-}CH_2{-}C(CH_3){=}C(H){-}CH_2{-}CH_2{-}C(CH_3){=}C(H){-}CH_2\sim$$

Isoprene is a volatile liquid. Rubber is a semisolid, elastic material.

Vulcanization: Cross-Linking

The long-chain molecules that make up rubber can be coiled and twisted and intertwined with one another. The stretching of rubber corresponds to the straightening of the coiled molecules. Natural rubber is soft and tacky when hot. It can be made harder by reaction with sulfur. This process, called **vulcanization**, cross-links the hydrocarbon chains with sulfur atoms (Figure 10.8).

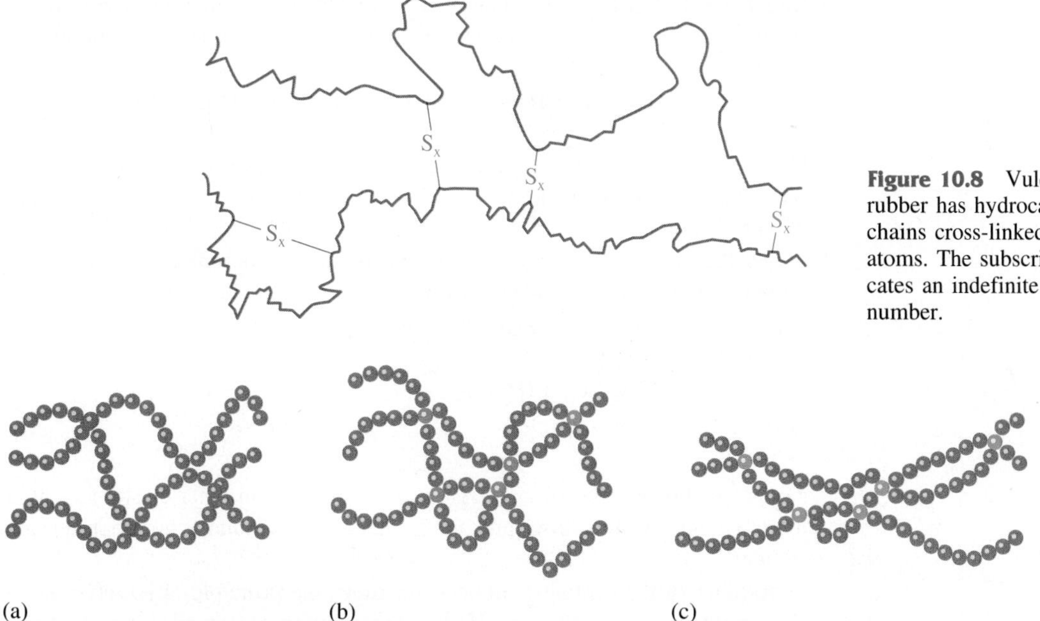

Figure 10.8 Vulcanized rubber has hydrocarbon chains cross-linked by sulfur atoms. The subscript x indicates an indefinite large number.

Figure 10.9 Vulcanization of rubber cross-links the molecular chains. (a) In unvulcanized rubber, the chains slip past one another when the rubber is stretched. (b) Vulcanization involves the addition of sulfur cross-linkages between the chains. (c) When the vulcanized rubber is stretched, the sulfur cross-links prevent the chains from slipping past one another. Vulcanized rubber is stronger and can be more elastic than unvulcanized rubber.

Vulcanization was discovered by Charles Goodyear, who was issued U.S. Patent No. 3633 in 1844.

Its three-dimensional cross-linked structure makes vulcanized rubber a harder, stronger substance that is suitable for automobile tires. Surprisingly, cross-linking also improves the elasticity of rubber. With just the right degree of cross-linking, the individual chains are still relatively free to uncoil and stretch. And when a stretched piece of this material is released, the cross-links pull the chains back to their original arrangement (Figure 10.9). Rubber bands owe their snap to this sort of molecular structure. Materials that act in this stretchable way are called **elastomers.**

Synthetic Rubber

Natural rubber is a polymer of isoprene. Some synthetic elastomers are closely related to it. For example, polybutadiene is made from the monomer butadiene.

$$CH_2{=}CH{-}CH{=}CH_2$$

Butadiene

Butadiene differs from isoprene only in that it lacks a methyl group on the second carbon atom. Polybutadiene is made rather easily from the monomer.

$$n\ \mathrm{CH_2{=}CH{-}CH{=}CH_2} \longrightarrow \mathrm{+_n CH_2{-}CH{=}CH{-}CH_2 +_n}$$

However, it has only fair tensile strength and poor resistance to gasoline and oils. These properties make it of limited value for automobile tires, the main use of elastomers.

Another synthetic elastomer, polychloroprene (Neoprene), is made from a monomer that is quite similar to isoprene, but this monomer has a chlorine in place of isoprene's methyl group.

$$n\ \mathrm{CH_2{=}\underset{\displaystyle Cl}{\underset{|}{C}}{-}CH{=}CH_2} \longrightarrow \mathrm{+CH_2{-}\underset{\displaystyle Cl}{\underset{|}{C}}{=}\underset{\displaystyle H}{\underset{|}{C}}{-}CH_2 +_n}$$

Neoprene shows better oil and gasoline resistance than other elastomers. It is used to make gasoline hoses and similar items used around automobile service stations.

Another of the synthetic rubbers illustrates the principle of *copolymerization.* In this process, a mixture of two monomers forms a product in which the chain contains units of both building blocks. Such a material is called a **copolymer**. Styrene–butadiene rubber (SBR) is a copolymer of styrene (25%) and butadiene (75%). A segment of an SBR molecule might look something like this.

$$\mathrm{{\sim}CH_2CH{=}CHCH_2{-}CH_2\underset{\displaystyle C_6H_5}{\underset{|}{C}H}{-}CH_2CH{=}CHCH_2{-}CH_2CH{=}CHCH_2{\sim}}$$

Butadiene unit — Styrene unit — Butadiene unit — Butadiene unit

This synthetic is more resistant to oxidation and abrasion than natural rubber, but it has less satisfactory mechanical properties.

Like natural rubber, the SBR molecules contain double bonds and can be cross-linked by vulcanization. SBR accounts for more than half of the total production of elastomers. It is mainly used for making tires. Chemists have even learned to make polyisoprene, a substance identical in every way to natural rubber except that it comes from petroleum refineries rather than plantations of rubber trees.

In addition to elastomers (mainly SBR), tires for automobiles contain many substances. Carbon black is added before vulcanization. It acts in some as yet unknown way to strengthen the elastomer. The tire is further strengthened mechanically by the use of nylon cord, fiberglass, or steel belts.

One of the surprising uses for elastomers is in paints. The substance in a paint that hardens to form a continuous surface coating is called a *binder.* In

Figure 10.10 Synthetic polymers serve as binders in paints, which provide us with a more colorful environment. [*Courtesy of Sherwin-Williams.*]

traditional oil-based paints the binder is usually linseed oil. Modern water-based latex paints contain a binder that is a synthetic polymer (Figure 10.10), usually one with with rubberlike properties. Various kinds of polymers can be used as binders, depending on the specific qualities desired in the paint.

SECTION 10.7

Condensation Polymers: Splitting Out Water

The polymers considered so far are all addition polymers. All the atoms of the monomer molecules are incorporated into the polymer molecules. In making a **condensation polymer**, a portion of the monomer molecule is not incorporated in the final polymer. What usually splits out during condensation polymerization is water.

Nylon

As an example, let's consider the formation of **nylon**. (There are several different nylons, each prepared from a different monomer or set of monomers, but all share certain common structural features.) The monomer in one type of nylon is a carboxylic acid with an amino group on the sixth carbon atom, 6-aminohexanoic acid. The polymerization involves the reaction of a carboxyl group of one monomer molecule with the amine group of another. This reaction produces an amide bond that holds the building blocks together in the final polymer.

$$H_2NCH_2(CH_2)_3CH_2\overset{\overset{O}{\|}}{C}OH$$

6-Aminohexanoic acid

$$\ldots + HO{-}\overset{\overset{O}{\|}}{C}CH_2CH_2CH_2CH_2CH_2\overset{\overset{H}{|}}{N}{-}H + HO{-}\overset{\overset{O}{\|}}{C}CH_2CH_2CH_2CH_2CH_2\overset{\overset{H}{|}}{N}{-}H + \ldots \longrightarrow$$

$$\sim\overset{\overset{O}{\|}}{C}CH_2CH_2CH_2CH_2CH_2\overset{\overset{H}{|}}{N}{-}\overset{\overset{O}{\|}}{C}CH_2CH_2CH_2CH_2CH_2\overset{\overset{H}{|}}{N}\sim + n\,H_2O$$

Water molecules are formed as a by-product. It is this formation of a nonpolymeric by-product that distinguishes condensation polymerization from addition polymerization. Note that the formula of a segmer unit is *not* the same as that of the monomer.

The polymer formed in the reaction shown above is nylon 6. And since the linkages holding it together are amide bonds, nylon 6 is a polyamide. Another polyamide, nylon 66, is made by the condensation of two different monomers.

Nylons are named according to the number of carbon atoms in each monomer unit. There are six each in 1,6-hexanediamine and adipic acid; hence, the polymer is nylon 66. The nylon on page 295 is made from a single monomer that contains six carbon atoms, so it is called nylon 6.

$$n\,\text{H}{-}\overset{\text{H}}{\overset{|}{\text{N}}}\text{CH}_2\text{CH}_2\text{CH}_2\text{CH}_2\text{CH}_2\text{CH}_2\overset{\text{H}}{\overset{|}{\text{N}}}{-}\text{H} + n\,\text{HO}{-}\overset{\text{O}}{\overset{\|}{\text{C}}}(\text{CH}_2)_4\overset{\text{O}}{\overset{\|}{\text{C}}}{-}\text{OH} \longrightarrow$$

1,6-Hexanediamine Adipic acid

$$\left[\overset{\text{H}}{\overset{|}{\text{N}}}\text{CH}_2\text{CH}_2\text{CH}_2\text{CH}_2\text{CH}_2\text{CH}_2\overset{\text{H}}{\overset{|}{\text{N}}}{-}\overset{\text{O}}{\overset{\|}{\text{C}}}\text{CH}_2\text{CH}_2\text{CH}_2\text{CH}_2\overset{\text{O}}{\overset{\|}{\text{C}}}\right]_n + n\,\text{H}_2\text{O}$$

This was the original nylon polymer discovered in 1937 by Du Pont chemist Wallace Carothers. Note that one monomer has two amino groups and the other has two carboxyl groups, but the product is still a polyamide, very similar to nylon 6. Silk and wool, which are protein fibers, are natural polyamides.

Although nylon can be molded into various shapes, most nylon is made into fibers. Some is spun into fine thread to be woven into silklike fabrics, and some is made into yarn that is much like wool. Carpeting, which was once made primarily from wool, is now made largely from nylon. Less than 3% of today's carpets are made of wool—68% are nylon.

Dacron

Polyesters, such as **Dacron**, are also condensation polymers. Dacron polyester is made by the condensation of ethylene glycol with terephthalic acid.

$$n\,\text{HO}{-}\text{CH}_2\text{CH}_2{-}\text{OH} + n\,\text{HO}{-}\overset{\text{O}}{\overset{\|}{\text{C}}}{-}\text{C}_6\text{H}_4{-}\overset{\text{O}}{\overset{\|}{\text{C}}}{-}\text{OH} \longrightarrow$$

Ethylene glycol Terephthalic acid

$$\left[\text{O}{-}\text{CH}_2\text{CH}_2{-}\text{O}{-}\overset{\text{O}}{\overset{\|}{\text{C}}}{-}\text{C}_6\text{H}_4{-}\overset{\text{O}}{\overset{\|}{\text{C}}}\right]_n + n\,\text{H}_2\text{O}$$

Dacron

The hydroxyl groups in ethylene glycol react with the carboxylic acid groups in terephthalic acid to produce long chains held together by many ester linkages. The polyester molecules make excellent fibers that are widely used in wash-and-wear clothing. More than 50% of all synthetic fabrics contain Dacron polyester.

The same polyester, when formed as a film rather than fiber, is called Mylar. When magnetically coated, Mylar tape is used in audio and video cassettes.

Bakelite: Phenol–Formaldehyde Resins

Let us now go back to **Bakelite**, which was the original synthetic polymer. These phenol–formaldehyde resins were first synthesized by Leo Baekeland, who received U.S. Patent No. 942,699 for the process in 1909.

The phenol–formaldehyde resins are formed by splitting out water molecules, the hydrogen atoms coming from the benzene ring and the oxygen atoms from the aldehyde. The reaction proceeds stepwise, with formaldehyde adding first to the 2- and 4-positions of the phenol molecule.

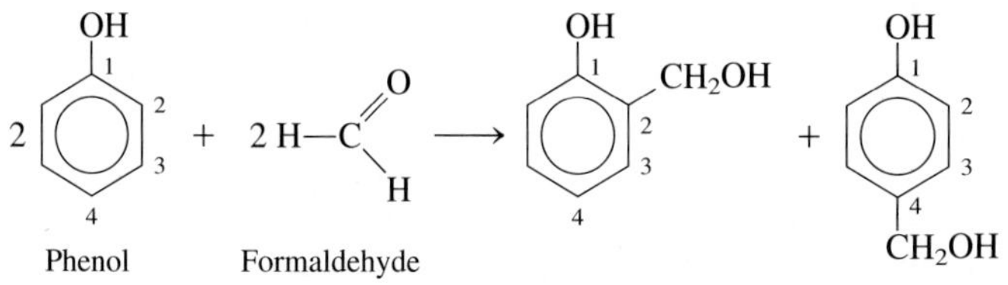

Phenol Formaldehyde

The substituted molecules then interact by splitting out water. (Remember that there are hydrogen atoms at all the unsubstituted corners of a benzene ring).

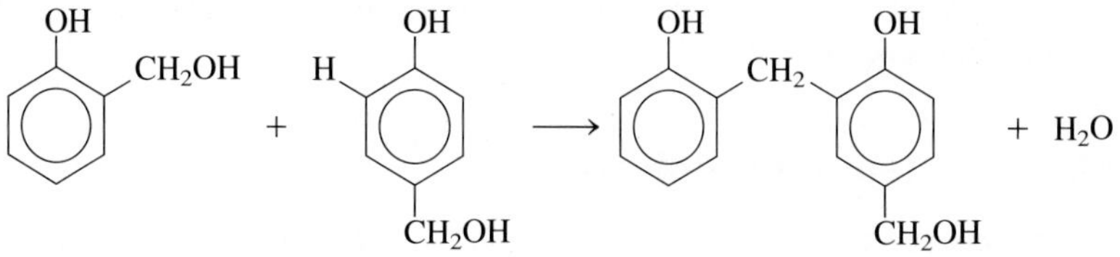

The hookup of molecules continues until an extensive network is achieved.

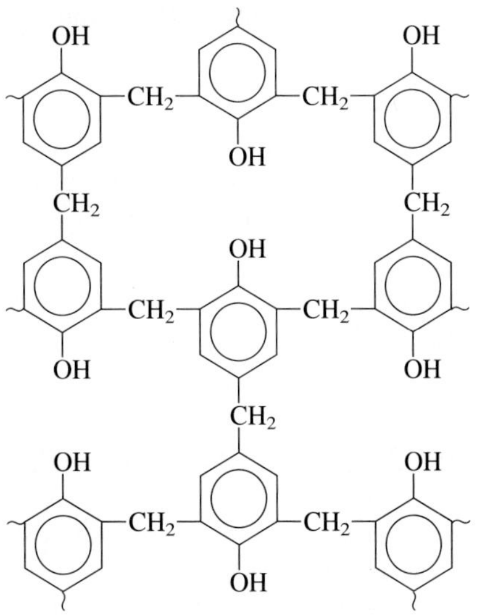

Phenol–formaldehyde resin

Water is driven off by heat as the polymer sets. The structure of the polymer is extremely complex, a three-dimensional network somewhat like the framework of a giant building. Note that the phenolic rings are joined together by CH_2 units from the formaldehyde. These resins cannot be melted and remolded. They are *thermoset resins.* Urea–formaldehyde and melamine–formaldehyde resins are similar to phenol–formaldehyde polymers.

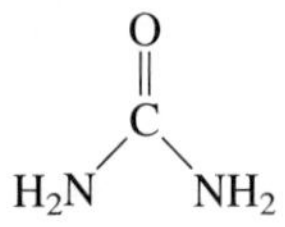

$$\underset{\text{Urea}}{H_2N{-}\overset{\overset{\displaystyle O}{\|}}{C}{-}NH_2}$$

$$\text{Melamine: } C_3N_3(NH_2)_3 \text{ (triazine ring with three } {-}NH_2 \text{ groups)}$$

Urea 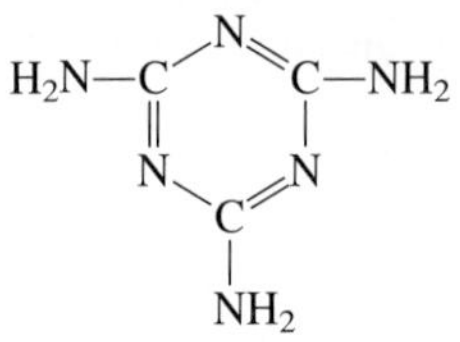 Melamine

(The melamine molecule is formed by condensation of three molecules of urea.) Both of these resins are complex three-dimensional networks formed by the splitting out of H_2O from formaldehyde ($H_2C{=}O$) molecules and amino ($-NH_2$) groups. Urea-formaldehyde resins are used to bind wood chips together in panels of particle board. Melamine–formaldehyde resins are used in plastic (Melmac) dinnerware and Formica counter tops. Like Bakelite, both of these are thermoset plastics.

Other Condensation Polymers

There are many other kinds of condensation polymers, but we will mention only a few.

Polycarbonates are tough, "clear as glass" polymers that are strong enough to be used as bullet-proof windows. They are also used in protective helmets. One polycarbonate, commonly called Lexan, has the following repeating unit.

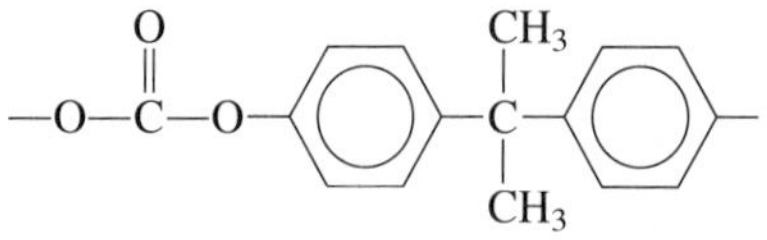

$$-O-\overset{\overset{\displaystyle O}{\|}}{C}-O-C_6H_4-\overset{\overset{\displaystyle CH_3}{|}}{\underset{\underset{\displaystyle CH_3}{|}}{C}}-C_6H_4-$$

Note the carbonatelike structure on the left.

Polyurethanes are elastomers that are similar to nylon in structure. The repeating unit in a common polyurethane is

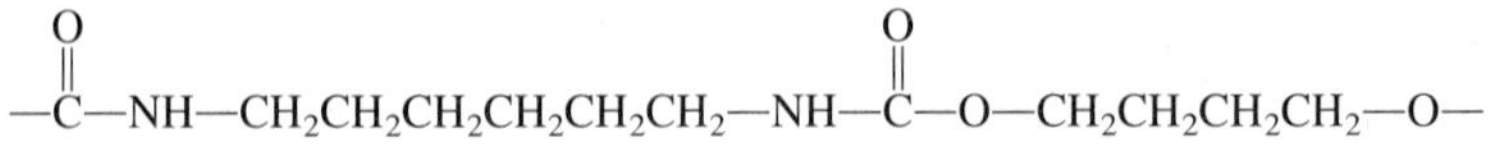

$$-\overset{\overset{\displaystyle O}{\|}}{C}-NH-CH_2CH_2CH_2CH_2CH_2CH_2-NH-\overset{\overset{\displaystyle O}{\|}}{C}-O-CH_2CH_2CH_2CH_2-O-$$

Polyurethanes are used especially as foamed padding ("foam rubber") in cushions, mattresses, and padded furniture.

Epoxy resins make excellent surface coatings, and they are powerful adhesives. Here is a typical repeating unit.

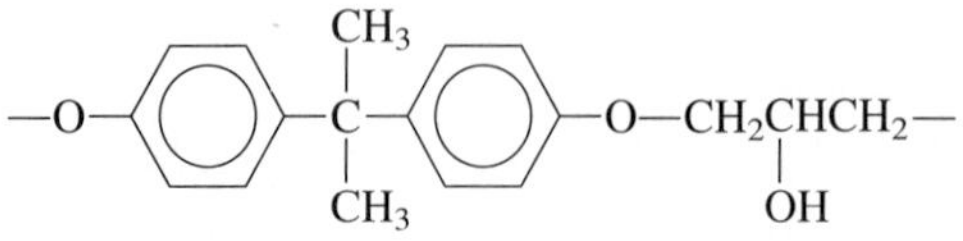

$$-O-C_6H_4-\overset{\overset{\displaystyle CH_3}{|}}{\underset{\underset{\displaystyle CH_3}{|}}{C}}-C_6H_4-O-CH_2\underset{\underset{\displaystyle OH}{|}}{C}HCH_2-$$

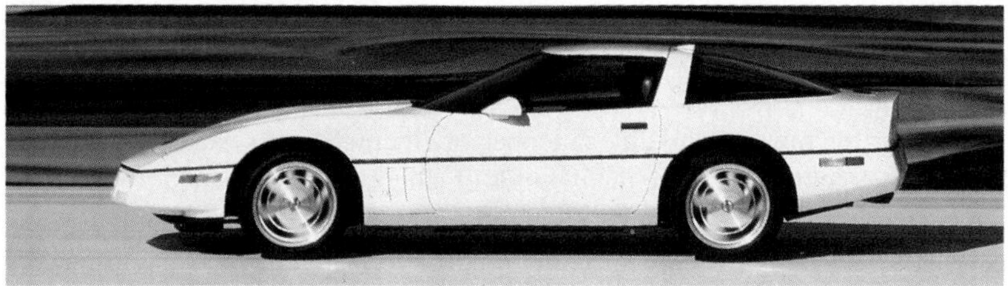

Figure 10.11 The body of this car is made of a plastic composite. [*Dick Reed/The Stock Market.*]

Epoxy adhesives have two components that are mixed just before they are used. The polymer chains become cross-linked, and the bonding becomes extremely strong.

Composites

Composite materials are made up of high-strength fibers (of glass, graphite, or ceramics) held together by a polymeric matrix, usually a thermoset condensation polymer. The fiber reinforcement provides the support, while the surrounding plastic acts to protect the fibers from breaking.

The most commonly used composite materials thus far have been polyester resins reinforced with glass fibers. These make up about 90% of reinforced plastic materials. They are widely used in sports gear (e.g., tennis rackets), boat hulls, molded chairs, and automobile panels. Various sports cars, including General Motors' Corvette and Fiero, have plastic composite bodies (Figure 10.11). Some composite materials have the strength and rigidity of steel, although they weigh only a fraction as much.

Silicones

Not all polymers are based on chains of carbon atoms. The silicones are a good example of a different type of polymer. A **silicone** is a polymer based on a series of alternating silicon and oxygen atoms.

$$
\begin{array}{ccccccccccccccc}
 & R & & R & & R & & R & & R & & R & & R & \\
 & | & & | & & | & & | & & | & & | & & | & \\
\sim & Si & -O- & Si & -O- & Si & -O- & Si & -O- & Si & -O- & Si & -O- & Si & \sim \\
 & | & & | & & | & & | & & | & & | & & | & \\
 & R & & R & & R & & R & & R & & R & & R &
\end{array}
$$

(R represents a hydrocarbon group, such as methyl, ethyl, or phenyl.) Silicones may be linear, cyclic, or cross-linked networks. They are very heat-stable and resistant to most chemicals, and they are excellent water-proofing materials. Some silicones are oils or greases, some are rubbery, and some are solid resins, depending on how long the chains are and how much cross-linking is present. Silicone oils are used as hydraulic fluids and lubricants, while other silicones are used in making such products as sealants, auto polish, shoe polish, and waterproof sheeting. Fabrics for raincoats and umbrellas are frequently treated with silicone (Figure 10.12).

Figure 10.12 The raincoat and umbrella are both waterproofed with silicone. [*Kim Robbie/The Stock Market.*]

An interesting silicone toy is the product known as "Silly Putty." It can be molded like clay or rolled up and bounced like a ball. On standing, it flows like a liquid.

Perhaps the most remarkable silicones of all are the ones used for synthetic human body parts. There are many kinds of silicone replacements, from finger joints to eye sockets. Artificial ears and noses are also made from silicone polymers. They can even be specially colored to match the surrounding skin.

For many years silicone gel was used for breast implants. In most cases these implants have remained perfectly stable over the years. However, because of a manufacturing error, some batches of the gel were not sufficiently cured. As a result, those gel implants later disintegrated, causing leakage into body tissues and leading to serious health problems.

SECTION 10.8

Properties of Polymers

There are many kinds of polymers, and no two are exactly alike. Thus there are many variations in the properties of polymers. We have already noted that some polymers are thermoplastic, which means that they can be melted down and remolded, while other plastics are thermosetting. Once a thermoset plastic has hardened, it cannot be softened by heat. Instead of softening when heated strongly, thermoset plastics become discolored and decompose.

Crystalline and Amorphous Polymers

Some polymers are *crystalline,* and their molecules line up neatly to form long fibers of high tensile strength. Other polymers are *amorphous,* composed of randomly oriented molecules that get tangled up with one another (Figure 10.13). Crystalline polymers tend to make good synthetic fibers, while amorphous polymers make good elastomers.

Sometimes the same polymer might be crystalline in one region and amorphous in another. For example, scientists have designed spandex fibers [used for stretch fabrics (Lycra) in ski pants, girdles, and bathing suits] so as to combine the tensile strength of crystalline fibers with the elasticity of amor-

(a) 

(b) 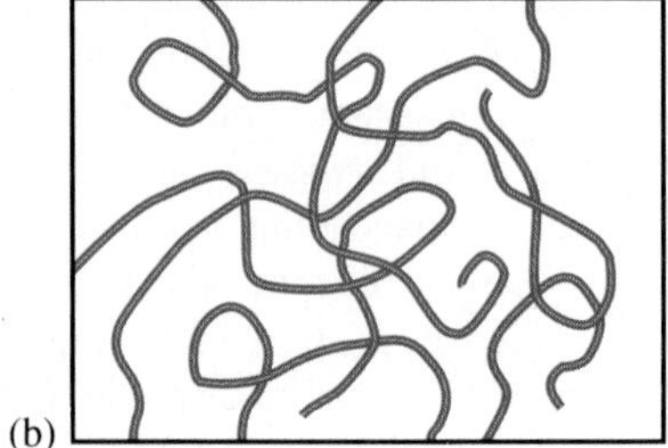

Figure 10.13 Organization of polymer molecules. (a) Crystalline arrangement. (b) Amorphous arrangement.

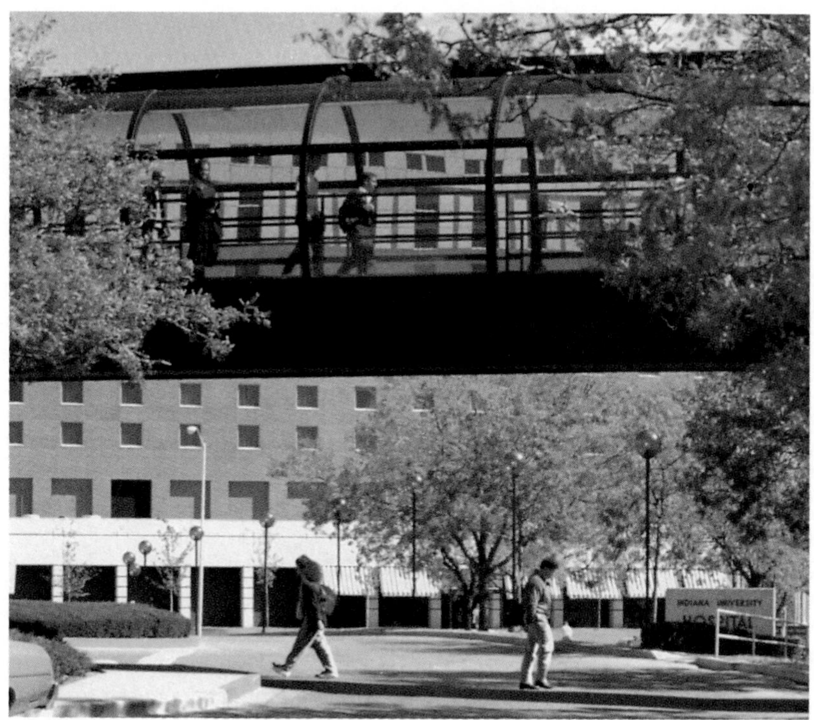

Figure 10.14 A high glass transition temperature (T_g) gives Plexiglas the glasslike properties well suited for this covered walkway at Indiana University at Indianapolis. Students have irreverently dubbed the walkway "The Gerbil Tube." [*Rick Baughn/Office of Learning Technologies/UPUI.*]

phous rubber. Two molecular structures are grafted onto one polymer chain, with blocks of crystalline character alternating with amorphous blocks. The resulting polymer exhibits both sets of properties—flexibility *and* rigidity!

The Glass Transition Temperature

An important parameter of most polymers is the *glass transition temperature* (T_g). Above this temperature, the polymer is rubbery and tough; below it, the polymer is like glass—hard, stiff, and brittle. Each polymer has a characteristic T_g. We want automobile tires to be tough and elastic, so we use materials with a low T_g. On the other hand, we want plastic substitutes for glass to be glassy (Figure 10.14). Thus, they have T_gs well above room temperature. We make use of the T_g concept in everyday life. To remove chewing gum from clothing, we apply a piece of ice to lower the temperature of the gum below the T_g of the polyvinyl acetate resin that makes up the bulk of the gum. The cold, brittle resin then crumbles readily and can be removed.

Figure 10.15 A liquid polymer is forced through the tiny holes of a device called a spinneret. The filaments solidify as they are cooled. [*Courtesy of Phillips Fibers Corporation.*]

Fiber-Forming Properties

Not all synthetic polymers can be converted into useful fibers, but those that can often have properties that are superior to those of natural fibers. That is why more than three-fourths of the fibers and fabrics used today in the United States are synthetic (Figure 10.15).

Silk fabrics are beautiful and have a luxurious "feel," but nylon fabrics are also beautiful and feel much like silk. Moreover, nylon fabrics wear longer, are easier to care for, and are less expensive than silk.

A Yale University study concluded that a cotton shirt, over its lifetime of use, requires 88% more energy than one that is 65% polyester and 35% cotton. (The cotton shirt requires more energy for washing and ironing than the one that is part polyester.)

Dacron (polyester) is a substitute for either cotton or silk, but it outperforms the natural fibers in many ways. It is not subject to mildew as cotton is, and polyester fabrics do not need ironing. Fabric made by blending Dacron with 35–50% cotton combines the comfort of cotton with the no-iron easy care of Dacron. The Japanese have developed a superior 50:50 blend of cotton and polyester by treating the fabric with formaldehyde (HCHO) in the vapor phase. After being washed and dried, the fabric looks as if it had been pressed. Even pure cotton, when given the formaldehyde treatment, becomes a "no-iron" fabric. The formaldehyde acts by cross-linking the strands of cellulose in the cotton.

Acrylic fibers, such as Orlon, Acrilan, and Creslan, can be spun into yarns that look like wool. Acrylic sweaters have the beauty and warmth of wool, but they do not shrink in hot water, are not attacked by moths, and do not bring on the allergic skin reactions that some people have with wool.

Some synthetic fibers have such phenomenal strength that they rival steel wire. One of the most interesting examples is Dyneema, a fiber developed in the Netherlands. It is made from extremely large molecules of polyethylene (about 100 times as large as ordinary polyethylene molecules). The fabric woven from these fibers is so amazingly strong that it is being used to make bullet-proof vests for policemen. The vests are much lighter and easier to wear than traditional metal vests. Dyneema fabric also has promise for use in crash helmets, sports gear, and specialized products in the fields of aerospace and medicine.

SECTION 10.9

Disposal of Plastics

Plastics make up about 8% by mass of our solid waste in the United States, but by volume they make up about 21%. This has created a serious problem because most of our solid waste goes into landfills, and we are rapidly running out of landfill space. One of the advantages of plastics is that they are durable and resistant to many things in the environment. It may be that they are too good in this respect. Some of them last almost forever! Once they are dumped, they do not go away. You see them littering our parks, our sidewalks, and our highways; and if you should go out to the middle of the ocean you would see them there, too. Small fish have been found dead with their digestive tracts clogged by bits of plastic foam ingested with their food.

Table 10.3 Code for Separating Plastics for Recycle

Plastic	Abbr.	Code
Polyethylene terephthalate	PET	1
High-density polyethylene	HDPE	2
Polyvinyl chloride	PVC	3
Low-density polyethylene	LDPE	4
Polypropylene	PP	5
Polystyrene	PS	6
All others		7

Degradable Plastics

About half of our waste plastic is from packaging. One approach to the plastics disposal problem is to make plastic packages that are biodegradable or photodegradable (broken down in the presence of bacteria or light). This has mainly been tried with plastic bags, six-pack ring connectors, and food cartons. Most biodegradable plastics are starch-based synthetic polymers. Photodegradable plastics usually contain a light-sensitive additive. Of course, it is important that the package remain intact and not start decomposing while it is still being used. So far many people seem reluctant to pay extra for garbage bags that are designed to fall apart.

Recycling

Perhaps the best way to dispose of discarded plastic is to recycle it. To do that the plastic must be collected, sorted, chopped, melted, and then remolded. Collection may be the hardest step in the process. It works best when there is strong community cooperation.

The separation step has been simplified by having code numbers (Table 10.3) stamped on most plastic bottles and other objects. Once the plastics are separated, they can be chopped into flakes, melted, and remolded or spun into fiber.

At present the only plastic items being recycled on a large scale are HDPE milk jugs, which are being remolded into detergent bottles, and PET soda bottles, which are being turned into fiber, mainly for carpets (Figure 10.16). Small companies are recycling such items as polystyrene containers or polyethylene bags. One recycler is turning polycarbonate water bottles into automobile bumpers, and others are converting milk jugs to artificial lumber.

Only about 2% of waste plastic is currently being recycled. However, there are many possible uses for recycled plastic, and the shortage of landfill space is becoming critical. Recycling of plastics is a field with great potential for growth.

Figure 10.16 Plastics can be recycled. Separated, the polymers can be melted and reshaped. Polyester bottles can be made into fibers for carpets and insulating fabrics. [*The Council for Solid Waste Solutions.*]

Figure 10.17 Accumulated old tires are an ideal breeding ground for disease-carrying insects and rats. [*Glenn McLaughlin/The Stock Market.*]

Incineration

Another way to dispose of discarded plastics is to burn them. Most plastics have a high fuel value. For example, a pound of polyethylene has about 20,000 Btu of energy, which is about the same as a pound of fuel oil. Some communities actually generate electricity with the heat from their garbage incinerators.

Various utility companies are looking into burning powdered coal mixed with a few percent of ground-up rubber tires. This would not only get extra energy from the tires but would also help solve the problem of tire disposal. Piles of discarded tires are not only unsightly but are breeding grounds for mosquitoes and rodents (Figure 10.17).

On the other hand, burning plastics and rubber can lead to some new problems. For example, polyvinyl chloride produces toxic hydrogen chloride gas when it burns, and burning automobile tires give off soot and a stinking smoke. Incinerators are corroded by acidic fumes and clogged by materials that are not readily burned.

SECTION 10.10

Plastics and Fire Hazards

The accidental ignition of fabrics, synthetic or otherwise, has caused untold human misery. The U.S. Department of Health and Human Services estimates that as many as 150,000 to 200,000 people are injured and several thousand are burned to death each year in accidents associated with flammable fabrics.

A good deal of research has been done to develop flame-retardant fabrics. One common approach involves the incorporation of chlorine and bromine atoms within the giant molecules of the polymeric fiber. Federal regulations require that children's sleepwear, in particular, be made of such flame-retardant materials. A major setback to the program occurred in 1977 when the Consumer Product Safety Commission banned a flame-retardant compound called Tris [tris(2,3-dibromopropyl)phosphate], which was used on sleepwear made from polyester and acetate fabrics. In laboratory tests, the compound had been shown to be both carcinogenic and mutagenic. The search for safe fire-retardants continues.

A *mutagen* is a substance that causes mutations, that is, induces changes in the DNA molecules that determine heredity. Provided they are not lethal, mutations are passed on to succeeding generations.

Another problem associated with plastics and fire is that of toxic gases. In 1972, lethal amounts of cyanide were found in the bodies of plane-crash victims in Chicago. The Illinois Department of Law Enforcement traced the cyanide to burned plastics in the plane's cabin. Firefighters often refuse to enter burning buildings without gas masks for fear of being overcome by fumes from burning plastics. Laboratory tests show that hydrogen cyanide is formed in large quantities by the burning of polyacrylonitrile (Orlon, Creslan, Acrilan, etc.) and other nitrogen-containing plastics. The increasing use of plastics in home furnishings and construction calls for more research in the area of identifying—and quantifying—the products of the combustion of plastics.

SECTION 10.11

Plasticizers and Pollution

It is difficult to process some plastics, particularly vinyl polymers. As formed, they can be rather hard and brittle. But they can be made more flexible by adding chemicals called plasticizers.

Ideally, **plasticizers** are liquids of low volatility that act by lowering the T_g of the plastic to make it more flexible and less brittle. Thin sheets of pure polyvinyl chloride (PVC) crack and break easily, but with proper plasticizers added, they are soft and pliable. Plastic raincoats, garden hoses, and seat covers for automobiles can be made from modified PVC. Plasticizers are generally lost by diffusion and evaporation as a plastic article ages. The plastic becomes brittle, then cracks and breaks.

Once used widely as plasticizers, but now banned, are the *polychlorinated biphenyls* (PCBs). The compounds are derived from a hydrocarbon called biphenyl ($C_{12}H_{10}$) by the replacement of hydrogen atoms with chlorine atoms. Some PCBs are shown in Figure 10.18. Note their structural similarity to the insecticide DDT.

Biphenyl is two benzene rings joined at a corner. As a substituent, the benzene ring is called a **phenyl group**. The phenyl group often is written as C_6H_5—.

PCB residues have been found in fish, birds, water, and sediments. The physiological effect of PCBs is similar to that of DDT. Poultry and eggs found to have concentrations greater than the 5 ppm allowed by the Food and Drug Administration (FDA) had to be destroyed.

Not all the PCBs manufactured were used as plasticizers. Some were used in electrical equipment. PCBs have a high electrical resistance, so they are useful as insulating materials in transformers, condensers, and other electrical apparatus.

The same properties that made the PCBs so desirable as industrial chemicals cause them to be an environmental hazard. They degrade slowly in na-

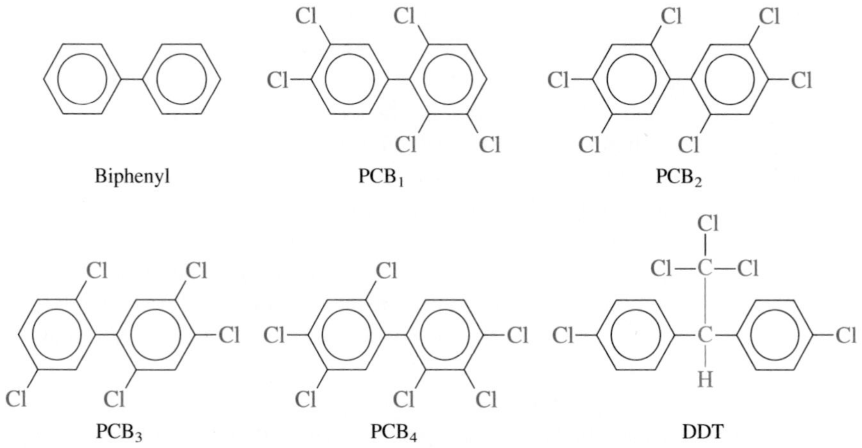

Biphenyl

PCB$_1$

PCB$_2$

PCB$_3$

PCB$_4$

DDT

Figure 10.18 Biphenyl and some of the PCBs derived from it. These are but a few of the hundreds of possible PCBs. DDT is shown for comparison.

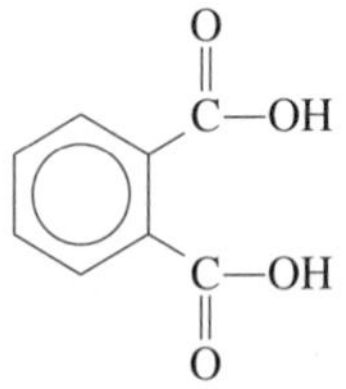

Phthalic acid: benzene ring with two adjacent $-C(=O)-OH$ groups

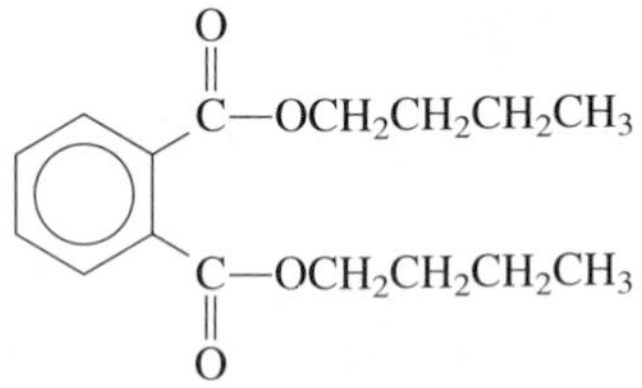

Dibutyl phthalate (DBP): benzene ring with two adjacent $-C(=O)-OCH_2CH_2CH_2CH_3$ groups

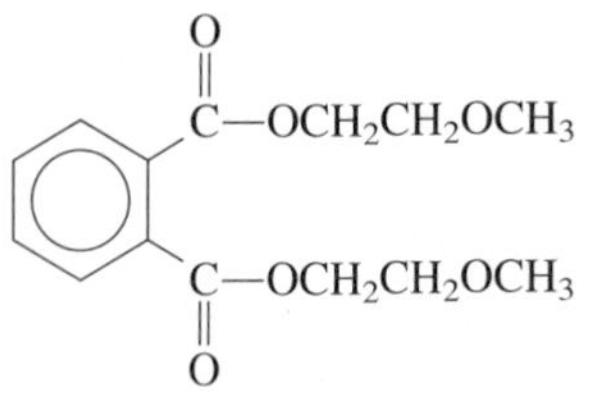

Dimethoxyethyl phthalate (DMEP): benzene ring with two adjacent $-C(=O)-OCH_2CH_2OCH_3$ groups

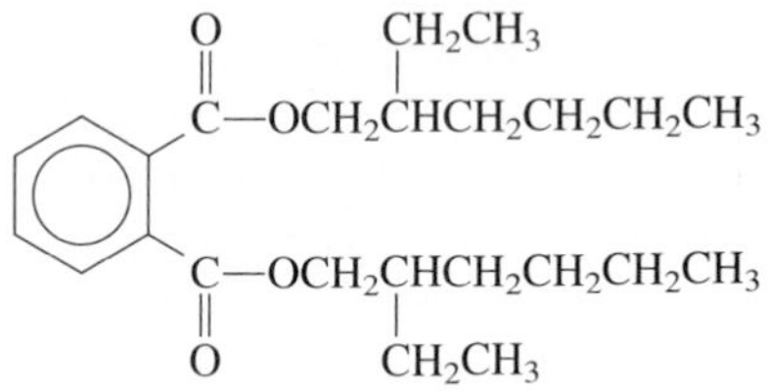

Dioctyl phthalate (DOP): benzene ring with two adjacent $-C(=O)-OCH_2CH(CH_2CH_3)CH_2CH_2CH_2CH_3$ groups

Figure 10.19 Phthalic acid and some derivatives. Dioctyl phthalate is also called di-2-ethylhexyl phthalate (DEHP).

ture, and their solubility in nonpolar media—animal fat as well as vinyl plastics—leads to their concentration in the food chain. Monsanto Corporation, the only company in the United States that produced PCBs, discontinued production in 1977, but PCBs will remain in the environment for years.

The most widely used plasticizers for flexible vinyl plastics are the phthalate esters, a group of diesters derived from phthalic acid (Figure 10.19). Phthalate plasticizers seem to have low acute toxicity, although their long-term effects are unknown.

SECTION 10.12

Plastics and the Future

In many applications, substitution of plastics for natural materials saves energy. Their use in place of metals in automobiles and airplanes saves weight and therefore saves fuel.

Synthetic polymers are the materials of the future. They are widely used today, but there is no doubt that there will be new kinds of polymers and even wider use in the future. We already have new polymers that conduct electricity, amazing new adhesives, and synthetic materials that are stronger than steel but much lighter in weight. Although plastics have presented us with a few problems, they have become such an important part of our daily lives that we would find it difficult to live without them.

In the field of medicine, body replacement parts made from polymers have become common, but in the future we will have artificial bones that can stimulate new bone growth so as to knit the new pieces more securely into place. We can also expect to have artificial lungs and livers, as well as artificial hearts.

Vinyl water pipes, siding, and window frames, plastic foam insulation, and polymeric surface coatings are all widely used in home construction today. Tomorrow's homes may also contain lumber and wall panels of artificial wood, made perhaps from recycled plastic.

Synthetic polymers are already used extensively in the interior of airplanes, but future planes will have entire bodies and wings made of light-weight composite materials.

There are presently several kinds of automobiles with plastic bodies. Within the near future it is possible that many cars will have bodies made from plastic composites, and some may also have frames built from high strength composite materials.

Electrically conducting polymers will help to make possible lighter weight batteries for electric automobiles. The electronics industry will certainly be using increasing amounts of electrically conducting thermoplastics in their miniaturized circuits.

There will probably be many more exciting polymers developed during the twenty-first century. Consider photosynthesis (Chapter 8), the process through which we get all the food we eat and all the oxygen we breathe. It is a process whereby energy from the sun is harnessed to carry out a sequence of life-sustaining chemical reactions. Many laboratories are trying to design polymers that would mimic nature in being able to convert sunlight directly to useful chemical energy.

But here is something to think about. Most synthetic polymers are made from petroleum or natural gas. Both of these natural resources are nonrenewable, and our supplies are limited. Especially in the case of petroleum, we are likely to run out during the twenty-first century. You might suppose that we would be actively conserving this valuable resource. Unfortunately, that is not the case. We are taking petroleum out of the ground at a rapid rate, converting most of it to gasoline and other fuels, and then simply burning it. There are other sources of energy, but is there anything that can replace petroleum as the raw material for making plastics?

Petroleum

That viscous, tarry liquid nature laid beneath the ground
  A hundred million years before man came upon the scene
Can be transformed to marvelous new products we have found—
  Like nylon, orlon, polyesters, polyethylene,
Synthetic rubber, plastics, films, adhesives, drugs, and dyes,
  And other things, some that we now can only dream about.
Who knows what wondrous products man might some day synthesize
  From oil! Except, alas, that our supplies are running out.
The time is near when Earth's prodigious flow of oil may stop,
  (We've taken so much from the ground, with no way to return it).
Meanwhile, we strive to find and draw out every precious drop,
  And then . . . incredibly . . . we take the bulk of it and burn it.

Summary

1. *Polymers* are giant molecules made up of many small units called *monomers.*
2. There are many polymers in nature, including starch, and cellulose, large numbers of proteins, and rubber.
3. The first synthetic polymer was called *celluloid.* It was actually natural cellulose that had simply been nitrated.
4. The simplest, least expensive, and biggest volume synthetic polymer is *polyethylene.*
5. *Addition polymers,* such as polyethylene, polypropylene, polystyrene, and polyvinyl chloride, are made from monomers containing double bonds.
6. *Condensation polymers* are formed by splitting out small molecules, usually water, from monomers containing at least two functional groups per molecule.
7. *Nylon, Dacron,* and *Bakelite* are all condensation polymers. Nylon and Dacron are both thermoplastic, but Bakelite is thermosetting.
8. Phenol–formaldehyde resin (Bakelite), made in 1909, was the first truly synthetic polymer.
9. *Thermoplastic* polymers can be softened by heating and then remolded; *thermoset* polymers cannot.
10. *Rubber* materials are *elastomers,* cross-linked amorphous polymers that are free to coil and stretch.
11. Strong fibers are made from crystalline polymers with molecules that are neatly aligned with one another.
12. Some plastics, especially vinyl polymers, need *plasticizers* in order to remain pliable. Polychlorinated biphenyls (PCBs), once used as plasticizers, are now banned.
13. Disposal of waste plastics is a problem that can probably best be solved by recycling.
14. Most of our synthetic polymers come from petroleum, which is a very limited and nonrenewable natural resource.

Key Terms

addition polymers 10.5
Bakelite 10.7
celluloid 10.3
condensation polymer 10.7
copolymer 10.6
Dacron 10.7
elastomer 10.6
macromolecule 10.1
monomer 10.1
nylon 10.7
plasticizer 10.11
polyester 10.7
polyethylene 10.4
polymer 10.1
polymerization 10.1
polystyrene 10.5
recycling 10.9
rubber 10.6
segmer 10.5
silicone 10.7
thermoplastic 10.4
thermoset 10.4
vinyl polymers 10.5
vulcanization 10.6

Review Questions

1. Define these terms.
   a. macromolecule b. monomer c. polymer
   d. segmer e. elastomer f. copolymer
   g. plasticizer
2. What does the word *plastic* mean in the scientific sense?
3. What is celluloid? How is it made?
4. Why is celluloid no longer used to make movie film?
5. Describe the structure of high-density polyethylene. How does this structure explain the properties of this polymer?
6. Describe the structure of low-density polyethylene. How does this structure explain the properties of this polymer?
7. What is a thermoplastic polymer?
8. How does the structure of polyvinyl chloride differ from that of polyethylene?
9. List several uses of polyvinyl chloride.
10. What is addition polymerization?
11. What structural feature usually characterizes molecules used as monomers in addition polymerization?

12. What is Teflon? What unique property does it have?
13. What inexpensive plastic is used to make clear, brittle, disposable drinking cups?
14. From what monomer are disposable foamed plastic coffee cups made?
15. What plastic is used in making gallon milk jugs?
16. What plastic is used to make 2-liter bottles for carbonated beverages?
17. Explain the elasticity of rubber.
18. Describe the process of vulcanization.
19. How does vulcanization change the properties of rubber?
20. Name three synthetic elastomers.
21. How does polybutadiene differ from natural rubber in properties? In structure?
22. How does polychloropropene (Neoprene) differ from natural rubber in properties? In structure?
23. How is styrene–butadiene rubber (SBR) made?
24. What substances other than elastomers are used in automobile tires?
25. What is condensation polymerization?
26. What is Bakelite? From what monomers is it made?
27. What is a thermosetting polymer?
28. What proportion of the fibers used in the United States are synthetic?
29. What is the glass transition temperature (T_g) of a polymer?
30. For what uses do we want polymers with a low T_g? With a high T_g?
31. How do plasticizers act to make polymers less brittle?
32. What are silicones?
33. What are PCBs? Why are they no longer used as plasticizers?
34. What problems arise when plastics are discarded into the environment?
35. What problems arise when plastics are disposed of in landfills?
36. What problems arise when plastics are disposed of by incineration?
37. What steps must be taken in order to recycle plastics?
38. Discuss plastics as fire hazards.
39. What is Tris? Why was it banned?
40. Dacron has a high T_g. How can the T_g be lowered so that a manufacturer can permanently crease a pair of Dacron slacks?
41. How can the use of plastics save energy?
42. Where will plastics come from when the Earth's supplies of coal and petroleum are exhausted?

Problems

Addition Polymerization

43. Give the structure of the monomer from which polyvinyl chloride is made.
44. Give the structure of a segment of polyvinyl chloride that is at least five segmer units long.
45. Give the structure of a segment of polystyrene in which the carbon chain is at least ten carbon atoms long.
46. Give the structure of the monomer from which polystyrene is made.
47. Give the structure of the polymer made from acrylonitrile ($CH_2{=}CH{-}C{\equiv}N$). Show at least four segmer units.
48. Give the structure of the polymer made from vinyl acetate ($CH_2{=}CH{-}O\overset{O}{\overset{\|}{C}}CH_3$). Show at least four segmer units.
49. Give the structure of the polymer made from 1-hexene ($CH_2{=}CHCH_2CH_2CH_2CH_3$). Show at least four segmer units.
50. Give the structure of the polymer made from styrene ($CH_2{=}CH{-}C_6H_5$, with the phenyl group drawn as a benzene ring). Show at least four segmer units.
51. Give the structure of the polymer made from methyl methacrylate ($CH_2{=}\overset{CH_3}{\overset{|}{C}}{-}\overset{O}{\overset{\|}{C}}OCH_3$). Show at least four segmer units.

52. One type of Saran has the structure

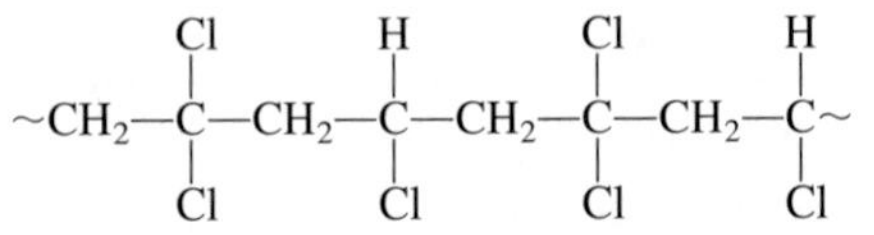

$$\sim CH_2-\underset{Cl}{\overset{Cl}{C}}-CH_2-\underset{Cl}{\overset{H}{C}}-CH_2-\underset{Cl}{\overset{Cl}{C}}-CH_2-\underset{Cl}{\overset{H}{C}}\sim$$

Give the structure of the two monomers from which this Saran is made.

53. Give the structure of the monomer from which polytetrafluoroethylene (Teflon) is made.

54. Give the structure of polytetrafluoroethylene. Show at least four segmer units.

55. Give the structure of isoprene, the ''monomer'' of natural rubber.

56. Give the structure of butadiene.

Condensation Polymerization

57. Nylon 46 is made from the monomers

$$H_2NCH_2CH_2CH_2CH_2NH_2$$

and

$$HO-\overset{O}{\overset{\|}{C}}CH_2CH_2CH_2CH_2\overset{O}{\overset{\|}{C}}-OH$$

Give the structure of nylon 46 showing at least two units from each monomer.

58. Give the structure of a polymer made from glycolic acid ($HO-CH_2\overset{O}{\overset{\|}{C}}-OH$). Show at least four segmer units. [*Hint:* Compare with nylon 6 in Section 10.7.]

59. Kodel is a polyester fiber. The monomers are terephthalic acid and 1,4-cyclohexanedimethanol. Write

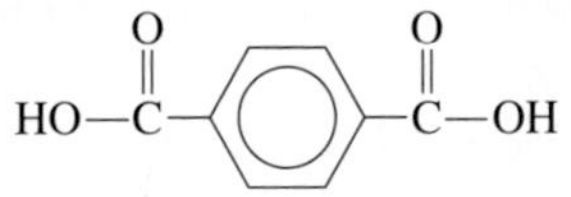

$$HO-\overset{O}{\overset{\|}{C}}-C_6H_4-\overset{O}{\overset{\|}{C}}-OH$$

Terephthalic acid

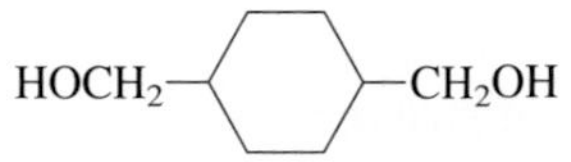

$$HOCH_2-C_6H_{10}-CH_2OH$$

1,4-Cyclohexanedimethanol

the structure of a segment of Kodel containing at least two of each monomer unit.

60. Kevlar, a polyamide used to make bulletproof vests, is made from terephthalic acid (Problem 59) and phenylenediamine. Write the structure for a segment of the Kevlar molecule.

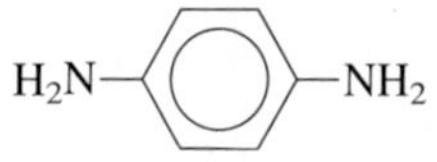

$$H_2N-C_6H_4-NH_2$$

Phenylenediamine

Additional Problems

61. In the following equation, identify the parts labeled a, b, and c as monomer, polymer, and segmer.

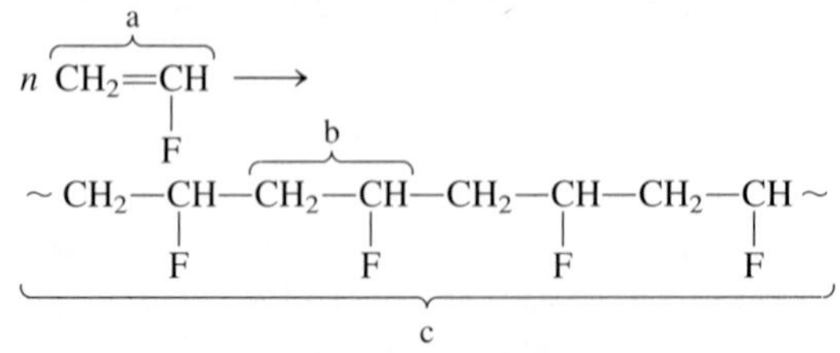

$$n\ \overbrace{CH_2=\underset{F}{CH}}^{a} \longrightarrow$$

$$\underbrace{\sim CH_2-\underset{F}{CH}-\overbrace{CH_2-\underset{F}{CH}}^{b}-CH_2-\underset{F}{CH}-CH_2-\underset{F}{CH}\sim}_{c}$$

62. What type of polymerization (addition or condensation) is represented in Problem 61?

63. Is nylon 46 (Problem 57) an addition polymer or a condensation polymer? Explain.

64. From what monomers might the following copolymer be made?

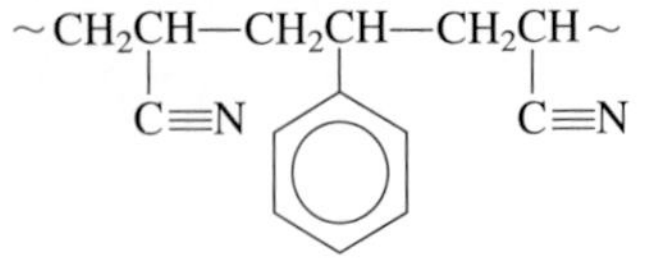

$$\sim CH_2\underset{C\equiv N}{CH}-CH_2\underset{C_6H_5}{CH}-CH_2\underset{C\equiv N}{CH}\sim$$

65. Isobutylene ($CH_2=\underset{CH_3}{C}CH_3$) polymerizes to form polyisobutylene, a sticky polymer used as an adhesive. Give the structure of polyisobutylene. Show at least four segmer units.

66. Copolymerized with isoprene, isobutylene (Problem 65) forms butyl rubber. Give the structure of butyl rubber. Show at least three isobutylene segmers and one isoprene segmer.

67. The bacteria *Alcalgenes eutrophus* produce a polymer called polyhydroxybutyrate with the structure

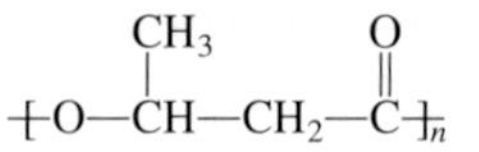

$$+O-\overset{CH_3}{\overset{|}{C}H}-CH_2-\overset{O}{\overset{\|}{C}}+_n$$

Give the structure of the hydroxy acid from which this polymer could be made.

Projects

68. Make a list of plastic objects (or parts of objects) that you encounter in your daily life. Try to identify a few of the kinds of polymers used in making the items. Compare your list with those of some of your classmates.

69. Do a risk–benefit analysis (Section 1.6) for the use of synthetic polymers as one or more of the following (as directed by your instructor).
a. grocery bags
b. building materials
c. clothing
d. carpets
e. food packaging
f. picnic coolers
g. automobile tires
h. artificial hip sockets

70. To what extent are plastics a litter problem in your community? Survey one city block (or other area as directed by your instructor) and inventory the litter found. (You might as well pick it up while you are at it.) What proportion of the trash is plastics? What proportion of it is fast-food containers?

71. To what extent are plastics recycled in your community? What factors limit further recycling?

References and Readings

1. Agoos, Alice. ''Hard as Horn: Plastics Development.'' *Today's Chemist,* December 1989, pp. 19–21.
2. Alper, Joseph, and Gordon L. Nelson. *Polymeric Materials.* Washington: American Chemical Society, 1989.
3. Artandi, Charles. ''Fibers in Medicine.'' *ChemTech,* August 1981, pp. 476–479.
4. Bell, John. ''Plastics: Waste Not, Want Not,'' *New Scientist,* Dec. 1, 1990, pp. 44–47.
5. Crawford, Mark. ''There's (Plastic) Gold in Them Thar Landfills.'' *Science,* 22 July 1988, pp. 411–412.
6. Deanin, Rudolph D. ''The Chemistry of Plastics.'' Journal of Chemical Education, Jan. 1987, pp. 45–47.
7. Harris, Frank W. ''Introduction to Polymer Chemistry.'' *Journal of Chemical Education,* November 1981, pp. 837–843.
8. Kauffman, George B., and Raymond B. Seymour. ''Elastomers I: Natural Rubber.'' *Journal of Chemical Education,* May 1990, pp. 422–425.
9. Letcher, Trevor M., and Nothando S. Lutseke. ''A Closer Look at Cotton, Rayon, and Polyester Fibers.'' *Journal of Chemical Education,* May 1990, pp. 361–363. Shows electron microscope photographs.
10. Marinelli, Janet. ''Packaging.'' *Garbage,* May–June 1990, pp. 28–35. Evaluates environment-friendly packaging.
11. Marvel, C. S. ''The Development of Polymer Chemistry in America—The Early Days.'' *Journal of Chemical Education,* July 1981, pp. 535–539.
12. Seymour, Raymond B., and C. E. Carraher. *Giant Molecules.* New York: John Wiley, 1989.
13. Seymour, Raymond B., and George B. Kauffman. ''Elastomers II: Synthetic Rubbers.'' *Journal of Chemical Education,* Mar. 1991, pp. 217–220.
14. Stone, R. F., A. D. Sagar, and N. A. Ashford. ''Recycling the Plastic Package.'' *Technology Review,* July 1992, pp. 48–56.
15. Thayer, Ann M. ''Degradable Plastics Generate Controversy in Solid Waste Issues.'' *Chemical and Engineering News,* 25 June 1990, pp. 7–14.
16. Wolfe, George. ''Pack it Right.'' *Nutrition Action Healthletter,* April 1990, pp. 5–7. Evaluates plastic versus paper as wrapping for food.
17. Wright, Karen. ''The Shape of Things to Go.'' *Scientific American,* May 1990, pp. 92–101. Discusses new materials and fuels for automobiles.

11 Chemistry of the Earth

Metals and Minerals

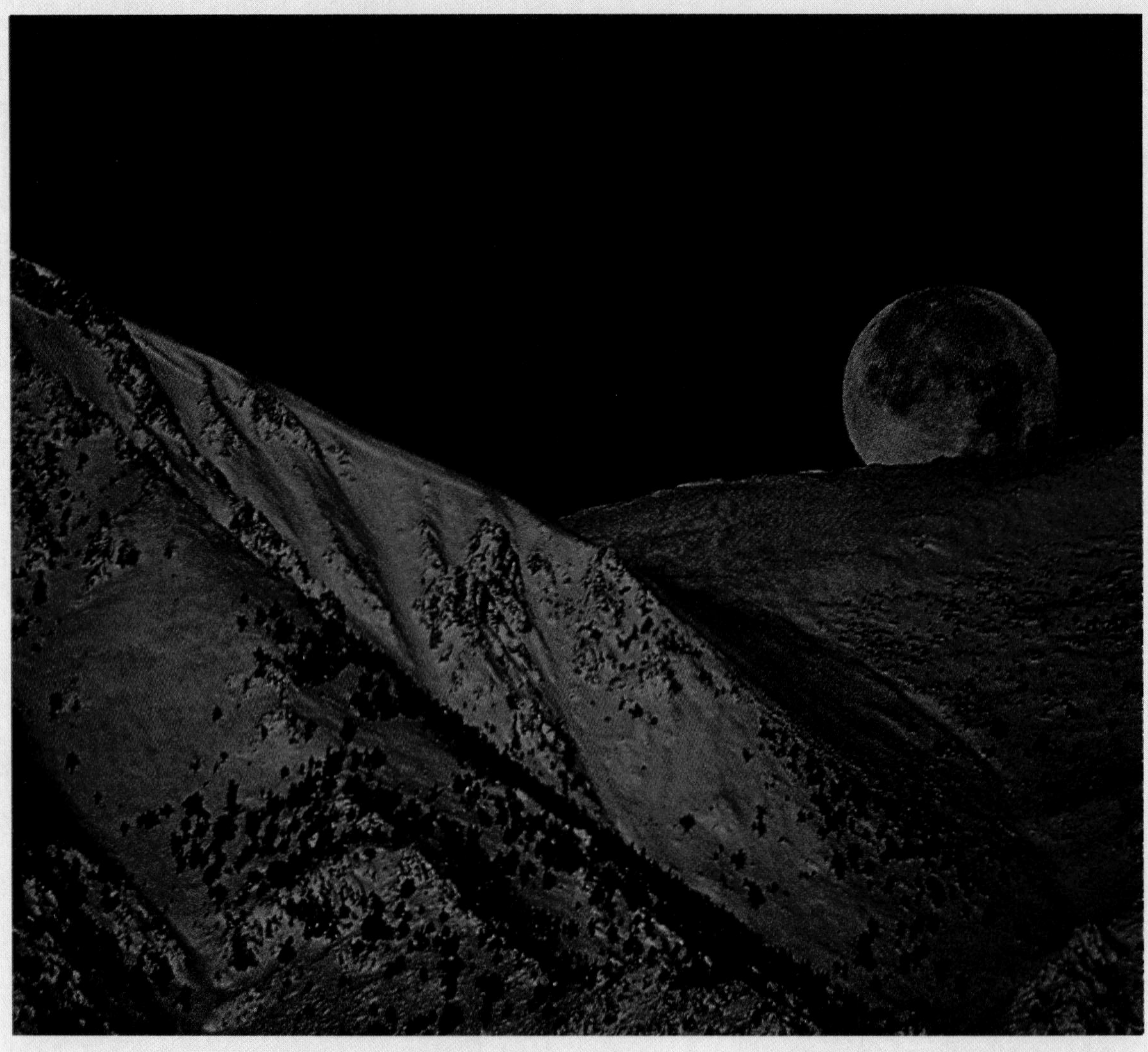

Hills and valleys all around,

Minerals beneath the ground,

Mountains, beaches, open space . . .

Earth is such a wondrous place!

This wondrous world of ours is a fertile sphere blanketed in air, with about three-fourths of its surface covered by water. Although it is but a tiny blue-green jewel in the vastness of space (Figure 11.1), our Spaceship Earth is about 40,000 km in circumference, with a surface area of half a billion square kilometers.

Our astronauts have walked on the barren surface of Earth's airless moon. Our space probes have explored the desolation of Mars and the crushing pressure and hellish heat of Venus, with its hurricane clouds of sulfuric acid. They have also given us close-up portraits of dry, pockmarked Mercury, as well as Jupiter and Saturn with their horrendous lightning storms and their turbulent atmospheres of hydrogen and helium. Earth is but a small island in the inhospitable immensity of space, a tiny oasis uniquely suited to the life that inhabits it.

More than 5 billion passengers are being carried aboard Spaceship Earth, but their numbers are increasing and will be at least 6 billion by the turn of the century. What kinds of materials do we have aboard this spaceship? Are they sufficient for this enormous load of passengers? Let us begin by looking at the composition of the Earth.

Figure 11.1 Spaceship Earth with the desolate moon in the foreground. [*NASA photo.*]

SECTION 11.1

Spaceship Earth: What It Is Made of

Earth is divided into three main regions: the *core,* the *mantle,* and the *crust* (Figure 11.2). The **core** is thought to consist largely of iron (Fe) and nickel (Ni). Since the core is not accessible, and since it does not seem likely that it will become so, we won't consider it as a source of materials.

The **mantle** is believed to be mostly silicates—compounds of silicon and oxygen with a variety of metals. Although the mantle might be reached even-

The Earth is a storehouse of chemicals—minerals, metals, and much more. Chemists modify many of these materials to make them more useful. [*Galen Rowell.*]

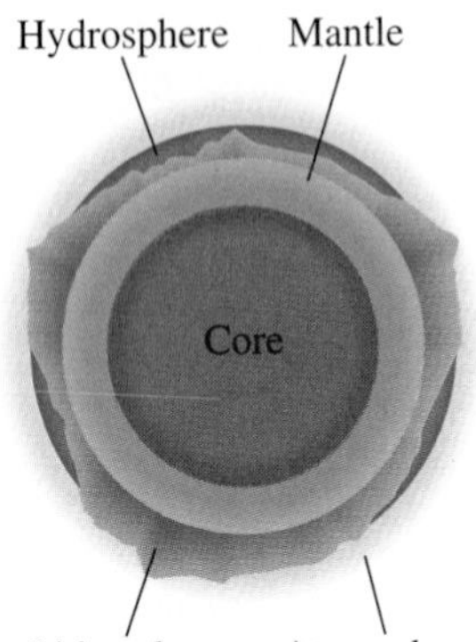

Figure 11.2 Schematic diagram of the Earth (not to scale).

tually, it probably has few useful materials that are not available in the more accessible crust.

The **crust** is the outer shell of the Earth. It has three parts: the **lithosphere** (solid part), the **hydrosphere** (oceans, seas, lakes, rivers, etc.), and the **atmosphere** (air). The lithosphere is about 35 km thick under the continents and about 10 km thick under the oceans. Extensive sampling of the crust has made it possible for scientists to estimate its elemental composition. Let's consider a random sample of 10,000 atoms from the crust. Of these, more than half (5330 atoms) would be oxygen. This element occurs in the atmosphere as molecular oxygen (O_2), in the hydrosphere in combination with hydrogen as water, and in the lithosphere in combination with silicon (pure sand is largely SiO_2) and various other elements.

The second most abundant element in the Earth's crust is silicon (Si). Of those 10,000 atoms in our sample, 1590 would be silicon. Hydrogen (H) would come in third with 1510 atoms, most of which would be in combination with oxygen in water. Hydrogen is such a light element—the lightest of all—that it makes up only 0.9% of the Earth's crust by weight. The nine most abundant elements and the number of their atoms in our sample of 10,000 are listed in Table 11.1. These nine elements would account for 9630 of the atoms, leaving only 370 atoms of all the other elements. However, some very important elements, such as carbon, nitrogen, and phosphorus, are among these elements that are rather scarce on Earth.

This chapter is mainly concerned with the solid portion of the crust—the lithosphere. We discuss the atmosphere (Chapter 12) and the hydrosphere (Chapter 13) later.

Table 11.1 Composition of the Earth's Surface: Atoms of Each Element in a Sample of 10,000 Atoms and Percent by Mass

Element	Symbol	Number of Atoms	Percent by Mass
Oxygen	O	5,330	49.5
Silicon	Si	1,590	25.7
Hydrogen	H	1,510	0.9
Aluminum	Al	480	7.5
Sodium	Na	180	2.6
Iron	Fe	150	4.7
Calcium	Ca	150	3.4
Magnesium	Mg	140	1.9
Potassium	K	100	2.4
All Others	—	370	1.4
Total		10,000	100.0

SECTION 11.2

The Lithosphere: Organic and Inorganic

The lithosphere is mainly rocks and minerals. Prominent among these are **silicates** (compounds of metals with silicon and oxygen), **carbonates** (metals combined with carbon and oxygen), **oxides** (metals combined with oxygen), and **sulfides** (metals combined with sulfur only). There are thousands of these mineral compounds making up the *inorganic* portion of the solid crust.

Though much, much smaller in quantity, the *organic* portion of the lithosphere includes all living creatures, their waste and decomposition products, and fossilized materials (such as coal, natural gas, petroleum, and oil shale) that once were living organisms. This organic material *always* contains the element carbon, nearly always has combined hydrogen, and often contains oxygen, nitrogen, and other elements.

The organic portion of the lithosphere is discussed in later chapters, petroleum under Energy (Chapter 14), living organisms under Biochemistry (Chapter 15), and edible products from nature under Food (Chapter 16).

In this chapter we concentrate on the *inorganic* portion of the lithosphere.

One of the more common rocks on Earth is limestone, which is calcium carbonate ($CaCO_3$). In nature calcium carbonate is found in many different physical forms. Marble is calcium carbonate, and so are seashells, egg shells, calcite, travertine, aragonite, coral, chalk, and the stalagmite and stalactite formations in caves (Figure 11.3).

Here is something to think about.

The Taj Mahal, resplendent in its solemn majesty;
Gibraltar's famous rock, a symbol of stability;
The snow white Cliffs of Dover in their bleak austerity;
Australia's mighty Barrier Reef that spans the Coral Sea;
The Roman Colosseum with its savage history;
The enigmatic pyramids enduring silently;
And Mammoth Cave's formations with their rich variety—
The formula, *in every case, is* $CaCO_3$.

Figure 11.3 Minerals from limestone. Carbon dioxide from the atmosphere dissolves in rainwater to form an acidic solution. As this acidic rainwater seeps through limestone formations, insoluble calcium carbonate is converted to soluble calcium bicarbonate.

$$CaCO_3 + H_2O + CO_2 \rightleftharpoons Ca(HCO_3)_2$$

Over the centuries, this action can produce a large cave in the limestone formation. The reaction is reversible, however, and as dripping water, saturated with calcium bicarbonate, evaporates, stalactites and stalagmites are formed. This photograph shows Temple of the Sun, Carlsbad Cavern, New Mexico. [*Fred E. Mang, Jr./courtesy of National Park Service.*]

SECTION 11.3

Meeting Our Needs: From Sticks to Bricks

The needs of primitive people were few. Food was obtained by hunting and gathering. Clothing, when it was needed, was provided by the skins of animals. Shelter was found in a convenient cave or constructed from available sticks and stones and mud.

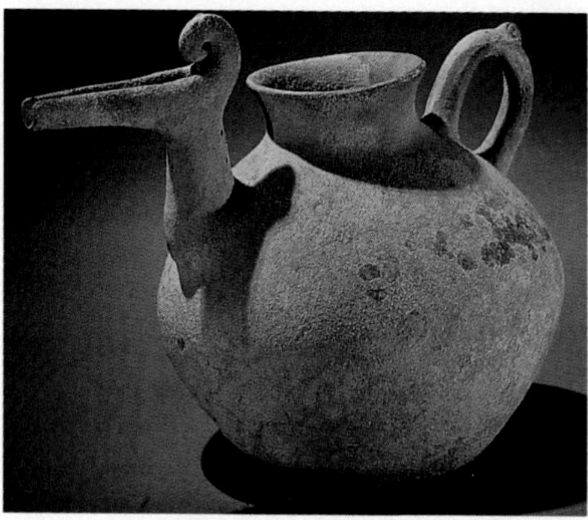

Figure 11.4 A prehistoric ceramic pot from Afghanistan. [*American Museum of Natural History (#K9927).*]

After the agricultural revolution (about 10,000 years ago), people no longer were forced to search for food. Domesticated animals and plants supplied their needs for food and clothing. A reasonably assured food supply enabled them to live in villages and created a demand for more sophisticated building materials. People learned to convert natural materials into other products with superior properties. Adobe bricks, which serve well in arid areas, can be made simply by drying mud in the sun.

Along the way people discovered fire, and they learned quite early that cooking meat and grains improved their flavor. Fire was one of the earliest agents of chemical change. With it people learned to heat their adobe bricks to make much stronger ceramic bricks. A similar process of firing clay at high temperatures produced ceramic pots (Figure 11.4), which made the cooking and storage of food much easier than using a spit for roasting it and gourds or wooden vessels for storing it.

Clay is a complex mixture of silicates. *Ceramics* are inorganic materials made by heating clay or other mineral matter to a high temperature at which the particles partially melt and fuse together.

With access to high temperatures, people later learned to make glass and to get metals from their ores. They were able to make tools from bronze, and later from iron. People were learning that there were some useful minerals lying in the ground.

SECTION 11.4

Silicates and the Shape of Things

There are thousands of different minerals in the lithosphere. We will consider only a few representative ones here. First, let's look at some of the **silicates**. The basic unit of silicate structure is the SiO_4 tetrahedron.

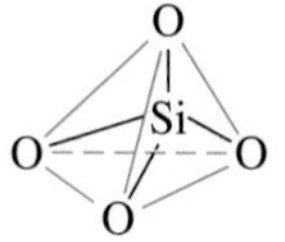

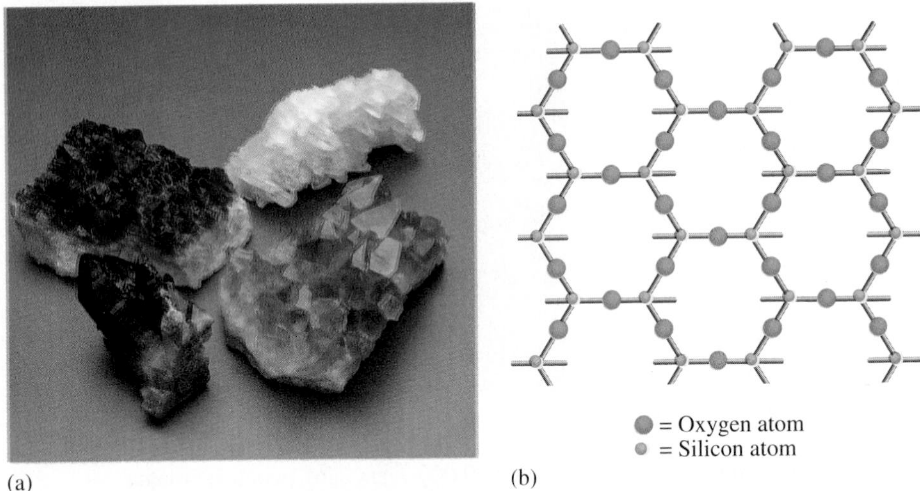

Figure 11.5 (a) A variety of quartz crystals. From the upper left: citrine (yellowish quartz), colorless quartz, amethyst (purple quartz), and smoky quartz. (b) The chemical structure of quartz. [*(a) Macmillan Publishing/Geoscience Resources.*]

Agate, jasper, and onyx are variously colored forms of chalcedony, a kind of quartz with a waxy luster.

Silicate tetrahedra can be arranged in a linear manner to form fibers (asbestos), in planar sheets (mica), or in complex three-dimensional arrays (quartz, feldspar).

Quartz is pure silicon dioxide (SiO_2). The ratio of silicon to oxygen atoms is 1:2, but since each silicon atom is surrounded by four oxygen atoms, the

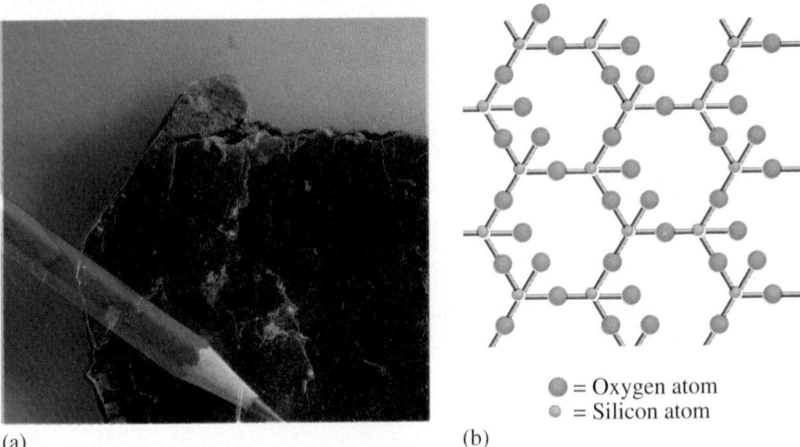

Figure 11.6 (a) A sample of mica, showing cleavage into thin transparent sheets. (b) The chemical structure of mica shows sheets of SiO_4 tetrahedra. These sheets are bound together by cations, principally Al^{3+} (not shown). [*(a) Fundamental Photographs.*]

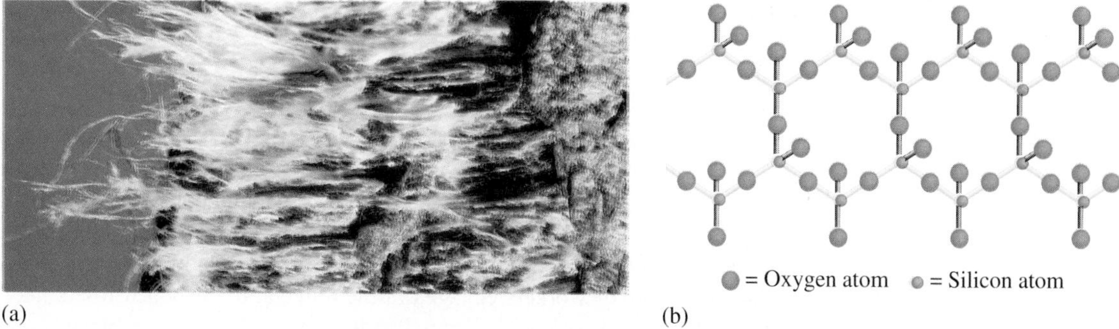

Figure 11.7 (a) A sample of chrysotile asbestos. (b) The chemical structure of chrysotile shows double chains of SiO_4 tetrahedra. The two chains are joined to each other through oxygen atoms. The double chains in turn are bound to each other by cations, principally Mg^{2+} (not shown). [*(a) Courtesy of U.S. Bureau of Mines.*]

basic unit of quartz is the SiO_4 tetrahedron. Crystals of pure quartz (rock crystal) are colorless, but various impurities produce amethyst (purple), citrine (yellow), rose quartz (pink), and smoky quartz (grey to black), some of the quartz crystals used as gems (Figure 11.5).

Micas are composed of SiO_4 tetrahedra arranged in two-dimensional sheet-like arrays. Micas are easily cleaved into thin, transparent sheets. Pieces of mica were once used as panels for lanterns and as windows in the doors of stoves. Figure 11.6 shows the transparency of a sheet of mica.

Asbestos is a generic term for a variety of fibrous silicates. Perhaps the best known of these is *chrysotile,* a magnesium silicate. A sample of this mineral is shown in Figure 11.7. Note that the chrysotile is a double chain of SiO_4 tetrahedra. The oxygen atoms that have only one covalent bond also bear a negative charge. It is with these that the magnesium ions (Mg^{2+}) are associated.

Figure 11.8 Clothing made of asbestos fibers was once widely used to protect workers exposed to high temperatures and hot materials. Asbestos for this purpose has been replaced to a large extent by synthetic polymers (Chapter 10). A worker is shown here pouring gold in a refinery. [*Mark Snyder/The Stock Market.*]

Asbestos has been used widely. It is an excellent thermal insulator. Great quantities of it have been used to insulate furnaces, heating ducts, and steam pipes. It has been used to make protective clothing for firefighters and others who are exposed to flames and high temperatures (Figure 11.8). It is used in brake linings for automobiles.

The health hazards to those who work with asbestos are well known. Inhalation of fibers 5 to 50 μm long over a period of 10 to 20 years causes *asbestosis.* After 30 to 45 years, some asbestos workers contract lung cancer. Others get mesothelioma, a rare and incurable cancer of the linings of body cavities.

Long-term occupational exposure to asbestos increases the risk of lung cancer by a factor of two. Cigarette smoking causes a tenfold increase in the risk of lung cancer. We would expect asbestos workers who smoke to have a risk 20 times that of nonsmokers who do not work with asbestos, but instead their risk is increased by a factor of 90! Cigarette smoke and asbestos fibers act in such a way that each enhances the action of the other. Such a joint action is called a **synergistic effect**.

The harmful effects of asbestos are due mainly to a relatively rare form called *crocidolite.* Chrysotile, which makes up 95% of the asbestos used in the United States, appears not to be nearly so dangerous. Government regulations do not distinguish between the two types.

Public fears of cancer from asbestos have led to regulations that require its removal from schools and other public buildings. Costs of removal will total \$50 to \$150 billion. Benefits from this expenditure are doubtful (reference 8). In fact, exposure of workers engaged in the removal to dislodged fibers may increase the overall risk as compared to leaving the asbestos in place.

SECTION 11.5

Modified Silicates: Ceramics, Glass, and Cement

Manufacture of ceramic pots, glass making, and formulation of cement were among the earliest technologies to be developed. Natural materials such as sand, clay, and limestone were modified, mainly by mixing and heating, to make much more useful products.

Ceramics

Early potters used natural clays, which they merely hardened by heat (Figure 11.9). Clays are exceedingly complex and their composition varies widely, but they are basically aluminum silicates. When clay is mixed with water, it can be molded into any shape. Firing leaves a hard, resistant (but porous) product. Bricks and tile are made in this manner. When porosity is not desirable—as in a cooking pot or a water jug—the pottery can be glazed by adding various salts to the surface. Heat then converts the entire surface to a glasslike matrix.

Figure 11.9 This ancient pottery jar is from San Jacinto, El Salvador. [*American Museum of Natural History (#278).*]

People long have been able to modify nature's materials to more nearly satisfy their needs and wants. For centuries, technological developments were based largely on trial and error. Modern science has, by developing an understanding of the structure of materials, greatly increased the human capacity for modification of natural materials. Thus, technological advances have been greatly accelerated by scientific knowledge.

Modern bone china, used to make fine dinnerware, is actually 50% bone (mainly calcium phosphate from animal bones), 25% kaolin (white clay), and 25% petuntse (decomposed granite). The mixture is blended with water, molded into shape, fired in a kiln, and then glazed.

Ceramic research has led to the development of some fantastic new **ceramic** materials. Some have such high heat resistance that they can withstand the extreme temperatures of atmospheric rocket re-entry. They are used to make rocket nose cones, surface tiles, and exhaust nozzles. Some have magnetic properties and can serve as memory elements in computers. And some have such exceptional electrical conductivity at the temperature of liquid nitrogen that they are known as "superconductors." Such superconducting materials are needed to build the powerful electromagnets used in particle accelerators and magnetically levitated ("maglev") trains.

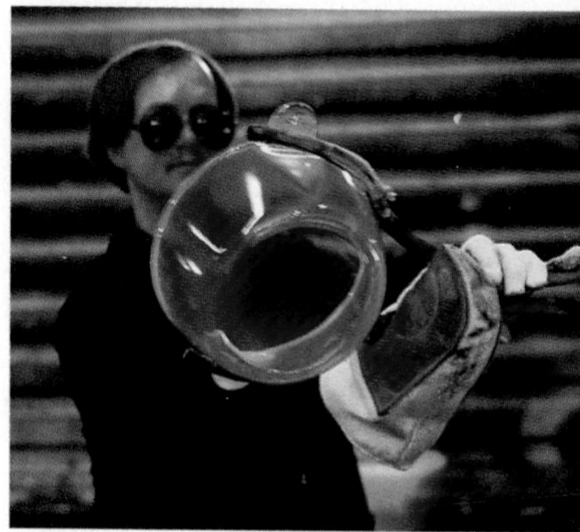

Figure 11.10 This heat-softened glass-ceramic material is shaped into a piece of Visions cookware. [*George Kemper Productions/Corning, Inc.*]

Glass

Glass is another technological development of ancient times. The first glass probably was made in ancient Egypt about 5000 years ago by heating a mixture of sand, sodium carbonate (Na_2CO_3), and limestone (calcium carbonate, $CaCO_3$). As the mixture melts, it becomes a homogeneous liquid. When the liquid cools, it becomes a hard, transparent material, **glass**.

When crystalline materials are heated, they melt over a narrow temperature range. Glass is different; when heated, it gradually softens. While soft, it can be blown (Figure 11.10), rolled, pressed or molded into almost any shape. The properties of glass are interpreted in terms of an *irregular* arrangement, in three dimensions, of SiO_4 tetrahedra. The chemical bonds in this arrangement are not all equivalent. Thus, when glass is heated, the weaker bonds break first and the glass softens gradually.

The basic ingredients in glass can be used in different proportions. Oxides of various metals can be substituted in whole or in part for the lime, soda, or sand. Thus, many special glasses can be made. Some examples are given in Table 11.2.

Table 11.2 Composition and Properties of Various Glasses

Type	Composition	Special Properties and Uses
Soda–lime glass	Sodium and calcium silicates (sand + soda + lime)	Ordinary glass (for windows, bottles, etc.)
Borosilicate glass (Pyrex, Kimax)	Boron oxide (instead of lime)	Heat resistant (for laboratory ware and ovenware)
Aluminosilicate glass	Aluminum oxide (instead of soda)	More highly heat resistant (for top-of-stove cookware and fiberglass)
Lead glass	Lead oxide (instead of lime)	Highly refractive (for optical glass, art glass, table crystal)
Alabaster glass	Sodium chloride (salt) added	White, opaque ("milk glass")
Colored glass	Selenium compounds added	Red color (ruby glass)
	Cobalt compounds added	Blue color (cobalt glass)
	Chromium compounds added	Green color
	Manganese compounds added	Violet color
	Cadmium sulfide added	Yellow color
	Carbon and iron oxide added	Brown color (amber glass)
Photochromic glass	Silver chloride or bromide added	Light sensitive, darkens when exposed to light (for sunglasses, hospital windows)
Laser glass	Contains neodymium	Strong laser
Frosted glass	Etched with hydrofluoric acid (HF)	Satiny frosted surface

Modern glass research has resulted in many wonderful new products, but one of the most remarkable developments is a product that is practically invisible. It consists of very pure threads of glass about as thick as a human hair. Called fiber optics, or optical wave guides, these fibers are already replacing many old telephone lines. A bundle of glass fibers can carry several hundred times as many messages as a copper cable the same size.

Fiber optics actually carry *light.* The messages they carry are intermittent bursts of light. When you speak into a telephone, the sound must be translated into an electrical signal, which then pulses a laser to give off light messages that are carried by the glass fibers. At the end of the line the light pulses must be converted back to sound waves.

These amazing glass fibers are not used only for telephone lines. They are already being used in computer interfacing and various internal but nonsurgical medical procedures. Who knows what we might be able to do with them tomorrow!

No vital raw materials are used in the manufacture of ordinary glass, but a furnace is required to melt and shape glass. Disposal presents a problem, since glass is one of the most permanent materials known. It doesn't rot when discarded. Glass is easily recycled, however, as long as different kinds are separated. It can be melted and formed into new objects at considerable energy savings compared to the manufacture of new glass. In fact, it is standard practice to add ''cullet'' (broken glass) to each batch of sand, soda, and limestone to be melted. The cullet is usually scrap glass from previous melts. This not only gets rid of waste glass, but it also actually improves the melting process.

Cement and Concrete

Cement is still another ancient technological development. The Romans used cement to construct roads, aqueducts, and the famous Roman baths. The raw materials for the production of **cement** are limestone (calcium carbonate, $CaCO_3$) and clay (aluminum silicates). The materials are finely ground, mixed, and roasted at about 1500 °C (in a kiln heated by burning natural gas or powdered coal) (Figure 11.11). The finished product is mixed with sand and gravel to form concrete.

Our understanding of the complex chemistry of cement is still imperfect. Nevertheless, several varieties of cement with special properties are available. There is a cement that sets quickly, with high early strength; a white cement; a waterproof cement; one that sets at high temperatures; and many other varieties.

(a)

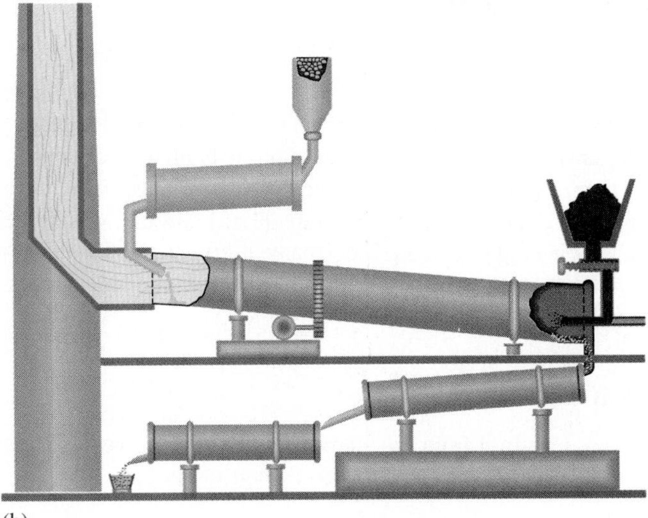

(b)

Figure 11.11 (a) A cement kiln. (b) Diagram of a cement kiln. [*(a) Courtesy of Portland Cement Association.*]

The production of cement involves extensive mining, with whole mountains being torn down for limestone rocks. The rotary kiln process consumes large quantities of irreplaceable fossil fuels. Particulate matter from the crushing operations and smoke and sulfur dioxide from the burning of fossil fuels make air pollution from cement plants especially serious. In use, concrete covers what once were green acres. In cities, the amount of paving with concrete and asphalt is enough to change the climate (temperature and rainfall). Runoff of rainfall from paved areas is especially rapid and contributes to flash flooding. But cement itself doesn't seem to be a bad pollutant. It can be broken up and used as rock fill.

Figure 11.12 This ancient bronze drum was made in Southeast Asia. [*American Museum of Natural History (#3121).*]

SECTION 11.6

Metals and Ores

Human progress through the ages often is described in terms of the materials that were used for making tools. We speak of the Stone Age, the Bronze Age, and the Iron Age. Ancient peoples knew about gold and silver because they are often found in the free state in nature. But these metals are too soft to use as tools. Metal tools were not possible until artisans learned to win certain metals from their ores by smelting.

Copper and Bronze

Copper was probably the first metal to be freed from its ore by early smelting techniques. We know from ancient Egyptian records that copper was known as early as 3500 B.C. It was likely isolated by heating copper carbonate ($CuCO_3$) or copper sulfide (Cu_2S) ore.

Today copper is important mainly because of its excellent electrical conductivity. It is used primarily for electrical wiring.

$$Cu_2S + O_2 \longrightarrow 2\,Cu + SO_2$$

Many copper ores are blue to green in color and easy to identify in the ground. The ancient Egyptians produced thousands of tons of copper metal.

Even more important than copper was **bronze**, which is a copper **alloy** containing around 10% tin (Figure 11.12). Bronze was harder than copper, and so it could be made into many useful tools. This was the first metal technology that really changed society. Bronze remained the most widely used metal for about 2000 years.

An *alloy* is a mixture of two or more elements, at least one of which is a metal. Alloys have metallic properties.

Iron and Steel

To produce iron from ore, carbon (coke) is employed as the reducing agent. First, the carbon is converted to carbon monoxide.

$$2\,C + O_2 \longrightarrow 2\,CO$$

Then the carbon monoxide reduces the iron oxide to iron metal.

$$Fe_2O_3 + 3\,CO \longrightarrow 2\,Fe + 3\,CO_2$$

Iron is produced in a huge chimney-like vessel called a blast furnace (Figure 11.13). The raw materials fed into the furnace are iron ore, coke, and limestone. The coke (which is mainly carbon) is made by heating coal in the absence of air to drive off coal oil and tar. The limestone is added to combine with silicate impurities to form a molten **slag** that floats on top of the iron and is drawn off.

$$\underset{\text{Limestone}}{CaCO_3} + \underset{\text{Silica}}{SiO_2} \longrightarrow \underset{\text{Slag}}{CaSiO_3} + CO_2$$

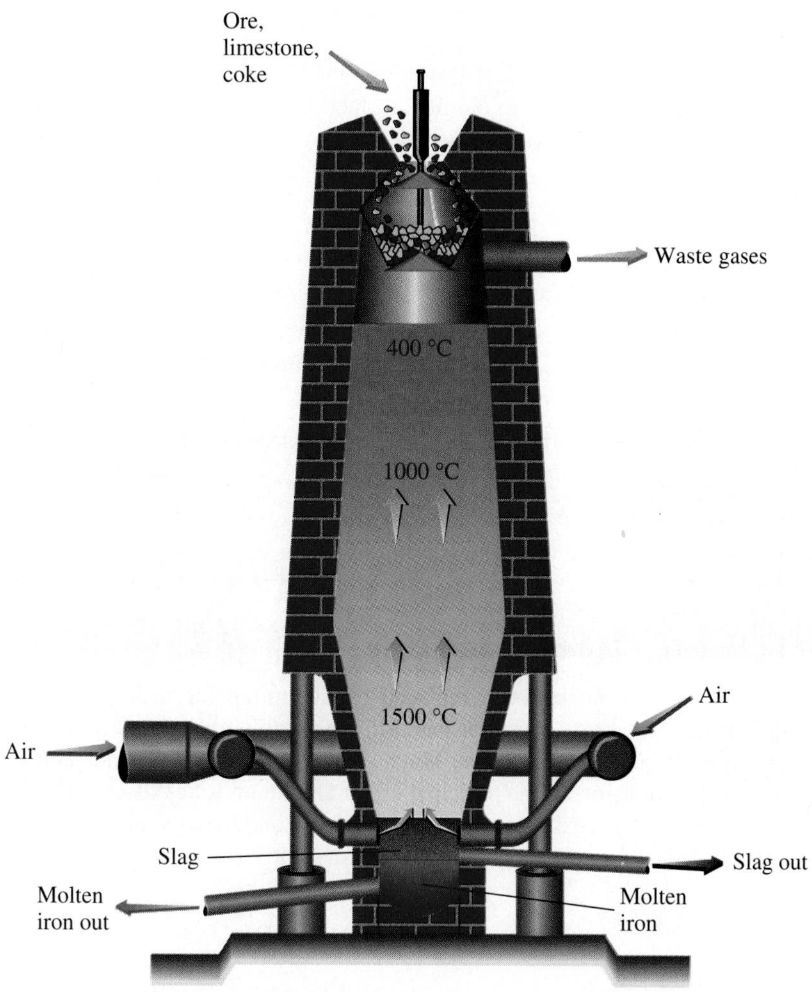

Figure 11.13 A blast furnace for the production of iron.

Molten iron is drawn off at the bottom of the furnace. The product of the blast furnace, called pig iron, has a fairly low melting point, so it is easily cast into molds. Hence, it is also known as cast iron.

Cast iron is brittle, and it has many impurities, so most iron from the blast furnace is converted to **steel**. In the steel furnace pressurized oxygen reacts with impurities, such as phosphorus, silicon, and excess carbon. Properties of steel can be varied over a wide range by adjusting the amount of carbon in it. High-carbon steel is hard and strong, whereas low-carbon steel is ductile and malleable.

Steel is an alloy of iron and carbon. It also may contain other elements.

Steel is commonly alloyed with other metals to give it special properties. For example, manganese imparts hardness, tungsten gives high temperature strength, and added chromium and nickel produce stainless steel. Because it is so abundant and can be made into so many different alloys, iron is the most useful of the metals.

Within recent years the steel industry has undergone some major changes. In the first place, small "minimills," which make steel from scrap iron instead of ore, now account for more than 50% of steel production in the United States. Although the blast furnace remains the primary method for producing iron from ore, that may not be true for long. The reason has to do with the need for coke in the blast furnace. In removing tar from coal, the coking operation produces huge amounts of environmental pollution, much of it carcinogenic. No new coking facilities have been built in the United States in the past half century. Present coking furnaces are old and dirty, and new ones are prohibitively expensive.

There are several new methods that might replace the old blast furnace. One involves direct reduction of iron ore with coal. Another reduces iron ore with natural gas to make iron carbide. However, a special ore from Australia or South America must be used in this case.

Aluminum: Abundant and Light

In recent years, iron has been replaced by aluminum for many purposes. **Aluminum** is the most abundant metal in the Earth's crust, but it is tightly bound in its compounds in nature. Much energy is required to extract aluminum metal from its ores. The principal ore of aluminum is bauxite, aluminum oxide. Impurities are extracted with a strong base, and then the melted oxide, Al_2O_3, is reduced by passing electricity through it.

$$2\ Al_2O_3 \xrightarrow{\text{electricity}} 4\ Al + 3\ O_2$$

It takes 2 t of aluminum oxide and 17,000 kilowatt hours (kWh) of electricity to produce 1 t of aluminum.

Aluminum is light and strong. A piece of aluminum weighs only one-third as much as a piece of steel the same size. Although it is considerably more active than iron, aluminum corrodes much more slowly. Freshly prepared

An interesting fact about aluminum is that the method we use to produce it was discovered more than 100 years ago by a college student named Charles Martin Hall. Even more fascinating is the fact that the same process was discovered almost simultaneously in France by another college student named Paul Héroult. Both were born in 1863, and it was in 1886 that they developed the process for manufacturing aluminum. But the coincidence doesn't end there. Both men died in 1914 at the age of 51.

aluminum metal reacts with oxygen to form a hard, transparent film of aluminum oxide (Al_2O_3) over its surface. The thin film protects the metal from further oxidation. Iron, on the other hand, forms an oxide coating that is porous and flaky. Instead of protecting the metal, this coating encourages further oxidation. Iron rusts; aluminum does not.

A tin can—really steel with a thin coating of tin—rusts when you throw it away, eventually disintegrating. Scientists at Pennsylvania State University estimate that it would take 500 years for an aluminum can to degrade completely.

The Environmental Cost of Iron and Aluminum

Steel and aluminum play vital roles in the modern industrial world. Their economic value cannot be overemphasized. But shouldn't we include the *environmental cost* of producing, using, and discarding these materials?

Steel mills and aluminum plants both create considerable pollution. Steel mills discharge lime, acids, grease, oil, and iron salts into the water. They also discharge particulate matter, carbon monoxide, nitrogen oxides, and other air pollutants. The manufacture of aluminum results in the discharge of iron, aluminum, and other metal oxides into waterways and the release of particulate matter, fluorides, and other pollutants into the air.

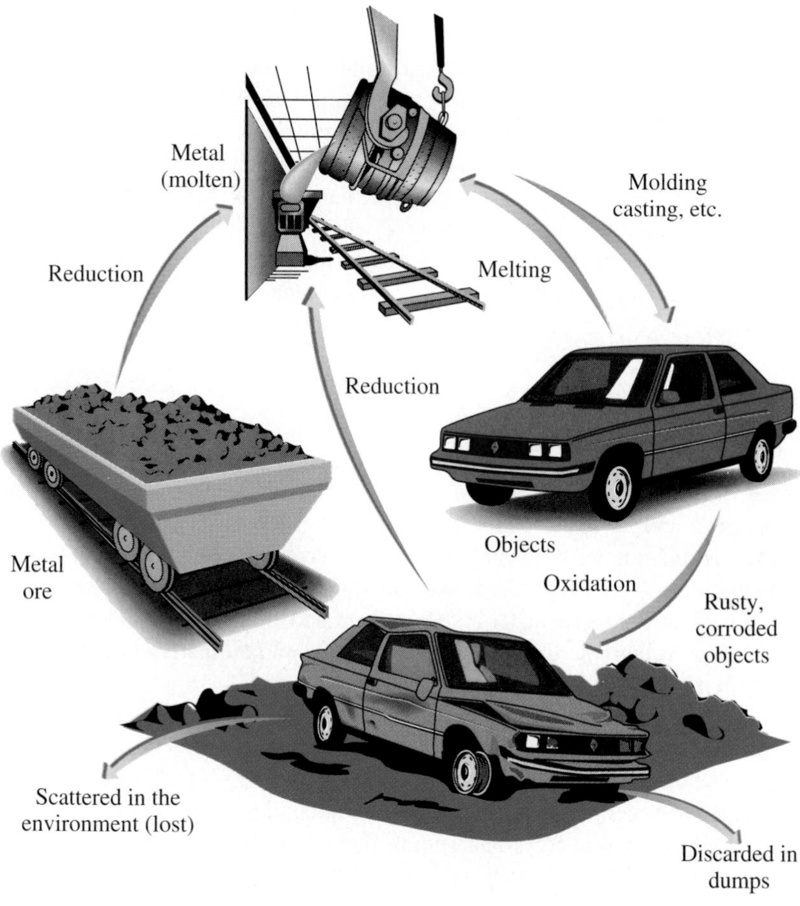

Figure 11.14 A metal cycle. The metal is obtained from its ore by reduction and then is slowly corroded (oxidized) back to the combined state (rust). Rust could be reduced back to metal and reused, but this is seldom done. Often rust is scattered in the environment and, for all practical purposes, lost. Many metal objects are melted down, the rust removed or reduced, and new objects molded from the molten metal.

Table 11.3 Some Other Important Metals

Metal	Symbol	Important Ore	Selected Properties	Typical Uses
Chromium	Cr	$FeCr_2O_4$	Shiny, resists corrosion	Chrome plating
Gold	Au	Au	Yellow metal, soft, dense	Coinage, jewelry, dentistry
Lead	Pb	PbS	Low melting, dense, soft	Plumbing, batteries
Magnesium	Mg	$MgCl_2$	Light and strong	Auto wheels, luggage
Mercury	Hg	HgS	Dense liquid	Thermometers, barometers
Nickel	Ni	NiS	Resists corrosion	Coinage, alloy for stainless steel
Platinum	Pt	Pt	Inert, high melting	Catalyst, instruments
Silver	Ag	Ag	Excellent electrical conductor	Electrical contacts, mirrors, jewelry, coins
Sodium	Na	NaCl	Reactive, soft	Heat transfer medium, reducing agent
Tin	Sn	SnO_2	Resists corrosion	Coating for steel cans
Tungsten	W	$CaWO_4$	Very high melting	Light-bulb filaments
Uranium	U	U_3O_8	Fissionable	Energy source
Zinc	Zn	ZnS	Forms protective coating	Galvanizing coating

One further comparison of steel and aluminum: Barry Commoner estimates that it takes 15 times as much fuel energy to produce aluminum as it does to produce a comparable weight of steel. Aluminum is a lighter (less dense) metal than steel, but making an aluminum can requires 6.3 times as much energy as making a steel can. On the other hand, when aluminum is recycled, the energy needed is only a fraction of what is required to make the metal from ore. Also consider the fact that a car made from aluminum would be so much lighter than a similar car made from steel that it would take less energy to operate it.

Other Important Metals

Several metals (as their ions) and minerals are essential to life. These are discussed in Chapter 16.

There are many other important metals. Some of them, with typical ores, properties, and uses, are listed in Table 11.3.

Metals and minerals are vital to a vigorous economy. The United States has depleted much of its high-grade ores and become increasingly dependent on other nations for these materials (Figure 11.15). It appears that the United States is as vulnerable to embargoes of these important materials as it was to the oil embargoes of the 1970s.

How can we ever run short of a metal? Aren't atoms conserved? Yes, there is as much iron on Earth as there was 100 years ago. But through use we scatter the metal throughout the environment. Gathering it back to a factory to make new objects requires energy, just as obtaining metals from low-grade ores requires more energy than getting them from high-grade ores.

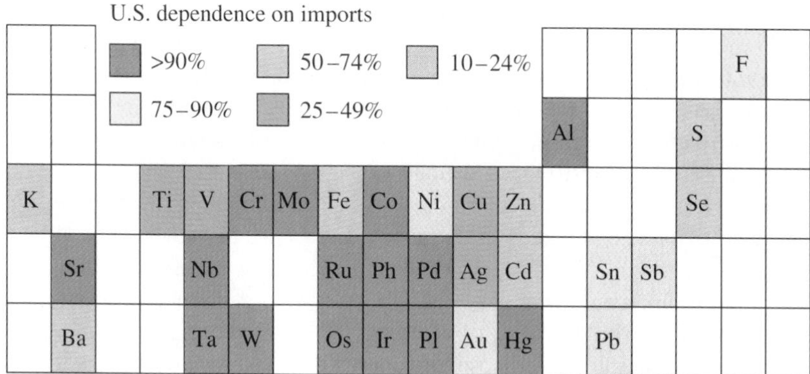

Figure 11.15 The United States imports a variety of strategic metals, minerals, and ores. It depends heavily on imports for some of the elements in the periodic table. (The actual dependence can vary considerably as changing economic conditions result in the opening and closing of mines.)

SECTION 11.7

Running Out of Everything: Earth's Dwindling Resources

Dramatic fuel shortages during the 1970s brought important changes in government policies and in industrial and individual practices. But now many of the conservation efforts that were begun at that time have largely been forgotten. We remain vulnerable to future fuel shortages that could be much worse than those of two decades ago.

We also face potential metals and minerals shortages. When the Europeans first came to North America, the native Americans were mining nearly pure copper in Michigan (Figure 11.16). Now we mine ore with less than half of

Figure 11.16 These crystals of native copper are from Houghton, Michigan. [*Breck P. Kent/Earth Scenes.*]

Figure 11.17 Nodules rich in manganese on the ocean floor. [*Frank Manheim/U.S. Geological Survey/Woods Hole Oceanographic Institution.*]

1% copper content. For much of the nineteenth century, we mined high-grade iron ore from the Mesabi Range in Minnesota. Now we mine taconite, a hard rock with small amounts of iron oxide dispersed through it. The same is true for gold and silver; the glory holes of western North America are gone. Indeed, there are few high-grade deposits of any metal ores left that are readily accessible to the industrialized nations.

What's wrong with low-grade ores? Considerable energy is required to concentrate them, and energy costs money. Let's consider an analogy. A bag of popcorn is useful. You can pop it and eat it. The same popcorn, scattered all over your room, would be less useful. Oh, you could gather it all up and then pop it, but that would take a lot of energy. Perhaps more energy than you would care to put into it. Perhaps more energy than you would get back if you popped the corn and ate it.

Not only are we running out of high-grade ores for metals such as copper and iron, but we also have the problem that we must import quite a few metals because their ores are not found in the United States. For example, we get tin from Bolivia, chromium from Rhodesia, and platinum from South Africa and Russia. But the metal reserves in other countries are being depleted, too.

Where will we get metals in the future? The sea is one possible source. Nodules rich in manganese cover vast areas of the ocean floor (Figure 11.17). These nodules also contain copper, nickel, and cobalt. Questions of who owns them and how to mine them without major environmental disruption remain to be resolved.

SECTION 11.8

Land Pollution: Solid Wastes

Productivity and creativity have enabled people in industrialized nations to have a fantastic variety of consumer goods in enormous quantities. These goods often are packaged in paper or plastic. Their wrappers litter the environment and fill our disposal sites. As the goods are broken, worn out, or merely become obsolete, they too add to our disposal problems. The approximate composition of solid wastes in the United States is given in Figure 11.18.

In the past, most solid wastes were simply discarded in *open dumps.* This led to infestation by rats, flies, and other pests that often spread to nearby areas. Open burning led to monstrously offensive air pollution. These open dumps are being rapidly phased out, but they still account for a proportion of the total land pollution, especially in rural areas.

The primary method of disposal today is the **sanitary landfill**. Garbage and trash are piled into a trench, compacted, and covered over. This eliminates the problem of rats, flies, and odors. Landfills leak, however, leading to groundwater contamination. And materials in landfills decompose much more slowly than previously thought. Newspapers are still readable after being entombed for 20 years. Without water and oxygen, microorganisms are unable to carry out the normal decay processes.

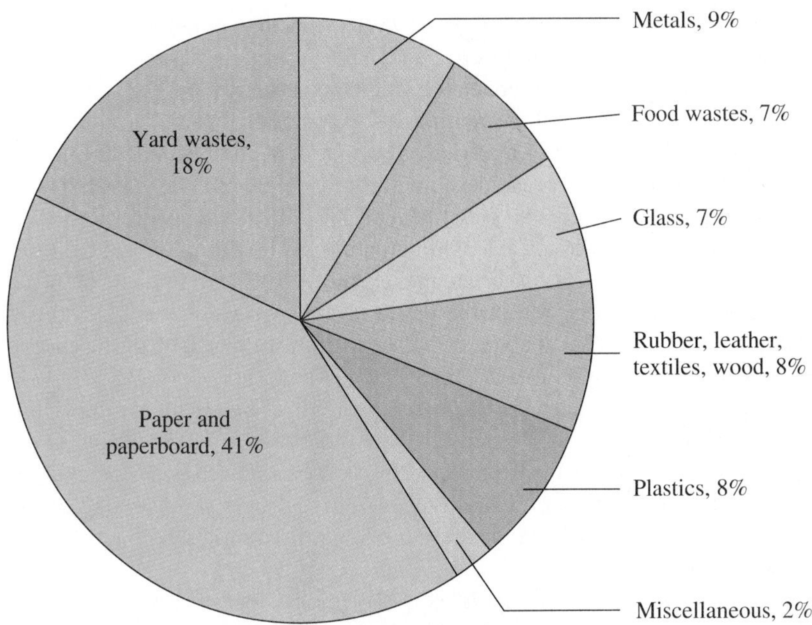

Figure 11.18 Municipal wastes in the United States are largely paper and cardboard and yard wastes. [*Data from U.S. Environmental Protection Agency, Office of Solid Wastes.*]

Another method of solid waste disposal is **incineration**. If this is carried out in properly designed incinerators, air pollution is minimal. However, many incinerators are old, or poorly designed, and they generate considerable smoke and odor. Building new incinerators tends to be extremely unpopular with local citizens.

Some cities not only burn their combustible solid wastes, but they also use the heat from the incinerators to warm buildings or to generate power, or both. A pound of polyethylene contains about 20,000 Btu of energy, about the same as a pound of #2 fuel oil.

Research in solid waste disposal has led to several new processes. The U.S. Bureau of Mines has a method for converting garbage to oil with an overall yield of 25%. Ground up rubber tires can be mixed with powdered coal and used as fuel. Waste glass and rubber tires have both been added to road-paving mixtures for highway construction.

About 8 million automobiles are junked each year in North America, and 250 million tires are thrown away.

SECTION 11.9

The Three Rs of Garbage: Reduce, Reuse, Recycle

There are several ways to deal with our garbage problem. Not all are equally desirable. We must choose wisely among them in order best to save energy and protect the environment.

Every day in the United States each person discards an average of about 1.5 kg of trash.

We throw away enough aluminum every three months to rebuild our entire commercial air fleet.

We discard so many glass jars and bottles each week that they would fill one of the towers at the New York Trade Center.

Every month we throw away a million tons of newspapers.

The best way to deal with our solid waste problem is to *reduce* the amount of throwaway materials produced. Don't buy unnecessary items. Don't use excessive packaging. If we reduce the volume of materials produced, we save resources and energy, and we minimize the disposal problem.

Next best is the *reuse* of materials. Make an item durable enough to withstand repeated use, rather than designing it for a single use after which it is to be discarded. Consider the glass beverage bottle. It takes about 6300 J (1500 kcal) of energy to make a nonreturnable half-liter bottle. It takes a little more energy, 8300 J (2000 kcal), to make a more durable, returnable bottle of the same capacity. The returnable bottle is used an average of 12.5 times before it is broken. That is about 660 J per use, only about 10% of the energy used to make the nonreturnable bottle for its single use.

The third way to reduce the volume of wastes is to *recycle* them. Recycling requires energy, and some material is unavoidably lost, but recycling saves materials. Over half of the copper and lead produced each year comes from recycled materials. About one-third of the iron and steel and over half of the aluminum also are salvaged. In the future, as raw materials become scarcer, these proportions are bound to increase.

Recycling also saves energy. It takes only 5% as much energy to make new aluminum cans from old ones as it does to make the cans from aluminum ore. Using a ton of scrap to make new iron and steel saves 1.5 tons of iron ore and

Figure 11.19 Waste materials separated for curbside collection. Once separated, they are ready to recycle. [*Steve Elmore/The Stock Market.*]

0.3 ton of coal. It also results in a 74% savings in energy, an 86% reduction in air pollution, and a 76% reduction in water pollution.

Recycling of household wastes requires that materials be separated. Many communities have curbside pickup of separated paper, glass, aluminum, and other materials (Figure 11.19). This separation depends on renewable human energy. Production of new materials requires nonrenewable energy from fossil fuels. Disposal of trash costs U.S. residents $4 billion each year. By following the three Rs, we can protect the environment at the same time that we save materials, energy, and money.

During 1992 in Maine recycling added more than 2000 jobs and about $300 million to the state's economy. Massachusetts reported that recycling added $600 million to that state's economy. By the year 2000 California expects to create as many as 65,000 new jobs in recycling.

Recycling a 1-m stack of newspapers saves the equivalent of a 10-m pine tree. Making paper from scrap instead of virgin wood pulp yields 50% savings of water and energy.

SECTION 11.10

How Crowded Is Our Spaceship?

Every day people die and other people are born, but the number born is about 250,000 more than the number that die. Every 4 days our world population goes up by a million people. In 1992 world population was 5.4 billion. It should reach 6.4 billion by the year 2000. A billion more people in just 8 years!

Figure 11.20 is a graphic picture of our population problem. For thousands of years human population had never been more than a few hundred million. It did not reach the 1 billion mark until about 1800. At the present rate of growth, world population should be 10 billion by 2025, and one century from now it could reach 40 billion!

How many people can Spaceship Earth accommodate? Many experts believe that we have already gone way beyond the optimum population for this planet. With our resources running out and our waste products piling up, we cannot afford to keep bringing so many more people onto this planet. One can argue that many of our other major problems—poverty, crime, starvation, war, pollution, and the rising number of jobless and homeless people—are linked directly to the population problem.

Why did that curve in Figure 11.20 start bending a few centuries ago? Did people suddenly start having more children? Of course not! Population growth depends on both the birth rate and the death rate. If they are equal, population growth is zero. They had been almost equal for thousands of years, but during the seventeenth and eighteenth centuries scientists began doing things that greatly lowered the death rate. Antiseptics, vaccines, antibiotics, and other medical advances have cut back our death rate enormously, but we have done very little to decrease the birth rate accordingly (especially in underdeveloped parts of the world). Thanks to our wonderful progress in medical science and technology, our birth and death rates are completely out of balance.

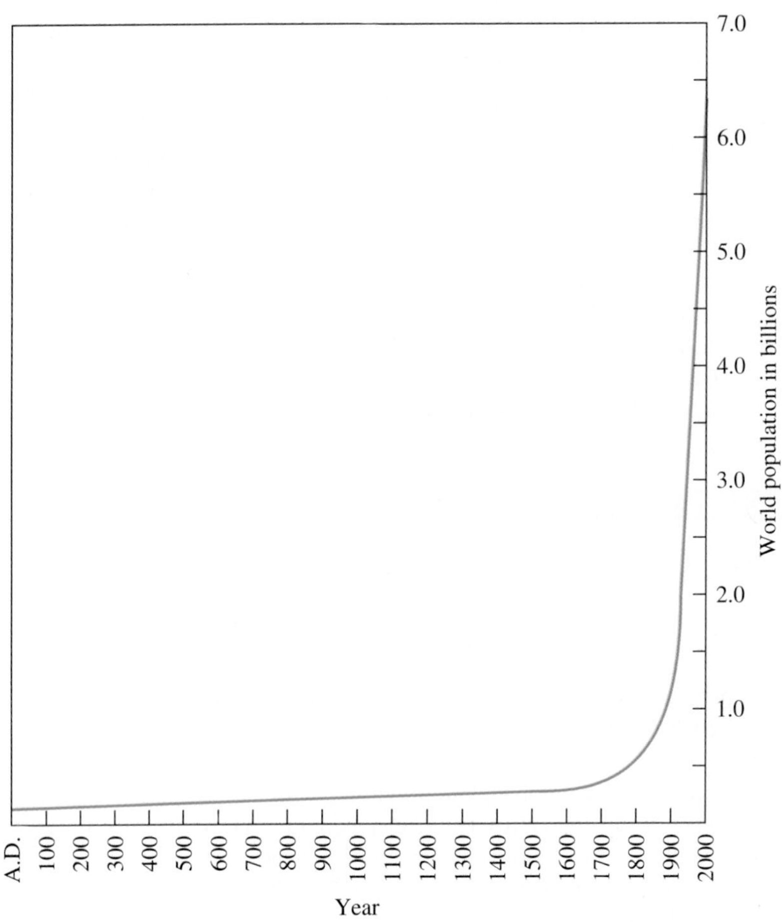

Figure 11.20 The growth of world population.

One thing is sure. That curve cannot keep rising indefinitely. It must bend sooner or later. If we do not make that happen by curbing our uncontrolled birth rate, then nature will take care of the problem by increasing our death rate. Let us hope it doesn't come to that. Nature's methods are not always kind.

Summary

1. The Earth is divided into three main regions: the *core,* the *mantle,* and the *crust.*
2. The *core* is thought to be made up largely of iron and nickel; the *mantle* is believed to be mostly silicates.
3. The *crust* (outer shell) includes the *lithosphere* (solid part), the *hydrosphere* (water), and the *atmosphere* (air).
4. The *lithosphere* is made up largely of rocks and minerals, such as silicates, carbonates, oxides, and sulfides.
5. *Mica* and *asbestos* are *silicates,* and *quartz* is pure silicon dioxide. Sand is also mainly silicon dioxide.
6. *Ceramics, glass,* and *cement* are modified silicates.
7. Metals must be won from their ores before they can be made into tools. Ancient peoples used *bronze,*

and later iron; modern societies use many metals, but especially *steel* and *aluminum.*

8. *Alloys* are metallic mixtures, such as bronze (copper and tin) and stainless steel (iron, carbon, chromium, and nickel).
9. Earth's resources are being used up at such a rate that we are bound to face shortages within the near future.
10. Our throw-away society generates mountains of solid waste each year, and we are running out of *landfill* space.
11. Three ways to deal with our mounting garbage problems are to reduce our volume of disposables, to reuse items when we can, and to recycle our waste whenever possible.

Key Terms

alloy 11.6
aluminum 11.6
asbestos 11.4
atmosphere 11.1
bronze 11.6
carbonates 11.2
cement 11.5
ceramics 11.5
coke 11.6
core 11.1
crust 11.1
glass 11.5
hydrosphere 11.1
incineration 11.8
lithosphere 11.1
mantle 11.1
mica 11.4
oxides 11.2
quartz 11.4
sanitary landfill 11.8
silicates 11.4
slag 11.6
steel 11.6
sulfides 11.2
synergistic effect 11.4

Review Questions

1. What are the three main regions of the Earth?
2. Name the three parts of the crust of the Earth.
3. What two elements are thought to make up most of the core of the Earth?
4. What is the most abundant element in the crust of the Earth?
5. What are silicates?
6. What materials make up the organic portion of the lithosphere?
7. What is inorganic chemistry?
8. What is the chemical composition of quartz?
9. Why does mica occur in sheets?
10. Why does asbestos occur as fibers?
11. What is a synergistic effect?
12. How does the structure of glass differ from that of crystalline silicates?
13. How do the properties of glass differ from those of crystalline substances?
14. What are the three principal raw materials for making glass?
15. How is the basic recipe for glass modified to make glasses with special properties?
16. What are the two basic raw materials for making cement?
17. What environmental problems are associated with the manufacture of cement?
18. What is concrete?
19. What are the principal raw materials for modern iron production? What is the purpose of each?
20. What is coke?
21. What is slag?
22. By what kind of chemical process is a metal obtained from its ore?
23. By what chemical process does a metal corrode?
24. What is an alloy?
25. What is steel?
26. What is the most abundant metal in the crust of the Earth?
27. Which metal is the most widely used? Why is the most abundant metal not the most widely used?

28. What environmental problems are associated with steel mills?

29. What environmental problems are associated with aluminum production?

30. If matter is conserved, how can we ever run out of a metal?

31. List three methods of solid waste disposal. Give the advantages and the disadvantages of each.

32. Explain why metals can be recycled fairly easily. What factors limit the recycling of metals?

Problems

Ores and Mining

33. An ore body at Crandon, Wisconsin, is estimated at 75 million tons and assays at 5.0% zinc, 0.4% lead, and 1.1% copper. What mass of each metal is contained in the deposit?

34. The Parc mine at Llanwrst, North Wales, in the United Kingdom, was closed in 1954, leaving behind a mound of about 250,000 tons of tailings. The tailings assay at 3.22% Zn, 0.82% Pb, and 0.023% Cd. Erosion has washed away 13,000 tons of the tailings. How much of each of the three elements has been carried away, presumably to enter the river ways?

35. Close to 81,000 tons of gold has been mined throughout history. What size cube of gold would this make if it was all in one piece? The density of gold is 19.3 g/cm^3.

36. Ten million carats of gem-quality rough diamonds are mined each year. What is the mass of these diamonds in kilograms? (1 carat = 200 mg.)

37. To obtain 1 carat of gem-quality diamonds, approximately 115 tons of earth has to be mined. How much earth is mined to produce 10 million carats of diamonds each year?

38. Gold ore in the United States today assays about 0.35 troy ounce of gold per ton of ore. How much earth is mined to produce the 0.7 million troy ounces of gold produced by mining each year? (Another 0.3 million troy ounces is obtained as a by-product of mining other metal ores.)

39. An aluminum plant produces 65 million kg of aluminum per year. How much aluminum oxide is required? (It takes 2.1 kg of crude bauxite to produce 1 kg of aluminum oxide.) How much bauxite is required?

40. It takes 17 kWh of electricity to produce 1 kg of aluminum. How much electricity does the plant (Problem 39) use for aluminum production in 1 year?

Oxidation and Reduction

41. The Hunter method for the production of titanium uses the reaction

$$TiCl_4 + 4\ Na \rightarrow Ti + 4\ NaCl$$

What substance is reduced? What is the reducing agent? What substance is oxidized? What is the oxidizing agent?

42. Pure chromium is prepared by reacting chromium(III) oxide with aluminum.

$$Cr_2O_3 + 2\ Al \rightarrow 2\ Cr + Al_2O_3$$

What substance is reduced? What is the reducing agent? What substance is oxidized? What is the oxidizing agent?

43. Thorium metal is prepared by reacting thorium(IV) oxide with calcium.

$$ThO_2 + 2\ Ca \rightarrow Th + 2\ CaO$$

What substance is reduced? What is the reducing agent? What substance is oxidized? What is the oxidizing agent?

44. Potassium metal is prepared by reacting molten potassium chloride with sodium metal.

$$KCl + Na \rightarrow NaCl + K$$

What substance is reduced? What is the reducing agent? What substance is oxidized? What is the oxidizing agent?

45. Aluminum chloride reacts with manganese metal to form aluminum metal and manganese(II) chloride.

$$2\ AlCl_3 + 3\ Mn \rightarrow 2\ Al + 3\ MnCl_2$$

What substance is reduced? What is the reducing agent? What substance is oxidized? What is the oxidizing agent?

46. The Toth Aluminum Company produces aluminum from kaolin. One step converts aluminum oxide to aluminum chloride by the following reaction.

$$Al_2O_3 + 3\ C + 3\ Cl_2 \rightarrow 2\ AlCl_3 + 3\ CO$$

What substance is reduced? What is the reducing agent? What substance is oxidized? What is the oxidizing agent?

Mass Relationships

47. Copper metal is prepared by blowing air through molten copper(I) sulfide.

$$Cu_2S + O_2 \rightarrow 2\ Cu + SO_2$$

How much copper is obtained from 143 g of Cu_2S?

48. How much sodium is required to react with 1.73 g of KCl? Use the equation in Problem 44.

49. How much cesium chloride must be reduced to produce 0.529 g of cesium? The reaction is

$$2\ CsCl + Ca \rightarrow 2\ Cs + CaCl_2$$

50. How much aluminum is required to reduce 41.7 g of calcium oxide? The reaction is

$$3\ CaO + 2\ Al \rightarrow 3\ Ca + Al_2O_3$$

References and Readings

1. Chenier, Philip J. ''A Summary Chart of the Manufacture of Important Inorganic Chemicals.'' *Journal of Chemical Education,* May 1983, pp. 411–413.
2. Commoner, Barry. *The Closing Circle.* New York: Bantam, 1972.
3. Denio, Allen A. ''Chemistry for Potters.'' *Journal of Chemical Education,* April 1980, pp. 272–275.
4. Ehrlich, Paul R., Anne H. Ehrlich, and John P. Holdren. *Ecoscience: Population, Resources, Environment.* San Francisco: W. H. Freeman, 1977.
5. Gore, Rick. ''The Planets: Between Fire and Ice.'' *National Geographic,* January 1985, pp. 4–51.
6. Kolb, Doris, and Kenneth E. Kolb. *Glass: Its Many Facets.* Hillside, NJ: Enslow, 1988.
7. Lal, R., and B. A. Stewart, Eds. *Soil Degradation.* New York: Springer-Verlag, 1990.
8. Mossman, B. T., et al. ''Asbestos: Scientific Developments and Implications for Public Policy.'' *Science,* 19 January 1990, pp. 294–301.
9. Ohashi, Nobuo. ''Modern Steelmaking.'' *American Scientist,* November–December 1992, pp. 540–555.
10. O'Leary, Philip R., Patrick W. Walsh, and Robert K. Ham. ''Managing Solid Wastes.'' *Scientific American,* December 1988, pp. 36–42.
11. Ponte, Lowell. ''Dawn of the New Stone Age.'' *Readers Digest,* July 1987, pp. 129–133. Describes space-age ceramics.
12. Roberts, Willard Lincoln, Thomas J. Campbell, and George Robert Rapp, Jr. *Encyclopedia of Minerals,* 2nd ed. New York: Van Nostrand Reinhold, 1989.
13. Stanglin, Douglas. ''Toxic Wasteland.'' *U.S. News & World Report,* 13 April 1992, pp. 40–49.
14. Studt, Tim. ''New Advances Revive Interest in Cement-Based Materials.'' *R & D Magazine,* November 1992, pp. 74–78.
15. Turner, Kenneth. ''Precious Metals.'' *Chemistry and Industry,* 19 July 1980, pp. 551–556.
16. Vandiver, Pamela B. ''Ancient Glazes.'' *Scientific American,* April 1990, pp. 106–113.
17. Young, John E. ''Aluminum's Real Tab.'' *World-Watch,* March–April, 1992, pp. 26–33.
18. Zurer, Pamela S. ''Asbestos: The Fiber That's Panicking America.'' *Chemical and Engineering News,* 4 March 1985, pp. 28–41.

12 Air

The Breath of Life

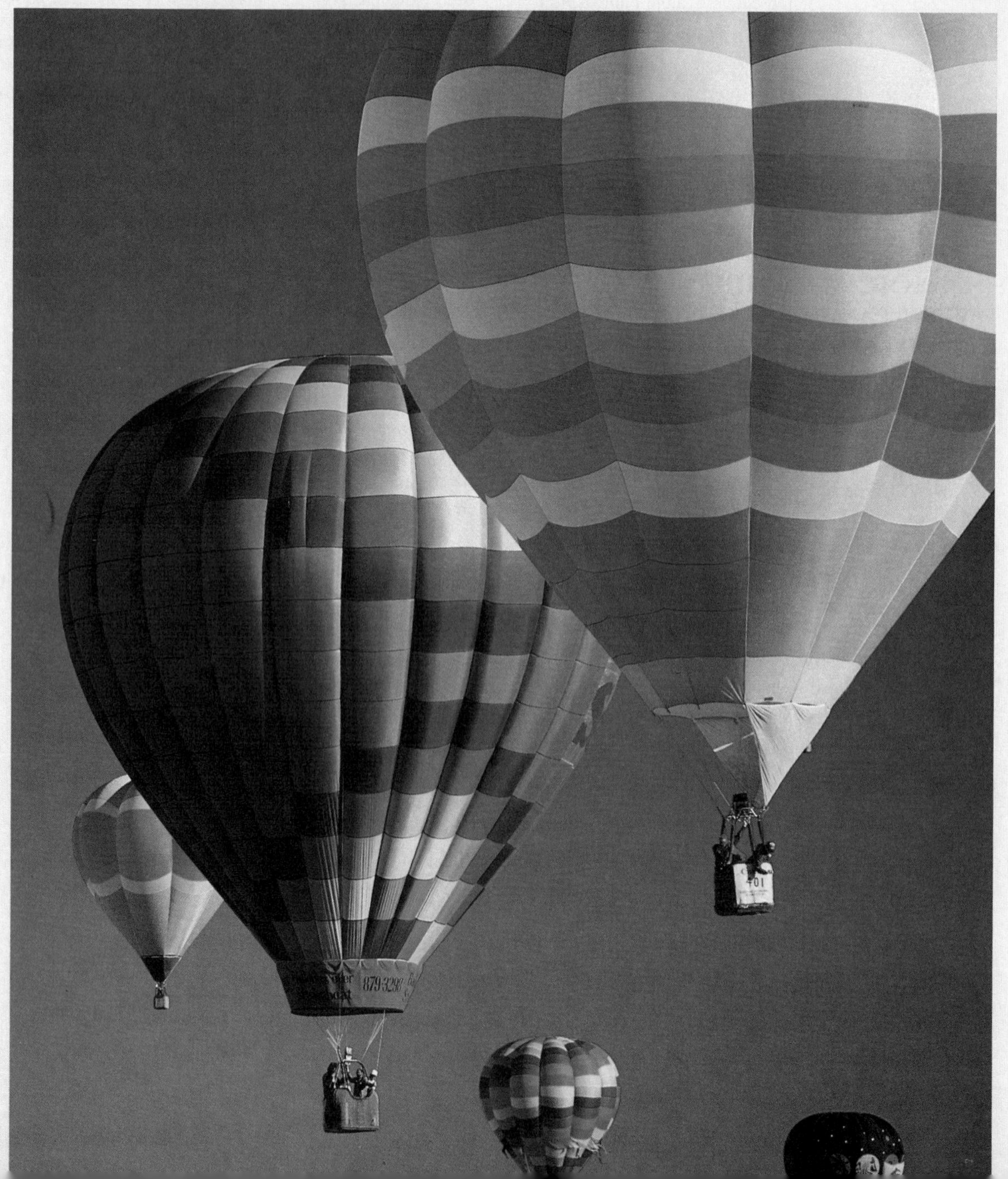

Although you cannot see it,

It's around you everywhere.

You cannot live without it.

It's the breath of life. It's AIR!

As passengers on Spaceship Earth, we live under a thin blanket of air called the atmosphere. It is difficult to measure exactly how deep the atmosphere is. It does not end abruptly but gradually gets thinner as it fades into space. But we know that 99% of the atmosphere lies within 30 km of Earth's surface. This is a very thin air layer indeed (a bit like the skin on a very large apple, only much thinner, relatively speaking).

Although other planets in our solar system have atmospheres, Earth's atmosphere appears to be unique in its ability to support life. We could live about a month without food. We could even live live for several days without water. But without air we cannot live more than a few minutes.

We may run short of food, and we may run short of fresh water, but we are not likely to run out of air. On the other hand, we might foul the air so badly that it could become unfit to breathe. In some areas of the world the air is so bad that people have become sick from breathing it, and it may eventually get so bad that many people could die because of it.

That won't really happen, you are probably thinking. We all certainly hope it won't happen, but we need to do much more than just hope!

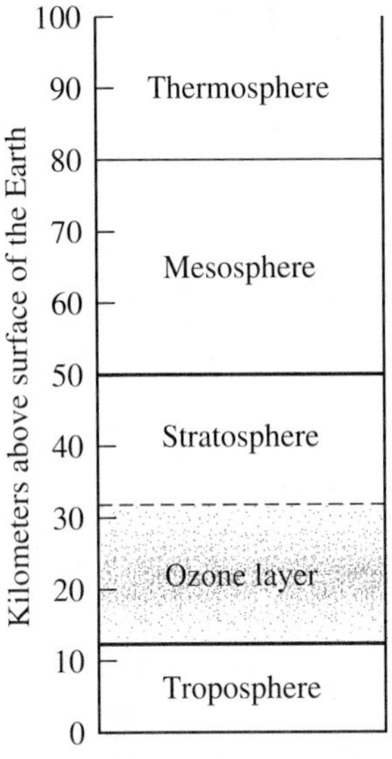

SECTION 12.1

The Atmosphere: Divisions and Composition

The **atmosphere** is divided into layers (Figure 12.1). The layer nearest Earth, the **troposphere**, harbors nearly all living things and nearly all human activity. The next region, the **stratosphere**, is where we find the **ozone layer** that shields living creatures from deadly ultraviolet radiation. Most of this chapter focuses on the troposphere, but we also examine some threats to the ozone

Figure 12.1 Divisions of the atmosphere and approximate location of the ozone layer (color).

Although we don't think about its presence most of the time, air is so essential that we could not live more than a few minutes without it. [*E. Gebhardt/ Mauritius.*]

Table 12.1 Composition of Earth's Atmosphere (Dry): Molecules of Each Substance per 100,000 Molecules of Air		
Substance	Formula	Number of Molecules
Nitrogen	N_2	78,083
Oxygen	O_2	20,945
Argon	Ar	934
Carbon dioxide	CO_2	35
All others	—	3
Total		100,000

layer in the stratosphere. We will not be concerned with higher regions of the atmosphere.

Air is a mixture of gases. Dry air is (by volume) about 78% nitrogen (N_2), 21% oxygen (O_2), and 1% argon (Ar). Water vapor varies from 0% up to about 4%. There are a number of minor constituents, the most important of which is carbon dioxide (CO_2). The concentration of carbon dioxide in the atmosphere increased from about 280 parts per million (ppm) in the preindustrial world to 296 ppm in 1900 to its present value of 356 ppm. It most likely will continue to rise as we burn more and more fossil fuels (coal, oil, and gas). The composition of the atmosphere is summarized in Table 12.1.

SECTION 12.2

The Nitrogen Cycle

Although nitrogen makes up 78% of the atmosphere, the N_2 molecules can't be used directly by higher plants or by animals. They first have to be *fixed*—that is, combined with another element.

Certain types of bacteria convert atmospheric nitrogen (N_2) to nitrates. Other bacteria convert the nitrogen in compounds back to N_2. Thus, a **nitrogen cycle** (Figure 12.2) is established. Lightning also serves to fix nitrogen by causing it to combine with oxygen. Nitric oxide (NO) and nitrogen dioxide (NO_2) are formed. The equations are

$$N_2 + O_2 \xrightarrow{\text{lightning}} 2\,NO$$

$$2\,NO + O_2 \longrightarrow 2\,NO_2$$

Nitrogen dioxide reacts with water to form nitric acid (HNO_3).

$$3\,NO_2 + H_2O \longrightarrow 2\,HNO_3 + NO$$

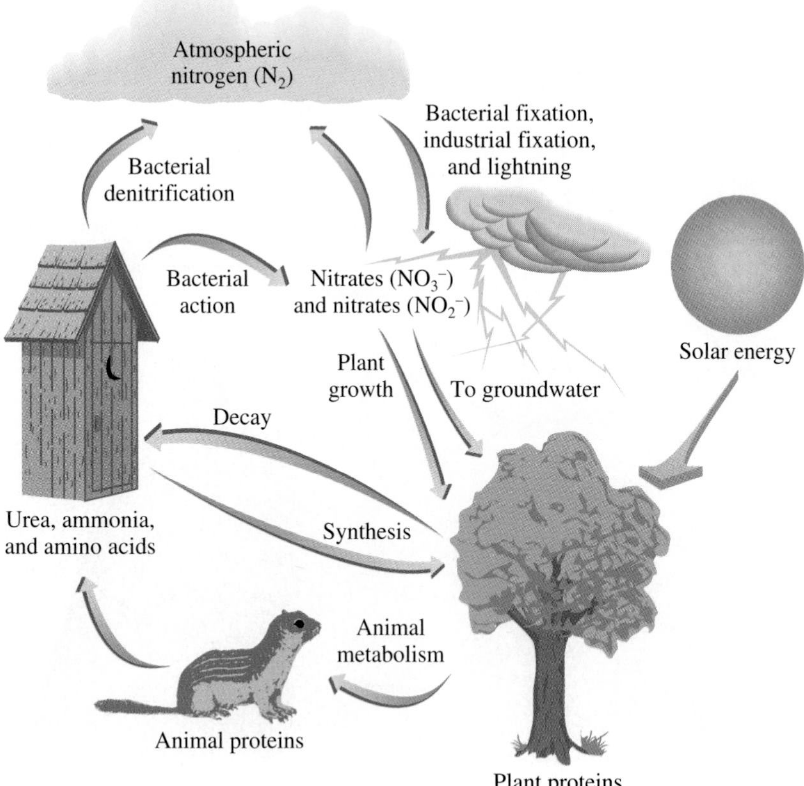

Figure 12.2 The nitrogen cycle.

The nitric acid falls in rainwater, adding to the supply of available nitrates in the oceans and the soil.

Humans have undertaken substantial intervention in the nitrogen cycle by fixing nitrogen industrially in the manufacture of nitrogen fertilizers. This intervention has greatly increased our food supply, since the availability of fixed nitrogen is often the limiting factor in the production of food (Chapter 16). Not all the consequences of this intervention have been favorable, however; excessive runoff of nitrogen fertilizer has led to serious water-pollution problems in some areas (Chapter 13).

SECTION 12.3

The Oxygen Cycle

The element oxygen makes up 21% of the Earth's atmosphere. The oxygen supply is constantly being replenished by green plants, including one-celled

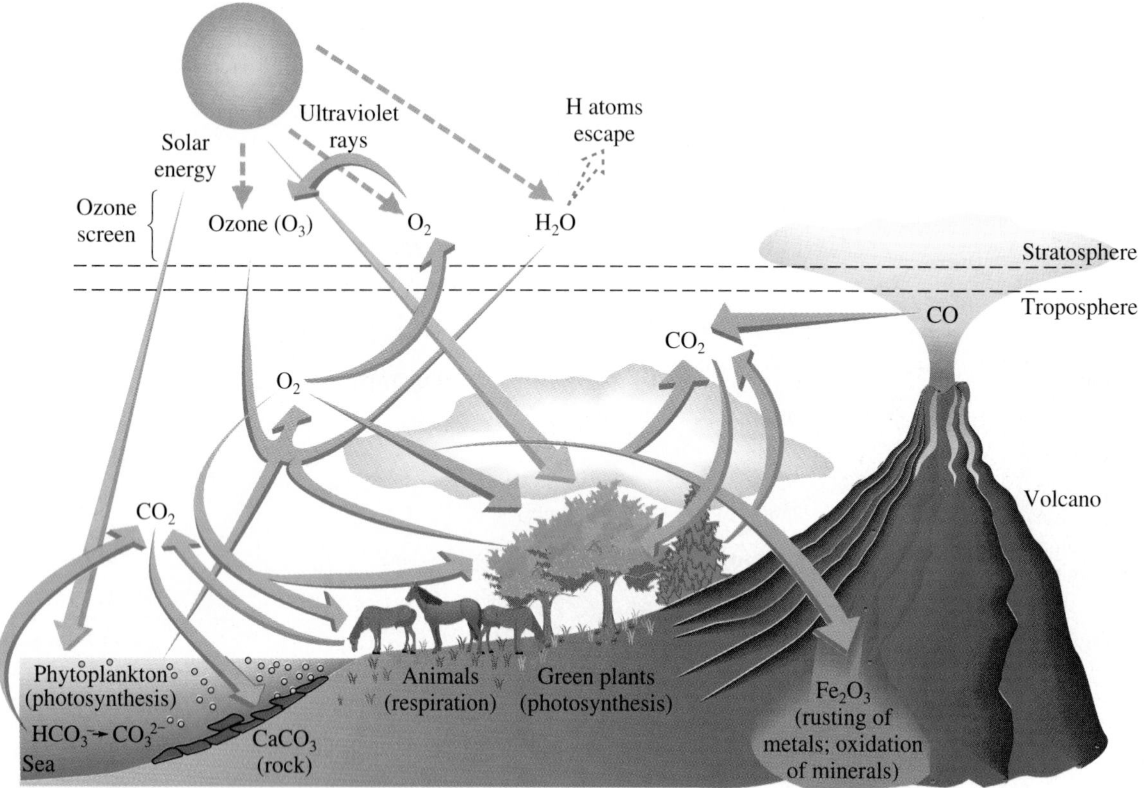

Figure 12.3 The oxygen cycle.

organisms, called phytoplankton, in the sea. Oxygen is used by both plants and animals in the metabolism of foods. It also is used in the decay and the combustion of plant and animal materials. Some oxygen is consumed in the rusting of metals and the weathering of rocks. A simplified **oxygen cycle** is illustrated in Figure 12.3.

In the stratosphere, some oxygen is formed by the action of ultraviolet rays on water. Perhaps more importantly, some oxygen is converted to ozone (Section 12.14).

$$3\ O_2 \xrightarrow{\text{ultraviolet rays}} 2\ O_3$$

The ozone, in turn, absorbs short-wavelength ultraviolet radiation that might otherwise make any sort of life on Earth impossible.

We discuss the atmosphere in more detail in subsequent sections, paying particular attention to changes wrought by human activities.

SECTION 12.4

Temperature Inversion

Normally, the temperature of the atmosphere decreases with altitude. The air is warmest near the ground. On clear nights, the ground cools rapidly, but the wind usually mixes the cooler air near the ground with the warmer air above it.

The scene is a small industrial town at the bottom of a river valley. The time is late October. A dense smog settles over the valley. The air is acrid with sulfur dioxide. Dust (probably fly ash) settles over the land in a layer so thick that footprints and tire tracks are visible in it. A Halloween parade passes through the streets, the marchers holding handkerchiefs over their faces in an attempt to keep the smog out of their lungs. People become seriously ill. Before the smog lifts, 5 days later, 17 people are dead. Four more, who became ill in October, die by Christmas Eve. Nearly half the population is made ill by smog. One in 10 is seriously ill.

What is this? A scene from a futuristic, nightmarish science-fiction story? No! It is history. The scene actually took place in Donora, Pennsylvania, in 1948. A zinc production plant and a huge iron and steel mill contributed to the smog. So did trains, automobiles, and even barges on the Monongahela River. An *atmospheric inversion*—a still, warm upper layer of air over a cool lower layer—kept the smog socked in on the valley (Figure 12.4).

(a)

(b)

Figure 12.4 Ordinarily, the air gets cooler as the altitude increases (a). During an atmospheric (thermal) inversion (b), a cold layer near the surface lies beneath a warmer layer.

Once in a while, the air is still, and the lower layer of cold air becomes trapped by the layer of warm air above it. This condition is known as an **atmospheric inversion** (or a thermal inversion). Any pollutants in the cooler air are also trapped near the ground, and the air can become quite seriously polluted if it takes several days for the wind to pick up again (Figure 12.4).

A temperature inversion can also occur when a warm front collides with a cold front. The warm air mass, being less dense, slides over the cold air mass, producing a condition of atmospheric stability. Because there is no vertical air movement, air stagnation results, and air pollutants accumulate.

SECTION 12.5

Natural Pollution

Even before there were people, there were cases of air **pollution**. Volcanoes erupted, spewing ash and poisonous gases into the atmosphere. They still do so. The eruption of Mount St. Helens in 1980 (Figure 12.5) ejected ash that blanketed areas over several states. It spewed out 1000 metric tons (t) of sulfur dioxide per day, and it emitted 160 t per day for several years following the eruption. El Chichon in Mexico in 1982 ejected 3.3 million t of sulfur dioxide into the stratosphere where it was converted to sulfuric acid mist. Kilauea in Hawaii emits 200 to 300 t of sulfur dioxide per day. Acid rain downwind from this volcano has created a barren region called the Kau desert.

Dust storms, especially in arid regions, add massive amounts of particulate matter to the atmosphere. Dust from the Sahara often reaches the Caribbean and South America. Swamps and marshes emit noxious gases. Nature isn't always benign.

Figure 12.5 An eruption of Mount St. Helens on 18 May 1980 ejected at least 1 km^3 of material as ash. [*Courtesy of U.S. Geological Survey.*]

The Air Our Ancestors Breathed

People always have altered their environment. The discovery of tools and the use of fire forever changed the balance between us and our environment. The fires early people built filled the air with smoke. They cleared land; this made possible even larger dust storms. They built cities, and the soot from their hearths and the stench from their wastes filled the air. The Roman author Seneca wrote in A.D. 61 of the "stink," soot, and "heavy air" of the imperial city. In 1257 the queen of England was forced to move away from the city of Nottingham because the heavy smoke was unendurable. The industrial revolution brought even worse air pollution. Coal was burned to power the factories and to heat the homes. Soot, smoke, and sulfur dioxide filled the air. The good old days? Not in the factory towns. But there were large rural areas relatively unaffected by air pollution.

Today we still have air pollution, but it is much more complex than the pollution our ancestors worried about. When society changes its activities, it often changes the nature of its waste materials, including the ones it dumps into the atmosphere. The world has gone through many changes. Change is not new, but the *rate* of change is. The pace of change in today's world has reached a level that our environment may not be able to absorb.

SECTION 12.7

Pollution Goes Global

With increasing population, the entire world is becoming more urban. It is the huge megalopolises that are most afflicted by air pollution. But rural areas are not unaffected. In the neighborhoods around smoky factories, there is evidence of increased rates of spontaneous abortion and of poor wool quality in sheep, decreased egg production and high mortality in chickens, and increased feed and care required for cattle. Plants are stunted, deformed, and even killed. The giant Ponderosa pines are dying over a hundred miles from the smog-plagued Los Angeles basin. Orbiting astronauts visually traced drifting blobs of Los Angeles smog as far east as western Colorado. Other astronauts, over 100 km up, were able to see the plume of smoke from the Four Corners power plant near Farmington, New Mexico. This was the only evidence they could see from that distance that Earth is inhabited!

The U.S. Environmental Protection Agency estimates that 100 million Americans live in places where the air is so polluted that breathing is hazardous to their health. Not quite so hazardous, however, as not breathing.

Snow in Norway is contaminated by pollutants from England and Germany. Traffic police in Tokyo have had to wear gas masks and take "oxygen breaks"—breathing occasionally from tanks of oxygen. Smog in Athens at times has forced factory closings and traffic restrictions. Acid rain in Canada is spawned by air pollution originating in the United States, contributing to strained relationships between the two countries. Buenos Aires, Sydney, Rome, Tehran, Ankara, Mexico City, and many other cities of the world have had frightening episodes of air pollution (Figure 12.6).

Probably the most terrible example of what can happen when people mistreat their environment is in the former Soviet Union. Since the 1920s, and especially since World War II, the Soviet Union had been aiming for rapid industrial growth—at any cost. The cost in environmental damage has been enormous. Gas flares in the oil fields spew out tons of black smoke, and their constant emission of sulfur dioxide has ruined thousands of square miles of timber in northern Siberia. It is claimed that oil spills approach a million barrels of oil every day, about 10% of their production. One single pool of spilled oil from a Siberian pipeline is about 7 miles long, 4 miles wide, and 6 feet deep. Some 35% of the people in the former Soviet Union are breathing air that is polluted with at least five times the limit of dangerous chemicals

Figure 12.6 Polluted air of Beijing, China, is conspicuous at sunrise. [*Larry Mulvehill/Photo Researchers, Inc.*]

allowed in the United States. There are at least 103 cities in Russia and other former Soviet countries that have serious air pollution problems.

The main difference between air-pollution problems today and problems in the past is that it is difficult to get away from air pollution now. The problem is worldwide. The air does purify itself, but we are pouring more pollutants into it than it can handle readily.

What is a **pollutant**? It is merely a chemical in the wrong place. For example, ozone is a natural and important constituent of the stratosphere, where it shields the Earth from life-destroying ultraviolet radiation. In the troposphere, however, ozone is a dangerous pollutant (Section 12.14).

SECTION 12.8

Coal + Fire → London Smog

There are two basic types of smog. One type—consisting mainly of smoke, fog, sulfur dioxide, sulfuric acid, ash, and soot—is called **London smog** (Figure 12.7). Indeed, the word **smog** is thought to have originated in England in 1905 as a contraction of the words *smoke* and *fog*.

London smog often is called *industrial smog* because most of its components come from industries such as smelters and coal-burning power plants.

Probably the most notorious case of smog in history started in London on Thursday, 4 December 1952. A large cold-air mass moved into the valley of the Thames River. A temperature inversion placed a blanket of warm air over the cold air. With nightfall, a dense fog and below-freezing temperatures caused the people of London to heap coal into their small stoves. Millions of

Figure 12.7 The London smog shown here is derived from the burning of coal in electric power plants and other industrial operations. It is characterized by high levels of sulfur dioxide, particulate matter, and (often) carbon monoxide. This type of smog can be minimized by the use of electrostatic precipitators, scrubbers, and other devices. [*Courtesy of Environmental Protection Agency.*]

these fires burned through the night, pouring sulfur dioxide and smoke into the air. The next day (Friday), the people continued to burn coal when the temperature remained below freezing. The factories added their smoke and chemical fumes to the atmosphere.

Saturday was a day of darkness. For 20 miles around London, no light came through the smog. The air was cold and still. And the coal fires continued to burn throughout the weekend. On Monday, 8 December, more than 100 people died of heart attacks while desperately trying to breathe. People tried to sleep sitting up in chairs in order to breathe a little easier. The city's hospitals were overflowing with people with respiratory diseases.

By the time a breeze cleared the air on Tuesday, 9 December, more than 4000 deaths had been attributed to the smog. Other people afflicted at the time died later. The total death count has been estimated at 8000. That is more people than were ever killed in any single tornado, mine disaster, shipwreck, or airplane crash. That is more people than were killed in the Japanese attack on Pearl Harbor in 1941. Air-pollution episodes may not be as dramatic as other disasters, but they can be just as deadly. (The smog at Donora was essentially London smog, with a few added ingredients—chlorides, fluorides, arsenic, lead, and zinc.)

SECTION 12.9

The Chemistry of London Smog

The chemistry of London smog is fairly simple. Coal is mainly carbon, but it contains as much as 3% sulfur. Coal also contains varying amounts of mineral matter. When burned, the carbon in coal is oxidized to carbon dioxide.

Figure 12.8 Scanning electromicrograph of particulate matter from a coal-burning power plant. [*Courtesy of Dr. Wojciech Jozewicz, Acurex Corp.*]

$$C + O_2 \longrightarrow CO_2$$

Heat is given off in this process. Not all the carbon is completely oxidized. Some of it winds up as carbon monoxide.

$$2\,C + O_2 \longrightarrow 2\,CO$$

Still other carbon, essentially unburned, ends up as soot.

Sulfur Oxides

The sulfur in coal also burns, forming sulfur dioxide, a choking, acrid gas.

$$S + O_2 \longrightarrow SO_2$$

Sulfur dioxide is readily absorbed in the respiratory system. It is a powerful irritant and is known to aggravate the symptoms of people who suffer from asthma, bronchitis, emphysema, and other lung diseases.

As if sulfur dioxide were not bad enough, things get worse. Some of the sulfur dioxide reacts further with oxygen in the air to form sulfur trioxide.

$$2\,SO_2 + O_2 \longrightarrow 2\,SO_3$$

The sulfur trioxide then reacts with water to form sulfuric acid.

$$SO_3 + H_2O \longrightarrow H_2SO_4$$

Sulfuric acid is much more irritating to the respiratory tract than sulfur dioxide.

Particulate Matter

The greatest contributor of particulates to the atmosphere is the sea. Wave action causes salt particles to be suspended in the air. The largest component (by mass) of all the particulate matter in the atmosphere is ordinary salt (NaCl), and it comes from a perfectly natural source.

Usually, London smog also is characterized by high levels of **particulate matter**, solid and liquid particles of greater than molecular size (Figure 12.8). The large particles often are visible in the air as dust and smoke. Smaller particles of 1 μm or less in diameter—called **aerosols**—are often invisible.

Particulate matter consists in part of soot (unburned carbon). A larger portion is made up of the mineral matter that occurs in coal. These minerals do not burn. Some are left behind as *clinkers*. In the roaring fire of a huge factory or power plant, however, much of this solid material is carried aloft in the tremendous draft. This *fly ash* settles over the surrounding area, covering everything with dust. It is also inhaled, thus contributing to respiratory problems in animals and humans.

Health Effects of London Smog

Perhaps a more insidious form of particulate matter is the sulfates. Some of the sulfuric acid in smog reacts with ammonia to form a solid material, ammonium sulfate.

$$2\,NH_3 + H_2SO_4 \longrightarrow (NH_4)_2SO_4$$

Sulfuric acid, in the form of minute liquid droplets, and the solid ammonium sulfate are easily trapped in the lungs. The harmful effect of these pollutants may be considerably magnified by their interaction. A certain level of sulfur dioxide, without the presence of particulate matter, might be reasonably safe. A certain level of particulate matter, without sulfur dioxide around, might be fairly harmless. But take the same levels of the two together, and the effect might well be deadly. Synergistic effects such as this are quite common whenever chemicals get together.

When the pollutants in London smog come into contact with the alveoli of the lungs, the cells are broken down. The alveoli lose their resilience, making it difficult for them to expel carbon dioxide. Such lung damage leads to—or at least contributes to—pulmonary emphysema, a condition characterized by an increasing shortness of breath. Emphysema is the fastest growing cause of death in the United States. The principal factor in the rise of emphysema is cigarette smoking. However, air pollution is known to be a factor, too. For instance, the incidence of the disease among smokers is three times as great in St. Louis, where air pollution is rather heavy, as in Winnipeg, Manitoba, where air pollution is rather mild.

The oxides of sulfur and the aerosol mists of sulfuric acid are damaging to plants. Leaves become bleached and splotchy when exposed to sulfur oxides. The yield and quality of farm crops can be severely affected. These compounds are also major ingredients for the production of acid rain.

SECTION 12.10

What to Do About London Smog

Much research has gone into the prevention and alleviation of London smog. Soot and smog can be removed from smokestack gases in several ways. One method uses **electrostatic precipitators** (Figure 12.9). These devices induce electrical charges on the particles, which are then attracted to oppositely charged plates and deposited (Figure 12.10).

Another method uses **bag filtration**, a system that works much like the bag in a vacuum cleaner. The filters are placed in a bag house (Figure 12.11),

Figure 12.9 Northern States Power Company's High Bridge power plant in St. Paul, Minnesota, after installation of an electrostatic precipitator. [*Jerry Mill/Northern States Power Company.*]

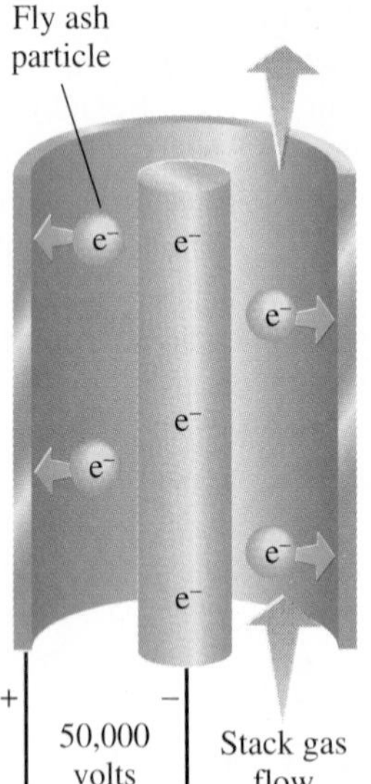

Figure 12.10 Cross section of a cylindrical electrostatic precipitator. Electrons from the negatively charged discharge electrode (in the center) attach themselves to the particles of fly ash and give them a negative charge. The charged particles then are attracted to the positively charged collector plate, on which they are deposited.

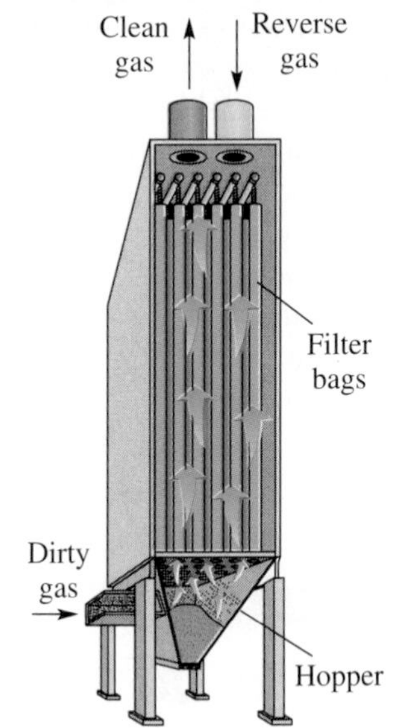

Figure 12.11 In this bag house, vibration of the bags shakes loose the particles, which then collect in the dust hopper at the bottom. Air normally flows upward, but the flow can be reversed to remove more adhering particulates. [*Adapted with permission from* EPRI Journal, *September 1983, p. 17.*]

arranged in a way that allows the filters to be shaken and air blown through periodically in the opposite direction to clean them.

A third device, called a **cyclone separator**, is arranged so that the stack gases spiral upward with a circular motion. The particles hit the outer walls, settle out, and are collected at the bottom.

Wet scrubbers are also used. These devices remove the particles by passing the stack gases through water. The water is usually sprayed in as a fine mist. The wastewater has to be treated to remove the particulates, and that adds to the cost of this method.

The exact device used depends on the type of coal being burned, the size of the power plant, and other factors. All require energy—electrostatic precipitators use 10% of the plant's output—and the collected ash has to be put somewhere. Ash production in the United States is about 70 million t per year. About 16 million t of this is used. Some replaces a part of the clay in making cement, and some is melted and blown in air to make mineral wool for insulation. The rest has to be stored; 70% of it goes into ponds and the rest into landfills.

It is harder to remove sulfur dioxide than particulates. Sulfur can be removed from the coal before burning, but both the flotation method (Chapter 14) and gasification or liquefaction processes are expensive. Another way to get rid of sulfur is to scrub sulfur dioxide out of stack gases after the coal has been burned. Perhaps the most promising of the scrubbers is the limestone–dolomite process. Limestone ($CaCO_3$) and dolomite (a mixed calcium–magnesium carbonate) are pulverized and heated. Heat drives off carbon dioxide to form calcium oxide (lime).

$$CaCO_3 + \text{heat} \longrightarrow CaO + CO_2$$

This basic oxide reacts with sulfur dioxide to form solid calcium sulfite.

$$CaO + SO_2 \longrightarrow CaSO_3$$

This by-product presents a sizable disposal problem. Removal of 1 t of sulfur dioxide produces almost 2 t of solids. Calcium sulfate ($CaSO_4$) is a much more useful product than calcium sulfite ($CaSO_3$), since calcium sulfate is used to make commercial products such as plasterboard. Some scrubber operations have been modified so as to oxidize the calcium sulfite produced to calcium sulfate.

$$2\,CaSO_3 + O_2 \longrightarrow 2\,CaSO_4$$

In nature calcium sulfate occurs as gypsum ($CaSO_4 \cdot 2H_2O$), which has the same composition as hardened plaster.

Figure 12.12 Downtown Los Angeles on a smoggy day. [*Ellis Herwig/Stock Boston.*]

SECTION 12.11

Photochemical Smog: Making Haze While the Sun Shines

The other main type of smog is **Los Angeles smog,** or, more properly, **photochemical smog**. Unlike London smog, which accompanies cold, damp air, photochemical smog usually occurs during dry, sunny weather. The principal culprits are unburned hydrocarbons and nitrogen oxides from automobiles. The warm, sunny climate that has drawn so many people to the Los Angeles area is also the perfect setting for photochemical smog (Figure 12.12).

The chemistry of Los Angeles smog is exceedingly complex. Let's look at the stuff that comes out of an automobile's exhaust pipe and examine the pollutants one at a time.

SECTION 12.12

Carbon Monoxide: The Quiet Killer

When a hydrocarbon burns in sufficient oxygen, the products are carbon dioxide and water. Since both of these substances are normal constituents of air, they are not generally considered to be pollutants. Let's illustrate the combustion process with an octane, one of the hundreds of hydrocarbons that make up the mixture we call gasoline.

$$2\,C_8H_{18} + 25\,O_2 \longrightarrow 18\,H_2O + 16\,CO_2$$

When insufficient oxygen is present, another oxide of carbon, **carbon monoxide**, is formed. Millions of metric tons of this invisible but deadly gas are poured into the atmosphere each year, about 75% of it from automobile exhausts. The U.S. goverment has set danger levels of 9 ppm carbon monoxide (average) over 8 h and 35 ppm (average) over 1 h. Even in off-street urban areas, levels often average 7 to 8 ppm. On streets, danger levels are exceeded much of the time. Such levels do not cause immediate death, but, over a long period, exposure can cause physical and mental impairment.

One-third of the U.S. population is exposed to excessive CO from auto emissions. Americans drive as many miles as all the rest of the world combined.

Carbon monoxide is an invisible, odorless, tasteless gas. There is no way for a person to tell that it is around (without using test reagents or instruments). Drowsiness is usually the only symptom, and drowsiness is not always unpleasant. How many auto accidents are caused by drowsiness or sleep induced by carbon monoxide? No one knows for sure. Cigarette smoke also contains a fairly high concentration of carbon monoxide.

Carbon monoxide is not an irritant. It exerts its insidious effect by tying up the hemoglobin in the blood. The normal function of hemoglobin is to transport oxygen (Figure 12.13). Carbon monoxide binds to hemoglobin so strongly that the hemoglobin is prevented from transporting oxygen. Therefore, the symptoms of carbon monoxide poisoning are those of oxygen deprivation. All except the most severe cases of carbon monoxide poisoning are reversible. The best antidote is the administration of pure oxygen. Artificial respiration may help if a tank of oxygen is not available.

Carbon monoxide poisoning impairs the ability of the blood to transport oxygen, and the heart has to work harder to supply oxygen to the tissues. Chronic exposure to even low levels of carbon monoxide may put an added strain on the heart and lead to an increased chance of a heart attack. In 1992 there were 10,595 cases of carbon monoxide poisoning reported in the United States. About 200 of those people died.

Carbon monoxide is a local pollution problem. It is a severe threat in urban areas with heavy traffic, but it does not appear to be a global threat. In laboratory tests, carbon monoxide survives about 3 years in contact with air. But nature is somehow able to prevent its buildup.

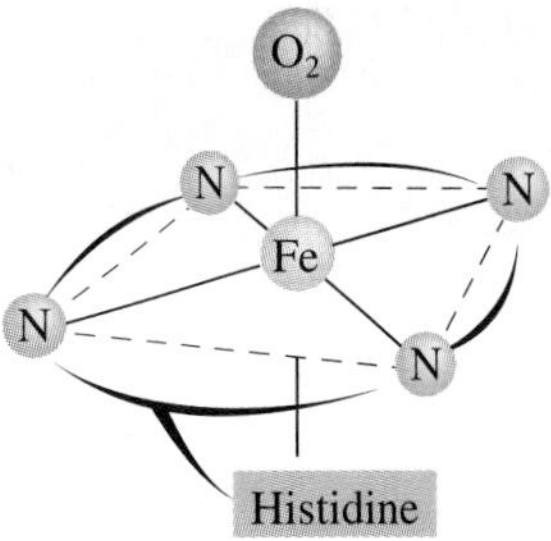

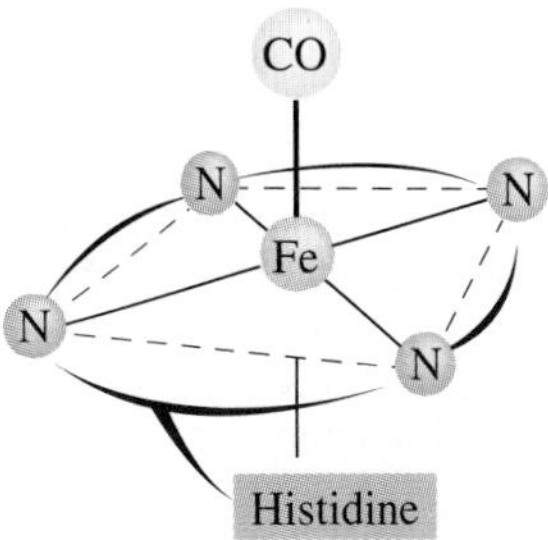

Figure 12.13 Schematic representations of a portion of the hemoglobin molecule. Histidine is an amino acid. Carbon monoxide bonds much more tightly than oxygen, as indicated by the heavier bond line.

SECTION 12.13

Nitrogen Oxides: Brown Is the Color of Los Angeles Air

In addition to carbon dioxide, carbon monoxide, and unburned hydrocarbons, automobile exhaust contains oxides of nitrogen. Power plants that burn fossil fuels are another major source of nitrogen oxides. In a reaction similar to the one that occurs in the atmosphere during electrical storms, nitrogen and oxygen are made to combine in combustion chambers; the main product is nitric oxide (NO).

$$N_2 + O_2 \longrightarrow 2\,NO$$

Figure 12.14 Photochemical smog. This type of air pollution results from the action of sunlight on oxides of nitrogen that are emitted from automobiles and other high-temperature combustion sources. Photochemical smog, often called Los Angeles smog, is characterized by an amber haze like that shown here. [*Sepp Seitz/Woodfin Camp & Associates.*]

Figure 12.15 A summary of some of the principal reactions in the formation of photochemical smog. Most reactive intermediates have been omitted from this simplified scheme.

The initiating reaction:

$$\underset{\text{Nitrogen dioxide}}{NO_2} + \text{sunlight} \longrightarrow \underset{\text{Nitric oxide}}{NO} + \underset{\text{Oxygen atom (highly reactive)}}{O}$$

Secondary reactions:

$$\underset{\text{Oxygen atom}}{O} + \underset{\text{Oxygen molecule}}{O_2} \longrightarrow \underset{\text{Ozone}}{O_3}$$

$$O + \text{hydrocarbons} \longrightarrow \text{aldehydes} \left(R{-}\overset{\overset{\large O}{\|}}{C}{-}H \right)$$

Tertiary reactions:

$$O_3 + \text{hydrocarbons} \longrightarrow \text{aldehydes}$$

$$\text{Hydrocarbons} + O_2 + NO_2 \longrightarrow \text{PAN} \left(R{-}\overset{\overset{\large O}{\|}}{C}{-}O{-}O{-}NO_2 \right)$$

Oxygen in the atmosphere oxidizes the nitric oxide to nitrogen dioxide (NO_2).

$$2\,NO + O_2 \longrightarrow 2\,NO_2$$

Nitrogen dioxide is an amber-colored gas. Smarting eyes and a brownish haze are excellent indicators of Los Angeles smog (Figure 12.14). It is this nitrogen dioxide that plays a vital (villain's) role in photochemical smog. It absorbs a photon of sunlight and breaks down into nitric oxide and reactive oxygen *atoms.*

$$NO_2 + \text{sunlight} \longrightarrow NO + O$$

The oxygen atoms react with other components of automobile exhaust and the atmosphere to produce a variety of irritating and toxic chemicals (Figure 12.15).

At present concentrations, nitrogen oxides don't seem particularly dangerous in themselves. Nitric oxide at high concentrations reacts with hemoglobin; as with carbon monoxide poisoning, this leads to oxygen deprivation. Such high levels seldom, if ever, result from ordinary air pollution, but they might be reached in areas close to industrial sources. Nitrogen dioxide is an irritant to the eyes and the respiratory system. Tests with laboratory animals indicate that chronic exposure in the range of 10 to 25 ppm might lead to emphysema and other degenerative diseases of the lungs.

The most serious environmental effect of the nitrogen oxides is that they produce smog. The gases also contribute to the fading and discoloration of fabrics, and by forming nitric acid, nitrogen oxides contribute to the acidity of rainwater (Section 12.18). They also contribute to crop damage, although their specific effects are difficult to separate from those of sulfur dioxide and other pollutants.

SECTION 12.14

Ozone: Protector and Pollutant

The ordinary oxygen that we breathe is made up of O_2 molecules. **Ozone** is a form of oxygen that consists of O_3 molecules. It is a natural component of the stratosphere, where it shields the Earth from life-destroying ultraviolet radiation. Ozone is also a familiar constituent of photochemical smog. Inhaled, it is a toxic, dangerous chemical. Ozone is a good example of just what a *pollutant* is—a chemical substance out of place in the environment. In the stratosphere, it helps make life possible. In the lower troposphere—the part we breathe—it makes life difficult.

Ozone (O_3) and oxygen (O_2) are **allotropes**: two different forms of the same element.

In the mesophere (see Figure 12.1), some ordinary oxygen molecules are split into oxygen atoms by short-wavelength, high-energy ultraviolet radiation.

$$O_2 \xrightarrow[\text{radiation}]{\text{ultraviolet}} 2\,O$$

Some of these highly reactive atoms diffuse down to the stratosphere where they react with oxygen molecules to form ozone.

$$\underset{\text{Oxygen atom}}{O} + \underset{\text{Oxygen molecule}}{O_2} \longrightarrow \underset{\text{Ozone}}{O_3}$$

Ultraviolet light sometimes is used to sterilize tools in the hairdresser's shop and safety glasses in laboratories to prevent the transfer of infectious microorganisms from one person to another. We do not want the entire planet rendered sterile, however, for we are a part of the life that would be harmed.

The ozone in turn absorbs shorter wavelength, but still lethal, ultraviolet rays, thus shielding us from this harmful radiation. In absorbing the rays, ozone is converted back to oxygen molecules and oxygen atoms, reversing the previous reaction.

$$O_3 \xrightarrow[\text{radiation}]{\text{ultraviolet}} O_2 + O$$

Undisturbed, the concentration of ozone is kept fairly constant by these cyclic processes. Human activity, however, threatens the existence of the protective shield. Before examining that problem, let's take a look at the effects of ozone down here where we live.

Los Angeles isn't the only place with an ozone problem. The Northeast, the Texas Gulf Coast, New York City, Denver, Salt Lake City, and Chicago also are afflicted. In all, some 94 urban areas in the United States failed to meet federal health-based ozone standards in 1988.

Ozone as an Air Pollutant

Ozone is a powerful oxidizing agent. At low levels, it causes eye irritation. At high levels, it can cause pulmonary edema, hemorrhage, and even death. The long-term effects of exposure to low levels of ozone are more difficult to evaluate. Inhalation of ozone is particularly dangerous during vigorous physical activity. Members of a New Jersey high school football team had to be hospitalized after collapsing during a severe pollution episode. School children in Los Angeles are not allowed to play outside when ozone reaches dangerous levels, as it often does. Exposure of animals to 1 ppm of ozone for 8 hr/day for 1 year has produced bronchial inflammation and irritation of fi-

Figure 12.16 Ozone causes rubber to harden and crack. Tire manufacturers incorporate paraffin wax into the rubber to protect tires from ozone. The wax migrates to the surface during use, replacing that worn off. Paraffins (alkanes) are among the very few substances resistant to attack by ozone. [*Courtesy of Smithers Scientific Services, Inc.*]

brous tissues. It is not known if the same thing occurs in humans. It is known that ozone levels have occasionally reached 0.5 ppm in southern California, and levels of 0.15 ppm are exceeded frequently.

In addition to adversely affecting health, ozone causes economic damage. It causes rubber to harden and crack, shortening the life of automobile tires (Figure 12.16) and other rubber items. Ozone also causes extensive damage to crops. Tobacco and tomatoes are particularly susceptible.

Exposure of a plant to as little as 0.06 ppm of ozone can cut photosynthesis in half. Plant growth is reduced, and leaves often become spotted.

SECTION 12.15

Chlorofluorocarbons and Other Threats to the Ozone Shield

A class of compounds called **chlorofluorocarbons** (Section 9.6) have been used as the dispersing gases in aerosol cans, as foaming agents for plastics, and as refrigerants. At room temperature, the chlorofluorocarbons are gases or liquids with low boiling temperatures. They are essentially insoluble in water and inert toward most other substances. These properties make them ideal propellants for use in aerosol cans of deodorants, hair sprays, and food products. Unfortunately, the inertness of these compounds allows them to persist in the environment.

Chlorofluorocarbons diffuse into the stratosphere, where they are broken down by ultraviolet radiation. Chlorine atoms formed in this process break down the ozone that protects the Earth from harmful ultraviolet radiation.

$$CF_2Cl_2 + \text{ultraviolet light} \longrightarrow CF_2Cl\cdot + Cl\cdot$$

$$Cl\cdot + O_3 \longrightarrow ClO\cdot + O_2$$

$$\cdot ClO + O \longrightarrow Cl\cdot + O_2$$

Note that the last step results in the formation of another chlorine atom that can break down another molecule of ozone. The second and third steps are repeated many times; thus, the decomposition of one molecule of chlorofluorocarbon can result in the destruction of many molecules of ozone.

The U.S. National Research Council predicts a 2 to 5% increase in skin cancer for each 1% depletion of the ozone layer. Even a 1% increase in ultraviolet radiation can lead to a 5% increase in melanoma, a deadly form of skin cancer.

In 1984, the NRC predicted a 2 to 4% depletion by late in the 21st century. Worries escalated in 1986 with the discovery that ozone concentrations over Antarctica plummet each September. This thinning or ''hole'' in the ozone layer (Figure 12.17) has been getting worse since 1979. A similar but less dramatic thinning has been detected in the Arctic. Winter and early spring stratospheric ozone depletion is estimated to be 2 to 10% over the last 20 years. Overall satellite data show that the ozone layer has been thinning at a

Why are there ozone holes at the poles? In the polar winter a vortex of intensely cold air with ice crystals develops. Reactions occur on the surface of the ice crystals where hydrogen chloride (HCl) and chlorine nitrate ($ClONO_2$) are adsorbed. These molecules react to form chlorine and nitric acid.

$$HCl + ClONO_2 \rightarrow Cl_2 + HNO_3$$

The Cl_2 molecules are readily dissociated by light into chlorine atoms.

$$Cl_2 \rightarrow 2\,Cl\cdot$$

These *free radicals* (Section 5.12) then enter into the ozone-destroying cycle.

Susan Solomon is a chemist at the Oceanic and Atmospheric Administration in Boulder, Colorado. In 1986 she proposed a mechanism for the action of chlorofluorocarbons on ozone that would explain the hole in the ozone layer. Later she confirmed her theory with on-site experiments in Antarctica. Her work led to international bans on the use of chlorofluorocarbons. [*Courtesy of Dr. Susan Solomon.*]

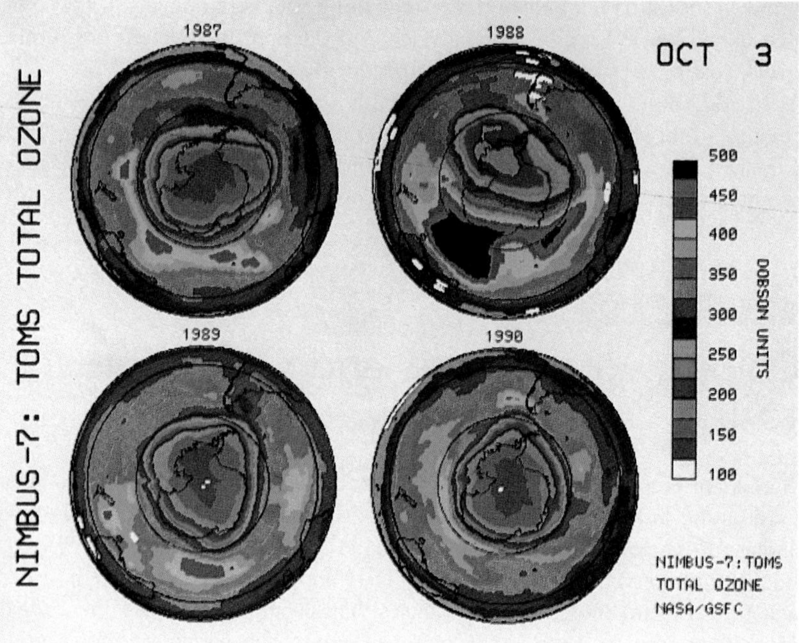

Figure 12.17 Maps of stratospheric ozone levels over Antarctica 1987–1990. The lowest amounts are over the South Pole. [*NASA photos.*]

rate of 0.5% per year since 1978. Evidence is now compelling that the thinning is caused by chlorofluorocarbons.

Chlorofluorocarbons have been banned in the United States from most aerosol preparations, but they are still used as refrigerants. They still escape into the atmosphere from refrigerators and air conditioners through leaks and when the appliances are discarded.

Most of the nations of the world have adopted the Montreal Protocol of 1987, updated in 1990 and 1992, which calls for the complete phaseout of the original CFCs. The goal is to limit production of CFCs in 1994 to 25% of the 1986 production, with a complete phaseout by January 1996. About half of 1994 model cars are using HCFCs in their air conditioners. These, too, are scheduled for phaseout between 2020 and 2040.

Chlorofluorocarbons are often referred to as CFCs, and hydrochlorofluorocarbons as HCFCs. A typical HCFC is dichlorotrifluoroethane.

$$Cl_2CH{-}CF_3$$

SECTION 12.16

Hydrocarbons: Another Culprit

Hydrocarbons are released from a variety of natural sources. Of all hydrocarbons found in the atmosphere, only about 15% are put there by people. In most urban areas, though, the processing and use of gasoline is the major source of hydrocarbons in the environment. Gasoline can evaporate anywhere

Hydrocarbons from autos come mainly from those without pollution control devices. Most hydrocarbon vapor comes from exhaust gas (65%), with some coming from crankcase (20%) and carburetor (15%) vents.

along the line. This simple process contributes substantially to the total amount of hydrocarbons in urban air. The automobile's internal combustion engine also contributes by exhausting unburned and partially burned hydrocarbons.

Certain hydrocarbons, particularly those that contain a double bond, combine with oxygen atoms or ozone molecules to form aldehydes. As a class, the aldehydes have foul, irritating odors. Another series of reactions involving hydrocarbons, oxygen, and nitrogen dioxide leads to the formation of peroxyacetyl nitrate (PAN). Ozone, the aldehydes, and PAN are responsible for much of the destruction wrought by smog. They make breathing difficult and make the eyes smart and itch. Those who already have respiratory ailments may be severely affected. The very young and the very old are particularly vulnerable.

$$CH_3\overset{\displaystyle O}{\overset{\|}{C}}—O—ONO_2$$

PAN

SECTION 12.17

What to Do About Photochemical Smog

The principal culprits in photochemical smog are nitrogen oxides, hydrocarbons, and sunlight. Reducing the amount of any of these would diminish the amount of smog. It isn't likely that we could—or would want to—reduce the amount of sunlight, so let's focus on the other two.

Hydrocarbons have many important uses as solvents. If we could reduce the amount of hydrocarbons entering the atmosphere, the amount of aldehydes and PAN formed would also be reduced. Many industries using hydrocarbons have substantially reduced emissions or switched to other solvents, and new storage and dispensing systems have reduced hydrocarbon emissions from gasoline filling stations. Similarly, modified gas tanks and crankcase ventilation systems have reduced evaporative emissions from automobiles. The main attack, however, has been through the use of **catalytic converters** to reduce hydrocarbon and carbon monoxide emissions in automotive exhausts. The catalyst in these converters is a precious metal such as platinum (Pt). Hydrocarbons and carbon monoxide react rapidly with oxygen on the surface of these metals. With no catalyst, the pollutants are more likely to reach the atmosphere unreacted.

Reducing the amount of nitrogen oxides has proven more difficult. Whereas carbon monoxide and hydrocarbons are removed by *oxidation*, NO must be *reduced* to N_2. Lowering the operating temperature of an engine helps, but this renders the engine less efficient. Running the engine on a richer (more fuel, less air) mixture lowers nitrogen oxide emissions but tends to raise carbon monoxide and hydrocarbon emissions. With a separate catalyst, some of the carbon monoxide can be used to reduce the nitric oxide.

$$2\,NO + 2\,CO \longrightarrow N_2 + 2\,CO_2$$

A one-car-one-person transportation system is inherently inefficient. The best way to reduce emissions from automobiles is to drive less. Walk, ride a bicycle, or use public transportation whenever possible. And if you must drive, use a smaller car.

The remaining carbon monoxide and the hydrocarbons are then oxidized over the platinum catalyst.

The state of California in 1990 passed a law requiring that by 1998 at least 2% of all cars sold in that state must be zero-emission vehicles. This means that 2% must be electric cars. The percentage increases to 5% in 2001 and 10% in 2003. Whether or not a practical electric car can be produced by 1998 remains to be seen. Currently, the projected price for electric cars is double that of conventional cars with gasoline engines. They are powered by more than two dozen heavy batteries weighing half a ton. But the driving range is only 70 miles: It will be interesting to see whether auto makers can succeed in making practical electric cars by 1998, or whether California will have to modify its law.

SECTION 12.18

Acid Rain: Air Pollution → Water Pollution

Normal rainwater, which is slightly acidic because of dissolved carbon dioxide, has a pH of 5.6. During thunderstorms, the pH of rainwater can be much lower due to nitric acid formed by lightning.

We have seen (Section 12.9) how sulfur oxides are converted to sulfuric acid. Similarly, nitrogen oxides become nitric acid (Section 12.2). These acids fall upon Earth as acid rain or acid snow or are deposited from acid fog or absorbed on particulate matter. **Acid rain** is defined as rain having a pH less than 5.6. Rain with a pH as low as 2.1 and fog with a pH of 1.8 have been reported (Figure 12.18). These values are lower than the pH of vinegar or lemon juice.

The best evidence indicates acid rain comes mainly from sulfur oxides emitted from power plants and smelters and from nitrogen oxides mostly from automobiles. These acids are carried for hundreds of kilometers before falling as rain or snow (Figure 12.19).

Keep in mind that a solution with a pH one unit lower than another is 10 times as acidic. A decrease in pH from 5.6 to 4.6, for example, means an increase in acidity by a factor of 10. Two pH units lower means 100 times as acidic; and three pH units lower means 1000 times as acidic.

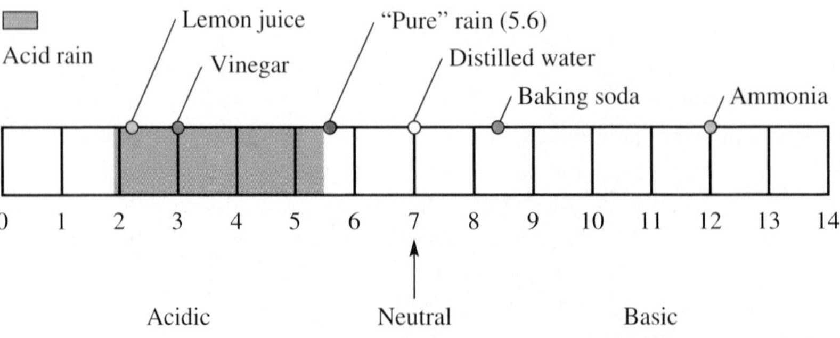

Figure 12.18 The pH of acid rain in relation to the pH of other familiar substances.

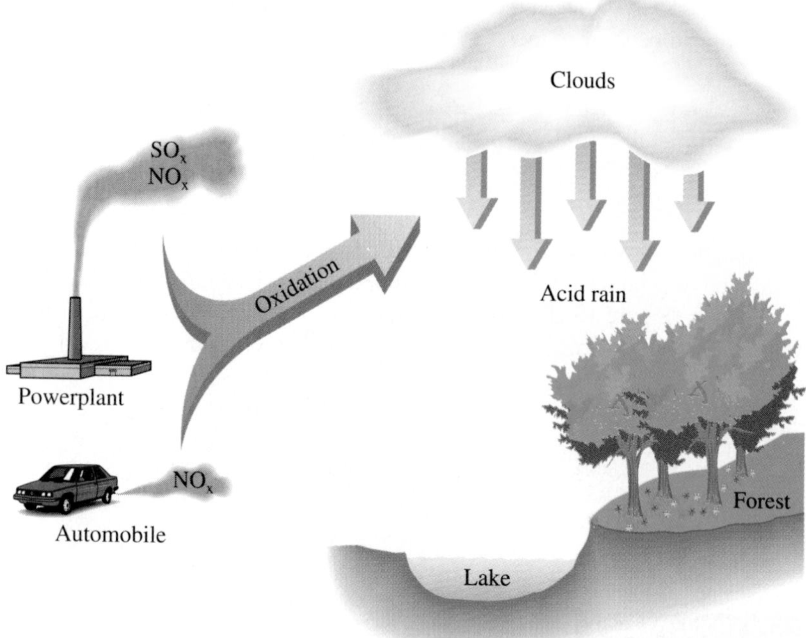

Figure 12.19 Acid rain usually falls far from the site where the acidic oxides are generated.

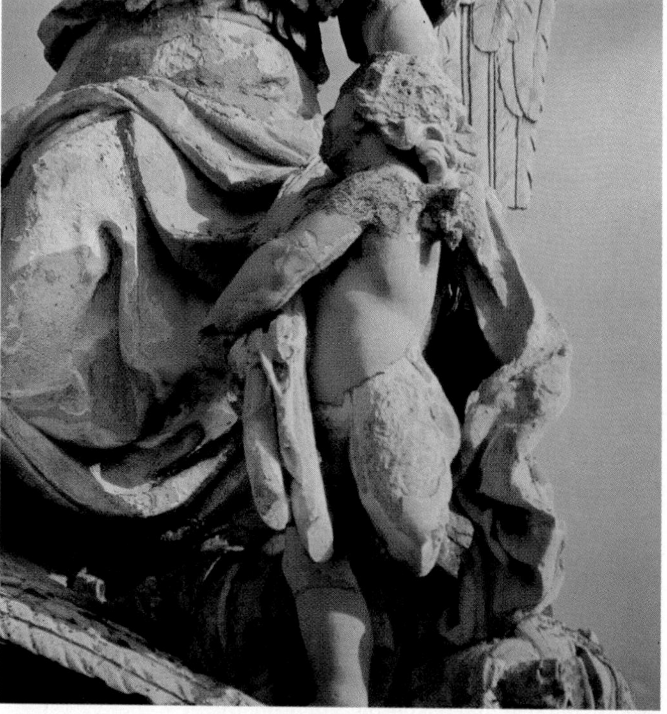

Figure 12.20 A particularly striking effect of air pollution is the action of acidic air pollutants on marble, as this statue shows. [*Courtesy of CNRI/French Information Services.*]

The effect of acidic waters on plant and animal life is discussed in Chapter 13.

Acids corrode metals and can even decompose stone buildings and statues. Sulfuric acid eats away metal to form a soluble salt and hydrogen gas.

$$\underset{\substack{\text{Iron}\\\text{(or steel)}}}{Fe} + H_2SO_4 \longrightarrow \underset{\substack{\text{Iron(II)}\\\text{sulfate}\\\text{(soluble)}}}{FeSO_4} + H_2$$

The reaction shown here is oversimplified. For example, in the presence of water and oxygen (air), the iron is converted to rust (Fe_2O_3). Marble buildings and statues (Figure 12.20) are disintegrated by sulfuric acid in a similar reaction that forms calcium sulfate, a slightly soluble, crumbly compound.

$$\underset{\substack{\text{Marble}\\\text{or limestone}}}{CaCO_3} + H_2SO_4 \longrightarrow \underset{\substack{\text{Calcium}\\\text{sulfate}}}{CaSO_4} + H_2O + CO_2$$

Catalytic converters and lead-free gasolines are far-from-perfect answers to air pollution problems. Cars with pollution-control devices often are harder to start and run less efficiently than cars without them. These devices work well only if cars are properly tuned (and many aren't).

SECTION 12.19

Get the Lead Out

Lead is a heavy-metal poison that affects the functioning of the blood, liver, kidneys, and brain (Chapter 20). For years, a lead compound, tetraethyllead, has been used to improve the antiknock qualities of gasoline. Modern high-compression engines will not run on straight-run gasoline. As it comes from the distillation tower of a petroleum refinery, gasoline has an octane rating of about 60. Science and technology, through the use of tetraethyllead, have done a fine job of providing high-octane fuel. Unfortunately, the solution to one problem may be the source of another. Large amounts of lead have been spewed into the environment by automobiles. Fairly high concentrations of lead are now found along heavily traveled streets, freeways, and other roads.

CH_3CH_2—Pb(CH_2CH_3)$_3$ (central Pb bonded to four CH_2CH_3 groups)

Tetraethyllead

In addition to its toxicity, lead fouls the catalysts in pollution-control devices. We have two reasons we should "get the lead out." The United States has led the world in getting rid of leaded gasoline. Beginning in 1974 all new cars in this country have used lead-free gasoline. By 1994 almost no leaded fuel was available in the United States. The only exceptions have been fuels for some farm and off-highway vehicles. Europe has been much slower in its phaseout of leaded motor fuel.

To achieve the same octane rating in lead-free gasoline, more branched-chain and aromatic hydrocarbons (Chapter 9) are needed. Even without these extra aromatic hydrocarbons, the burning of gasoline produces 3,4-benzpyrene, a chemical that induces cancer in laboratory animals. Increasing the aromatic hydrocarbon content of gasoline probably increases the concentration of this carcinogen in the atmosphere.

3,4-Benzpyrene

SECTION 12.20

The Inside Story: Indoor Air Pollution

You can't necessarily escape air pollution by staying indoors. In high-traffic areas, air indoors has been found to have the same carbon monoxide level as air outside. In some places, office buildings, airport terminals, and apartments have indoor levels that exceed government safety standards. It is recommended that such buildings be tightly sealed on the lower floors and that they be spaced in such a way that the wind can disperse pollutants around them.

The Home: No Haven from Air Pollution

As we try to save energy by installing better insulation and by sealing air leaks around windows and doors, we often make indoor pollution worse. We trap pollutants inside. A kitchen with a gas range often has levels of nitrogen oxides above United States government standards. Those who try to save on the fuel bill by using free-standing kerosene stoves produce levels of nitrogen oxides up to 20 times greater than those permitted by federal regulations for outdoor air. (These regulations do not apply to indoor air.) Another problem, the release of toxic, irritating formaldehyde gas from foamed insulation, led in 1982 to a ban on the use of urea–formaldehyde polymer foams for insulation. (This ban was lifted by the courts in 1983.) Similar polymers are used as adhesives in plywood and particle board; these uses have not been banned.

Cigarettes and Secondhand Smoke

Perhaps the prevalent indoor polluter is the cigarette. More than 40 carcinogens have been identified among the 4000 or so chemical compounds found in cigarette smoke.

The carcinogens in cigarette smoke include aromatic hydrocarbons, aromatic amines, nitrosamines, and radioactive polonium-210.

The health effects of smoking on the smoker are widely known. Smokers face twice the risk of bladder cancer, stroke, and heart attack as those who do not smoke; 4 times the risk of cancer of the esophagus; 7 times the risk of death from chronic bronchitis or emphysema; 8 times the risk of cancer of the larynx; and 10 to 20 times the risk of lung cancer. These risks have been widely publicized in a series of reports released by the Surgeon General of the United States. The first report, published in 1964, led to a ban on television advertisements for cigarettes. Subsequent reports have dealt with the effects of tobacco smoke as an air pollutant. Studies have shown that the air quality in smoke-filled rooms is poor. The level of carbon monoxide in a room, even a well-ventilated one, where people are smoking is often equal to, or greater than, the legal limits for ambient air. High levels of carbon monoxide have been shown to impair time-interval discrimination. In severe cases, performance on psychomotor tests is impaired. People already suffering from heart or lung disease might be severely affected. Smoke from just one burning cigarette can raise the level of particulate matter above government standards.

There is a linear relationship between cigarettes smoked per day and levels of damage to DNA found in smokers' lungs. This damaged DNA perhaps is involved in the production of lung cancer.

The nonsmoker also is exposed to significant levels of tar and nicotine. In fact, the average nonsmoker has 5% as much nicotine in the blood as a smoker. Workers at Cornell University Medical College found a glycoprotein (a protein molecule with sugar units attached) in tobacco that causes allergic reactions in many smokers and nonsmokers. The reaction is so severe that it may damage small blood vessels and lead to strokes and heart disease. The risk to nonsmokers from cigarette smoke is well established but of uncertain magnitude. Several states and municipalities have already banned or restricted smoking in many public places.

Radon and Her Dirty Daughters

A most enigmatic indoor air pollutant is radon. A noble gas, **radon** is colorless, odorless, tasteless, and unreactive chemically. Radon is, however, radioactive. It is the decay products of radon, called **daughter isotopes**, that are the problem. Radon itself decays by alpha emission (Chapter 4) with a half-life of 3.8 days.

$$^{222}_{86}\text{Rn} \longrightarrow \, ^{218}_{84}\text{Po} + \, ^{4}_{2}\text{He}$$

When radon is inhaled, the polonium-218 and other daughters, including lead-214 and bismuth-214, are trapped in the lungs where their decay damages the tissues.

Radon is released naturally from soils and rocks, particularly granite and shale, and from minerals such as phosphate ores and pitchblend. The ultimate source is uranium atoms found in these materials. Radon is only one of several radioactive materials formed during the multistep decay of uranium. However, radon is unique in that it is a gas and escapes. The others are solid and remain in the soil and rock.

Outdoors, the radon dissipates and presents no problems. Trapped inside a house, however, levels build up, sometimes reaching quantities several times the maximum safe level established by the U.S. Environmental Protection Agency (0.15 disintegration per second per liter of air). At five times that level, the hazard is thought to equal that of smoking two packs of cigarettes a day. Some scientists say that radon threatens 8 million homes in the United States and may cause 20,000 lung cancer deaths a year. The exact threat isn't known, however. The "safe level" was set for people who work in uranium mines. These workers do have high rates of lung cancer, but they also breathe radioactive dust and many smoke cigarettes. These risk factors are difficult to separate. We need careful studies in order to estimate the risk of radon alone.

SECTION 12.21

Who Pollutes? How Much?

Air pollution causes material damage by dirtying and destroying buildings, clothing, and other material objects. It increases health hazards, especially for

Table 12.2 The Seven Major Air Pollutants

Pollutant	Symbol	Major Sources	Health Effects	Environmental Effects
Carbon monoxide	CO	Motor vehicles	Interferes with oxygen transport, causing dizziness, death; possibly contributes to heart disease	Slight
Hydrocarbons	C_nH_m	Motor vehicles, industry, solvents	Narcotic at high concentrations; some aromatics are carcinogens	Precursor to aldehydes, PAN
Sulfur oxides	SO_x	Power plants, smelters	Irritate respiratory system; aggravate lung and heart diseases	Reduce crop yields; precursor to acid rain, SO_4^{2-} particulates
Nitrogen oxides	NO_x	Power plants, motor vehicles	Irritate respiratory system	Reduce crop yields; precursor to ozone and acid rain; produce brown haze
Particulate matter	—	Industry, power plants, dust from farms and construction sites	Irritates respiratory system; synergistic with SO_2; contains adsorbed carcinogens, toxic metals	Impairs visibility
Ozone	O_3	Secondary pollutant from NO_2	Irritates respiratory system; aggravates lung and heart diseases	Reduces crop yields; kills trees (synergistic with SO_2); destroys rubber, paint, etc.
Lead	Pb	Motor vehicles, smelters	Toxic to nervous system and blood-forming system	Toxic to all living things

the very young, the old, and those already ill. It causes crop damage by stunting or killing green plants. Air pollution reduces visibility, thus increasing auto and air-traffic accidents. With its ugly smoke plumes and unpleasant odors, it is even an aesthetic problem.

Who causes all that pollution? Table 12.2 summarizes the seven major pollutants, their main sources, and their health and environment effects. Note that motor vehicles are a prominent source. Indeed, they are the source of about one-half (by mass) of all air pollutants. Our transportation system accounts for about 80% of carbon monoxide emissions, 40% of hydrocarbon emissions, and 40% of nitrogen oxide emissions. Since most transportation in the United States is by private automobile, we can conclude that the car goes a long way toward doing us in.

On the other hand, most particulate matter comes from power plants (about 40%) and industrial processes (about 45%). Similarly, over 80% of sulfur oxide emissions come from power plants, with an additional 15% coming from other industries. Power plants alone contribute about 55% of nitrogen oxide emissions. Who uses electricity from power plants? You do. A 100-W bulb burning for 1 year uses the electricity generated by burning 275 kg of coal.

What is the *worst* pollutant? Carbon monoxide is produced in huge amounts and is quite toxic. Yet it is deadly only in concentrations approaching 4000 ppm. Its contribution to cardiovascular disease, by increasing stress on the heart, is difficult to measure.

The World Health Organization (WHO) rates sulfur oxides as the worst pollutants. Sulfur oxides are powerful irritants, and, according to WHO, people with respiratory illnesses are more likely to die from exposure to sulfur oxides than to any other kind of pollutant. Sulfur oxides and sulfuric acid formed from them have been linked to as many as 53,000 deaths per year in the United States.

Who pollutes? We all do. In the words of Walt Kelley's comic-strip character Pogo, "We have met the enemy, and he is us." Don't despair, however. You can be a part of the solution by conserving fuel and electricity. Many utility companies now offer suggestions for saving energy.

Overall, air quality is improving. Lead emissions are down 96% from 1970. Sulfur oxide emissions have been reduced 28% in the last 20 years. Particulate emissions are down 61%, with especially dramatic reductions in urban areas. Carbon monoxide levels are down 38%, but many cities still exceed health standards. Levels of nitrogen oxides have increased 8%, however. Despite improvements the air is still polluted at unsafe levels for much of the population at times. We still have a long way to go before we all have clear air to breathe.

SECTION 12.22

Carbon Dioxide

No matter how clean an engine or a factory is, as long as it burns coal or petroleum products it produces carbon dioxide. The concentration of carbon dioxide in the atmosphere has increased 18% in this century. It continues to increase at an expanding rate because of the increased burning of carbon fuels. We generally don't even consider carbon dioxide to be a pollutant. It is a

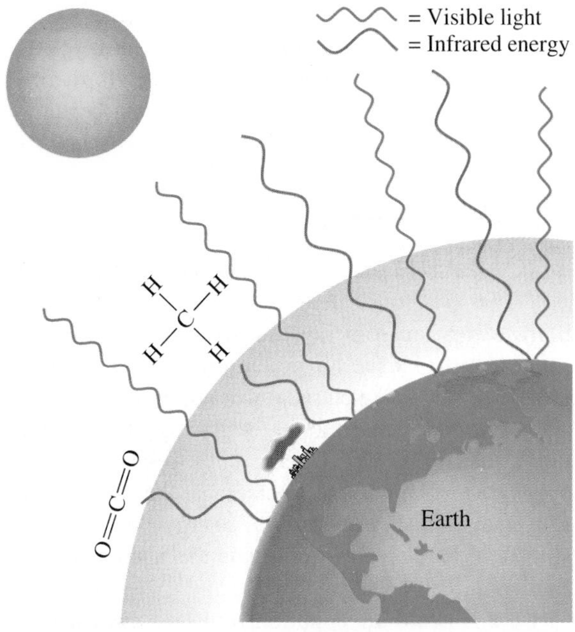

Figure 12.21 The greenhouse effect. Molecules such as CO_2 and CH_4 are transparent to visible light from the sun, but they trap infrared energy as Earth attempts to radiate heat back into space.

natural component of the environment. Certainly its immediate effect upon us is slight. But what about the long-term effects?

The Greenhouse Effect: Planet with a Fever

Carbon dioxide and other gases produce the **greenhouse effect**. These chemicals let the sun's rays (visible light) in to warm the surface of the Earth. When the Earth tries to radiate this heat (infrared energy) back out into space, however, the energy is trapped by molecules of carbon dioxide and other gases (Figure 12.21).

The amount of carbon dioxide in the atmosphere could double by 2030 if we continue our present practices.

Human activities add 25 billion t of carbon dioxide to the atmosphere each year, with 22 billion t coming from the burning of fossil fuels. About 15 billion t is removed by plants, the soil, the oceans, leaving a net addition of 10 billion t per year. The concentration of carbon dioxide is therefore increasing at a rate of 1 ppm per year.

Methane, chlorofluorocarbons, and other trace gases also contribute to the greenhouse effect. The concentration of methane in the atmosphere has been increasing since 1977. Although present in much smaller amounts than carbon dioxide, these trace gases are much more efficient at trapping heat. Methane is 20 to 30 times and chlorofluorocarbons 20,000 times as effective as carbon dioxide at holding heat in Earth's atmosphere.

Water vapor also acts as a greenhouse gas. When released into the atmosphere, however, water soon falls back to Earth as rain. It therefore affects the climate mostly at the local level.

Many scientists predict a warming trend, but they often differ in their estimates of its magnitude and effect. Some fear that global warming will melt the polar ice caps and flood coastal cities. The oceans are now rising at 3 mm per year. That doesn't seem like much, but even slight rises increase tides and

Estimates of sea level rise by 2030 vary from 20 cm to 14 m, with a most probable rise of 28 cm. Differences arise from the use of various models of the atmosphere and from different estimates of the parameters that go into the model. Most predictions are made with the use of powerful computers.

result in higher, more damaging storm surges. The fears have increased in recent years. The 1980s saw the four warmest years in the last 110, and heat waves and drought scorched much of the Northern Hemisphere. Then 1990 was the warmest year since record keeping began.

SECTION 12.23

The Ultimate Pollutant: Heat

Electric cars are sometimes advocated as a solution to air pollution. They could help to reduce air pollution in urban areas. But electric cars require electric power, and electric power requires power plants. Conventional power plants burn coal, gas, and oil. Replacing cars that run on fossil fuel with ones run on electricity would serve only to change the *site* of the pollution and perhaps spread it out a bit. Remember that there are losses at every step in the conversion and transmission of energy. More fossil fuels would have to be burned at power plants to provide the same amount of energy for propelling automobiles that we now use.

We can avoid air pollution by using nuclear power plants (Chapter 14). They do not spew soot, smoke, and poisonous chemicals into the atmosphere. But public fears of nuclear accidents seem to render nuclear power an unlikely choice.

There is one pollutant we cannot avoid: heat. According to the second law of thermodynamics, in any conversion some of the energy winds up as heat. All power plants dump vast quantities of residual heat into the environment as they produce electrical energy. This heat may be the ultimate pollutant. Urban areas are already warmed by human activities, with cities often 4 or 5 °C warmer than surrounding rural areas.

We use fantastic amounts of energy—in cars, factories, homes, schools, and hospitals. We use it everywhere, for everything. The release of energy heats up the environment. Eventually, this heat will change the climate of the Earth, and the ecology of the planet will be affected. As Rene Dubos, world-renowned biologist, put it: "We will destroy our lives by producing more useless, destructive energy to make more and more needless things that do not increase the happiness of people."

Figure 12.22 This statue at Valley Forge, Pennsylvania, has been soiled by air pollution. [*Nicholas Veloz/ courtesy of National Park Service.*]

SECTION 12.24

Paying the Price

Air pollution costs us billions of dollars each year. It wrecks our health by causing or aggravating bronchitis, asthma, emphysema, and lung cancer. It destroys our crops and sickens and kills our livestock. It corrodes our machines, blights our buildings, and even destroys our works of art (Figure 12.22).

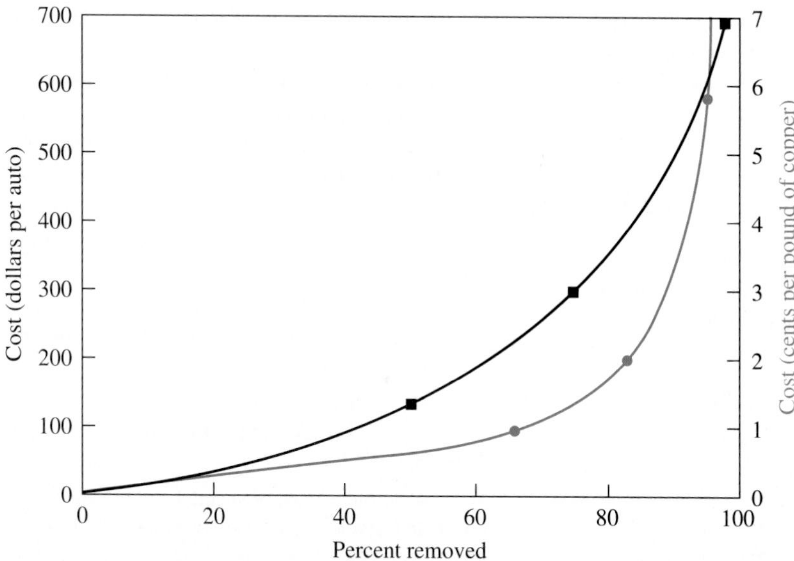

Figure 12.23 The cost of removing pollutants increases exponentially with the percentage of material removed. The color line shows the cost of removing sulfur dioxide from copper smelter fumes. The black line shows the cost of reducing automobile emissions.

Elimination of air pollution would not be cheap or easy. It is especially expensive to remove the last fractions of pollutants. Figure 12.23 shows how costs increase as you try to remove a larger percent of pollutants. Note that the curves are exponential: costs soar to infinity as pollutants are reduced to zero. But we could have cleaner air. How clean depends on how much we are willing to pay.

What would we gain by getting rid of air pollution? How much is it worth to see the clear blue sky or to see the stars at night? How much is it worth to breathe clean, fresh air?

Summary

1. The *atmosphere* is made up of layers: the *troposphere* (nearest Earth), the *stratosphere* (which includes the ozone layer), the mesosphere, and the thermosphere.
2. Clean, dry air contains about 78% nitrogen, 21% oxygen, and 1% argon (by volume), plus very small amounts of other gases.
3. Although the air is 78% nitrogen, animals and most plants cannot use the nitrogen unless it has been "fixed" (combined with other elements).
4. Nitrogen goes from the air into plants and animals and eventually back to the air via the *nitrogen cycle*.
5. Oxygen from the air is involved in oxidation (of plant and animal materials) and is eventually returned to the air via the *oxygen cycle*.
6. Oxygen is converted to *ozone* and then changed back to oxygen in the stratosphere.
7. Natural air *pollution* includes dust storms, noxious gases from swamps, and ash from erupting volcanoes.

8. The primary air *pollutants* due to human activity are the smoke and gases produced by the burning of fuels and the fumes and particulates emitted by factories.

9. *Atmospheric inversion* is a condition in which cool, dirty air is held down near the ground by a stable warm upper layer of air.

10. *London smog* is a combination of smoke and fog with sulfur dioxide, sulfuric acid, and soot, the kind of smog produced in cold, damp air by excessive burning of coal.

11. *Photochemical* (or *Los Angeles*) *smog* is produced in dry, sunny weather, mainly from auto exhaust fumes.

12. *Carbon monoxide* (which is almost always produced when organic fuels are burned) is a poisonous gas that ties up hemoglobin in the blood so that it cannot transport oxygen to the cells.

13. Nitrogen oxides are present in auto exhaust gases and emissions from power plants that burn fossil fuels. The brown color of nitrogen dioxide causes the brownish haze often seen over certain large cities.

14. *Ozone*, an allotrope of oxygen, is a harmful air pollutant in the troposphere but a protective layer in the stratosphere that shields Earth against ultraviolet radiation from the sun.

15. *Chlorofluorocarbons* have been linked to the "hole" in the *ozone layer* and are currently being phased out.

16. *Acid rain* is caused by sulfur oxides (mainly from power plants) and nitrogen oxides (mainly from automobiles).

17. Tetraethyllead, once added to gasoline as an antiknock agent, is being phased out because of lead pollution.

18. Because of the effects of secondhand smoke, many cities and states are restricting smoking in public places.

19. *Radon* gas, an indoor pollutant in some places, is a problem because it is radioactive.

20. Gases such as carbon dioxide and methane can produce a *greenhouse effect* and lead to global warming.

▸ Key Terms

acid rain 12.18
aerosols 12.9
air 12.1
allotropes 12.14
atmosphere 12.1
atmospheric inversion 12.4
bag filtration 12.10
carbon monoxide 12.12
catalytic converter 12.17
chlorofluorocarbons 12.15
cyclone separator 12.10
daughter isotopes 12.20
electrostatic precipitator 12.10
greenhouse effect 12.22
London smog 12.8
Los Angeles smog 12.11
nitrogen cycle 12.2
oxygen cycle 12.3
ozone 12.14
ozone layer 12.1
particulate matter 12.9
photochemical smog 12.11
pollutant 12.7
pollution 12.5
radon 12.20
smog 12.8
stratosphere 12.1
troposphere 12.1
wet scrubber 12.10

▸ Review Questions

1. Which layer of the atmosphere lies nearest the Earth?

2. Which layer of the atmosphere contains the ozone layer?

3. List the three major components of dry air and give the approximate (nearest whole number) percentage by volume of each.
4. List the two most important variable components of the atmosphere and give the approximate concentration range of each.
5. What is meant by nitrogen fixation? Why is it important?
6. How has industrial fixation of nitrogen to make fertilizers affected the nitrogen cycle?
7. How is the oxygen supply of Earth's atmosphere replenished?
8. How does the ozone layer protect the inhabitants of Earth's surface?
9. What is an atmospheric (thermal) inversion?
10. How did the atmospheric inversion contribute to the air-pollution episode in Donora, Pennsylvania, in 1948?
11. Is all air pollution the result of human activity? Explain.
12. What is a pollutant?
13. What is smog?
14. What are the chemical components of London (industrial) smog?
15. What weather conditions characterize London smog?
16. What is particulate matter?
17. What is an aerosol?
18. What are clinkers?
19. What is fly ash?
20. Describe the synergistic action of sulfur dioxide and particulate matter.
21. What is an electrostatic precipitator? How does it work?
22. Describe how a bag filter removes particulate matter from stack gases.
23. How does a wet scrubber remove particulate matter? What is a major disadvantage of wet scrubbers?
24. Describe how a limestone scrubber removes sulfur dioxide from stack gases.
25. Give two uses for fly ash.
26. Give the equation for the reaction of sulfur dioxide with hydrogen sulfide to form elemental sulfur.
27. What are the chemical components of photochemical smog?
28. What weather conditions characterize photochemical smog?
29. What is the main source of carbon monoxide in polluted air?
30. What are the health effects of carbon monoxide?
31. How might exposure to carbon monoxide contribute to heart disease?
32. What are the health effects of ozone in polluted air?
33. List two uses of chlorofluorocarbons.
34. How might chlorofluorocarbons be involved in the depletion of the ozone layer?
35. What health effect might result from depletion of the ozone layer?
36. How might nuclear war bring about an "ultra-violet summer"?
37. What is the main source of hydrocarbons in polluted air?
38. What is PAN? From what is it formed? What are its health effects?
39. What are catalytic converters? How do they work?
40. Give two ways the level of nitrogen oxide emissions from an automobile can be reduced.
41. What is acid rain?
42. How is acid rain formed?
43. How might acid rain be alleviated?
44. What is tetraethyllead? Why was it used in gasoline?
45. Why has leaded gasoline been phased out in the United States?
46. List several indoor air pollutants. What is the source of each?
47. List some risks associated with cigarette smoking.
48. What is the greenhouse effect?
49. Why is zero pollution not possible?
50. Is the electric car a solution to air pollution on a nationwide basis? Within a given city?
51. Should students be allowed to smoke in school buildings? On school grounds? Defend your position.
52. Should smoking be allowed in restaurants? In meeting rooms? On buses, trains, and airplanes?

Problems

Chemical Equations

53. Give the equation for the conversion of oxygen to ozone in the ozone layer.

54. Give the equations by which lightning fixes nitrogen.

55. Give the equation for the oxidation of sulfur to sulfur dioxide.

56. Give the equation for oxidation of sulfur dioxide to sulfur trioxide.

57. Give the equation for the reaction of sulfur trioxide with water to form sulfuric acid.

58. What compound is produced when sulfuric acid reacts with ammonia? Give the equation for the reaction.

59. Under what conditions do nitrogen and oxygen combine? Give the equation.

60. Give the equation for the reaction of nitric oxide with oxygen to form nitrogen dioxide.

61. What happens to nitrogen dioxide in sunlight? Give the equation.

62. Give the equation by which carbon monoxide reduces nitric oxide to nitrogen gas.

63. What is the effect of acid rain on iron? Give an equation.

64. What is the effect of acid rain on marble? Give the equation.

Additional Problems

65. The average person breathes about 20 m^3 of air a day. What weight of particulates would a person breathe in a day if the particulate level were 400 $\mu g/m^3$.

66. The atmosphere contains 5.2×10^{15} t of air. How much carbon dioxide (CO_2) is in the atmosphere if the concentration is 350 ppm?

67. The world's termite population is estimated to be 2.4×10^{17}. These termites produce an estimated 4.6×10^{16} g of carbon dioxide, 1.5×10^{14} g of methane, and 2×10^{14} g of hydrogen. How much of each gas is produced by each termite? What percent increase in the total amount of carbon dioxide in the atmosphere (see Problem 66) would the termites cause in 1 year?

Projects

68. What are the average and peak concentrations of each of the following in your community? How can you find out?
a. carbon monoxide
b. ozone
c. nitrogen oxides
d. sulfur dioxide
e. particulate matter

69. Does your community have an air-pollution problem? If so, describe it. How could it be solved?

References and Readings

1. Bernarde, Melvin A. *Our Precarious Habitat: Fifteen Years Later*. New York: Wiley, 1989. Chapter 10 discusses air pollution.
2. Brimblecombe, Peter. *Air: Composition and Chemistry*. Cambridge: Cambridge University Press, 1986.
3. Christopherson, Robert W. *Geosystems: An Introduction to Physical Geography*, 2nd ed. New York: Macmillan, 1994.
4. Graedel, Thomas E., and Paul J. Crutzen. "The Changing Atmosphere." *Scientific American*, September 1989, pp. 58–68.

5. ''How Tobacco Smoke Kills Nonsmokers.'' *Smoking and Health Review*, July 1989, pp 3–8. Includes extensive references to original studies.

6. Legge, Alan H. *Acidic Deposition: Sulfur and Nitrogen Oxides*. Chelsea, MI: Lewis Publishers, 1989.

7. Lyman, Francesca. *The Greenhouse Trap*. Boston: Beacon Press, 1990.

8. Manahan, Stanley E. *Environmental Chemistry*, 5th ed. Chelsea, MI: Lewis Publishers, 1991.

9. Matthews, Samuel W., and James A. Sugar. ''Under the Sun: Is Our World Warming?'' *National Geographic*, October 1990, pp. 66–99. Contains excellent photographs and graphics.

10. McLoughlin, Merrill. ''Our Dirty Air.'' *U.S. News and World Report*, 12 June 1989, pp. 48–54. Presents a good overview of our air pollution problems.

11. Monastersky, R. ''Antarctic Ozone Bottoms at Record Low.'' *Science News*, 13 October 1990, p. 228.

12. Moran, Joseph M., and Michael D. Morgan. *Meteorology—The Atmosphere and the Science of Weather*, 4th ed. New York: Macmillan, 1994.

13. Office of Public Awareness. ''Air Pollution and Your Health.'' Washington, DC: U.S. Environmental Protection Agency, June 1979.

14. Rahn, Kenneth A. ''Who's Polluting the Arctic?'' *Natural History*, May 1984, pp. 30–38.

15. Spengler, John D., and Ken Sexton. ''Indoor Air Pollution: A Public Health Perspective.'' *Science*, 1 July 1983, pp. 9–16.

16. Trefl, James. ''Modeling Earth's Future Climate Requires Both Science and Guesswork.'' *Smithsonian*, December 1990, pp. 28–37.

17. Wallach, Paul. ''Murky Water.'' *Scientific American*, May 1990, pp. 25–26. Role of oceans in removing CO_2 is uncertain.

18. Washington, Warren M. ''Where's the Heat?'' *Natural History*, March 1990, pp. 67–72. The greenhouse gases have increased, but the effect on Earth's temperature is less certain.

19. White, Robert M. ''The Great Climate Debate.'' *Scientific American*, July 1990, pp. 36–43. Discusses uncertainties in computer modeling of future climate changes.

13 Water

To Drink and to Dump Our Wastes In

You drink it. You pour it on flowers you've planted.

You wash your clothes in it. You take it for granted.

"Earth has so much *water!" you probably think;*

But most of that water is not fit to drink!

Earth is a watery planet, with most of its surface covered with oceans and seas. People who live on this planet contain a lot of water, too. Did you know that about two-thirds of your body weight is water? The water in your blood is quite similar to water in the ocean. In fact, you might say that we are walking sacks of sea water!

The presence of large amounts of water makes our planet unique in the solar system, probably the only one capable of supporting life. The nature of water makes it essential to life; and the nature of life makes it dependent on water. If we should someday discover life outside our solar system, it will likely be on another watery planet similar to this one.

Those properties of water that make it able to support life, however, also make it easy to pollute. Many chemical substances are soluble in water. They are easily dispersed and eventually are scattered to nearly infinite dilution in the ocean. Removing these chemicals from our water supplies, once they are there, is not easy. In this chapter we discuss some of the unusual properties of water, as well as water pollution and water treatment. Don't underestimate the importance of good, clean drinking water. We could live for several weeks without food, but without water we would last only a few days.

SECTION 13.1

Water: Some Unusual Properties

Water is a very familiar substance. Even so, it is a most unusual compound. It is the only common liquid on the surface of our planet. The solid form of water (ice) is less dense than the liquid because water expands when it freezes. The consequences of this peculiar characteristic are immense for life on this planet. Ice forms on the surface of lakes and insulates the lower layers of water. This enables fish and other aquatic organisms to survive winters in the temperate zones. If ice were more dense than liquid water, it would sink to the

Water is the only substance on Earth to exist in large amounts in all three physical states: solid (ice caps); liquid (oceans); and gas (steam from geysers).

This sparkling dewdrop might remind us of how very precious water is. Three-fourths of Earth's surface is covered with water, but so little of it is fresh and clean. [*S. J. Krasemann/Photo Researchers, Inc.*]

Figure 13.1 Oil coats the water of Prince William Sound, Alaska, where the *Exxon Valdez* ran aground on Bligh Reef, 24 March 1989, spilling 42 million L of oil. One liter of oil can create a slick 2.5 hectares in size (1 hectare = 2.47 acres). [*Michael Baytoff/ Black Star.*]

bottom as it formed, and even the deeper lakes of the northern United States would freeze solid in winter.

The same property—ice being less dense than liquid water—has dangerous consequences for living cells. When living tissues freeze, ice crystals are formed, and the expansion ruptures and kills cells. The slower the cooling, the larger the crystals of ice and the more damage there is to the cell. Frozen-food manufacturers take advantage of the properties of water when they freeze foods so rapidly that the ice crystals are kept very small (and thus do minimum damage to the cellular structure of the food); the industry calls such foods "flash frozen."

Water also has a higher density than most other familiar liquids. As a consequence, liquids that are less dense than water and insoluble in water float on its surface. A familiar problem in recent years has been the gigantic oil spills that occur when a tanker ruptures or when an offshore well gets out of control. The oil, floating on the surface of the water, often is washed onto beaches, where it does considerable ecological and aesthetic damage (Figure 13.1). If water were less dense than oil, the problem would certainly be different, though not necessarily less serious.

Another unusual property of water is its high heat capacity. The **heat capacity** of a substance is the quantity of heat that must flow into it in order to raise its temperature by 1 °C. (**Specific heat** is the amount of heat required to raise the temperature of 1 gram of a substance by 1 °C.) Table 13.1 gives the specific heat of a number of familiar substances. Note that it takes almost 10 times as much heat to raise the temperature of 1 g of water by 1 °C as to raise the temperature of 1 g of iron by the same amount. Conversely, much heat is given off by water for even a small drop in temperature. The vast amounts of water on the surface of the Earth thus act as a giant thermostat to moderate

Table 13.1 Specific Heats of Some Familiar Substances

Substance	Specific Heat $\left(\frac{\text{cal}}{\text{g °C}}\right)$
Aluminum	0.22
Carbon (graphite)	0.17
Copper	0.092
Ethanol	0.59
Diethyl ether	0.54
Iron	0.11
Lead	0.031
Water (ice)	0.50
Water (liquid)	1.00

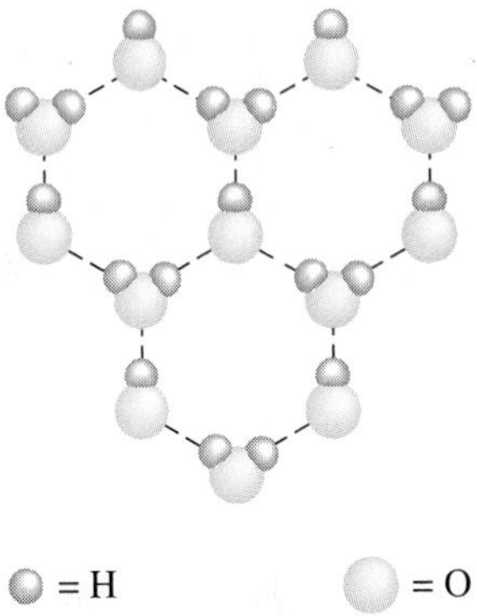

Figure 13.2 The structure of ice has large hexagonal holes. (In this drawing some of the hydrogen atoms are hidden behind the larger oxygen atoms.)

daily temperature variations. You need only consider the extreme temperature changes on the surface of the waterless moon to appreciate this important property of water.

Still another way in which water is unique is that it has a high **heat of vaporization**; that is, a large amount of heat is required to evaporate a small amount of water. This is of enormous importance to us, for large amounts of body heat can be dissipated by the evaporation of small amounts of water (perspiration) from the skin. This effect also accounts for the climate-modifying property of lakes and oceans. A large portion of the heat that would otherwise heat up the land is used to vaporize water from the surface of the lake or the sea. Thus, in summer it is cooler near a large body of water than in interior land areas.

All these fascinating properties of water depend on the unique structure of the water molecule. Recall (Chapter 5) that the water molecule is highly polar

Figure 13.3 The hexagonal symmetry of the ice crystals in these snowflakes reflects the arrangement of molecules in the crystal structure. [*American Museum of Natural History (#K5300).*]

and that in the liquid and solid states, water molecules are strongly associated through hydrogen bonds. In liquid water, the association of molecules is random, and the molecules are close together. When water freezes, its molecules take on a more ordered arrangement with large hexagonal holes (Figure 13.2). This three-dimensional structure extends out for billions and billions of molecules. The holes account for the fact that ice is less dense than liquid water. The hexagonal arrangement allows water to assume forms of exquisite beauty as snowflakes (Figure 13.3).

SECTION 13.2

Water, Water, Everywhere

About 10 million tons/year of salt is used during winter months in the northern part of the United States.

At the present rate of accumulation, the seas will become saturated with salt (about 36%) in another 3.5 billion years. If our descendants are still around in that distant age, they will find our oceans much like today's Dead Sea.

Three fourths of the surface of the Earth is covered with water (Figure 13.4). Nearly 98%, though, is seawater—unfit for drinking and not even suitable for most industrial purposes. Seawater is very salty. Because of its polar nature, water tends to dissolve ionic substances. (See Figure 5.17.) This solvent power of water accounts for the saltiness of the sea. Rainwater dissolves minerals (ionic in nature), and these are carried by streams and rivers to the sea. There the heat of the sun evaporates part of the water, leaving the "salts" behind. The oceans grow more salty as the years go by. (But this is a very slow process. The increase in salt content is not noticeable in one lifetime.)

Rain is the purest kind of natural water, and it falls upon the Earth in enormous amounts; however, most rainwater falls into the sea or on areas otherwise inaccessible. Something less than 2% of Earth's water is frozen in the polar ice caps, leaving less than 1% available as fresh water (Figure 13.5). The average use of water per person (including that used for industrial, agricultural, and other purposes) in the United States is 8 million L per year, and the rate of use is increasing rapidly. Moreover, the available water is not always where the people are. Quite often, too, freshwater supplies are so polluted that they are unfit for human use.

When you gaze out across an ocean, the world appears to be mainly water. In fact, most of Earth's surface *is* covered with water. Actually, we might look at the surface of the Earth as one single ocean with all the continents very large islands. On the other hand, only about 1/4000 of Earth's mass is water. If Earth were the size of a basketball and you could hold it in your hands, you would notice that it was wet on the surface, but the sphere would seem quite solid otherwise. When we discuss our water resources on this planet, it is like talking about the dampness on the surface of a wet basketball.

Figure 13.4 Three-fourths of the Earth's surface is covered with water, making the planet mostly blue when viewed from space. [*NASA photo.*]

Figure 13.5 Life becomes difficult without an adequate supply of water. This photograph was taken in drought-stricken Chad in 1984. [*Photo by John Isaac/ courtesy of UN Photos.*]

SECTION 13.3

The Water Cycle

The distribution of water between the sea, the ice caps, and freshwater rivers, lakes, and streams is fairly constant. There is, however, a dynamic cycle of water between these various components (Figure 13.6). Water is constantly evaporating from both water and land surfaces. This water vapor condenses into clouds and returns to Earth in the form of rain, sleet, and snow. This fresh water becomes part of the ice caps, runs off in streams and rivers, and fills lakes and underground reservoirs.

The cycling of water serves to replenish our supply of fresh water. When water evaporates from the sea, salts are left behind. When water moves through the ground, impurities are trapped in the rock, gravel, sand, and clay. This capacity to purify is not infinite, however.

According to the Biblical story of Moses, getting a dependable supply of fresh water has long been a problem (Figure 13.7). When Moses led the Israelites out of Egypt into the wilderness, he encountered a desert area where potable water was scarce. While nearly everyone knows the Biblical account of how Moses struck the rock to bring forth water, an incident at Marah is less well known. At Marah, the Israelites couldn't drink the water because it was bitter. According to the account in Exodus, God commanded Moses to throw a tree into the water to purify it.

Recall from Chapter 7 that one property of basic solutions is that they are bitter. The water at Marah was probably basic, or *alkaline*. Such waters are common in desert areas. Attempts have been made to give a chemical explanation of Moses's purification of the brackish water. The tree was probably a dead one, bleached by the desert sun. Such bleaching oxidizes the alcohol groups in cellulose to carboxylic acid groups.

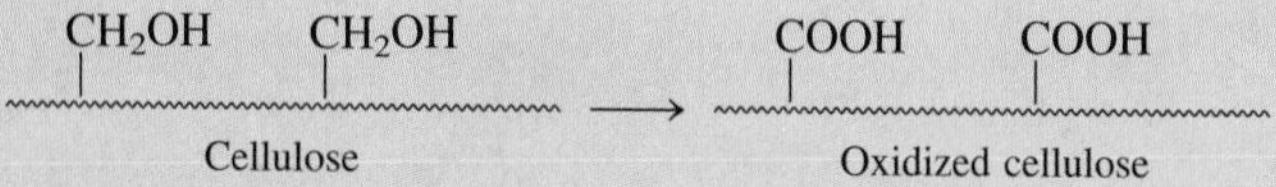

These acidic groups serve to neutralize the alkali in the water. Moses didn't have to understand the science of water purification in order to apply the appropriate technology.

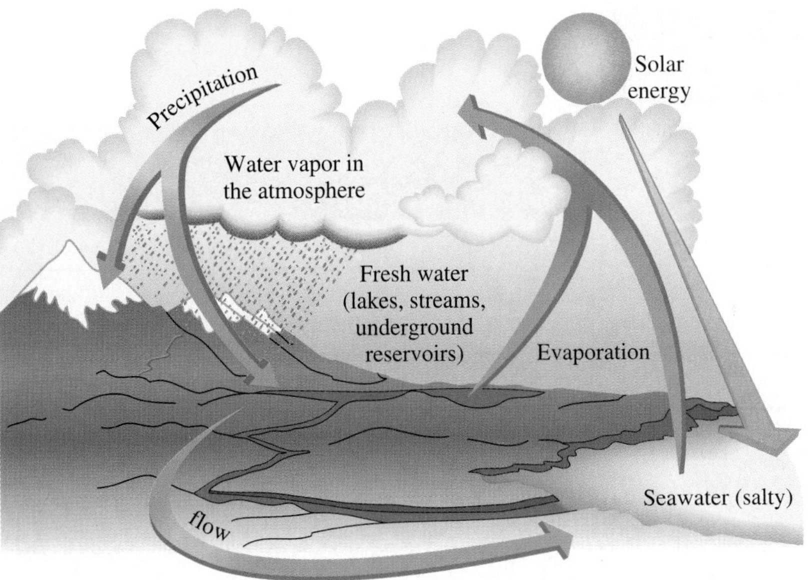

Figure 13.6 The water cycle.

Figure 13.7 An adequate supply of water is essential to life. This photograph shows a community water tap in Asmar, Ethiopia. [*Photo by John Isaac/courtesy of UN Photos.*]

SECTION 13.4

Biological Contamination: The Need for Clean Water

The pollutants in the waters of Marah were natural ones. Early people did little to pollute the water and the air, if only because their numbers were so few. It was with the coming of the agricultural revolution and the rise of cities that there were enough *Homo sapiens* to pollute the environment seriously. Even then, the pollution was mostly local and largely biological. Human wastes were dumped on the ground or into the nearest stream. Disease organisms were transmitted through food, water, and direct contact.

Contamination of water supplies by microorganisms from human wastes was a severe problem throughout the world until about a hundred years ago. During the 1830s, severe epidemics of cholera swept the Western World. Typhoid fever and dysentery were common. In 1900, for example, there were over 35,000 deaths from typhoid in the United States. Today, as a result of chemical treatment, municipal water supplies in the advanced nations are generally safe. However, waterborne diseases are still quite common in much of Asia, Africa, and Latin America. Indeed, it is estimated that 80% of all the world's sickness is caused by contaminated water. People with waterborne diseases fill half the world's hospital beds and die at a rate of 25,000 people a day. Less than 10% of the people of the world have access to sufficient clean

Table 13.2 Average Direct Daily Use of Water in the United States per Person

Use	Amount (L)
Drinking and cooking	7
Flushing toilets	80
Swimming pools and lawns	85
Dishwashing	14
Bathing	70
Laundry	35
Miscellaneous	9
Total	300

Table 13.3 Average Total Daily Use of Water in the United States per Person

Use	Amount (L)
Industrial	3800
Irrigation (agriculture)	2150
Municipal water (nonindustrial)	550
Total	6500

water. We still have epidemics of cholera, typhoid, and dysentery in many parts of the world.

How much water does one really need? Only about 1.5 L a day for drinking. In the United States each day, we use about 7 L per person for drinking and cooking; 120 L for cleanliness (bathing, dish washing, laundering, and housecleaning); 80 L for flushing the toilet; and 85 L for swimming pools and lawns (Table 13.2). We use much more *indirectly* in agriculture and industry to produce food and other materials; it takes 800 L of water to produce 1 kg of vegetables and 13,000 L of water to produce a steak (Table 13.3). We also use water for recreation (for example, swimming, boating, and fishing). For most of these purposes, we need water free from bacteria, viruses, and parasitic organisms.

Figure 13.8 Contaminated beaches are closed to swimming and other recreational activities. [*SYGMA*.]

The threat of biological contamination has not been totally eliminated from the developed nations. It is estimated that 30 million people in the United States are at risk because of bacterial contamination of drinking water. Hepatitis, a viral disease occasionally spread through drinking water, at times threatens to reach epidemic proportions, even in the most advanced nations. Biological contamination also lessens the recreational value of water. Swimming is forbidden in many areas (Figure 13.8).

Biological contamination may have reached a peak in the 1960s. Lake Erie was described as ''America's Dead Sea'' and ''the world's largest cesspool.'' Polluted waters were not unusual; unpolluted waters were. The St. Croix River (Figure 13.9), which forms part of the border between Minnesota and Wisconsin, was described as the *only* major unpolluted river near a sizable metropolitan area (Minneapolis–St. Paul).

Much improvement has been made in recent years. Fish thrive again in Lake Erie. Beaches in Lake Michigan near Chicago, closed in 1969 because of pollution, were reopened in 1975. The Mississippi below Minneapolis–St. Paul has been reopened to recreation and swimming after being closed for many years. Shellfish are being harvested again in Maine's Belfast Bay. Shrimp and oysters are returning to Escambia Bay off Pensacola, Florida.

Figure 13.9 The St. Croix River, as yet a relatively unpolluted river. [*Richard Frear/courtesy of National Park Service.*]

Atlantic salmon have returned to the Connecticut River for the first time in 100 years. And, in England, fish once again inhabit the Thames. Despite all the good news, however, pollution from domestic sewage remains the gravest threat to the water supply in the United States. We still have a long way to go.

SECTION 13.5

Chemical Contamination: From Farm, Factory, and Home

The industrial revolution added a new dimension to our water pollution problems. Factories often were built on the banks of streams, and wastes were dumped into the streams to be carried away. The rise of modern agriculture has led to increased contamination as fertilizers and pesticides have found their way into the water system. Transportation of petroleum results in oil spills in oceans, estuaries, and rivers. Acids enter waterways from mines and factories and from acid precipitation. Household chemicals also contribute to water pollution when detergents, solvents, and other chemicals are dumped down drains.

Before we consider water pollution, let's first look at water as it occurs in nature.

Table 13.4 Some Substances Found in Natural Waters

Substance	Formula	Source
Carbon dioxide	CO_2	Atmosphere
Dust	—	Atmosphere
Nitrogen	N_2	Atmosphere
Oxygen	O_2	Atmosphere
Nitric acid (thunderstorms)	HNO_3	Atmosphere
Sand and soil particles	—	Soil and rocks
Sodium ions	Na^+	Soil and rocks
Potassium ions	K^+	Soil and rocks
Calcium ions	Ca^{2+}	Limestone rocks
Magnesium ions	Mg^{2+}	Dolomite rocks
Iron(II) ions	Fe^{2+}	Soil and rocks
Chloride ions	Cl^-	Soil and rocks
Sulfate ions	SO_4^{2-}	Soil and rocks
Bicarbonate ions	HCO_3^-	Soil and rocks

SECTION 13.6

Natural Water Isn't All H_2O

Rainwater carries dust particles from the atmosphere to the ground. Rainwater also dissolves a little oxygen, nitrogen, and carbon dioxide as it falls through the atmosphere. During electrical storms, lightning causes nitrogen, oxygen, and water vapor to combine to form nitric acid. Traces of this, too, are found in rainwater.

As water moves along or beneath the surface of the Earth, it dissolves minerals from rocks and soil. It also dissolves matter from decaying plants and animals. Recall that minerals (salts) are ionic, and that ions have either positive or negative charges. The principal positive ions (cations) in natural water are sodium (Na^+), potassium (K^+), calcium (Ca^{2+}), magnesium (Mg^{2+}), and sometimes iron (Fe^{2+} or Fe^{3+}). The negative ions (anions) are usually sulfate (SO_4^{2-}), bicarbonate (HCO_3^-), and chloride (Cl^-). Table 13.4 provides a summary of substances found in natural water.

Soft water may contain ions, such as Na^+ or K^+, but they do not form insoluble scum with soap.

Water containing calcium, magnesium, or iron salts is called **hard water**. The positive ions react with the negative ions in soap to form a scum that clings to clothes and leaves them dingy looking. The same scum is responsible for the familiar bathtub ring.

SECTION 13.7

Sewage: Some Chemistry and Biology

Pathogenic (disease-causing) *microorganisms* are not the only problem caused by dumping human sewage into our waterways. The breakdown of

Figure 13.10 Common sewage from homes and businesses depletes the dissolved oxygen in water. [*Larry Lee/Westlight.*]

organic matter by bacteria depletes **dissolved oxygen** in the water and enriches the water with plant nutrients. A stream can handle a small amount of waste without difficulty, but when massive amounts of raw sewage are dumped into a waterway, undesirable changes occur (Figure 13.10).

Most organic material can be broken down (degraded) by microorganisms. This biodegradation can be either **aerobic** or **anaerobic**. Aerobic oxidation occurs in the presence of dissolved oxygen. A measure of the amount of oxygen needed for this degradation is the **biochemical oxygen demand** (**BOD**). The greater the quantity of degradable organic wastes, the higher the BOD. If the BOD is high enough, oxygen is depleted and no life (other than odor-producing anaerobic microorganisms) can survive in the lake or the stream. Flowing streams can regenerate themselves. Rapid ones soon come alive again as oxygen is dissolved by the moving water. Lakes with little or no flow can remain dead for decades.

With adequate dissolved oxygen, aerobic bacteria (those that require oxygen) oxidize the organic matter to carbon dioxide, water, and a variety of inorganic ions (Table 13.5). The water is relatively clean, but the ions, particularly the nitrates and phosphates, may serve as nutrients for the growth of algae, which also cause problems. When the algae die, they become organic waste and increase the BOD. This process is called **eutrophication** (Figure 13.11). Algal bloom and die-off are stimulated by the runoff of agricultural fertilizers (Chapter 16). This combination leads to dead and dying streams and lakes that nature cannot purify nearly as quickly as we can pollute.

When the dissolved oxygen in a body of water is depleted by too much organic matter—whether from sewage, dying algae, or other sources—anaerobic bacteria thrive. Instead of oxidizing the organic matter, they reduce it. Methane (CH_4) is formed. Sulfur is converted to hydrogen sulfide (H_2S)

Table 13.5 Some Substances Added to Water by the Breakdown of Organic Matter

Substance	Formula
Aerobic conditions	
Carbon dioxide	CO_2
Nitrate ions	NO_3^-
Phosphate ions	PO_4^{3-}
Sulfate ions	SO_4^{2-}
Bicarbonate ions	HCO_3^-
Anaerobic conditions	
Methane	CH_4
Ammonia	NH_3
Amines	*
Hydrogen sulfide	H_2S
Methanethiol	CH_3SH

*See Chapter 9.

and other foul-smelling organic compounds. Nitrogen is reduced to ammonia and odorous amines (Chapter 9). The foul odors are a good indication that the water is overloaded with organic wastes. No life, other than the anaerobic microorganisms, can survive in such water.

Figure 13.11 The eutrophication of a lake is a natural process, but the effect can be accelerated greatly by human effect such as agricultural runoff of fertilizers. [*Tim McCage/courtesy of Soil Conservation Service.*]

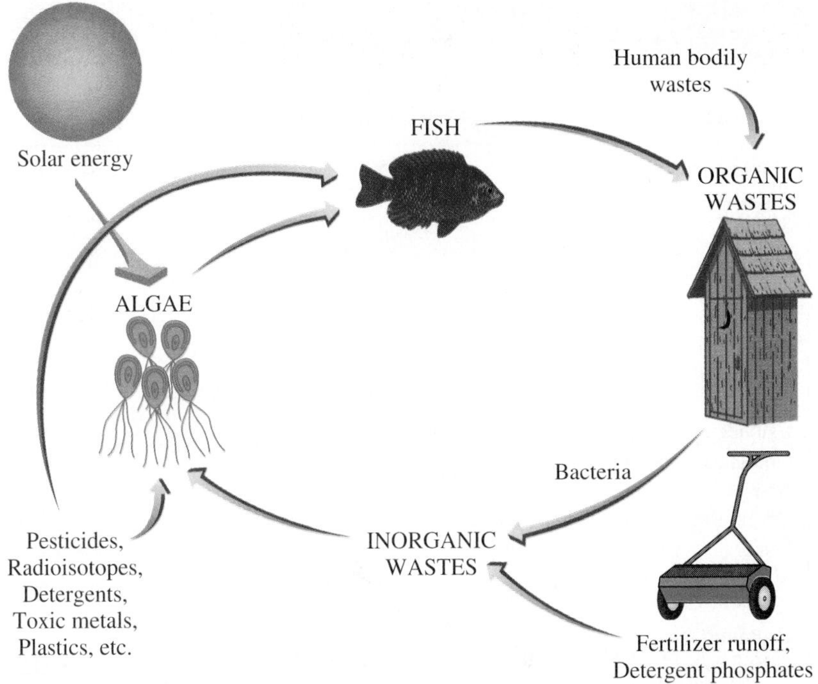

Figure 13.12 A simple ecological cycle and some ways humans disrupt it.

SECTION 13.8

Ecological Cycles

Let's consider water pollution from the point of view of a fish. A simplified ecological cycle (Figure 13.12) could take place in a small lake. Fish in the water produce organic wastes. Bacteria break down these wastes into organic materials that serve as nutrients for the growth of algae. Fish eat the algae, balance is established, and the cycle is complete.

Now let's look at some of the ways people can disrupt the cycle. We can increase the organic wastes by dumping sewage into the water. In breaking down these wastes, bacteria consume all the dissolved oxygen and the fish die. Chemists can monitor the BOD of the water, thus identifying the problem and gauging its severity. Chemists also can contribute to the development of better methods of wastewater treatment.

Dumping our sewage into the water isn't the only way we can foul up an ecological cycle. Fertilizer runoff and seepage from feedlots add inorganic nutrients to the cycle, and an algal bloom can lead to oxygen depletion and death for the fish. Perhaps the most enigmatic influence of all comes through the introduction of new substances into the ecological water cycle: pesticides, radioisotopes, detergents, toxic metals, and plastics.

Table 13.6 Maximum Levels of Substances Allowed in Drinking Water by the U.S. Public Health Service

Substance	Maximum Concentration (mg/L)*	
Total dissolved solids	500	Less toxic
Chloride	250	
Sulfate	250	
Nitrate	45	
Zinc	5	
Fluoride	2	
Barium	1	
Copper	1	
Iron	0.3	
Cyanide	0.2	
Organic compounds	0.2	
Arsenic	0.05	
Chromium	0.05	
Lead	0.05	
Manganese	0.05	
Silver	0.05	
Cadmium	0.01	
Selenium	0.01	More toxic

*1 mg/L often is called 1 part per million (1 ppm).

By creating new materials, chemists have brought on a number of ecological problems. But chemists and chemistry are also necessary to any understanding of our pollution problems. It is only through the development of sophisticated analytical methods and instruments that we are even aware of some of our ecological problems. A stinking lake would be obvious to everyone, but only a person trained in chemistry could determine the level of dangerous, yet invisible, materials in the water we drink (Table 13.6).

SECTION 13.9

Groundwater Contamination: Tainted Tap Water

Groundwater is the main source of drinking water in 32 of the 50 United States, and 34 of our largest cities rely on it.

Among our rural populations, 97% drink groundwater.

About half the people in the United States drink surface water (from streams and lakes). The other half get their drinking water from **groundwater**. Toxic chemicals have been found in the groundwater in many areas. People living around the Rocky Mountain Arsenal, near Denver, have found their wells contaminated by wastes from the production of pesticides. Wells near Minneapolis, Minnesota, are contaminated with creosote, a chemical used as a wood preservative. Wells in Wisconsin and on Long Island have been contaminated with aldicarb, a pesticide used on potato crops. Community water supplies in New Jersey have been shut down because of contamination with industrial wastes.

Collectively, these compounds are called *volatile organic chemicals (VOC)*. They evaporate readily and add an undesirable odor to water. VOCs are used in homes and factories as solvents, cleaners, and fuels. They are components of gasoline, spot removers, oil-based paints, thinners, and some drain cleaners. When spilled or discarded, they enter the soil and eventually get into the groundwater.

Volatile Organic Chemicals in Groundwater

Chemicals buried in dumps—often years ago, before there was much awareness of environmental problems—have now infiltrated groundwater supplies. Often, as at the Love Canal site in Niagara Falls, New York, schools and houses were built on or near old dump sites (Figure 13.13). Common contaminants are hydrocarbon solvents, such as benzene and toluene, and chlorinated hydrocarbons, such as carbon tetrachloride, chloroform, and methylene chloride (Chapter 9). Especially common is trichloroethylene ($CCl_2{=}CHCl$), widely used as a dry-cleaning solvent and as a degreasing compound. Most of these compounds are suspected carcinogens. These organic compounds do not dissolve in water. However, they do have a slight solubility, often in the parts per million or parts per billion range. It is these trace amounts that are found in groundwater. These compounds are also noted for their lack of reactivity. They react so slowly that they are likely to be around for a long time.

Figure 13.13 Homes and a school were built upon the site of an old chemical dump at the Love Canal section of Niagara Falls, New York. Now a contaminated area near the old dump is fenced to keep people out of the area. [*AP/Wide World Photos.*]

Leaking Underground Storage Tanks: LUST

Another major source of groundwater contamination is LUST—an attention-grabbing acronym for leaking underground storage tanks. Gasoline at service stations is usually stored in buried steel tanks. There are perhaps 2.5 million such tanks in the United States. The tanks last an average of about 15 years before they rust through and begin to leak. As many as 200,000 of these tanks may be leaking. Many of them are at stations that went out of business during the fuel shortages of the 1970s. Gasoline is being discovered in wells near these tanks in many areas of the country, but the full extent of the problem is not yet known. There are now laws regarding replacement of old gasoline tanks and proper cleanup of any contaminated ground.

Trace Toxics and Public Perception

Groundwater contamination is particularly alarming because, once contaminated, an underground aquifer may remain unusable for decades or longer. There is no easy way to remove the contaminants. Pumping out the water and purifying it could take years and cost billions of dollars.

The problem of groundwater pollution is serious, but it is often overdramatized by the news media. The amounts of contaminants are often minute. Chemists now can measure parts per million (ppm) and parts per billion (ppb) on a routine basis. Some contaminants can even be detected at parts per trillion (ppt) or lesser amounts. Without sophisticated chemical techniques, we wouldn't even know about some problems. For example, the U.S. Environmental Protection Agency has set a safety limit of 10 ppb for aldicarb—that's 10 mg of aldicarb in 1000 L of water. You would have to drink 32,000 L of water to get as much aldicarb as there is aspirin in one tablet! Contamination of groundwater by toxic substances is serious and should not be taken lightly, but we must keep in mind the fantastically small amounts of toxic materials usually involved.

Toxic substances are discussed further in Chapter 20.

SECTION 13.10

Acid Waters: Dead Lakes

Acids pour down upon us from the heavens as **acid rain**, fog, and snow. They corrode metals, limestone, and marble, and even ruin the finishes on our automobiles. Acids also flow into streams from abandoned mines (Figure 13.14).

Acid water is detrimental to life in lakes and streams. Over 1000 bodies of water in the eastern United States are already acidified, and 11,000 others have only a limited ability to neutralize the acids that enter them. Already 140 lakes in Ontario are devoid of any life. Another 48,000 lakes in Ontario alone are threatened. The threat to these areas is presumed to be acid rain originating mainly in the Ohio Valley and Great Lakes regions. The sulfur oxides and nitrogen oxides from power plants, industries, and automobiles travel hundreds of kilometers downwind and fall as sulfuric and nitric acids. Acid rain has been linked to declining crop and forest yields.

The effects of acid waters on living organisms are hard to pin down pre-

Figure 13.14 Acidic water draining from an abandoned mine. [*Kent and Donna Dannen/Photo Researchers, Inc.*]

cisely. Probably the greatest effect of acidity is that it causes the release of toxic ions from rocks and soil. For example, aluminum ions, which are tightly bound in clays and other minerals, are released by acid. Aluminum ions have low toxicity to humans, but they seem to be deadly to young or small fish. Many of the dying lakes have only old fish; none of the young survive. Ironically, lakes destroyed by excess acidity are often quite beautiful. The water is clear and sparkling—quite a contrast to those in which fish are killed by algal blooms.

Acids are no threat to lakes and streams in areas where the rock is limestone (calcium carbonate), which can neutralize excess acid.

$$\underset{\text{Limestone}}{CaCO_3} + \underset{\text{Acid}}{2\,H_3O^+} \longrightarrow Ca^{2+} + CO_2 + 3\,H_2O$$

Where rock is principally granite, however, no such neutralization occurs.

Acidic waters can be treated with pulverized limestone (or other basic substances). A few such attempts have been carried out, but the process is costly and the results last only a few years. An obvious way to alleviate the problem is to remove the sulfur from the coal before combustion or to scrub the sulfur oxides from the smokestack gases. But all such solutions are expensive, and they add to the cost of electricity.

SECTION 13.11

Industrial Water Pollution

Industries in the United States have eliminated a considerable proportion of the water pollution they once perpetrated. Most are in compliance with the Water Pollution Control Act, which requires that they use the best practicable technology. We examine here only a few examples of industrial pollution and some ways to alleviate it.

Automobiles and Water Pollution

It takes several hundred kilograms of steel to make an automobile. To make a metric ton of steel requires about 100 t of water. About 4 t of water are lost through evaporation. The remainder is contaminated with acids, grease and oil, lime, and iron salts. This polluted water can be cleaned up, and most of it is recycled.

Chrome plating on bumpers, grills, and ornaments is also a source of pollution. Waste chromium [in the form of chromate ions (CrO_4^{2-})] and cyanide ions (CN^-) are products of this process. In the past, these toxic substances were dumped into waterways. Nowadays, they generally are removed to a large extent by chemical treatment.

Table 13.7 Water Required to Produce Various Materials	
Material	Water Required*
Steel	100
Paper	20
Copper	400
Rayon	800
Aluminum	1280
Synthetic rubber	2400

*In cubic meters per metric ton (t). A cubic meter of water weighs 1000 kg, or 1 t.

Cyanide is treated with chlorine to form nitrogen gas, bicarbonate ions, and chloride ions.

$$10\,OH^- + 2\,CN^- + 5\,Cl_2 \longrightarrow N_2 + 2\,HCO_3^- + 10\,Cl^- + 4\,H_2O$$

The products of this treatment, bicarbonate and chloride, are still pollutants, but they are much less toxic than cyanide. Chromate is removed by reduction with sulfur dioxide. The chromium winds up as Cr^{3+} ion, the sulfur as sulfate.

$$2\,CrO_4^{2-} + 3\,SO_2 + 2\,H_2O \longrightarrow 2\,Cr^{3+} + 3\,SO_4^{2-} + 4\,OH^-$$

Sulfate is still a pollutant, but it is generally not a serious one. Cr^{3+} is relatively insoluble in alkaline solution but is soluble enough in acidic media to constitute a problem.

Now, add in the cost (environmental and economic) of elastomers for tires, fabrics for upholstery, glass for windows, 520 kWh of electricity—the list goes on and on—and it is easy to see that the private automobile is an ecological problem even before it hits the road. And once on the road, it is a major contributor to air pollution (Chapter 12).

Most other industries also contribute to water pollution. Table 13.7 lists the water required (per metric ton) for the production of a variety of materials. Most of this water is cleaned up and recycled, but the need for clean water still is enormous.

Water Pollutants from Other Industries

Wastes from the textile industries include conditioners, dyes, bleaches, oils, dirt, and other organic debris. Most of these can be removed by conventional sewage treatment. Wastes from meatpacking plants include blood and various animal parts. Other plants generate fruit and vegetable waste. Food industry wastes also are usually treated by regular sewage plants.

Oil refineries produce wastes that include dyes, oils, acids, brines, and sulfur compounds. Chemical plants produce a variety of waste materials. Most industries have made substantial reductions in the amount of wastes produced and are committed to further reductions.

Figure 13.15 So many dams and factories have been built along the Volga River over past decades that it is no longer the mighty, vigorous waterway that it once was. [*Wolfgang Kaehler.*]

One of the most famous rivers in the world is the mighty Volga in Russia. It must be heartbreaking for the Russian people to see what has happened to it. The secrecy that prevailed in the Soviet Union for such a long time covered up the environmental problems of eastern Europe. Even the nuclear accident at Chernobyl was not admitted until it could not be hidden any longer. We now know that there were more than 130 nuclear explosions carried out for the purpose of moving earth in order to build dams and other structures. So many dams have been built along the Volga that its flow has been slowed to a crawl. It used to take 50 days for water to travel the 2300 miles from the source to the mouth of the river. Now it takes 1 1/2 years. The sluggish flow plus the pollution from all the power plants and factories along the way have created an ecological catastrophe. There are tons of industrial waste (cleaning fluids, fertilizers, pesticides, heavy metals, toxic chemicals, radioactive waste, and waste from pulp and paper mills) pouring into the Caspian Sea, which appears to be dying. In some places you can see yellow, red, and black streams of water carrying sulfur, iron oxide, and various oils. The Volga once teemed with caviar-producing sturgeon, but now it is so polluted that the sturgeon are threatened with extinction.

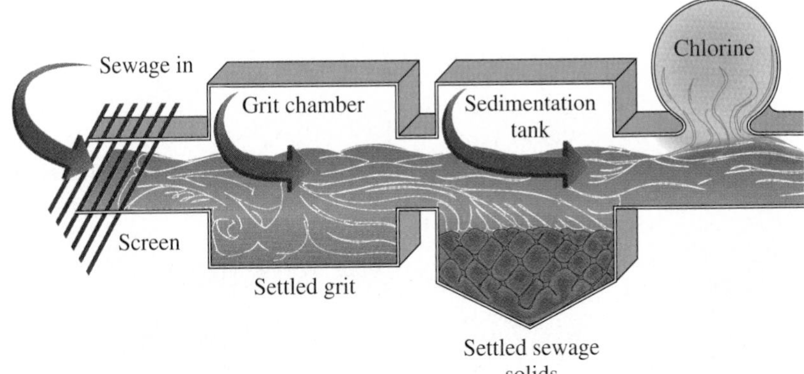

Figure 13.16 A diagram of a primary sewage treatment plant.

SECTION 13.12

Wastewater Treatment Plants

For many years most communities simply held sewage in settling ponds for a while before discharging it into a stream, lake, or ocean. This form of facility (Figure 13.16) provides what now is called **primary sewage treatment**. Primary treatment removes some of the solids as *sludge*. The effluent has a huge BOD. Often all the dissolved oxygen in the pond is used up, and anaerobic decomposition—with its resulting odors—takes over. Effluent from a primary treatment plant contains a lot of dissolved and suspended organic matter. A **secondary sewage treatment** plant passes effluent from the primary treatment through sand and gravel filters. There is some aeration in this step, and aerobic bacteria convert most of the organic matter to stable inorganic materials.

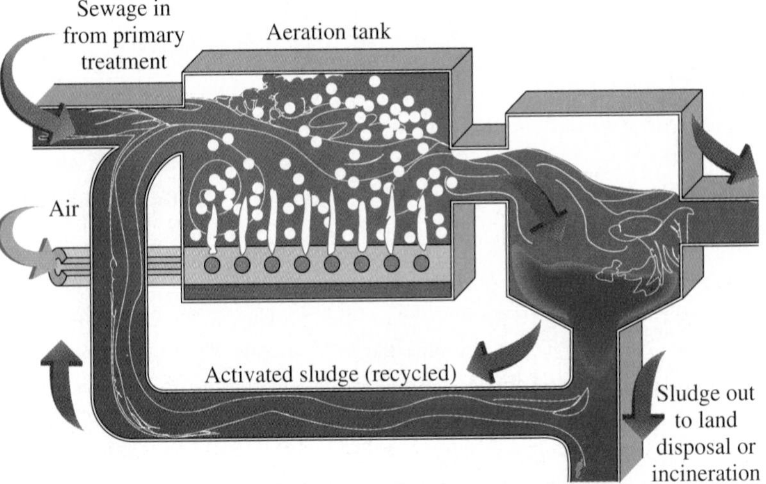

Figure 13.17 A diagram of a secondary wastewater treatment plant that uses the activated sludge method.

A combination of primary and secondary treatment methods, known as the **activated sludge method** (Figure 13.17), is frequently employed. The sewage is placed in tanks and aerated with large blowers. This causes the formation of large, porous clumps called *flocs.* These serve to filter and absorb contaminants. The aerobic bacteria further convert the organic material to sludge. A part of the sludge is recycled to keep the process going, but huge quantities must be removed for disposal. This sludge is stored on land (where it requires large areas), dumped at sea (where it pollutes the ocean), or burned in incinerators (where it requires energy—such as natural gas—and contributes to air pollution). Sometimes the sludge is used as fertilizer.

Chlorination is widely used to treat drinking water. **Ozone** also has been used. Over 1000 cities, mostly in Europe, use ozone to treat their drinking water. Ozone is more expensive than chlorine, but less of it is needed. An added advantage is that ozone kills viruses on which chlorine has little, if any, effect. Tests have shown ozone to be 100 times as effective as chlorine for killing polio viruses.

Ozone (O_3) acts by transferring its "extra" oxygen to the contaminant. The oxidized contaminants are generally less toxic than the chlorinated ones. In addition, ozone imparts no chemical taste to the water. Ozone does have the disadvantage that it provides no residual protection against microorganisms. If the problems with chlorine get worse, though, we may see a shift from chlorine to ozone in the treatment of our drinking water and wastewater. Perhaps ozone could be used to disinfect the water, then just enough chlorine added to provide residual protection.

In many areas, secondary treatment of wastewater is inadequate. **Advanced treatment** (sometimes called **tertiary treatment**) has become increasingly important. A number of advanced processes are in use. One process, charcoal filtration, promises to become increasingly important. Charcoal absorbs organic molecules that are difficult to remove by any other method. There are federal mandates for more communities to treat their water with charcoal, but many cities are not yet complying because of the expense.

Charcoal filtration is especially effective for removing organic compounds from water. The organic molecules are *adsorbed* on the surface of the charcoal and thus removed from the water. After a period of time, the charcoal becomes saturated and is no longer effective. It can be regenerated by heating to drive off the adsorbed substances.

Even more costly processes, such as **reverse osmosis** (pressure filtration through a semipermeable membrane), may be needed in some cases. Finding the money to finance adequate sewage treatment will be a major political problem for years to come. (See Table 13.8 for a summary of wastewater treatment methods.)

The effluent from sewage plants usually is treated with chlorine before it is returned to a waterway. Chlorine is added in an attempt to kill any pathogenic microorganisms that might remain. Such treatment has been quite effective in preventing the spread of waterborne infectious diseases such as typhoid fever. Further, some chlorine remains in the water, providing residual protection against pathogenic bacteria. However, chlorination is not effective against viruses, such as those that cause hepatitis.

According to the World Health Organization, about 40,000 children around the world die each day for want of pure water.

Use of chlorine is questioned by some people. Chlorination converts dissolved organic compounds into chlorinated hydrocarbons. Many chlorinated hydrocarbons, including such known carcinogens as chloroform and carbon tetrachloride, have been found in the drinking water of several cities that take

Table 13.8 Summary of Wastewater Treatment Methods

Method	Cost	Material Removed	Percent Removal
Primary			
Sedimentation	Low	Dissolved organics	25–40
		Suspended solids	40–70
Secondary			
Trickling filters	Moderate	Dissolved organics	80–95
		Suspended solids	70–92
Activated sludge	Moderate	Dissolved organics	85–95
		Suspended solids	85–95
Advanced (tertiary)			
Carbon bed with regeneration	Moderate	Dissolved organics	90–98
Ion exchange	High	Nitrates and phosphates	80–92
Chemical precipitation	Moderate	Phosphates	88–95
Filtration	Low	Suspended solids	50–90
Reverse osmosis	Very high	Dissolved solids	65–95
Electrodialysis	Very high	Dissolved solids	10–40
Distillation	Extremely high	Dissolved solids	90–98

*See the references for discussion of methods not covered in the text.

their water from rivers. The concentration is in the parts-per-billion range. The threat they pose is probably small, but it's worrisome, nonetheless. (It is not nearly so worrisome, however, as the waterborne diseases that prevail in much of the world where adequate water treatment is not available.)

SECTION 13.13

Back to the Soil: An Alternative Solution

A few communities use plants to treat their sewage. The sewage is pumped into a marshy area where plants use nutrients from the sewage as fertilizer. Properly selected, the plants even remove toxic metals. Effluent from the marsh often is as clean as that in reservoirs (Section 13.14).

Why do we dump our wastes in the water in the first place? Dumping fouls the water with wastes that contain nutrients that could fertilize the land. Why not return the nutrients to the soil? In many societies both human and animal wastes always have been used as fertilizer. It is only in the "overdeveloped" countries that these vital resources are dumped into the waterways on a large scale.

Several communities already return sludge to the soil. The sludge can be dried, sterilized, and transported to farmlands. Wet, suspended sludge also can be pumped directly to the fields. The water in the mixture irrigates the crops; the sludge is a source of nutrients and humus. One concern is that the pathogens in the sludge will survive and spread disease. The biggest problem, though, seems to be that much of the sludge is contaminated with toxic metals. These metals could be taken up by the plants and eventually end up in our food.

Other solutions are possible. Toilets that compost wastes and use no energy or water have been developed. The mild heat of composting drives off water from the wastes. The system is ventilated to keep the process aerobic. No odors enter the house from a properly installed system. Dried waste is removed about once a year. The initial cost is much higher than that of a flush toilet, however. In the United States, we flush about 50,000 L each of drinking-quality water per person annually. Perhaps we should consider an alternative to flushing our wastes into the water we drink.

SECTION 13.14

A Drop to Drink

The water we drink often comes from reservoirs, lakes, and rivers. A number of cities use water that has been used by other cities upstream. Such water may be badly polluted with chemicals and pathogenic microorganisms. To make it safe and palatable involves several steps of chemical and physical treatment (Figure 13.18).

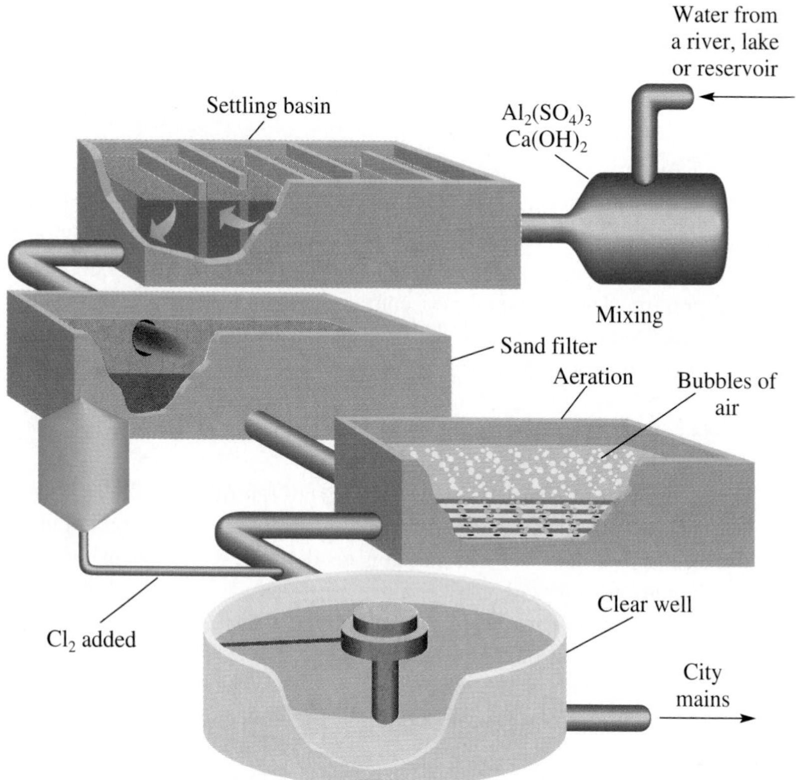

Figure 13.18 A schematic diagram of a municipal water purification plant.

The water to be purified usually is placed in a settling basin, where it is treated with slaked lime and a flocculant such as aluminum sulfate. These materials react to form a gelatinous mass, aluminum hydroxide.

$$\underset{\text{Slaked lime}}{3\,Ca(OH)_2} + \underset{\text{Aluminum sulfate}}{Al_2(SO_4)_3} \longrightarrow \underset{\text{Aluminum hydroxide}}{2\,Al(OH)_3} + \underset{\text{Calcium sulfate}}{3\,CaSO_4}$$

The aluminum hydroxide carries down dirt particles and bacteria. The water is then filtered through sand and gravel.

Sometimes water is treated by **aeration**; it is sprayed into the air to remove odors and improve its taste (water without dissolved air tastes flat). Sometimes the water is filtered through charcoal to remove colored and odorous compounds. In the final step, chlorine is added to kill any remaining bacteria. In some communities that use river water, a lot of chlorine is needed to kill all the bacteria, and you can taste the chlorine in the water.

Over 400,000 people were made ill in 1993 when *Cryptosporidium,* a microorganism common in barnyard runoff, contaminated the water supply in the Milwaukee area.

We generally take our drinking water for granted. Perhaps we shouldn't. There are at least 4000, and perhaps as many as 40,000, cases of waterborne illnesses in the United States each year. Twenty million people have no running water at all, and many of these people get water from suspect sources. Another 30 million tap individual wells or springs. Many of these sources are of uncontrolled and unknown quality. Much remains to be done before we can all be assured of safe drinking water.

Fluorides

Dental caries (tooth decay) was once considered to be "the leading chronic disease of childhood." That is no longer true, thanks mainly to fluoride. Up to a point, the hardness of our tooth enamel can be correlated with the amount of fluoride present. Tooth enamel is a complex calcium phosphate called hydroxyapatite. Fluoride ions replace some of the hydroxide ions, forming a harder mineral called fluorapatite.

$$Ca_5(PO_4)_3OH + F^- \longrightarrow Ca_5(PO_4)_3F + OH^-$$

Exhaustive studies of fluoride as a carcinogen were concluded in 1990. The results were equivocal; we still don't know whether or not fluoride causes cancer. It does seem unlikely that levels found in fluoridated water cause cancer in humans.

Concentrations of 0.7 to 1.0 ppm (by weight) of fluoride (usually as H_2SiF_6 or Na_2SiF_6) have been added to the drinking water of many communities. Evidence indicates that such fluoridation results in a reduction in the incidence of dental caries (cavities) by as much as two-thirds in some areas. While only 29% of the nine-year-olds in the United States were cavity-free in 1971, over 65% of that age group were cavity-free in 1987.

There are some who object to having fluoride added to their water. Fluoride salts are acute poisons in moderate to high concentrations. Indeed, sodium fluoride (NaF) is used as a poison for roaches and rats. Small amounts of fluoride ion, however, do seem to contribute to our well-being.

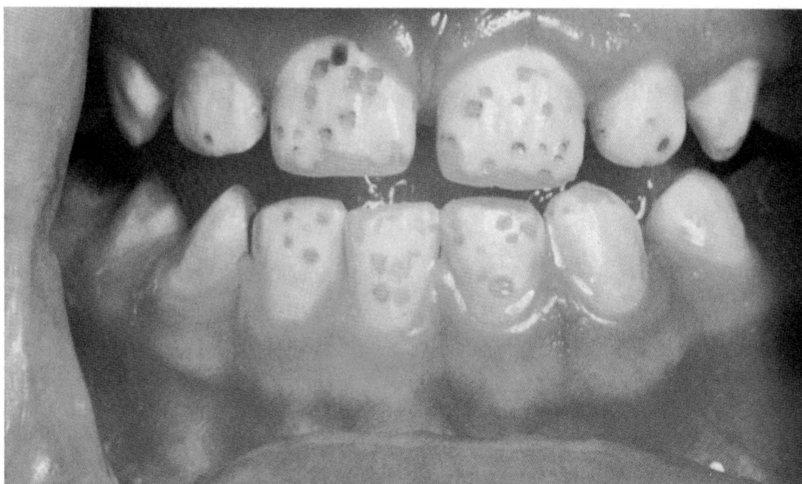

Figure 13.19 Excessive fluoride consumption in early childhood can cause mottling of tooth enamel. A severe case caused by continuous use during childhood of a water supply with an excessive natural concentraiton of fluoride is shown here. [*Courtesy of National Institute of Dental Research, Bethesda, MD.*]

There is some concern about cumulative effects of consuming fluorides in drinking water, in the diet, in toothpaste, and from other sources. Excessive fluoride consumption during early childhood can cause mottling of the tooth enamel (Figure 13.19). The enamel becomes brittle in certain areas and gradually discolors. Fluorides in high doses also interfere with calcium metabolism, with kidney action, with thyroid function, and with the action of other glands and organs. Although there is little or no evidence that optimal fluoridation causes problems such as these, fluoridation of public water supplies will most likely remain a subject of controversy.

Nitrates

People in rural areas often obtain water from wells. Because the water table is dropping in most parts of the country, ever deeper wells are required. And the well water can be contaminated. In many agricultural areas, the concentration of nitrates (NO_3^-) in well water is above the maximum safe level (for infants) of 10 ppm. For adults, the maximum safe level is 45 ppm. Even this level is exceeded in some areas.

Excessive nitrates (which are reduced to nitrites in the infant's digestive tract) cause methemoglobinemia, the blue baby syndrome. In parts of Illinois, baby pigs turn blue and die after drinking the water. In California's Imperial Valley some parents have to buy bottled water for their babies.

Nitrates in the groundwater come from agricultural fertilizers, from decomposition of organic wastes in sewage treatment, and from runoff from animal feedlots. They are highly soluble and thus difficult to remove from water.

Only expensive treatment can remove these compounds from water once they are there.

SECTION 13.15

You're the Solution to Water Pollution

The U.S. Congress has passed laws intended to drastically reduce water pollution. It can't be eliminated entirely, however. To use water is to pollute it. You can do your share by conserving water and by minimizing your use of products that require vast amounts of water for production. And you, the citizen, must be prepared to foot the bill for waste treatment. As our population grows, it will cost plenty just to maintain the present (and often inadequate) water quality. To clean our water up, and then keep it clean, will cost even more. Remember that the cost of unclean water is even higher—discomfort, loss of recreation, illness, even death.

Summary

1. Water covers three-fourths of Earth's surface, but only about 1% of it is available as fresh water.
2. Water has a higher density than most other liquids.
3. It also has an unusually high *heat capacity* and *heat of vaporization.*
4. Unlike most liquids, water expands when it freezes.
5. Seawater is very salty because water is a good solvent for ionic compounds.
6. Water evaporated from oceans and lakes condenses into clouds and then returns to Earth as rain or snow. This is the water cycle.
7. Fresh water can readily become contaminated by various kinds of chemicals and microorganisms, some from natural processes and others from human activities.
8. *Hard water* contains salts of calcium, magnesium, or iron.
9. Dumping sewage into water increases its *biological oxygen demand* (BOD).
10. Water pollutants that are nutrients for the growth of algae can lead to *eutrophication* of a lake or stream.
11. *Acid rain* can cause lakes and streams to become so acidic that they damage fish populations and other marine life.
12. *Groundwater* obtained from wells provides drinking water for many cities and most rural households.
13. Wastewater treatment usually includes sludge removal, sand and gravel filtration, *aeration,* and *chlorination.*
14. *Advanced treatment* methods include charcoal filtration, ion exchange, and *reverse osmosis.*
15. Although fluoride is poisonous in large concentrations, addition of 1 ppm to municipal water supplies has greatly reduced the incidence of dental caries.
16. Maintaining water quality is expensive, and will become more so; but not maintaining the quality of our water will cost much more in terms of our health and comfort.

Key Terms

acid rain 13.10
activated sludge method 13.12
advanced treatment 13.12
aeration 13.14
aerobic 13.7
anaerobic 13.7
biochemical oxygen demand 13.7
chlorination 13.12
dissolved oxygen 13.7
eutrophication 13.7

groundwater 13.9
hard water 13.6
heat capacity 13.1
heat of vaporization 13.1
ozone 13.12
primary sewage treatment 13.12
reverse osmosis 13.12
secondary sewage treatment 13.12
specific heat 13.1
tertiary treatment 13.12

Review Questions

1. Ice is less dense than liquid water. What consequences does this property have for life in northern lakes?
2. Why should foods be flash-frozen rather than frozen slowly?
3. A barge filled with gasoline sinks and breaks open. Would the gasoline dissolve in the water? Would it float or sink?
4. Define heat capacity. Why is the high heat capacity of water important to planet Earth?
5. Why is the high heat of vaporization of water important to our bodies?
6. Why is it cooler near a lake than inland during the summer?
7. Why is ice less dense than liquid water?
8. Why are seas salty?
9. What proportion of Earth's water is seawater?
10. How is our supply of fresh water replenished by natural processes?
11. List some water-borne diseases. Why are these no longer common in developed countries?
12. What impurities are present in rainwater?
13. List four cations present in groundwater.
14. List three anions present in groundwater.
15. What is hard water? Why is it sometimes undesirable?
16. What is BOD? Why is a high BOD undesirable?
17. What are pathogenic microorganisms?
18. What are the products of the breakdown of organic matter by aerobic bacteria?
19. What is eutrophication?
20. List some of the products of anaerobic decay.
21. List some ways in which groundwater is contaminated.
22. What problems are caused by LUST?
23. List some common industrial contaminants of groundwater.
24. Why do chlorinated hydrocarbons remain in groundwater for such a long time?
25. List two ways by which lakes and streams have become acidic.
26. Why is acidic water especially harmful to fish?
27. List several ways by which the acidity of rain can be reduced.
28. What kind of rocks tend to neutralize acidic waters?
29. How can we restore (at least temporarily) lakes that are too acidic?
30. List two toxic compounds found in wastes from the chrome plating process. How is each removed?
31. Describe a primary sewage treatment plant.
32. What impurities are removed by primary sewage treatment?
33. Describe a secondary sewage treatment plant.
34. What impurities are removed by secondary sewage treatment?
35. Describe the activated sludge method of sewage treatment.
36. What substances remain in wastewater after effective secondary treatment?
37. Why is wastewater chlorinated before it is returned to a waterway?
38. List the advantages and disadvantages of chlorination of wastewater.
39. List the advantages and disadvantages of ozone as a disinfectant of wastewater.
40. What is meant by advanced (or tertiary) treatment of wastewater?
41. What kinds of substances are removed from wastewater by charcoal filtration?
42. How does alum (aluminum sulfate) remove phosphates from water?

43. What are the advantages and disadvantages of spreading sewage sludge on farmlands?
44. Describe an alternative to the flush toilet.
45. List two methods of advanced sewage treatment.
46. Why are municipal water supplies aerated?
47. How does fluoride strengthen tooth enamel?
48. What are some health effects of too much fluoride in the diet?
49. What is the source of nitrate ions in well water?
50. What is methemoglobinemia? How is it caused?

▸ Problems

Chemical Equations

51. Give the equation for the neutralization of acidic water (H_3O^+) by limestone (calcium carbonate).
52. Give the equation for the reaction of slaked lime (calcium hydroxide) with alum (aluminum sulfate) to form aluminum hydroxide. What is the purpose of this reaction in water purification?
53. Give the equation for the precipitation of phosphate ions by aluminum ions (from alum). What is the purpose of this reaction in wastewater treatment?

▸ Additional Problems

54. Wastewater disinfected with chlorine must be dechlorinated before it is returned to sensitive bodies of water. The dechlorinating agent often is sulfur dioxide. The reaction is

$$Cl_2 + SO_2 + 2\,H_2O \rightarrow 2\,Cl^- + SO_4^{2-} + 4\,H^+$$

Is the chlorine oxidized or reduced? Identify the oxidizing agent and reducing agent in the reaction.
55. Radioactive aluminum-26 is used to study the mobilization of aluminum by acidic waters. When it decays, the isotope is transmuted into magnesium-26. What kind of particle is emitted by aluminum-26?

▸ Projects

56. Consult the references or other sources and prepare a report on the desalination of seawater by one of the following methods.
    a. distillation
    b. freezing
    c. electrodialysis
    d. reverse osmosis
    e. ion exchange
57. What sort of wastewater treatment is used in your community? Is it adequate?
58. Where does your drinking water come from? What steps are used in purifying it?
59. Call your water utility office and obtain a chemical analysis of your drinking water. What substances are tested for? Are any of those substances considered to be problems? (If you get water from a private well, has the water been analyzed? If so, what were the results?)

▸ References and Readings

1. "An Integrated Approach to Industrial Wastewater Management." *The National Environmental Journal,* September–October 1993, pp. 42–47.
2. Bernarde, Melvin A. *Our Precarious Habitat: Fifteen Years Later.* New York: Wiley, 1989. Chapters 11 and 12 discuss water pollution.

3. Boutacoff, David. ''Working with the Watershed.'' *EPRI Journal,* January–February 1990, pp. 28–33. Liming the watershed keeps a lake neutral longer than liming the lake directly.

4. ''Fit to Drink.'' *Consumer Reports,* January 1990, pp. 27–43. Describes problems with our drinking water and evaluates home treatment devices.

5. La Riviere, J. W. Maurits. ''Threats to the World's Water.'' *Scientific American,* September 1989, pp. 80–94.

6. Manahan, Stanley E. *Environmental Chemistry,* 5th ed., chs. 2–8. Chelsea, MI: Lewis Publishers, 1991.

7. Morganthau, Tom. ''Don't Go Near the Water.'' *Newsweek,* 1 August 1988, pp. 42–48. Describes the plight of our coastal waters.

8. Noether, Dorit L. ''Water Is . . .'' *ChemTech,* February 1982, pp. 84–90.

9. Ouellette, Robert P. ''A Perspective on Water Pollution,'' *The National Environmental Journal,* September–October 1991, pp. 20–24.

10. Page, George W., Ed. *Planning for Groundwater Protection.* New York: Academic Press, 1987.

11. Pope, Victoria. ''Poisoning Russia's River of Plenty.'' *U.S. News & World Report,* 13 April 1992, pp. 49–51.

12. Postel, Sandra. *Last Oasis: Facing Water Scarcity.* New York: W. W. Norton, 1992.

13. Stanglin, Douglas. ''Toxic Wasteland.'' *U.S. News & World Report,* 13 April 1992, pp. 40–49.

14. Sun, Marjorie, ''Ground Water Ills: Many Diagnoses, Few Remedies.'' *Science,* 20 June 1986, pp. 1490–1493.

15. Thayer, Ann M. ''Water Treatment Chemicals: Tighter Rules Drive Demand.'' *Chemical and Engineering News,* 26 March 1990, pp. 17–34.

16. Van der Leeden, F., F. L. Troise, and D. K. Todd. *The Water Encyclopedia,* 2nd ed. Chelsea, MI: Lewis Publishers, 1990.

17. Wehle, D. H. S., and Felicia C. Coleman. ''Plastics at Sea.'' *Natural History,* February 1983, pp. 20–26.

14 Energy

A Fuels Paradise

Though we burn gas, and coal, and oil,

When all is said and done,

We get most of our energy

As light rays from the sun.

We were tempted to call this chapter *Fire.* Then the titles of Chapters 11, 12, 13, and 14 would have been *Earth, Air, Water,* and *Fire*—the four elements of the ancient world!

Actually, that would not have been such an inappropriate name for this chapter. The subject is energy, and most of our energy is obtained by burning fuels. And the burning of fuels certainly does involve fire.

On the other hand, there are many sources of energy that do not involve burning anything. Furthermore, we will be depending more heavily on those other sources as our reserves of fuel get used up.

Energy is the ability to do work. It takes energy to make an automobile move, or an airplane fly, or to make electrons flow through a copper wire. The United States, with less than 5% of the world's population, uses almost one-fourth of all the energy presently being generated. Abundant energy has enabled the United States to build its industrial base and provide its people with one of the highest standards of living in the world.

Industry uses about 38% of all energy produced in the United States. This energy is used to convert raw materials into the many products our society seems to demand. Utilities use about 35% of the nation's energy production, primarily to generate electricity. Transportation uses about 27%, to power private automobiles, trucks, trains, airplanes, and buses. Private homes and commercial spaces use about 26%.

These figures do not add up to 100% because some of the energy is counted twice. For example, electricity generated by utilities is then used in industries, homes, and commercial spaces.

Energy lights our homes, heats and cools our living spaces, and makes us the most mobile society in the history of the human race. It powers the factories that provide us with abundant material goods. Indeed, energy is the basis of our modern society.

SECTION 14.1

Heavenly Sunlight Flooding the Earth with Energy

Did you know that you and all other living things on Earth depend on nuclear energy for survival? Most of the energy available to us on this planet comes

◀ Petroleum, natural gas, and coal are the principal sources of the energy that sustains our modern civilization. Petrochemical plants also provide the chemicals from which many modern materials are made. [*David Frazier/Tony Stone Worldwide.*]

Figure 14.1 The sun is fueled by nuclear reactions. Nearly all the energy on Earth originated in the sun. [*NASA photo.*]

from that giant nuclear reactor we call the sun (Figure 14.1). Although it is about 150 million km away, the sun has been supplying Earth with most of its energy for billions of years, and it will likely continue to do so for another 10 to 15 billion years (Table 14.1).

The SI unit of energy is the joule (J). A watt (W), the SI unit for power, is 1 joule per second (J/s). The sun is a nuclear fusion reactor that steadily converts hydrogen to helium (Chapter 4) with a power output of 4×10^{26} watts (W). Although the Earth receives only about 1 part in 50 billion of that energy, it is abundantly bathed in energy, receiving about 1.73×10^{17} W

Table 14.1 The Earth's Energy Ledger (Rough Estimates)

Item	Energy (TW)	Approximate Percent
In		
Solar radiation	173,000	99+
Internal heat	32	0.02
Tides	3	0.002
Out		
Direct reflection	52,000	30
Direct heating*	81,000	47
Water cycle*	40,000	23
Winds*	370	0.2
Photosynthesis*	40	0.02

*This energy is eventually returned to space via long-wave radiation (heat).

Source: M. King Hubbert, "The Energy Resources of the Earth," *Scientific American*, September 1971.

[173,000 terawatts (TW)] from the sun—an amount equivalent to the output of 115 million nuclear power plants. In three days, the Earth receives energy from the sun equivalent to all fossil fuel reserves.

$1 \text{ TW} = 10^{12} \text{ W}$

▶ **EXAMPLE 14.1**

How many joules of electrical energy are consumed by a 100-W bulb burning for 1 hr (3600 s)?

SOLUTION

$$100 \text{ W} = 100 \text{ J/s}$$

$$\frac{100 \text{ J}}{1 \cancel{\text{s}}} \times 3600 \cancel{\text{s}} = 360{,}000 \text{ J}$$

Exercise 14.1

How many joules of electrical energy are consumed by a 600-W microwave oven running for 10 min?

Energy and the Life Support System

Only a small fraction of the energy that the biosphere receives is used to support life. In fact, about 30% of the incident radiation is immediately reflected back into space as short-wave radiation (ultraviolet and visible light). Nearly half is converted to heat, making the third planet a warm and habitable place. About 23% of the solar radiation is used to power the water cycle (Chapter 13), evaporating water from land and seas. The radiant energy of the sun is converted into the potential energy of water vapor, water droplets, and ice crystals in the atmosphere. This potential energy is converted to the kinetic energy of falling rain and snow and of flowing rivers.

The biosphere is the thin film of air, water, and soil in which all life exists. It is only about 15 km thick.

A tiny—but most important—fraction of the solar energy is absorbed by green plants, which use it to power **photosynthesis**. In the presence of green pigments, called *chlorophylls,* this energy is used to convert carbon dioxide and water into glucose, a simple sugar rich in energy.

$$6\,CO_2 + 6\,H_2O \xrightarrow[\text{chlorophyll}]{\text{sunlight}} \underset{\text{Glucose}}{C_6H_{12}O_6} + 6\,O_2$$

This reaction also serves to replenish the oxygen in the atmosphere. The glucose can be stored or it can be converted into more complex foods and structural materials. All animals depend on the stored energy of green plants for survival.

SECTION 14.2

Energy and Chemical Reactions

We make use of our knowledge of the temperature effect on chemical reactions in our daily lives. For example, we freeze foods to retard those chemical reactions that lead to spoilage. When we want to speed up the reactions involved in cooking food, we turn up the heat.

How fast—or how slowly—chemical reactions take place depends on a number of factors. One such factor is *temperature.* Reactions generally take place faster at a higher temperature. For example, coal (carbon) reacts so slowly with oxygen (from the air) at room temperature that the change is imperceptible. However, if coal is heated to several hundred degrees, it reacts at a rapid rate. The heat evolved in the reaction keeps the coal burning smoothly.

The effect of temperature on the rates of chemical reactions is explained in terms of the kinetic-molecular theory. At high temperatures, molecules move more rapidly. Thus, they collide more frequently, creating an increased chance for reaction. The increase in temperature also supplies more energy for the breaking of chemical bonds—a condition necessary for most reactions.

Another factor affecting the rate of a chemical reaction is the *concentration* of reactants. The more molecules there are in a given volume of space, the more likely a collision. The more collisions there are, the more reactions that occur. For example, you could light a wood splint and then blow out the flame. The splint would continue to glow as the wood reacted slowly with the oxygen of the air. If the glowing splint were placed in pure oxygen, the splint would burst into flame, indicating a much more rapid reaction. This factor can be interpreted in terms of the concentration of oxygen: air is about 21% oxygen, so the concentration of O_2 molecules in pure oxygen is about five times as great as in air.

Catalysts are even more important in living organisms. Biological catalysts, called *enzymes,* mediate nearly all the chemical reactions that take place in living systems (Chapter 15).

Catalysts (Chapter 8) also affect the rate of chemical reactions (Figure 14.2). Catalysts are of great importance in the chemical industry. Reactions that otherwise would be so slow as to be impractical can be made to proceed much faster using appropriate catalysts.

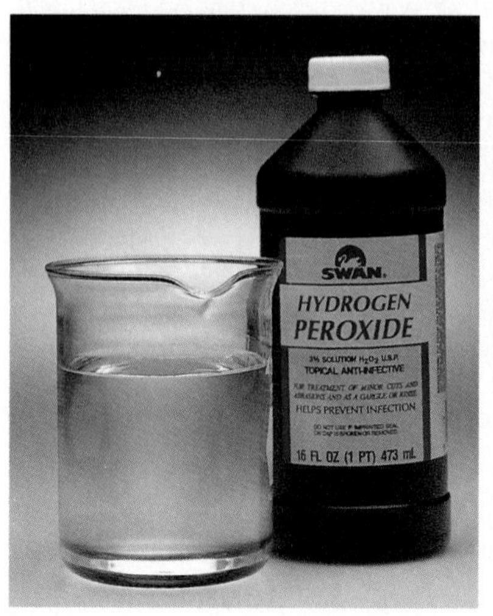

(a)

(b)

Figure 14.2 Hydrogen peroxide decomposes slowly to water and oxygen (a). When platinum metal is inserted into a solution of hydrogen peroxide (b), the reaction proceeds rapidly. The exothermic process produces steam, and the evolving oxygen gas causes frothing. [*(a) Richard Megna/ Fundamental Photographs; (b) Carey B. Van Loon.*]

Energy Changes and Chemical Reactions

The energy changes associated with chemical reactions are quantitatively related to the *amounts* of chemicals that are changed. For example, burning 1 mol (16.0 g) of methane to form carbon dioxide and water releases 192 kcal of energy as heat.

$$CH_4 + 2\,O_2 \longrightarrow CO_2 + 2\,H_2O + 192\text{ kcal}$$

Burning 2 mol (32.0 g) of methane produces twice as much heat, 384 kcal.

Figure 14.3 Coal burns with a highly exothermic reaction. The heat released can be used to convert water to steam that can turn a turbine to produce electricity. [*Barry L. Runk/ Grant Heilman Photography.*]

▶ EXAMPLE 14.2

Burning 1.00 mol of propane releases 526 kcal of energy.

$$C_3H_8 + 5\,O_2 \longrightarrow 3\,CO_2 + 4\,H_2O + 526\text{ kcal}$$

How much energy is released when 15.0 mol of propane is burned?

SOLUTION

$$\frac{526\text{ kcal}}{1\text{ }\cancel{\text{mol}}} \times 15.0\text{ }\cancel{\text{mol}} = 7890\text{ kcal}$$

Exercise 14.2

The reaction of nitrogen and oxygen to form nitrogen monoxide (nitric oxide) requires an input of energy.

$$N_2 + O_2 + 4.32\text{ kcal} \longrightarrow 2\,NO$$

How much energy is absorbed when 5.00 mol of N_2 reacts with oxygen to form NO?

Chemical reactions that result in the release of heat are said to be **exothermic**. The burning of methane, gasoline, and coal (Figure 14.3) are all exothermic reactions. In each case, chemical energy is converted into heat energy. There are other reactions, such as the decomposition of water, in which energy must be supplied.

$$2\,H_2O + 137\text{ kcal} \longrightarrow 2\,H_2 + O_2$$

If energy is supplied as heat, such reactions are said to be **endothermic**. It takes 137 kcal of energy to decompose 36.0 g (2 mol) of water into hydrogen and oxygen. It should be noted that the same amount of energy is released when enough hydrogen is burned to form 36.0 g of water.

$$2\,H_2 + O_2 \longrightarrow 2\,H_2O + 137\text{ kcal}$$

When heat energy is released during a chemical reaction, heat often is listed as a product in the chemical equation. When energy is supplied to keep a chemical reaction going, heat can be listed as a reactant in the equation.

EXAMPLE 14.3

How much energy is released when 220 g of propane (Example 14.2) is burned?

SOLUTION

First calculate the formula weight of propane.

$$(3 \times \text{C}) + (8 \times \text{H}) = (3 \times 12.0) + (8 \times 1.0) = 36.0 + 8.0 = 44.0$$

One mole of propane weighs 44.0 g. Therefore,

$$220\ \cancel{\text{g}} \times \frac{1\ \cancel{\text{mol}}}{44.0\ \cancel{\text{g}}} \times \frac{526\ \text{kcal}}{1\ \cancel{\text{mol}}} = 2630\ \text{kcal}$$

Exercise 14.3

How much energy is absorbed when 0.528 g of N_2 is converted to NO (Exercise 14.2)?

SECTION 14.3

Energy and the First Law: Energy Is Conserved, Yet We're Running Out

We take a look at our present energy sources and our future energy prospects shortly. To do so scientifically, however, let's first look at some of the laws of nature. You should recall that natural laws merely summarize the results of many experiments. We won't trouble you with a recounting of all those experiments here; we will merely state the laws and some of their consequences.

Can we somehow get around the law of conservation of energy, as some people claim? The law has been verified experimentally to one part per quadrillion (1 part in 10^{15}). Any exception to the law would have to be less than that. How small is the possible deviation? A typist with that rate of precision could type at 100 words per minute for 30 million years and make only one mistake. For all practical purposes, the law is absolute.

The **first law of thermodynamics** grew out of a variety of experiments during the early 1800s. By 1840, it was clear that, although it can be changed from one form to another, energy is neither created nor destroyed. This law (also called the **law of conservation of energy**) has been restated in a number of ways, including "you can't get something for nothing" and "there is no such thing as a free lunch." Energy can't be made from nothing. Neither does it just disappear, although it may go someplace else.

From the first law of thermodynamics we can conclude that we can't win—in the sense that we can't make a machine that produces energy from nothing. From that law alone, however, we might conclude that we can't possibly run out of energy because energy is conserved. That is true enough, but it doesn't mean that we don't have problems. There is another law with a long arm from which we cannot escape.

SECTION 14.4

Energy and the Second Law: Things Are Going to Get Worse

Figure 14.4 Energy always flows spontaneously from hot to cold, never the reverse.

Despite innumerable attempts, no one has ever built a successful ''perpetual-motion'' machine. You can't make a machine that will produce more energy than it consumes. Even if an engine isn't doing any work, it loses energy (as heat) due to the friction of its moving parts. In fact, in any real engine, you can't get as much useful energy out as you put in. You can't even break even!

If energy is neither created nor destroyed, why do we always need more? Will what we have now not last forever? The answer lies in the fact that energy can be changed from one form or another and not all forms are equal; and high-grade forms of energy are constantly being degraded into low-grade forms. Energy flows downhill. Mechanical energy eventually is changed into heat energy. Hot objects cool off by transferring their heat to cooler objects. There is a tendency toward an even distribution of energy. Energy always flows from a hot object to a cooler one (Figure 14.4). The reverse never occurs spontaneously.

Observations of heat flow led to the formulation of the **second law of thermodynamics**. In one form (of many), the law states that energy does not flow spontaneously from a cold object to a hot one. Now it is true that we can *make* energy flow from a cold region to a hot one—that's what refrigerators are all about—but we cannot do so without producing changes elsewhere. This sort of reversal of a natural process can only be done at a price. The price, in the case of refrigerators, is the consumption of electricity; that is, you must use energy to make energy flow from a cold space to a warmer one.

The second law tells us that it is easy to scatter materials because this increases the entropy of the universe. That means it is easy to pollute; a barrel of some toxic substance can pollute a river or lake. On the other hand, once a substance is scattered, it takes a lot of energy to concentrate it. It costs a lot to clean up polluted water, soil, or air. Prevention of pollution is by far the better alternative.

Another way to look at the second law is in terms of disorder. Scientists use the term **entropy** to indicate the randomness of a system. The more mixed up a system is, the higher its entropy. This concept is illustrated in Figure 14.5. Natural processes tend toward a greater entropy—toward disorder.

Actually, we sometimes *can* reverse the tendency toward randomness—but only through the expenditure of energy. For example, a new deck of playing cards could be arranged by suit and by rank, a situation of high order (low entropy). If the cards were dropped and scattered randomly about the floor (higher entropy), they could be returned to their ordered state only by the expenditure of energy—bending, stooping, and stacking. Your body would

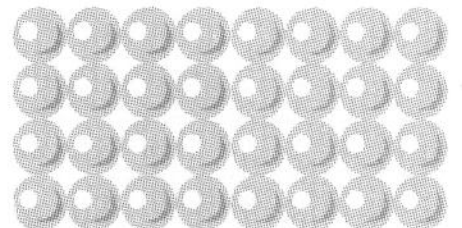

High order
low entropy

Less order
more entropy

Figure 14.5 Any spontaneous change is toward a less ordered state—toward greater entropy.

have to convert food energy to muscular energy to do the task. It is quite unlikely that the cards would pick themselves up off the floor and resume their ordered state.

Thus, the second law places restrictions on what can be done. It also has something to say about what we should do about many of our environmental problems, and it points out—rather firmly—a number of things we cannot do.

SECTION 14.5

People Power: Early Uses of Energy

Primitive peoples obtained their energy (food and fuel) by collecting wild plants and hunting wild animals. They expended the energy so obtained in hunting and gathering. Domestication of the horse and the ox increased the availability of energy only slightly. The raw materials used by these work animals were natural—and replaceable—plant materials.

Plants were also the first fuels. Tree branches and other combustible materials from dead trees and bushes kept primitive fires burning. Even today, wood remains the primary fuel in some parts of the world.

Probably the first device used to convert energy into useful work was the waterwheel (Figure 14.6). The Egyptians first used waterpower about 2000 years ago, primarily for grinding grain. Later on, waterpower was used for sawmills, textile mills, and other small factories. Windmills were introduced into western Europe during the Middle Ages, primarily for pumping water and grinding grain. More recently, wind power has been used to generate electricity.

Windmills and waterwheels are fairly simple devices for converting the kinetic energy of blowing wind and flowing water into mechanical energy. They were sufficient to power the early part of the industrial revolution, but it was the development of the steam engine that freed the factories from loca-

Figure 14.6 Waterwheels were described by the Roman engineer Vitruvius in 14 B.C. They were seldom used in Europe, however, until the Middle Ages. The Romans used human slaves instead.

tions along waterways. Since 1850, the water turbine, the internal combustion engine, the steam turbine, the gas turbine, and a variety of other energy-conversion devices have been added to the arsenal. In fact, it has been estimated that the power output of all these conversion devices has increased the energy available for use by a factor of 10,000.

Let's turn our attention now to the fossils that fueled the industrial revolution and still serve as the basis for modern civilization.

SECTION 14.6

Fossil Fuels

Our modern industrial civilization is fueled by fossils. Almost 90% of the energy used to sustain our way of life comes from the **fossil fuels**—coal, petroleum, and natural gas. In the next few sections, we consider the origin and chemical nature of these fuels and how they are burned to release energy that plants of ages past captured from rays of sunlight.

Fuels are substances that burn readily with the release of significant amounts of energy. Fuels are *reduced* forms of matter, and the burning process is oxidation (Chapter 8). If an atom already has its maximum number of bonds to oxygen (or to other electronegative atoms such as chlorine or bromine), the atom cannot serve as a fuel. Indeed, such substances can be used to put *out* fires. Figure 14.7 shows some representative fuels and nonfuels.

We are so accustomed to having plenty of fuel that we tend to forget that Earth has only a limited supply of fossil fuels. Within your lifetime, it is likely that natural gas and petroleum will become so scarce and so expensive that people won't be able to afford them as fuels. At the current rate of production, U.S. reserves of petroleum will be running out sometime during the twenty-first century. More than half of the world's petroleum reserves are in the troubled lands of the Middle East. Natural gas reserves are also expected to

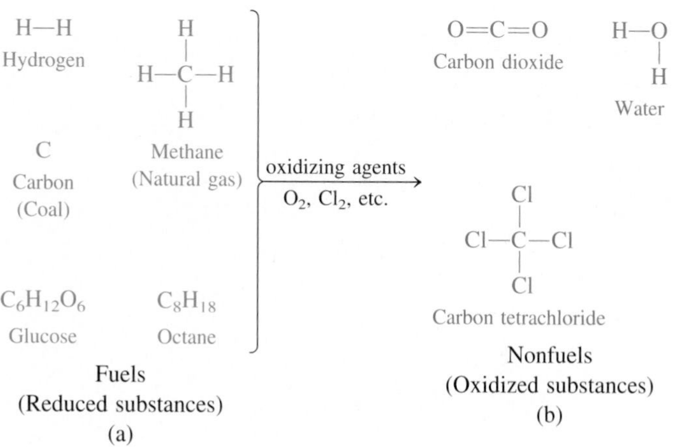

Figure 14.7 (a) Some fuels (reduced forms of matter). (b) Some nonfuels (oxidized forms of matter).

Table 14.2 Approximate Composition (Percent by Mass) of Various Typical Grades of Coal (Dry Basis)				
Grades of Coal	Carbon	Hydrogen	Oxygen	Nitrogen
Wood (for comparison)	50	6	43	1
Peat	59	6	33	2
Lignite (brown coal)	69	5	25	1
Bituminous (soft coal)	88–89	5	5–15	1
Anthracite (hard coal)	95	2–3	2–3	Trace

Source: M. G. Zabetakis and L. D. Phillips, *Coal Mining*. Washington, DC: U.S. Government Printing Office, 1975.

run out during the twenty-first century. Coal reserves should last at least 300 years at the present rate of use, but the rate of use is increasing, and it will increase even more dramatically as the other two fossil fuels become depleted.

SECTION 14.7

Coal: The Carbon Rock of Ages

Coal is a complex material. It is composed principally of carbon, but it also contains smaller percentages of other elements. The quality of coal as an energy source is based on its carbon content; the rankings run from low-grade peat and lignite to high-grade anthracite (Table 14.2). Soft (bituminous) coal is much more plentiful than hard coal (anthracite). Lignite and peat have

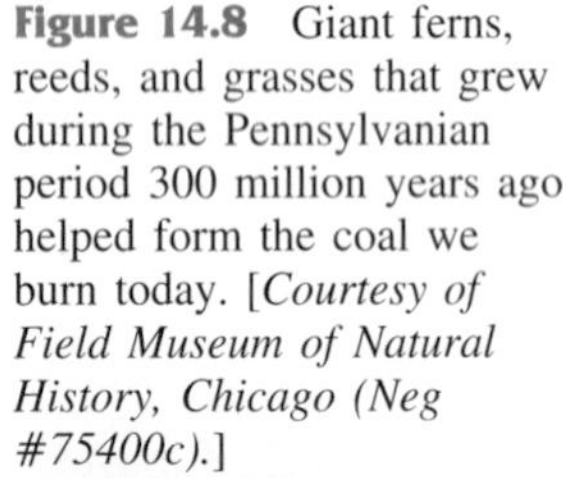

Figure 14.8 Giant ferns, reeds, and grasses that grew during the Pennsylvanian period 300 million years ago helped form the coal we burn today. [*Courtesy of Field Museum of Natural History, Chicago (Neg #75400c).*]

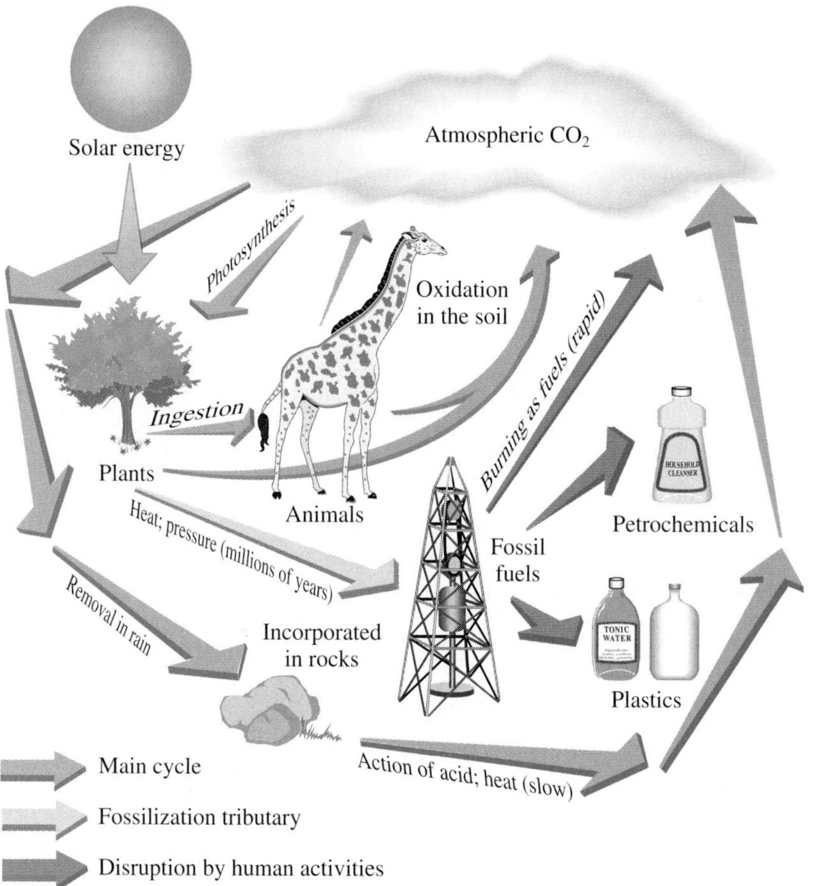

Figure 14.9 The carbon cycle.

become increasingly important as the supply of higher grades of coal is depleted.

The supply of coal is limited. Deposits that exist today are less than 600 million years old. For a part of that time, the Earth was much warmer than it is now, and plant life flourished (Figure 14.8). Most plants lived, died, and decayed—playing their normal role in the carbon cycle (Figure 14.9). But a small portion of the plant material became buried under mud and water. There, in the absence of oxygen, it could decay only partially. The structural material of plants is largely cellulose, a compound of carbon, hydrogen, and oxygen. Under the action of increasing pressure, as the material was buried more deeply, the cellulose molecules broke down. Small molecules rich in hydrogen and oxygen escaped, leaving behind a material increasingly rich in carbon. Thus, peat is a young coal, only partly converted. Anthracite, on the other hand, has been almost completely carbonized.

Presumably, fossil fuels are still being formed in nature. This process is perhaps most evident in peat bogs. The rate of formation is extremely slow, however. In fact, it has been estimated that we are using fuels 50,000 times as fast as they are being formed.

Even in 1850 the burning of wood still accounted for more than 90% of the energy produced in the United States.

People probably have used small amounts of coal since prehistoric times, but, as late as 1760, wood was—for practical purposes—the only fuel being used. In only a few hundred more years, we will have removed from the Earth and burned virtually all the remaining recoverable fossil fuels. Of all that ever existed, we will use about 90% in a period of 300 years. On a geological time scale, that is but a moment. Even on a historical basis, the fossil-fuel feast is but a brief interlude.

Inconvenient Fuel

Coal is an inconvenient fuel. As a solid, it can't be pumped through pipes to get it out of the ground. It must be removed by dangerous deep mining or devastating strip mining. It may be possible to transport coal by grinding it to a powder, mixing it with water, and pumping the slurry through a pipeline. At the present time, though, coal has to be hauled by trucks, trains, and barges. A good deal of energy is consumed in getting coal from the mine to the power plant or the factory where it is to be used.

Abundant Fuel

Coal is our most plentiful fossil fuel. The United States is estimated to have over 40% of the world's reserves of coal. There is probably enough easily recoverable coal to last us several hundred years. Electric utilities in the United States burn nearly 700 million t of coal each year, generating 1300 TW of electricity (55% of the total). Unfortunately, this coal use is associated with some of our worst environmental problems.

(a)

(b)

Figure 14.10 The site of a strip mine in Morgan County, Tennessee. (a) The abandoned mine in 1963. (b) The same mine after it was reclaimed in a demonstration project, 1971. [*Courtesy of U.S. Department of the Interior, Office of Surface Mines.*]

Coal can be obtained most inexpensively by strip mining. In the process, however, vast areas are ripped bare of all vegetation and the exposed soil washes away, filling streams with mud and silt (Figure 14.10). Stripped areas sometimes can be restored, but restoration is expensive. Laws in the United States now require the restoration of most stripped areas and ban stripping on steep slopes where restoration is not feasible. Unfortunately, a lot of damage was done before the laws were passed, and many conservationists consider the present laws inadequate.

The fuel value of coal depends largely on its carbon content. Complete combustion of the carbon produces carbon dioxide.

$$C + O_2 \longrightarrow CO_2$$

In limited amounts of air, however, large quantities of carbon monoxide and soot are formed.

$$2\,C + O_2 \longrightarrow 2\,CO$$

The soot is mostly unburned carbon.

Source of Pollution

Coal, a solid, contains many impurities. Unlike natural gas and petroleum, it often contains a lot minerals. When the coal is burned, the minerals are left as ash, presenting a major problem in solid-waste disposal. Some minerals enter the air as particulate matter, constituting a major pollution problem.

Perhaps even worse, most of our remaining coal is high in sulfur. When coal burns, choking sulfur dioxide is poured into the atmosphere. Reacting with oxygen and moisture in the air, the sulfur dioxide is converted into sulfuric acid. This acid slowly damages steel and aluminum structures, marble buildings and statues, and human lungs.

There *are* ways to clean coal before it is burned. A *flotation method* makes use of the different densities of coal and its major impurities. Coal has a density of about 1.3 g/cm^3. Shale, a rock formed from hardened clay, has a density of about 2.5 g/cm^3. Pyrite (FeS_2), the major source of sulfur in the coal, has a density of about 5.0 g/cm^3. By using a solution of the proper density and adding detergents to cause the coal to float, the coal can be floated off, leaving the heavier minerals behind. Unfortunately, the process adds to the cost of the coal. Coal also can be converted to more convenient gaseous and liquid fuels.

The flotation method does a fairly good job of removing inorganic sulfur compounds such as pyrite from coal. It does not remove sulfur atoms covalently bonded to the carbon atoms in the coal. This bound sulfur still contributes to our pollution problems.

Source of Chemicals

But coal is not just a fuel. When coal is heated in the absence of air, volatile material is driven off, leaving behind a mostly carbon product called coke, used in the production of iron and steel. The more volatile material is condensed to liquid coal oil and a gooey mixture called coal tar, both of which are good sources of organic chemicals.

Figure 14.11 Natural gas burns with a relatively clean flame. [*John Kapreilian/ Photo Researchers, Inc.*]

SECTION 14.8

Natural Gas: Mostly Methane

Natural gas is a fossil fuel. It is composed principally of methane, but its composition varies greatly. In North America, natural gas typically contains 60 to 80% methane, 5 to 9% ethane, 3 to 18% propane, and 2 to 14% butane and pentane. The gas, as it comes from the ground, often contains sulfur and nitrogen compounds as impurities.

Natural gas was most likely formed—ages ago—by the action of heat, pressure, and perhaps bacteria on buried organic matter. The gas is trapped in geological formations capped by impermeable rock. It is removed through wells drilled into the gas-bearing formations.

Most natural gas is used as fuel, but it is also an important raw material. In North America, some of the higher alkanes are separated out of the natural gas. Ethane and propane are **cracked**—decomposed by heat in the absence of air—to form ethylene and propylene. These alkenes are intermediates in the synthesis of plastics (Chapter 10) and many other useful commodities. Natural gas is also the raw material from which methanol and many other one-carbon compounds are made.

Composed mainly of methane, natural gas burns with a relatively clean flame (Figure 14.11), and its products are mainly carbon dioxide and water.

$$CH_4 + 2\,O_2 \longrightarrow CO_2 + 2\,H_2O$$

As in any combustion in air, some nitrogen oxides are formed. The role of these compounds in air pollution is discussed in Chapter 12. When natural gas is burned in limited amounts of air, carbon monoxide or even elemental carbon (soot) can be major products.

$$2\,CH_4 + 3\,O_2 \longrightarrow 2\,CO + 4\,H_2O$$

$$CH_4 + O_2 \longrightarrow C + 2\,H_2O$$

Even so, natural gas is the cleanest of the fossil fuels. Unfortunately, it is also in short supply (Table 14.3). Our reserves are expected to dwindle over the next 30 years or so, especially if our rate of use continues to increase.

SECTION 14.9

Petroleum: Liquid Hydrocarbons

Petroleum is an exceedingly complicated liquid mixture of organic compounds. Petroleum deposits nearly always have associated natural gas, part or all of which is sent through a separation process. The main components of petroleum are hydrocarbons. These are chiefly alkanes, but some are cyclic

Table 14.3 Estimated United States and World Reserves of Economically Recoverable Fuels*		
Fuel	United States	World
Coal	264,000	926,000
Petroleum	5,400	199,000
Natural gas	6,110	143,000

*Expressed in thousands of metric tons of coal equivalents for ready comparison of energy content. See the following source for assumptions made in making these estimates.
Source: Energy Information Administration, *International Energy Outlook 1990.* Washington, DC: U.S. Department of Energy, 1990.

compounds. Petroleum also has varying proportions of sulfur-, nitrogen-, and oxygen-containing compounds.

We have seen that coal is primarily of plant origin. Recent evidence indicates that petroleum is of animal origin. Most likely it is formed primarily from the fats (Chapter 15) of ocean-dwelling, microscopic animals, for it is

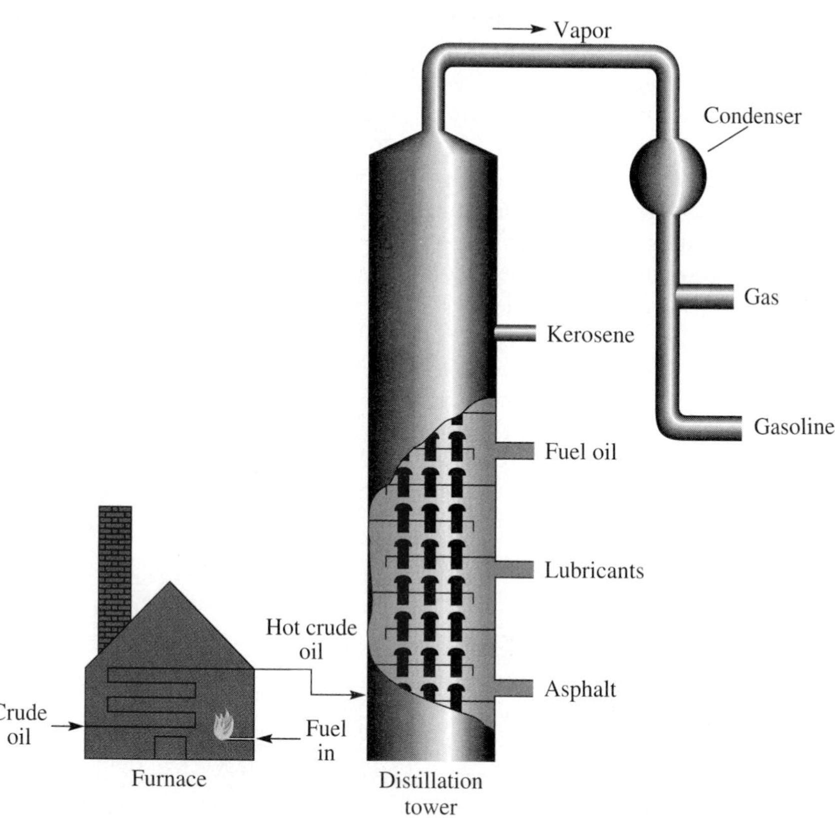

Figure 14.12 The fractional distillation of petroleum.

Table 14.4 Typical Petroleum Fractions

Fraction	Typical Range of Hydrocarbons	Approximate Range of Boiling Points (°C)	Typical Uses
Gas	CH_4 to C_4H_{10}	Less than 40	Fuel, starting materials for plastics
Gasoline	C_5H_{12} to $C_{12}H_{26}$	40–200	Fuel, solvents
Kerosene	$C_{12}H_{26}$ to $C_{16}H_{34}$	175–275	Diesel fuel, jet fuel, home heating; cracking to gasoline
Heating oil	$C_{15}H_{32}$ to $C_{18}H_{38}$	250–400	Industrial heating, cracking to gasoline
Lubricating oil	$C_{17}H_{36}$ and up	Above 300	Lubricants
Residue	$C_{20}H_{42}$ and up	Above 350 (some decomposition)	Paraffin, asphalt

nearly always found in rocks of oceanic origin. Fats are made up mainly of compounds of carbon and hydrogen, with a little oxygen (for example, $C_{57}H_{110}O_6$). The removal of the oxygen and the slight rearrangement of the carbon and hydrogen atoms in these fats produce typical hydrocarbon molecules as they occur in petroleum.

Petroleum, as it comes from the ground, is of limited use. To make it better suit our needs, we separate it into fractions by boiling it in a distillation column (Figure 14.12). The lighter molecules come off at the top of the column, the heavier at the bottom. (The various fractions are described in Table 14.4.)

Gasoline usually is the fraction of petroleum most in demand, and fractions that boil at higher temperatures are often in excess supply. The latter are often converted to gasoline by heating in the absence of air. The **cracking process** breaks down the bigger molecules into smaller ones. The effect of the process, with $C_{14}H_{30}$ as an example, is illustrated in Figure 14.13. Not only does

$C_{14}H_{30}$ —heat→

Smaller saturated hydrocarbons (having from 1 to 13 carbon atoms)		Unsaturated hydrocarbons (having from 2 to 13 carbon atoms)		
$C_{13}H_{28}$		$C_{13}H_{26}$	$C_{13}H_{24}$	
$C_{12}H_{26}$		$C_{12}H_{24}$	$C_{12}H_{22}$	
$C_{11}H_{24}$		$C_{11}H_{22}$	$C_{11}H_{20}$	
$C_{10}H_{22}$	+	$C_{10}H_{20}$ +	$C_{10}H_{18}$	+ etc.
C_9H_{20}		C_9H_{18}	C_9H_{16}	
CH_4		C_2H_4	C_2H_2	

Figure 14.13 Formulas of some of the products formed when $C_{14}H_{30}$ (a typical molecule in kerosene) is cracked. Note the great variety of hydrocarbons with fewer atoms that are formed.

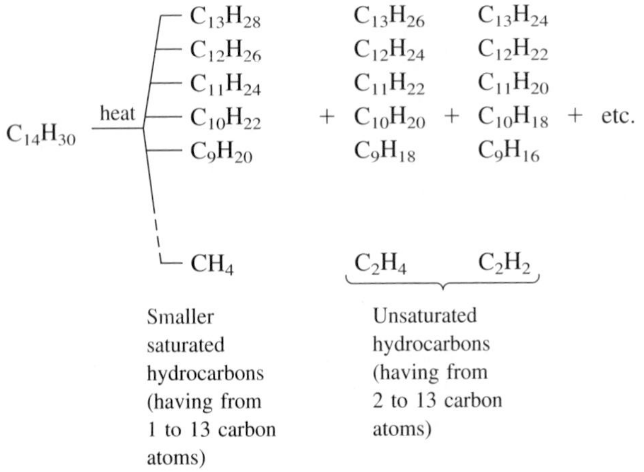

cracking convert some molecules into those in the gasoline range (C_5H_{12} through $C_{12}H_{26}$), it produces a variety of useful by-products. The unsaturated hydrocarbons are starting materials for the manufacture of a whole host of petrochemicals.

The cracking process described here is a crude, yet illustrative, example of how chemists modify nature's materials to meet human needs and desires. Starting with coal tar or petroleum, a chemist can create a dazzling array of substances with a wide variety of properties. A few of these substances are plastics, pesticides, herbicides, perfumes, preservatives, pain killers, antibiotics, stimulants, depressants, and detergents. Many of these substances are discussed in other chapters.

Petroleum is largely a mixture of hydrocarbons, and hydrocarbons burn readily. As with natural gas, complete combustion yields mainly carbon dioxide and water. A representative reaction is that of an octane.

$$2\,C_8H_{18} + 25\,O_2 \longrightarrow 16\,CO_2 + 18\,H_2O$$

Combustion in air also leads to the formation of nitrogen oxides. Incomplete burning produces carbon monoxide and soot. Petroleum usually contains small amounts of sulfur compounds that produce sulfur dioxide when burned.

Efficiently burned, petroleum products are rather clean fuels. Fuel oil, used for heating homes or to produce electricity, can be burned efficiently and thus contributes only moderately to air pollution. Gasoline, however, the major fraction of petroleum, is used to power automobiles; since the internal combustion engines in most automobiles are rather inefficient, this inefficient combustion of gasoline contributes heavily to air pollution (Chapter 12).

Air pollution isn't the only problem: burning up our petroleum reserves will leave us without a ready source of many familiar materials. Nearly all industrial organic chemicals come from petroleum (smaller amounts are derived from coal and natural gas); and plastics, synthetic fibers, solvents, and many other consumer products are made from these chemicals.

Petroleum appears to be somewhat more abundant than natural gas (see Table 14.3). It also contains more impurities. Some of the impurities can be removed by processing, but petroleum-derived fuels are generally dirtier than natural gas. Crude oil is liquid, a convenient form for its transportation. Petroleum is pumped easily through pipelines or hauled across the oceans in giant tankers. Little energy is required to pump petroleum from the ground.

As domestic supplies diminish, however, we will have to expend an increasing fraction of the energy content of petroleum to transport it to where it is needed. Also, as petroleum is pumped out of a well, that which remains behind is increasingly difficult to get out. This is a natural consequence of the second law of thermodynamics: that which is left behind is more scattered, and more energy is required to collect it.

United States petroleum reserves are expected to decline rapidly. Even the massive field on the North Slope of Alaska would last only 2 or 3 years if it were our only source. Offshore deposits along the Atlantic coast of the United

The invasion of Kuwait by Iraq in 1990 and the subsequent defeat of Iraq by the U.S.-led coalition in early 1991 provide examples of economic and political problems that arise (at least in part) from our dependence on imported oil. The burning of Kuwaiti oil wells by retreating Iraqi forces has created problems that could last for years.

States would produce only a few years' supply, and offshore drilling requires a great deal more energy (and money) than drilling on land.

The developed countries of North America, western Europe, and Japan depend heavily on oil imports from developing nations, some of which are politically unstable. Such dependence makes the industrial nations vulnerable economically and politically.

Gasoline

Gasoline, like the petroleum from which it is derived, is a mixture of hydrocarbons. The typical alkanes in gasoline have formulas ranging from C_5H_{12} to $C_{12}H_{26}$. Since there are many isomeric forms, particularly for the higher members of the group, we see that gasoline is an exceedingly complex mixture of alkanes. There are also small amounts of other kinds of hydrocarbons present, and even some sulfur- and nitrogen-containing compounds. The gasoline fraction of petroleum as it comes from the distillation column is called **straight-run gasoline**. It doesn't burn very well in modern, high-compression automobile engines. Chemists have learned how to modify it in a variety of ways to make it burn more smoothly.

The Octane Rating of Gasolines

Early in the development of the automobile engine, scientists learned that some types of hydrocarbons burned more evenly and were less likely to ignite prematurely than others. Ignition before the piston was in proper position led to a knocking in the engine. Scientists soon were able to correlate good performance with a branched-chain structure in hydrocarbon molecules. An arbitrary performance standard, called the **octane rating**, was established in 1927. Isooctane was assigned a value of 100 octane. An unbranched-chain compound, heptane, was given an octane rating of 0. A gasoline rated 90 octane was one that performed the same as a mixture that was 90% isooctane and 10% heptane.

$$CH_3-\underset{\underset{CH_3}{|}}{\overset{\overset{CH_3}{|}}{C}}-CH_2-\overset{\overset{CH_3}{|}}{CH}-CH_3$$

Isooctane

$$CH_3CH_2CH_2CH_2CH_2CH_2CH_3$$

Heptane

During the 1930s, chemists discovered that the octane rating of gasoline could be improved by heating gasoline in the presence of catalysts such as sulfuric acid (H_2SO_4) or aluminum chloride ($AlCl_3$). This increase in octane rating was attributed to a conversion (**isomerization**) of part of the unbranched structures to highly branched molecules. Heptane molecules can be isomerized to branched structures, for example.

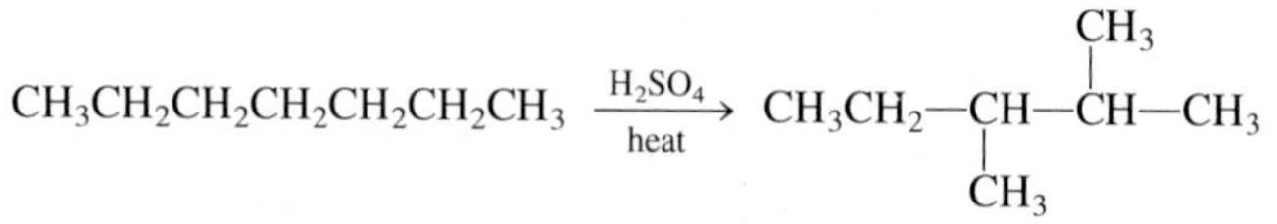

$$CH_3CH_2CH_2CH_2CH_2CH_2CH_3 \xrightarrow[\text{heat}]{H_2SO_4} CH_3CH_2{-}CH(CH_3){-}CH(CH_3){-}CH_3$$

Chemists also are able to combine small hydrocarbon molecules (below the gasoline range) into larger ones more suitable for use as fuel. This process is called **alkylation.** In a typical alkylation reaction, shown below, isobutylene is reacted with propane.

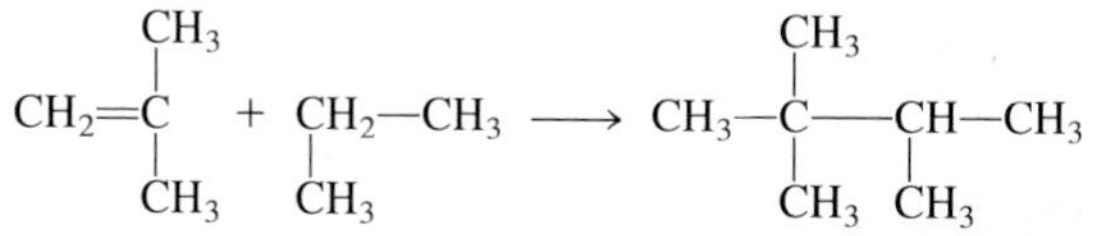

$$CH_2{=}C(CH_3)_2 + CH_3{-}CH_2{-}CH_3 \longrightarrow CH_3{-}C(CH_3)_2{-}CH(CH_3){-}CH_3$$

The product molecules are in the right size range for gasoline, and they are also highly branched, and therefore high in octane number.

Octane Boosters

Certain chemical substances also were discovered that, when added in small amounts, substantially improved the antiknock quality of gasoline. Chief among these additives was tetraethyllead. This compound, when added in amounts as small as 1 mL per liter of gasoline (about 1 part per thousand), would increase the octane rating by 10 or more.

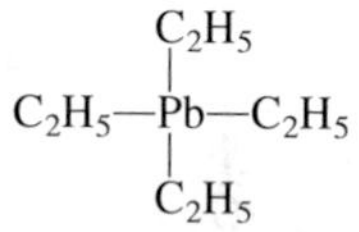

$$Pb(C_2H_5)_4$$

Tetraethyllead

Lead is toxic, however. Large amounts of it have entered the environment through the combustion of leaded gasoline in automobiles. Further, lead fouls the catalytic converters used in modern automobiles. In the United States, because lead would destroy the catalytic converter, unleaded gasoline became available in 1974. Today almost all of our gasoline is unleaded. In Europe leaded gasoline is still being used, but it is being phased out.

Lead is especially toxic to the brain. Even small amounts lead to learning disabilities in children.

Scientists have developed a number of ways to get high octane ratings in unleaded fuels. For example, petroleum refineries use **catalytic reforming** to convert low-octane alkanes to high-octane aromatic compounds. Hexane (with an octane number of 25) is converted to benzene (with an octane number of 106).

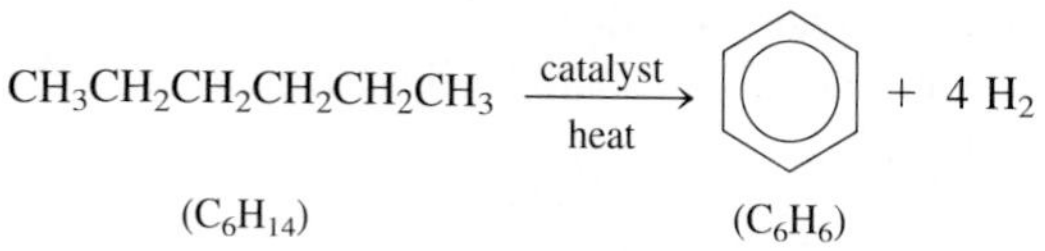

Tetraethyllead was added to gasoline along with ethylene dibromide in order to keep lead metal from fouling the spark plugs in the engine. The lead was converted to lead bromide, which was able to leave the engine in the exhaust gas. Instead of fouling the spark plugs, the lead just fouled up the atmosphere.

$$\underset{(C_6H_{14})}{CH_3CH_2CH_2CH_2CH_2CH_3} \xrightarrow[\text{heat}]{\text{catalyst}} \underset{(C_6H_6)}{\text{benzene}} + 4\,H_2$$

Catalytic reforming was an extremely important oil refining process for half a century. In the decade of the 1990s, however, we are trying to reduce the aromatics in gasoline, and especially benzene, because of health concerns.

Octane boosters to replace tetraethyllead have been developed. Methyl *tert*-butyl ether (Section 9.10) is perhaps the most important. Methanol, ethanol, and *tert*-butyl alcohol also are used. None of these is nearly as effective as tetraethyllead in boosting the octane rating. They must therefore be used in fairly large quantities. The amount that can be used in gasoline is limited by solubility problems. Methanol in excess of 5% and ethanol in excess of 10% tend to separate from the gasoline, especially if moisture gets into the fuel.

Unlike gasoline, which is made up of hydrocarbons, methyl *tert*-butyl ether and the various alcohols and their derivatives all contain oxygen. They are sometimes referred to as "oxygenates." These oxygenates are added to gasoline not only for their octane improvement, but also, and perhaps more importantly, for their ability to decrease carbon monoxide in the auto exhaust gas.

SECTION 14.10

The Fossil Fuel Feast: A Brief Interlude

After the steam engine came into widespread use (by about 1850), the industrial revolution was powered largely by coal. By 1900, about 95% of the world's energy production came from burning coal. With the development of the internal combustion engine, petroleum became increasingly important and by 1950 replaced coal as the principal fuel.

The supply of fossil fuels on the Earth is limited. Estimated United States and world reserves are given in Table 14.3 and shown graphically in Figure 14.14. We burn these materials at a rapid rate. Annual U.S. and world production is given in Table 14.5. Estimates of reserves vary greatly, depending on the assumptions made. Even the most optimistic estimates, however, lead to the conclusion that our nonrenewable energy resources are being depleted rapidly. Indeed, in just a century, we will have used up over half the fossil fuels that were formed over the ages. As far as energy sources are concerned, we are in the midst of the "chemical century."

The ability to modify hydrocarbon molecules enables the petroleum industry to produce increased amounts of whatever fraction they desire. They can, on demand, increase the proportion of gasoline or of fuel oil from a given supply of petroleum. They can even make gasoline from coal. They can't, however, increase the amount of fossil fuels aboard Spaceship Earth. Let's hope that scientists soon develop new sources of energy that do not depend on petroleum, so that we can stop what is, in many ways, a profligate waste of resources. Spaceship Earth has aboard it all the supplies it will ever have. We must use them wisely.

SECTION 14.11

Convenient Energy: Electricity

The convenience of a fossil fuel depends on its physical state: gases are most convenient, liquids next, solids least convenient. Perhaps the most convenient

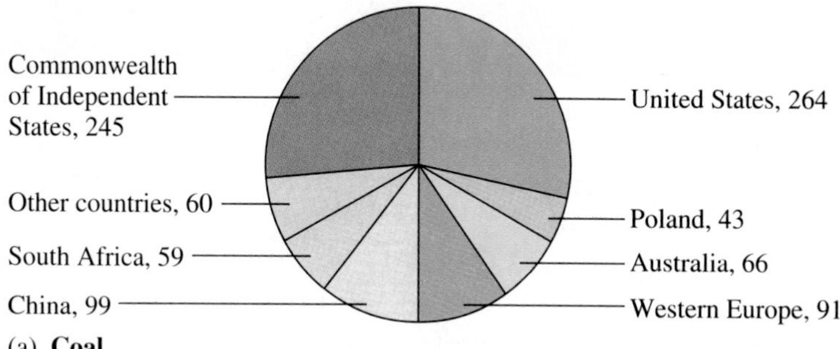

(a) **Coal**

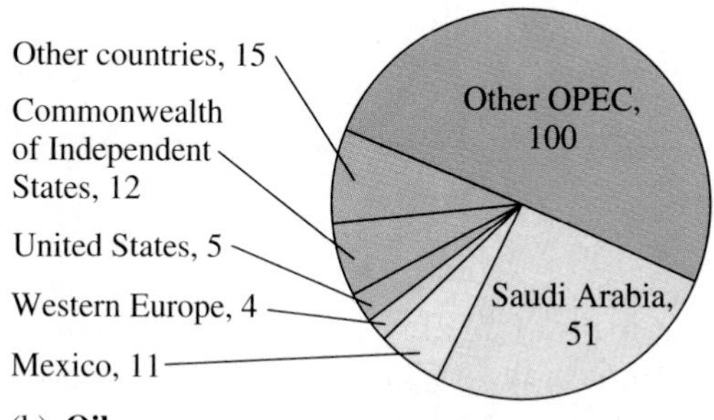

(b) **Oil**

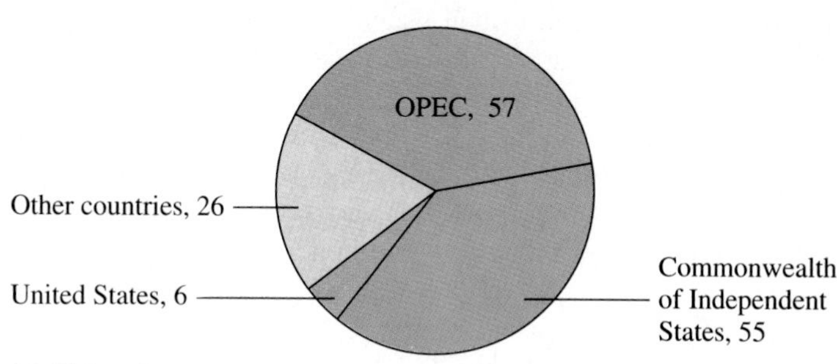

(c) **Natural gas**

Figure 14.14 Estimates of world reserves of coal, petroleum, and natural gas (in millions of metric tons of coal equivalents). Note that the U.S. shares of petroleum and natural gas are meager; coal is plentiful. (Commonwealth of Independent States is the former Soviet Union.)

Table 14.5 Annual United States and World Production of Fossil Fuels*

Fuel	United States	World
Coal	788,752	3,301,831
Petroleum	528,113	4,308,141
Natural gas	598,546	2,499,832

*Expressed in thousands of metric tons of coal equivalents for ready comparison of energy content.

Source: Department of Economic and Social Information and Public Affairs, *Statistical Yearbook 1990–1991*. New York: United Nations, 1993.

Table 14.6 Extraction, Transportation, Distribution, and Convenience of Fuels in Various Physical States

Physical State	Extraction	Transportation to Cities	Distribution Within a City	In Use: Convenience	In Use: Cleanliness
Solids (coal, wood)	Shovels, borers, blasting	Trucks, trains, barges (Slurry with water in pipe)	Trucks, buckets	Least ↓	Dirtiest ↓
Liquids (gasoline, fuel oil)	Pumps	Pipelines, tankers, barges, trucks	Trucks		
Gases (natural gas)	Pumps	Pipelines	Pipes		
Electricity* (electron flow)	—	Wires	Wires	Most	Cleanest

*Produced by burning any of the primary fuels, and included for comparison.

Electricity is a *secondary* energy source. A primary source, such as coal, must be used to produce it.

form of energy of all is electricity (Table 14.6). With it we can have light, and hot water, and can run motors of all sorts. We can use it to heat or cool our homes and workplaces. When looking at future energy sources, then, we look mainly at ways of generating electricity.

Any fuel can be burned to boil water, and the steam produced (in great enough quantities) can turn a turbine to generate electricity. Figure 14.15 shows a coal-fired steam power plant. At present, 56% of U.S. electrical energy comes from coal-burning plants (Figure 14.16). Such facilities are at best only about 40% efficient; 60% of the energy of the fossil fuel is wasted as heat. Today the idea of *cogeneration* is very popular. Power installations are using the waste heat generated to warm buildings.

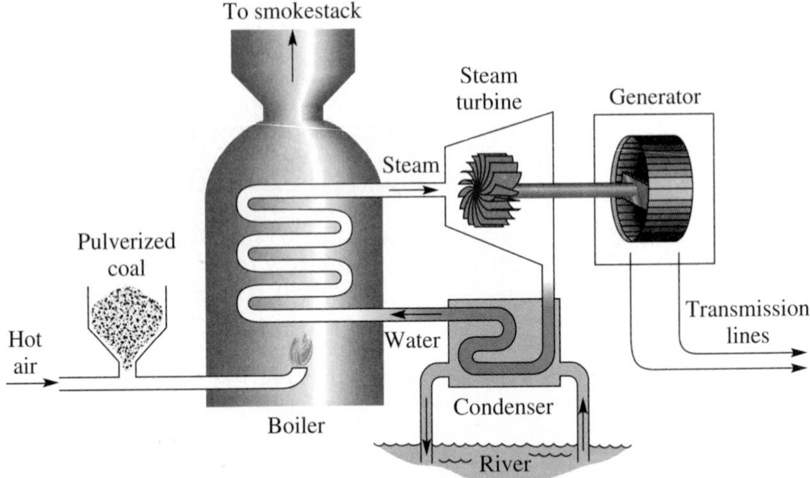

Figure 14.15 A coal-fired power plant for generating electricity.

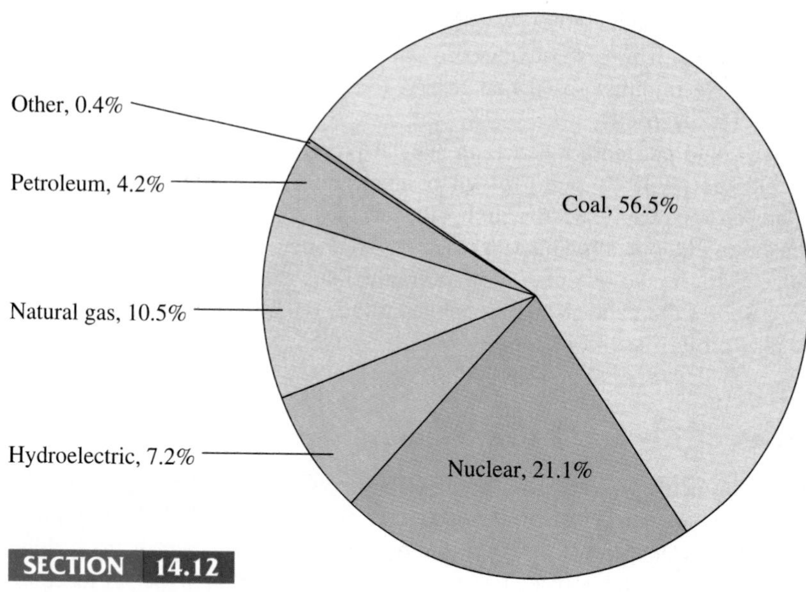

Figure 14.16 Percentages of electric power generation in the United States in 1993 from various energy sources. [*Data from Energy Information Administration,* Monthly Energy Review, *December 1993.*]

SECTION 14.12

Nuclear Power

The **fission reactions** that power nuclear bombs can be controlled in **nuclear reactors**. The energy released during fission can be used to generate steam, and the steam can turn a turbine to generate electricity (Figure 14.17).

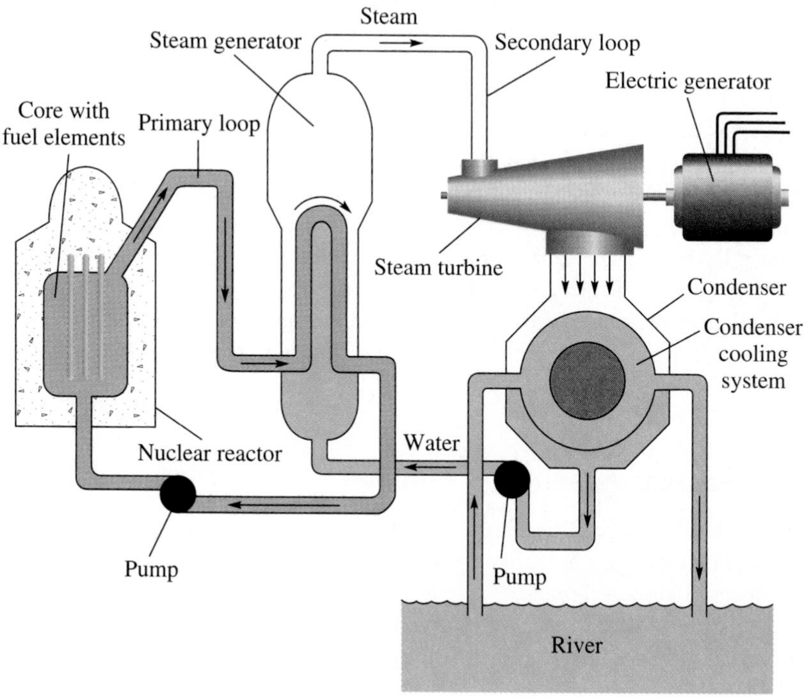

Figure 14.17 Schematic diagram of a nuclear power plant used to generate electricity.

Table 14.7 Percentage of Total Electricity Generated by Nuclear Power Plants

Country*	Percent
Belgium	59.9
Bulgaria	32.5
Finland	33.2
France	72.9
Germany	30.1
Hungary	48.4
Japan	27.7
S. Korea	43.2
Lithuania	80.0†
Spain	36.4
Sweden	43.2
Switzerland	39.6
Ukraine	25.0
United Kingdom	23.2
United States	22.3

*Only those countries generating more than 20% of their electricity by nuclear power are included in this list.
†Estimated.
Source: International Atomic Energy Agency, December 1992.

At present, about 20% of United States electricity comes from nuclear power plants. The eastern seaboard and upper midwestern states, many of which have minimal fossil fuel reserves, are heavily dependent on nuclear power for electricity.

We could put more reliance on nuclear power. Other nations have (Table 14.7). The public is quite fearful of nuclear power plants, however. That apprehension was exacerbated by the accident at Chernobyl. The United States has 109 operating nuclear reactors, but no new plants have been ordered since 1978. It takes about 10 years to build a nuclear power plant. There will have to be a dramatic change in public attitude quite soon if nuclear power is to play a big role in our future.

Types of Nuclear Power Plants

There are actually several types of nuclear power plants. The one in Figure 14.17 is a pressurized water reactor. Earlier models were mainly boiling water reactors in which the steam from the reactor was used to power the turbine directly. No attempt is made here to discuss all the possible types of reactors.

At the dawn of the nuclear age, nuclear power was envisioned by some to be destined to fulfill the Biblical prophecy of a fiery end to our world. Indeed, as the cold war between the United States and the Union of Soviet Socialist Republics intensified during the 1950s, it was difficult to see how nuclear war could be avoided. If it had come, it could well have been the end of life on Earth.

While some were predicting doom, others saw nuclear power as a source of unlimited energy. During the late 1940s, some predicted that electricity from nuclear plants would become so cheap that it eventually would not have to be metered. Nuclear power has not yet brought us either paradise or perdition, but it has become one of the most controversial issues of our time. Our great need for energy indicates that the controversy will continue for years to come.

Nuclear power plants use the same fission reactions as nuclear bombs. A **moderator** is used to slow down the fission neutrons. The reaction is also controlled by the insertion of boron steel or cadmium **control rods**. Boron and cadmium absorb neutrons readily, so the rods prevent the neutrons from participating in the chain reaction. These rods are installed when the reactor is being built. Removing them part way starts the chain reaction; the reaction is stopped when the rods are pushed in all the way.

The United States has 109 operating nuclear power plants. If they were to be replaced by coal-burning plants, airborne pollutants would increase by 19,000 tons per day.

The Nuclear Advantage: Minimal Air Pollution

The main advantage of nuclear power plants over those that burn fossil fuels is in what they do *not* do. Unlike the fossil-burning plants, nuclear power plants produce no carbon dioxide to add to the greenhouse effect, and they add no soot, fly ash, sulfur oxides, or nitrogen oxides to our atmosphere. They contribute almost nothing to global warming, air pollution, or acid rain (Figure 14.18).

Figure 14.18 The cloud above this power plant is just that—a cloud composed of condensed droplets of water. Nuclear power plants emit a minuscule amount of radioactivity, but none of the noxious pollutants that come from coal-burning power plants. [*Sinclair Stammers/SPL/Photo Researchers, Inc.*]

If used in place of oil-burning plants, nuclear power plants also reduce our dependence on foreign oil and lessen our trade deficit.

Problems with Nuclear Power

Nuclear power plants have a number of disadvantages. Elaborate and expensive safety precautions must be taken to protect plant workers and the inhabitants of surrounding areas from radiation.

The reactor itself must be heavily shielded and housed inside a containment building of metal and reinforced concrete (Figure 14.19). Because loss of coolant water can result in meltdown of the reactor core, backup emergency cooling systems have to be constructed. Despite what proponents call the utmost precautions, some opponents of nuclear power still fear a runaway nuclear reaction in which the containment building would be breached and massive amounts of radioactivity would escape into the environment. The chance for such an accident is probably exceedingly small, but if it did occur, thousands could be killed and large areas rendered uninhabitable for centuries.

Figure 14.19 Nuclear reactors are housed in containment buildings made of steel and reinforced concrete. These structures are designed to withstand nuclear accidents and prevent the escape of radioactive materials into the environment. [*Richard Megna/Fundamental Photographs.*]

The production of nuclear weapons also results in radioactive wastes. Whether or not we ever had used nuclear reactions to produce electricity, we still would have nuclear wastes to store.

The United States tentatively has decided to store its nuclear wastes deep within Yucca Mountain in a remote area of Nevada. Scientific evaluation of the site, lawsuits, and other considerations likely will delay the construction for years. Its use is unlikely before 2010, if then. In the meantime, wastes are stored under water on the sites of the plants. (It was intended that on-site storage be temporary, just long enough for short-lived radioisotopes to decay.) These temporary tanks are rapidly being filled; the permanent site doesn't seem to get any closer to realization.

The benefit of nuclear power—abundant electrical energy—is clear, but the small probability of an accident causes scientists and others to debate endlessly the desirability of nuclear power.

Another problem with nuclear power is that the fission products are highly radioactive and must be isolated from the environment for centuries. Again, scientists disagree about the feasibility of nuclear waste disposal. Proponents of nuclear power say that such wastes can be safely stored in old salt mines or other geologic formations. Opponents fear that the wastes may arise from their ''graves'' and eventually contaminate the groundwater. It is impossible to do a million-year experiment in a few years to prove who is right.

There also are wastes from the mining of uranium ore. Over 200 million tons of *tailings* (wastes left from uranium ore processing) now plague ten western states (Figure 14.20). These tailings are mildly radioactive, giving off radon gas and gamma radiation. Dust from these tailings carries problems to surrounding areas.

Still another problem, **thermal pollution**, is unavoidable. As the energy from any material is converted to heat to generate electricity, some of the energy is released into the environment as ''waste'' heat. Nuclear power plants do generate more thermal pollution than plants that burn fossil fuels, but the difference is not as important, perhaps, as the other problems we have mentioned.

Nuclear Accidents: Real and Imagined Risks

In 1979, a loss-of-coolant accident at the Three Mile Island nuclear power plant near Harrisburg, Pennsylvania, released a very small amount of radioac-

Figure 14.20 Mildly radioactive tailings are a by-product of uranium processing. [*Courtesy of U.S. Department of Energy.*]

Figure 14.21 An accident at this nuclear power plant at Chernobyl, Ukraine, increased public fears of nuclear power. The reactor lacked a reinforced containment building, so considerable radioactivity was released. The plant used graphite as a moderator (U.S. plants use water). The burning graphite hindered efforts to tame the runaway nuclear reaction. [*Sovfoto/Eastfoto.*]

tivity into the environment. Although no one was killed or seriously injured, this accident whetted public fear of nuclear power. The accident at Chernobyl was much more frightening. The reactor core meltdown killed several people outright. Others died from radiation sickness in the following weeks and months. Thousands were evacuated. A large area will remain contaminated for decades. Radioactive fallout spread across much of Europe. Thousands, particularly those close to the accident, are at greatly increased risk of cancer from exposure to the radiation. At Three Mile Island, a containment building kept most of the radioactive material inside. The Chernobyl plant (Figure 14.21) had no such protective structure.

An unfounded fear is that nuclear power plants might blow up like the bombs that devastated Hiroshima and Nagasaki—but uranium-fueled plants can't explode like nuclear bombs. The uranium is enriched to, at most, 3 or 4% uranium-235. To make a bomb, you need about 90% uranium-235.

There is considerable controversy over most aspects of nuclear power, and there are scientists on both sides. While they may be able to agree on the results of laboratory experiments, scientists obviously do not agree on what is best for society.

A study of nuclear shipyard workers tracked over 70,000 workers for about 20 years. They had 24% lower mortality than nonnuclear workers in the general population. Those with the highest chronic radiation exposure had the lowest mortality from all causes—including cancer. Maybe small doses of radiation might be beneficial.

Breeder Reactors: Making More Fuel Than They Burn

Less than 1% of naturally occurring uranium is the fissionable uranium-235 isotope. Large quantities of uranium-238 are available as a by-product of the production of uranium-235. Uranium-238 can be converted to fissile plutonium-239 by bombardment with neutrons. Uranium-239 is formed initially,

$$^{238}_{92}U + ^{1}_{0}n \longrightarrow ^{239}_{92}U$$

$$^{239}_{92}U \longrightarrow ^{239}_{93}Np + ^{0}_{-1}e$$

$$^{239}_{93}Np \longrightarrow ^{239}_{94}Pu + ^{0}_{-1}e$$

Plutonium-breeding reactions

(a)

$$^{232}_{90}Th + ^{1}_{0}n \longrightarrow ^{233}_{90}Th$$

$$^{233}_{90}Th \longrightarrow ^{233}_{91}Pa + ^{0}_{-1}e$$

$$^{233}_{91}Pa \longrightarrow ^{233}_{92}U + ^{0}_{-1}e$$

Uranium-233-breeding reactions

(b)

Figure 14.22 Reactions for breeding fissile fuels from nonfissile isotopes. In each case a nonfissile isotope is converted by neutron bombardment into an unstable isotope. That isotope decays into another unstable isotope, which in turn decays into a fissile isotope.

but it rapidly decays to neptunium-239, and the neptunium quickly decays to plutonium. The reactions are shown in Figure 14.22a.

A reactor can be built with a core of fissionable plutonium surrounded by uranium-238. As the plutonium fissions, neutrons convert the uranium-238 shield to more plutonium; thus, the reactor breeds more fuel than it consumes. There is enough uranium-238 to last several centuries, so one of the disadvantages of nuclear plants could be overcome by the use of **breeder reactors**.

Breeder reactors have some problems of their own, however. Plutonium is fairly low melting (640 °C), and the plant is limited to fairly cool—and inefficient—operation. Water is not adequate as a coolant; it is necessary to use molten sodium metal in the primary loop. These reactors often are called liquid-metal fast breeder reactors (LMFBR). If an accident occurred in such a breeder, the sodium could react violently with both the water and air.

Since plutonium is low melting, a failure of the cooling system could cause the reactor's core to melt. All reactors are required to have an emergency backup core-cooling system. Whether these systems work or not is a principal area of controversy.

Plutonium is highly toxic. It emits alpha particles, a property that makes it especially dangerous when it is ingested. It is estimated that 1 microgram (μg) in the lungs of a human is enough to induce lung cancer. A further hazard is that reactor-grade plutonium, unlike the uranium used in nuclear reactors, could be readily converted into a nuclear bomb. There is even the possibility that terrorist groups could fashion a crude weapon from stolen plutonium and threaten to destroy an entire city to force authorities to meet their demands.

Plutonium is produced for nuclear weapons, some in ordinary (nonbreeder) reactors. Plutonium has a half-life of about 25,000 years. Assuming that it will be essentially gone after 10 half-lives, we can estimate that our descendants would have to contend with plutonium for 250,000 years.

Another possible breeder reaction is that in which thorium-232 is converted to fissile uranium-233 (Figure 14.27b). It is thought that it would be rather difficult to make bombs from reactor-grade uranium-233. However, uranium-233 is like plutonium in that it emits damaging alpha particles.

No breeder reactors are operating in the United States, but they are being used in France and several other countries.

SECTION 14.13

Nuclear Fusion: The Sun in a Magnetic Bottle

In Chapter 4, we discussed the **thermonuclear reactions** that power the sun and that have been adapted by scientists and engineers to make hydrogen bombs. If we could find some way to control these fusion reactions and use them to produce electricity, we would have a nearly unlimited source of power. To date, the **fusion reactions** are useful only for making bombs, al-

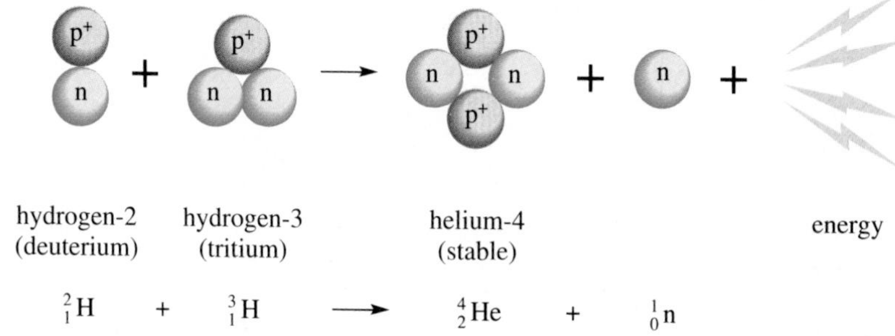

Figure 14.23 The fusion reaction that is most promising is the deuterium-tritium (DT) reaction. A hydrogen-2 (deuterium) isotope fuses with a hydrogen-3 (tritium) nucleus to form a helium nucleus plus a neutron with the release of considerable energy.

though research in the control of nuclear fusion is progressing (Figure 14.23). Controlled fusion would have several advantages over nuclear fission reactors. The principal fuel, deuterium ($^{2}_{1}H$), is plentiful and is obtained by the fractional electrolysis (splitting apart by means of electricity) of water. (Only 1 hydrogen atom in 5000 is a deuterium atom, but we have oceans full of water to work with.) The problem of radioactive wastes would be minimized. The end product—helium—is stable and biologically inert. Escape of tritium might be a problem, because this hydrogen isotope would be readily incorporated into organisms. Tritium ($^{3}_{1}H$) undergoes beta decay, with a half-life of 12.3 years. And there is one other problem associated with any production and use of energy: the unavoidable loss of part of the energy as heat. We would still be concerned with thermal pollution.

Great technical difficulties have to be overcome before a controlled fusion reaction can be used to produce energy. Temperatures of 50,000,000 °C would have to be attained, and no material on Earth could withstand more than a few thousand degrees. At a temperature of 50,000,000 °C, no molecule could hold together, nor could the atoms from which the molecule is made. All atoms would be stripped of their electrons, and the nuclei and free electrons would form a mixture called a **plasma**. There is hope that this plasma could be contained by a strong magnetic field (Figure 14.24), however. Scientists in

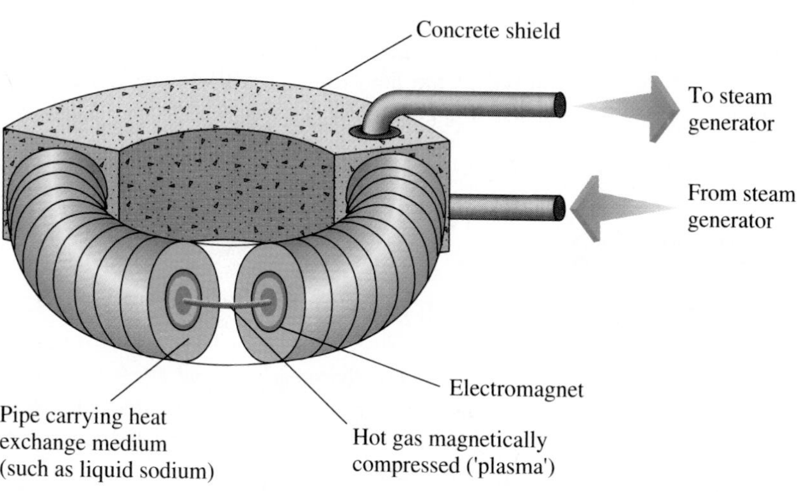

Figure 14.24 A giant donut-shaped electromagnet called a *tokamak* is designed to confine plasma at the exceedingly high temperatures and pressures required for nuclear fusion.

Perhaps you have seen headlines or read articles referring to *cold* fusion. In the spring of 1989 two chemists named Stanley Pons (University of Utah) and Martin Fleischman (University of Southampton, England) announced that they had achieved nuclear fusion in an ordinary laboratory at ordinary temperatures. The process immediately became known as "cold fusion."

What they had done was to carry out an electrolysis of heavy water using palladium electrodes. Deuterium, like hydrogen, is quite soluble in palladium. If the electrolysis is allowed to run long enough, the palladium becomes "saturated" with deuterium, so that the deuterium atoms are pushed very close together within the palladium crystal structure. Pons and Fleischman claimed that much more energy was produced by the cathode than was used for the electrolysis. (In one experiment, the cathode actually melted.)

Many other scientists have tried to repeat this work. Some have been at least somewhat successful, but others have not. Part of the problem is the unpredictability of the system. Excess heat is not given off every time. Even when extra heat is given off, it can stop quite abruptly. Sometimes a cell can run for months with no sign of excess energy, and then it may start giving off large amounts of energy for several days before stopping as suddenly as it began.

Something unusual certainly seems to be happening with this system, but whether or not it actually involves nuclear fusion remains to be seen.

several nations are getting closer to the development of the environment necessary for controlled fusion.

Nuclear fusion may well be our best hope for relatively clean, abundant energy in the future. Much work remains to be done, however. Even when controlled fusion is achieved in the laboratory, it will still be a long time before it becomes a practical source of energy.

SECTION 14.14

Harnessing the Sun: Solar Energy

At the beginning of the chapter, we saw that nearly all the energy on Earth comes from the sun. With all that energy from our celestial power plant, why do we face an energy crisis? The answer lies in the fact that this energy is thinly spread and difficult to concentrate.

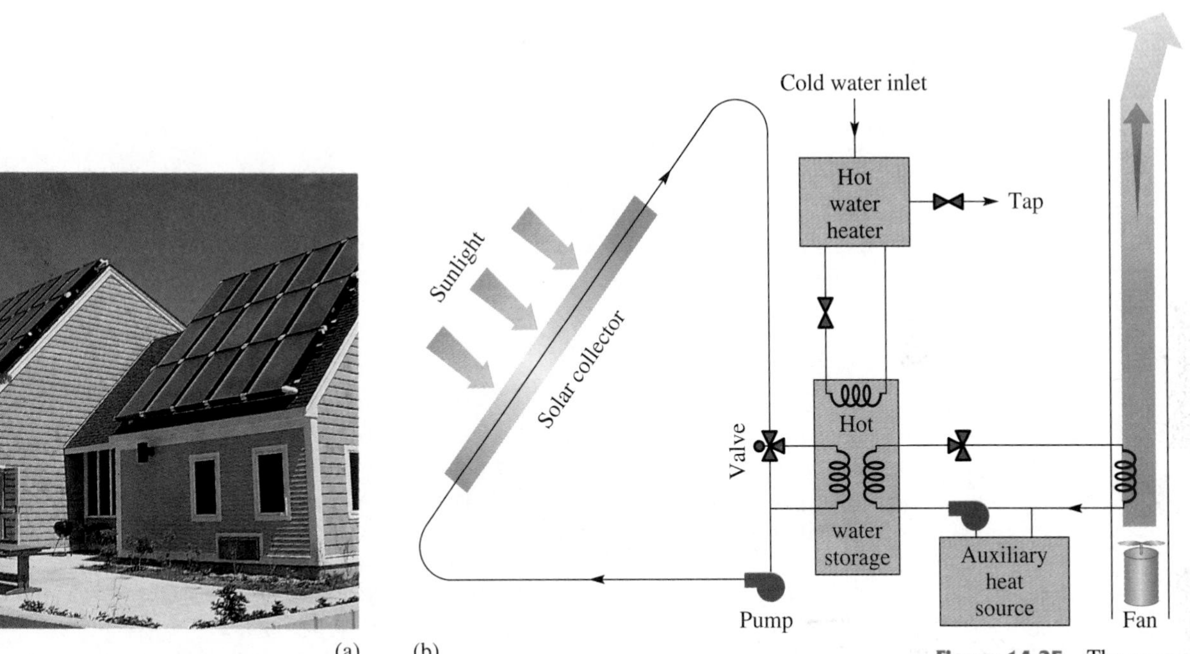

Figure 14.25 The energy in sunlight can be absorbed by solar collectors and used to heat water. The hot water can be used directly, or it can be used to (partially) heat a building. The photograph (a) shows solar collectors on a rooftop. The diagram (b) shows how a solar collector can furnish hot water and warm air heat. [*(a) Bruce W. Wellman/Stock Boston.*]

Solar Heating

Diffuse energy is not very useful. As it arrives on the surface of the Earth, about half of the solar energy is converted to heat. Another 30% is simply reflected back into space. We can increase the efficiency of this conversion rather easily. A black surface absorbs radiation better than a colored one. A simple solar collector can be made by covering a metal surface, painted black, with a glass plate. The glass is transparent to the incoming solar radiation, but it partially prevents the heat from escaping back into space. The hot surface is used to heat water or other fluids, and the hot fluids usually are stored in an insulated tank.

Water heated in this manner can be used directly for bathing, dish washing, and laundry. To heat a building, air can be warmed by being passed around the reservoir and then circulated through the building (Figure 14.25). Even in the cold northern climates, about 50% of home heating requirements could be met by solar collectors. These installations are expensive but could pay for themselves, through fuel savings, in a few years.

Solar Cells: Electricity from Sunlight

Sunlight also can be converted directly to electricity by devices called **photovoltaic cells** or, more simply, **solar cells**. These devices can be made from a variety of substances, but most are made from elemental silicon. In a crystal of

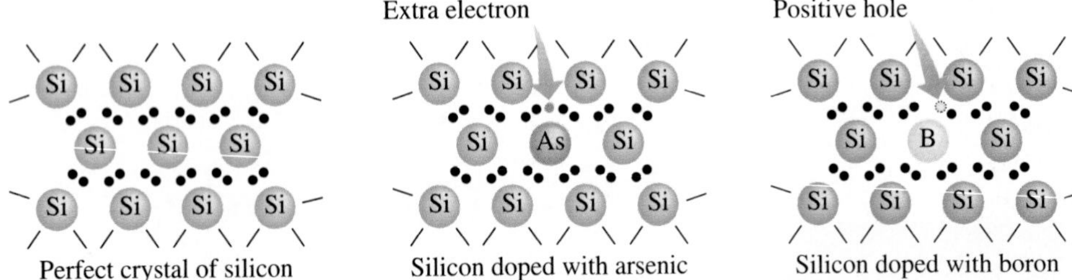

Figure 14.26 Models of silicon crystals. Crystals doped with impurities are used in solar cells.

pure silicon, each silicon atom has four valence electrons and is covalently bonded to four other silicon atoms (Figure 14.26). In the manufacture of a solar cell, extremely pure silicon is doped with small amounts of specific impurities and formed into single crystals. One type of crystal has about 1 part per million (ppm) of arsenic added. Arsenic atoms have five valence electrons, four of which are used to form bonds to silicon atoms. The fifth electron is relatively free to move around. This type of crystal, with extra electrons, is called a *donor crystal.* Another crystal is made by adding about 1 ppm of boron. Boron atoms have three valence electrons. These three electrons are used to bond silicon atoms, but there is a shortage of one electron, leaving a *positive hole* in the crystal. This boron-doped crystal is called an *acceptor crystal.*

When the two types of crystals (acceptors and donors) are joined, there is a strong tendency for electrons to flow from the donor to the acceptor. However, the holes near the junction are quickly filled by nearby mobile electrons, and the flow ceases. When sunlight strikes the cell, more electrons are dislodged, creating more mobile electrons and more positive holes. When the two crystals are connected by an external circuit, electrons flow from the donor to the acceptor (Figure 14.27).

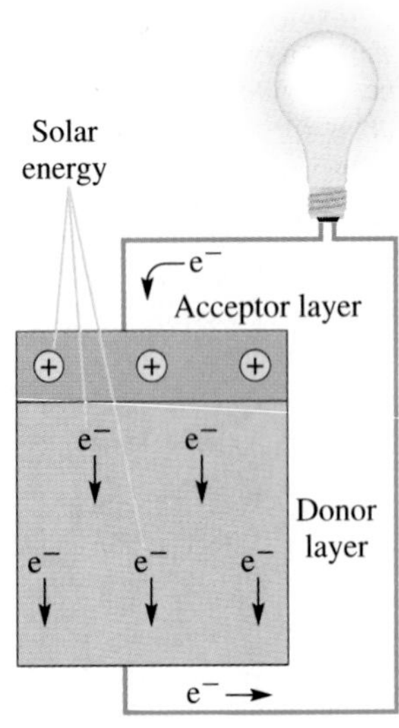

Figure 14.27 Schematic diagram of a solar cell. Electrons flow from the donor crystal to the acceptor crystal through the external circuit.

An array of solar cells can be combined to form a solar battery. Such solar batteries can produce about 100 W per square meter of surface; it takes a square meter of cells to power one 100-W light bulb. Solar batteries have been used for years to power spaceships. They are now being used to provide electricity for weather instruments in remote areas. Prices are decreasing, and solar batteries are now widely used to power small devices such as electronic calculators.

Solar cells are not very efficient. Much of the sunlight striking them is reflected back into space. Their present efficiency of conversion is only about 10%. Generation of enough energy to meet a significant portion of our demands would require covering vast areas of desert land with solar cells. It would require 2000 hectares (ha), or about 5000 acres, of cloud-free desert land to produce as much energy as one nuclear power plant. Research is under way to increase the efficiency of solar cells. It is hoped that an efficiency of 20 to 30% can be attained, thus reducing the land area required to 50% of that now needed.

Over 10% efficiency has now been achieved using amorphous silicon, which is much easier to produce than silicon crystals.

Using solar energy would require the storage of energy for use at night and on cloudy days. One scheme for this involves storage of energy as heat. While the sun shines, energy would be transferred to tanks of molten salts. Then, heat from these salts would, as needed, be used to generate steam to run a turbine, which would generate electricity.

The technology is available now for the use of solar energy for space heating and for providing hot water. Electricity from solar energy is probably several years away, however, and more research and development are needed.

SECTION 14.15

Biomass: Photosynthesis for Fuel

Why bother with solar collectors and photovoltaic cells to capture energy from the sun when green plants do it every day? Indeed, the idea of growing plants and burning them for energy has been explored. Dry plant material burns quite well. It could be used to fuel a power plant for the generation of electricity. We could start "energy plantations" to grow plants for use as fuel. Burning this plant **biomass** has some nice advantages: it is a *renewable resource* whose production is powered by the sun (Figure 14.28).

Hemp, the fiber crop from which rope is made, is said to be a good energy source. It has woody stalks that are 77% cellulose, and they can produce 10 tons of biomass per acre in just 4 months. The oil from hemp seeds can also be used as diesel fuel. There is just one problem: it is illegal to raise hemp in the United States. (The leaves and flowers can be used to make the drug marijuana.)

Unfortunately, there are a number of disadvantages to this scheme, too. Most of the available land is needed for the production of food. Even where productive land is available, there are problems. The plants have to be planted, harvested, and hauled to the power plant. Often, the land is far from where the energy is needed, and the overall efficiency is even less than that of solar cells—only about 3% at best.

Nevertheless, there is considerable research activity in the area of biomass. Scientists in upstate New York, for example, are actively investigating fast-growing trees as fuel for utility companies.

Plant material does not have to be burned directly. Plants high in starches and sugars can be fermented to form ethyl alcohol. Wood can be distilled in the absence of air to produce methyl alcohol. Both alcohols are liquids and convenient to transport. Both are also excellent fuels that burn relatively cleanly. Bacterial breakdown of plant material produces methane. Under proper conditions, this process can be controlled to produce a clean-burning fuel similar to natural gas. It should be noted, however, that any of these conversions must result in the loss of a portion of the useful energy. The laws of thermodynamics tell us that we would get the most energy by burning the biomass directly rather than converting it to a more convenient liquid or gaseous fuel.

Tumbleweed has been pressed into logs, peanut shells have been formed into briquettes, and mutton fat has been added to fuel oil.

The shortage of land probably means that we will never obtain a major portion of our energy from biomass. We could, however, supplement our other

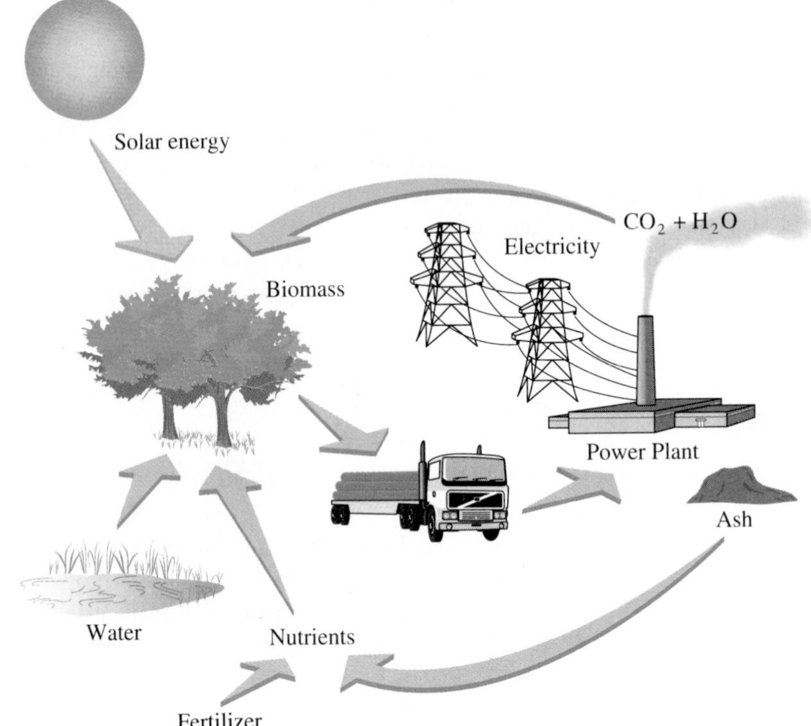

Figure 14.28 Energy from the sun can be converted to electrical energy by way of green plants. If the ash from the burned plants was not returned to the soil, the soil would become depleted of nutrients and have to be fertilized. Indeed, most of the land being considered for biomass production is too poor to be productive without massive applications of fertilizer.

sources by burning agricultural wastes, fermenting some to ethyl alcohol, producing methyl alcohol from wood where wood is plentiful, and fermenting human and animal wastes to produce methane. The technology for all these processes is readily available. Each has been used in the past. Several of the processes now are being used on a limited scale or are being investigated for use in the future.

SECTION 14.16

Other Energy Sources

Modern civilization requires a lot of energy. Our reserves of fossil fuels are running out, and nuclear energy has some serious problems. Using the sun's energy sounds like a good idea, but solar energy cannot fill all our energy needs. It is clear that we are going to need some alternative energy sources, and many different possibilities are being examined.

In this section we look at various other kinds of energy sources. Some of them are only potential sources, but some are already being used. Indeed, some have been used for centuries.

Figure 14.29 The Grand Coulee Dam on the Columbia River in Washington State produces over 5000 MW of electrical energy. [*James Hanley/Photo Researchers, Inc.*]

Wind Power and Water Power

The sun, by heating the Earth, causes the winds to blow and the water to evaporate and rise in the air, later to fall as rain. The blowing wind and the flowing water can be used as sources of energy. Indeed, they have been used that way for centuries. Waterwheels were used to lift water into the irrigation ditches of ancient Egypt. Windmills were used in tenth-century Persia to grind grain.

What is the connection between waterpower and wind power and chemistry? Chemists are involved in the production of new materials such as metal alloys, plastics, and lubricants that are vital to the construction and operation of dams, windmills, and turbines. Chemists also monitor water quality above and below hydroelectric dams, and they participate in many other activities vital to the protection of the environment around generating facilities.

Why not use wind power and waterpower to solve the energy crisis? We do use waterpower. Nearly 10% of our current electricity production is hydroelectric, most of it in the mountainous western United States. In the modern hydroelectric plant, water is held behind huge dams (Figure 14.29). Some of this stored water is released against the blades of water turbines. The potential energy of the stored water is converted to the kinetic energy of flowing water. The moving water imparts mechanical energy to the turbine. The turbine drives a generator that converts mechanical energy to electrical energy.

Hydroelectric plants are relatively clean, but most of the best dam sites in the United States have already been used. To obtain more hydroelectric energy, we would have to dam up scenic rivers and flood valuable cropland and recreational areas. Reservoirs silt up over the years, and sometimes dams break, causing catastrophic floods. Even hydroelectric power has its problems.

We could use more wind power. The kinetic energy of moving air is readily converted to mechanical energy to pump water and grind grain. Windmills have been so used for centuries. Wind also can be used to turn turbines and generate electricity. Giant windmills have been built to test and develop this potential (Figure 14.30). Smaller units are available for use on farms and for rural dwellings.

Interest in wind power continues to be strong. In 1994 utility companies in New England, the Midwest, and the Southwest United States announced plans to build some experimental wind turbine generators for producing electricity.

Properly developed, wind power could supply about 10% of our energy

Figure 14.30 Wind turbine generators at Altamont Pass, California, that produce electricity. [*Dan Cabe/Photo Researchers, Inc.*]

needs. Wind is clean, free, and abundant. However, one serious drawback is that the wind does not always blow. Some means of energy storage or an alternate source of energy must be available. Land use might become a problem if wind power was used widely. Presumably, land under windmills could be used for farming or grazing, but giant complexes with power lines tend to be ugly.

The Tides: Moon Power

The power of the tides is another potential source of energy. One tide-powered generating plant is now in operation in France. Although the tides produce great amounts of energy, there are few appropriate sites available in the United States. Some sites that could be used, such as Passamaquoddy Bay, are prized for their beauty; any attempt to construct a power plant would be strongly opposed by environmentalists (Figure 14.31). Another drawback is that electricity could be generated only when the tide is coming in or going out. There would be breaks in the production of electricity and a system of energy storage would be required.

Geothermal Energy

The geysers in northern California supply about 7% of the state's electric power. This geothermal power is low in cost (second only to hydropower).

The interior of the Earth is heated by immense gravitational forces and by natural radioactivity. This heat comes to the surface in some areas through geysers and volcanoes. **Geothermal energy** has long been used in Iceland, New Zealand, and Italy. It has some potential in the United States—indeed, it is being used now in California (Figure 14.32)—but in the near future this potential could only be realized in areas where steam or hot water is at or near

Figure 14.31 The tide comes in on a rocky coast. Are we willing to give up the view for a tide-powered electricity-generating plant? [*Michael Neveux/Westlight.*]

the surface. One drawback of geothermal energy is that the wastewater is quite salty; its disposal could be a major problem.

Oil Shale and Tar Sands

There are fossil carbon compounds other than coal, petroleum, and natural gas. For example, there are immense reserves of oil shale in Colorado, Utah, and Wyoming and of tar sands in Alberta, Canada. The problem with both resources is that their extraction is difficult and environmental disruption can be severe.

(a)

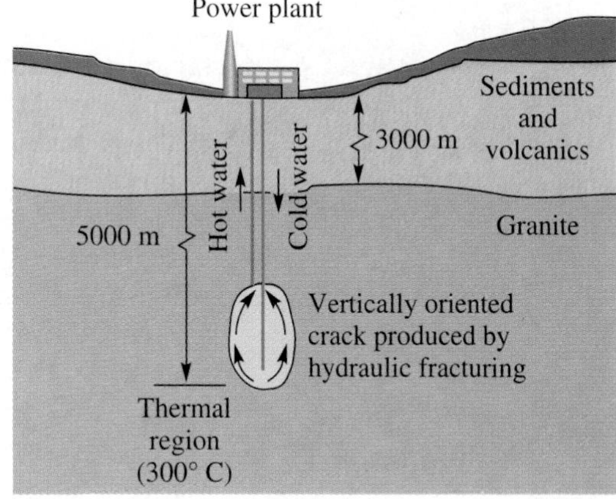

(b)

Figure 14.32 Geothermal energy has the potential of contributing to our energy supply. (a) A commercial geothermal plant at the Geysers in Sonoma County, California. (b) A scheme for extracting energy from dry hot rock 5 km below the surface of the Earth. [*(a) Courtesy of Sweinn Einareson/UN Photos.*]

Figure 14.33 Oil shale is a rock rich in kerogen. Pyrolysis (heating in the absence of air) of oil shale produces a liquid from which fuels can be made. [*Courtesy of American Petroleum Institute Photographic and Film Services.*]

Actually, **oil shale** (Figure 14.33) contains no oil. The organic matter is present as a complex material called **kerogen**, which has an approximate composition of $(C_6H_8O)_n$, where n is a large number. When heated in the absence of air, kerogen breaks down, forming hydrocarbon oils similar to petroleum. The problem is that most of the kerogen is thinly distributed through rock. Although it does not seem to be covalently bonded to the rock, it can't be pumped out in the way petroleum can be pumped from the porous rock in which it occurs. It takes a lot of energy to get oil from oil shale, and there is a lot of waste rock left to dispose of. The energy content of the shale oil—by the time it is refined into gasoline and fuel oil—is not a whole lot greater than the energy put into producing it.

Tar sands (Figure 14.34) are quite different from oil shale, but the problems associated with extracting fuel from them are similar. The organic matter of **tar sands** is present as **bitumen**, a hydrocarbon mixture. Separating bitumen from sand is difficult and requires a lot of energy. As in oil shale, the organic matter is thinly distributed in a lot of inorganic waste.

It may be that we will someday get a significant portion of our energy from oil shale and tar sands. It will be expensive energy, though. The thinly distributed, intimately mixed organic matter is high-entropy stuff. To get it into a useful, low entropy (purified) form will require a lot of energy. The return on the energy invested will probably be rather low.

Oil from Seeds

Many plants produce oil-bearing seeds. Cotton, peanuts, soybeans, and sunflowers are a few examples. Usually the oils are extracted and bottled as salad

Figure 14.34 Tar sands are composed of bitumen, a tarry hydrocarbon mixture, and sand. [*Courtesy of American Petroleum Institute Photographic and Film Services.*]

or cooking oil, but the oils also have value as fuels. Some farmers in North Dakota have used sunflower oil to power their diesel tractors.

In Brazil there has already been considerable use of soybean oil as diesel fuel.

Since plants are a renewable resource, fast-growing plants with high oil yields might be developed for producing oil-seed fuels. If the plants could also be made to grow in areas and climates unsuitable for most food crops, perhaps they might really help to solve part of our energy problem.

Coal Gasification and Liquefaction

Quite often, in discussions of future energy supplies, we hear about converting coal to gas or oil. We are running short of gas and petroleum. Why not make them out of coal? The technology has been around for years. Why not get on with it? Gasification and liquefaction do have some advantages: gases and liquids are easy to transport, and the process of conversion leaves the sulfur behind, thus overcoming one of the serious disadvantages of coal as a fuel.

There are a variety of experimental projects under way in which coal is being converted to a synthetic gaseous fuel. The basic process is one of reduction of carbon by hydrogen.

$$C + 2\,H_2 \longrightarrow CH_4$$

The hydrogen can be produced by passing steam over hot charcoal.

$$C + H_2O \longrightarrow CO + H_2$$

Coal also can be reacted with hydrogen to form petroleumlike liquids from which gasoline and fuel oil can be made.

Coal can also be converted to methanol, which can be used directly as fuel or converted, in turn, to gasolinelike hydrocarbons. In the first step, the coal reacts with extremely hot steam to form carbon monoxide and hydrogen.

$$C + H_2O \longrightarrow CO + H_2$$

Part of the carbon monoxide is then reacted with water to form more hydrogen.

$$CO + H_2O \longrightarrow CO_2 + H_2$$

The carbon monoxide is combined with hydrogen to form methanol.

$$CO + 2\,H_2 \longrightarrow CH_3OH$$

Finally, the methanol is converted to a hydrocarbon mixture that we represent as C_nH_m.

$$n\,CH_3OH \longrightarrow C_nH_m + x\,H_2O$$

The Mobil process is actually being used in New Zealand. However, the fuel is very expensive.

This process was invented by Mobil Oil Corporation. Each step requires an appropriate catalyst.

But both gasification and liquefaction of coal require a lot of energy. The process is inherently wasteful: up to one-third of the energy content of the coal is lost in the conversion. Liquid fuels from coal are high in unsaturated hydrocarbons and in sulfur, nitrogen, and arsenic compounds. Their combustion products are high in particulate matter. Coal conversions also require large amounts of water, yet the large coal deposits on which they would be based are in arid regions. Further, the conversions are messy. Without stringent safeguards, the plants would seriously pollute both air and water. To increase our supply of more convenient fossil fuels—oil and gas—we would deplete our third fossil fuel—coal—even more rapidly.

Other methods that have been used to make liquid fuels from coal are the Bergius and Fischer-Tropsch methods. In the Bergius process the coal is reacted with hydrogen. In the Fischer-Tropsch method the coal is reacted with steam to make a mixture of carbon monoxide (CO) and hydrogen (H_2), which is then reacted over an iron catalyst to produce a mixture of hydrocarbons. Both processes were used in Germany during World War II. Since 1955 the SASOL process, which uses the Fischer-Tropsch method, has provided a significant portion of the liquid fuels for South Africa.

Hydrogen as Fuel

Just as natural gas is sent through pipes to wherever it is needed, so other fuel gases could also be sent through those same pipes. One such gas is hydrogen (H_2).

When hydrogen burns, it produces water and gives off energy.

$$2\,H_2 + O_2 \longrightarrow 2\,H_2O + 572\text{ kJ}$$

Several automobile companies have built experimental cars that operate on hydrogen fuel.

On a weight basis, hydrogen has more energy than any other fuel (except nuclear fuel). It is also a very clean fuel, yielding no chemical products other than water. Since hydrogen can be made from seawater, the supply is virtually inexhaustible.

It should be noted, however, that hydrogen is strictly a secondary energy source, since hydrogen does not exist in significant amounts on Earth. It must be produced from a hydrogen compound (usually water), and that requires energy. Hydrogen, therefore, is not really a basic energy source but just a convenient energy delivery system.

One possibility is that solar cells might be used to produce hydrogen from water by electrolysis. The hydrogen gas could then be sent by pipelines to various destinations. It costs only about only about one-fourth as much to pipe hydrogen across a long distance at is does to transmit high voltage electricity.

Some people are afraid of using hydrogen as fuel for home heating because of the possibility of explosion, such as occurred in the *Hindenburg* disaster. (As shown in Figure 8.6, the hydrogen-filled German dirigible exploded in flames on landing at Lakehurst, N.J., in May 1937.) It might be pointed out that another large dirigible, the *Graf Zeppelin,* which was also filled with hydrogen, crossed the Atlantic many times on a regular basis from 1928 until 1937, when passenger airships were discontinued. It might also be mentioned that serious explosions can result from natural gas leaks, too.

Hydrogen escapes more readily than natural gas or gasoline because of its smaller molecular size, but it also dissipates more readily, so that it could never form a flammable low-lying vapor as gasoline sometimes does.

Alcohol as Fuel

Ethanol (ethyl alcohol or grain alcohol) can be mixed with gasoline and sold under the name ''gasohol.'' Up to 10% alcohol can be added to gasoline without any need to adjust the auto engines using it. If the engines are adjusted, however, automobiles can run on alcohol alone. About half of Brazil's cars run on alcohol made from sugar cane. (The alcohol they use is 95%, rather than the more expensive absolute alcohol that is used in making gasohol.) In the United States the ethanol is usually made by fermentation of corn.

Methanol (methyl alcohol or wood alcohol) can also be used as automobile fuel. It is less popular than ethanol, however, because it is much more toxic and it promotes rusting of engine parts.

Fuel Cells

A **fuel cell** is a device in which fuel is oxidized in an electrochemical cell (Chapter 8) so as to produce electricity directly. These fuel cells differ from the usual electrochemical cell in two ways. First, the fuel and oxygen are fed

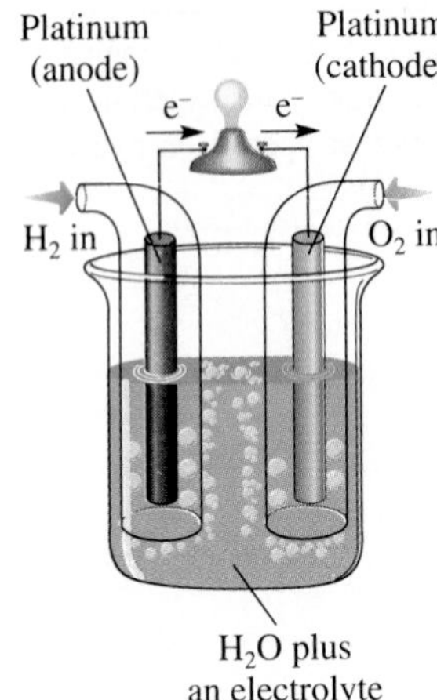

Figure 14.35 A hydrogen-oxygen fuel cell.

in continuously. Second, the electrodes are made of an inert material such as platinum that does not react during the process. The electrodes serve to conduct the electrons to and from the solution.

Consider a fuel cell using hydrogen and oxygen (Figure 14.35). Hydrogen is oxidized at the anode; oxygen is reduced at the cathode. The overall reaction is similar to combustion, but not all the chemical energy is converted to heat.

$$2\,H_2 + O_2 \longrightarrow 2\,H_2O$$

Rather, about 40 to 55% of the chemical energy is converted to electricity.

Fuel cells are used on space craft to produce electricity. The water produced can be used for drinking, so less water need be taken aboard. Fuel cells overall have a weight advantage over storage batteries, an important consideration when launching a spaceship.

On Earth, more research is needed to reduce cost and design long-lasting cells. Perhaps someday fuel cells will provide electricity to meet peak needs in large power plants. Unlike huge boilers and nuclear reactors, they can be started and stopped simply by turning the fuel on or off.

SECTION 14.17

Energy: How Much Is Too Much?

Although there are many problems, science and technology probably will be able to provide us with a plentiful supply of energy for the foreseeable future. This energy will not—indeed, cannot—be pollution free. Any production and use of energy will be accompanied by pollution.

According to a study by the World Energy Council, the global demand for energy in 2020 will be 30 to 100% higher than it was in 1990 because of population increase and improved living conditions.

How do we choose the best method of energy production? The task is certainly difficult. The choice should be made by informed citizens who have examined the process from beginning to end. We must know what is involved in the construction of power plants, the production of fuels, and the ultimate use of energy in our homes and factories. We must know that energy is wasted (as heat) at every step in the process. Plentiful power usually involves substantial thermal pollution at the site of the power plant. We cannot escape the consequences of the second law of thermodynamics.

Will our profligate consumption of energy affect the Earth's climate? Our activities have already modified the climate in and around metropolitan areas. The worldwide effects of our expanding energy consumption are harder to estimate.

What can we do as individuals? We can conserve. We can walk more and use cars less. We can reduce our wasteful use of electricity. We can buy more efficient appliances, and we can avoid purchasing energy-intensive products.

Over the past quarter century we have made significant progress in the area of energy conservation. There is a new awareness regarding energy efficiency.

New furnaces are 90 to 95% efficient, as compared with the 60% efficiency of 20-year-old furnaces. New refrigerators and air conditioners use about 35% less energy than earlier models. The average gasoline mileage of automobiles is now about twice what it was in 1970. There is also greater home use of fluorescent lamps, which are at least four times as energy efficient as incandescent lamps. Overall American industry has reduced its energy consumption per product by almost 30%.

One thing that could greatly reduce energy consumption in the United States is greater use of public transportation. In Europe, where there is much wider use of public transport systems, there is considerably less energy use per capita. It will not be easy to get people who are used to the convenience of personal cars to start using busses and trains, but it may become necessary. In Europe the average price for gasoline approaches $4.00 per gallon. If gasoline were that expensive in the United States, perhaps public transportation would seem more attractive.

The simple facts are that our population on Spaceship Earth is going up, and our fuel resources are going down. We clearly need to conserve energy and to look for new energy sources. But we also need to think more seriously about trying to stabilize the number of energy users.

A bicycle not only has no fuel tank to fill but it also provides wonderful exercise.

Summary

1. The United States with only 5% of the world's population uses almost one-quarter of all the energy being generated.
2. Most of our energy on Earth originated in the sun.
3. *Photosynthesis* is a process whereby energy is stored in plants. Animals depend on this stored energy for their survival.
4. A chemical reaction is either *exothermic* (giving off energy) or *endothermic* (requiring energy).
5. The *first law of thermodynamics* deals with conservation of energy: energy can be neither created not destroyed.
6. The *second law of thermodynamics* states that natural processes tend toward greater *entropy* (disorder).
7. Ancient societies used people power and animal power for doing work. Wood was the fuel burned.
8. Water wheels have been used for about 2000 years, and windmills about half that long.
9. ''Fossil fuels'' are natural fuels that were once living plants. They include coal, natural gas, and petroleum.
10. *Coal* is a black burnable rock that is mainly carbon. It is very slowly made from peat.
11. *Natural gas* is mainly methane. It is the cleanest of the fossil fuels.
12. *Petroleum* is a black liquid that is a complex mixture of organic compounds, mainly hydrocarbons.
13. *Gasoline* is a mixture of hydrocarbons, from about C_5 to C_{12}, obtained from petroleum.
14. Petroleum is fractionally distilled to separate gasoline from higher boiling heating and lubricating oils. Then the higher boiling oils are *cracked* to increase the yield of gasoline.
15. In order to raise the *octane rating* (antiknock quality) of gasoline, oil refiners use *isomerization* (to increase branching), *catalytic reforming* (to convert straight-chain hydrocarbons to aromatic rings), and *alkylation* (to convert gases to highly branched gasoline molecules).
16. Tetraethyllead was added to gasoline to improve its octane rating for more than 50 years (from the late 1920s until the mid-1980s), but its use has been largely discontinued.
17. Burning *fossil fuels* produces mainly carbon dioxide and water, but it also produces other compounds that pollute the air.

18. Most electric power is generated by boiling water to produce steam, which turns the turbines that produce electricity. Most power plants are fired with coal.

19. *Nuclear reactors* can also provide the heat to turn the steam turbines in power plants. They create very little air pollution.

20. Nuclear reactors produce energy by nuclear *fission.* They must be well shielded so as to contain all radioactivity.

21. *Breeder reactors* make more fissile fuel (plutonium-239) than they use (uranium-235).

22. Nuclear *fusion* generates energy by fusing small atoms to form larger ones. Thus far, attempts to control fusion for commercial purposes have not been successful.

23. With *solar cells* (*photovoltaic cells*) sunlight can be converted directly to electricity.

24. Burning *biomass* to generate heat might involve growing plants specifically for use as fuel, or just burning paper and other combustible garbage.

25. Energy can also be obtained as hydroelectric power from waterfalls or as windpower using giant windmills.

26. Other natural energy sources include *geothermal* power (geysers and volcanoes), the tides, and the fossil fuels in *oil shale* and *tar sands.*

27. Synthetic fuels can be made by liquefying or gasifying coal.

28. *Fuel cells* generate electricity by reacting a fuel, such as hydrogen, with oxygen in an electrochemical cell.

Key Terms

alkylation 14.9
biomass 14.15
bitumen 14.16
catalytic reforming 14.9
coal 14.7
control rods 14.12
cracked 14.8
cracking process 14.9
endothermic 14.2
entropy 14.4
exothermic 14.2
first law of thermodynamics 14.3
fission reactions 14.12
fossil fuels 14.6
fuel cell 14.16
fuels 14.6
fusion reactions 14.13
gasoline 14.9
geothermal energy 14.16
isomerization 14.9
kerogen 14.16
law of conservation of energy 14.3
moderator 14.12
natural gas 14.8
nuclear reactor 14.12
octane rating 14.9
oil shale 14.16
petroleum 14.9
photosynthesis 14.1
photovoltaic cells 14.14
plasma 14.13
second law of thermodynamics 14.4
solar cells 14.14
straight-run gasoline 14.9
tar sands 14.16
thermal pollution 14.12
thermonuclear reactions 14.13

Review Questions

1. What is a fuel?
2. What kind of reaction powers the sun?
3. How does temperature affect the rate of a chemical reaction?
4. Why does wood burn more rapidly in pure oxygen than in air?
5. What is an exothermic reaction? Give an example.
6. What is an endothermic reaction? Give an example.
7. State the first law of thermodynamics.
8. State the second law of thermodynamics in terms of energy flow and in terms of order and disorder.
9. What is entropy?
10. What are the advantages and disadvantages of coal as a fuel?
11. Which of the fossil fuels is most plentiful in the United States?

12. What is the main component of natural gas?
13. What are the advantages and disadvantages of natural gas as a fuel?
14. What is the physical state of each of the three fossil fuels?
15. What is thought to be the origin of petroleum?
16. What are the advantages and disadvantages of petroleum as a source of fuels?
17. What is meant by the octane rating of a gasoline?
18. What are the advantages and disadvantages of tetraethyllead as an octane booster?
19. What was the principal fuel used in the United States before 1800?
20. What was the principal fuel used in the United States from about 1850 to 1950?
21. What has been our principal fuel since 1950?
22. Energy is conserved. How can we ever run out of energy?
23. Is entropy increased or decreased when a fossil fuel is burned?
24. How is crude petroleum modified to better meet our needs and wants?
25. What is the ultimate source of nearly all the energy on Earth?
26. Why did the United States shift from coal to petroleum and natural gas when we have much larger reserves of coal than of the two hydrocarbon fuels?
27. What proportion of United States electricity is generated by nuclear power plants?
28. What proportion of the electricity in France is generated by nuclear power plants?
29. Can a nuclear power plant explode like a nuclear bomb? Explain your answer.
30. How does a breeder reactor produce more fuel than it consumes?
31. Can nuclear bombs be made from reactor-grade uranium? Explain your answer.
32. List some advantages of nuclear power plants over coal-fired plants.
33. What are some of the disadvantages of breeder reactors?
34. Can a nuclear bomb be made from reactor-grade plutonium?
35. List some possible advantages of a nuclear fusion reactor over a fissin reactor.
36. List some possible problems with nuclear fusion reactors.
37. What is plasma?
38. What are some problems associated with the use of solar energy?
39. What is a photovoltaic cell?
40. What is plant biomass?
41. List some advantages and disadvantages of the use of biomass as a source of energy.
42. List two ways that fuel cells differ from electrochemical cells.
43. List the advantages, disadvantages, and limitations of each of the following as an energy source.
    a. wind power
    b. geothermal power
    c. power from the tides
    d. hydroelectric power
44. Compare a nuclear power plant with a coal-burning plant. Which would you rather have in your neighborhood? Why?
45. Which of the following is the best fuel for heating your home?
    a. natural gas
    b. electricity
    c. coal
    d. fuel oil

    What problems with supply, use, and waste products are involved with each fuel?

Problems

Fuels and Combustion

46. Which of the following are fuels?
    a. C_3H_8
    b. CCl_4
    c. C_8H_{18}
    d. CO_2
47. Which of the following are fuels?
    a. H_2
    b. C_2H_2
    c. C_3F_8
    d. H_2O
48. Give the equation for the complete combustion of coal. Assume the coal is carbon (C).

49. Give the equation for the incomplete combustion of coal to form carbon monoxide. Assume the coal is carbon (C).

50. Give the equation for the complete combustion of methane.

51. Give the equation for the incomplete combustion of methane to form carbon monoxide.

Energy Calculations from Chemical Equations

52. Burning 1.00 mol of methane releases 192 kcal of energy.

$$CH_4 + 2\,O_2 \rightarrow CO_2 + 2\,H_2O + 192\text{ kcal}$$

How much energy is released by burning 5.00 mol of methane?

53. It takes 137 kcal of energy to decompose 2 mol of water.

$$2\,H_2O + 137\text{ kcal} \rightarrow 2\,H_2 + O_2$$

How much energy does it take to decompose 55.5 mol of water?

Fossil Fuel Reserves

54. The estimated world oil reserves in 1992 were 989 billion barrels. The world rate of annual use was 23.4 billion barrels. How long would these reserves last if that rate of use continued?

55. The estimated United States oil reserves in 1992 were 25 billion barrels. How long would those reserves last if there were no imports or exports and if the U.S. annual rate of use of 5.4 billion barrels were to continue?

56. The estimated world natural gas reserves in 1992 were 4376 trillion ft^3. The world rate of annual use was 68.4 trillion ft^3. How long would these reserves last if that rate of use continued?

57. The estimated United States natural gas reserves in 1992 were 167 trillion ft^3. How long would those reserves last if there were no imports or exports and if the U.S. annual rate of use of 18.0 trillion ft^3 continued?

Nuclear Reactions

58. Give nuclear equations to show how thorium-232 is converted to fissile uranium-233.

59. Give nuclear equations to show how uranium-238 is converted to fissile plutonium-239.

60. Give the nuclear equation that shows how deuterium and tritium fuse to form helium and a neutron.

61. Give the nuclear equation that shows how four protons fuse to form helium and two positrons.

Synthetic and Converted Fuels

62. Give the chemical equation for the basic process by which coal is converted to methane.

63. Give the chemical equation for the conversion of coal (carbon) to carbon monoxide and hydrogen.

64. Give the chemical equation for the conversion of carbon monoxide and hydrogen to methanol.

65. Give the chemical equation for the reaction that occurs in the hydrogen-oxygen fuel cell.

Additional Problems

66. A woman uses 1000 kcal in running 10 km in 40 minutes. Calculate her average power output in watts. (1.000 kcal = 4184 J.)

67. Energy requirements of the human brain are about 20% of the total body metabolism. Calculate the power output (in watts) of your brain if you use 2000 kcal of energy per day.

68. What weight of carbon dioxide is formed by the combustion of 1250 kg of coal when the coal is 40% carbon? The equation is

$$C + O_2 \rightarrow CO_2$$

69. What mass of methanol can be made from 437 g of carbon monoxide if sufficient hydrogen is available?

References and Readings

1. Asimov, Isaac. ''In Dancing Flames a Greek Saw the Basis of the Universe.'' *Smithsonian,* November 1971, pp. 52–57. Discusses heat, from the Greek ''element'' fire to the first law of thermodynamics.
2. Bailey, Maurice E. ''The Chemistry of Coal and Its Constituents.'' *Journal of Chemical Education,* July 1974, pp. 446–448.
3. Bent, Henry A. ''Haste Makes Waste: Pollution and Entropy.'' *Chemistry,* October 1971, pp. 6–15. This article should be required reading for all science—and nonscience—students.
4. Bozak, Richard E., and Manuel Garcia, Jr. ''Chemistry in the Oil Shales.'' *Journal of Chemical Education,* March 1976, pp. 154–155.
5. Brown, Lester R., Christopher Flavin, and Sandra Postel. ''Earth Day 2030.'' *WorldWatch,* March 1990, pp. 12–21. Sustainable practices are needed to ensure a healthy environment by 2030.
6. Burnett, W. M., and S. D. Ban. ''Changing Prospects for Natural Gas in the United States.'' *Science,* 21 April 1989, pp. 305–310.
7. Commoner, Barry. *The Poverty of Power: Energy and the Economic Crisis.* New York: Knopf, 1976.
8. Davis, Ged R. ''Energy for Planet Earth.'' *Scientific American,* September 1990, pp. 54–62. Part of a special issue on energy.
9. Fischetti, M. ''Here Comes the Electric Car—It's Sporty, Aggressive and Clean.'' *Smithsonian,* April 1992, pp. 34–43.
10. Gibbons, John H., Peter D. Blair, and Holly L. Gwin. ''Strategies for Energy Use.'' *Scientific American,* September 1989, pp. 136–143. Part of a special issue on managing planet Earth.
11. Golay, Michael W., and Neil E. Todreas. ''Advanced Light-Water Reactors.'' *Scientific American,* April 1990, pp. 82–89.
12. Hubbard, H. M. ''Photovoltaics Today and Tomorrow.'' *Science,* 21 April 1989, pp. 297–304.
13. Kolb, Doris, and Kenneth E. Kolb. ''Chemical Principles Revisited: Petroleum Chemistry.'' *Journal of Chemical Education,* July 1979, pp. 465–469.
14. Krieger, James. ''Energy.'' *Chemical and Engineering News,* 17 June 1991, pp. 18–46.
15. Lapp, Ralph E. ''The Chemical Century.'' *Bulletin of the Atomic Scientists,* September 1973, pp. 8–14. Use of fossil fuels will begin to decline after the year 2000, so the twentieth century will be unique in the use of chemical energy.
16. Lenssen, Nicholas, and John E. Young. ''Filling Up in the Future.'' *WorldWatch,* May–June 1990, pp. 18–26.
17. Lihach, Nadine. ''Fuel Cells for the 90's.'' *EPRI Journal,* September 1984, pp. 6–13.
18. Maize, Kennedy. ''Water to Burn: The Coming Hydrogen Age?'' *Public Power,* September–October 1992, pp. 20–27.
19. Miller, Peter. ''Our Electric Future.'' *National Geographic,* August 1991, pp. 62–87. Considers the building of new nuclear power plants as the answer to supplying electric power in the future.
20. Morton, Lawrence, Norman Hunter, and Hyman Gesser. ''Methanol: A Fuel for Today and Tomorrow.'' *Chemistry and Industry,* 16 July 1990, pp. 457–462.
21. Owsley, Dennis C., and Jordan J. Bloomfield. ''Energy Facts: A Basis for Decision.'' *ChemTech,* February 1985, pp. 94–98.
22. Stevenson, Kenneth L. ''Brief Introduction to the Three Laws of Thermodynamics.'' *Journal of Chemical Education,* May 1975, pp. 330–331.
23. Westbrook, Charles K. ''The Chemistry Behind Engine Knock.'' *Chemistry and Industry,* 3 August 1992, pp. 562–566.
24. ''Which Gasoline for Your Car?'' *Consumer Reports,* January 1990, pp. 8–15. Includes a section on fuels of the future.

15 Biochemistry

A Molecular View of Life

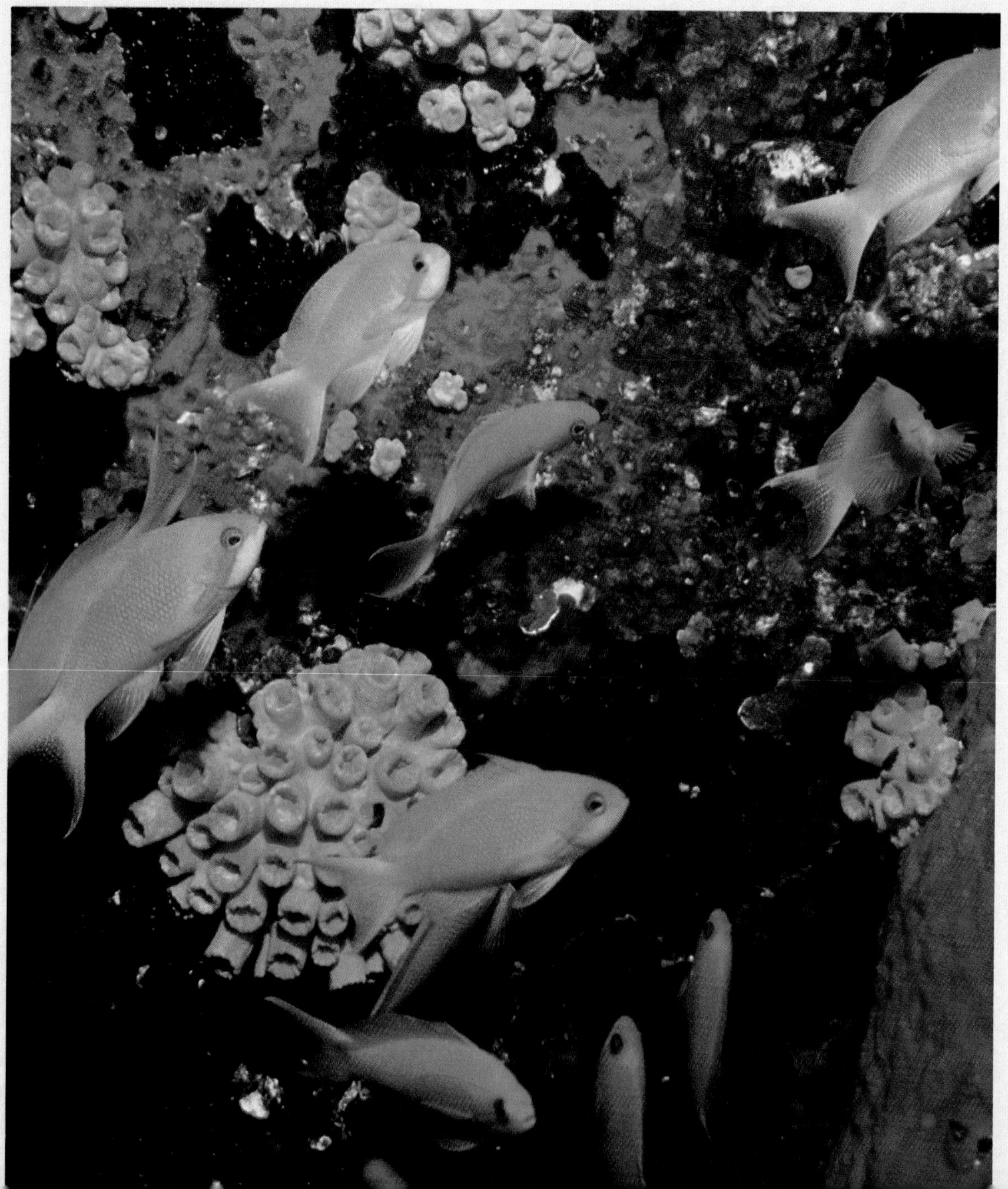

A myriad different forms of life
In dazzling array!
Vast multitudes of cells, each one
Programmed by DNA.

Your body is an incredible chemical factory. It is also a living machine that can do many different kinds of jobs. To stay in good condition it needs many specific chemical compounds, and it manufactures most of them all by itself in an exquisitely organized network of chemical production lines.

Your body is made up of millions of tiny cells, and thousands of chemical reactions are taking place within each of those cells every minute of every day. The study of these reactions and the chemicals that they produce is called biochemistry.

SECTION 15.1

The Cell

Biochemistry is the chemistry of living things and life processes. The structural unit of all living things is the **cell**. Every cell is enclosed in a membrane, and plant cells also have walls made of cellulose. Every kind of tissue is made up of cells specific to the function of that particular tissue. Muscle cells are different from nerve cells, and red blood cells are different from skin cells.

Not many years ago scientists thought that a cell was a fairly simple entity, with an outer membrane and a central nucleus. All the space in between was thought to be filled with a substance called cytoplasm. With the development of the electron microscope came the discovery of many more cell parts (Figure 15.1). We can consider only a few of them here.

Each human cell is enclosed in a **cell membrane** through which it gains nutrients and gets rid of wastes. Inside are various structures that serve a multiplicity of functions. The largest structure is usually the **cell nucleus**, which contains the material that controls heredity. Protein synthesis takes place in the **ribosomes**. The **mitrochondria** are the cell ''batteries.'' These are the sites where energy is produced.

◀ Earth has so many different forms of life, each with its own distinctive set of DNA ''blueprints.'' [*Norbert Wu.*]

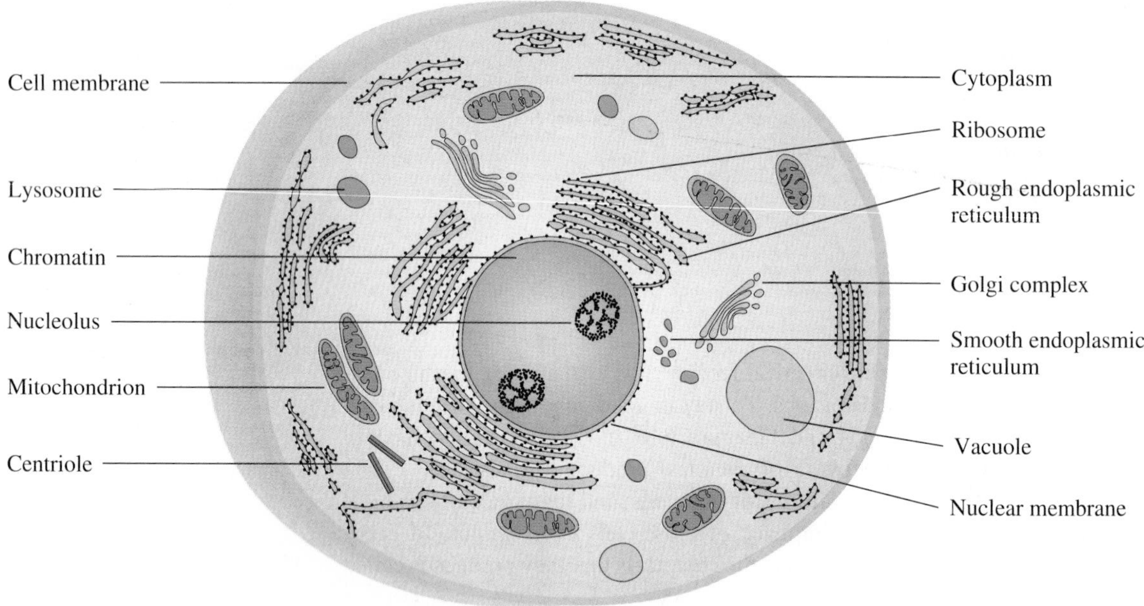

Figure 15.1 An idealized "typical" animal cell.

Plant cells (but not animal cells) also contain some parts called chloroplasts (Figure 15.2). It is in the **chloroplasts** that energy from the sun is converted to chemical energy, which is stored in the plant in the form of carbohydrates.

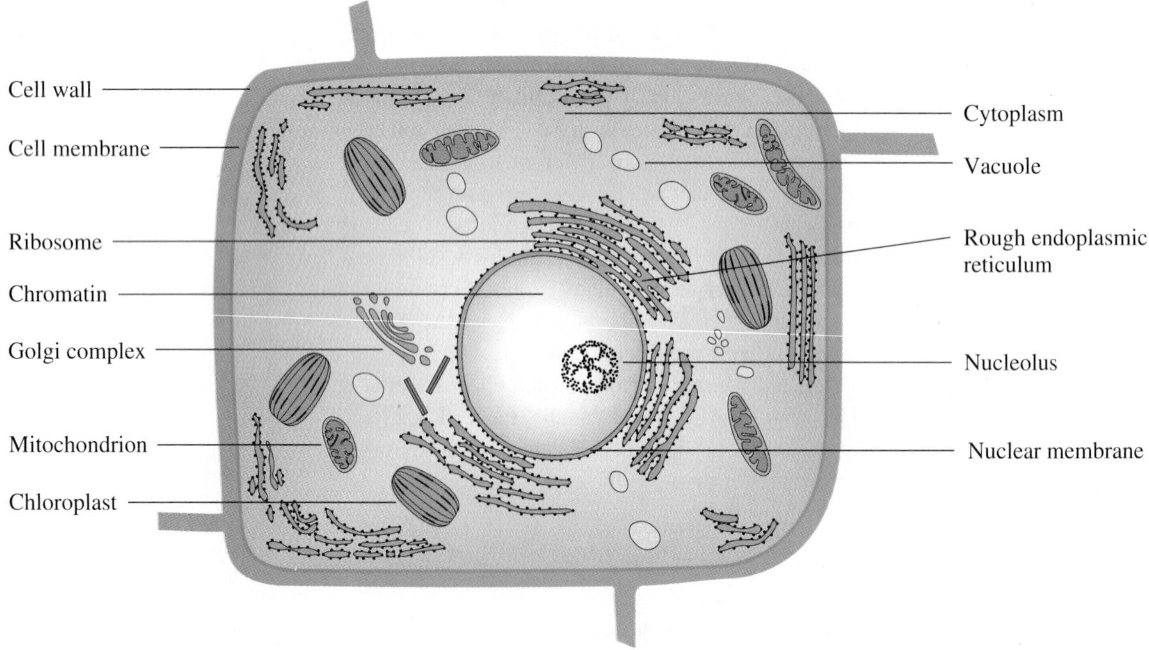

Figure 15.2 An idealized "typical" plant cell.

SECTION 15.2

Energy in Biological Systems

Life requires energy. Living cells are inherently unstable and avoid falling apart only because of a continued input of energy. Living organisms are restricted to using certain forms of energy. Supplying a plant with heat energy by holding it in a flame will do little to prolong its life. On the other hand, a green plant is uniquely able to tap sunlight, the richest source of energy on Earth. Chloroplasts in green plant cells capture the radiant energy of the sun and convert it into chemical energy, which is then stored in carbohydrate molecules. The photosynthesis of glucose is represented by the equation

$$6\,CO_2 + 6\,H_2O \longrightarrow C_6H_{12}O_6 + 6\,O_2$$

The plant cell also can convert those carbohydrate molecules into fat molecules and, with the proper inorganic nutrients, into protein molecules (Figure 15.3).

Carbohydrates, fats, and proteins as foods are discussed in Chapter 16. In this chapter, we deal mainly with the synthesis and structure of these vital materials.

Animals cannot directly use the energy of sunlight. They must get their energy by eating plants, or by eating other animals that eat plants. There are three major types of foods from which animals obtain energy: carbohydrates, fats, and proteins.

Once digested and transported to the cell, a food molecule can be used in

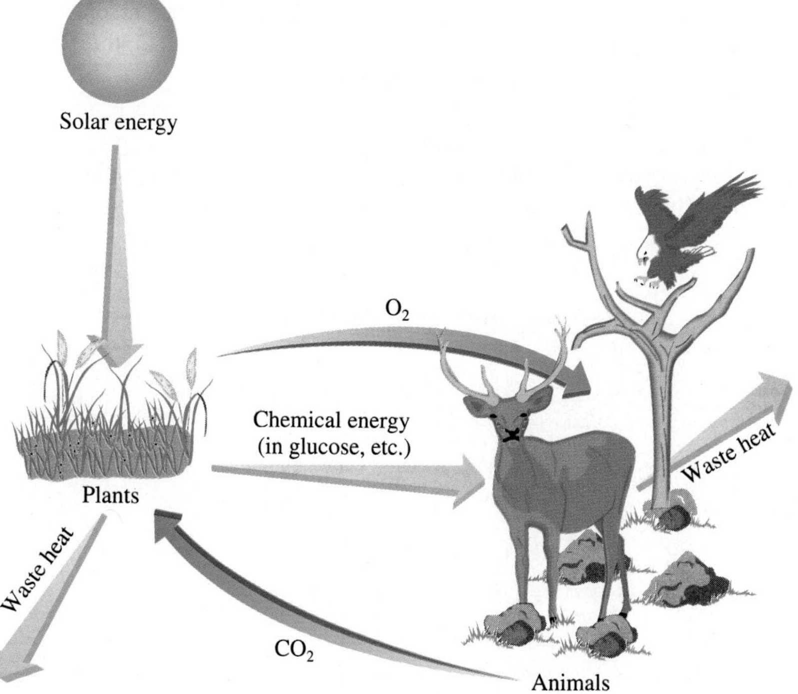

Figure 15.3 Some energy transformations in living systems.

either of two ways. It can be used as a building block to make new cell parts or to repair old ones, or it can be "burned" for energy. The entire series of coordinated chemical reactions that keep cells alive is called **metabolism**. In general, metabolic reactions are divided into two classes. The degrading of molecules to provide energy is called **catabolism**. The process of building up or synthesizing the molecules of living systems is called **anabolism**.

SECTION 15.3

Carbohydrates: A Storehouse of Energy

A formal definition of carbohydrates is difficult. Chemically, **carbohydrates** are polyhydroxy aldehydes or ketones or compounds that can be hydrolyzed to form those compounds. (Hydrolyzed means split by water.) Note the hydroxyl groups and aldehyde or ketone functions in the compounds in Figure 15.4.

Carbohydrates are compounds of the elements carbon, hydrogen, and oxygen. Usually, the atoms of these elements are present in a ratio expressed by the formula $C_x(H_2O)_y$. Glucose, a simple sugar that has the formula $C_6H_{12}O_6$, could be written $C_6(H_2O)_6$. It is from formulas such as these that the term *carbohydrate* is derived. We use it even though carbohydrates are not hydrates of carbon. The carbohydrates include the sugars, starches, and cellulose.

Some Simple Sugars

Sugars are sweet-tasting carbohydrates. The simplest of these are the monosaccharides, carbohydrates that cannot be further split by water (hydrolyzed). Familiar monosaccharides include glucose (also called dextrose), fructose (fruit sugar), and galactose (a component of milk sugar). Structures of these are given in Figure 15.4. Although represented in the figure as open-chain compounds (to show the aldehyde or ketone functional groups), these sugars exist mainly as cyclic molecules (Figure 15.5).

Sucrose (cane or beet sugar) and lactose (milk sugar) are examples of **disaccharides**, carbohydrates consisting of molecules that can be hydrolyzed to two monosaccharide units (Figure 15.6). Sucrose is split into glucose and fructose. Hydrolysis of lactose gives glucose and galactose

$$\text{Sucrose} + H_2O \longrightarrow \text{Glucose} + \text{Fructose}$$

$$\text{Lactose} + H_2O \longrightarrow \text{Glucose} + \text{Galactose}$$

Glucose	Galactose	Fructose
H—C=O	H—C=O	CH_2OH
H—C—OH	H—C—OH	C=O
HO—C—H	HO—C—H	HO—C—H
H—C—OH	HO—C—H	H—C—OH
H—C—OH	H—C—OH	H—C—OH
CH_2OH	CH_2OH	CH_2OH

Figure 15.4 Three common monosaccharides. All have hydroxyl groups. Glucose and galactose have aldehyde functions. Fructose has a ketone group. Glucose and galactose differ only in the arrangement of the H and OH on the fourth carbon from the top.

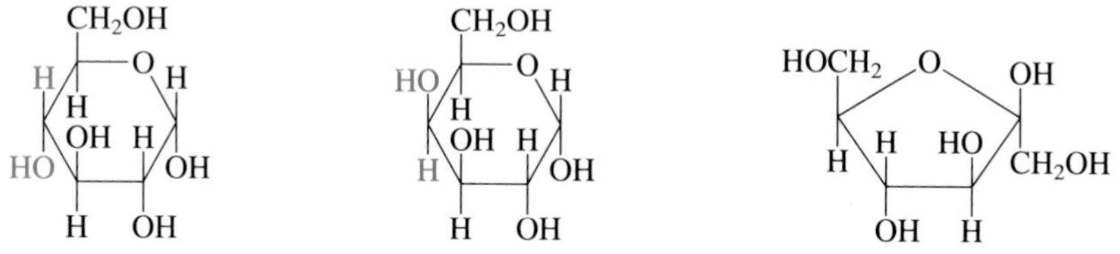

Glucose Galactose Fructose

Figure 15.5 Cyclic structures for glucose, galactose, and fructose. A corner not otherwise occupied represents a carbon atom. Glucose and galactose are represented as six-membered rings. They differ only in the arrangement of the H and OH on the lower left corner. Fructose is shown as a five-membered ring.

Some sugars exist in more than one cyclic form. For simplicity, we show only one form of each here.

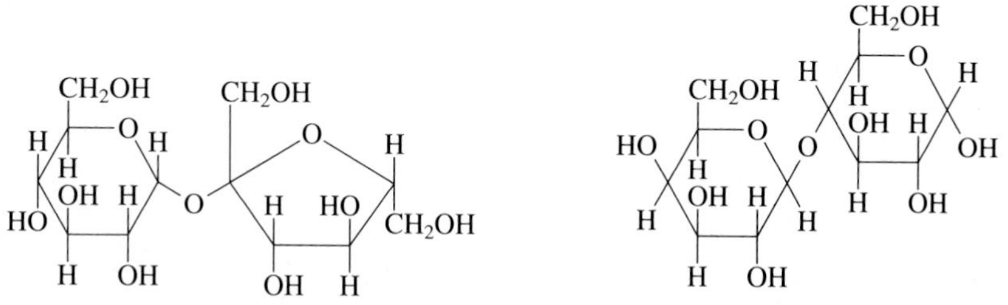

Sucrose Lactose

Figure 15.6 Sucrose and lactose are disaccharides. Sucrose yields glucose and fructose upon hydrolysis, while lactose gives glucose and galactose.

SECTION 15.4

Polysaccharides: Starch and Cellulose

Polysaccharides are composed of molecules that yield many monosaccharide units upon hydrolysis. Polysaccharides include the starches, the main energy storage system of many plants, and cellulose, which is the structural material of plants. Figure 15.7 shows short segments of starch and cellulose molecules. Notice that both are polymers of glucose. **Starch** molecules generally have from 100 to about 6000 glucose units. **Cellulose** molecules are composed of from 1800 to 3000 or more glucose units.

A crucial structural difference between starch and cellulose is in the way the glucose units are hooked together. Consider starch first. Using the CH_2OH as a reference, the oxygen atom joining the glucose units is pointed down. This arrangement is called an *alpha linkage.* In cellulose, again with the CH_2OH as a point of reference, the oxygen atom connecting the glucose segmers is pointed up, an arrangement called a *beta linkage.*. This difference in linkages may seem to be a minor one, but it determines whether or not the materials can be digested (Chapter 16). The different linkages also result in

As foods (Chapter 16), the polysaccharides are called complex carbohydrates. Starches serve as foods, and cellulose serves as dietary fiber. The simple sugars are rapidly absorbed into the body from the digestive tract. The starches must be digested before absorption. Cellulose cannot be digested by humans.

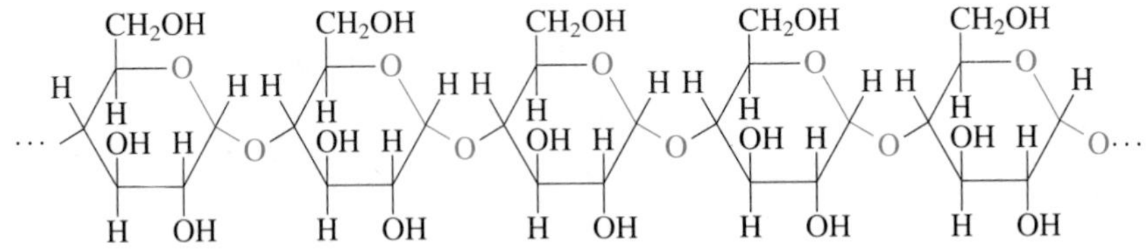

Starch

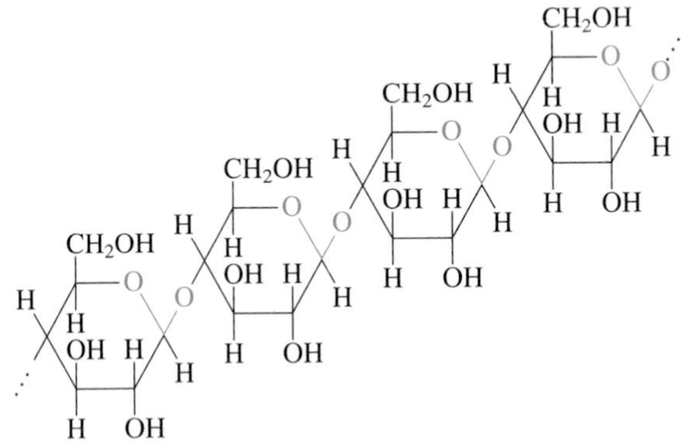

Cellulose

Figure 15.7 Carbohydrates. Both starch and cellulose are polymers of glucose. Note the different linkages joining the glucose units in the two structures. In starch the glucose segmers are joined by alpha linkages. In cellulose the linkages are beta.

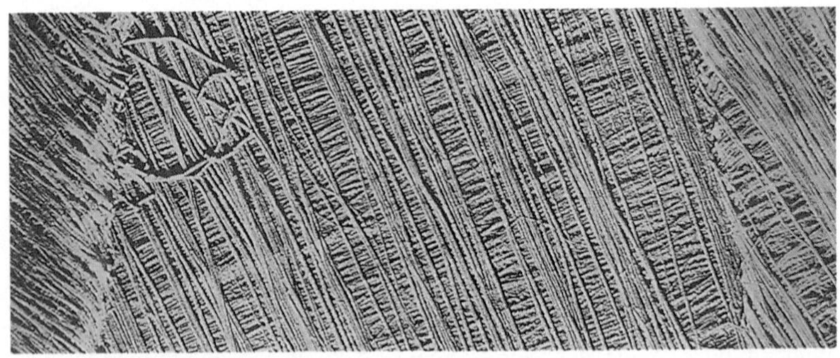

Figure 15.8 Electron micrograph of the cell wall of an alga. The wall consists of successive layers of cellulose fibers in parallel arrangement. [*Fred Bavendam/Peter Arnold, Inc.*]

```
. . . Glc-Glc-Glc-Glc-Glc-Glc-Glc-Glc-Glc-Glc-Glc-Glc-Glc-Glc . . .
                         Amylose

 . . . Glc-Glc
             \
. . . . Glc-Glc-Glc-Glc
                       \
     . . . Glc-Glc-Glc-Glc-Glc-Glc-Glc-Glc-Glc
                                            \
    . . . Glc-Glc-Glc-Glc-Glc-Glc-Glc-Glc-Glc-Glc-Glc-Glc
                                                       \
                         . . . . Glc-Glc-Glc-Glc-Glc-Glc-Glc-Glc
                         Amylopectin
```

Figure 15.9 Schematic representations of amylose and amylopectin. Glc represents a glucose unit. The structure of glycogen is similar to that of amylopectin.

different three-dimensional forms for cellulose and starch. For example, cellulose in the cell walls of plants is arranged in fibrils, bundles of parallel chains. The fibrils, in turn, lie parallel to each other in each layer of the cell wall (Figure 15.8). In alternate layers, the fibrils are perpendicular; this imparts great strength to the wall.

Actually there are two kinds of plant starch. One, called amylose, has the glucose units joined in a continuous chain, like beads on a string. The other kind, amylopectin, has branched chains of glucose units. These are perhaps best represented schematically, as in Figure 15.9 (each glucose unit is represented by the letters Glc).

Animal starch is called **glycogen**. Like amylopectin, it is composed of branched chains of glucose units. Glycogen in muscle and liver tissue is arranged in granules (Figure 15.10), clusters of small particles. Plant starch, on the other hand, forms large granules. Plant-starch granules rupture in boiling water to form a paste. On cooling, the paste gels. Potatoes and cereal grains form this type of starchy broth. All forms of starch are hydrolyzed to glucose during digestion.

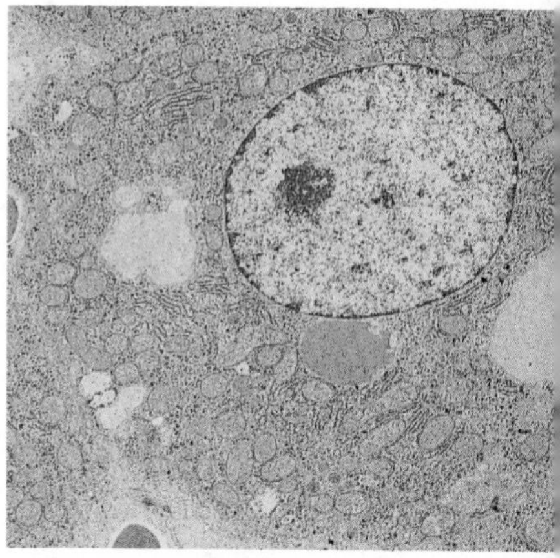

Figure 15.10 Electron micrograph of glycogen granules in a liver cell of a rat. [*Biophoto Associates/ Photo Researchers, Inc.*]

SECTION 15.5

Fats and Other Lipids

Fats are the predominant form of a class of compounds called lipids. These substances are not defined structurally; rather they have common solubility properties. A **lipid** is a cellular constituent that is soluble in organic solvents of low polarity such as hexane, diethyl ether, or carbon tetrachloride. Lipids are insoluble in water. In addition to the fats, the lipid family contains **fatty acids** (long-chain carboxylic acids), steroids such as cholesterol and the sex hormones (Chapter 18), the fat-soluble vitamins (Chapter 16), and other substances.

Fats are esters of fatty acids and the trihydroxy alcohol glycerol. A fat may contain up to three fatty acid chains joined to glycerol through ester linkages. Fats are classified according to the number of fatty acid chains they contain. A monoglyceride has one fatty acid chain, a diglyceride has two, and a triglyceride has three fatty acid chains. Figure 15.11 gives an illustration of the formation of a triglyceride from glycerol and three fatty acid molecules.

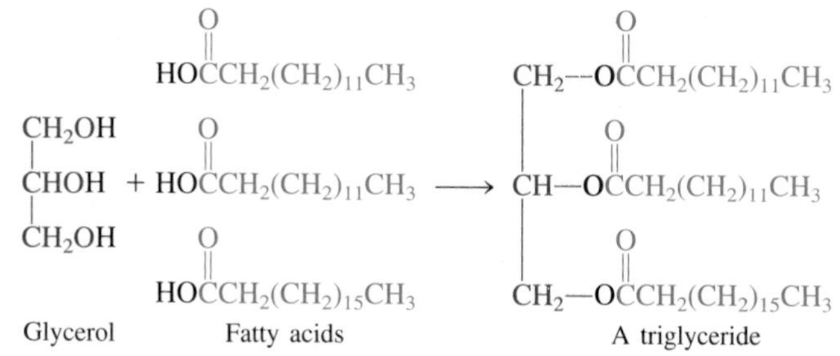

$$\begin{array}{lclcl} & & HO\overset{O}{\overset{\|}{C}}CH_2(CH_2)_{11}CH_3 & & CH_2{-}O\overset{O}{\overset{\|}{C}}CH_2(CH_2)_{11}CH_3 \\ CH_2OH & & & & | \\ | & & & & | \\ CHOH & + & HO\overset{O}{\overset{\|}{C}}CH_2(CH_2)_{11}CH_3 & \longrightarrow & CH{-}O\overset{O}{\overset{\|}{C}}CH_2(CH_2)_{11}CH_3 \\ | & & & & | \\ CH_2OH & & & & | \\ & & HO\overset{O}{\overset{\|}{C}}CH_2(CH_2)_{15}CH_3 & & CH_2{-}O\overset{O}{\overset{\|}{C}}CH_2(CH_2)_{15}CH_3 \\ \text{Glycerol} & & \text{Fatty acids} & & \text{A triglyceride} \end{array}$$

Figure 15.11 Triglycerides are esters in which the alcohol glycerol, with three hydroxyl groups, is esterified with three fatty acid groups.

Table 15.1 Some Fatty Acids in Natural Fats

Number of Carbon Atoms	Condensed Structure	Name	Source
4	$CH_3CH_2CH_2COOH$	Butyric acid	Butter
6	$CH_3(CH_2)_4COOH$	Caproic acid	Butter
8	$CH_3(CH_2)_6COOH$	Caprylic acid	Coconut oil
10	$CH_3(CH_2)_8COOH$	Capric acid	Coconut oil
12	$CH_3(CH_2)_{10}COOH$	Lauric acid	Palm kernel oil
14	$CH_3(CH_2)_{12}COOH$	Myristic acid	Oil of nutmeg
16	$CH_3(CH_2)_{14}COOH$	Palmitic acid	Palm oil
18	$CH_3(CH_2)_{16}COOH$	Stearic acid	Beef tallow
18	$CH_3(CH_2)_7CH{=}CH(CH_2)_7COOH$	Oleic acid	Olive oil
18	$CH_3(CH_2)_4CH{=}CHCH_2CH{=}CH(CH_2)_7COOH$	Linoleic acid	Soybean oil
18	$CH_3CH_2(CH{=}CHCH_2)_3(CH_2)_6COOH$	Linolenic acid	Fish oils
20	$CH_3(CH_2)_4(CH{=}CHCH_2)_4CH_2CH_2COOH$	Arachidonic acid	Liver

Naturally occurring fatty acids nearly always have an even number of carbon atoms. Representative ones are listed in Table 15.1. Animal fats generally are rich in saturated fatty acids with a smaller proportion of unsaturated fatty acids. At room temperature, most animal fats are solids. Liquid fats, called **oils**, are obtained principally from vegetable sources. Structurally, oils are identical to fats except that they incorporate a higher proportion of unsaturated fatty acid units.

Fats often are classified according to the degree of unsaturation of the fatty acids they incorporate. Saturated fatty acids contain no carbon-to-carbon double bonds, monounsaturated fatty acids contain one carbon-to-carbon double bond per molecule, and polyunsaturated fatty acids are those that have two or more carbon-to-carbon double bonds. Saturated fats contain a high proportion of saturated fatty acids; the fat molecules have relatively few double bonds. Polyunsaturated fats (oils) incorporate mainly unsaturated fatty acids; these fat molecules have many double bonds.

The degree of unsaturation of a fat or an oil is usually measured by the iodine number. Chlorine and bromine add readily to carbon-to-carbon double bonds.

$$\text{>C=C<} + Br_2 \longrightarrow -\underset{Br}{\overset{|}{\underset{|}{C}}}-\underset{Br}{\overset{|}{\underset{|}{C}}}-$$

Iodine also adds, but less readily. The **iodine number** is the number of grams of iodine that will be consumed by 100 g of fat or oil. The more double bonds a fat contains, the more iodine is required for the addition reaction; thus, a high iodine number means a high degree of unsaturation. Representative iodine numbers are listed in Table 15.2. Note the generally lower values for the animal fats (butter, tallow, lard) compared to those for the vegetable oils.

Table 15.2 Typical Iodine Numbers for Some Fats and Oils*

Fat or Oil	Iodine Number	Fat or Oil	Iodine Number
Coconut oil	8–10	Cottonseed oil	100–117
Butter	25–40	Corn oil	115–130
Beef tallow	30–45	Fish oils	120–180
Palm oil	37–54	Soybean oil	125–140
Lard	45–70	Safflower oil	130–140
Olive oil	75–95	Sunflower oil	130–145
Peanut oil	85–100	Linseed oil	170–205

*Oils shown in green are from plant sources. Three fats and one oil come from animals.

Coconut oil, which is highly saturated, and fish oils, which are relatively unsaturated, are notable exceptions to the general rule.

Fats and oils feel greasy. They are insoluble in water but soluble in organic solvents such as carbon tetrachloride, gasoline, and diethyl ether. Fats and oils are lighter than water; consequently, they float on top of water.

SECTION 15.6

Proteins: Polymers of Amino Acids

Proteins are vital components of all life. No living part of the human body—or of any other organism, for that matter—is completely without protein. There is protein in the blood, the muscles, the brain, and even the tooth enamel. The smallest cellular organisms—the bacteria—contain protein. Viruses, so small that they make bacteria look like giants, are nothing but proteins and nucleic acids. This combination of nucleic acids and proteins is found in all cells and is the stuff of life itself.

Each type of cell makes its own kinds of proteins. The proteins serve as the structural material of animals, much as cellulose does for plants. Muscle tissue is largely protein; so are skin and hair. Silk, wool, nails, claws, feathers, horns, and hooves are proteins. All proteins contain the elements carbon, hydrogen, oxygen, and nitrogen. Most also contain sulfur. The structure of a short segment of a typical protein molecule is shown in Figure 15.12.

Proteins are copolymers of about 20 different amino acids (Table 15.3). **Amino acids** have two functional groups. The amino group (—NH_2) is on the carbon (called the alpha carbon) next to the carbon of the carboxyl group (—COOH).

$$H_2N-\underset{|}{\overset{|}{C}}-\overset{O}{\overset{\|}{C}}-OH$$

This partial formula indicates the proper placement of these groups, but the structure is not really correct. Acids react with bases to form salts; the car-

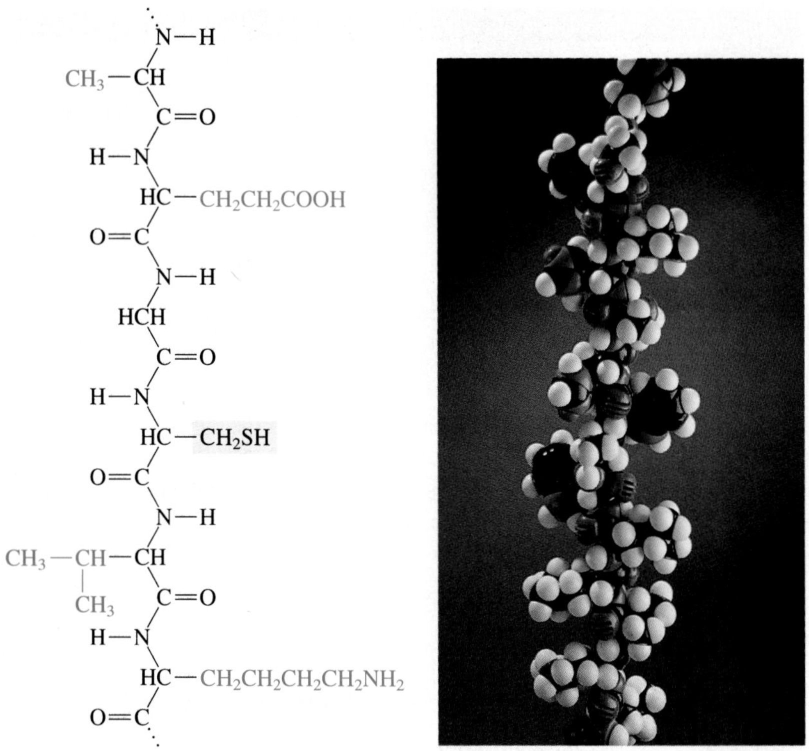

N—H
CH_3—CH
C=O
H—N
HC—CH_2CH_2COOH
O=C
N—H
HCH
C=O
H—N
HC—CH_2SH
O=C
N—H
CH_3—CH—CH
CH_3
C=O
H—N
HC—$CH_2CH_2CH_2CH_2NH_2$
O=C

(a) (b)

Figure 15.12 Structural formula (a) and space-filling model (b) of a short segment of a protein molecule. In the structural formula hydrocarbon side chains (green), an acidic side chain (magenta), a basic side chain (blue), and a sulfur-containing side chain are highlighted. [*(b) Fundamental Photographs/Macmillan Science Files.*]

boxyl group is acidic and the amino group is basic. Therefore, these two functional groups interact, the acid transferring a proton to the base. The resulting product is an inner salt, or **zwitterion**, a compound in which the anion and the cation are parts of the same molecule. The amino acids differ in that a variety of other groups (symbolized by the —R) may be attached to the carbon atom that bears the amino group.

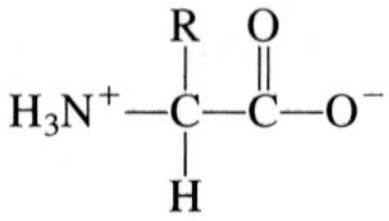

$$H_3N^+\text{—}\overset{\displaystyle R}{\underset{\displaystyle H}{C}}\text{—}\overset{\displaystyle O}{\overset{\|}{C}}\text{—}O^-$$

A zwitterion

Plants are able to synthesize proteins from carbon dioxide, water, and minerals [such as nitrates (NO_3^-) and sulfates (SO_4^{2-})]. Animals require proteins as one of the three major classes of foods.

Table 15.3 Some Representative Amino Acids

Name	Abbreviation	Essential	Structure
Glycine	Gly	No	$^{+}NH_3-CH_2-COO^-$
Alanine	Ala	No	$CH_3-CH(^{+}NH_3)-COO^-$
Phenylalanine	Phe	Yes	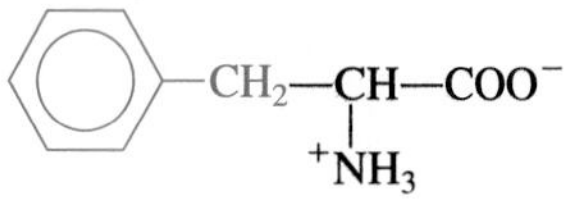 $C_6H_5-CH_2-CH(^{+}NH_3)-COO^-$
Valine	Val	Yes	$CH_3-CH(CH_3)-CH(^{+}NH_3)-COO^-$
Leucine	Leu	Yes	$CH_3CH(CH_3)CH_2-CH(^{+}NH_3)-COO^-$
Isoleucine	Ile	Yes	$CH_3CH_2CH(CH_3)-CH(^{+}NH_3)-COO^-$
Proline	Pro	No	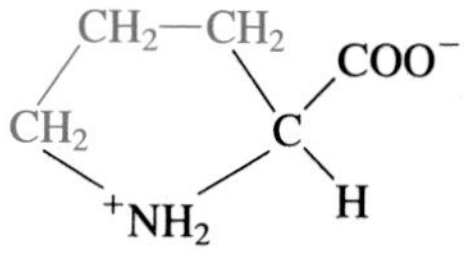 ring: $CH_2-CH_2-C(H)(COO^-)-{}^{+}NH_2-CH_2$ (five-membered ring)
Methionine	Met	Yes	$CH_3-S-CH_2CH_2-CH(^{+}NH_3)-COO^-$
Serine	Ser	No	$HO-CH_2-CH(^{+}NH_3)-COO^-$
Threonine	Thr	Yes	$CH_3CH(OH)-CH(^{+}NH_3)-COO^-$
Asparagine	Asn	No	$H_2N-C(=O)-CH_2-CH(^{+}NH_3)-COO^-$
Glutamine	Gln	No	$H_2N-C(=O)-CH_2CH_2-CH(^{+}NH_3)-COO^-$
Cysteine	Cys	No	$HS-CH_2-CH(^{+}NH_3)-COO^-$

Table 15.3 Some Representative Amino Acids (continued)

Name	Abbreviation	Essential	Structure
Tyrosine	Tyr	No	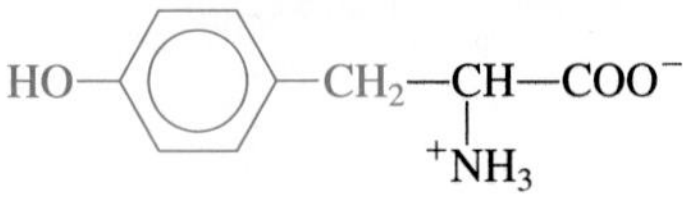 $HO-C_6H_4-CH_2-CH(NH_3^+)-COO^-$
Tryptophan	Trp	Yes	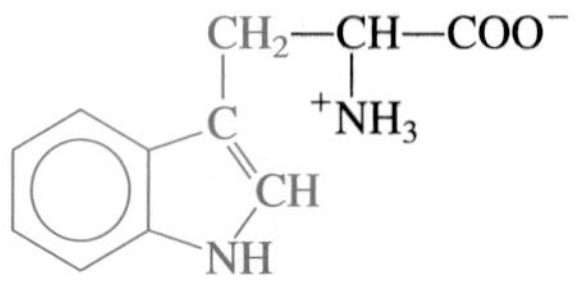 (indole ring: benzene fused to C=CH–NH) $-C-CH_2-CH(NH_3^+)-COO^-$
Lysine	Lys	Yes	$H_3\overset{+}{N}CH_2CH_2CH_2CH_2-CH(NH_2)-COO^-$
Arginine	Arg	—*	$H_2N-C(=\overset{+}{N}H_2)-NHCH_2CH_2CH_2-CH(NH_2)-COO^-$
Histidine	His	—†	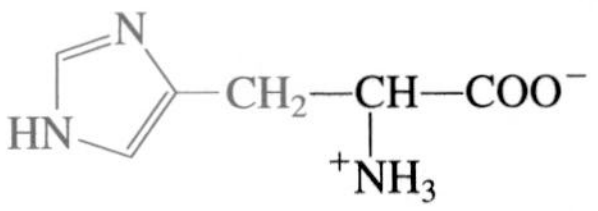 (imidazole ring: HN, =N) $-CH_2-CH(NH_3^+)-COO^-$
Aspartic acid	Asp	No	$HOOC-CH_2-CH(NH_3^+)-COO^-$
Glutamic acid	Glu	No	$HOOC-CH_2CH_2-CH(NH_3^+)-COO^-$

*Essential to growing children but not to adult humans.
†Essential to human infants.

SECTION 15.7

The Peptide Bond: Peptides and Proteins

The end of the protein molecule with a free carboxyl group (—COOH) is called the C-terminal end. The end with a free amino group (—NH_2) is designated the N-terminal end.

The human body contains about 30,000 different proteins. Each of us has his or her own tailor-made set. Proteins are polyamides. The amide linkage (—CONH—) is called a peptide bond when it joins two amino acid units.

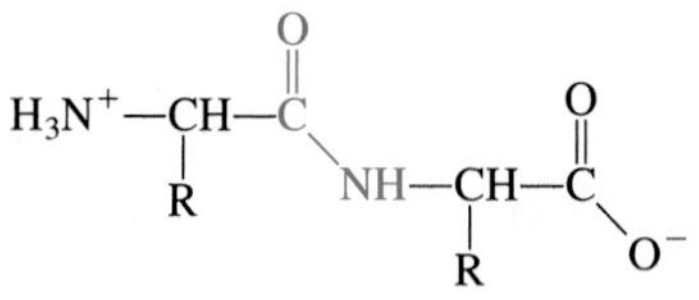

$$H_3N^+-\underset{R}{CH}-\overset{O}{\overset{\|}{C}}-NH-\underset{R}{CH}-\overset{O}{\overset{\|}{C}}-O^-$$

Peptide bond

Note that there is still a reactive amino group on the left and a carboxyl group on the right. Each of these can react further to join more amino acid units. This process can continue until thousands of units have joined to form a giant molecule—a polymer called a protein.

$$H_3N^+CH(R)C(=O)\text{—}NHCH(R)C(=O)\left(NHCH(R)C(=O)\right)_n NHCH(R)C(=O)\text{—}NHCH(R)C(=O)\text{—}O^-$$

N-terminal C-terminal

When only two amino acids are joined, the product is a **dipeptide**.

$$H_3N^+\text{—}CH_2\text{—}C(=O)\text{—}NH\text{—}CH(CH_2C_6H_5)\text{—}C(=O)\text{—}O^-$$

Glycylphenylalanine
(a dipeptide)

When three amino acids are combined, the substance is a *tripeptide*.

$$H_3N^+\text{—}CH(CH_2OH)\text{—}C(=O)\text{—}NH\text{—}CH(CH_3)\text{—}C(=O)\text{—}NH\text{—}CH(CH_2SH)\text{—}C(=O)\text{—}O^-$$

Serylalanylcysteine
(a tripeptide)

Note how these substances are named. Other combinations are named in a similar manner. A molecule with more than 10 amino acid units often is simply called a **polypeptide**. When the molecular weight of a compound exceeds 10,000, it is called a **protein**. The distinction between polypeptides and proteins is an arbitrary one, and it is not always precisely applied.

The Sequence of Amino Acids

For peptides and proteins to be physiologically active, it is not enough that they incorporate certain amounts of specific amino acids. The order or *sequence* in which the amino acids are connected is also of critical importance. Glycylalanine is different from alanylglycine, for example. Although the difference seems minor, the two substances behave differently in the body.

When chemists describe peptides (and proteins), they find it much simpler to indicate the amino acid sequence by using the abbreviations for the amino acids (see Table 15.3). The sequence for glycylalanine is written Gly-Ala, and that for alanylglycine is Ala-Gly. It is understood from this shorthand that the peptide is arranged with the free amino group (N-terminal) to the left and the free carboxyl group (C-terminal) to the right.

As the length of a peptide chain increases, the possible sequential variations become very large. Just as we can make millions of different words with our 26-letter English alphabet, we can make millions of different proteins with the 20 or so amino acids. Just as one can write gibberish with the English alphabet, one can make nonfunctioning proteins by putting together the wrong sequence of amino acids.

Yet while the correct sequence is ordinarily of utmost importance, it is not always absolutely required. Just as you can sometimes make sense of incorrectly spelled English words, a protein with a small percentage of ''incorrect'' amino acids may continue to function. It may not function as well, however. And sometimes a seemingly minor change can have a disastrous effect. Some people have hemoglobin with one incorrect amino acid unit in about 300. That ''minor'' error is responsible for sickle cell anemia, an inherited condition that ordinarily proves fatal.

Some protein molecules are enormous, with molecular weights in the tens of thousands. The molecular formula for hemoglobin, the oxygen-carrying protein in red blood cells, is $C_{3032}H_{4816}O_{780}N_{780}S_8Fe_4$, corresponding to a molecular weight of 64,450. Although they are huge compared to ordinary molecules, a billion average-sized protein molecules still would fit on the head of a pin.

SECTION 15.8

Structure of Proteins

The structure of proteins is generally discussed at four organizational levels. The **primary structure** of a protein molecule is its amino acid sequence. To specify the primary structure, one merely writes out the sequence of amino acids in the long-chain molecule. The primary structure is held together by the peptide links between the amino acid units.

Figure 15.13 Pleated sheet conformation of polypeptide chains. (a) Ball-and-stick model. (b) Schematic drawing emphasizing the pleats. [*Redrawn from C. David Gutsche and Daniel J. Pasto,* Fundamentals of Organic Chemistry, 1975. *Reprinted by permission of Prentice-Hall, Inc., Englewood Cliffs, NJ.*]

Intermolecular hydrogen bonds

Top view

(a)

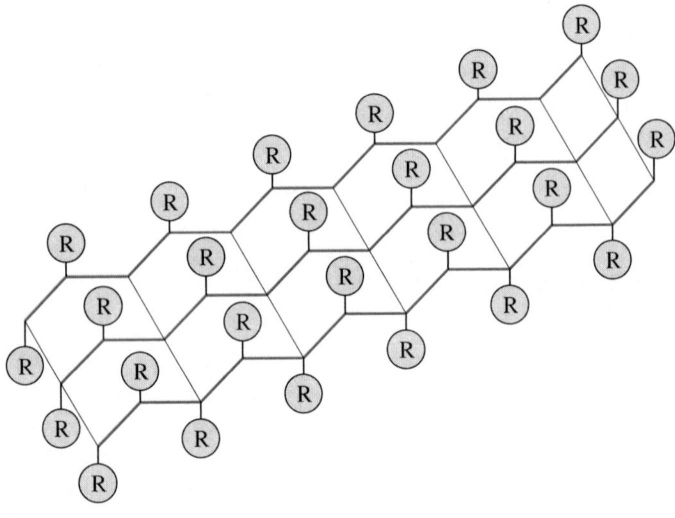

(b)

Molecules of proteins aren't just arranged at random as tangled threads. The chains are held together in unique configurations. The **secondary structure** of a protein refers to the arrangement of chains about an axis. This arrangement may be a pleated sheet, as in silk (Figure 15.13), a helix, as in wool (Figure 15.14), or whatever.

The primary structure of angiotensin II, a peptide produced in the kidneys, is

Asp-Arg-Val-Tyr-Ile-
Phe-Pro-His

This sequence specifies an octapeptide with aspartic acid at the N-terminal, joined to arginine, valine, tyrosine, isoleucine, histidine, proline, and ending with phenylalanine at the C-terminal. Angiotensin II causes powerful constrictions of blood vessels.

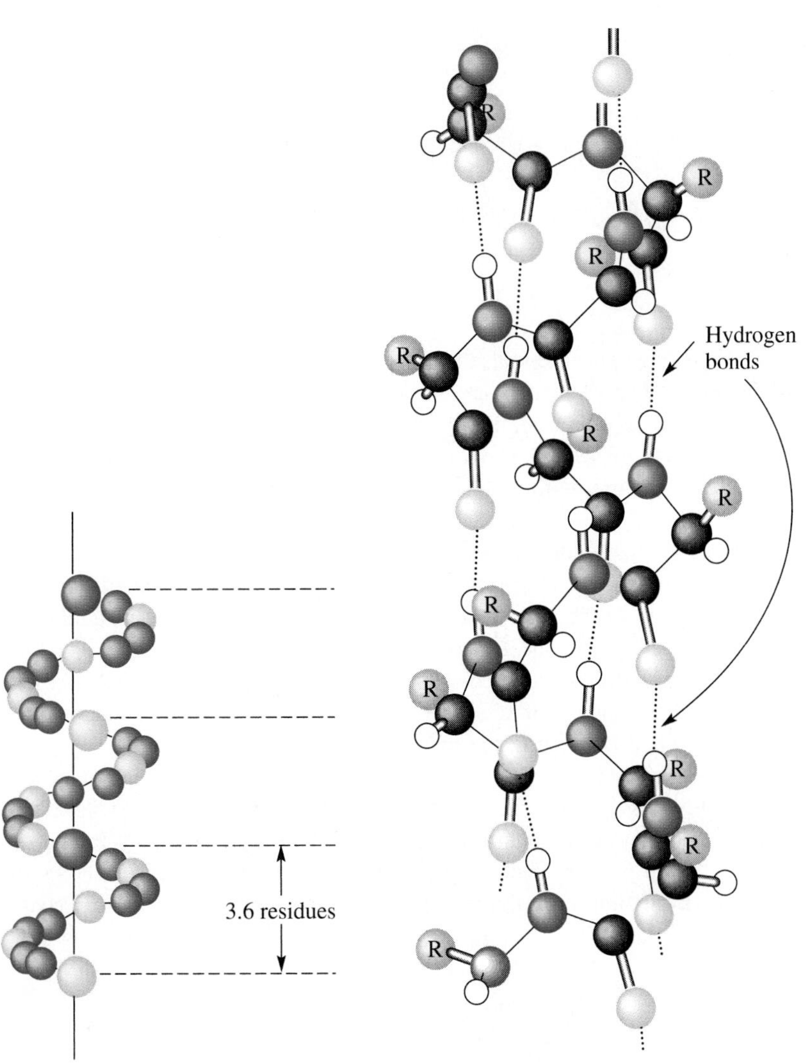

Figure 15.14 Two representations of the alpha-helical conformation of a polypeptide chain. (a) The skeletal representation best shows the helix. (b) Hydrogen bonding between turns of the helix is shown in the ball-and-stick model. [*Redrawn from C. David Gutsche and Daniel J. Pasto,* Fundamentals of Organic Chemistry, 1975. *Reprinted by permission of Prentice-Hall, Inc., Englewood Cliffs, NJ.*]

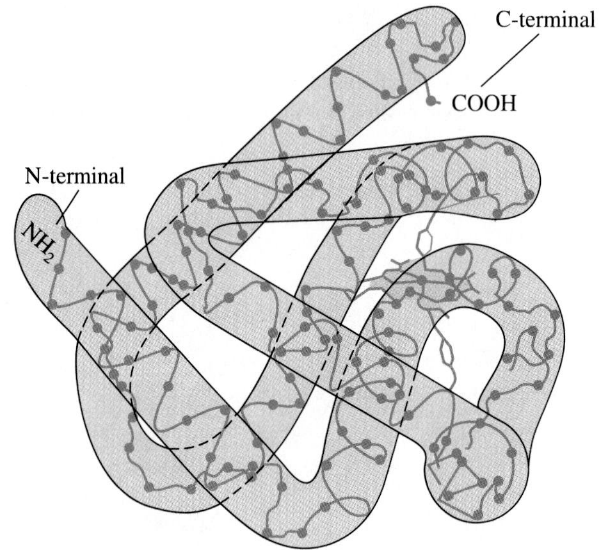

Figure 15.15 A skeletal representation of the tertiary structure of myoglobin. The protein chain is folded to form a globular structure, much as string can be folded into a ball. [*Reprinted with permission from H. Neurath (Ed.),* The Proteins, *vol. 2. Academic Press, 1964. Copyright 1964 by Academic Press, Inc.*]

We can relate the three levels of protein organization to a more familiar object. Think of the coiled cord on a telephone receiver. The cord starts out as a long, straight wire (Figure 15.16). We'll call that the primary structure. The wire is coiled into a helical arrangement. That is its secondary structure. When the receiver is hung up, the coiled cord folds in a particular pattern. That is its tertiary structure.

In the pleated sheet layout, the protein chains exist in an extended zigzag arrangement. The molecules are stacked in extended arrays, with hydrogen bonds holding adjacent chains together. The appearance (see Figure 15.13) gives this type of secondary structure its name, the **pleated sheet** arrangement. This structure, with its multitude of hydrogen bonds, makes silk strong and flexible.

The protein molecules in wool, hair, and muscle contain large segments arranged in the form of a right-handed helix, or as it is usually called, an **alpha helix**. Each turn of the helix (see Figure 15.14) requires 3.6 amino acid units. The N—H groups in one turn form hydrogen bonds to carbonyl (C=O) groups in the next. These helices wrap around one another in threes or sevens like strands of a rope. Unlike silk, wool can be stretched, much like a spring can be stretched by pulling the coils apart.

The **tertiary structure** of a protein refers to the spatial relationships of amino acid units that are relatively far apart in the protein chain. In describing tertiary structure, we frequently will talk about how the molecule is folded. An example is the protein chain in **globular proteins**. Figure 15.15 shows the structure of myoglobin, which is folded into a compact, spherical shape.

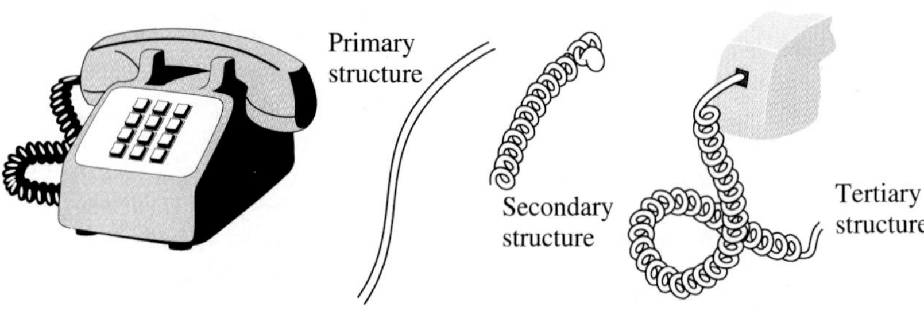

Figure 15.16 Three levels of structure of a telephone cord.

Some proteins contain more than one polypeptide chain; that is, a protein molecule can be an aggregate of subunits. Hemoglobin is the most familiar example. A single hemoglobin molecule contains four polypeptide units. Each unit is roughly comparable to a myoglobin molecule. The four units are arranged in a specific pattern (Figure 15.17). When we describe the *quaternary structure* of hemoglobin, we describe in detail the way in which the four units are packed together in the hemoglobin molecule.

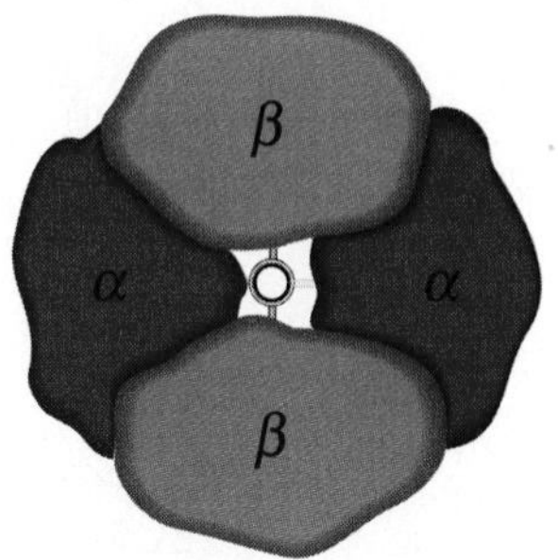

Figure 15.17 The quaternary structure of hemoglobin has the four coiled chains, each analogous to myoglobin (Figure 15.15), stacked in a nearly tetrahedral arrangement.

Four Ways to Link Protein Chains

Peptide bonds fix the primary structure of proteins. What forces hold proteins in the structural arrangements we have referred to as secondary, tertiary, and quaternary? There are four kinds of forces—hydrogen bonds, ionic forces called salt bridges, covalent disulfide linkages, and dispersion forces. The last are the only forces operating between nonpolar side chains; these are called hydrophobic interactions. The various forces are illustrated in Figure 15.18.

A number of the amino acids have side chains that can and do participate in *hydrogen bonding* (the hydroxyl group of serine is one example). Nonetheless, the hydrogen bonds of greatest importance are those that involve an interaction between the atoms of one peptide bond and those of another. Thus, the carbonyl (C=O) oxygen of one peptide link may form a hydrogen bond to an amide hydrogen (N—H) located some distance away on the same chain or located on an entirely different chain. The secondary structure of wool (alpha helix) illustrates the former situation; the secondary structure of silk (pleated sheet) illustrates the latter. In both cases you will notice a pattern of such interactions. Because the peptide links are regularly spaced along the chain, there is a regular repetition of hydrogen bonds, both intramolecular (in wool) and intermolecular (in silk). Such patterns are a fairly common occurrence.

Intramolecular means within the same molecule. Intermolecular means between two molecules.

Salt bridges occur when an amino acid with a basic side chain appears opposite one with an acidic side chain. Proton transfer results in opposite charges, which then attract one another. These interactions can occur between relatively distant groups that happen to come into contact because of some folding or coiling of a single chain. They also occur between chains.

Disulfide linkages are formed when two cysteine units (whether on the same chain or two different chains) are oxidized. The disulfide bond is a

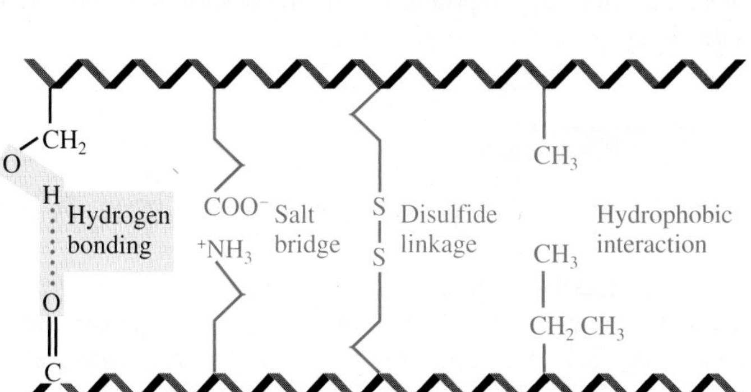

Figure 15.18 The tertiary structure of proteins is maintained by four different types of interactions.

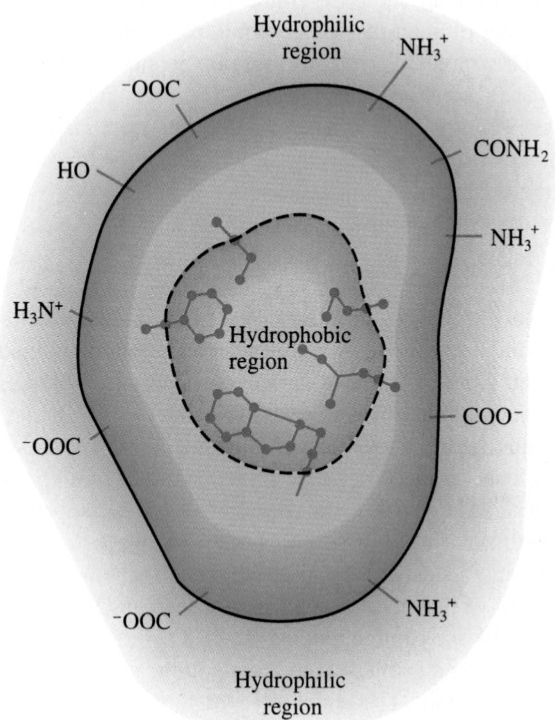

Figure 15.19 A folded protein chain often has a hydrophobic region on the inside (away from water) and a hydrophilic region on the outside with polar groups extending into the water.

covalent bond. The strength of this bond is much greater than that of a hydrogen bond. The much larger number of hydrogen bonds does compensate somewhat for their individual weakness.

Still weaker are the **hydrophobic interactions** between nonpolar side chains. These can be important, however, when other types of interactions are missing or are minimized. The hydrophobic interactions are made stronger by the cohesiveness of the water molecules surrounding the protein. Nonpolar side chains minimize their exposure to water by clustering together on the inside folds of the protein in close contact with one another (Figure 15.19). Hydrophobic interactions become fairly significant in structures such as that of silk, in which a high proportion of amino acids in the protein have nonpolar side chains.

We have emphasized protein structure and function in this chapter. Proteins as foods are discussed in Chapter 16.

SECTION 15.9

Enzymes: Exquisite Precision Machines

A highly specialized class of proteins, **enzymes** are biological catalysts produced by the cell. They have enormous catalytic power and are ordinarily highly specific, catalyzing only one reaction or a closely related group of reactions. Nearly all known enzymes are proteins.

Enzymes enable reactions to occur at much more rapid rates and lower temperatures than they would otherwise. They do this by changing the reaction path. The enzyme reacts with a compound, called the **substrate**, and bonds to it. This newly formed complex then separates into products and the regenerated enzyme. Schematically, this may be written

$$\text{Enzyme} + \text{Substrate} \rightleftharpoons \text{Enzyme–substrate complex} \rightleftharpoons \text{Enzyme} + \text{Products}$$

The double arrows indicate that the steps are reversible; the enzyme catalyzes both the forward and reverse reactions.

Enzyme action is often explained by the analogy of a lock and key. The substrate must fit a portion of the enzyme, called the **active site**, quite precisely, just as a key must fit a tumbler lock in order to open it (Figure 15.20). Not only must the enzyme and substrate fit precisely, but they likely are held together by electrical attraction. This requires that certain charged groups on the enzyme complement certain charged or partially charged groups on the substrate. Formation of new bonds to the enzyme by the substrate weakens bonds within the substrate. These weakened bonds can then be more easily broken to form products.

Portions of the enzyme molecule other than the active site may be involved

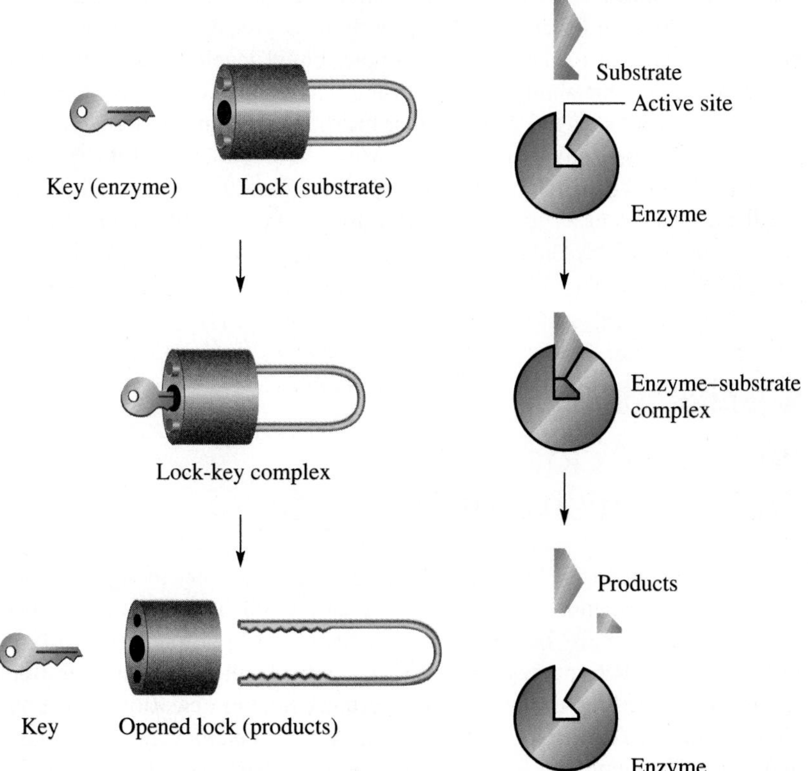

Figure 15.20 The lock-and-key model for enzyme action.

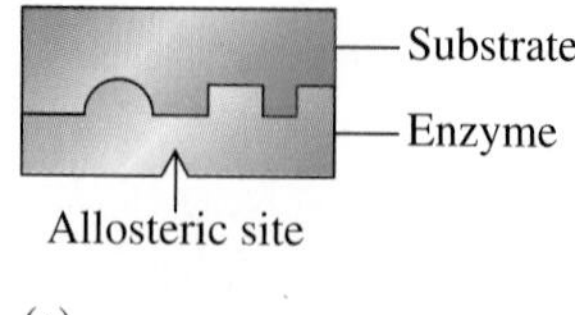

(a)

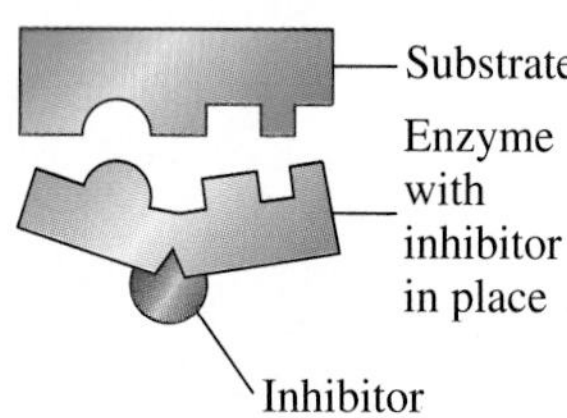

(b)

Figure 15.21 A model for the inhibition of an enzyme. (a) The enzyme and its substrate fit like a lock and key. (The allosteric site is the place where the inhibitor binds.) (b) With the inhibitor bound to the enzyme, the active site of the enzyme is distorted and the enzyme will not bind its substrate.

in the catalytic process. Some interaction at a position remote from the active site can change the shape of the enzyme and thus change its effectiveness as a catalyst. In this way it is possible to slow or stop the catalytic action. Figure 15.21 offers a model for the inhibition of enzyme catalysis. An inhibitor molecule attaches to the enzyme at a position remote from the active site where the substrate is bound. The enzyme changes shape as it accommodates the inhibitor, and the substrate is no longer able to bind to the enzyme. This is one of the mechanisms by which cells ''turn off'' enzymes when their work is done.

Some enzymes consist entirely of protein chains. In others, another chemical component is necessary for the proper function of the enzyme. Such a component is called a **cofactor**. The cofactor may be a metal ion such as zinc (Zn^{2+}), manganese (Mn^{2+}), magnesium (Mg^{2+}), iron(II) (Fe^{2+}), or copper(II) (Cu^{2+}). If the cofactor is organic in nature, it is called a **coenzyme**. By definition, coenzymes are nonprotein. The pure protein part of an enzyme is called the **apoenzyme**. Both the coenzyme and the apoenzyme must be present for enzymatic activity to take place.

$$\underset{\substack{\text{Nonprotein}\\ \text{(inactive)}}}{\text{Coenzyme}} + \underset{\substack{\text{Protein}\\ \text{(inactive)}}}{\text{Apoenzyme}} \longrightarrow \underset{\text{(Active)}}{\text{Enzyme}}$$

Many coenzymes are vitamins or are derived from vitamin molecules.

Clinical analysis for enzymes in body fluids or tissues is a common diagnostic technique in medicine. For example, blood levels of an enzyme involved in muscle metabolism rise in case of a heart attack in which the heart muscle has been damaged. On the other hand, many forms of strenuous (and healthful) physical activity also result in elevated blood levels of the enzyme. (Indeed, it is even possible for the level to increase in a person who hates needles and has tensed up while waiting for the blood sample to be taken!) Despite these variables, analysis for specific enzymes is an increasingly valuable diagnostic tool.

SECTION 15.10

Nucleic Acids: The Chemistry of Heredity

Complex compounds called **nucleic acids** are found in every living cell. They serve as the information and control centers of the cell. There are actually two kinds of nucleic acids. **Deoxyribonucleic acid (DNA)** is found primarily in the cell nucleus. **Ribonucleic acid (RNA)** is found in all parts of the cell. Both nucleic acids are long chains of repeating units called **nucleotides**. Each nucleotide, in turn, consists of three parts: a sugar, a heterocyclic amine base, and a phosphate unit. The sugar is either ribose or deoxyribose (Figure 15.22).

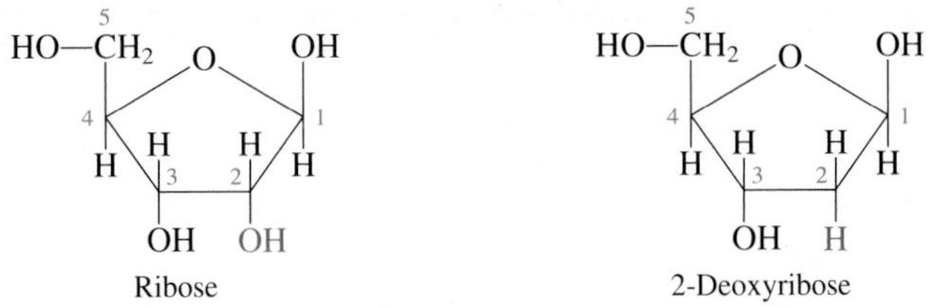

Ribose

2-Deoxyribose

Figure 15.22 The sugars found in nucleic acid. 2-Deoxyribose is a component of DNA, and ribose is found in RNA. The two differ in that 2-deoxyribose lacks an oxygen atom ("deoxy") on the second carbon atom; that is, ribose has an OH group there and 2-deoxyribose does not.

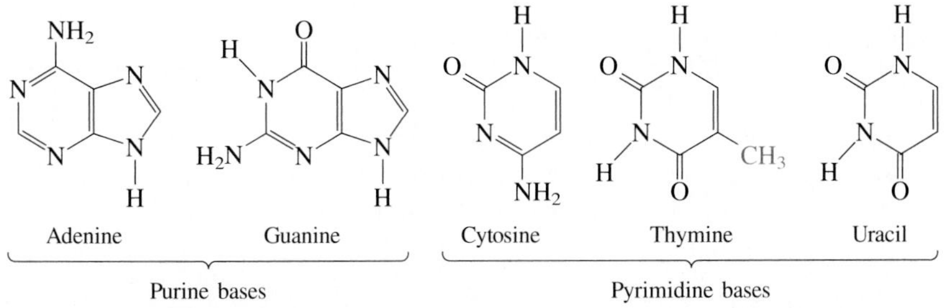

Adenine Guanine Cytosine Thymine Uracil

Purine bases Pyrimidine bases

Figure 15.23 Heterocyclic bases found in nucleic acids. Adenine, guanine, and cytosine are found in both DNA and RNA. Thymine is a component of DNA only, and uracil is found only in RNA. Note that thymine has a methyl group that is lacking in uracil.

Note that deoxyribose differs from ribose in that it lacks an oxygen atom on the second carbon atom. Ribose is found in ribonucleic acid (RNA), and deoxyribose is found in deoxyribonucleic acid. When either is incorporated into a nucleic acid, the hydroxyl group on the first carbon atom is replaced by one of five heterocyclic amine bases (Figure 15.23). The bases with two fused rings, adenine and guanine, are classified as **purines**. Cytosine, thymine, and uracil have only one ring; they are called **pyrimidines**.

The hydroxyl group on the fifth carbon of the sugar unit is converted into a phosphate ester group. Adenosine monophosphate (AMP) is a representative nucleotide. In AMP, the base is adenine and the sugar is ribose.

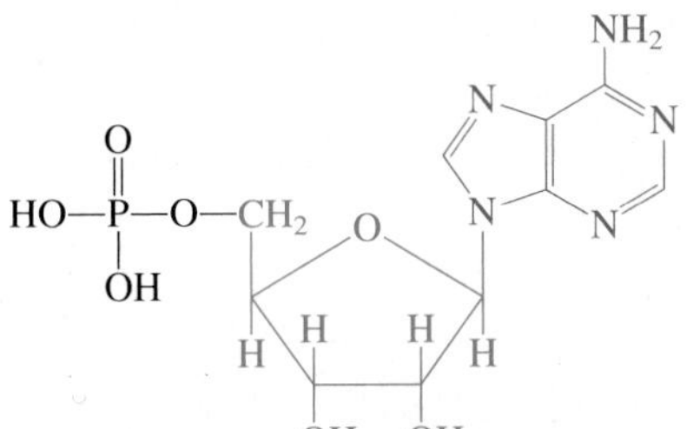

Adenosine monophosphate

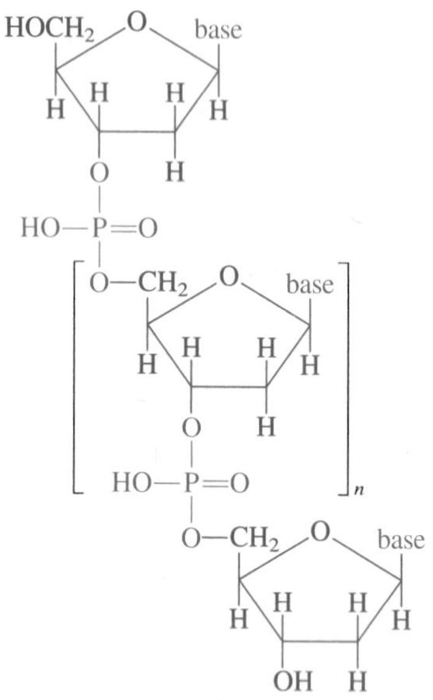

Figure 15.24 The backbone of a deoxyribonucleic acid molecule. The *n* indicates that the unit is repeated many times. Each repeating unit is composed of a base, phosphate unit, and sugar.

The phosphate units often are abbreviated as P_i, a compact designation of "inorganic phosphate."

Table 15.4 Components of DNA and RNA

	DNA	RNA
Purine bases	Adenine	Adenine
	Guanine	Guanine
Pyrimidine bases	Cytosine	Cytosine
	Thymine	Uracil
Pentose sugar	Deoxyribose	Ribose
Inorganic acid	Phosphoric acid	Phosphoric acid

Nucleotides are joined to one another through the phosphate group to form nucleic acid chains. The phosphate unit on one nucleotide forms an ester linkage to the hydroxyl group on the third carbon atom of the sugar unit in a second nucleotide. This unit is in turn joined to another nucleotide, and the process repeated to build up the long nucleic acid chain (Figure 15.24). The backbone of the chain consists of alternating phosphate and sugar units. The heterocyclic bases are branched off this backbone.

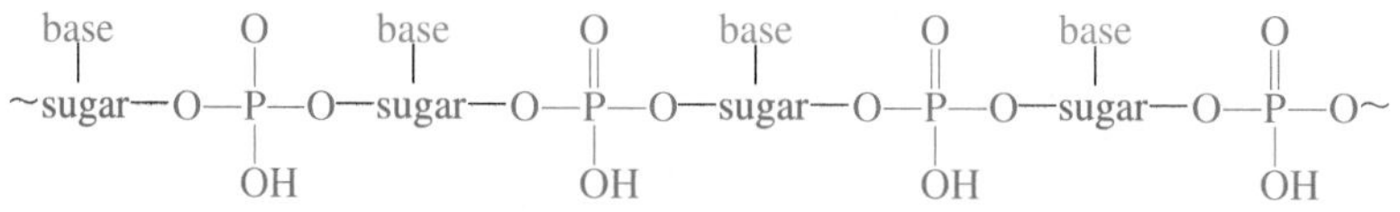

As we have seen, the sugar in DNA is deoxyribose, and that in RNA is ribose. The bases in DNA are adenine, guanine, cytosine, and thymine. Those in RNA are adenine, guanine, cytosine, and uracil (Table 15.4).

EXAMPLE 15.1

Identify the sugar and the base in the following nucleotide.

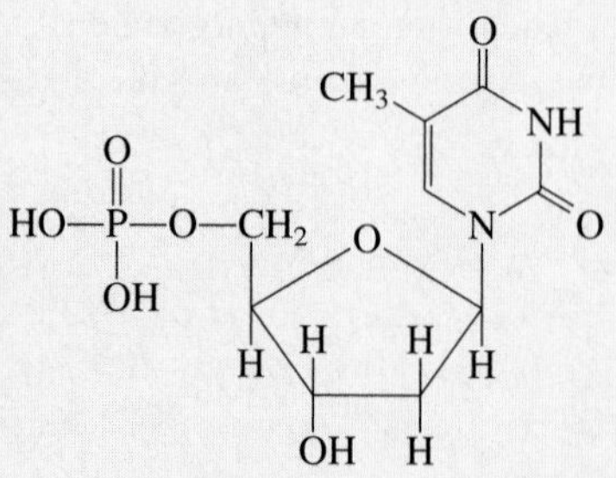

SOLUTION

The sugar is deoxyribose, and the base is thymine.

Exercise 15.1
Identify the sugar and the base in the following nucleotide.

Base Sequence in Nucleic Acids

A most important feature of the nucleic acid molecule is the sequence of the four bases along the strand. The molecules are huge, with molecular weights ranging into the billions for mammalian DNA. Along these great chains, the four bases may be arranged in essentially infinite variations. That is a crucial feature of these molecules, because it is the base sequence that is used to store the multitude of information needed to build living organisms. Before we examine that aspect of nucleic acid chemistry, let us consider one more important feature of nucleic acid structure.

The Double Helix

In an experiment designed to probe the structure of DNA, it was determined that the molar amount of adenine (A) in DNA corresponded to the molar amount of thymine (T). Similarly, the molar amount of guanine (G) is essentially the same as that of cytosine (C). To maintain this balance, the bases in DNA must be paired, A to T and G to C. But how? At the midpoint of this century, it was quite clear that the answer to this question would bring with it a Nobel prize. Although many illustrious scientists worked on the problem, two scientists who were relatively unknown in the world of science announced in 1953 that they had worked out the structure of DNA. Using data that involved quite sophisticated chemistry, physics, and mathematics, and working with models not unlike a child's construction set, James D. Watson and Francis Crick determined that DNA must be composed of two helixes wound about one another. The phosphate and sugar backbone of the polymer chains form the outside of the structure, which is rather like a spiral staircase. The heterocyclic amines are paired on the inside—with guanine always opposite cytosine and adenine always opposite thymine. In our staircase analogy, these base pairs are the steps (Figure 15.25).

Why do the bases pair in this precise pattern, always A to T and T to A, always G to C and C to G? The answer is hydrogen bonding and a truly

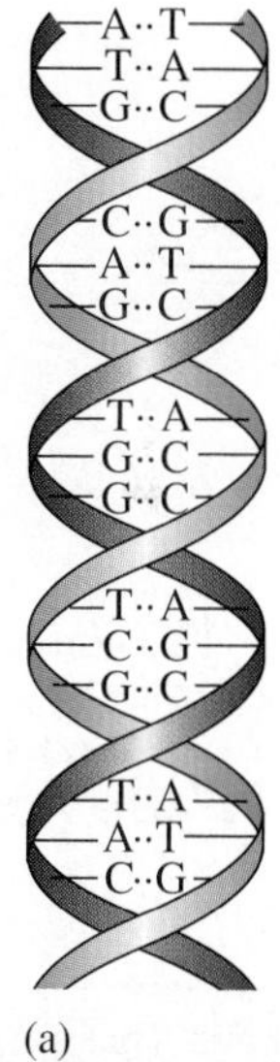

(a)

(b)

Figure 15.25 (a) A schematic representation of the DNA double helix. (b) A model of a portion of the DNA molecule. [*(b) Fundamental Photographs/Macmillan Science Files.*]

James D. Watson and Francis Crick, discoverers of the double helix model of DNA. [*Courtesy of Harvard University Biological Laboratories.*]

Watson and Crick received the Nobel prize in 1962 for discovering, as Crick put it, "the secret of life."

elegant molecular design. Figure 15.26 shows the two sets of base pairs. You should notice two things. First, a pyrimidine is paired with a purine in each case, and the long dimensions of both pairs are identical (1.085 nm).

The second thing you should notice in Figure 15.26 is the hydrogen bonding between the bases in each pair. When guanine is paired with cytosine, three hydrogen bonds can be drawn between the bases. No other pyrimidine–purine pairing will permit such extensive interaction. Indeed, in the combination shown in the figure, both pairs of bases fit like lock and key.

The Watson–Crick structure was accepted almost immediately by other scientists around the world because it answers so many crucial questions. It can explain how cells are able to divide and go on functioning, how genetic data are passed on to new generations, and even how proteins are built to required specification. It all depends on the base-pairing.

Structure of RNA

The molecules of RNA consist of single strands of the nucleic acid. Some internal (intramolecular) base-pairing may occur in sections where the molecule folds back on itself. Portions of the molecule may exist in double-helical form (Figure 15.27).

SECTION 15.11

DNA: Self-replication

Cats have kittens that grow up to be cats. Bears have cubs that grow up to be bears. How is it that each species reproduces after its own kind? How does a fertilized egg "know" that it should develop into a kangaroo and not a koala?

Thymine
Adenine
To chain
To chain
1.085 nm
Cytosine
Guanine
To chain
To chain
1.085 nm

Figure 15.26 The pairing of bases in the DNA double helix.

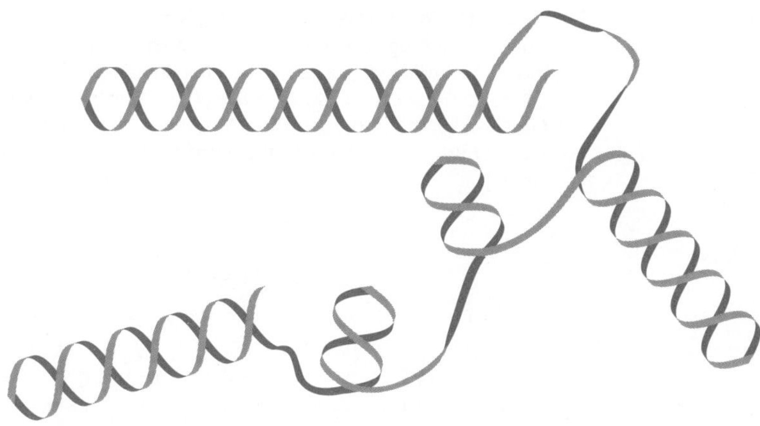

Figure 15.27 RNA occurs as single strands that can form double-helical portions by internal base pairing.

The physical basis of heredity has been known for a long time. Most higher organisms reproduce sexually. A sperm cell from the male unites with an egg cell from the female. The fertilized egg so formed must carry all the information needed to make the various cells, tissues, and organs necessary for the functioning of a new individual. For human beings, that single cell must carry the information for the making of legs, liver, lungs, heart, head, hair, and hands—in short, all the instruction ever needed for growth and maintenance of the individual. In addition, if the species is to survive, information must be set aside in germ cells—both sperms and eggs—for the production of new individuals.

Chromosomes and Genes

The hereditary material is found in the nuclei of all cells, concentrated in elongated, threadlike bodies called **chromosomes**. The number of chromosomes varies with the species. Human body cells have 46 chromosomes. Germ cells carry only half that number. Thus, in sexual reproduction, the entire complement of chromosomes is achieved only when the egg and sperm combine; a new individual receives half its hereditary material from each parent.

Humans have about 50,000 active genes. Scientists have embarked on an ambitious endeavor, called the *human genome project* to map the location of each of these genes on the chromosomes. (The *genome* of an organism is its complete set of genes.)

Chromosomes are made of nucleic acids and proteins. The nucleic acid in chromosomes is DNA, and it is the DNA that is the primary hereditary material (Figure 15.28). Arranged along the chromosomes are the basic units of heredity, the genes. Structurally, **genes** are sections of the DNA molecule (some viral genes contain only RNA). When cell division occurs, each chromosome produces an exact duplicate of itself. Transmission of genetic information therefore requires the **replication** (copying or duplication) of DNA molecules.

Genes have functional regions called *exons* interspersed with inactive portions called *introns.* During protein synthesis, introns are snipped out and not translated.

The Watson–Crick double helix provides a ready model for the process of replication. If the two chains of the double helix are pulled apart, then each

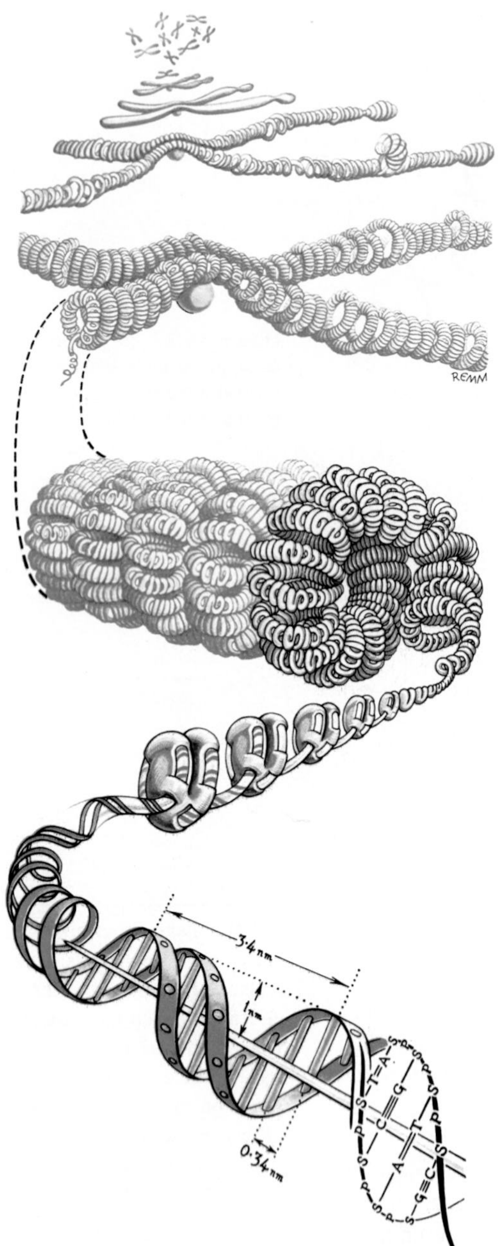

Figure 15.28 Chromosomes and DNA. Each chromosome (top) is made up of two chromatids united at the centromere. A chromatid is a protein-coated strand of multicoiled DNA. [*Adapted from Roger Warwick and Peter L. Williams,* Gray's Anatomy, *36th British ed. (London: Longmans Co. Ltd., 1980), p. 15, with permission of Churchill Livingstone.*]

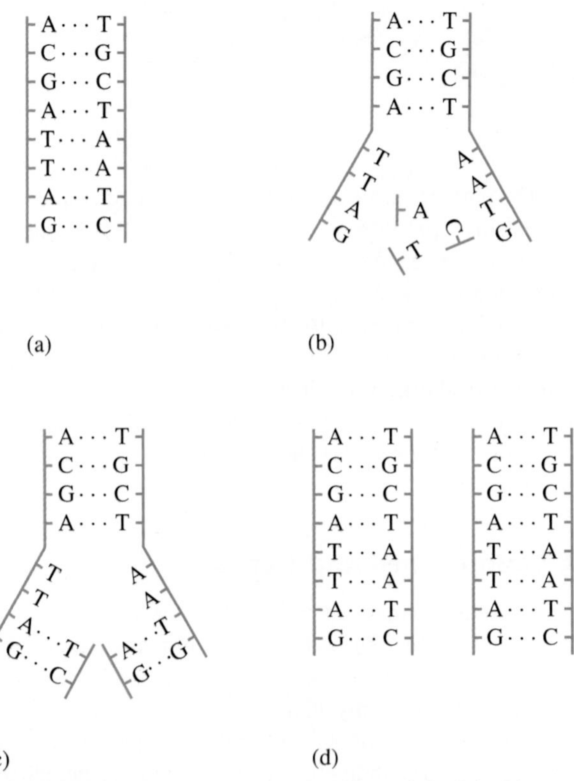

Figure 15.29 DNA replication. (a) The original double helix, flattened out here for clarity. (b) The helix beginning to split. Some free nucleotides from the cell are shown. (c) Nucleotides from the cell beginning to pair with bases on each original strand. (d) The two new double helixes, each identical to the original.

chain can direct the synthesis of a new DNA chain. In the cellular fluid surrounding the DNA are all the necessary nucleotides. Synthesis begins with the base on a nucleotide pairing with its complementary base on the DNA strand (Figure 15.29). Keep in mind that adenine can pair only with thymine and guanine only with cytosine. Each base unit in the separated strand can pick up only a unit identical to the one it had paired with before. Each of the separating chains serves as a template, or pattern, for the formation of a new complementary chain.

As the nucleotides align, enzymes connect them to form the sugar–phosphate backbone of the new chain. In this way, each strand of the original DNA molecule forms a duplicate of its former partner. Whatever information was encoded in the original DNA double helix is now contained in each of the replicates. When the cell divides, each daughter cell gets one of the DNA molecules and all the information that was available to the parent cell.

DNA can be compared to a book containing directions for putting together a model airplane or knitting a sweater. Knitting directions store information as words on paper. Letters of the alphabet are arranged in a certain way (for example, "knit one, purl two"), and these words direct the knitter to carry out a particular operation with needles and yarn. If all the directions are correctly followed, the ball of yarn becomes a sweater.

How is information stored in DNA? The sequence of bases along the DNA chain encodes the directions for building an organism. Just as *saw* means one thing in English and *was* means another, the sequence of bases C-G-T means one thing, and G-C-T means something else. Although there are only four "letters"—the four bases—in the genetic codes of DNA their sequence along the long strands can vary so widely that essentially unlimited information storage is available. Each cell carries in its DNA all the information it needs to determine all its hereditary characteristics.

SECTION 15.12

RNA: Protein Synthesis and the Genetic Code

DNA carries a message that must somehow be relayed and acted upon in the cell. Because DNA is seldom identified outside of the cell nucleus, its information, or "blueprint," must be transported by something else. It is! In the first step, called **transcription**, DNA transfers its information to a special RNA molecule called **messenger RNA (mRNA)**. The base sequence of DNA specifies the base sequence of mRNA. Thymine in DNA calls for adenine in mRNA, cytosine specifies guanine, guanine calls for cytosine, and adenine requires uracil (Table 15.5). Remember that in RNA molecules, uracil is used in place of DNA's thymine. Notice the similarity in structure of these two bases (see Figure 15.23).

The next step in protein synthesis involves interpretation of the code that has been copied by mRNA and **translation** of that code into a protein structure. mRNA travels from the nucleus to structures called ribosomes in the cytoplasm of the cell. Ribosomes also are constructed of nucleic acids and proteins. mRNA becomes attached to the ribosome, and it is here that the

Table 15.5	DNA Bases and Their Complementary RNA Bases
DNA Base	Complementary RNA Base
Adenine	Uracil
Thymine	Adenine
Cytosine	Guanine
Guanine	Cytosine

genetic code is deciphered. Each mRNA contains all the information necessary to make one particular protein.

There is another type of RNA molecule, called **transfer RNA (tRNA)**, located in the cytoplasm of cells. The tRNA is responsible for translating the specific base sequence of mRNA into a specific amino acid sequence in a protein. (The function of proteins is dependent on the sequence of their amino acid building blocks.) The tRNA molecule has a looped structure (Figure 15.30) that contains a critical set of three bases, called a **base triplet**; the base triplet determines which amino acid will be attached to an end of tRNA. Each molecule of tRNA can carry only one amino acid. For example, tRNA molecules with the triplet configuration of CCU (the complement of the mRNA triplet GGA) always carry the amino acid glycine, and a tRNA molecule with the base triplet GCG (the complement of the mRNA triplet CGC) always carries arginine.

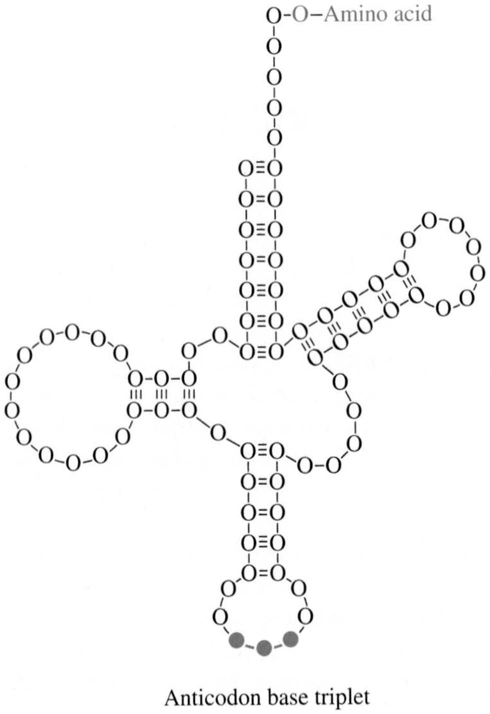

Figure 15.30 A given transfer RNA can carry only one kind of amino acid, which is specified by the base triplet in the anticodon.

Amino acid

Anticodon base triplet

We are now ready to build proteins. Strung out along some ribosomes in the cytoplasm of a cell is an mRNA molecule. The sequence of bases along this molecule was determined by the DNA of a gene in the cell nucleus. Floating around in the cytoplasm surrounding the mRNA are tRNA molecules, each carrying its own amino acid. Let us suppose that a portion of the mRNA base sequence reads.

~C—G—C—G—G—A—G—G—C~

The first three bases (CGC) could pair, through hydrogen bonding, with a nucleic acid that had a GCG sequence. There, floating by, is just such a nucleic acid, a tRNA molecule with the amino acid arginine attached. The triplet of bases of tRNA pairs up with the first three bases of the mRNA.

```
~C—G—C—G—G—A—G—G—C~     (mRNA strand)
 :  :  :
 G—C—G                   (tRNA)
   |
  Arg
```

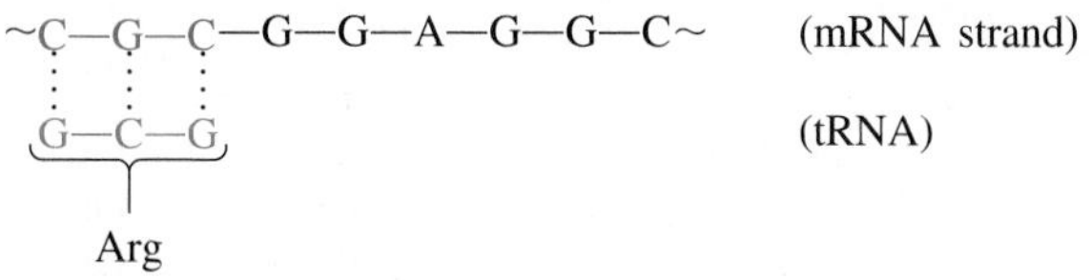

The next three bases of the mRNA, GGA, pair up with a CCU on a tRNA molecule, which carries the amino acid glycine.

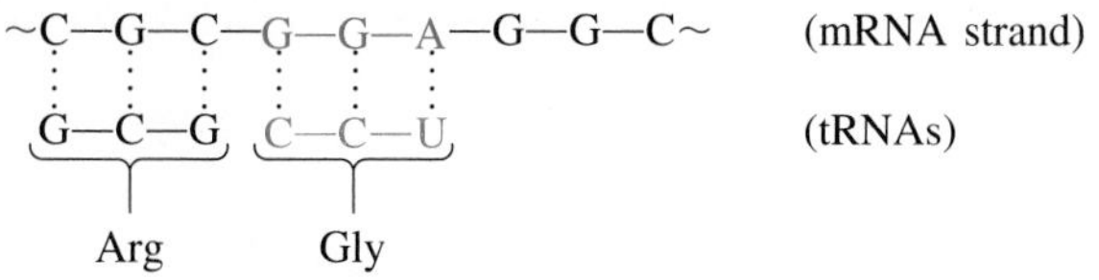

```
~C—G—C—G—G—A—G—G—C~     (mRNA strand)
 :  :  :  :  :  :
 G—C—G  C—C—U            (tRNAs)
   |      |
  Arg    Gly
```

When the two tRNA molecules are appropriately lined up along the mRNA, a bond forms between the two amino acids.

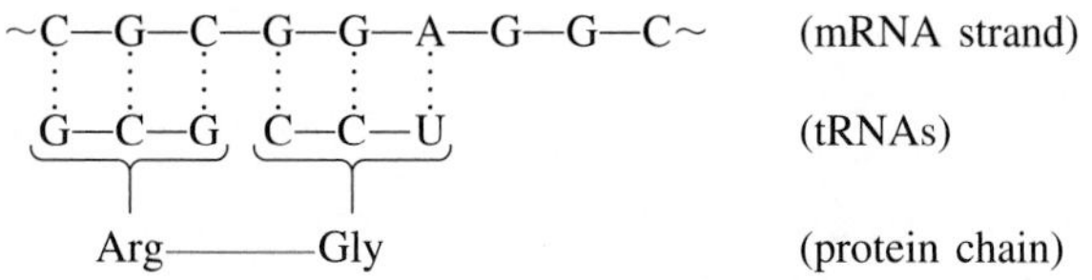

```
~C—G—C—G—G—A—G—G—C~     (mRNA strand)
 :  :  :  :  :  :
 G—C—G  C—C—U            (tRNAs)
   |      |
  Arg————Gly             (protein chain)
```

After the bond is formed the tRNA with the GCG triplet returns to the cytoplasm, and the next three bases on the mRNA (GGC) pair with a CCG tRNA molecule.

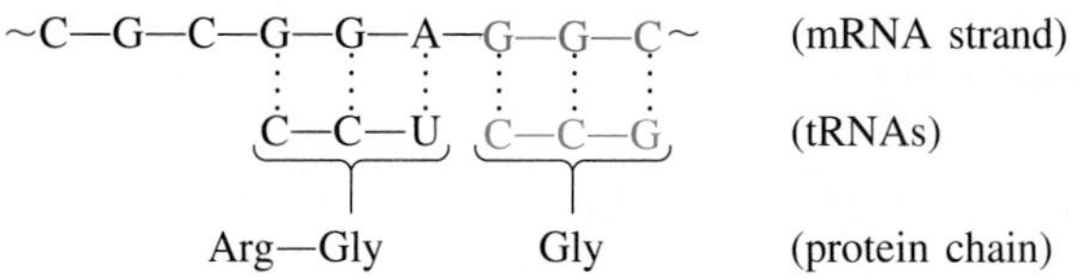

```
~C—G—C—G—G—A—G—G—C~     (mRNA strand)
          :  :  :  :  :  :
          C—C—U  C—C—G   (tRNAs)
            |      |
       Arg—Gly    Gly    (protein chain)
```

The amino acid carried by the tRNA is then bonded to the previously linked amino acids.

Each step in the replication-transcription-translation process is subject to error. In replication alone, each time a human cell divides, a copy is made of 4 billion bases to make a new strand of DNA. There are perhaps 2000 errors each time replication occurs. Most such errors are unimportant, but some have terrible consequences: genetic disease or even death may result.

Table 15.6 The Genetic Code for Protein Synthesis

First base	Second base U		Second base C		Second base A		Second base G		Third base
U	UUU	Phe	UCU	Ser	UAU	Tyr	UGU	Cys	U
U	UUC	Phe	UCC	Ser	UAC	Tyr	UGC	Cys	C
U	UUA	Leu	UCA	Ser	UAA	Stop	UGA	Stop	A
U	UUG	Leu	UCG	Ser	UAG	Stop	UGG	Trp	G
C	CUU	Leu	CCU	Pro	CAU	His	CGU	Arg	U
C	CUC	Leu	CCC	Pro	CAC	His	CGC	Arg	C
C	CUA	Leu	CCA	Pro	CAA	Gln	CGA	Arg	A
C	CUG	Leu	CCG	Pro	CAG	Gln	CGG	Arg	G
A	AUU	Ile	ACU	Thr	AAU	Asn	AGU	Ser	U
A	AUC	Ile	ACC	Thr	AAC	Asn	AGC	Ser	C
A	AUA	Ile	ACA	Thr	AAA	Lys	AGA	Arg	A
A	AUG	Met	ACG	Thr	AAG	Lys	AGG	Arg	G
G	GUU	Val	GCU	Ala	GAU	Asp	GGU	Gly	U
G	GUC	Val	GCC	Ala	GAC	Asp	GGC	Gly	C
G	GUA	Val	GCA	Ala	GAA	Glu	GGA	Gly	A
G	GUG	Val	GCG	Ala	GAG	Glu	GGG	Gly	G

~C—G—C—G—G—A—G—G—C~ (mRNA strand)

G—C—U C—C—G (tRNAs)

Arg—Gly—Gly (protein chain)

In that way the protein chain is gradually built up. The chain is released from the tRNA and mRNA as it is formed.

Each base triplet on the mRNA molecule is called a **codon**. The base triplets of the tRNA are called **anticodons**. A codon on mRNA always pairs with its complementary tRNA anticodon. A complete dictionary of the genetic code has been compiled. Table 15.6 shows which amino acids are called for by all the possible mRNA codons. There are 64 possible codons and only 20 amino acids; there is considerable redundancy in the code. Two of the amino acids (serine and leucine) are each specified by six different codons. Two others (tryptophan and methionine) have only one codon each. Three codons are stop signals, calling for termination of the protein chain.

SECTION 15.13

Genetic Engineering

Over 3000 human diseases have a genetic component. Over the last decade or so, researchers have linked specific genes to specific diseases. Now the ability

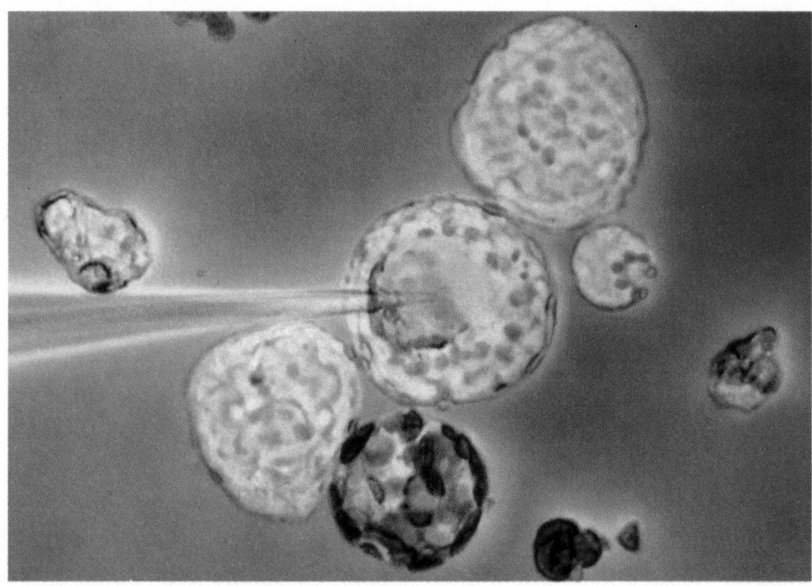

Figure 15.31 Under a powerful microscope, a tiny pipet is used to introduce foreign genes into a cell nucleus. [*Courtesy of Bayer AG, Germany.*]

to use this information to diagnose and cure genetic diseases appears to be within our grasp. By determining the location of a gene on the DNA molecule, scientists have been able to identify and isolate genes with specific functions.

RFLPs: Predicting Genetic Disease

A gene is a rather elusive substance. There are approximately 10,000 genes on each human chromosome, and isolating the one defective gene that causes a particular genetic disease is a monumental task. One approach to gene location is to treat the DNA with enzymes called *restriction endonucleases.* This method yields segments of genetic material called *restriction fragment length polymorphisms (RFLPs;* pronounced ''rif lips''). These segments are much easier to work with because they contain fewer genes. RFLP patterns characteristic of certain families can be isolated. If the pattern of a relative matches that of a person with a genetic disease, that relative will probably develop the disease. Thus it is possible to predict the occurrence of a genetic disease.

A further hope of genetic engineering is that we will be able to introduce a functioning gene into a person's cells (Figure 15.31), thus correcting the action of a defective gene.

SECTION 15.14

Recombinant DNA

All living organisms (except some viruses) have DNA as their hereditary material. It should be possible, then, to place a gene from one organism into

Many viruses readily penetrate human cells where they use the cell machinery to replicate themselves. A human gene can be cloned into a modified virus that will then transport that gene into the human cells where it produces multiple copies of the human gene. The first human gene transplant was carried out in 1989. It was a demonstration project that proved the technique works, although it was not used to treat disease. The first human gene therapy was attempted in 1990. Modified human white blood cells were infused into a young girl with severe immune deficiency (reference 19).

the genetic material of another. **Recombinant DNA** technology does just that.

By working backward from the amino acid sequence of the protein, scientists can work out the base sequence of the gene that codes for the protein. When isolated by the RFLP process, the gene is spliced into a special kind of bacteria DNA called a **plasmid**. The recombined plasmid is then inserted into the host organism (Figure 15.31). Once inside the host, the plasmid replicates, making multiple exact copies (or **clones**) of itself. As the engineered bacteria multiply, they become virtual factories for producing the desired protein.

Many valuable materials, difficult to obtain in any other way, are now made using recombinant DNA technology. People with diabetes formerly had to use insulin from pigs or cattle. Now human insulin, a protein coded for by human DNA, is being produced by the cell machinery of bacteria. All newly diagnosed insulin-dependent diabetics in the United States are now treated with human insulin produced by recombinant DNA technology. The hope for the future is that a functioning gene for insulin can be incorporated into the cells of insulin-dependent diabetics.

Human growth hormone, used to treat children who fail to grow properly, was formerly available only in tiny amounts obtained from cadavers. Now it is readily available through recombinant DNA technology. This technology also yields interferon, a promising anticancer agent. The gene for epidermal growth factor, which stimulates the growth of skin cells, has been cloned. It has been used to speed the healing of burns and other skin wounds. Scientists have even designed bacteria that will "eat" the oil released in an oil spill, although their success in actual spills has been minimal.

Concern over the potential for disaster in this type of research has lessened somewhat in recent years. Initially, scientists worried about the possibility of producing a deadly "artificial" organism. What if a gene that causes cancer was spliced into the DNA of a bacterium that normally inhabits our intestine? To protect against such a development, strict guidelines for recombinant-DNA research have been instituted.

One of the products of recombinant DNA technology is a growth hormone for cows, bovine somatotropin (BST). When injected into cows, this hormone increases their milk production by 10 to 20%. The Food and Drug Administration, after 10 years of study, finally gave approval for the use of BST starting 4 February 1994. Many milk producers are very positive about using BST, but some consumers are unhappy at the thought of having someone tamper with their milk. There is much controversy as to whether or not a hormone such as this should actually be used.

The new molecular genetics has already resulted in some impressive achievements. Its possibilities are mind boggling—elimination of genetic defects, a cure for cancer, a race of geniuses, and who knows that else? Knowledge gives power; it does not necessarily give wisdom. Who will decide what sort of creature the human species should be? The greatest problem we shall likely face in our use of bioengineering is that of choosing who is to play God with the new "secret of life."

Summary

1. The human body is made up of millions of *cells,* each one with its own *nucleus, mitochondria, ribosomes,* and other parts.
2. Cell *metabolism* includes many processes of both *anabolism* (building up of molecules) and *catabolism* (breaking down of molecules).
3. The three basic foodstuffs are *carbohydrates, fats,* and *proteins.*
4. Sugars are simple carbohydrates that have a sweet taste.
5. *Starches* are polymers of glucose held together by alpha linkages; *cellulose* is a glucose polymer held together by beta linkages.
6. *Lipids* include solid *fats* and liquid *oils,* which are both triglycerides, esters of fatty acids with glycerol.
7. *Iodine number* measures the degree of unsaturation (double bonds) in a fat or oil. Oils tend to have more, and fats fewer, double bonds.
8. *Proteins* are polymers of amino acids, with 20 amino acids in nature forming thousands of different proteins.
9. The *primary structure* of a protein is its sequence of amino acids, which are held together by peptide bonds (amide linkages).
10. The *secondary structure* of a protein is its conformation, (e.g., *alpha helix* or *pleated sheet*) held in place by hydrogen bonds.
11. The *tertiary structure* of a protein is its folding pattern, determined by the formation of *salt bridges, disulfide linkages,* and *hydrophobic interactions.*
12. An *enzyme* is a biological catalyst, made up of an *apoenzyme* (protein) and a *coenzyme* (often a vitamin), and sometimes containing a *cofactor* (such as a metal ion).
13. *Nucleic acids* are nucleotide polymers.
14. A *nucleotide* is a compound made from phosphoric acid, a pentose sugar, and a nitrogen base (purine or pyrimidine).
15. The nitrogen bases in nucleic acids are adenine and guanine (purines) and thymine, cytosine, and uracil (pyrimidines).
16. *DNA (deoxyribonucleic acid)* is a nucleic acid in which the pentose sugar is deoxyribose and the nitrogen bases are adenine, thymine, guanine, and cytosine.
17. *RNA (ribonucleic acid)* is a nucleic acid in which the pentose is ribose and the nitrogen bases are the same as those in DNA, except that uracil replaces thymine.
18. The sequence of nitrogen bases on the sugar–phosphate stem carries the message on the DNA molecule, each triplet of bases acting as code for an amino acid.
19. The nitrogen bases always pair up the same way, guanine with cytosine and adenine with thymine (in DNA) or with uracil (in RNA).
20. DNA has a double helix structure that is able to unwind during *replication* or *transcription* to RNA, so that a new strand of nucleic acid can be produced.
21. RNA carries out protein synthesis, with *mRNA* carrying the ''blueprint'' from DNA to the ribosomes and *tRNA* delivering amino acids to the growing protein chain.
22. Each *base triplet* on mRNA is a *codon* for a specific amino acid. The base triplets on tRNA are called *anticodons.*
23. Through genetic engineering, using *recombinant DNA* technology, it may be possible to eliminate hundreds of hereditary diseases.

Key Terms

active site 15.9
alpha helix 15.8
amino acids 15.6
anabolism 15.2
anticodon 15.12
apoenzyme 15.9
base triplet 15.12
biochemistry 15.1
carbohydrate 15.3
catabolism 15.2
cell 15.1
cell membrane 15.1
cell nucleus 15.1
cellulose 15.4
chloroplasts 15.1
chromosomes 15.11

clone 15.14
codon 15.12
coenzyme 15.9
cofactor 15.9
deoxyribonucleic acid (DNA) 15.10
dipeptide 15.7
disaccharides 15.3
disulfide linkage 15.8
enzymes 15.9
fats 15.5
fatty acids 15.5
genes 15.11
globular proteins 15.8
glycogen 15.4
hydrophobic interactions 15.8
iodine number 15.5
lipid 15.5
messenger RNA (mRNA) 15.12
metabolism 15.2
mitochondria 15.1
nucleic acids 15.10
nucleotides 15.10
oils 15.5
plasmid 15.14
pleated sheet 15.8
polypeptide 15.7
polysaccharides 15.4
primary structure 15.8
protein 15.7
purines 15.10
pyrimidines 15.10
recombinant DNA 15.14
replication 15.11
ribonucleic acid (RNA) 15.10
ribosomes 15.1
salt bridge 15.8
secondary structure 15.8
starch 15.4
substrate 15.9
tertiary structure 15.8
transcription 15.12
transfer RNA (tRNA) 15.12
translation 15.12
zwitterion 15.6

Review Questions

1. What is biochemistry?

2. Briefly identify and state a function of each of the following parts of a cell.
a. cell membrane b. cell nucleus
c. chloroplasts d. mitochondria
e. ribosomes

3. How do green plants obtain food?

4. What are the three major types of foods from which animals obtain energy?

5. Define each of the following.
a. anabolism b. catabolism
c. metabolism

6. What are carbohydrates?

7. What are monosaccharides?

8. In what way are amylose and cellulose similar?

9. What is the main structural difference in starch and cellulose?

10. What are the two kinds of plant starch? How do they differ?

11. What is glycogen?

12. What is a lipid? Name at least three kinds of lipids.

13. What are fatty acids?

14. What is the chemical nature of fats?

15. Define monoglyceride, diglyceride, and triglyceride.

16. How do fats and oils differ in structure? In properties?

17. What is meant by the iodine number of a fat?

18. List some of the properties of fats and oils.

19. In the determination of the iodine number of a fat, what part of the fat molecule reacts with the reagent?

20. Define saturated fat. Define polyunsaturated fat (oil).

21. In what parts of the body are proteins found?

22. What tissues are largely protein?

23. What is the chemical nature of proteins?

24. How does the elemental composition of proteins differ from that of carbohydrates and fats?

25. What functional groups are found on amino acid molecules?

26. How many different amino acids are incorporated into proteins?

27. What is a zwitterion?

28. About how many different proteins are there in the human body?

29. What is a peptide bond?

30. What is a dipeptide? A tripeptide? A polypeptide?

31. Of what importance is the sequence of amino acids in a protein molecule?

32. What is the difference between a polypeptide and a protein?

33. What is meant by the primary structure of a protein?

34. What is meant by the secondary structure of a protein?
35. What is meant by the tertiary structure of a protein?
36. What is meant by the quaternary structure of a protein?
37. What is meant by the pleated sheet arrangement in a protein?
38. What is meant by the alpha helix arrangement in a protein?
39. What is a globular protein?
40. List the four different ways protein chains are bonded to one another.
41. Describe the lock-and-key model of enzyme action.
42. Describe how an inhibitor deactivates an enzyme.
43. Name the two kinds of nucleic acids. Which is found primarily in the nucleus of the cell?
44. What is the sugar unit in RNA? What sugar is found in DNA?
45. List the heterocyclic bases found in DNA. List those found in RNA.
46. How do DNA and RNA differ in structure?
47. What kind of intermolecular force is involved in base-pairing?
48. Describe the process of replication.
49. In replication, a parent DNA molecule produces two daughter molecules. What is the fate of each strand of the parent DNA double helix?
50. We say DNA controls protein synthesis, yet most DNA resides within the cell nucleus while protein synthesis occurs outside the nucleus. How does DNA exercise its control?
51. Explain the role of messenger RNA in protein synthesis.
52. Explain the role of transfer RNA in protein synthesis.
53. Which nucleic acid or acids is (are) involved in the process referred to as transcription?
54. Which nucleic acid or acids is (are) involved in the process referred to as translation?
55. Which nucleic acid contains the codon?
56. Which nucleic acid contains the anticodon?
57. What is the relationship between the cell parts called chromosomes, the units of heredity called genes, and the nucleic acid DNA?
58. What are RFLPs? How do RFLP patterns indicate that a person probably will develop a genetic disease?
59. List the four steps in recombinant DNA technology.
60. What is a plasmid?
61. How can a virus be used to replace a defective gene in a human?
62. Discuss some applications of genetic engineering.

Problems

Carbohydrates

63. Which of the following are monosaccharides?
    a. amylose
    b. cellulose
    c. fructose
    d. galactose
64. Which of the following are monosaccharides?
    a. glucose
    b. glycogen
    c. lactose
    d. sucrose
65. Which of the carbohydrates in Problems 63 and 64 are disaccharides?
66. Which of the carbohydrates in Problems 63 and 64 are polysaccharides?
67. What are the hydrolysis products of each of the following?
    a. amylose
    b. lactose
    c. glycogen
68. What are the hydrolysis products of each of the following?
    a. amylopectin
    b. sucrose
    c. cellulose
69. Give the formula for the open-chain form of glucose. What functional groups are present?
70. Give the formula for the open-chain form of fructose. What functional groups are present?

71. Give the formula for the open-chain form of galactose. What functional groups are present?

72. Mannose differs from glucose only in that the H and OH on the second carbon atom are reversed. Give the formula for the open-chain form of mannose. What functional groups are present?

Fats and Oils

73. Which of the following fatty acids are saturated?
a. linolenic acid b. linoleic acid
c. oleic acid d. palmitic acid
e. stearic acid

74. Which of the fatty acids in Problem 73 are unsaturated?

75. Which of the fatty acids in Problem 73 are monounsaturated?

76. Which of the fatty acids in Problem 73 are polyunsaturated?

77. How many carbon atoms are there in a molecule of each of the following fatty acids?
a. linolenic acid b. palmitic acid
c. stearic acid

78. How many carbon atoms are there in a molecule of each of the following fatty acids?
a. linoleic acid b. oleic acid
c. butyric acid

79. Which would you expect to have the higher iodine number, corn oil or beef tallow? Explain your reasoning.

80. Which would you expect to have the higher iodine number, lard or liquid margarine? Explain your reasoning.

Proteins

81. Is the dipeptide represented by Ser-Ala the same as the one represented by Ala-Ser? Explain.

82. A chemist is asked to make some Phe-Ile-Leu. He makes some Leu-Ile-Phe. Evaluate the chemist for possible job advancement.

83. Give structural formulas for each of the following amino acids.
a. alanine b. serine

84. Give structural formulas for each of the following amino acids.
a. glycine b. phenylalanine

85. Give structural formulas for each of the following dipeptides.
a. glycylalanine b. alanylserine

86. Give structural formulas for each of the following dipeptides.
a. alanylglycine b. phenylalanylserine

Nucleic Acids

87. Which of the following are nucleotides?

a. $HOCH_2$ O OH
H H H H
OH OH

b. $HOCH_2$ O adenine
H H H H
OH OH

c. OH
O=P—O—CH_2 O guanine
OH
H H H H
OH H

88. Which of the following are nucleotides?

a. OH
O=P—O—CH_2 O uracil
OH
H H H H
OH OH

b. OH
O=P—O—CH_2 O cytosine
OH
H H H H
OH H

c. $HOCH_2$ O adenine
H H H H
OH H

89. For each of the structures in Problem 87, identify the sugar unit as ribose or deoxyribose.

90. For each of the structures in Problem 88, identify the sugar unit as ribose or deoxyribose.

91. Identify the sugar and the base in the following nucleotide.

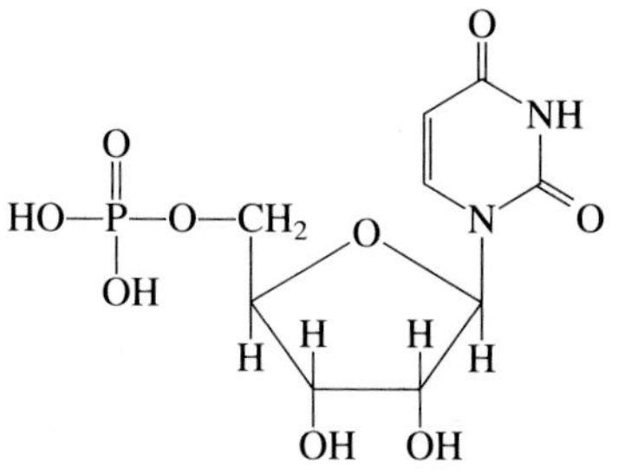

92. Identify the sugar and the base in the following nucleotide.

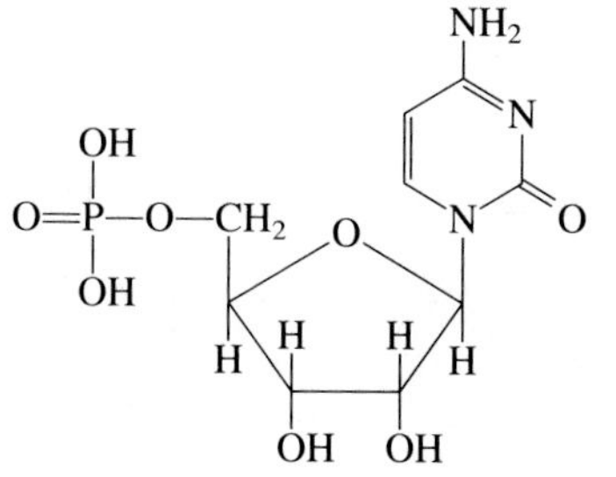

93. In DNA, which base would be paired with the base listed?

a. cytosine b. adenine
c. guanine d. thymine

94. In an RNA molecule, which base would pair with the base listed?

a. adenine b. guanine
c. uracil d. cytosine

95. The base sequence along one strand of DNA is AATTCG. What would be the sequence of the complementary strand of DNA?

96. What sequence of bases would appear in the messenger RNA molecule copied from the original DNA strand shown in Problem 95?

97. If the sequence of bases along a messenger RNA strand is UCCGAU, what was the sequence along the DNA template?

98. What are the complementary triplets on tRNA for the following triplets on mRNA?

a. UUU b. CAU
c. AGC d. CCG

99. What are the complementary codons on mRNA for the following anticodons on tRNA molecules?

a. UUG b. GAA

100. What are the complementary codons on mRNA for the following anticodons on tRNA molecules?

a. UCC b. CAC

Additional Problems

101. Which base is purine and which is pyrimidine?

a. 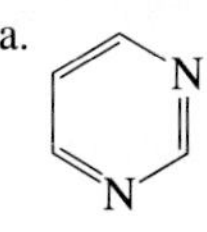

b.

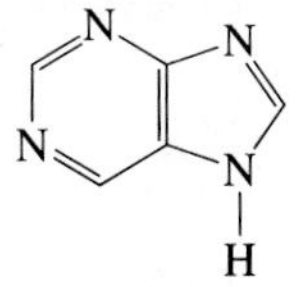

102. Answer the questions for the molecule shown.

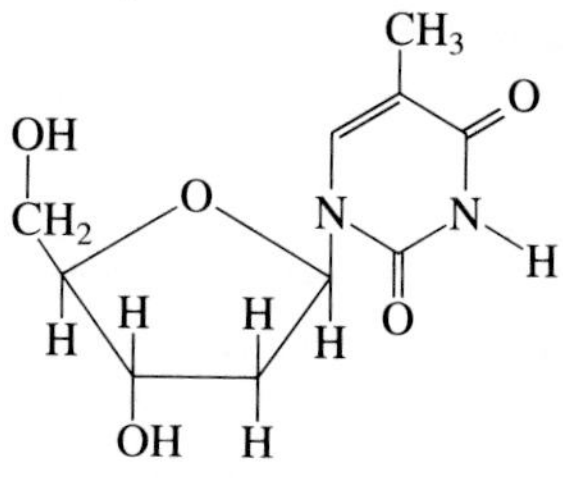

a. Is the base a purine or a pyrimidine?
b. Would the compound be incorporated in DNA or in RNA?

103. Answer the questions posed in Problem 102 for the following compound.

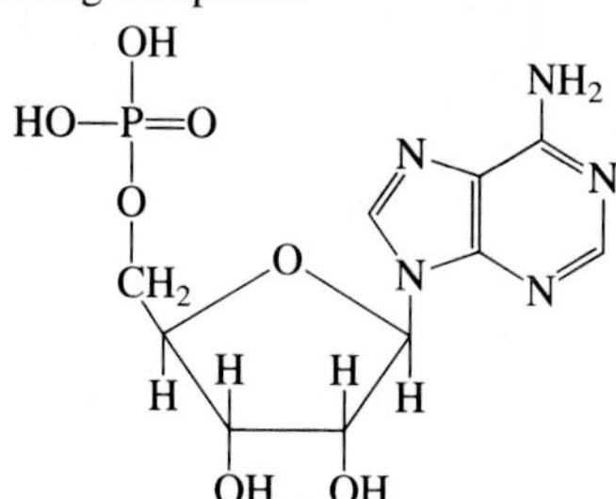

104. In the schematic below, *Glc* represents glucose. What substance is indicated?

```
. . . Glc-Glc
           \
. . . Glc-Glc-Glc-Glc
                 \
  . . . Glc-Glc-Glc-Glc-Glc-Glc-Glc-Glc-Glc
                                       \
     . . . Glc-Glc-Glc-Glc-Glc-Glc-Glc-Glc-Glc-Glc-Glc-Glc
                                                  \
                      . . . Glc-Glc-Glc-Glc-Glc-Glc-Glc
```

105. Which of the following is a lipid?

a. $H_3\overset{+}{N}\underset{\underset{CH_3}{|}}{C}H\overset{\overset{O}{\|}}{C}—NH\underset{\underset{CH_2SH}{|}}{C}H\overset{\overset{O}{\|}}{C}—NHCH_2COOH$

b. $CH_2—O—\overset{\overset{O}{\|}}{C}CH_2(CH_2)_9CH_3$
$\quad|$
$CH—O—\overset{\overset{O}{\|}}{C}CH_2(CH_2)_9CH_3$
$\quad|$
$CH_2—O—\overset{\overset{O}{\|}}{C}CH_2(CH_2)_9CH_3$

c. CH_2OH $\quad CH_2OH$
O $\quad$ O
H $\quad$ H H $\quad$ H
H $\quad$ H
HO $\quad$ OH H $\quad$ O $\quad$ OH H $\quad$ OH
H OH $\quad$ H OH

106. Write the abbreviated version of the following structural formula.

$$H_3\overset{+}{N}—\overset{\overset{CH_3}{|}}{C}H—\overset{\overset{O}{\|}}{C}—NH—\overset{\overset{\overset{OH}{|}}{CH_2}}{|}\!\!\!\!CH—\overset{\overset{O}{\|}}{C}—NH—\overset{\overset{\overset{SH}{|}}{CH_2}}{|}\!\!\!\!CH—\overset{\overset{O}{\|}}{C}—NH—\overset{\overset{\overset{C_6H_5}{|}}{CH_2}}{|}\!\!\!\!CH—\overset{\overset{O}{\|}}{C}—NH—\overset{\overset{H}{|}}{C}H—\overset{\overset{O}{\|}}{C}—NH—\overset{\overset{H}{|}}{C}H—\overset{\overset{O}{\|}}{C}—O^-$$

▸ References and Readings

1. Brenchley, Jean E. "For a Better World." *Today's Chemist,* April 1990, pp. 10–12. Discusses biotechnology.
2. Brownlee, Shannon, and Joanne Silberner. "The Age of Genes." *U.S. News and World Report,* 4 November 1991, pp. 64–76.
3. Diamond, Jared. "The Cruel Logic of Our Genes." *Discover,* November 1989, pp. 72–78. Discusses genetic diseases.
4. Friedmann, Theodore. "Progress Toward Human Gene Therapy." *Science,* 16 June 1989, pp. 1275–1281. Part of a special issue on genetic engineering.
5. Gasser, Charles S., and Robert T. Fraley. "Transgenic Crops." *Scientific American,* June 1992, pp. 62–69.
6. Grierson, Don, ed. *Plant Genetic Engineering.* New York: Chapman and Hall, 1991.
7. Good, Mary L. *Biotechnology and Materials Science.* Washington, DC: American Chemical Society, 1988.
8. Hill, John W., Dorothy M. Feigl, and Stuart J. Baum. *Chemistry and Life,* 4th ed. New York: Macmillan, 1991.
9. Marks, Dawn B. *Biochemistry.* Baltimore: Williams and Wilkins, 1990.
10. Mathews, Christopher K., and K. E. van Holde. *Biochemistry.* Redwood City, CA: Benjamin/Cummings, 1990.
11. Miller, J. A. "First Live Gene-Splice Release: It's Already History." *Science News,* 12 April 1986, p. 228.
12. Montgomery, Geoffrey. "The Ultimate Medicine." *Discover,* March 1990, pp. 60–68. Viruses can carry replacements for defective genes into human cells.

13. Moore, Michael. ''Within Our Grasp: The New Hope of Genetic Treatment.'' *Health Sciences,* Summer 1986, pp. 1–6.

14. Neufeld, Peter J., and Neville Colman. ''When Science Takes the Witness Stand.'' *Scientific American,* May 1990, pp. 46–53. Discusses DNA ''fingerprinting.''

15. O'Neill, Luke, Michael Murphy, and Richard B. Gallagher. ''What Are We? Where Did We Come From? Where Are We Going?'' *Science,* 14 January 1994, pp. 181–183.

16. Richards, Frederic M. ''The Protein Folding Problem.'' *Scientific American,* January 1991, pp. 54–63. Discusses how proteins fold to form biologically active molecules.

17. ''The Telltale Gene.'' *Consumer Reports,* July 1990, pp. 483–488. Discusses genetic screening for abnormal genes.

18. Watson, J. D. *The Double Helix.* New York: New American Library, 1968. Irreverent, gossipy account of the discovery of the structure of DNA.

19. ''4-Year-Old Infused After Human Gene Therapy Approval.'' *FDA Consumer,* December 1990, p. 3.

20. Weiss, Jiri, Jr. ''Weed on Parole.'' *Discover,* March 1992. p. 28. Discusses advantages of tobacco leaves in biotechnology.

16 Food

Those Incredible Edible Chemicals

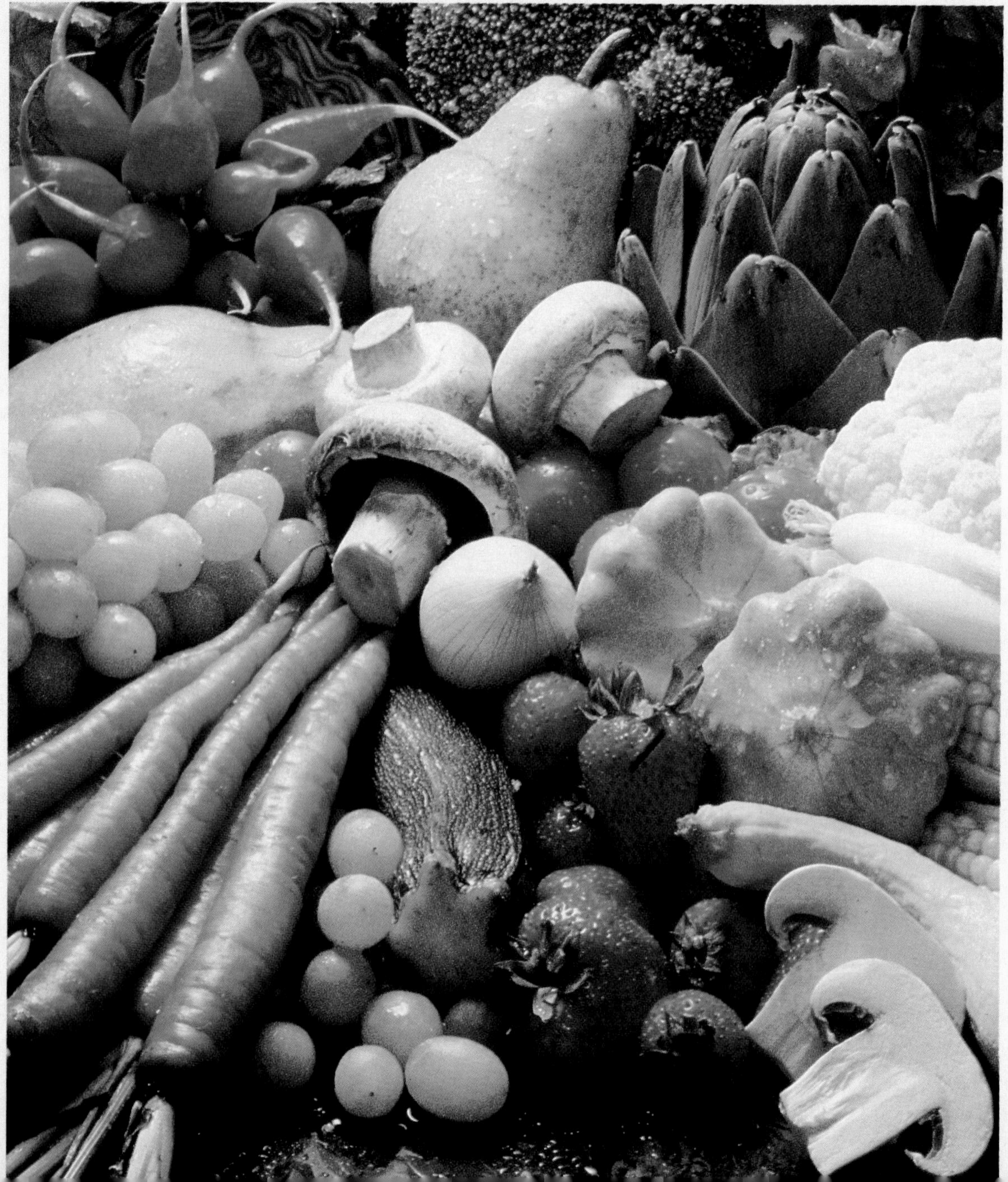

▼

Apples, peanuts, corn, tomatoes,

Peppers, olive oil,

Lettuce, mustard, beans, potatoes—

All come from the soil.

FOOD! The word conjures up images of holiday feasting and midnight snacking. It may make you think about fields ready for harvest, or orchards ready for picking. Perhaps it makes you think about some personal favorite—apple pie, or chocolate cake, or extra-cheese pizza.

There are some places in the world where people never have enough food. They are always hungry. Meanwhile, there are other places with surplus food. In the Western World we have such an overabundance of food that we often eat too much. And frequently we eat the wrong kinds of food. It is estimated that more than 30% of the adults in the United States are overweight. Obesity contributes to poor health, increased risk of diabetes and heart attacks, and increased susceptibility to other diseases as well. Our diet (too rich in fat, sugar, alcohol, and cholesterol) has been linked to five of the ten leading causes of death in the United States.

The human body is a conglomeration of chemicals. If we did a chemical analysis of the human body, we would find that it is largely—about two-thirds—water. There are also a number of minerals present and thousands of other chemicals that play vital roles in the life processes of the body.

Our foods are also made up of chemicals. They enable children to grow. They provide people of all ages with energy, and they supply the chemicals needed for repair and replacement of body tissues.

The three main classes of foods are carbohydrates, fats, and proteins. These, too, are chemicals. For proper nutrition our diet should include balanced proportions of these three foodstuffs, plus water, vitamins, minerals, and fiber.

In this chapter we discuss food—what it's made of and where it comes from.

SECTION 16.1

Carbohydrates: The Preferred Fuels

Dietary **carbohydrates** include the sugars and starches. The sugars are monosaccharides and disaccharides; the starches are polysaccharides.

◀ The foods from which we get all our chemical building blocks and all our energy obtain their energy from the sun through the remarkable process of photosynthesis. [*Richard Fukuhara/Westlight.*]

Some of the chemistry of carbohydrates is discussed in Section 15.3.

Sweet Chemicals: The Sugars

Sugars have been used for ages to make certain foods sweeter and more palatable. Two monosaccharides, **glucose** (also called **dextrose**) and **fructose** (fruit sugar), and the disaccharide **sucrose** are the most common dietary sugars (Figure 16.1). Glucose is the sugar used by the cells of our bodies for energy. It is the sugar that is circulated in the bloodstream; that is why medical workers call it **blood sugar**.

Glucose and fructose are isomers; both have the molecular formula $C_6H_{12}O_6$. Sucrose has the molecular formula $C_{12}H_{22}O_{11}$. Structures of these and other sugars are given in Section 15.3.

Common table sugar is the disaccharide sucrose, derived principally by crystallization from the juice of sugar cane and sugar beets. Corn syrup is mainly glucose. It is made by the hydrolysis of starch. *Fructose* is found in some fruits, but much of it is made from glucose. *High-fructose corn syrup* is made by using enzymes to convert much of the glucose in the syrup to fructose. Fructose is sweeter than glucose or sucrose. Foods sweetened to the same degree with high-fructose corn syrup have somewhat fewer calories than those sweetened with sucrose.

Per capita consumption of sugars in the United States is about 61 kg per year. Most are consumed in soft drinks, presweetened cereals, candy, and other highly processed foods with little or no other nutritive value. The empty calories from sweetened foods contribute to tooth decay and obesity.

Hydrolysis means splitting by water. By "sucrose is hydrolyzed" we mean that sucrose reacts with water and is broken down to glucose and fructose.

Digestion and Metabolism of Carbohydrates

Glucose and fructose are absorbed directly into the blood stream from the digestive tract. *Sucrose* is hydrolyzed during digestion to glucose and fructose.

$$\text{Sucrose} + H_2O \longrightarrow \text{Glucose} + \text{Fructose}$$

Figure 16.1 The principal sugars in our diets are sucrose (cane or beet sugar), glucose (corn syrup), and fructose (fruit sugar, often in the form of high-fructose corn syrup). [*Richard Megna/Fundamental Photographs.*]

The disaccharide **lactose** occurs in milk. During digestion, it is hydrolyzed to two simpler sugars, glucose and galactose.

$$\text{Lactose} + H_2O \longrightarrow \text{Glucose} + \text{Galactose}$$

Nearly all human babies have the enzyme necessary to accomplish this breakdown, but many adults do not. People who lack the enzyme get digestive upsets from drinking milk. The condition is called *lactose intolerance.* When milk is cooked or fermented, the lactose is at least partially hydrolyzed. People with lactose intolerance may still be able to enjoy cheese, yogurt, or cooked foods containing milk with little or no problem.

All monosaccharides are converted during metabolism to glucose. Some babies are born deficient in the enzyme that catalyzes the conversion of galactose to glucose, a condition known as *galactosemia.* A synthetic formula (Figure 16.2) to replace milk is used to provide the babies with proper nutrition.

Figure 16.2 Infants born with galactosemia can thrive on a milk-free substitute formula. [*Richard Megna/ Fundamental Photographs.*]

The enzyme that acts on lactose is now readily available in drug stores.

Complex Carbohydrates: Starches

As foods, complex carbohydrates (starch and cellulose) have a somewhat better reputation than the simple sugars. Starch is an important part of any balanced diet (Figure 16.3). **Cellulose** is an important component of dietary fiber.

Starch is hydrolyzed to glucose when it is digested. The body then metabolizes the glucose, using it as a source of energy. Glucose is broken down

Figure 16.3 Bread, flour, cereals, and pasta, as well as potatoes, are rich in starches. [*Robert Mathena/ Fundamental Photographs.*]

All animals can digest and metabolize starch, while humans and most other animals get no food value from cellulose. Bacteria in the digestive tract of grazing animals and termites have enzymes that break down cellulose to glucose. The animals then use part of the glucose and other bacterial materials as sources of energy.

through a complex set of more than 50 chemical reactions to produce carbon dioxide and water, with the release of energy.

$$C_6H_{12}O_6 + 6\,O_2 \longrightarrow 6\,CO_2 + 6\,H_2O + \text{Energy}$$

These reactions are essentially the reverse of photosynthesis. In this way, animal organisms are able to make use of the energy from the sun that was captured by plants in the process of photosynthesis.

Carbohydrates supply 4 kcal of energy per gram. When we eat more than we can use, a small amount of carbohydrate can be stored in our liver and muscle tissues as **glycogen** (animal starch). Large excesses, however, are converted to fat for storage.

There is no set dietary recommendation for carbohydrates, but at least 65%, and perhaps as much as 80%, of our calories should come from them—for these are our bodies' preferred fuels. The majority of our carbohydrate intake should be the complex carbohydrates found in cereal grains, rather than the simple sugars found in many prepared foods.

When violently shaken together with water, fats are broken into tiny, submicroscopic particles and dispersed through the water. Such a mixture is called an *emulsion.* Unless a third substance has been added, the emulsion breaks down rapidly. The droplets then recombine and float to the surface of the water. Soap, certain types of gum, or protein can stabilize the emulsion by forming protective coatings on the fat droplets that prevent them from coming together. Compounds called bile salts keep tiny fat droplets suspended in the aqueous digestive media during digestion.

Many foods are emulsions. Milk is an emulsion of butterfat in water. The stabilizing agent is a protein called casein. Mayonnaise is an emulsion of salad oil in water, stabilized by egg yolk.

SECTION 16.2

Fats: Energy Reserves, Cholesterol, and Cardiovascular Disease

When most people diet, they reduce their intake of such foods as bread and potatoes. But it usually is not these carbohydrate foods that make us fat; rather it is the high proportion of fat in our diets that leads to obesity. We eat too much fat-laden snack foods, fast foods cooked in fat, and too much cream, butter, margarine, salad oils, and the like (Figure 16.4).

Digestion and Metabolism of Fats

Dietary **fats** are mainly **triglycerides**; that is, they are esters of glycerol and fatty acids (Chapter 15). Fats are digested by enzymes called lipases, to fatty acids, glycerol, soaps (salts of fatty acids), and monoglycerides and diglycerides (Figure 16.5). The products of fat digestion are reassembled into triglycerides, which are attached to proteins for transportation through the bloodstream.

Fats are stored throughout the body, principally in **adipose tissue**. The storage places are called **fat depots**. Considerable fat is stored around vital organs, such as the heart, kidneys, and spleen. There it serves as a protective cushion, helping to prevent injury to the organs. Fat is also stored under the skin, where it helps insulate against sudden temperature changes.

When the fat reserves are called upon for energy, the fat molecules are hydrolyzed once more to glycerol and fatty acids. The glycerol can be "burned" for energy or it can be converted to glucose. The fatty acids can be

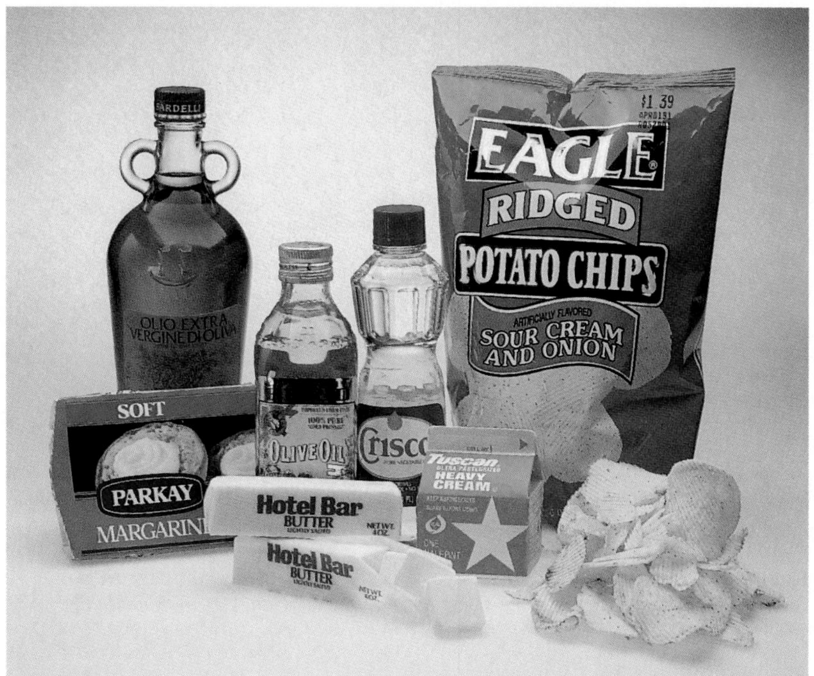

Figure 16.4 Cream, butter, margarine, cooking oils, and foods fried in fat are rich in fats. [*Richard Megna/Fundamental Photographs.*]

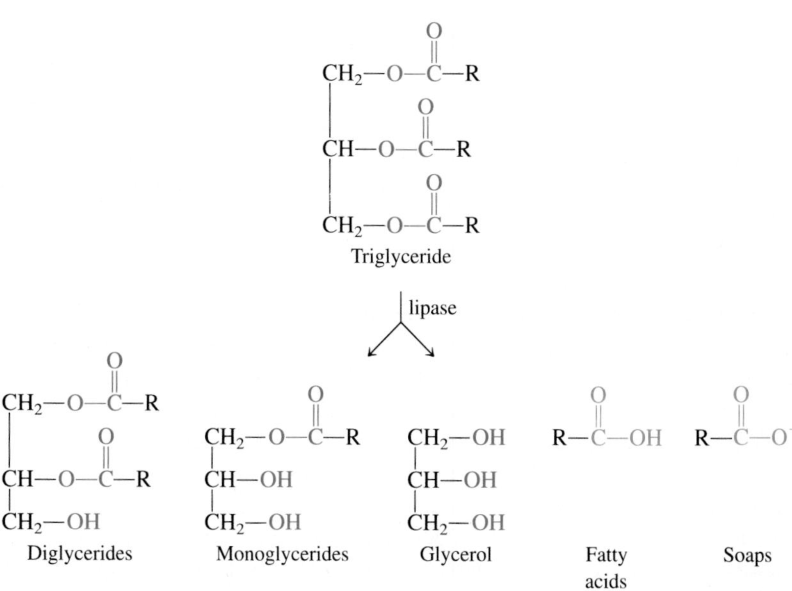

Figure 16.5 Products formed in the digestion of fats. Triglyceride (fat) is hydrolyzed into diglycerides, monoglycerides, glycerol, fatty acids, and soaps. The R symbolizes the hydrocarbon portion of the fatty acids.

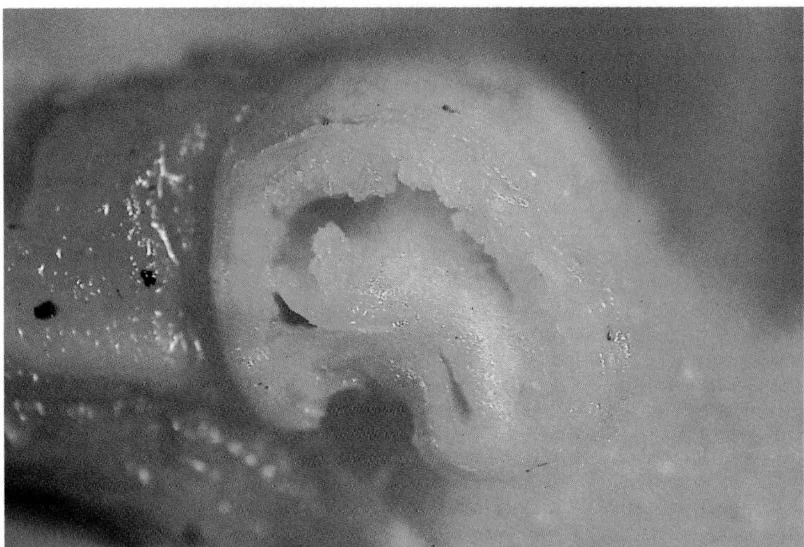

Figure 16.6 Photomicrograph of a cross section of a hardened artery showing deposits of plaque. The deposits contain cholesterol. [*Dan McCoy/Rainbow.*]

used for energy in a process called the fatty acid spiral that chops off the carbon atoms two at a time. These two-carbon fragments can be used for energy or used to synthesize new fatty acids.

Saturated fats are those that are made of a large proportion of saturated fatty acids such as palmitic acid and stearic acid. Polyunsaturated fats (oils) have a high proportion of unsaturated fatty acids such as linoleic acid and linolenic acid. (Structures of these compounds are given in Chapter 15.)

A **lipoprotein** is any of a group of proteins combined with a lipid (Section 15.5) such as a triglyceride or cholesterol.

Fats, Cholesterol, and Human Health

Dietary **saturated fats** have been implicated, along with cholesterol, in *arteriosclerosis* (hardening of the arteries). As this condition develops, deposits form on the walls of arteries. Eventually these deposits become calcified (harden), robbing the vessels of their elasticity (Figure 16.6). Blood clots tend to lodge in the narrowed arteries, leading to heart attacks (if the blocked artery is in heart muscle) and strokes (if the blockage occurs in an artery that supplies the brain).

The plaque in clogged arteries is rich in **cholesterol**. High blood levels of cholesterol, like those of triglycerides, correlate closely with the risk of cardiovascular disease. Also like the fats, cholesterol is insoluble in water. It is transported in blood by water-soluble proteins. They usually are classified according to their density (Table 16.1). Very-low-density lipoproteins (VLDL) serve mainly to transport triglycerides. Low-density lipoproteins (LDL) are the main carriers of cholesterol. The LDL carry cholesterol to the cells for use. It is these lipoproteins that are thought to deposit cholesterol in arteries, leading to cardiovascular disease. The high-density lipoproteins (HDL) also carry cholesterol, but they carry it to the liver for processing and excretion. Exercise is thought to increase the levels of HDL, the lipoprotein

Table 16.1 Lipoproteins in the Blood

Class	Abbreviation	Percent Protein	Density (g/mL)	Main Function
Very-low-density	VLDL	5	1.006–1.019	Transport triglycerides
Low-density	LDL	25	1.019–1.063	Transport cholesterol to the cells for use
High-density	HDL	50	1.063–1.210	Transport cholesterol to the liver for processing for excretion

sometimes called "good cholesterol." High levels of LDL, called "bad cholesterol," put one at increased risk of heart attack and stroke.

There is also statistical evidence that fish oils can prevent heart disease. Researchers at the University of Leiden in the Netherlands have found that Greenlanders who eat a lot of fish have a low risk of heart disease, despite a diet that is high in total fat and cholesterol. The probable effective agents are the **polyunsaturated** fatty acids such as eicosapentaenoic acid.

One large egg contains 213 mg of cholesterol. Most health authorities recommend a maximum of 300 mg/day of cholesterol in the diet.

$$CH_3CH_2CH{=}CHCH_2CH{=}CHCH_2CH{=}CHCH_2CH{=}CHCH_2CH{=}CHCH_2CH_2CH_2\overset{\overset{\displaystyle O}{\|}}{C}{-}OH$$

Other studies have shown that diets with added fish oil lead to lower cholesterol and triglyceride levels in the blood.

Fats: Useful, But Harmful in Excess

Fats are high-energy foods. They yield about 9 kcal of energy per gram. Some fats are "burned" as fuel for our activities. Others are used to build and maintain important constituents of our cells, such as the cells of brain and nerve tissue and the membranes that surround each cell.

Fat in our diet comes from butter, cream, margarine, cheese, vegetable oils and shortenings, meat products, and some seeds and nuts. The American Heart Association and most other health authorities recommend that not more than 30% of our calories should come from fat. No more than 10% of total calories should come from saturated fats, with 10% from monounsaturated and 10% from polyunsaturated fats. The average American diet contains too much fat; about 34% of our calories come from fat, with about 13% of calories from saturated fats.

We get half our total fats, three-fourths of our saturated fats, and *all* our cholesterol from animal products such as meat, milk, cheese, and eggs. Advertising claims that a vegetable oil (for example) contains no cholesterol are simply silly; no vegetable product contains cholesterol.

EXAMPLE 16.1

One of McDonald's Big Mac sandwiches furnishes a total of 541 kcal of which 279 kcal comes from fat. What percentage of total calories is from fat?

SOLUTION

Simply divide the calories from fat by the total calories. Then multiply by 100 to get percentage (parts per 100).

$$\% \text{ calories from fat} = \frac{279 \text{ kcal}}{541 \text{ kcal}} \times 100 = 51.6\%$$

Exercise 16.1

One of Burger King's Whopper sandwiches furnishes a total of 606 kcal of which 288 kcal comes from fat. What percentage of total calories is from fat?

▶ EXAMPLE 16.2

You can calculate your maximum fat intake in grams by multiplying your daily calorie intake by 0.30 (30%), and then by 1 g fat/9 kcal. What is the maximum fat and maximum saturated fat that should be included in the diet of a person who needs 1800 kcal/day to maintain proper weight?

SOLUTION

$$1800 \cancel{\text{kcal}} \times 0.30 \times \frac{1 \text{ g fat}}{9 \cancel{\text{kcal}}} = 60 \text{ g fat}$$

Only one-third of total fat should be saturated fat. The person should eat not more than 60 g of total fat of which 20 g is saturated fat each day.

Exercise 16.2

What is the maximum fat and maximum saturated fat that should be included in the diet of a person who needs 2500 kcal/day to maintain proper weight?

▶ EXAMPLE 16.3

The label on a macaroni and cheese dinner indicates that each $\frac{3}{4}$-cup serving (as prepared) furnishes 290 kcal, 9 g protein, 34 g carbohydrate, and 13 g of fat. Calculate the percentage of calories from fat.

SOLUTION

First, calculate the calories from each kind of nutrient.

$$9 \cancel{\text{g}} \times 4 \text{ kcal}/\cancel{\text{g}} = 36 \text{ kcal (from protein)}$$

$$34 \cancel{\text{g}} \times 4 \text{ kcal}/\cancel{\text{g}} = 136 \text{ kcal (from carbohydrate)}$$

$$13 \cancel{\text{g}} \times 9 \text{ kcal}/\cancel{\text{g}} = 117 \text{ kcal (from fat)}$$

Then calculate the percent calories from fat.

$$\frac{117\ \cancel{\text{kcal}}}{290\ \cancel{\text{kcal}}} \times 100 = 40\%$$

Exercise 16.3

The label on a can of cream-style corn indicates that each $\frac{1}{2}$-cup serving furnishes 90 kcal, 2 g protein, 22 g carbohydrate, and 1 g of fat. Calculate the percentage of calories from fat.

SECTION 16.3

Proteins: Muscle and Much More

Proteins (Chapter 15) are polymers of **amino acids**. Each gene carries the blueprint for a specific protein, and each protein serves a particular purpose. We need protein in our diet in order to make muscles and hair and enzymes and many other cellular components vital to life.

Protein Metabolism: Essential Amino Acids

Proteins are broken down in the digestive tract into the component amino acids. From these our bodies synthesize proteins for growth and for the repair of tissues. When a diet contains more protein than is needed for the body's growth and repair, the leftover protein is used as a source of energy.

The adult human body can synthesize all but eight of the amino acids needed for making proteins. These eight (see Table 15.3 for structures)

Isoleucine	Lysine	Phenylalanine	Tryptophan
Leucine	Methionine	Threonine	Valine

are called **essential amino acids**. They must be included in our diet. We eat proteins, break them down in our bodies to their constituent amino acids, and then use some of these amino acids to build other proteins essential for our health. Each of the essential amino acids is a **limiting reagent**. When the body runs out of one of them, it can't make proper proteins.

An *adequate protein* supplies all the essential amino acids in the quantities needed for the growth and repair of body tissues. Most proteins from plant sources are deficient in one or more amino acids. Corn protein is lacking in lysine and tryptophan. People whose diet consists chiefly of corn may suffer from malnutrition even though the calories supplied by the food are adequate. Protein from rice is short of lysine and threonine. Wheat protein is lacking in lysine. Even soy protein, probably the best nonanimal protein, is lacking in the essential amino acid methionine.

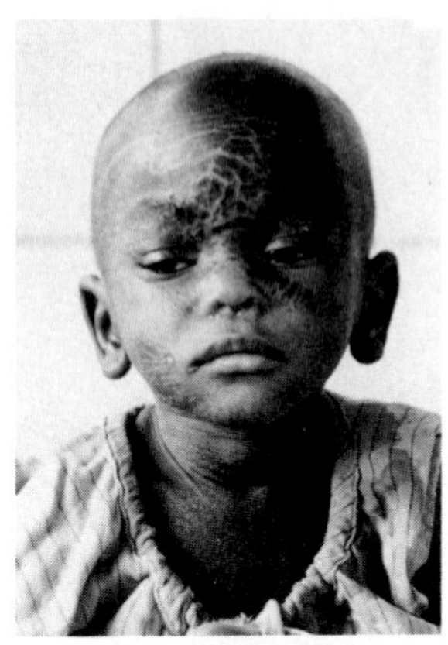

Figure 16.7 A lack of proteins and vitamins causes a deficiency disease known as kwashiorkor. The symptoms include retarded growth, discoloration of skin and hair, bloating, swollen belly, and mental apathy. [*Courtesy of Paul Almasy, World Health Organization.*]

Most proteins from animal sources contain all the essential amino acids in adequate amounts. Lean meat, milk, fish, eggs, and cheese supply adequate proteins. In fact, gelatin is one of the few inadequate animal proteins. It contains almost no tryptophan and has only small amounts of threonine, methionine, and isoleucine.

Protein Deficiency: Kwashiorkor

Our requirement for protein is about 0.8 g per kilogram of body weight. Diets with inadequate protein are common in some parts of the world. A protein-deficiency disease called *kwashiorkor* (Figure 16.7) is common at times in parts of Africa where corn is the major food.

Nutrition is especially important in a child's early years. Protein deficiency leads to both physical and mental retardation. The effect on a human's mental capacity is readily apparent from a consideration of Figure 16.8, which shows that the human brain reaches nearly full size by the age of 2 years.

Vegetarian Diets

Green plants such as grasses and grains trap a small fraction of the energy of the sun that falls upon them. They use this energy to convert carbon dioxide, water, and mineral nutrients (including nitrates, phosphates, and sulfates) into proteins. Cattle eat the plant protein, digest it, and convert a small portion of it into animal protein. People eat this animal protein, digest it, and reassemble some of the amino acids into human protein. Some of the energy originally trapped by the green plants is lost as heat at every step of the food chain (Table 16.2). If people ate the plant protein directly, one highly inefficient step could be skipped. Vegetarianism is in harmony with energy conservation.

It is interesting to note that a variety of ethnic dishes supply relatively good protein by combining a cereal grain with a legume (beans, peas, peanuts, etc.) The grain is deficient in tryptophan and lysine, but it has sufficient methio-

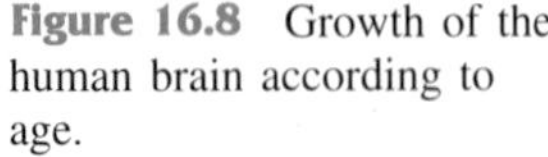

Figure 16.8 Growth of the human brain according to age.

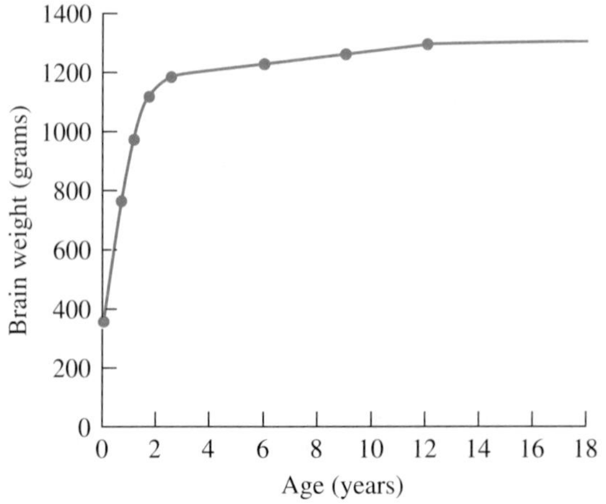

Table 16.2 Efficiencies of Protein Conversions

Food	Efficiency of Production (%)*
Beef or veal	4.7
Pork	12.1
Chicken or turkey	18.2
Milk	22.7
Eggs	23.3

*Calculated by dividing the weight of edible protein by the weight of the protein feed required to produce it, then multiplying the result by 100.

Source: President's Science Advisory Committee. *The World Food Problem,* vol. 2, 1967.

Table 16.3 Ethnic Foods That Combine a Cereal Grain with a Legume

Ethnic Group	Food*
Mexicans	Corn tortillas and refried beans
Japanese	Rice and soybean curds (tofu)
English working classes	Baked beans on toasted bread
American Indians	Corn and beans (succotash)
Children (especially in the United States)	Peanut butter on bread (sandwiches)
Western Africans	Rice and peanuts (ground nuts)
Cajuns (Louisiana)	Red beans and rice

*Cereal grains are in red, legumes in green.

nine. The legume is deficient in methionine, but it has enough tryptophan and lysine. A few such combinations are listed in Table 16.3.

Although complete proteins can be obtained by a careful mixture of vegetable foods, extreme vegetarianism is dangerous. Even when the diet includes a wide variety of plant materials, an all-vegetable diet is likely to lack vitamin B_{12}, because this nutrient is not found in plants. Other nutrients scarce in all-plant diets include calcium, iron, riboflavin, and (for children not exposed to sunlight) vitamin D. A modified vegetarian diet that includes milk, eggs, cheese, and fish can provide excellent nutrition, with red meat totally excluded.

SECTION 16.4

Minerals: Important Inorganic Chemicals in Our Lives

As might be expected, many of the chemicals in food are organic. But there are also inorganic chemicals that are vital to sustaining life. These inorganic

Table 16.4 Elements Essential to Life

Element	Symbol	Form Used	Element	Symbol	Form Used
Bulk Structural Elements			**Ultratrace Elements**		
Hydrogen	H	Covalent	Manganese	Mn	Mn^{2+}
Carbon	C	Covalent	Molybdenum	Mo	Mo^{2+}
Oxygen	O	Covalent	Chromium	Cr	?
Nitrogen	N	Covalent	Cobalt	Co	Co^{2+}
Phosphorus	P	Covalent	Vanadium	V	?
Sulfur	S	Covalent	Nickel	Ni	Ni^{2+}
			Cadmium	Cd	Cd^{2+}
Macrominerals			Tin	Sn	Sn^{2+}
			Lead	Pb	Pb^{2+}
Sodium	Na	Na^+	Lithium	Li	Li^+
Potassium	K	K^+	Fluorine	F	F^-
Calcium	Ca	Ca^{2+}	Iodine	I	I^-
Magnesium	Mg	Mg^{2+}	Selenium	Se	SeO_4^{2-}?
Chlorine	Cl	Cl^-	Silicon	Si	?
Phosphorus	P	$H_2PO_4^-$	Arsenic	As	?
Sulfur	S	SO_4^{2-}	Boron	B	H_3BO_3
Trace Elements					
Iron	Fe	Fe^{2+}			
Copper	Cu	Cu^{2+}			
Zinc	Zn	Zn^{2+}			

nutrients are called **dietary minerals.** It is estimated that minerals represent about 4% of the weight of a human body. Some of these, such as the chlorides (Cl^-), phosphates (PO_4^{3-}), bicarbonates (HCO_3^-), and sulfates (SO_4^{2-}), occur in the blood and the other body fluids. Others, such as iron (as Fe^{2+}) in the hemoglobin and phosphorus in the nucleic acids (DNA and RNA), are constituents of complex organic compounds.

Note that phosphorus and sulfur appear twice each in Table 16.4. They are structural elements and also components of the macrominerals phosphate and sulfate.

Table 16.4 lists the 30 elements known to be essential to one or more living organisms. The 11 that comprise the structural elements and the macrominerals make up more than 99% of all the atoms in the human body. The 19 trace elements include iron, zinc, copper, and 16 others called ultratrace elements.

Minerals serve a variety of functions. The function of iodine is perhaps the most dramatic. A small amount of iodine is necessary for the thyroid gland to function properly. A deficiency of iodine has dire effects, of which goiter is perhaps the best known (Figure 16.9). Iodine is available naturally in seafood, but, to guard against iodine deficiency, a small amount of potassium iodide (KI) is added to table salt. The use of iodized salt has greatly reduced the incidence of goiter.

Iron(II) ions (Fe^{2+}) are necessary for the proper function of the oxygen-transporting compound hemoglobin. Without sufficient iron, not enough oxygen is supplied to the body tissues. Anemia, a general weakening of the body, results. Foods especially rich in iron compounds include red meat and liver. It appears that most adult males need very little dietary iron, since iron seems to be retained by the body and lost mainly through bleeding.

Calcium and phosphorus are necessary for the proper development of bones and teeth. Growing children need about 1.5 g of each per day. These elements are available from milk. The needs of adults for calcium and phosphorus are less widely known but are very real just the same. For example, calcium ions are necessary for the coagulation of blood (to stop bleeding) and for maintenance of the rhythm of the heartbeat. Phosphorus is necessary in order for the body to obtain energy from carbohydrates. Without phosphorus compounds, we couldn't get *any* energy from our main energy foods.

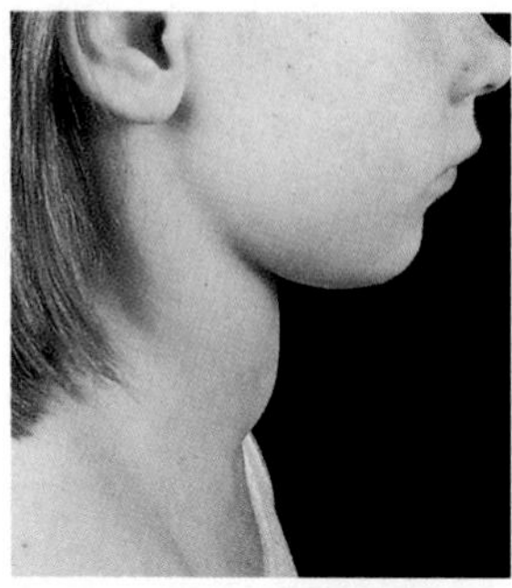

Figure 16.9 A person affected by goiter. The swollen thyroid gland in the neck results from a lack of the trace element iodine in the diet. [*Biophoto Associates/Photo Researchers, Inc.*]

Sodium chloride—in moderate amounts—is essential to life. It is important in the exchange of fluids between cells and plasma, for example. The presence of salt increases water retention, though, and a high volume of retained fluids can cause swelling and high blood pressure (hypertension). An estimated 36 million people in the United States suffer from hypertension. Most physicians agree that our diets generally contain too much salt.

Iron, copper, zinc, cobalt, manganese, molybdenum, calcium, and magnesium are essential to the proper functioning of metalloenzymes, which are life sustaining. A great deal remains to be learned about the role of inorganic chemicals in our bodies. Bioinorganic chemistry is a flourishing area of research.

SECTION 16.5

The Vitamins: Vital, but Not All Are Amines

Why are British sailors called ''limeys''? And what does that have to do with food? Sailors have been plagued since early times by *scurvy*. In 1747, Scottish navy surgeon James Lind showed that scurvy could be prevented by the inclusion of fresh fruit and vegetables in the diet. Convenient fresh fruits to carry on long voyages (they didn't have refrigeration) were limes, lemons, and oranges. British ships put to sea with barrels of limes aboard, and the sailors ate a lime or two every day. That is how they came to be known as ''lime eaters,'' or simply ''limeys.''

In 1897, the Dutch scientist Christiaan Eijkman showed that polished rice lacked something found in the hull. Lack of that something caused the disease *beriberi,* which was quite a problem in the Dutch East Indies at that time. A British scientist, F. G. Hopkins, fed a synthetic diet of carbohydrates, fats, proteins, and minerals to a group of rats. The rats were unable to sustain healthy growth. Again, something was missing.

Table 16.5 Some of the Vitamins (Polar groups are in red.)

Vitamin	Structure and Name	Sources	Deficiency Symptoms
	Fat-Soluble Vitamins		
A	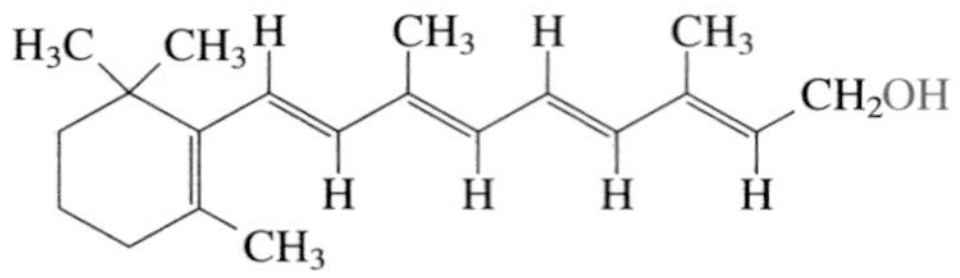 Retinol	Fish, liver, eggs, butter, cheese; also a vitamin precursor in carrots and other vegetables	Night blindness
D	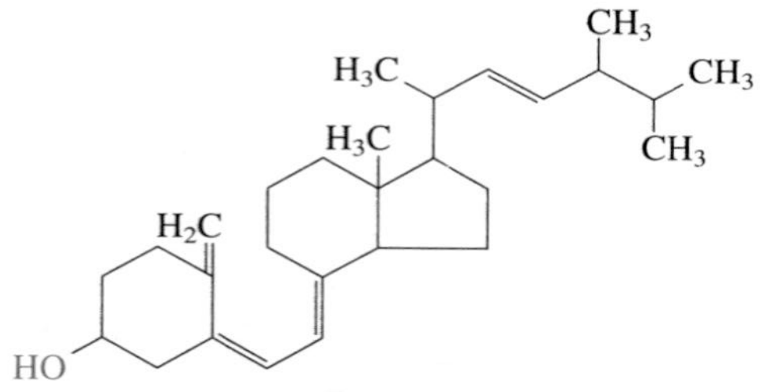 Vitamin D_2 (calciferol)	Cod liver oil, irradiated ergosterol (milk supplement)	Rickets
E	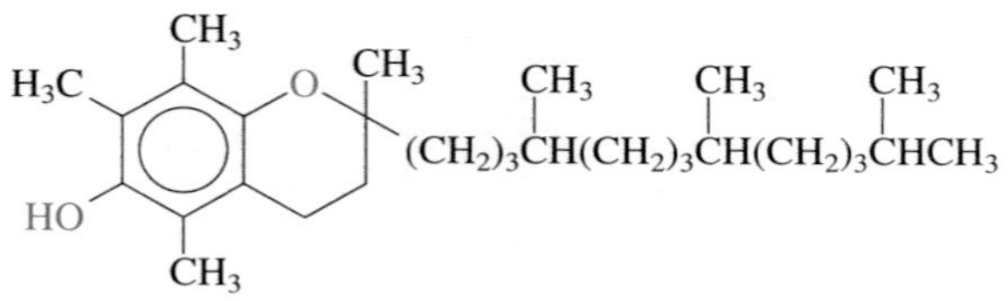 α-Tocopherol	Wheat germ oil, green vegetables, egg yolks, meat	Sterility, muscular dystrophy
K	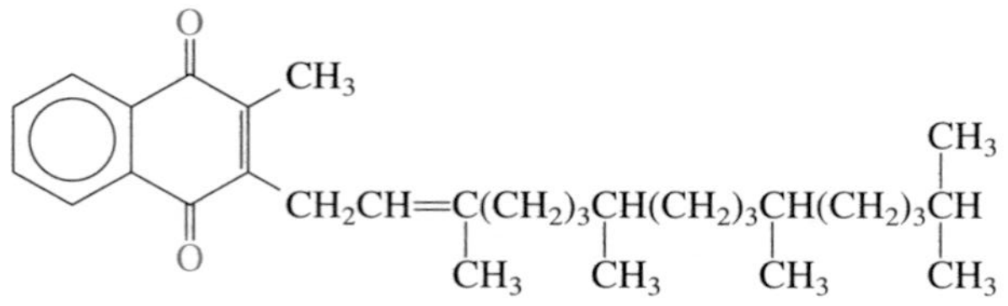 Vitamin K_1 (phylloquinone)	Spinach, other green leafy vegetables	Hemorrhage
	Water-Soluble Vitamins		
The B Complex	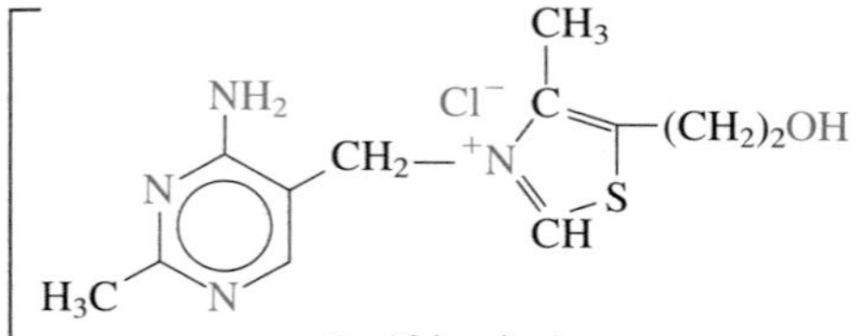 B (thiamine)	Germ of cereal grains, legumes, nuts, milk, and brewer's yeast	Beriberi—polyneuritis resulting in muscle paralysis, enlargement of heart, and ultimately heart failure

	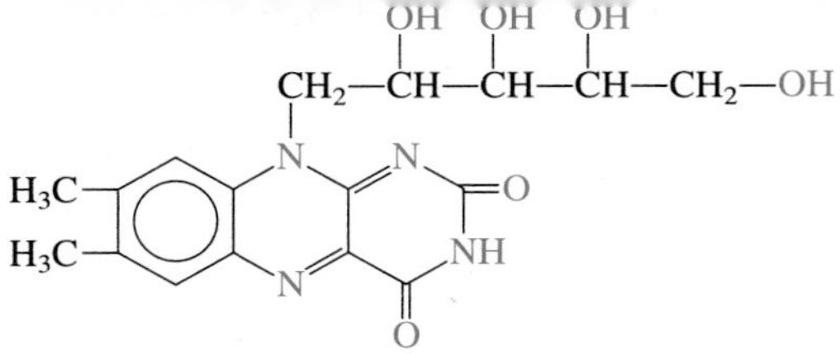 B_2 (riboflavin)	Milk, red meat, liver, egg white, green vegetables, whole wheat flour (or fortified white flour), and fish	Dermatitis, glossitis (tongue inflammation)
	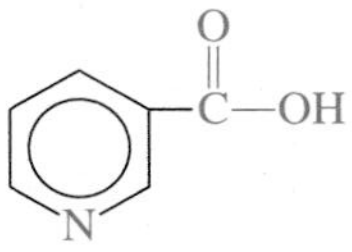 Nicotinic acid, 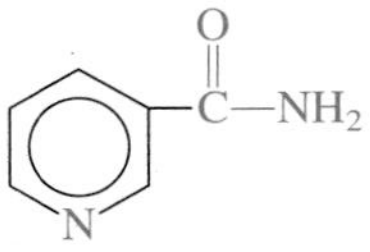 Nicotinamide — Niacin	Red meat, liver, collards, turnip greens, yeast, and tomato juice	Pellagra—skin lesions, swollen and discolored tongue, loss of appetite, diarrhea, various mental disorders (Figure 16.11)
	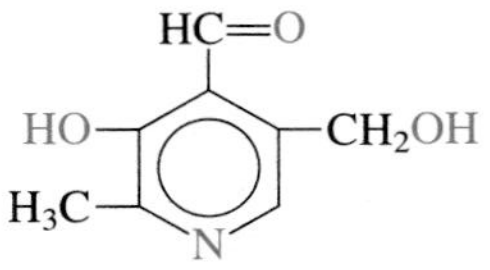 Pyridoxal, 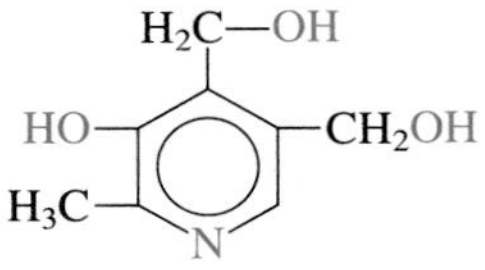 Pyridoxol, 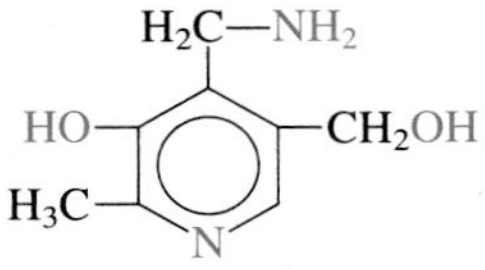 Pyridoxamine — B_6	Eggs, liver, yeast, peas, beans, and milk	Dermatitis, apathy, irritability, and increased susceptibility to infections; convulsions in infants
	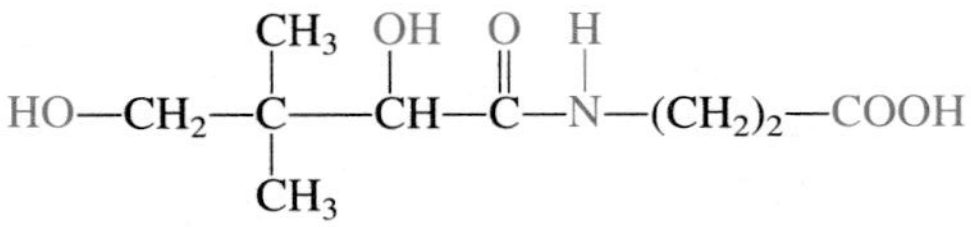 Pantothenic acid	Liver, eggs, yeast, and milk	(Possibly) emotional problems and gastrointestinal disturbances
	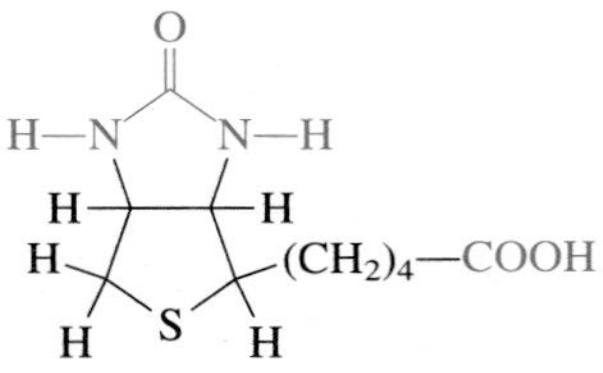 Biotin	Beef liver, yeast, peanuts, chocolate, and eggs (although this vitamin cannot be synthesized by humans, it is a product of their intestinal bacteria)	Dermatitis
	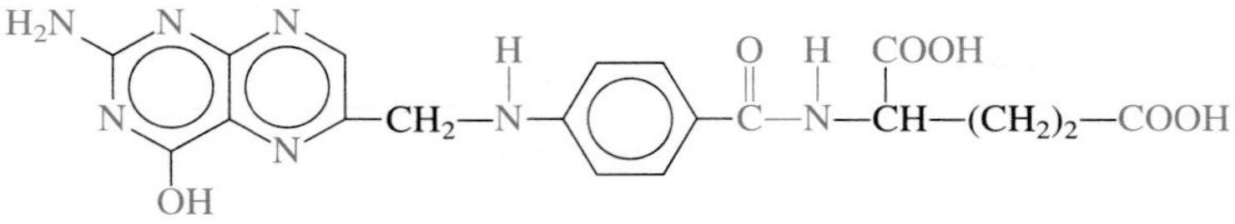 Folic acid	Liver, kidney, mushrooms, yeast, and green leafy vegetables	Anemias (folic acid is used in the treatment of megaloblastic anemia, a condition characterized by giant red blood cells)

Table 16.5 Some of the Vitamins (Polar groups are in red.) (cont.)

Vitamin	Structure and Name	Sources	Deficiency Symptoms
	B_{12} (cyanocobalamine)	Liver, meat, eggs, and fish (not found in plants)	Pernicious anemia
C	Ascorbic acid	Citrus fruits, tomatoes, green peppers	Scurvy

In 1912, Casimir Funk, a Polish biochemist, coined the word *vitamine* (from the Latin *vita*: ''life'') for these missing factors. Funk thought all factors contained the amine group. In the United States, the final *e* was dropped after it was found that not all the factors were amines. The generic term became *vitamin*. Eijkman and Hopkins shared the 1929 Nobel prize in physiology and medicine for their important discoveries relating to vitamins.

Vitamins are specific organic compounds that are required in the diet (in addition to the usual proteins, fats, carbohydrates, and minerals) to prevent specific diseases. Some of the vitamins, their structures, sources, and deficiency symptoms are shown in Table 16.5. The role of vitamins in the prevention of deficiency diseases (Figure 16.10) has been well established.

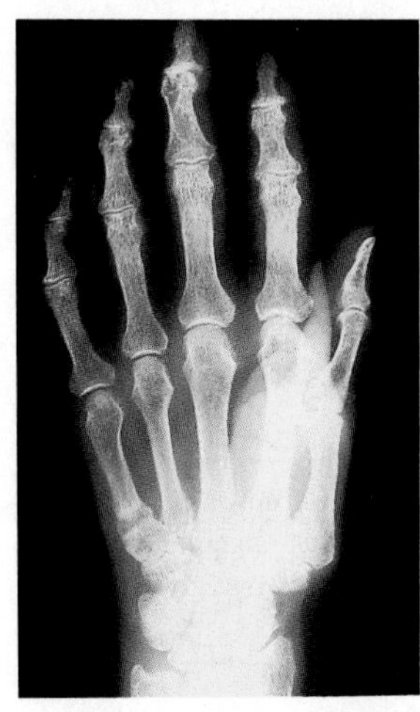

(a)

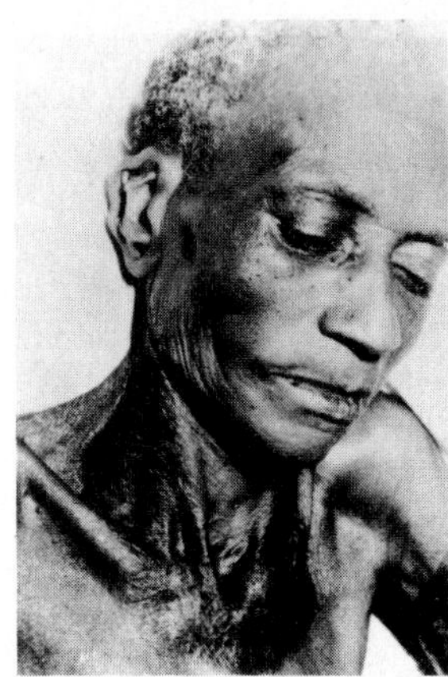

(b)

Figure 16.10 (a) This X-ray shows softened bones caused by a deficiency of vitamin D. (b) Inflammation and abnormal pigmentation characterize pellagra, caused by niacin deficiency. [*(a) Science Photo Library/ Photo Researchers, Inc.; (b) courtesy of World Health Organization.*]

As you can see from Table 16.5, the vitamins do not share a common chemical structure. They can, however, be divided into two broad categories: the *fat-soluble vitamins*—including A, D, E, and K—and the *water-soluble vitamins*—made up of the B complex and vitamin C. All the fat-soluble vitamins incorporate a high proportion of hydrocarbon structural elements. There are one or two oxygen atoms present, but the compounds as a whole are nonpolar. In contrast, a water-soluble vitamin contains a high proportion of the electronegative atoms oxygen and nitrogen, which can form hydrogen bonds to water; therefore, the molecule as a whole is soluble in water.

Fat-soluble vitamins dissolve in the fatty tissue of the body, and reserves of these vitamins can be stored there for future use. For example, an adult can store several years' supply of vitamin A. If the diet becomes deficient in vitamin A, these reserves are mobilized, and the adult remains free of the deficiency disease for quite a while. On the other hand, a small child who has not had the opportunity to build up a store of the vitamin soon exhibits symptoms of the deficiency. Many children in developing countries are permanently blinded by vitamin A deficiency.

Because the fat-soluble vitamins are efficiently stored in the body, overdoses can result in adverse effects. Large excesses of vitamin A cause irritability, dry skin, and a feeling of pressure inside the head. Massive doses administered to pregnant rats result in malformed offspring. Vitamin D, like vitamin A, is fat soluble. Too much vitamin D can cause pain in the bones, nausea, diarrhea, and weight loss. Bonelike material may be deposited in joints; kidney tubules; blood vessels; and heart, stomach, and lung tissue. The amounts of both vitamin D and A in nonprescription capsules are regulated by the U.S. Food and Drug Administration.

Even though vitamins E and K also are fat soluble, they are metabolized and excreted. They are not stored to the extent vitamins A and D are, and excesses seldom cause problems.

The body has a limited capacity to store water-soluble vitamins. It excretes anything over the amount that can be immediately used. Water-soluble vitamins must be taken at frequent intervals, whereas a single dose of a fat-soluble vitamin can be used by the body over several weeks. The vitamin content of some foods can be lost when the food is cooked in water and then drained. The water-soluble vitamins go down the drain with the water.

SECTION 16.6

Other Essentials

In addition to carbohydrates, proteins, fats, minerals, and vitamins, there are other things in our food that are important. One of them is *fiber.*

Dietary Fiber

What we now call dietary fiber was once called *roughage* or referred to as *bulk in the diet.*

The importance of fiber in the diet often was ignored or overlooked until the mid-1970s. Since then, several studies have been published that have the public excited about fiber.

First, it was noted that people in the developed countries are much more likely to get cancer of the colon than are those in underdeveloped nations. Also, people in the developed countries eat diets rich in highly processed, low-fiber foods, while those in more "primitive" areas eat high-fiber diets. High-fiber diets lead to frequent and robust bowel movements. Cellulose molecules (see Figure 15.7) have lots of —OH groups. These readily form hydrogen bonds to water molecules. Cellulose therefore absorbs a lot of water, leading to softer stools. Low-fiber diets result in less frequent bowel action, with high retention times for feces in the colon.

Wheat bran is about 90% cellulose. It and related materials are not dissolved by water or digestive juices. This type of fiber, called *insoluble fiber,* passes unchanged into the colon.

Bacteria act upon the materials in the colon. (Indeed, bacteria, both living and dead, make up about one-third of the dry weight of feces.) With a high-fiber diet, the materials seldom remain in the colon for more than 1 day. With low-fiber diet, the retention time can be as long as 3 days, allowing for prolonged bacterial activity that produces a high level of mutagenic chemicals. Chemicals that are mutagenic often are also carcinogenic. Thus, a possible chemical link between low-fiber diets and cancer of the colon has been established.

Water

Much of the food we eat is mainly water. It should come as no surprise that tomatoes are 90% water, or that melons, oranges, and grapes are largely water. But water is one of the main ingredients in practically all foods, from roast beef and seafood to potatoes and onions.

In addition to the water we get in our food, it is recommended that we drink about 1.0 to 1.5 L of water each day. We could satisfy this need by drinking plain water, but often we choose other beverages. We drink coffee, tea, milk, fruit juice, and soft drinks. Almost all beverages are mostly water. In 1985 carbonated soft drinks replaced water as the most consumed kind of beverage in the United States.

Many people also drink beverages that contain ethyl alcohol. These are made by fermenting grains or fruit juices. Beer is usually made from malted barley and wine from grape juice. Even these alcoholic beverages are mainly water.

SECTION 16.7

Starvation, Fasting, and Fad Diets

When the human body is totally deprived of food, whether voluntarily or involuntarily, the condition is known as **starvation**. During total fasting, the body's glycogen stores are depleted rapidly and the body calls on its fat reserves. Fat is first taken from around the kidneys and the heart. Then it is removed from other parts of the body. Ultimately, even the bone marrow (which is also a fat-storage depot) is depleted, and it becomes red and jellylike (it is usually white and firm).

Increased dependence on stored fats as an energy source leads to *ketosis,* a condition characterized by the appearance of *ketones* in the urine. One of the ketones is acetone. Two others are acetoacetic acid and β-hydroxybutyric acid. Collectively, these three compounds are called *ketone bodies,* even though one is not a ketone (Figure 16.11). Note that two of the ketone bodies are acids. Ketosis rapidly develops into *acidosis;* the blood pH drops and oxygen transport is hindered. Oxygen deprivation leads to depression and lethargy.

In the more extreme low-carbohydrate diets (for weight loss), ketosis is deliberately induced. Understandably, two of the side effects of these diets are often depression and lethargy. In the early stages of a diet deficient in carbohydrates, the body converts amino acids to glucose. The brain must have glucose; it can't obtain sufficient energy from fats. If there are enough adequate proteins in the diet, tissue proteins are spared. (The liquid protein diets, from which several people died, were based on the inadequate proteins obtained from beef hides and connective tissues.)

In the early stages of a *total* fast, body protein is metabolized at a relatively rapid rate. After several weeks, the rate of protein breakdown slows considerably, as the brain adjusts to using the breakdown products of fatty acid metabolism for its energy source. When fat reserves are substantially depleted, the body must again draw heavily on its structural proteins for its energy requirements. The emaciated appearance of a starving individual is due to the depletion of muscle proteins.

Even low-carbohydrate diets that are high in adequate proteins are hard on the body, which must rid itself of the nitrogen compounds—ammonia and urea—formed by the breakdown of proteins. This puts an increased stress on the liver, where the waste products are formed.

It is interesting to note that contrary to a popular notion, fasting does not cleanse the body. Indeed, quite the reverse occurs. A shift to fat metabolism produces the ketone bodies, and protein breakdown produces ammonia, urea, and other wastes. You can lose weight by fasting, but the process should be carefully monitored by a physician.

Involuntary starvation is a serious problem in much of the world. Even so, starvation is seldom the sole cause of death. Weakened by starvation, the victims of starvation and malnutrition succumb to disease. Even usually minor

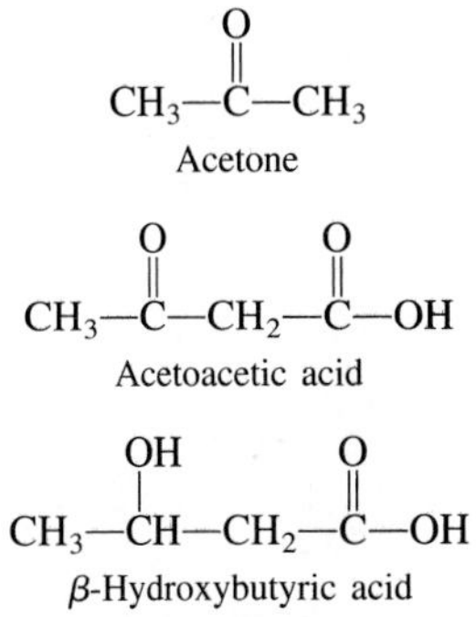

$$CH_3-\overset{\overset{\displaystyle O}{\|}}{C}-CH_3$$

Acetone

$$CH_3-\overset{\overset{\displaystyle O}{\|}}{C}-CH_2-\overset{\overset{\displaystyle O}{\|}}{C}-OH$$

Acetoacetic acid

$$CH_3-\overset{\overset{\displaystyle OH}{|}}{CH}-CH_2-\overset{\overset{\displaystyle O}{\|}}{C}-OH$$

β-Hydroxybutyric acid

Figure 16.11 The three ketone bodies. Only two are ketones; the other has an alcohol function. Two of the compounds are acids.

Ellen Richards, the first woman graduate of the Massachusetts Institute of Technology (1873), was among the first to apply the laws of science to nutrition. [*The MIT Museum.*]

diseases, such as chickenpox and measles, become life-threatening disorders. Barring disease, starvation alone would lead eventually to death from circulatory failure when the heart muscle became too weak to pump blood.

Weight loss through diet and exercise is discussed in Chapter 18.

Processed Food: Less Nutrition

Malnutrition need not be due to starvation or dieting. It can be the result of eating too much highly processed food.

Whole wheat is an excellent source of vitamin B_1 and other vitamins. To make white flour, the wheat germ and bran are removed from the grain. The remaining material has few minerals and no vitamins. We eat the starch and use the germ and bran for animal food. At least our cattle and hogs get good nutrition. Similarly, polished rice has had most of its protein and minerals removed, and it has virtually no vitamins. The disease beriberi became prevalent when polished rice was introduced to Southeast Asia.

When fruits and vegetables are peeled, most of their vitamins, minerals, and fibers are lost. The peels often are dumped (directly or through a garbage disposal) into a nearby stream, where they contribute to water pollution (Chapter 13). The heat used to cook food also destroys some vitamins. If water is used in cooking, part of the water-soluble vitamins (C and B complex) and some of the minerals often are drained off and discarded with the water.

The methods we use for processing and cooking food often are chosen for aesthetic reasons, but sometimes cooking methods help preserve the food. For example, highly milled white flour does keep better than whole wheat flour. Cooked tomatoes keep better than fresh tomatoes, although their vitamin C is partly destroyed by cooking. But often tomatoes, potatoes, and apples are peeled because they look better that way.

It is estimated that over half of the diet of the average person in the United States consists of processed foods. The "teenager's" diet of hamburgers, potato chips, and colas is lacking in many essential nutrients. Highly processed convenience foods threaten to leave the people of the developed nations poorly nourished despite their abundance of food.

SECTION 16.8

Food Additives

The label reads "egg whites, vegetable oils, nonfat dry milk, lecithin, mono- and diglycerides, propylene glycol monostearate, xanthan gums, sodium citrate, aluminum sulfate, artificial flavor, iron phosphate, niacin, riboflavin, and irradiated ergosterol." You recognize only a few of the ingredients. Most of the substances listed are **food additives**, substances other than basic foodstuffs that are present in food as a result of some aspect of production, processing, packaging, or storage.

Since food processing results in the removal or destruction of certain essential food substances, it is sometimes necessary to include additives in prepared food to increase its nutritional value. Other chemicals are added to enhance color and flavor, to retard spoilage, to provide texture, to sanitize, to bleach, to ripen (or to prevent ripening), to control moisture or dryness, and to prevent (or enhance) foaming. There are several thousand different additives. Sugar, salt, and corn syrup are used in the greatest amounts. These three, plus citric acid, baking soda, vegetable colors, mustard, and pepper, make up over 98% (by weight) of all additives.

In the United States, food additives are regulated by the Food and Drug Administration (FDA). The original Food, Drug, and Cosmetic Act was passed by Congress in 1938. Under that act, the FDA had to prove that an additive was unsafe before its use could be prevented. The Food Additives Amendment of 1958 shifted the burden of proof to industry. Now, a company that wishes to use a food additive must first furnish proof to the FDA that the additive is safe for the intended use. The FDA can also regulate the amount of chemicals that can be added.

Food additives are not a recent development. Salt has been used to preserve meat and fish since the time when people lived in caves. Spices have been used since earliest recorded history to flavor and preserve foods. Other additives have been used throughout the centuries. The movement of the population from farms to cities in recent years has increased the necessity of using additives to preserve foods. An increased desire for convenience foods also has led to greater use of additives.

Our increased consumption and concern about some food additives have led to a lot of worry, often expressed as concern about "chemicals in our food." Food itself is chemical, however. Table 16.6 shows the chemical composition of a typical breakfast. Many of the chemicals in the breakfast would be harmful in large amounts but are harmless in the trace amounts that occur naturally in foods. Indeed, some make important contributions to the flavors and aromas that make food so delightful.

Our bodies are also collections of chemicals. Broken down to its elements (Table 16.7), your body would be worth only a few dollars as chemicals in their elemental form. It is the unique combination of the elements in every human body that makes you different from everyone else and makes each individual's value beyond measure. Since food is chemical and we are chemical, we shouldn't have to worry, in general, about chemicals in our food. Perhaps we should be concerned, however, about *some* of the specific chemicals in our food.

Additives That Improve Nutrition

Only a small amount of iodine is necessary for the proper functioning of the thyroid gland. Iodine is plentiful in seafood, but people in inland areas often

Table 16.6 Your Breakfast—As Seen by a Chemist*

Chilled Melon

Starches	Anisyl propionate
Sugars	Amyl acetate
Cellulose	Ascorbic acid
Pectin	Vitamin A
Malic acid	Riboflavin
Citric acid	Thiamine
Succinic acid	

Scrambled Eggs

Ovalbumin	Lecithin
Conalbumin	Lipids (fats)
Ovomucoid	Fatty acids
Mucin	Butyric acid
Globulins	Acetic acid
Amino acids	Sodium chloride
Lipovitellin	Lutein
Livetin	Zeazanthine
Cholesterol	Vitamin A

Sugar-Cured Ham

Actomyosin	Adenosine triphosphate (ATP)
Myogen	Glucose
Nucleoproteins	Collagen
Peptides	Elastin
Amino acids	Creatine
Myoglobin	Pyroligneous acid
Lipids (fats)	Sodium chloride
Linoleic acid	Sodium nitrate
Oleic acid	Sodium nitrite
Lecithin	Sodium phosphate
Cholesterol	
Sucrose	

Coffee

Caffeine	Acetone
Essential oils	Methyl acetate
Methanol	Furan
Acetaldehyde	Diacetyl
Methyl formate	Butanol
Ethanol	Methylfuran
Dimethyl sulfide	Isoprene
Propionaldehyde	Methylbutanol

Cinnamon Apple Chips

Pectin	Propanol
Cellulose	Butanol
Starches	Pentanol
Sucrose	Hexanol
Glucose	Acetaldehyde
Fructose	Propionaldehyde
Malic acid	Acetone
Lactic acid	Methyl formate
Citric acid	Ethyl formate
Succinic acid	Ethyl acetate
Ascorbic acid	Butyl acetate
Cinnamyl alcohol	Butyl propionate
Cinnamic aldehyde	Amyl acetate
Ethanol	

Toast and Coffee Cake

Gluten	Mono- and diglycerides
Amylose	Methyl ethyl ketone
Amino acids	Niacin
Starches	Pantothenic acid
Dextrins	Vitamin D
Sucrose	Acetic acid
Pentosans	Propionic acid
Hexosans	Butyric acid
Triglycerides	Valeric acid
Sodium chloride	Caproic acid
Phosphates	Acetone
Calcium	Diacetyl
Iron	Maltol
Thiamine	Ethyl acetate
Riboflavin	Ethyl lactate

Tea

Caffeine	Phenyl ethyl alcohol
Tannin	Benzyl alcohol
Essential oils	Geraniol
Butyl alcohol	Hexyl alcohol
Isoamyl alcohol	

*The chemicals listed are those found normally in the foods. No food additives are itemized, and the chemical listings are not necessarily complete.

Source: Manufacturing Chemists Association, Washington, DC.

Table 16.7 Approximate Elemental Analysis of the Human Body	
Element	Percent by Weight in Human Body
Oxygen	65
Carbon	18
Hydrogen	10
Nitrogen	3
Calcium	1.5
Phosphorus	1
Potassium	0.35
Sulfur	0.25
Chlorine	0.15
Sodium	0.15
Magnesium	0.05
Iron	0.004
Trace elements to make 100%	

suffer from iodine deficiency. A particularly striking result is goiter, a swelling of the thyroid often symptomized by an unsightly swollen neck (see Figure 16.9). The first nutrient supplement approved by the Bureau of Chemistry of the U.S. Department of Agriculture, potassium iodide (KI), was added to table salt in 1924 to reduce the incidence of goiter. (The Bureau of Chemistry later became the FDA.)

A number of other chemicals have been added to foods specifically to prevent deficiency diseases. The addition of vitamin B_1 (thiamine) to polished rice is essential in the Far East, where beriberi still is a problem. The replacement of the B-complex vitamins thiamine, riboflavin, and niacin (which are removed in processing) and the addition of iron (usually ferrous carbonate [$FeCO_3$]) to flour is called **enrichment**. Enriched bread made from this flour still isn't as good a food as bread made from whole wheat. It lacks vitamin B_6, pantothenic acid, folic acid, zinc, and magnesium, nutrients usually provided by whole grain flour.

An investigator in Texas fed enriched white bread to rats. Most of them died of malnutrition in fewer than 60 days. The staff of life, in its modern form, is unable to sustain life. However, the enrichment of bread, corn meal, and cereals has served to virtually eliminate pellagra, a disease that once plagued the southern United States.

Vitamin C (ascorbic acid) is added frequently to fruit juices, flavored drinks, and beverages. Although our diets generally contain enough ascorbic acid to prevent scurvy, a number of scientists recommend a much larger intake than basic minimum requirements.

Vitamin D is added to milk in developed countries. This use of fortified milk has lead to the virtual elimination of rickets. Similarly, vitamin A is added to margarine. (This vitamin occurs naturally in butter; it is added to margarine so that the substitute more nearly matches the nutritional quality of butter.)

If we ate a balanced diet of foods fresh from the farm, we probably wouldn't need nutritional supplements. With our usual diets rich in highly processed foods, however, we need the nutrients provided by vitamin and mineral food additives.

Chemicals That Taste Good

If you like spice cake, soda pop, gingerbread, and sausage, you like food additives. These and many other foods depend almost totally on spices and other flavorings for their flavor. Cloves, ginger, cinnamon, and nutmeg are examples of natural spices. Natural flavors also can be found in herbs or extracted from fruits and other plant materials. Vanilla extract is a familiar example.

Chemists sometimes analyze a natural flavor and determine its components. Then they synthesize the components and make a mixture that may closely resemble the natural product. The major components of natural and artificial flavors are often identical. For example, both vanilla extract and imitation vanilla owe their flavor mainly to vanillin (Figure 16.12). The natural flavor is usually more complex than the imitation because the natural product contains a wider variety of chemicals than the imitation. Flavors, whether natural or synthetic, probably present little hazard when used in moderation, and they contribute considerably to our enjoyment of food.

Figure 16.12 Some molecular flavorings.

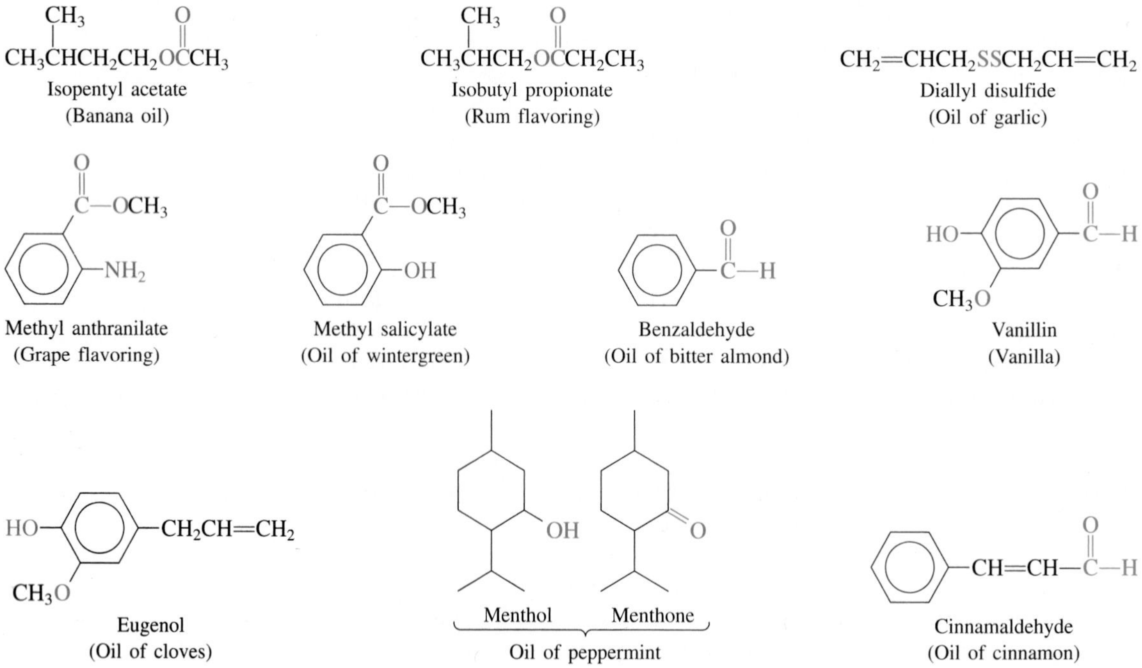

$CH_3CH(CH_3)CH_2CH_2OC(=O)CH_3$
Isopentyl acetate
(Banana oil)

$CH_3CH(CH_3)CH_2OC(=O)CH_2CH_3$
Isobutyl propionate
(Rum flavoring)

$CH_2{=}CHCH_2SSCH_2CH{=}CH_2$
Diallyl disulfide
(Oil of garlic)

Methyl anthranilate
(Grape flavoring)

Methyl salicylate
(Oil of wintergreen)

Benzaldehyde
(Oil of bitter almond)

Vanillin
(Vanilla)

Eugenol
(Oil of cloves)

Menthol Menthone
Oil of peppermint

Cinnamaldehyde
(Oil of cinnamon)

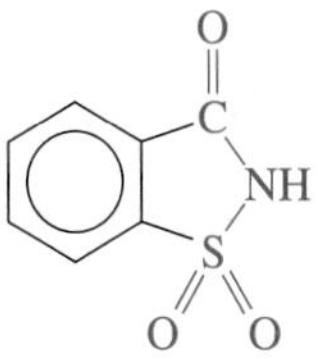

Figure 16.13 Four artificial sweeteners.

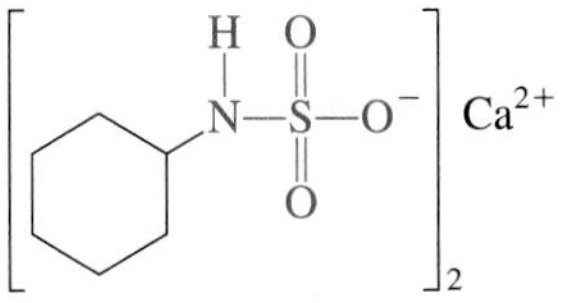

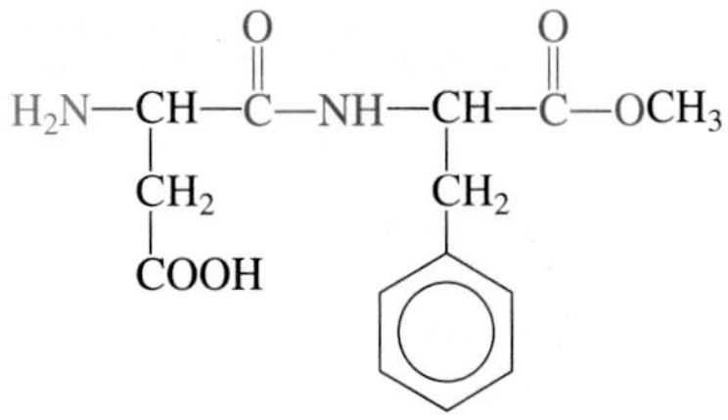

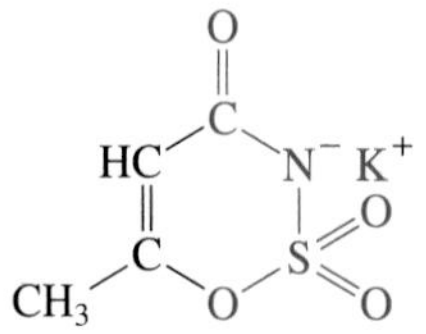

Artificial Sweeteners

Obesity is a major problem in most of the developed countries. Presumably, we could reduce the intake of calories by replacing sugars with noncaloric sweeteners. Although there is little evidence that artificial sweeteners are of any value in controlling obesity, they have still become part of our culture.

For many years, the major artificial sweeteners were saccharin and cyclamates. Cyclamates were banned in the United States in 1970 after studies showed they caused cancer in laboratory animals. Subsequent studies have failed to confirm that finding. Nevertheless, the FDA has not lifted the ban.

In 1977 saccharin was shown to cause bladder cancer in laboratory animals. The FDA's move to ban it (as required by the Delaney amendment) was blocked by Congress because at that time saccharin was the only approved artificial sweetener. Its ban would have meant the end of diet soft drinks and low-calorie products.

In 1981 the FDA approved aspartame, the methyl ester of the dipeptide aspartylphenylalanine, as an artificial sweetener. **Aspartame** is about 160 times sweeter than sucrose. Aspartame has largely replaced saccharin as the artificial sweetener of choice. There are anecdotal reports of problems with aspartame, but repeated studies have shown it to be safe, at least in moderate amounts. An exception is that it is not safe for people with phenylketonuria, an inherited condition in which phenylalanine cannot be metabolized properly. Acesulfame K also is now approved for use in the United States. It can survive the high temperatures of cooking processes, whereas aspartame is broken down by heat. Structures of the artificial sweeteners are given in Figure 16.13.

The popular commercial products Nutrasweet and Equal are both aspartame.

Table 16.8 compares the sweetness of a variety of substances. What makes a compound sweet? There is little structural similarity among the compounds. Saccharin, the cyclamates, and compound P-4000 bear little resemblance to the sugars. Even more baffling are the tastes of two compounds that closely resemble P-4000: compound I, which has the NO_2 and NH_2 groups reversed, is tasteless; compound II, which has two NO_2 groups, is extremely bitter (Figure 16.14). P-4000 is one of the sweetest substances known. In 1974, its use in the United States was banned because of possible toxic effects.

Recall from Chapter 15 that sugars are polyhydroxy compounds. Indeed, many compounds with hydroxyl groups on adjacent carbon atoms are sweet.

Table 16.8 Sweetness of Some Compounds	
Compound	Relative Sweetness*
Acesulfame K	20,000
Glucose	74
Fructose	173
Lactose	16
Sucrose	100
Maltose	33
P-4000	400,000
Saccharin	50,000
Aspartame	16,000

*Sweetness is relative to sucrose at a value of 100.

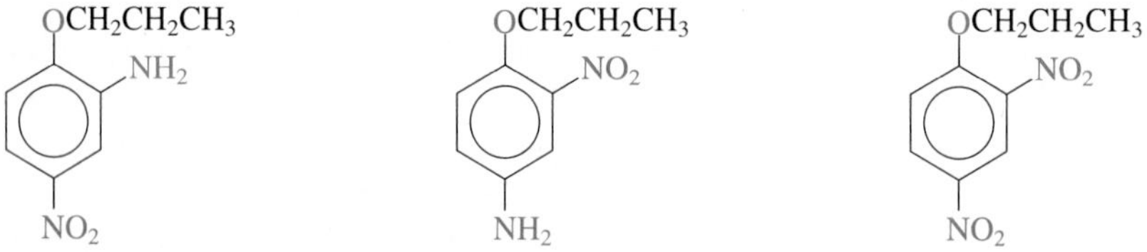

$OCH_2CH_2CH_3$, NH_2, NO_2 — P-4000 (extremely sweet)

$OCH_2CH_2CH_3$, NO_2, NH_2 — Compound I (tasteless)

$OCH_2CH_2CH_3$, NO_2, NO_2 — Compound II (bitter)

Figure 16.14 Taste responses to these three compounds are dramatically different.

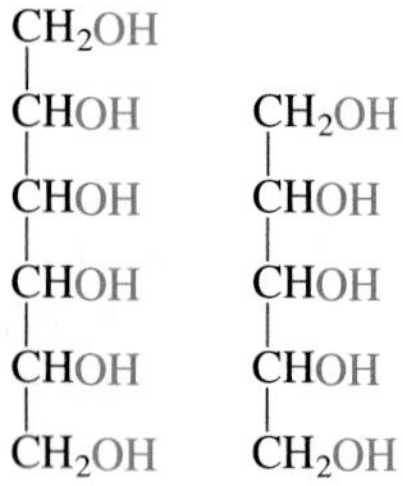

CH_2OH–CHOH–CHOH–CHOH–CHOH–CH_2OH
Sorbitol

CH_2OH–CHOH–CHOH–CHOH–CH_2OH
Xylitol

Even ethylene glycol ($HOCH_2CH_2OH$) is sweet, though it is quite toxic. Glycerol, obtained from the hydrolysis of fats (Section 16.2), is also sweet. It is used as a food additive, principally for its properties as a **humectant** (moistening agent), however, not as a sweetener.

Other polyhydroxy alcohols used as sweeteners are *sorbitol,* made by the reduction of glucose, and *xylitol,* which has five carbon atoms with a hydroxyl group on each. These compounds have an advantage over sugars in that they are not broken down in the mouth and thus do not contribute to tooth decay. This makes these alcohols useful in sugar-free chewing gums.

MSG is used heavily in Chinese foods. For several years MSG was thought to be responsible for a set of symptoms (including headaches and a feeling of weakness) known as Chinese-restaurant syndrome. Experts who reviewed 230 studies concluded that MSG was not the cause of the syndrome. This "not guilty" finding has been confirmed by further studies.

Flavor Enhancers

Some chemical substances, though not particularly flavorful themselves, are used to enhance other flavors. Common table salt (sodium chloride) is a familiar example. Salt seems to increase sweetness but helps to mask bitterness and sourness.

Salt is a necessary nutrient in moderate amounts. Many physicians and scientists contend that the average person's diet in the United States contains too much salt. Snack foods (crackers, potato chips, pretzels) are particularly heavy in salt. There is evidence that too much salt contributes to high blood pressure (hypertension); three out of four people between 65 and 80 years old suffer from this condition.

Another popular flavor enhancer is monosodium glutamate (MSG). Glutamic acid is one of about 20 amino acids that occur naturally in proteins.

MSG is the sodium salt of glutamic acid. It is used in many convenience foods. It also is found in many foods for which the FDA has established *standards of identity*. Such foods must contain certain ingredients in established proportions. They may, however, contain certain other substances as well. Mayonnaise, for example, is defined by a standard of identity. It may contain MSG without listing it as an ingredient.

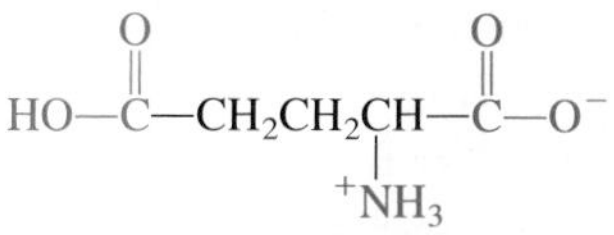

$$HO{-}C({=}O){-}CH_2CH_2CH({-}^{+}NH_3){-}C({=}O){-}O^-$$

Glutamic acid

$$HO{-}C({=}O){-}CH_2CH_2CH({-}NH_2){-}C({=}O){-}O^-\ Na^+$$

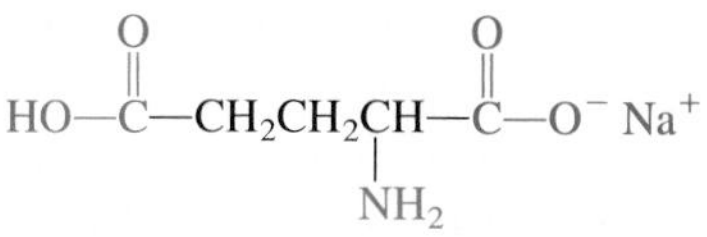

MSG

Although glutamates are found naturally in proteins, there is evidence that huge excesses can be harmful. MSG can numb portions of the brains of laboratory animals. It also may be teratogenic; that is, it may cause birth defects when administered in large amounts.

Spoilage Inhibitors

Food spoilage can result from the growth of molds (fungi) or bacteria. Propionic acid and its sodium and calcium salts are added to bread and cheese to act as mold and yeast inhibitors. Sorbic acid, benzoic acid, and their salts are also used. The structures of some of these inhibitors are shown in Figure 16.15.

$$CH_3CH_2C({=}O){-}OH$$

Propionic acid

$$CH_3CH_2C({=}O){-}O^-\ Na^+$$

Sodium propionate

$$CH_3CH{=}CHCH{=}CHC({=}O){-}OH$$

Sorbic acid

$$CH_3CH{=}CHCH{=}CHC({=}O){-}O^-\ K^+$$

Potassium sorbate

$$C_6H_5{-}C({=}O){-}OH$$

Benzoic acid

$$C_6H_5{-}C({=}O){-}O^-\ Na^+$$

Sodium benzoate

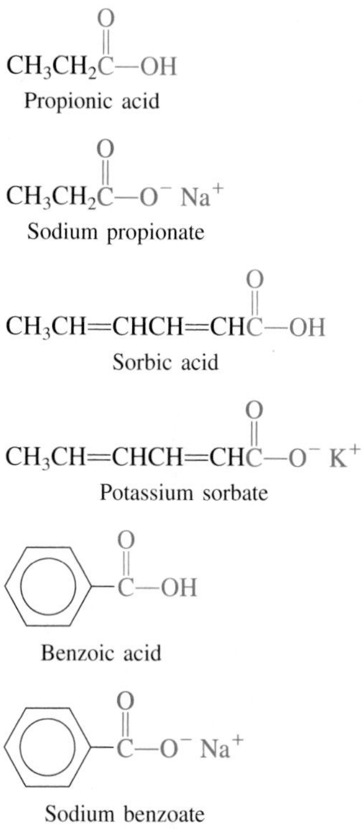

Figure 16.15 Some spoilage inhibitors.

Some inorganic compounds are also used as spoilage inhibitors. Sodium nitrite ($NaNO_2$) is used in meat curing and to maintain the pink color of smoked hams, frankfurters, and bologna. Nitrites also contribute to the tangy flavor of processed meat products.

Nitrites are particularly effective as inhibitors of *Clostridium botulinum,* the bacterium that produces botulism poisoning. However, only about 10% of the amount used to keep meat pink is needed to prevent botulism.

Nitrites have been investigated as possible causes of cancer of the stomach. In the presence of the hydrochloric acid (HCl) in the stomach, nitrites are converted to nitrous acid.

> Sodium nitrate ($NaNO_3$) also was once used in curing meat. The FDA banned the use of nitrates since they have no advantage over nitrites. At any rate, bacteria in our stomachs readily reduce nitrates to nitrites.

$$NaNO_2 + HCl \longrightarrow HNO_2 + NaCl$$

The nitrous acid then may react with secondary amines (amines with two alkyl groups on nitrogen) to form nitroso compounds.

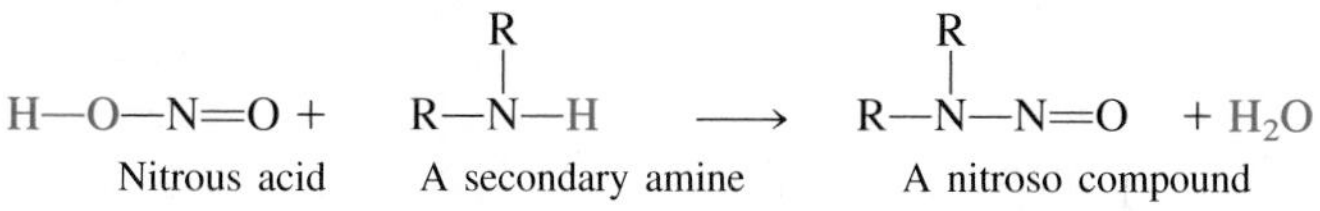

$$\underset{\text{Nitrous acid}}{H{-}O{-}N{=}O} + \underset{\text{A secondary amine}}{R{-}N(R){-}H} \longrightarrow \underset{\text{A nitroso compound}}{R{-}N(R){-}N{=}O} + H_2O$$

The R— groups may be alkyl groups such as methyl (CH_3—) or ethyl (CH_3CH_2—), or they may be more complex. In any case, these nitroso compounds are among the most potent carcinogens known. By eating foods containing nitrates and nitrites, we may be giving our stomachs the raw materials for cancer. It is a fact that the rate of stomach cancer is higher in countries that use prepared meats than in the developing nations where people eat little or no cured meat. The incidence of stomach cancer is *decreasing* in the United States, however, and we aren't quite sure why. We do know that the reaction between nitrous acid and amines to form nitrosamines is inhibited by ascorbic acid (vitamin C). Perhaps a similar inhibition occurs in our stomachs. Problems with nitrites have led the FDA to approve sodium hypophosphite (NaH_2PO_2) as a substitute.

Sulfur dioxide is another inorganic food additive. A gas at room temperature, sulfur dioxide serves as a disinfectant and preservative, particularly for dried fruits such as peaches, apricots, and raisins. It is also used as a bleach to prevent the browning of wines, corn syrup, jelly, dehydrated potatoes, and other foods. Sulfur dioxide seems safe for most people when ingested with food. However, it is a powerful respiratory irritant when inhaled, and is a damaging ingredient of polluted air in some areas (Chapter 12). Sulfite salts once were used in restaurants to keep salad vegetables appearing crisp and fresh. They were shown to cause severe allergic reactions in some people. The FDA banned this use of sulfites in 1986. Their use in other foods must be indicated on the label.

Antioxidants

Antioxidants are added to foods to prevent fats and oils from turning rancid and making the food unpalatable. Packaged foods that contain vegetable oils or animal fats (bread, potato chips, sausage, dry breakfast cereals) often have antioxidants added.

BHA and BHT: Free Radical Reactions

Two compounds often used as antioxidants are butylated hydroxytoluene (BHT) and butylated hydroxyanisole (BHA) (Figure 16.16).

Figure 16.16 Two common antioxidants. BHA is a mixture of two isomers.

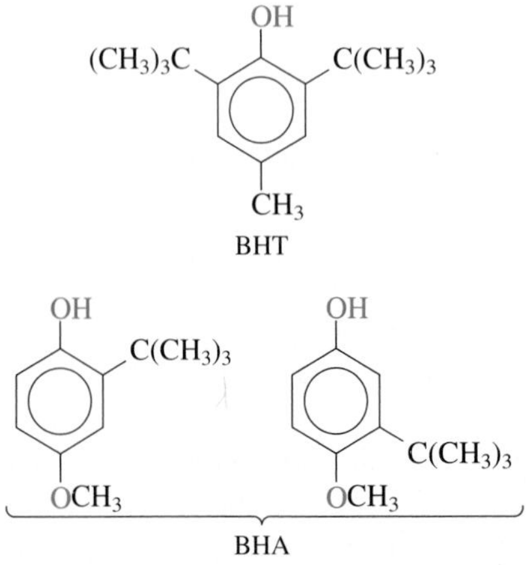

OH
$(CH_3)_3C$
$C(CH_3)_3$
CH_3
BHT

OH
$C(CH_3)_3$
OCH_3
OH
$C(CH_3)_3$
OCH_3
BHA

Fats turn rancid, in part, through oxidation. This process occurs through the formation of molecular fragments called **free radicals**, which have an unpaired electron as a distinguishing feature. (Recall from Chapter 5 that covalent bonds are shared *pairs* of electrons.) We need not concern ourselves with the details of the structures of radicals, but we can summarize the process. First, a fat molecule reacts with oxygen to form a free radical (which we will call Rad·).

$$\text{Fat} + O_2 \longrightarrow \text{Rad}\cdot$$

Then, the radical reacts with another fat molecule to form a new free radical that can repeat the process. A reaction such as this, in which intermediates are formed that keep the reaction going, is called a **chain reaction**. One molecule of oxygen can lead to the decomposition of many fat molecules.

To preserve foods containing fats, processors package the products to exclude air. It isn't possible to exclude air completely, however, so chemical antioxidants are used to stop the chain reaction by reacting with the free radicals.

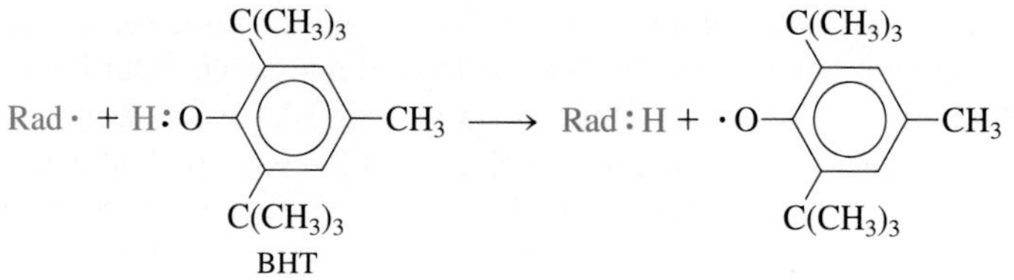

C(CH3)3 C(CH3)3

Rad· + H:O–(ring)–CH3 ⟶ Rad:H + ·O–(ring)–CH3

C(CH3)3 C(CH3)3

BHT

The new radical formed from BHT is rather stable. The unpaired electron doesn't have to stay on the oxygen atom but can move around in the electron cloud of the benzene ring. The BHT radical doesn't react with fat molecules, and the chain is broken.

Why the butyl groups? Without them, the phenols would simply couple when exposed to an oxidizing agent.

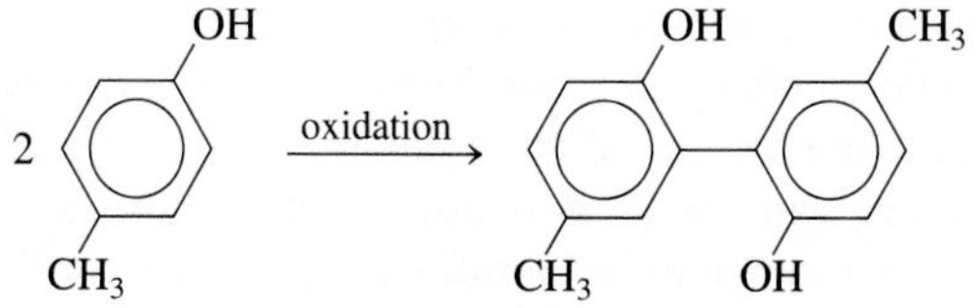

OH OH CH3

2 (ring) —oxidation→ (ring)–(ring)

CH3 CH3 OH

With the bulky butyl groups aboard, the rings can't get close enough together for coupling. They are free, then, to trap free radicals from the oxidation of fats.

Many food additives have been criticized as harmful, and BHA and BHT are no exceptions. They have been reported to cause allergic reactions in some people. Pregnant mice fed a diet containing 0.5% of either BHA or BHT give birth to offspring with abnormalities in their brains.

Not all reports about antioxidants have been unfavorable, however. When relatively large amounts of BHT were fed to rats daily, their life spans were increased by a human equivalent of 20 years. One theory about aging is that it is caused in part by the formation of free radicals. BHT retards this chemical breakdown in the cells in the same way it retards spoilage in foods. The discovery of the secret of aging might lead to longer life spans for humans. In the laboratory, BHT also has shown antitumor activity. Some scientists have speculated that the increased use of antioxidants as food additives has contributed to a decline in the incidence of stomach cancer in the United States.

Vitamin E: A Natural Antioxidant

BHA and BHT are synthetic chemicals, but there are natural antioxidants. Perhaps the most notable of these is vitamin E (Section 16.5). Note that vitamin E is also a phenol, with lots of substituents on the benzene ring. Presumably, its action as an antioxidant is quite similar to that of BHT.

Food Colors

β-Carotene is thought to serve as an anticarcinogen (Chapter 20).

Some foods are naturally colored. For example, the yellow compound β-carotene (read *beta*-carotene) occurs in carrots. β-Carotene is used as a color additive in other foods, such as butter and margarine. (Our bodies convert β-carotene to vitamin A; thus, it is a vitamin additive as well as a color additive.) Other natural food colors include beet juice, grape-hull extract, and saffron (from autumn-flowering crocus flowers).

β-Carotene

We have come to expect many foods to have characteristic colors. The food industry, to increase the attractiveness and acceptability of its products, has used synthetic food colors for decades. Since the Food and Drug Act of 1906, the FDA has regulated the use of these coloring chemicals and set limits on their concentrations. But the FDA is not infallible. Colors once on the approved list later were shown to be harmful and were removed from the list. In 1950, a candy company tried to duplicate in Halloween candy the bright orange color of pumpkins. They used a large amount of Food, Drug, and Cosmetic (FD&C) Orange No. 1. Although this color had been safe in amounts previously used, it caused severe gastrointestinal upset in a number of trick-or-treaters. Thus, it was banned by the FDA.

A few years later, two more dyes were banned. FD&C Yellow Nos. 3 and 4 were found to contain small amounts of β-naphthylamine, a carcinogen. Furthermore, the dyes reacted with acids in the stomach to produce more β-naphthylamine. This compound induces cancer of the bladder in laboratory

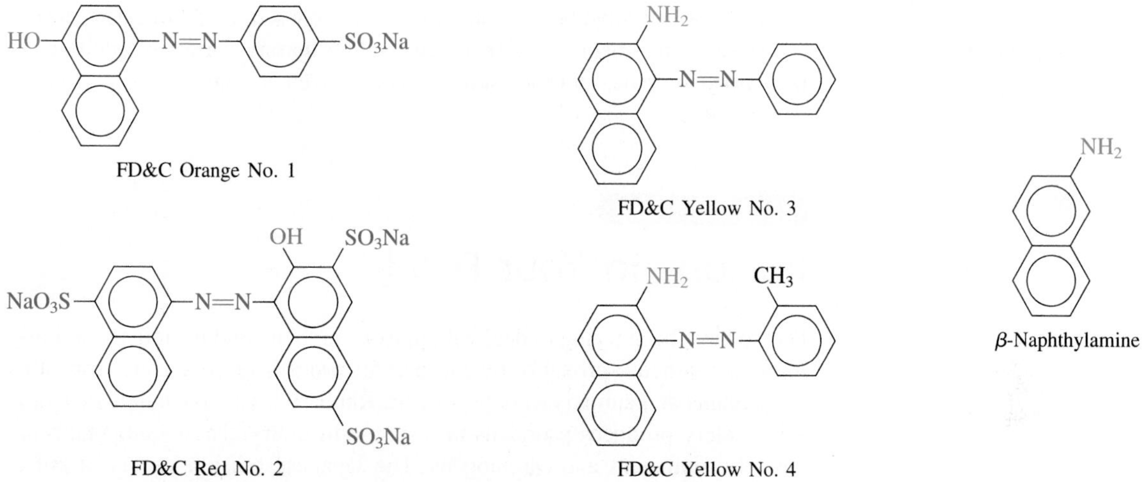

Figure 16.17 Four synthetic food colors that have been banned by the FDA. Note that the two yellow dyes are related to β-naphthylamine, a carcinogen.

animals. Any chemical shown to induce cancer in laboratory animals is automatically banned under the 1958 **Delaney Amendment** to the Food and Drug Act. Several other dyes have since been banned, including FD&C Red No. 2, which was ''delisted'' in 1976 after it was shown to cause cancer in laboratory animals. Structures of these synthetic food colors are shown in Figure 16.17.

It should be noted that food colorings, even those that have been banned, present an extremely low hazard. They have been used for years with apparent safety. Even if the hazard is low, however, there is essentially no benefit from food colors, other than an aesthetic one. Perhaps it would be best to ban these chemicals rather than to assume the risk, however small.

There are only a few artificial food colors still approved by the FDA. Several of these are being studied for possible harmful effects. Any foods that contain artificial colors are supposed to say so on the label, and you can avoid them if you want to.

The GRAS List

Some food additives have been used for many years without apparent harmful effects. In 1958, the United States Congress established a list of additives ''generally recognized as safe.'' This compilation became known as the **GRAS list**. Recent developments have brought to light some deficiencies in the original testing procedures. New research findings—and greater consumer awareness—have led the FDA to begin to reevaluate all the chemicals on the list.

Saccharin, the cyclamates, and several of the food colors now banned were once on the GRAS list, but improved instruments and better experimental designs have revealed possible harm where none was thought to exist. Most of the newer experiments have involved feeding massive doses of additives to laboratory animals, and they have been criticized in that regard.

SECTION 16.9

Poisons in Your Food

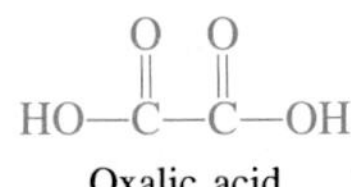

Oxalic acid

People have been trying to deal with poisons in their food for millennia. Early foragers learned—probably by the painful process of trial and error—that some plants and animals were poisonous. Rhubarb leaves contain toxic oxalic acid. Celery produces psoralens in response to injury. These compounds are powerful mutagens and carcinogens. The Japanese relish a variety of puffer fish that contains a deadly poison in its ovaries and liver. More than a hundred Japanese die each year from improperly prepared puffers.

The most toxic substance known is the toxin produced by the bacterium *Clostridium botulinum.* This organism grows in improperly canned food by a perfectly natural process. If the food isn't properly sterilized before it is sealed in jars or cans, the microorganism flourishes in anaerobic (without air) conditions. The poison it produces is so toxic that 1 g of it could kill more than a million people. The point is that a food is not inherently good simply because it is natural. Neither is it necessarily bad because a synthetic chemical substance has been added to it.

There's no doubt that we could get better nutrition if we ate nothing but fresh food. For people in large cities, however, that might be impossible; and few people anywhere want to spend all the time it takes to gather fresh food and prepare meals from scratch. Convenience foods are indeed convenient, and food additives make them easier to fix, more attractive, and (in some cases) more nutritious.

Carcinogens

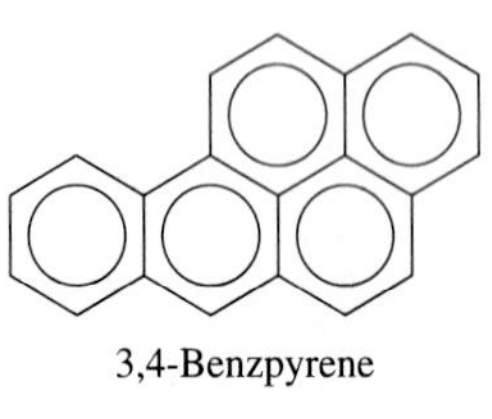

3,4-Benzpyrene

O O

$CH_2CH{=}CH_2$

Safrole

There is little chance that we will suffer acute poisoning from approved food additives. But what about cancer? Could all those chemicals in our food increase our risk of cancer? The possibility exists, even though the risk is low.

We should recognize, though, that carcinogens occur in food naturally. A charcoal-broiled steak contains 3,4-benzpyrene, a carcinogen also found in cigarette smoke and automobile exhaust fumes. Cinnamon and nutmeg contain safrole, a carcinogen that has been banned as a flavoring in root beer.

Among the most potent carcinogens are the **aflatoxins**, compounds produced by molds growing on stored peanuts and grains (Figure 16.18). Aflatoxin B_1 is estimated to be 10 million times as potent a carcinogen as saccharin, and there is no way to keep it completely out of our food. The FDA sets a tolerance of 20 ppb for aflatoxins.

Figure 16.18 Aflatoxins, highly toxic and carcinogenic compounds, are produced by molds that grow on peanuts and stored grains. [*Richard Megna/ Fundamental Photographs.*]

Aflatoxin B_1

It is estimated that we consume 10,000 times as much natural carcinogens as synthetic ones.

Should we ban steaks, spices, peanuts, and grains because they contain naturally occurring carcinogens? Probably not. The risk is slight, and life is filled with more serious risks. Should we ban additives that have been shown to be carcinogenic? Probably yes. Why take the risk, however slight, when it is easy to avoid?

Incidental Additives

There are two major categories of food additives: *intentional additives* are put in a product on purpose to perform a specific function; *incidental additives* get in accidentally during production, processing, packaging, or storage. Pesticide residues, insect parts, and antibiotics added to animal feeds are examples of the incidental additives that sometimes get into our food. There are about 2800 intentional additives. There are perhaps 10,000 incidental additives. These incidental additives often receive wide publicity.

In 1989 discovery of residues of daminozide (Alar) on apples caused great public concern. Daminozide is a plant growth regulator that causes apples to ripen at the same time and have an improved appearance compared to untreated apples.

$$HO-\overset{\overset{O}{\|}}{C}CH_2CH_2\overset{\overset{O}{\|}}{C}-NHN(CH_3)_2$$

Daminozide

The compound itself appears to be harmless, but it breaks down to give dimethylhydrazine, a suspected carcinogen.

$$H_2NN(CH_3)_2$$

Dimethylhydrazine

To get a dose equivalent to that fed to laboratory animals, a person would have to eat 13,000 kg of apples a day for 70 years. To most scientists, the risk seemed vanishingly small. Many consumers were unwilling to assume the risk, however small. Falling sales of apples led to a withdrawal of the chemical from use by apple growers and removal of the compound from the market by the manufacturer.

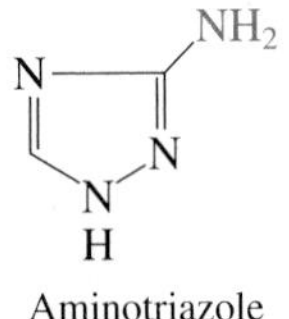

Aminotriazole

Alar was not the first incidental additive to cause concern. In 1959, the sale of cranberries was forbidden after some shipments were found to be contaminated by the herbicide aminotriazole, a compound shown to be carcinogenic in tests on laboratory animals. In 1969, coho salmon taken from Lake Michigan were shown to contain DDT above the tolerance level. Sale of these fish was banned, also.

Polychlorinated biphenyls (PCBs) have been found in poultry and eggs. These products, too, were seized and destroyed. Related compounds, polybrominated biphenyls (PBBs), meant to be used as fire-retardants, were accidentally mixed with animal feed in western Michigan. Many farm animals were destroyed, and still PBBs got into the food supply.

Br Br Br Br Br Br Br Br Br Br

Typical PBB compounds

Chemical substances in animal feed often show up in the meat we eat. Antibiotics are added to animal feed to promote weight gain. In fact, nearly 50% of the 9 million kg of antibiotics produced annually in the United States go into low-level dosages in animal feeds. Residues of these have been found in up to 25% of the animals slaughtered. These residues may result in sensitization of individuals who eat the meat, thus hastening the development of allergies. Use of antibiotics in animal feeds also may hasten the process by which bacteria become drug resistant.

Diethylstilbestrol (DES), a synthetic female hormone, also has been added to animal feeds to promote weight gain. It was banned after evidence showed that it caused vaginal cancer in the daughters of women who had taken DES during pregnancy. DES has also been shown to cause testicular cancer in sons of women who took DES while pregnant.

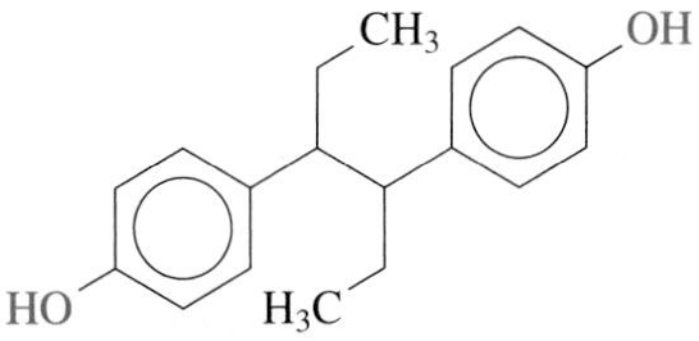

CH_3 OH

HO H_3C

Diethylstilbestrol

Note that the effect of DES did not show up for 15 years. Even then it appeared in the *offspring* of the women who took the drug. This points out some of the problems involved in evaluating a chemical for its possible harmful effects.

SECTION 16.10

A World Without Food Additives

Could we get along without food additives? Some of us could. But food spoilage might drastically reduce the food supply in an already hungry world. And diseases due to vitamin and mineral deficiencies might flourish again. Foods could cost more and be less nutritious. Food additives seem to be a necessary part of modern society. It is true that there are hazards associated with the use of some food additives, but the major problem with our food supply still is contamination by rodents, insects, and harmful microorganisms. Indeed, there are about 9000 deaths and 6 million illnesses annually in the United States from food poisoning caused by bacterial toxins. There are few, if any, deaths associated with the use of intentional food additives.

In 1993, improperly cooked hamburgers contaminated with a virulent strain of *E. coli* bacteria made several hundred people ill in Washington, Idaho, and Nevada. Two children died, one of whom ate a cheeseburger at a Jack-in-the-Box restaurant.

What should we do about food additives? We should be sure that the FDA is staffed with qualified personnel to ensure the adequate testing of proposed food additives. People trained in chemistry are necessary for control and monitoring of food additives and the detection of contaminants. Research in the analytical techniques necessary for the detection of trace quantities is vital to adequate consumer protection. We also should work for laws adequate to prevent the unnecessary and excessive use of pesticides and other agricultural chemicals that might contaminate our food. Above all, we should be alert and informed about these problems so vital to our health and well-being.

SECTION 16.11

Plants: Sun-Powered Food-Making Machines

All food comes ultimately from green plants. Although all organisms can transform one type of food into another, only green plants can harness sunlight and use it to convert carbon dioxide and water into the sugars that directly or indirectly fuel all living things.

$$6\,CO_2 + 6\,H_2O \xrightarrow{\text{photosynthesis}} C_6H_{12}O_6 + 6\,O_2$$

This reaction also replenishes the oxygen in the atmosphere (Chapter 12). With other nutrients, particularly compounds of nitrogen and phosphorus, plants can convert the sugars from photosynthesis into proteins, fats, and other chemicals that we use as food.

Soil is rock broken down by weathering plus decaying plant and animal matter. It takes many years to build even a few millimeters of soil. The United States loses 5 billion t of soil each year to erosion.

The structural elements of plants—carbon, hydrogen, and oxygen—are derived from air and water. Other plant nutrients are taken from the soil, and energy is supplied by the sun. In a primitive society, people grow plants for food and obtain energy from the food. Nearly all this energy is reinvested in the production of food, although a portion goes into making clothing and building shelter. Figure 16.19 shows a simplified diagram of the energy flow in a primitive society. Obviously, a real society would be much more complicated than the diagram. Some of the plants, for example, might be fed to animals, and human energy in turn would be obtained from animal flesh or animal products (such as milk and eggs).

In primitive societies, nearly all the energy comes from renewable resources. One unit of human work energy, supplemented liberally by energy

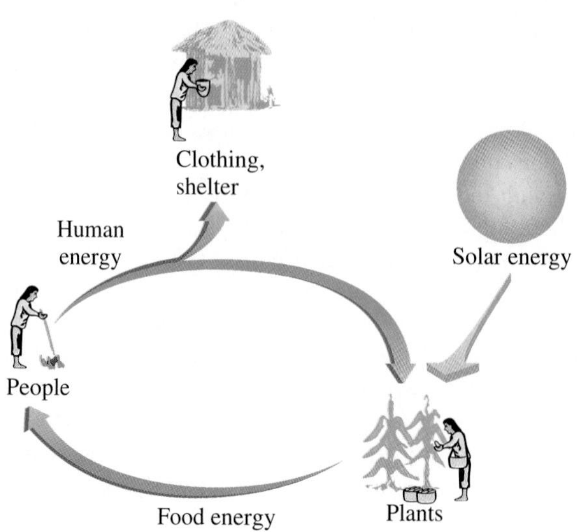

Figure 16.19 Energy flow in a primitive society.

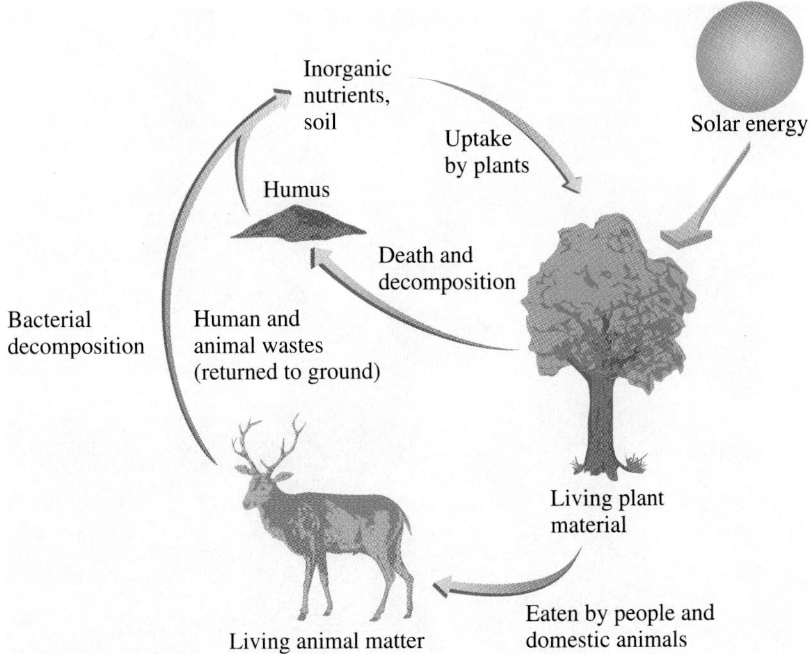

Figure 16.20 Flow of nutrients in a primitive system.

from the sun, might produce 10 units of food energy. The surplus energy might be used to make clothing or to provide shelter. It also might be used in games or cultural activities.

The flow of nutrients is also rather simple. Unused portions of plants and human and animal wastes are returned to the soil. These are broken down by bacteria to provide the nutrients for the growth of the new plants (Figure 16.20). The humus content of the soil is maintained. Properly practiced, this primitive agriculture could be continued for centuries without seriously depleting the soil. The only problem is that farming at that level doesn't support very many people.

SECTION 16.12

Farming with Chemicals: Fertilizers

To replace nutrients lost from the soil and to increase crop production, modern farmers use a variety of chemical fertilizers. The three **primary plant nutrients** are nitrogen, phosphorus, and potassium. Let's consider nitrogen first.

Nitrogen, though present in air in the elemental form (N_2), is not generally available to plants. There are some forms of bacteria that are able to fix nitrogen, that is, to convert it to a combined, soluble form. Colonies of bacteria that can perform this vital function grow in nodules on the roots of legumes

Figure 16.21 Bacteria in nodules on the roots of legumes fix atmospheric nitrogen by converting it to soluble compounds that plants can use as nutrients. [*Runk/Schoenberger/Grant Heilman Photography.*]

Figure 16.22 The intense heat in a lightning flash fixes nitrogen that falls to Earth as nitric acid, enriching the soil. [*William J. Warren/Westlight—Backgrounds.*]

(plants such as clovers and peas) (Figure 16.21). Thus, farmers are able to restore fertility to the soil by crop rotation. A nitrogen-fixing crop (such as clover) is alternated with a nitrogen-consuming crop (such as corn).

Legumes could still be used to supply nitrogen to the soil, and they still are to some extent. But the modern methods of high-yield production demand chemical fertilizers. Why get a corn crop every other year, when you can have one every year?

Plants usually take up nitrogen in the form of nitrate ions (NO_3^-) or ammonium ions (NH_4^+). These are combined with carbon compounds from photosynthesis to form amino acids, the building blocks of proteins (the polymeric compounds essential to all life processes). For years, farmers were dependent on manure as a source of nitrates. Discovery of deposits of sodium nitrate (called Chile saltpeter) in the deserts of northern Chile led to exploitation of this source as a supplemental source of nitrogen.

A rapid rise in population growth during the late nineteenth and early twentieth centuries led to increasing pressure on the available food supply. This pressure led to an increasing demand for nitrogen fertilizers. The atmosphere offered a seemingly inexhaustible supply—if only it could be converted to a form useful to humans. Every flash of lightning forms some nitric acid in the air (Figure 16.22). This process probably contributes about 9 kg of nitrogen per hectare of land per year.

Nitrates

The first real breakthrough in nitrogen fixation came in Germany on the eve of World War I. The process, developed by Fritz Haber, made possible the combination of nitrogen and hydrogen to make ammonia.

$$3\,H_2 + N_2 \longrightarrow 2\,NH_3$$

By 1913, one nitrogen-fixation plant was in production and several more were under construction. The Germans were able to make ammonium nitrate (NH_4NO_3), an explosive, by oxidizing part of the ammonia to nitric acid.

$$2\,NH_3 + 4\,O_2 \longrightarrow 2\,HNO_3 + 2\,H_2O$$

The nitric acid then was reacted with ammonia to produce ammonium nitrate.

$$HNO_3 + NH_3 \longrightarrow NH_4NO_3$$

The Germans were interested mainly in ammonium nitrate as an explosive, but it turned out to be a valuable nitrogen fertilizer as well.

A gas at room temperature, ammonia is easily compressed into a liquid that can be stored and transported in tanks. In this form, called anhydrous ammonia, it is applied directly to the soil as fertilizer (Figure 16.23).

Anhydrous means without water.

Some ammonia is reacted with carbon dioxide to form urea, an organic compound that releases nitrogen slowly to the soil (rather than all at once as the inorganic forms do).

$$2\,NH_3 + CO_2 \longrightarrow \underset{\text{Urea}}{NH_2{-}\overset{\overset{\displaystyle O}{\|}}{C}{-}NH_2} + H_2O$$

Figure 16.23 This farmer is applying anhydrous ammonia, a source of the plant nutrient nitrogen, to his fields. [*Grant Heilman/ Grant Heilman Photography.*]

Table 16.9 Various Forms of Nitrogen Fertilizers Made from Ammonia		
Reagent	Product Fertilizer	Formula
(none)	Anhydrous ammonia	NH_3
Carbon dioxide	Urea	NH_2CONH_2
Sulfuric acid	Ammonium sulfate	$(NH_4)_2SO_4$
Nitric acid	Ammonium nitrate	NH_4NO_3
Phosphoric acid	Ammonium hydrogen phosphate	$(NH_4)_2HPO_4$

Some ammonia, as we mentioned, is converted to ammonium nitrate, a crystalline solid. Some is converted to solid ammonium sulfate [$(NH_4)_2SO_4$] by reaction with sulfuric acid.

$$2\,NH_3 + H_2SO_4 \longrightarrow (NH_4)_2SO_4$$

These ammonia products (Table 16.9) can be applied separately or combined with other plant nutrients to make a more complete fertilizer.

In 1912 American farmers produced an average of 26 bushels of corn per acre. Today the yield per acre is almost 100 bushels. This four-fold increase is due in large part to the increased use of nitrogen fertilizers.

It is largely through the use of nitrogen fertilizers that farmers are able to feed the world. The atmosphere has vast amounts of nitrogen, but present industrial ways of fixing it require a lot of energy and use up nonrenewable resources. (The hydrogen for the synthesis of ammonia is made from natural gas.) Genes for nitrogen fixation have been transferred from bacteria into higher (nonlegume) plants. Perhaps some day soon corn and cotton will be able to produce their own nitrogen fertilizer the way clovers and peas do.

Fritz Haber, the German chemist who invented a process for manufacturing ammonia. [*Courtesy of Encyclopedia Britannica, Inc., Chicago.*]

Fritz Haber was awarded the Nobel prize for chemistry in 1918. There are a number of ironies in this. Alfred Nobel, the Swedish inventor and chemist who died in 1896, endowed the Nobel prize (including the peace prize) with a fortune derived from his own work with explosives. During his lifetime, he was bitterly disappointed by the fact that the explosives he had developed for excavation and mining were put to such destructive uses in war. And Haber, a man who helped his country during World War I to the extent of going to the front to supervise the release of chlorine in the first poison-gas attack, was exiled from his native land in 1933. He was a Jew, and Nazi racial laws forced him out of his position as director of the Kaiser Wilhelm Institute of Physical Chemistry. He accepted a post at Cambridge University in England, but his life was ended by a stroke less than a year later.

Phosphates

Phosphates have probably been used as fertilizer since ancient times—in the form of bone, guano (bird droppings), or fish meal. However, phosphates as such were not recognized as plant nutrients until 1800. Following that discovery, the great battlefields of Europe were dug up and bones were shipped to plants for processing into fertilizer.

Animal bones are rich in phosphorus, but the phosphorus is tightly bound and not readily available to plants. In 1831, the Austrian chemist Heinrich Wilhelm Köhler treated animal bones with sulfuric acid to convert them to a more soluble form, called superphosphate. In 1843, John Lawes applied the same treatment to phosphate rock. The essential reaction for forming superphosphate is

$$\underset{\text{Phosphate rock or bone (insoluble)}}{Ca_3(PO_4)_2} + 2\,H_2SO_4 \longrightarrow \underbrace{Ca(H_2PO_4)_2 + 2\,CaSO_4}_{\text{Superphosphate (more soluble)}}$$

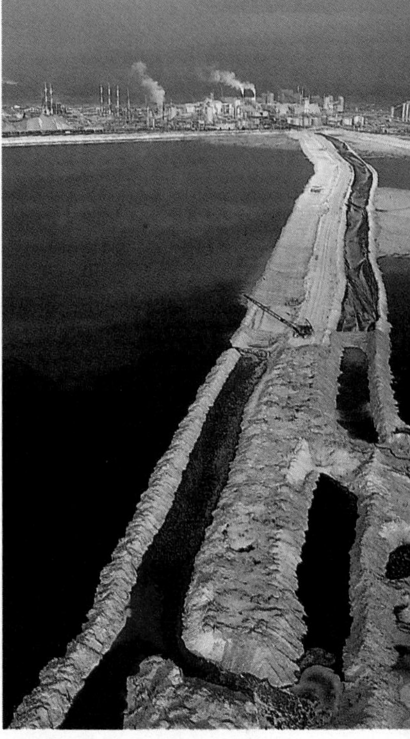

Figure 16.24 Phosphate fertilizers are obtained from ores (rock phosphate) that likely are the skeletal remains of ancient sea creatures. [*Nathan Benn/Stock Boston.*]

Modern phosphate fertilizers often are made by treating phosphate rock with phosphoric acid to make water-soluble calcium dihydrogen phosphate.

$$Ca_3(PO_4)_2 + 4\,H_3PO_4 \longrightarrow 3\,Ca(H_2PO_4)_2$$

Even more common today is the use of ammonium monohydrogen phosphate, $(NH_4)_2HPO_4$, which supplies both nitrogen and phosphorus.

Phosphate is quite common in the soil but available forms are often at concentrations too low for adequate support of plant growth. Fortunately, there are deposits that are more concentrated. The presence of bones and teeth from early fish and other animals in these ores indicates that the deposits are largely the skeletal remains of sea creatures of ages past (Figure 16.24). Unfortunately, fluorides often are associated with phosphate ores, and they constitute a serious pollution problem from the production of phosphate fertilizers.

Usually phosphorus is absorbed by a plant as $H_2PO_4^-$. In plants, phosphates are incorporated into DNA and RNA (Chapter 15). They also are constituents of compounds essential for the conversion of starches to sugars. Phosphates appear to play a vital role in photosynthesis, mainly by involvement in energy-transfer processes. Phosphorus compounds also are essential to the formation of fats and of some proteins and other cell constituents. The availability of phosphorus is often the limiting factor in plant growth.

About 90% of all phosphates produced are used in agriculture. The United States is the leading producer and user, but the rich phosphate deposits in the United States may be depleted in the 1990s. Large offshore deposits have been discovered in the Atlantic Ocean off North Carolina. Worldwide, Morocco has two-thirds of the reserves. All the high-grade phosphate reserves may be exhausted in 30 or 40 years. Our use of phosphates scatters them irretrievably throughout the environment.

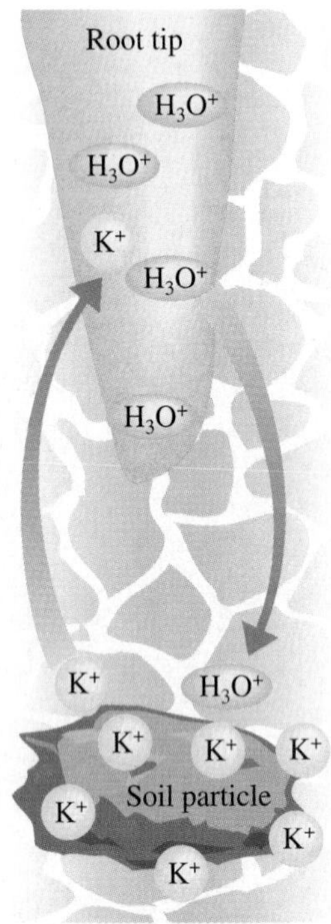

Figure 16.25 When a root tip takes up potassium ions from the soil, hydronium ions are transferred to the soil. Potassium uptake by plants tends to make the soil acidic.

Potassium

The third major element necessary for plant growth is potassium. Plants use it in the form of the simple ion K^+. Generally, potassium is abundant and there are no problems with solubility. The precise function of potassium in plant cells is difficult to determine. It seems to be involved in the formation and transport of carbohydrates. Also, it may be necessary for the buildup of proteins from amino acids. Uptake of potassium ions from the soil leaves the soil acidic; each time one potassium ion enters the root tip, a hydronium ion must leave in order for the plant to maintain electrical neutrality (Figure 16.25).

The usual chemical form of potassium in commercial fertilizers is potassium chloride (KCl). Vast deposits of this salt occur in Stassfurt, Germany. For years, this source supplied nearly all the world's potassium fertilizer. With the coming of World War I, the United States sought supplies within its own borders. Deposits at Searles Lake, California, and Carlsbad, New Mexico, now supply much of United States needs. Canada has vast deposits in Saskatchewan and Alberta. Beds of potassium chloride up to 200 m thick lie about 1.5 km below the Canadian prairies (Figure 16.26).

Although reserves are large, potassium salts are a nonrenewable resource. We should use them wisely.

Figure 16.26 Mining potassium chloride about 1.5 km below the prairies of Saskatchewan, Canada. [*Courtesy of Potash Corporation of Saskatchewan.*]

Other Essential Elements

In addition to the three major nutrients (nitrogen, phosphorus, and potassium), a variety of other elements are necessary for proper plant growth. Three **secondary plant nutrients**—magnesium, calcium, and sulfur—are needed in moderate amounts. Calcium, in the form of lime (calcium oxide), is used to neutralize acidic soils.

$$CaO + 2\,H_3O^+ \longrightarrow Ca^{2+} + 3\,H_2O$$

Calcium ions are also necessary plant nutrients. Magnesium ions (Mg^{2+}) are incorporated into chlorophyll molecules and therefore are necessary for photosynthesis. Sulfur is a constituent of several amino acids, and it is necessary for protein synthesis.

Seven other elements, called **micronutrients**, are needed in very small amounts. These are summarized in Table 16.10. Many soils contain these trace elements in sufficient quantity. However, some soils are deficient in one or more. Productivity of these soils can be markedly increased by the addition of small amounts of the needed elements.

It is possible that other elements also are required by plants. Some elements that may be needed include sodium, silicon, vanadium, chromium, selenium, cobalt, fluorine, and arsenic. Most of these elements are present in most soils,

Table 16.10 Seven Micronutrients Necessary for Proper Plant Growth

Element	Form Used by Plants	Function	Deficiency Symptoms
Boron	H_3BO_3	Required for protein synthesis; essential for reproduction and for carbohydrate metabolism	Death of growing points of stems, poor growth of roots, poor flower and seed production
Copper	Cu^{2+}	Constituent of enzymes; essential for reproduction and for chlorophyll production	Twig dieback, yellowing of newer leaves
Iron	Fe^{2+}	Constituent of enzymes; essential for chlorophyll production	Yellowing of leaves, particularly between veins
Manganese	Mn^{2+}	Essential for redox reactions and for the transformation of carbohydrates	Yellowing of leaves, brown streaks of dead tissue
Molybdenum	MoO_4^{2-}	Essential in nitrogen fixation by legumes and reduction of nitrates for protein synthesis	Stunting, pale green or yellow leaves
Zinc	Zn^{2+}	Essential for early plant growth and maturing	Stunting, seed and grain yields reduced
Chlorine	Cl^-	Increases water content of plant tissue; involved in carbohydrate metabolism	Shriveling

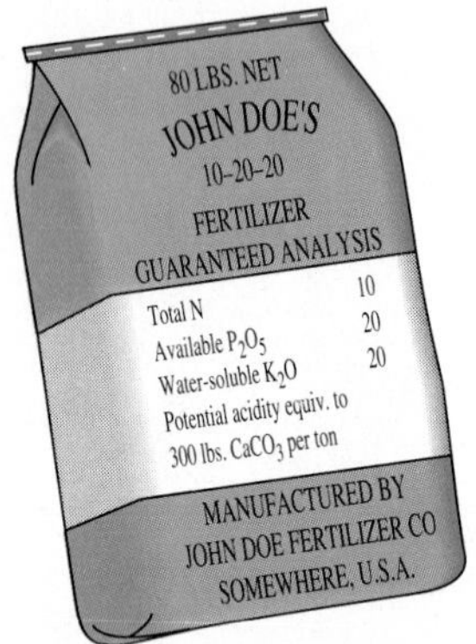

Figure 16.27 The numbers on this bag of fertilizer tell us that the fertilizer is 10% nitrogen, 20% phosphorus (as P_2O_5), and 20% potassium (as K_2O).

but it has not been established whether or not they are necessary for plant growth.

Fertilizers: A Mixed Bag

Most farmers buy *complete* **fertilizers**, which, despite the name, usually contain only the three main nutrients. There are usually three numbers on fertilizer bags (Figure 16.27). The first number represents the percent of nitrogen (N); the second, the percent of phosphorus (calculated as P_2O_5); and the third, the percent of potassium (calculated as K_2O). So 20-10-5 means that the fertilizer contains 20% N, 10% P_2O_5, and 5% K_2O. The rest is inert material.

Fertilizers must be water soluble to be used by plants. When it rains, the nutrients from the fertilizers are washed into streams and lakes, where they stimulate blooms of algae. These chemicals, particularly the nitrates, also penetrate the groundwater.

There really is no K_2O or P_2O_5 in fertilizer. These formulas merely are used as a basis for calculation. The actual form of potassium is nearly always KCl, although any potassium salt would furnish the needed K^+ ion. Phosphorus is supplied as one of several salts. Calculation as K_2O or P_2O_5 gives numbers that compare amounts of nutrient regardless of the compound that supplies them.

SECTION 16.13

The War Against Pests

Since the earliest days of recorded human history (and surely even before that), people have been plagued by insect pests. Three of the 10 plagues of Egypt (described in the Book of Exodus) were insect plagues—lice, flies, locusts. The decline of Roman civilization has been attributed in part to malaria, a disease carried by mosquitoes, which destroys vigor and vitality when it does not kill. Bubonic plague carried by rats (and by fleas from rats to humans) swept through the Western World repeatedly during the Middle Ages. One such plague (during the 1660s) is estimated to have killed 25 million people—25% of the population in Europe at that time. The first attempt to dig a Panama Canal (by the French during the 1880s) was defeated by an outbreak of malaria.

The use of modern chemical **pesticides** may be the only thing that stands between us and some of these insect-borne plagues. Pesticides also prevent the consumption of a major portion of our food supplies by insects and other pests. Crop losses to insects in the United States are estimated to be $4 billion a year. When losses to other pests—rodents, fungi, nematodes, and weeds—are included, the losses total $15 billion a year.

There are about 50,000 different pesticide products on the market based on 1400 active ingredients. Some are used in homes and industrial buildings, but 77% of all pesticides are used in agriculture.

In earlier days, people tried to control insect pests by draining swamps, pouring oil on ponds (to kill the mosquito larvae), and using a variety of chemicals. Most of these chemicals were compounds of arsenic. Lead arsenate [$Pb_3(AsO_4)_2$] is a particularly effective poison, since both the lead and the arsenic in it are toxic. A few pesticides, like pyrethrum (used in mosquito control) and nicotine sulfate (Black Leaf 40), are obtained from plant matter.

Only a few insect species are harmful. Many are beneficial, and others play

Table 16.11 Toxicity of Insecticidal Preparations Administered Orally to Rats

Pesticide	LD_{50}*		
Pyrethrins[(1)]	1200		Least toxic
Malathion	1000	(1375)	
Lead arsenate	825		
Diazinon	285	(250)	
Carbaryl	250		
Nicotine[(2)]	230		
DDT[(3)]	118	(113)	
Lindane	91	(88)	
Methyl parathion	14	(24)	
Parathion	3.6	(13)	
Carbofuran[(4)]	2		
Aldicarb	1		Most toxic

*Dose in milligrams per kilogram of body weight that will kill 50% of test population (see Section 20.13). Values in parentheses are for male rats.

Notes:

(1) Active ingredients of pyrethrum.

(2) In mice. Nicotine is much more toxic by injection.

(3) Estimated LD_{50} for humans is 500 mg/kg.

(4) In mice.

Source: Susan Budavari (Ed.), *The Merck Index.* Rahway, NJ: Merck and Co., 1989.

important roles in ecological systems and are indirectly beneficial. Most poisons are indiscriminate. They kill all insects, not just those we consider pests. Many are also toxic to humans and other animals. Some say we should call such poisons **biocides** (because they kill living things) rather than insecticides. Table 16.11 lists the toxicities of some insecticides.

EXAMPLE 16.4

What amount of the pesticide lindane could lead to the death of a 10-kg (22-lb) child if the lethal dose was the same as it is for rats (see Table 16.11)?

SOLUTION

The lethal dose is given as 88 mg of lindane per kilogram of body weight.

$$10\,\cancel{\text{kg}} \times \frac{88\text{ mg}}{1\,\cancel{\text{kg}}} = 880\text{ mg or }0.88\text{ g}$$

Exercise 16.4

How much parathion would it take to kill a 70-kg farm worker if the lethal dose was the same as it is for rats (see Table 16.11)?

DDT: The Dream Insecticide

$Cl-C_6H_4-CH(C_6H_4Cl)-CCl_3$

DDT
(Dichlorodiphenyl-
trichloroethane)

Shortly before World War II, a chlorinated hydrocarbon (DDT) was found to be a potent insecticide. The discovery was made in Switzerland by Paul Müller. **DDT** was tested and found effective against grapevine pests and against a particularly bad infestation of potato beetles.

When the war came, the supply of the pesticide pyrethrum was cut off by the Japanese occupation of Southeast Asia and the Dutch East Indies (now Indonesia). Lead, arsenic, and copper compounds, which had been used for insecticides, were needed for armaments and other military purposes. The British and the Americans found themselves in need of an insecticide that would protect soldiers from lice and ticks (bearers of typhus). Those who had to fight in the jungles of the South Pacific needed protection from mosquitoes and other disease-bearing pests.

A few kilograms of DDT were obtained by the Allies and hurriedly tested. Combined with talcum, it made a very effective delousing powder. Clothing was impregnated with DDT, and it seemed to have no deleterious effects on humans or other large animals. Allied soldiers were virtually free from lice. The Germans missed out on DDT, even though it was discovered just across their border. German troops were heavily infested with lice and many were sick with typhus. In wars before World War II, more soldiers probably died from typhus than from bullet wounds.

DDT is easily synthesized from cheap, readily available chemicals. Chlorobenzene and chloral hydrate are warmed in the presence of sulfuric acid. When the reaction is complete, the mixture is poured into water and the DDT separates out (like other chlorinated hydrocarbons, DDT is essentially insoluble in water).

$$2\,Cl-C_6H_5 + HO-CH(OH)-CCl_3 \xrightarrow{H_2SO_4} Cl-C_6H_4-CH(C_6H_4Cl)-CCl_3 + 2\,H_2O$$

Chlorobenzene   Chloral hydrate   DDT

A cheap **insecticide** effective against a variety of insect pests, DDT came into widespread use after the war (Figure 16.28). Other chlorinated hydrocarbons were synthesized, tested, and pressed into service in the war against insects. Although invaluable to farmers in the production of food and fiber, the chlorinated hydrocarbons won their most dramatic victories in the field of public health. According to the World Health Organization, approximately 25 million lives have been saved and hundreds of millions of illnesses prevented by the use of DDT and other chlorinated-hydrocarbon pesticides. In India alone, malaria cases were reduced from 75 million to 5 million per year. The average life span in India has been increased by 15 years since the mosquito-eradication program was begun in 1954.

Figure 16.28 Insecticides often are applied by aerial spraying. [*USDA Photo.*]

Striking evidence of the value of DDT in public health comes from Sri Lanka (formerly Ceylon), where there were 2.8 million cases of malaria in 1946. By 1963, spraying with DDT had reduced the incidence of malaria to only 17 cases. The spraying program was terminated in 1964, and by 1968, the number of cases had risen to over 1 million per year.

DDT was a dream come true. It seemed that the world would be free at last from insect plagues and insect-borne diseases. Crops would be protected from the ravages of insects, and food production would be increased. In recognition of his magnificent discovery, Paul Müller was awarded the Nobel prize in physiology and medicine in 1948.

The Decline and Fall of DDT

Before Müller even received his prize, however, there were warnings that all was not well. Houseflies resistant to DDT were reported as early as 1946. DDT's toxicity to fish was reported by 1947. By 1948, 11 additional resistant species of insects had been discovered. Such early warnings were largely ignored. It was also assumed that the toxicity would disappear soon after the chemical was discharged into the environment. DDT was used extensively to protect crops and control mosquitoes and to try to prevent Dutch elm disease. Bird populations began to decline in agricultural areas. By 1962, the year Rachel Carson's book *Silent Spring* appeared, United States production of DDT had reached 76 million kg per year.

Today, in the developed countries, DDT is known mostly for its harmful environmental effects. It interferes with calcium metabolism. Birds are threatened because egg shells are composed mainly of calcium compounds. Eggs of birds that have ingested DDT have thin shells that are poorly formed and easily broken. Even a few parts per billion (ppb) of DDT interfere with the growth of plankton and the reproduction of crustaceans such as shrimp. As bad as DDT sounds, though, it has probably saved the lives of more people than any other chemical substance.

Chlorinated hydrocarbons generally are unreactive. This lack of reactivity was a major advantage of DDT. Sprayed on a crop, DDT stayed there and killed insects for weeks. This *pesticide persistence* was also a major disadvantage: the substance did not break down readily in the environment. Although not very toxic to humans and other warm-blooded creatures, DDT is much more toxic to cold-blooded organisms. That includes insects, of course, and fish. DDT's lack of reactivity toward oxygen, water, and components of the soil led to its buildup in the environment, where it threatened fish, birds, and other wildlife.

Biological Magnification: Concentration in Fatty Tissues

Chlorinated hydrocarbons are good solvents for fats, and fats are good solvents for chlorinated hydrocarbons such as DDT. When these compounds are ingested as contaminants in food or water, they are concentrated in fatty tissues. Their fat-soluble nature causes chlorinated hydrocarbons to be con-

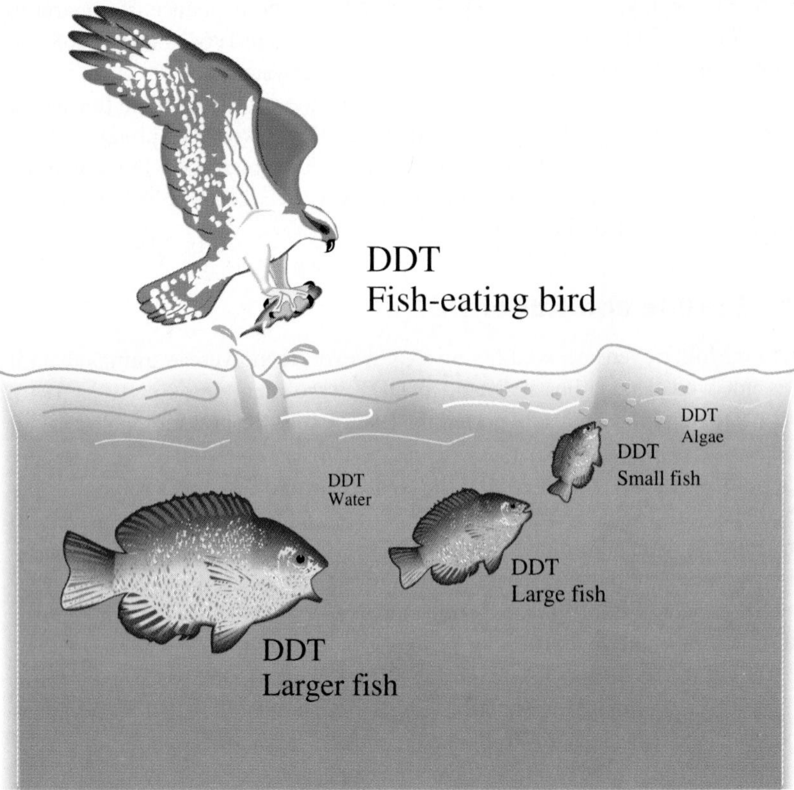

Figure 16.29 Concentration of DDT up the food chain. Animals at the top of the chain have the highest concentration of the insecticide.

centrated up the food chain. This *biological magnification* was graphically demonstrated in California in 1957. Clear Lake, about a hundred miles north of San Francisco was sprayed with DDT in an effort to control gnats. The water, after spraying, contained only 0.02 ppm of DDT. The microscopic plant and animal life contained 5 ppm—250 times as much. Fish feeding on these microorganisms contained up to 2000 ppm. Grebes, the diving birds that ate the fish, died by the hundreds (Figure 16.29).

DDT and other chlorinated hydrocarbons such as PCBs (Chapter 10) are nerve poisons. Concentrated in the fatlike compounds that make up nerve sheaths, they somehow interfere with the transmission of electrical impulses along these sheaths. DDT also interferes with calcium metabolism, essential to the formation of healthy bones and teeth. Although no harm to humans has ever been conclusively demonstrated, the disruption of calcium metabolism in birds has been disastrous for some species. The egg shells, mainly composed of calcium compounds, are thin and poorly formed. The bald eagle, peregrine falcon, and other birds became endangered species. All have made dramatic recoveries since DDT was banned in the United States in 1972.

Malaria still kills. Worldwide, 100 million people are ill with the disease. It causes 1 million deaths each year, mostly among children under 5 years old.

The ban was not total. DDT is still produced for export and is exported particularly to developing countries, where it still sometimes plays a role in malaria control. But the problem of getting a worldwide ban on DDT may solve itself as more and more insect species develop resistance. Its use may be discontinued simply because it is no longer effective.

Lindane, a chlorinated hydrocarbon pesticide, is used to kill head lice and crab lice on humans.

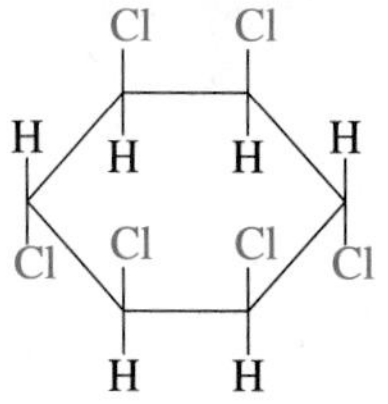

Cl Cl H H H H Cl Cl Cl Cl H H

Lindane

Most chlorinated hydrocarbon insecticides have been banned or severely restricted in developed countries. In following sections, we consider some alternatives—including some biological methods of insect control.

Organic Phosphorus Compounds

Bans and restrictions on chlorinated hydrocarbon insecticides have led to increased use of organic phosphorus compounds. Typical of these are malathion, diazinon, and parathion (Figure 16.30). Over two dozen of these insecti-

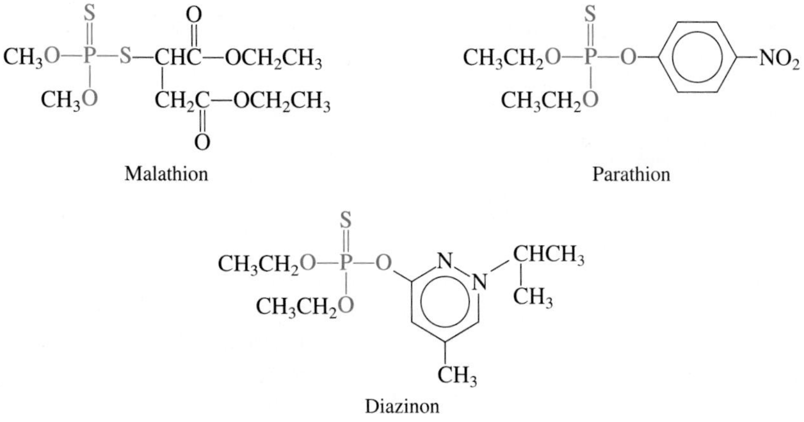

$CH_3O-P(=S)(OCH_3)-S-CH(CH_2C(=O)-OCH_2CH_3)C(=O)-OCH_2CH_3$

Malathion

$CH_3CH_2O-P(=S)(OCH_2CH_3)-O-C_6H_4-NO_2$

Parathion

$CH_3CH_2O-P(=S)(OCH_2CH_3)-O-$ (pyrimidine ring with CH_3 and $CHCH_3$ / CH_3)

Diazinon

Figure 16.30 Three organic phosphorus compounds that are used as insecticides.

About 45,000 pesticide poisonings of humans are recorded in the United States each year. About 200 of these cases are fatal.

In 1992 Americans spread 60 million pounds of pesticides and fertilizer on their lawns.

Both the carbamates and the organic phosphorus compounds are nerve poisons. Their mode of action is discussed in Chapter 20.

Plants produce their own pesticides to protect them from being eaten. These pesticides sometimes make up 5 to 10% of the dry weight of plants. Nicotine protects tobacco plants. The pyrethrins also are plant-produced pesticides. Physostigmine is a naturally occurring carbamate. We probably eat 10,000 times as much natural pesticides as we do synthetic ones.

It was carbamates that were being made at the chemical plant in Bhopal, India, in December 1984 when an explosion released large amounts of a very toxic gas called methyl isocyanate.

cides are available commercially. They have been extensively studied and evaluated for effectiveness against insects and for toxicity to people, laboratory animals, and farm animals.

The organic phosphorus compounds generally are more toxic to mammals than the chlorinated hydrocarbons (see Table 16.11). (Malathion, used to control fruit fly infestations in California, is a notable exception; it is less toxic than DDT.) Like the chlorinated hydrocarbons, the organic phosphorus pesticides are concentrated in fatty tissues. The phosphorus compounds are less persistent in the environment; they break down in days or weeks while the chlorinated hydrocarbons often persist for years. Residues of organic phosphorus compounds are seldom found in food.

Carbamates

Another group of compounds that have gained prominence since the ban on DDT are the carbamates. Examples are carbaryl (Sevin), carbofuran (Furadan), and aldicarb (Temik) (Figure 16.31). Carbaryl has a low toxicity to mammals, but carbofuran and aldicarb have toxicities similar to that of parathion. Most of the carbamates are directed specifically at one, or a few, insect pests. Thus, they are called *narrow spectrum insecticides,* in contrast to the chlorinated hydrocarbons and organic phosphorus compounds that kill many kinds of insects. The latter are called *broad spectrum insecticides.* Carbamates break down rapidly in the environment. Unfortunately, carbaryl, the most widely used, is particulary toxic to honeybees. Millions of these valuable insects have been wiped out by spraying crops with carbaryl.

The essential feature of the carbamate molecule is the group

$$\underset{|}{\text{O}}-\overset{\overset{\displaystyle\text{O}}{\|}}{\text{C}}-\underset{|}{\text{N}}-$$

Many variations are possible by changing the groups attached to the oxygen and the nitrogen. Generally, they break down rapidly in the environment, and they do not accumulate in fatty tissue, an advantage of these insecticides over the chlorinated hydrocarbons and organic phosphorus compounds.

SECTION 16.14

Biological Controls

A rapidly developing technique is the use of natural enemies to control pests. Praying mantises and ladybugs are sold commercially and are used to destroy garden pests. Biological controls also include bacteria and viruses, often specific for a given species of pest. *Bacillus thuringiensis,* sold under such trade names as Dipel, is widely used by home gardeners against cabbage loopers,

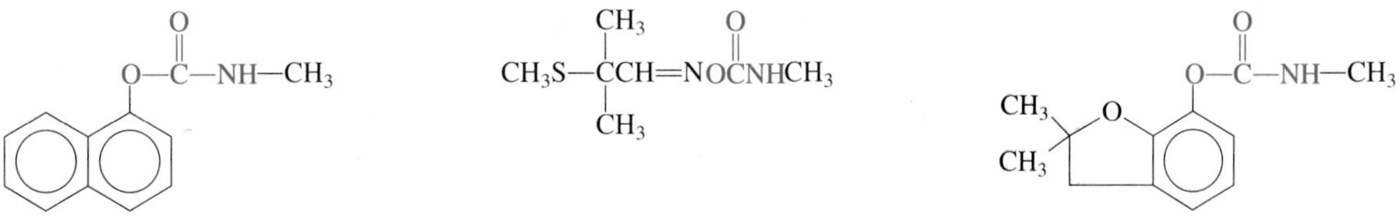

O—C(=O)—NH—CH_3

Carbaryl
(Sevin)

$CH_3S—C(CH_3)_2—CH{=}NOC(=O)NHCH_3$

Aldicarb
(Temik)

CH_3, CH_3, O, O—C(=O)—NH—CH_3

Carbofuran
(Furadan)

Figure 16.31 Three carbamate insecticides.

hornworms, and other moth larvae. Cost limits the agricultural use of this bacterial pesticide.

Viruses have been tested successfully against a number of insect pests. A virus used as a pesticide against the cotton bollworm is cultured in a "factory" that will be able to supply enough virus to treat a substantial portion of the United States cotton crop. Viral pesticides also show promise against grasshoppers and many other insect pests.

Production of viral pesticides is expensive. Nevertheless, this approach to insect control is intriguing because viral agents are highly specific, generally harmless to people and other animals, and completely biodegradable.

Another biological approach is the breeding of insect- and fungus-resistant plants. Such plants, particularly grain bearers, have contributed considerably to increased crop production in recent years. Scientists using genetic engineering have been able to do some remarkable things. They have been able to incorporate a gene from *Bacillus thuringiensis* into a cotton plant. The gene controls production of a protein that disrupts the digestive system of insect pests that eat the plant. Not all research is successful, however. Plant breeders have produced a potato that is insect-resistant, but it had to be taken off the market because it also is toxic to people.

Sterilization

A control technique effective on some insects is sterilization. Large numbers of males can be sterilized by radiation, chemicals, or cross-breeding. Then, enough of them can be released so that they far outnumber the local fertile males. If a female meets with a sterile male, no offspring are produced.

Radiation sterilization has virtually eliminated the screwworm fly, once a serious pest that affected cattle in the southern United States. Tropical fruit flies also have been eradicated by this method in some areas. (It is likely that the 1981 outbreak of Mediterranean fruit flies in California was caused in part by the accidental release of males that had not been sterilized effectively. Extensive spraying with malathion and the stripping of fruit from trees ended the outbreak.) Sterilization involves raising vast numbers of insects, separating the males from the females, then sterilizing and releasing the males. The great expense and limited applicability of this process probably means that it will not become the major method of insect control.

Pheromones: The Sex Trap

One intriguing area of research in insect control is the investigation of **pheromones**. These chemicals are secreted externally by insects and serve the function of marking a trail, sending an alarm, or attracting a mate. Perhaps the most interesting pheromones are the insect **sex attractants**, usually secreted by the female to attract males (Figure 16.32). Chemical research can identify the sex attractants and determine their structures. These compounds then can be synthesized and used to lure males into traps. Alternatively, the attractant can be used in quantities sufficient to confuse and disorient the males, who detect a female in every direction but can't find one to mate with.

Some sex attractants are amazingly simple. Others are complex. The sex attractant for the common housefly is a simple, unsaturated hydrocarbon.

$$CH_3CH_2CH_2CH_2CH_2CH_2CH_2CH_2CH{=}CHCH_2CH_2CH_2CH_2CH_2CH_2CH_2CH_2CH_2CH_2CH_2CH_2CH_3$$

Although this molecule contains 23 carbon atoms (with a double bond between the ninth and tenth), it is fairly easy to synthesize. Quantities of it are now available for testing and development.

Most research in sex attractants is not easy. Some have complicated structures and all are secreted in extremely tiny amounts. For example, a team of U.S. Department of Agriculture researchers had to use the tips of 87,000 female gypsy moths to isolate a minute amount of a powerful sex attractant. Field tests showed the synthetic attractant to be effective in concentrations as low as 1 picogram (1×10^{-12} g) in baited traps. (Other pheromones, too, are effective at extremely low concentrations. A male silkworm moth can detect as few as 40 molecules per second. If a female releases as little as 0.01 μg, she can attract every male within 1 km.)

Figure 16.32 A male gypsy moth detects odor from a female by using his large antennae. [*USDA Photo.*]

The gypsy moth larvae have defoliated great forests, mainly in the northeastern United States, and have now spread over much of the country. Its pheromone has been used mainly in traps to monitor insect populations.

Pheromones for a variety of insects have been isolated and identified. Most insects use two or three compounds. Some cases are quite unusual. An intriguing example is that of the male cotton bollworm moth. An unsaturated aldehyde called Z-9-tetradecenal causes this moth to try to mate with female tobacco budworm moths. Because their genitals don't fit, the two insects become locked in an amorous embrace and eventually die.

Much more research has to be done before pheromones will play a major role in insect control. The method is expensive, and research is painstaking and time-consuming. Workers must be careful not to get the attractants on their clothes. Who wants to be attacked on a warm summer night by a million sex-crazed gypsy moths?

Juvenile Hormones

Another approach to insect control is the use of juvenile hormones. Hormones are the chemical messengers that control many life functions in plants and animals. Minute quantities produce profound physiological changes. In the insect world, **juvenile hormones** control the rate of development of the young. Normally, production of the hormone is shut off at the appropriate time to allow proper maturation to the adult stage.

Chemists have been able to isolate insect juvenile hormones and determine their structure. With knowledge of the structure, they can synthesize the hormone. Application of this hormone to the ponds where mosquitoes breed keeps mosquitoes in the harmless pre-adult stage. Because only adult insects can reproduce, juvenile hormones appear to be a nearly perfect method of mosquito control.

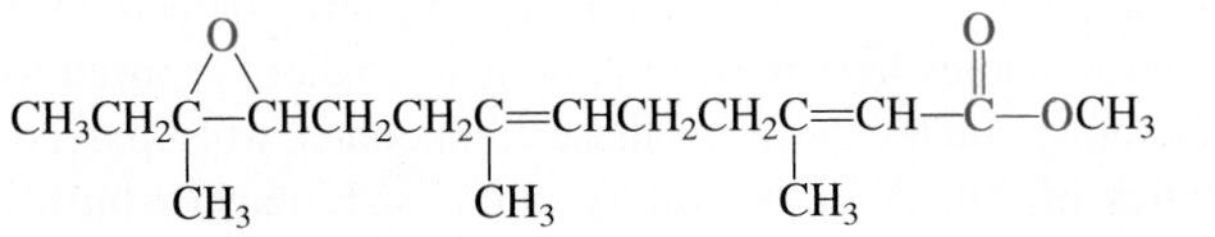

$$CH_3CH_2\underset{\underset{CH_3}{|}}{C}\overset{O}{\frown}CHCH_2CH_2\underset{\underset{CH_3}{|}}{C}{=}CHCH_2CH_2\underset{\underset{CH_3}{|}}{C}{=}CH{-}\overset{\overset{O}{\|}}{C}{-}OCH_3$$

A natural juvenile hormone

Chemists have synthesized analogs of juvenile hormones. One such analog, methoprene, is approved by the U.S. Environmental Protection Agency for use against mosquitoes and fleas.

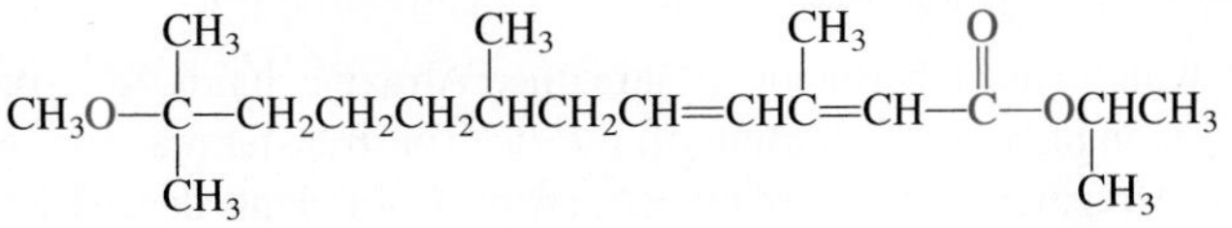

$$CH_3O{-}\underset{\underset{CH_3}{|}}{\overset{\overset{CH_3}{|}}{C}}{-}CH_2CH_2CH_2\overset{\overset{CH_3}{|}}{C}HCH_2CH{=}CH\overset{\overset{CH_3}{|}}{C}{=}CH{-}\overset{\overset{O}{\|}}{C}{-}O\underset{\underset{CH_3}{|}}{C}HCH_3$$

Methoprene

The synthesis of juvenile hormones is difficult (as you could probably guess from their complex structures) and, consequently, quite expensive. The hormones are for use against insects that are pests at the adult stage. Little would be gained by keeping a moth or a butterfly in the caterpillar stage for a longer period of time; caterpillars have voracious appetites and do a great deal of damage to crops.

SECTION 16.15

Herbicides and Defoliants

The United States produces about 700 million kg of pesticides annually. Of these, nearly 400 million kg are **herbicides** (chemicals used to kill weeds).

Herbicides kill weeds. *Defoliants* cause leaves to fall off plants.

2,4-D and 2,4,5-T

Crops produce more abundant harvests when they have no competition from weeds. Removing the weeds by hand is tedious, backbreaking work. Chemical herbicides have been used for a number of years to kill the unwanted plants.

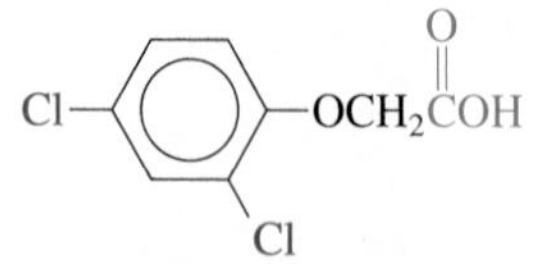

2,4-Dichlorophenoxyacetic acid (2,4-D)

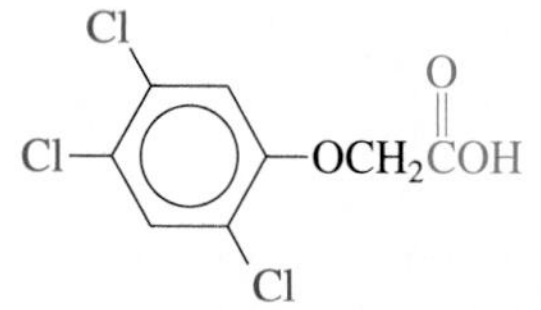

2,4,5-Trichlorophenoxyacetic acid (2,4,5-T)

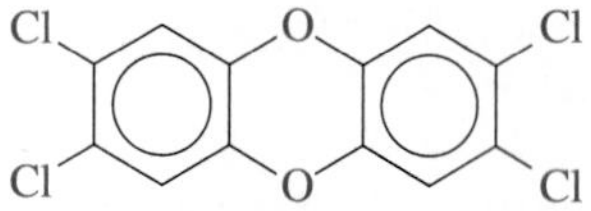

2,3,7,8-Tetrachlorodibenzo-para-dioxin (a "dioxin")

Atrazine

Solutions of copper salts, sulfuric acid, and sodium chlorate ($NaClO_3$) have been used, but it wasn't until the introduction of 2,4-D in 1945 that the use of herbicides became common. Chemically, 2,4-D is 2,4-dichlorophenoxyacetic acid, or one of its derivatives. These chemicals are growth-regulator herbicides and are especially effective against newly emerged, rapidly growing broad-leaved plants.

A relative of 2,4-D, called 2,4,5-T, is especially effective against woody plants; it works by causing the leaves to fall off the plants (**defoliation**). These two chemicals, combined in a formulation called **Agent Orange**, were used extensively in Vietnam to remove enemy cover and to destroy crops that maintained enemy armies. In addition to causing vast ecological damage, 2,4-D and 2,4,5-T were suspected of causing birth defects in Vietnamese children and in babies later born to American soldiers exposed to the herbicides. Laboratory studies show that these compounds, when pure, do not cause abnormalities in fetuses of laboratory animals. Extensive birth defects are caused, though, by contaminants called **dioxins**, once frequently found in the herbicides. These dioxins are also chlorinated compounds. Continuing concern over dioxin contamination led the U.S. Environmental Protection Agency to ban 2,4,5-T in 1985.

Atrazine and Glyphosate

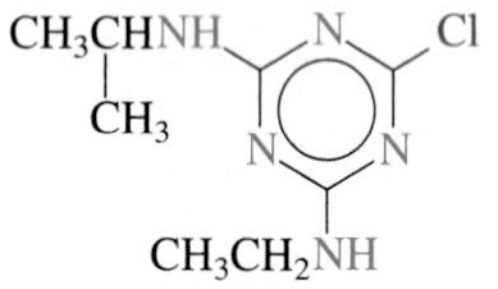

Another widely used herbicide is atrazine. Atrazine binds to a protein in chloroplasts in plant cells, shutting off the electron-transfer reactions of photosynthesis. Atrazine is often used on corn crops. Corn plants deactivate atrazine by removing the chlorine atom. Weeds cannot deactivate the compound and are killed.

Glyphosate, a derivative of the amino acid glycine, is used to control perennial grasses.

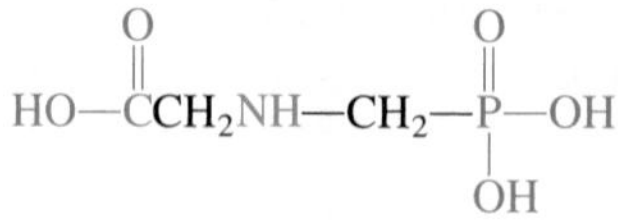

$$HO-\overset{O}{\overset{\|}{C}}CH_2NH-CH_2-\underset{OH}{\underset{|}{\overset{O}{\overset{\|}{P}}}}-OH$$

Glyphosate

It is not selective; it kills all vegetation.

The isopropylamine salt of glyphosate is sold by Monsanto Corporation under the trade name Round-up.

Paraquat: A Preemergent Herbicide

Another type is the **preemergent herbicides**, such as paraquat. This ionic compound is toxic to most plants, but it is rapidly broken down in the soil. Therefore, paraquat can be used to kill weed plants before the crop seedlings emerge. Paraquat inhibits photosynthesis by accepting the electrons that otherwise would reduce carbon dioxide. It has been used in the United States and Latin America to destroy marijuana crops. Some of the sprayed crop has been harvested anyway, and contaminated marijuana has appeared in United States cities. It is feared that paraquat inhaled with marijuana smoke can cause severe lung damage.

$$CH_3{-}^{+}N\langle\bigcirc\rangle{-}\langle\bigcirc\rangle N^{+}{-}CH_3 \quad Cl^- \quad Cl^-$$

Paraquat

Risks and Benefits

There is no doubt that the use of herbicides and defoliants has increased the value of agricultural crops by billions of dollars. Roadsides, vacant lots, and industrial areas also have been kept free of weeds. People who suffer from allergies caused by pollen from ragweed and other plant pests have been relieved, and poison ivy and other toxic weeds have been made less common.

With the exception of problems with dioxin contaminants in 2,4,5-T, herbicides generally have a better safety record than insecticides. We still don't know, though, the ultimate effect on the environment of the long-term use of herbicides.

Chemical defoliants facilitate the harvesting of crops. Calcium cyanamide (CaNCN), one of the earliest defoliants, causes cotton plants to lose their leaves when the bolls are mature. This makes it possible to use mechanical cotton pickers. If the leaves were left on, they would be crushed by the machinery and would stain the cotton.

SECTION 16.16

Alternative Agriculture

Problems with pesticides have led to calls for alternatives to conventional farming, which uses pesticides and fertilizers from chemical plants. This call was heightened in 1989 when the Board of Agriculture of the National Research Council urged farmers to consider other ways to deal with pests and provide plant nutrients. Among the Board's suggestions were crop diversification, integrated pest management (using a mixture of biological controls and synthetic chemicals), disease prevention by careful crop management, and the genetic improvement of crops. Much of the advice is similar to that given to organic farmers and gardeners for generations.

Modern agriculture is also energy intensive. Agriculture uses more petroleum than any other basic industry, accounting for about 13% of all our energy use. This nonrenewable energy is required for the production of fertilizers, pesticides, and farm machinery. Energy also is required to run the machinery used for tilling, harvesting, drying, and transporting the crops, and to process and package the food (Figure 16.33).

One-third the cost of food in the United States goes to transportation. We spend $6 million and use 3.6 million L of fuel each year just to transport broccoli from California to New York.

Organic farmers use manure from farm animals for fertilizer, and they rotate other crops with legumes to restore nitrogen to the soil. They control insects by planting a variety of crops, alternating the use of fields. (A corn pest has a hard time surviving for the year that its home field is planted in soy beans.) Organic farming also is less energy intensive. According to a study by the Center for the Biology of Natural Systems at Washington University,

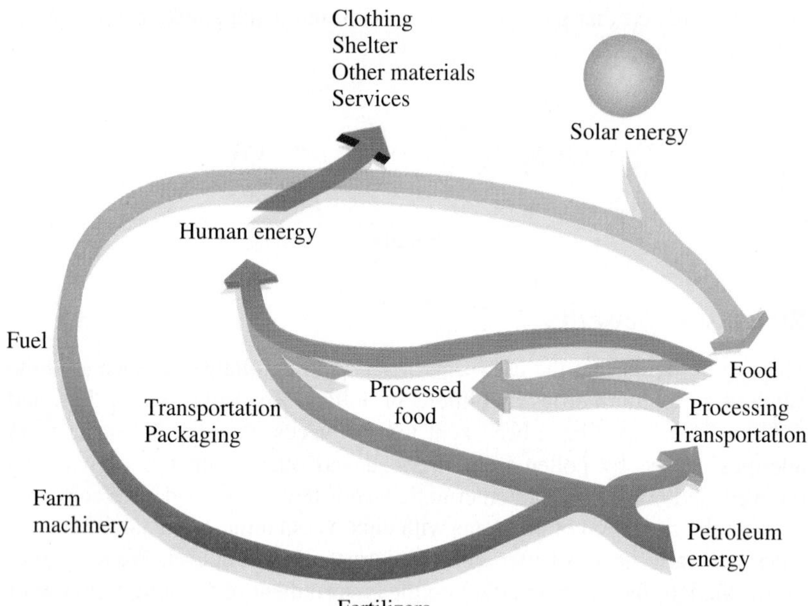

Figure 16.33 Energy flow in modern agriculture.

Organic farming is carried out without the use of synthetic fertilizers or pesticides.

Despite claims to the contrary, organically grown food has not been shown to be more nutritious than food grown with synthetic chemicals. Chemical analysis can detect no difference between food organically grown and food produced using chemical fertilizers. Anyway, organic fertilizers must be broken down to inorganic chemicals before they can be used by plants. Corn doesn't care if nitrate ions are formed by bacteria in manure or by chemists in factories.

comparable conventional farms used 2.3 times as much energy as organic farms. Production on the organic farms was 10% lower, but costs were lower by a comparable percent. Organic farms require 12% more labor than conventional ones, but human labor is a renewable resource, whereas petroleum energy is not.

Conventional agriculture also often results in severe soil erosion and is the source of considerable water pollution. No doubt we should practice organic farming to the limit of our ability to do so. But we should not delude ourselves. Banning synthetic fertilizers and pesticides likely would lead to a drastic drop in food production.

As far as human energy is concerned, United States agriculture is enormously efficient. Each farm worker produces enough food for about 80 people. But that productivity is based on fossil fuels. It is estimated that 10 units of petroleum energy are required to produce 1 unit of food energy. If we consider production per hectare, modern farming is marvelously efficient. If we consider the energy used in relation to the energy produced, it is remarkably inefficient. It should be noted, however, that in an energy-efficient, primitive society, nearly all human energy goes into food production. In modern societies, only about 10% of energy is devoted to producing food. The other 90% is used to provide the materials and services that are so much a part of our civilization. We should try to make our food production more energy efficient, but it is unlikely that we will want to return to a primitive way of life.

Table 16.12 Arithmetic and Geometric Growth Through Ten Periods (Starting with One Unit)

	Growth Period										
	0	1	2	3	4	5	6	7	8	9	10
Arithmetic growth	1	2	3	4	5	6	7	8	9	10	11
Geometric growth	1	2	4	8	16	32	64	128	256	512	1024

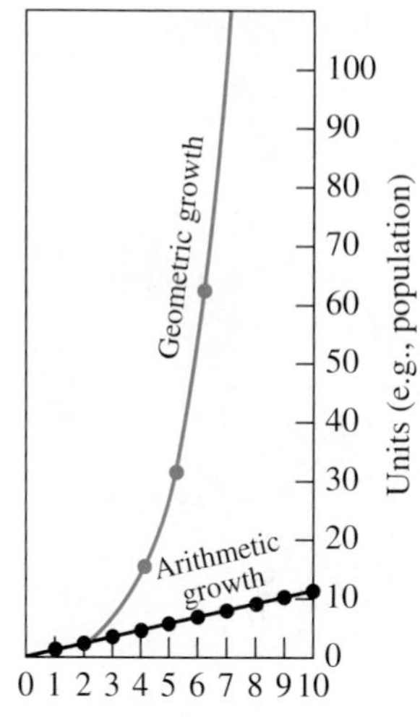

Figure 16.34 Arithmetic and geometric growth through 10 periods (starting with 1 unit).

SECTION 16.17

Some Malthusian Mathematics

In 1830, Thomas Robert Malthus, an English clergyman and political economist, made the statement that population increases faster than the food supply. Unless the birthrate was controlled, he said, poverty and war would have to serve as restrictions on the increase.

Malthus's predictions were based on simple mathematics. Population, he said, grows geometrically, while the food supply increases arithmetically. In **arithmetic growth**, a constant amount is added during each growth period. As an example, consider a cookie-jar savings account. The first week, little Lavinia puts in the 25¢ she received for her birthday. Each week thereafter, she adds 25¢. The growth of the savings is arithmetic; it increases by a constant amount (25¢) each week. At the end of the first week, she will have 25¢; at the end of the second, 50¢; the third, 75¢; the fourth, \$1.00; the fifth, \$1.25; the sixth, \$1.50; and so on.

Now let's consider an example of **geometric growth**, in which the increment increases in size for each growth period. Again, let's use little Lavinia's bank as an example. The first week she puts in 25¢. The second week she puts in another 25¢ to double the amount (to 50¢). The third week she puts in 50¢ to double this amount again. Each week, she puts in an amount equal to what is already there; she doubles the amount in the bank. At the end of the first week, she has 25¢; the second, 50¢; the third, \$1.00; the fourth, \$2.00; the fifth, \$4.00; the sixth, \$8.00; and so on. Before long, little Lavinia will have to start robbing banks to keep up her geometrically growing deposits.

Table 16.12 compares arithmetic and geometric growth through ten growth periods. These data are shown graphically in Figure 16.34. Note that arithmetic growth is slow and steady; geometric growth starts slowly, then shoots up like a rocket.

For a population growing geometrically, we can calculate the **doubling time** from the **Rule of 70**. Simply divide the percent of annual growth into 70. For example, the Earth's population is growing 1.7% per year. If it continues to grow at that rate, the population will double in 41 years.

The Rule of 70 is derived in the same way as the time for the doubling of savings. Divide 70 by the annual interest rate (compounded daily) to get the number of years in which your money will have doubled.

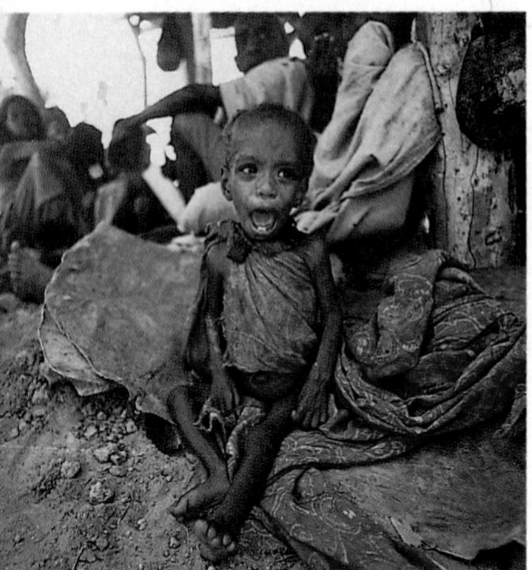
Figure 16.35 In the developed countries, science has thwarted the Malthusian prediction of hunger. But starvation is still a fact of life for many people in developing countries, particularly for those in areas wracked by war. [*J. Frank/courtesy of UN Photos.*]

▶ EXAMPLE 16.5

The population of Mexico City was estimated to be 19.4 million in 1989 and to be growing at a rate of 5.0% a year. If growth continues at the same rate, when will the population have doubled to 38.8 million?

SOLUTION

Divide the annual growth rate into 70.

$$\frac{70}{5.0} = 14$$

The population will have doubled to 38.8 million by the year 2003 if this growth rate continues.

Exercise 16.5

The population of Kenya was estimated to be 26.2 million in 1992 and to be growing at a rate of 3.6% a year. If growth continues at the same rate, when will the population have doubled to 52.4 million?

The population of planet Earth has grown enormously since Malthus's time, reaching 5 billion in 1986. Famine has brought death to millions in Ethiopia and other war-plagued areas (Figure 16.35). But farmers in developed nations have produced enormous surpluses of food, depressing prices and driving many farmers out of business. Even the developing nations have made great strides. In ten years (1971 to 1981) the population of India, the world's second most populous nation, grew by 136 million people. That increase is equal to the population of Brazil, the sixth most populous country. Yet India increased its food production even more rapidly, becoming a net exporter of grain after decades as an importer.

Despite great surpluses of food in some parts of the world, Spaceship Earth still carries 800 million seriously undernourished passengers. Half of these are children under five who will carry the physical and mental scars of this deprivation the rest of their lives.

Scientific developments—fertilizers, pesticides, and new genetic varieties of crop plants—have brought abundance to the developed countries and promise the same for developing nations. It was mainly through the increased application of fertilizers that India increased its food production so dramatically.

World food production over the next generation must equal or surpass that of all previous generations combined to ensure adequate food.

SECTION 16.18

Can We Feed a Hungry World?

The population of planet Earth reached 5 billion in 1986. At a growth rate of 1.7%, it will reach 10 billion in 2027 and 20 billion just 40 years later. Will we be able to feed all those people?

An alliance of science and agriculture has brought an increase in food supplies beyond the imagination of people of a few generations ago. Through the use of irrigation, synthetic fertilizers, pesticides, and improved genetic varieties of plants and animals, most of the people of the world have abundant food.

We may be able to increase food production even more through genetic engineering (Chapter 15). Scientists can design plants that produce more food, that are resistant to disease and insect pests, and that grow well in hostile environments. They can design animals that grow larger and produce more meat and milk. They may some day provide us with food in unimaginable abundance. But the hungry are with us now, and 100 million more come to dinner each year.

Even by quadrupling present food production, we could meet our needs only until 2050. Virtually all the world's available arable land is now under cultivation, and we lose farm land every day to housing, roads, erosion, encroaching deserts, and to the increasing salt content of irrigated soils. It seems unlikely that we can forever keep food production ahead of the high rate of population growth, particularly in developing countries.

We seem to be hooked on a high-energy form of agriculture that uses synthetic fertilizers, pesticides, and herbicides and depends on the power of machinery that burns fossil fuels. Ultimately, the only solution to the problem is population stabilization. Obviously, that will come someday. The only questions are when and how. Population control could come through decreasing the birthrate. This is happening in almost all the developed world and in many developing nations. Other nations still have populations growing at explosive rates that far outstrip their food supply. Birth control isn't the only way to limit populations. Another possibility is an increase in the death rate—through catastrophic war, famine, pestilence, or the poisoning of the environment by the wastes of an ever-expanding population. The ghost of Thomas Malthus haunts us yet.

The current farm population in the United States is less than 5 million. From 1915 to 1945 there were about 30 million people living on farms. Today fewer people produce more food by using fertilizers, insecticides, and herbicides.

Summary

1. The three main classes of foods are *carbohydrates, fats,* and *proteins.*
2. We also need to include in our diets *vitamins,* minerals, fiber, and water.
3. Carbohydrates include sugars, *starches,* and *cellulose.*
4. Dietary fats and oils are mainly triglycerides, esters of fatty acids and glycerol.
5. Solid fats (mostly from animals) are largely saturated; liquid oils (mostly from plants) are usually unsaturated or *polyunsaturated.*
6. Proteins are *amino acid* polymers. Dietary proteins supply all the amino acids for making body proteins and other nitrogen compounds.
7. There are 30 elements essential to life: some structural elements, some macrominerals, and some trace elements.
8. Vitamins are compounds that our bodies need but cannot make. Most are needed as coenzymes.
9. Vitamins B and C are water soluble, whereas A, D, E, and K are all fat soluble.
10. Dietary fiber appears to give protection against colon cancer.

11. Drinking enough water is very important, but it need not be plain water. Most drinks are mainly water.

12. *Food additives* are there to add nutritional enrichment, to inhibit spoilage, to add color or flavor, to add sweetness, to bleach, or to provide texture.

13. Toxic substances found in food are sometimes natural ingredients, but sometimes they are added inadvertently to our food in the form of pesticide residues or animal feed additives.

14. All food originates in plants. Photosynthesis occurring in green plants is the reaction that stores energy from the sun, providing fuel for all living things.

15. *Fertilizers* add the essential elements nitrogen, phosphorus, and potassium to the soil (in the form of ammonium, nitrate, phosphate, and potassium salts).

16. Plants also need calcium, magnesium, and sulfur, as well as a number of *micronutrients*.

17. Various toxic substances are added to farm crops in order to control insects, rodents, and weeds.

18. Although *DDT* saved millions of lives during World War II, its harmful effect on the environment has led to a government ban.

19. Certain organic phosphorus compounds and carbamates are nerve poisons currently used as *insecticides*.

20. *Pheromones (sex attractants), juvenile hormones,* viruses, radiation, and other kinds of insects (enemies) have been used to help control insect populations.

21. *Herbicides* are used to kill weeds, and *defoliants* are used to remove leaves.

22. According to Thomas Malthus, since population increases geometrically and food production arithmetically, food production cannot keep up with population growth.

Key Terms

adipose tissue 16.2
aflatoxins 16.9
Agent Orange 16.15
amino acids 16.2
antioxidants 16.8
arithmetic growth 16.17
aspartame 16.8
biocide 16.13
blood sugar 16.1
carbohydrates 16.1
cellulose 16.1
chain reaction 16.8
cholesterol 16.2
DDT 16.13
Delaney Amendment 16.8
dextrose 16.1
dietary minerals 16.4
defoliation 16.15
dioxins 16.15
doubling time 16.17
enrichment 16.8
essential amino acids 16.3
fat depots 16.2
fats 16.2
fertilizers 16.12
food additives 16.8
free radicals 16.8
fructose 16.1
geometric growth 16.17
glucose 16.1
glycogen 16.1
GRAS list 16.8
herbicide 16.15
humectant 16.8
insecticide 16.13
juvenile hormone 16.14
lactose 16.1
limiting reagent 16.3
lipoprotein 16.2
micronutrients 16.12
organic farming 16.16
pesticides 16.13
pheromones 16.14
polyunsaturated 16.2
preemergent herbicides 16.15
primary plant nutrients 16.12
protein 16.3
Rule of 70 16.17
secondary plant nutrients 16.12
sex attractant 16.14
starch 16.1
starvation 16.7
sucrose 16.1
triglycerides 16.2
vitamins 16.5

Review Questions

1. List the three major types of food.

2. What is the role of carbohydrates in the diet?

3. What is the chemical name of the following sugars?
   a. blood sugar b. table sugar
   c. fruit sugar

4. What is the principal sugar in corn syrup?
5. How is high-fructose corn syrup made?
6. What is the dietary difference in starch and cellulose?
7. What sugar is formed when starch is digested?
8. What are the ultimate products formed when cells metabolize glucose?
9. How much energy is supplied by 1 g of carbohydrates?
10. What proportion of our diet should be carbohydrates?
11. What functions do fats serve?
12. What are polyunsaturated fats?
13. How do animal fats differ from vegetable oils?
14. How much energy is supplied by 1 g of fat?
15. Where is fat stored in the body?
16. What is the maximum percentage of calories in our diet that should come from fats?
17. What is the maximum percentage of calories in our diet that should come from saturated fats?
18. What are essential amino acids?
19. What is a limiting reagent?
20. What is an adequate protein?
21. List some foods that contain adequate proteins.
22. Which essential amino acids are likely to be lacking in corn? In beans?
23. A new bread is made by adding pea flour to wheat flour. Would the bread provide adequate protein? Why or why not?
24. In general, what are the problems associated with a strict vegetarian diet?
25. A diet high in meat products makes less efficient use of the energy originally captured by plants through photosynthesis than a vegetarian diet does. Explain why.
26. Vitamins and minerals are discussed in this chapter. Which are organic and which are inorganic?
27. Indicate a biological function for each of the following.
    a. iodine b. iron
    c. calcium d. phosphorus
28. Which of the following minerals would you expect to find in relatively large amounts in the human body? Explain your reasoning.
    a. Ca b. Cl c. Co d. Mo
    e. Na f. P g. Zn
29. Match the compound with its designation as a vitamin.

Compound	Designation
Ascorbic acid	Vitamin A
Calciferol	Vitamin B_{12}
Cyanocobalamine	Vitamin C
Retinol	Vitamin D
Tocopherol	Vitamin E

30. Which of the following are B vitamins?
    a. folic acid
    b. riboflavin
    c. β-hydroxybutyric acid
    d. thiamine
    e. niacin
31. Identify the vitamin deficiency associated with these diseases.
    a. scurvy
    b. rickets
    c. night blindness
32. In each case, identify the deficiency disease associated with a diet lacking in the indicated vitamin.
    a. vitamin B_1 (thiamine)
    b. niacin
    c. vitamin B_{12} (cyanocobalamine)
33. Identify the following vitamins as water soluble or fat soluble.
    a. vitamin A b. vitamin B_6
    c. vitamin B_{12} d. vitamin C
    e. vitamin K
34. Identify the following vitamins as water soluble or fat soluble.
    a. calciferol b. niacin
    c. riboflavin d. tocopherol
35. Is an excess of a water-soluble vitamin or a fat-soluble vitamin more likely to be dangerous? Why?
36. What is starvation?
37. In fasting, which stores are depleted first, fats or glycogen?
38. What is a food additive?
39. What are the two major categories of food additives? Give an example of each.
40. What is the function of each of the following food additives?
    a. potassium iodide b. vanilla extract
    c. MSG d. sodium nitrite

41. What is the function of each of the following food additives?
a. $FeCO_3$ b. SO_2
c. potassium sorbate

42. What is the purpose of each of the following food additives?
a. BHA
b. FD&C Yellow No. 5
c. saccharin

43. What is the purpose of each of the following food additives?
a. aspartame
b. vitamin D
c. sodium hypophosphite

44. What is enriched bread? Is it equal in nutritional value to bread made from whole grain?

45. What is MSG?

46. What is botulism?

47. What are antioxidants?

48. What is a chain reaction?

49. What vitamin serves as a fat-soluble antioxidant?

50. Name three artificial sweeteners. Which are approved for current use in the United States?

51. What is the chemical nature of aspartame?

52. What is the GRAS list?

53. What are aflatoxins?

54. List some incidental additives that have been found in foods.

55. What is the Delaney Amendment to the Food and Drug Act?

56. What United States government agency regulates the use of food additives?

57. What must be done before a food company can use a new food additive?

58. List five functions of food additives.

59. Where does most of the matter of a growing plant come from?

60. List the three structural elements of a plant.

61. In what form is nitrogen used by plants?

62. How is ammonia made? What are the raw materials for ammonia synthesis?

63. List several ways that nitrogen can be fixed.

64. What is urea? From what is synthetic urea made?

65. What is anhydrous ammonia? What is its use?

66. What is the source of phosphate fertilizers?

67. What is the role of nitrogen in plant nutrition?

68. What is the role of phosphorus in plant nutrition?

69. Why does soil become acidic when potassium ions are absorbed by plants?

70. List the advantages and disadvantages of DDT as an insecticide.

71. Why is DDT especially harmful to birds?

72. Describe how chlorinated hydrocarbons become concentrated in a food chain.

73. What is a narrow spectrum insecticide? Name one.

74. What is a broad spectrum insecticide? Name two.

75. What are pheromones?

76. What are juvenile hormones? How are they used against insect pests?

77. Describe the technique of sterilization as a method for controlling insects. Why is it not more widely used?

78. What are herbicides?

79. What is a defoliant? Why are defoliants used on cotton crops?

80. What is a preemergent herbicide?

81. List the four main features of alternative agriculture.

82. Compare organic farming with conventional farming. Consider energy requirements, labor, profitability, and crop yields.

83. What did Thomas Malthus predict in his famous 1830 statement?

84. Why has Malthus's prediction not come true?

85. What is arithmetic growth? Give an example.

86. What is geometric growth? Give an example.

Problems

Chemical Equations

87. Give the chemical equation for the photosynthesis reaction.

88. Give the chemical equation for the synthesis of ammonia.

Calorie Calculations

89. The label on a can of white beans indicates that each half-cup portion supplies 80 kcal and has 6 g protein, 18 g carbohydrate, and 1 g fat. Calculate the percent of calories from fat.

90. The label on a can of milk substitute indicates that each half-cup serving supplies 150 kcal and has 8 g protein, 12 g carbohydrate, and 8 g fat. Calculate the percent of calories from fat.

Nutrient Calculations

91. Our requirement for protein is about 0.8 g per kilogram of body weight. What is the protein requirement of 50-kg (110-lb) person?

92. Graham crackers are 8% protein. Assume that your Recommended Daily Allowance (RDA) is 60 g of protein. What percent of your RDA for protein would you receive if you ate 150 g of graham crackers?

93. Assume that 1 cup of skim milk contains 225 g of milk. Skim milk is 3.6% protein. How many grams of protein are there in 4 cups of skim milk?

94. Campbell's Chunky Vegetable Beef Soup is available in regular and low-sodium versions. The regular has 935 mg of Na^+ and the low-sodium has 90 mg of Na^+ per 10.75-oz portion. What percent of the recommended daily maximum of 3300 mg of Na^+ would you get from one portion of each?

95. One cup of whole milk supplies 34 mg of cholesterol. The American Heart Association recommends a maximum of 300 mg of cholesterol per day. What percentage of the maximum recommendation is met by 1 cup of whole milk?

Toxicity

96. How much DDT would it take to kill a person weighing 60 kg if the lethal dose was 0.5 g per kilogram of body weight?

97. How much parathion would it take to kill the person in problem 96 if the lethal dose was 5 mg per kilogram of body weight?

Growth

98. You start a stamp collection, with a plan to buy two stamps a week. How many weeks would it take to acquire 100 stamps? Is the collection growing arithmetically or geometrically?

99. You raise rabbits. Starting with two, you have four at the end of the first month, eight at the end of the second month, and so on; the rabbit population doubles each month. How many rabbits would you have at the end of 12 months? Is the rabbit population growing arithmetically or geometrically?

100. The populations of some Latin American, African, and Asian nations are growing at 3.5% a year. At that rate, how many years of growth will it take for those populations to double?

101. The population of China, now 1.2 billion, is growing at an annual rate of 1.5%. At that rate, how many years of growth will it take for the population to double?

Additional Problems

102. Structural formulas for two vitamins are shown. Identify each of them as water soluble or fat soluble.

a. $HOCH_2C(CH_3)_2{-}CH(OH){-}C(=O){-}NHCH_2CH_2COOH$

b. $(\text{ring})\text{–}CH{=}CHC(CH_3){=}CHCH{=}CHC(CH_3){=}CHCH_2OH$ (ring bearing CH_3, CH_3, and CH_3 substituents)

103. Identify the functional group in the sex attractant for the common housefly (Section 16.14).

104. Identify the functional groups in the juvenile hormone molecule (Section 16.14).

105. Identify the functional groups in the methoprene molecule (Section 16.14).

▸ Projects

106. Consult a recent issue of the *FDA Consumer* and make a list of incidental food additives that are reported on in that issue. What are the most common causes of contamination?

107. Examine the label on a sample of each of the following.
a. a can of soft drink
b. a can of beer
c. a dried soup mix
d. a can of soup
e. a can of fruit drink
f. a cake mix
Make a list of the food additives in each. Try to determine the function of each additive.

108. Maraschino cherries are bleached (with sulfur dioxide) and then dyed with an organic food coloring. Should the dye be banned? Justify your conclusion.

109. When Russian refugee Alexander Jourjine escaped from the U.S.S.R. by hiking for 23 days across Finland to Sweden, he carried 9 lb of lard and cheese, several loaves of dried black bread, tea, sugar, and 12 chocolate bars. His diet was supplemented with berries picked along the way. Defend or criticize his choice of each food item.

110. Examine the label on a package of yard or garden pesticide. List its trade name and active ingredients.

111. Examine the label of a bag of fertilizer. What is its composition?

112. Use a reference such as *The Merck Index* to look up each of these pesticides.
What is the toxicity of each?
a. methoxychlor
b. aldrin
c. heptachlor

113. If you grew a garden, would you use chemical fertilizers? Pesticides? Why or why not?

▸ References and Readings

1. Adcock, Louis H. "Fredrick Christian Accum: A Chemist for All Seasonings." *Chemistry,* May 1973, pp. 16–18. Accum exposed hazards in the food industry—150 years ago.
2. Barrons, Keith C. *Are Pesticides Really Necessary?* Chicago: Regnery Gateway, 1981.
3. Beardsley, Tim. "The A Team." *Scientific American,* February 1991, pp. 16–19. Discusses how vitamin A and related compounds act at the cellular level.
4. Brody, Jane. *Jane Brody's Nutrition Book.* New York: Bantam Books, 1981. An excellent general source.
5. Committee on Chemistry and Public Affairs. *Chemistry and the Food System.* Washington, DC: American Chemical Society, 1980.
6. Crosson, Pierre R., and Norman J. Rosenberg. "Strategies for Agriculture." *Scientific American,* September 1989, pp. 128–135.
7. Eaton, S. Boyd, and Marjorie Shostak. "Fat Tooth Blues." *Natural History,* July 1986, pp. 6–15. What fats do to us.
8. Ehrlich, Paul R., Anne H. Ehrlich, and John P. Holdren. *Ecoscience: Population, Resources, Environment.* San Francisco: W. H. Freeman, 1977.
9. Feigl, Dorothy M., John W. Hill, and Erwin Boschmann. *Foundations of Chemistry.* 3rd ed. New York: Macmillan, 1991. Chapters 14–18 treat many of the topics in this chapter in more detail.
10. Frieden, Earl. "New Perspectives on the Essential Trace Elements." *Journal of Chemical Education,* November 1985, pp. 917–923.

11. Guild, Walter, Jr. ''The Theory of Sweet Taste.'' *Journal of Chemical Education,* March 1972, pp. 171–173.

12. Hassell, Kenneth A. *The Biochemistry and Uses of Pesticides,* 2nd ed. New York: VCH Publishers, 1991.

13. Heylin, Michael. ''Pesticides: Costs Versus Benefits.'' *Chemical and Engineering News,* 7 January 1991, p. 5. This editorial introduces and summarizes a news forum (pp. 27–55 of the same issue) on the risk assessment of pesticides.

14. Hileman, Bette. ''Alternative Agriculture.'' *Chemical and Engineering News,* 5 March 1990, pp. 26–40.

15. Holmes, Alan. ''Role of Food Additives.'' *Chemistry and Industry,* 6 February 1984, pp. 104–107.

16. Hui, Y. H. *Encyclopedia of Food Science and Technology.* New York: Wiley, 1991.

17. Keyfitz, Nathan. ''The Growing Human Population.'' *Scientific American,* September 1989, pp. 118–126.

18. Kourik, Robert. ''Combatting Household Pests Without Chemical Warfare.'' *Garbage,* March–April 1990, pp. 22–29.

19. Kourik, Robert. ''Noxious Naturals.'' *Garbage,* September–October 1990, pp. 54–57. Plants make their own pesticides.

20. Larkin, Tim. ''Herbs Are Often More Toxic Than Magical.'' *FDA Consumer,* October 1983, pp. 5–11.

21. LeBaron, Homer M., and Janis E. McFarland. ''Resistance to Herbicides.'' *ChemTech,* August 1990, pp. 508–511.

22. Lisansky, Stephen. ''Biopesticides: The Next Revolution?'' *Chemistry and Industry,* 7 August 1989, pp. 478–482.

23. McDermott, Jeanne. ''Some Heartland Farmers Just Say No to Chemicals.'' *Smithsonian,* April 1990, pp. 114–127.

24. McKone, Harold T. ''Copper in the Candy, Chromium in the Custard.'' *Today's Chemist,* October 1990, pp. 22–25. A history of food colors before the synthetic organic dyes.

25. Raloff, J. ''Colon Cancer: Clues to Fiber's Benefits.'' *Science News,* 4 August 1990, p. 69.

26. Reganold, John P., Robert I. Papendick, and James F. Parr. ''Sustainable Agriculture,'' *Scientific American,* June 1990, pp. 112–120.

27. Robinson, Corinne H., Marilyn R. Lawler, Wanda Chenoweth, and Ann Elizabeth Garwick. *Normal and Therapeutic Nutrition.* New York: Macmillan, 1990. Good reference for anyone interested in nutrition.

28. Rodricks, Joseph V. ''Aflatoxins: Hazards from Nature.'' *FDA Consumer,* May 1978, pp. 16–19.

29. ''Salt and Your Health.'' *Consumer Reports,* January 1984, pp. 17–22.

30. Sheldon, Richard P. ''Phosphate Rock.'' *Scientific American,* June 1982, pp. 45–51.

31. Silberner, Joanne. ''Food: What's in a Label?'' *U.S. News and World Report,* 18 June 1990, pp. 56–60. Part of a special ''Health Guide'' issue.

32. Silverstein, Robert M. ''Pheromones: Background and Potential for Use in Insect Pest Control.'' *Science,* 18 September 1981, pp. 1326–1332.

33. Simpson, Lance L. ''Deadly Botulism.'' *Natural History,* January 1980, pp. 12–24.

34. Skinner, Karen Joy. ''Nitrogen Fixation.'' *Chemical and Engineering News,* 4 October 1976, pp. 22–35.

35. Torey, John G. ''The Development of a Plant Biotechnology.'' *American Scientist,* July–August 1985, pp. 354–363.

36. Worthy, Ward. ''Evidence Mounts for Dietary Soluble Fiber Benefits.'' *Chemical and Engineering News,* 28 May 1990, pp. 23–24.

37. Ziporyn, Terra. ''The Food and Drug Administration: How 'Those Regulations' Came to Be.'' *Journal of the American Medical Association,* 18 October 1985, pp. 2037–2046.

17 Household Chemicals

Helps and Hazards

Car wax, toothpaste, bleach, detergent,

Lotion, paint, shampoo . . .

There are chemicals for almost

Everything we do.

Suppose that someone offered you a job in a place where poisonous chemicals were used every day. Toxic vapors and harmful dusts were a common hazard. Corrosive acids and alkalies were often used. Highly flammable liquids and vapors nearby posed a fire hazard. Would you take the job? You probably would. That was a typical American home we were describing, and the potential employer might have been one of your parents.

People tend to be very careful in a chemistry laboratory because they know the place is dangerous. The shelves are stocked with various kinds of chemicals, and those chemicals could be toxic or hazardous. However, people are much less cautious in their own homes, even though some of the chemicals on the shelves are just as toxic and hazardous as those in a laboratory.

There are perhaps half a million chemical products available for use in the American home. They include waxes, wax removers, paints, paint removers, bleaches, insecticides, rodenticides, spot removers, solvents, disinfectants, detergents, toothpaste, shampoo, perfumes, lotions, shaving cream, deodorants, hair spray, and many other products (Figure 17.1). Some of these products are quite harmless, but others contain corrosive or toxic chemicals, or they may present fire hazards. Some cause environmental problems when they are used or discarded.

The typical U.S. supermarket has 5000 different consumer products.

Extensive use of chemicals in the home has led to an increasing number of accidents. The chemistry laboratory is probably a safer place than many homes. In the laboratory, chemicals usually are used under carefully controlled conditions. By contrast, studies have shown that chemicals often are used in the home without regard to the directions or the precautions given on their labels. Indeed, all too frequently the labels aren't read at all! It is this *misuse* of household chemicals that often ends in tragedy.

We have already discussed the chemical composition of our food and the chemicals that are used in producing it (Chapter 16). Some of those agricultural chemicals are also used around the home, especially in the yard and garden. We have discussed polymers used in our clothing and home furnishings (Chapter 10) and fuels used in our furnaces and automobiles

The household chemicals that are sold in greatest volume are cleaning products—soaps, detergents, and various kinds of cleaning mixtures. [*Prisma/Westlight.*]

Figure 17.1 A modern home is stocked with a variety of chemical products. [*Courtesy of Vista Chemical Company.*]

(Chapter 14). We have even discussed the chemistry of our own bodies (Chapter 15). But there are many other chemicals around the house—cleaning materials, personal care products, and a miscellaneous assortment of other things. In this chapter we look at some of these various household chemicals. Let's begin with the ones that make up the largest volume—cleaning compounds.

SECTION 17.1

A History of Cleaning

In primitive societies, even today, clothes are cleaned by beating them with rocks in the nearest stream. Sometimes plants, such as the soapworts of Europe or the soapberries of tropical America, are used as cleansing agents. The leaves of the soapwort and soapberries contain **saponins**, chemical compounds that produce a soapy lather. These saponins were probably the first detergents.

Ashes of plants contain potassium carbonate (K_2CO_3) and sodium carbonate (Na_2CO_3). The carbonate ion, present in both of these compounds, reacts with water to form an alkaline solution.

$$\underset{\substack{\text{Carbonate}\\ \text{ion}}}{CO_3^{2-}} + H_2O \longrightarrow \underset{\substack{\text{Bicarbonate}\\ \text{ion}}}{HCO_3^-} + \underset{\substack{\text{Hydroxide}\\ \text{ion}\\ \text{(basic)}}}{OH^-}$$

The basic solution has detergent properties. These alkaline plant ashes were used as cleansing agents by the Babylonians at least 4000 years ago. Europeans were using plant ashes to wash their clothes as recently as 100 years ago. Sodium carbonate is still sold today as washing soda.

The Romans, with their great public baths, probably did not use any sort of soap. They covered their bodies with oil, worked up a sweat in a steam bath, and then had the oil wiped off by a slave. They finished by taking a dip in a pool of fresh water, And the slaves? They probably didn't bathe at all.

The history of cleanliness of body and clothes is rather spotty. During the Middle Ages (about 450 to 1450) cleanliness of body was prized if not always attained. In twelfth-century Paris, with a population of about 100,000, there were many public bathhouses. The Renaissance, which began in the fourteenth century and extended to the seventeenth, was noted for a revival of learning and art. However, it was not noted for cleanliness. Queen Elizabeth I of England (1558–1603) bathed once a month, a habit that caused many to think her peculiar for her fastidiousness. Queen Isabella of Castille (1474–1504), who supported Columbus's 1492 voyage to the New World, is reported to have bathed only twice in her life. People did sometimes use perfume to mask body odors, and wore clean outer clothes to hide the filth of underclothes. Though soap was known, it was used as a medicine—when it was used at all. The discovery of disease-causing microorganisms and subsequent public health practices brought about increased interest in cleanliness by the late eighteenth century. Soap was in common use by the middle of the nineteenth century.

SECTION 17.2

Soap: Fat + Lye

The first written record of soap is found in the writings of Pliny the Elder, the Roman who described the Phoenicians' synthesis of soap by using goat tallow and ashes. By the second century A.D., sodium carbonate (produced by the evaporation of alkaline water) was heated with lime (from limestone or seashells) to produce sodium hydroxide (lye).

$$Na_2CO_3 + Ca(OH)_2 \longrightarrow 2\,NaOH + CaCO_3$$

The sodium hydroxide was heated with animal fats or vegetable oils to produce soap (Figure 17.2). [Note that **soap** is a salt of a long-chain carboxylic acid (Chapter 9).] The American pioneers made soap in much the same manner. Lye was added to animal fat in a huge iron kettle. The mixture was cooked over a wood fire for several hours. The soap rose to the surface and upon cooling, solidified. The glycerol remained as a liquid on the bottom of the pot. Both the glycerol and the soap often contained unreacted alkali, which ate away the skin. Grandma's lye soap is not just a myth!

In modern commercial soapmaking, the fats and oils often are hydrolyzed with superheated steam. The fatty acids then are neutralized to make soap. Toilet soaps usually contain a number of additives such as dyes, perfumes, creams, and oils. Scouring soaps contain abrasives such as silica, pumice, and

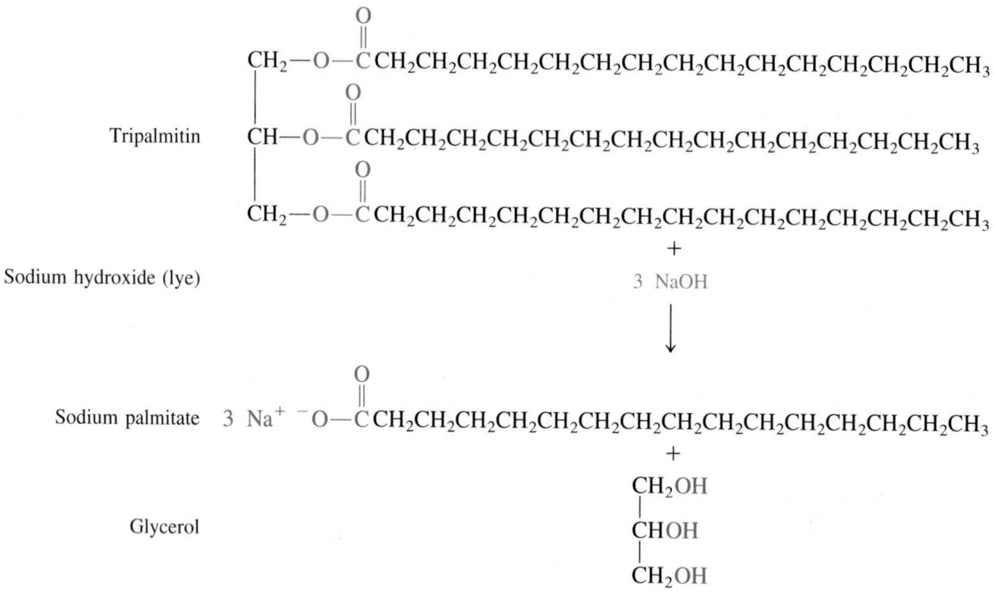

$$\begin{array}{ll} & CH_2-O-\overset{O}{\overset{\|}{C}}CH_2CH_2CH_2CH_2CH_2CH_2CH_2CH_2CH_2CH_2CH_2CH_2CH_2CH_2CH_3 \\ \text{Tripalmitin} & CH-O-\overset{O}{\overset{\|}{C}}CH_2CH_2CH_2CH_2CH_2CH_2CH_2CH_2CH_2CH_2CH_2CH_2CH_2CH_2CH_3 \\ & CH_2-O-\overset{O}{\overset{\|}{C}}CH_2CH_2CH_2CH_2CH_2CH_2CH_2CH_2CH_2CH_2CH_2CH_2CH_2CH_2CH_3 \\ & + \\ \text{Sodium hydroxide (lye)} & 3\ NaOH \\ & \downarrow \\ \text{Sodium palmitate} & 3\ Na^+\ {}^-O-\overset{O}{\overset{\|}{C}}CH_2CH_2CH_2CH_2CH_2CH_2CH_2CH_2CH_2CH_2CH_2CH_2CH_2CH_2CH_3 \\ & + \\ & CH_2OH \\ \text{Glycerol} & CHOH \\ & CH_2OH \end{array}$$

Figure 17.2 Soap is made by the reaction of animal fat or vegetable oil with sodium hydroxide. Animal fats yield hard soaps. Vegetable oils, with unsaturated carbon chains, produce soft soaps. Coconut oils, with shorter carbon chains, yield soaps that are more soluble in water.

oatmeal(!). Many soaps claim to have deodorant action, but few have any active deodorant other than soap itself. Some soaps have air blown in before they solidify to lower their density so that they float. Some bath bars contain synthetic detergents. Their action is similar to that of soap.

$CH_3(CH_2)_{14}COO^-\ K^+$
Potassium palmitate

$CH_3(CH_2)_{14}COO^-$
$(HOCH_2CH_2)_3\overset{+}{N}H$
Triethanolammonium palmitate

Potassium soaps (see margin) are softer than sodium soaps, and they produce a finer lather. They are used alone, or in combination with sodium soaps, in liquid soaps and shaving creams. Soaps also are made by reacting fatty acids with triethanolamine. These substances are used in shampoos and other cosmetics.

How Soap Works

Dirt and grime usually adhere to skin, clothing, and other surfaces, because they are combined with greases and oils—body oils, cooking fats, lubricating greases, and a variety of similar substances—that act a little like sticky glues. Since oils are not miscible with water, washing with water alone does little good.

A soap molecule has a dual nature. One end is ionic and dissolves in water. The other end is like a hydrocarbon and dissolves in oils (Figure 17.3). If we represent the ionic end of the molecule as a circle and the hydrocarbon end as a zigzag line, we can illustrate the cleansing action of soap schematically (Figure 17.4). The hydrocarbon ''tails'' stick into the oil. The ionic ''heads'' remain in the aqueous phase. In this manner, the oil is broken into tiny drop-

$$CH_3CH_2CH_2CH_2CH_2CH_2CH_2CH_2CH_2CH_2CH_2CH_2CH_2CH_2CH_2\overset{\overset{\displaystyle O}{\|}}{C}{-}O^-\ Na^+$$

Hydrocarbon end (dissolves in oils) — Ionic end (dissolves in water)

(a)

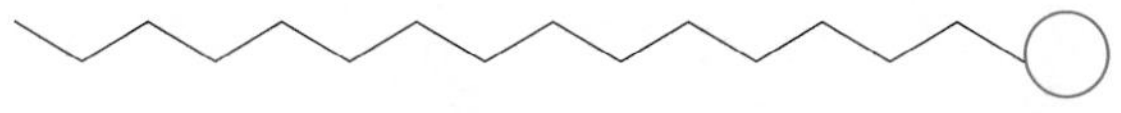

Hydrocarbon "tail" — Polar "head"

(b)

Figure 17.3 Sodium palmitate, a soap. (a) Structural formula. (b) A schematic representation.

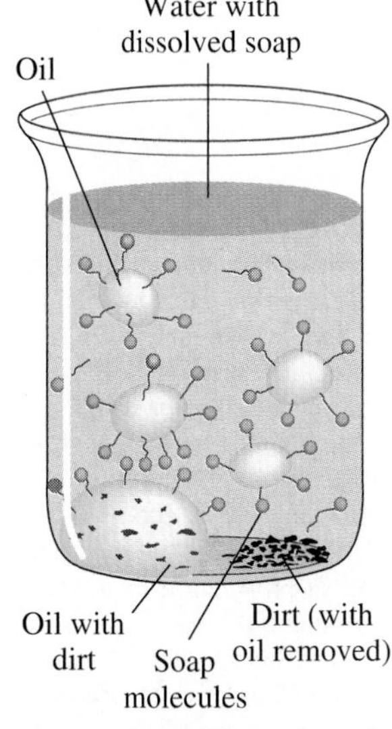

Figure 17.4 The action of soap in removing dirt.

lets and dispersed throughout the solution. The droplets don't coalesce because of the repulsion of the charged groups (the carboxyl anions) on their surfaces. The oil and water form an emulsion, with soap acting as the emulsifying agent. With the oil no longer "gluing" it to the surface, the dirt can be removed easily.

Disadvantages of Soap

For cleaning clothes and for many other purposes, soap has been largely replaced by synthetic detergents because soaps have two rather serious shortcomings. One of these is that, in acidic solutions, soaps are converted to free fatty acids (Figure 17.5a). The fatty acids, unlike soap, don't have an ionic end so they do not exhibit any detergent action. What's more, these fatty acids are insoluble in water and separate as a greasy scum.

$$\underset{\text{A soap}}{CH_3CH_2CH_2CH_2CH_2CH_2CH_2CH_2CH_2CH_2CH_2COO^-\ Na^+} + \underset{\text{An acid}}{H^+} \longrightarrow$$

$$\underset{\text{A fatty acid}}{CH_3CH_2CH_2CH_2CH_2CH_2CH_2CH_2CH_2CH_2CH_2COOH} + Na^+$$

(a)

$$2\ \underset{\text{Soap anion}}{CH_3CH_2CH_2CH_2CH_2CH_2CH_2CH_2CH_2CH_2CH_2COO^-} + Ca^{2+} \longrightarrow$$

$$\underset{\text{Bathtub ring (insoluble)}}{(CH_3CH_2CH_2CH_2CH_2CH_2CH_2CH_2CH_2CH_2CH_2COO^-)_2\ Ca^{2+}}$$

(b)

Figure 17.5 Soap suffers two disadvantages. (a) Acids convert soap anions to fatty acids. (b) Hard-water ions such as Ca^{2+} precipitate soap as insoluble curds. Neither the fatty acids nor the curds have detergent action.

Figure 17.6 Sudsing quality of hard water versus soft water. *From left:* Detergent in hard water, soap in hard water, detergent in soft water, and soap in soft water. Note that the sudsing of detergent differs little in hard and soft waters. Note also the absence of sudsing and the formation of insoluble material when soap is used in hard water. [*Richard Megna/Fundamental Photographs.*]

The second, and more serious, disadvantage of soap is that it doesn't work well in hard water (Figure 17.6). Hard water is water that contains certain metal ions, particularly magnesium, calcium, and iron ions. The soap anions react with these metal ions to form greasy, insoluble curds (Figure 17.5b). These deposits make up the familiar bathtub ring. They leave freshly washed hair sticky and are responsible for "tattletale gray" in the family wash.

SECTION 17.3

Water Softeners

To aid the action of soaps, a variety of water-softening agents and devices have been developed. An effective water softener is washing soda, sodium carbonate ($Na_2CO_3 \cdot 10H_2O$). It makes the water basic (preventing the precipitation of fatty acids) and removes the hard water ions calcium and magnesium. These jobs are performed by the carbonate ion. The ion reacts with water to raise the pH (that is, to make the solution more basic).

$$CO_3^{2-} + H_2O \longrightarrow HCO_3^- + OH^-$$

The carbonate ion also reacts with the ions that cause hard water and removes the ions as insoluble salts.

$$Mg^{2+} + CO_3^{2-} \longrightarrow MgCO_3$$

$$Ca^{2+} + CO_3^{2-} \longrightarrow CaCO_3$$

Trisodium phosphate (Na_3PO_4) is another water-softening agent. Like washing soda, it makes the wash basic and precipitates calcium and magnesium ions.

$$PO_4^{3-} + H_2O \longrightarrow HPO_4^{2-} + OH^-$$

$$2\,PO_4^{3-} + 3\,Mg^{2+} \longrightarrow Mg_3(PO_4)_2$$

In addition, phosphates seem to aid in the cleaning process in some other way that is not yet well understood.

Water-softening tanks are also available for use in homes and businesses. These tanks contain an insoluble polymeric material that attracts and holds the calcium, magnesium, and iron ions to its surface, thus softening the water (Figure 17.7). After a period of use, the polymer becomes saturated and must be discarded or regenerated.

Before leaving the subject of soap, let's mention that soap has some advantages. It is an excellent cleanser in soft water, it is relatively nontoxic, it is

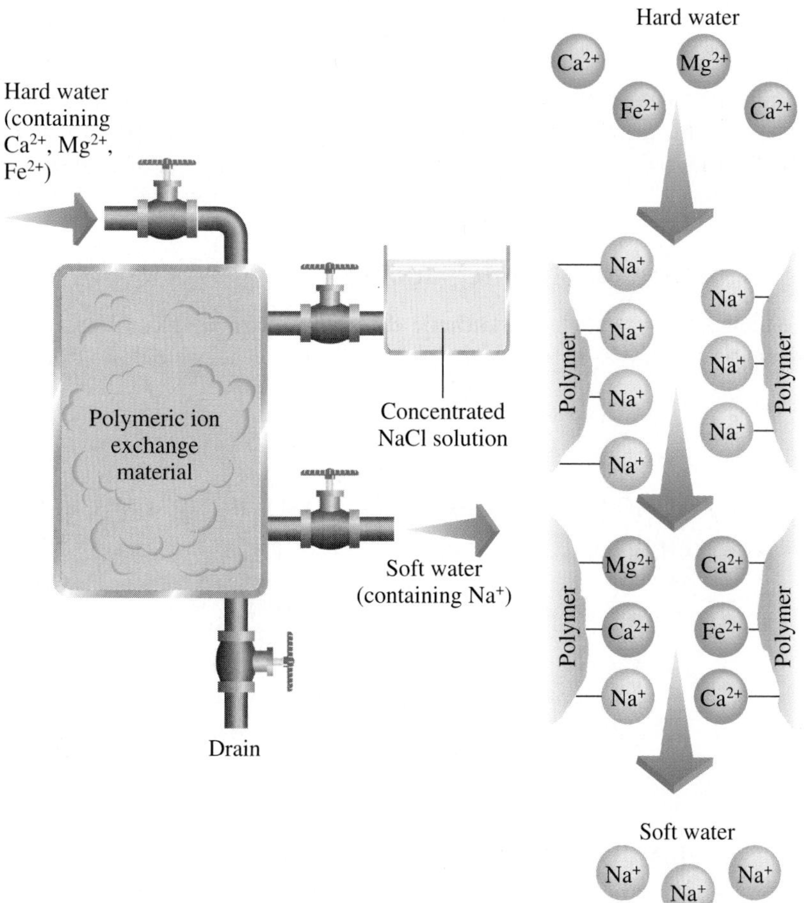

Figure 17.7 One type of water-softening tank contains a polymeric material on which hard-water ions are exchanged for sodium ions. When the material becomes saturated with calcium, magnesium, and/or iron ions, it is regenerated by flushing with a saturated salt solution.

derived from renewable resources (animal fats and vegetable oils), and it is biodegradable.

SECTION 17.4

Synthetic Detergents

A second technological approach to the problems with soap was to develop a new synthetic detergent. The molecules of the synthetic detergents were enough like those of soap to have the same cleaning action, but different enough to resist the effects of acids and hard water. Sodium lauryl sulfate is typical of the first (but fairly expensive) synthetic detergents.

$$CH_3CH_2CH_2CH_2CH_2CH_2CH_2CH_2CH_2CH_2CH_2CH_2O\overset{\overset{O}{\|}}{\underset{\underset{O}{\|}}{S}}O^- \ Na^+$$

Sodium lauryl sulfate
(sodium dodecyl sulfate)

These first detergents were derived from fats by reduction with hydrogen, followed by reaction with sulfuric acid, then neutralization (Figure 17.8). Sodium lauryl sulfate is still used in toothpastes, shampoos, and other cosmetics.

ABS Detergents: Nondegradable

Within a few years, cheap **synthetic detergents** were produced from petroleum products. Alkylbenzenesulfonate (ABS) detergents were made from the

(a) *Reduction:*

$$\begin{array}{l} CH_2OCO(CH_2)_{10}CH_3 \\ | \\ CHOCO(CH_2)_{10}CH_3 \\ | \\ CH_2OCO(CH_2)_{10}CH_3 \end{array} + 6\,H_2 \xrightarrow{\text{Ni (catalyst)}} 3\,CH_3(CH_2)_{10}CH_2OH + \begin{array}{l} CH_2OH \\ | \\ CHOH \\ | \\ CH_2OH \end{array}$$

Trilaurin — Lauryl alcohol (Dodecyl alcohol) — Glycerol

(b) *Reaction with sulfuric acid:* $CH_3(CH_2)_{10}CH_2OH + H_2SO_4 \longrightarrow CH_3(CH_2)_{10}CH_2OSO_3H + H_2O$

(c) *Neutralization:* $CH_3(CH_2)_{10}CH_2OSO_3H + NaOH \longrightarrow CH_3(CH_2)_{10}CH_2OSO_3^-\ Na^+ + H_2O$

Figure 17.8 Early synthetic detergents were made from fats by a three-step process.

Figure 17.9 Foaming rivers were quite common during early 1960s. The problem was solved in the United States by the development of biodegradable detergents. This photograph shows detergent foam on the Bogota River in Columbia. [*Viviane Holbrooke, courtesy of UN Photos.*]

alkene propylene ($CH_3CH{=}CH_2$), benzene, sulfuric acid, and a base (usually sodium carbonate).

$$CH_3\underset{\displaystyle CH_3}{\underset{|}{C}}HCH_2\underset{\displaystyle CH_3}{\underset{|}{C}}HCH_2\underset{\displaystyle CH_3}{\underset{|}{C}}HCH_2\underset{\displaystyle CH_3}{\underset{|}{C}}H{-}C_6H_4{-}SO_3^-\ Na^+$$

An ABS detergent

Sales of ABS detergents soared. For a decade or more, nearly everyone was happy. Then suds began to appear in sewage treatment plants. Foam piled high in the rivers (Figure 17.9). People in some areas even got a head of foam on their drinking water (Figure 17.10). It was found that the branched-chain structure of ABS molecules was not readily broken down by microorganisms in the sewage treatment plants. The whole supply of groundwater was threatened. Public outcries caused laws to be passed and industries to change their processes. Biodegradable detergents were quickly put on the market, and nondegradable detergents were banned.

Figure 17.10 A glass of suds. [*Courtesy of Bergwall Productions, New York.*]

LAS Detergents: Biodegradable

The degradable detergents (called linear alkylsulfonates, LAS) have linear chains of carbon atoms. Microorganisms can break down LAS molecules by

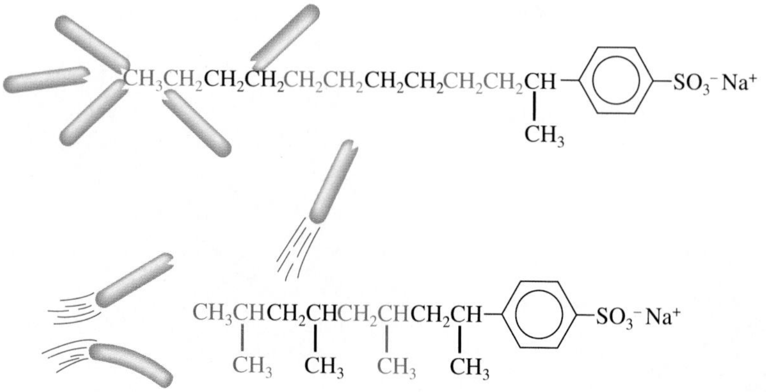

$CH_3CH_2CH_2CH_2CH_2CH_2CH_2CH_2CH_2CH_2CH(CH_3)\text{—}C_6H_4\text{—}SO_3^-\ Na^+$

$CH_3CH(CH_3)CH_2CH(CH_3)CH_2CH(CH_3)CH_2CH(CH_3)\text{—}C_6H_4\text{—}SO_3^-\ Na^+$

Figure 17.11 Microorganisms can readily digest LAS molecules, but it takes them much longer to break down the branched chains of ABS molecules.

producing enzymes that degrade the molecule two (and only two) carbon atoms at a time (Figures 17.11 and 17.12). The branched chain blocks this enzyme action, preventing the degradation of ABS molecules. Thus, technology has solved the problem of foaming rivers.

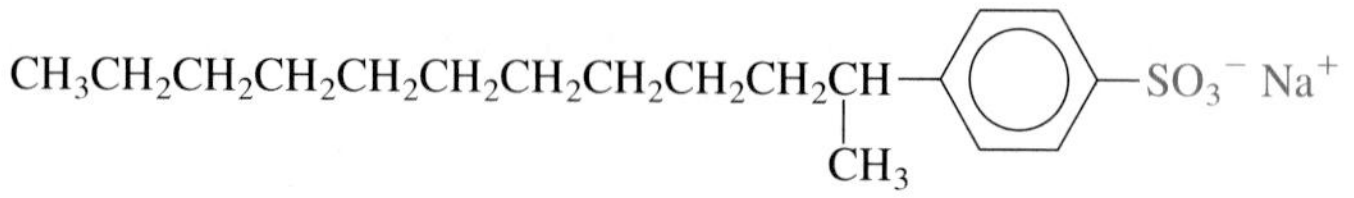

$$CH_3CH_2CH_2CH_2CH_2CH_2CH_2CH_2CH_2CH_2CH(CH_3)\text{—}C_6H_4\text{—}SO_3^-\ Na^+$$

An LAS detergent

It should be pointed out that just because something is degraded, it doesn't simply disappear. Matter is conserved. Everything has to go somewhere. A completely degraded LAS molecule winds up as carbon dioxide, water, and sulfate (SO_4^{2-}). Barry Commoner (reference 5) contends that LAS degradation leads to phenol, a toxic material, and that degradable detergents are more

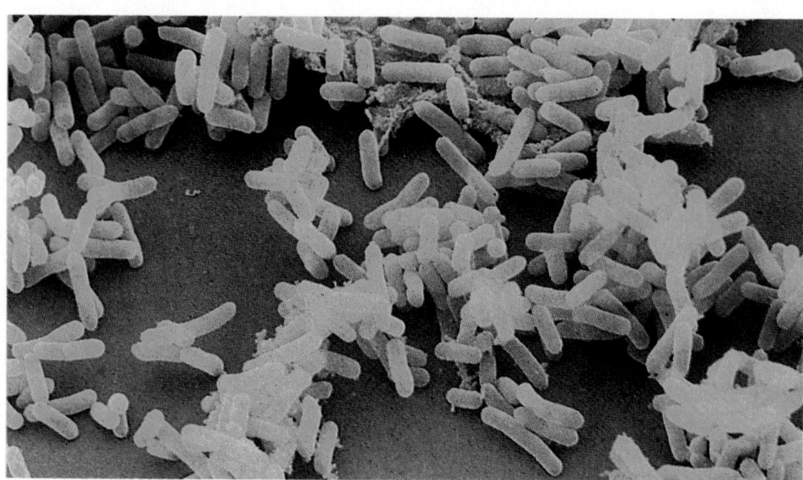

Figure 17.12 Microorganisms, such as the flagellated *Escherichia coli* shown here (magnified 42,500 times by scanning electron microscopy), are able to degrade LAS detergents. [*David M. Phillips/Visuals Unlimited.*]

likely to kill fish than are nondegradable ones. These allegations are denied by the detergent industry.

The cleansing action of synthetic detergents is quite similar to that of soaps. The synthetics work better in acidic solution and in hard water, though. Their calcium and magnesium salts, unlike those of soap, are soluble and do not separate out, even in extremely hard water (see Figure 17.6). Thus, the cleansing action of the synthetic detergent is little affected by hard water.

SECTION 17.5

Laundry Detergent Formulations: Builders

Even if Barry Commoner is wrong about the toxicity of the degradation products of biodegradable detergents, the laundry detergents we use do affect streams and lakes. The products for use in homes and commercial laundries usually contain much more than LAS molecules. The LAS is a surface-active agent, or surfactant. (Any agent, including soap, that stabilizes the suspension of nonpolar substances—such as oil and grease—in water is called a **surface-active agent**.) In addition to LAS, modern detergent formulations contain a number of other substances to improve detergency, to bleach, to lessen the redeposition of dirt, to brighten, or simply to reduce the cost of the formulation.

Any substance added to surfactant to increase its detergency is called a **builder**. Common builders, once widely used but now banned or restricted in many areas, are the phosphates. An example is sodium tripolyphosphate ($Na_5P_3O_{10}$). It ties up Ca^{2+} and Mg^{2+} in soluble complexes—thus softening the water—and produces a mild alkalinity, providing a favorable environment for detergent action.

Phosphates may totally disappear from detergents in the United States before the end of the century.

We have seen (in Chapter 13) how phosphates speed the eutrophication of lakes. In some areas, phosphates from detergents have been shown to contribute to this process. The degree of their contribution compared to phosphates that occur naturally in sewage remains open to question, however, in other locations.

Several state and local governments have banned the sale of detergents containing phosphates, and the detergent industry has offered a variety of replacements; the most prominent are sodium carbonate and complex aluminosilicates called *zeolites*. Sodium carbonate acts by precipitating the calcium ions, thus softening the water (Section 17.3).

$$Ca^{2+} + CO_3^{2-} \longrightarrow CaCO_3$$

The $CaCO_3$ precipitate appears to be harmful to automatic washing machines. Further, excess carbonate ions form strongly basic solutions, that is, solutions that contain an excess of OH^- ions.

Laundry detergents sometimes are eaten by small children. Those with sodium carbonate builders are more toxic than those with phosphate builders. Carbonate builders can cause death in children who eat them.

$$CO_3^{2-} + H_2O \longrightarrow HCO_3^- + OH^-$$

The zeolites are perhaps the most promising of the substitutes. The zeolite anions trap calcium ions by exchanging them for their own sodium ions.

$$Ca^{2+} + Na_2Al_2Si_2O_8 \longrightarrow 2\,Na^+ + CaAl_2Si_2O_8$$

The calcium ions are held in suspension by the zeolites (rather than being precipitated). Further, zeolite solutions are not strongly basic and are therefore less likely than sodium carbonate and sodium silicate solutions to irritate the skin and eyes.

Brighteners

A variety of other additives are used in detergent formulations. Many contain **optical brighteners**. These compounds, called blancophors (or colorless dyes), absorb the invisible ultraviolet component of sunlight and reemit it as visible light at the blue end of the spectrum. The fabric appears brighter and the blue light camouflages any yellowing. A diagram of this action, along with the structure of one such dye, is shown in Figure 17.13. Clothes treated with an optical brightener on the surface may be dirty underneath, but they look "whiter and brighter than new" (Figure 17.14). These brighteners are also used in cosmetics, paper, soap, plastics, and other products.

Optical brighteners have no known immediate toxic effect on humans. There is some possibility that these compounds might cause cancer or genetic disease. Bjorn Gillberg of the Royal Agricultural College of Sweden has

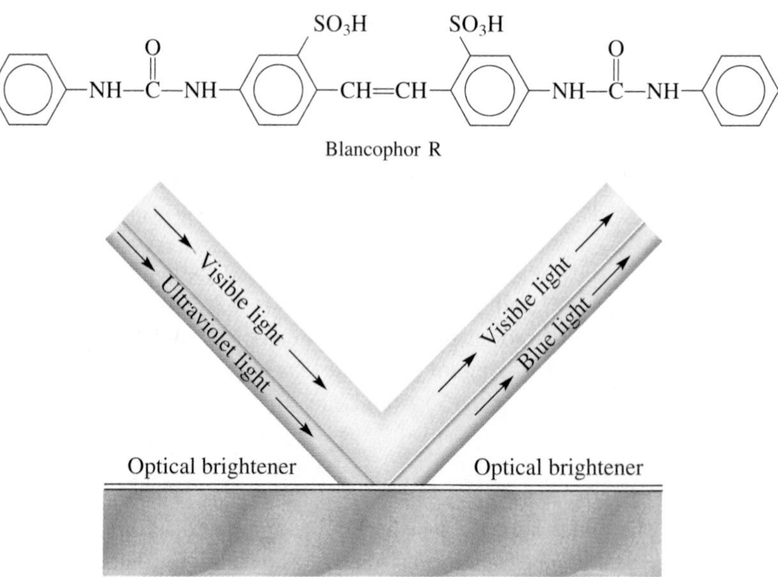

Figure 17.13 An optical brightener, such as the molecule illustrated here, converts invisible ultraviolet light to visible blue light, making the fabric look brighter and masking any yellowish color.

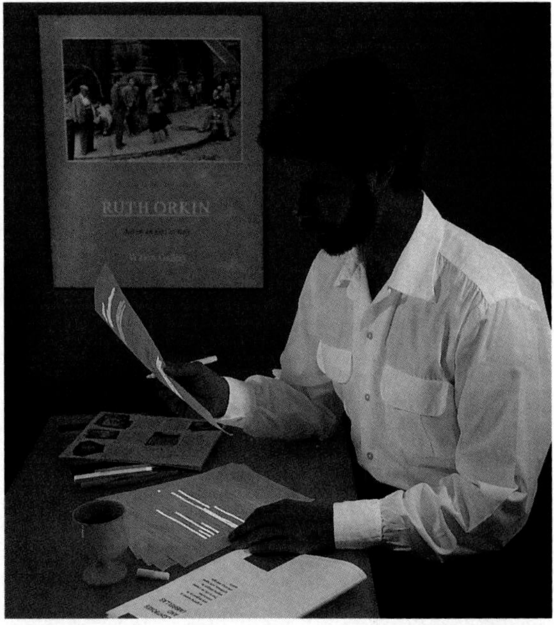

Figure 17.14 Clothes washed in a detergent formulation that contains an optical brightener glow brightly in the light from an ultraviolet lamp (black light). Paper, chalk, and other white materials often have optical brighteners added. [*Richard Megna/ Fundamental Photographs.*]

found that brighteners cause some mutations in microorganisms. Brighteners have been shown to cause skin rashes. Their effect on aquatic systems is largely unknown, yet large amounts are entering our waterways. The only possible benefit of these compounds is cosmetic: the appearance of our laundry may be more pleasing. Chemists can make optical brighteners, but it is up to the consumer to decide whether the benefit obtained by their use outweighs their monetary, environmental, and health costs.

SECTION 17.6

Liquid Laundry Detergents

Rapidly rising in the home laundry market are liquid detergent formulations. There are two basic types: those built with phosphates or other additives and unbuilt ones that are high in surfactant but contain no phosphates or other builders.

LAS is the cheapest surfactant in liquid laundry detergents. In unbuilt formulations, LAS usually is used as the sodium or triethanolamine salt. In built varieties, LAS often is present as the potassium salt. Other popular liquids are the alcohol ether sulfates. These contain a hydrocarbon portion derived from an alcohol (or alkylphenol), a polar portion derived from ethylene oxide, and a sulfate salt portion. These surfactants are efficient but more expensive.

O
CH_2—CH_2
Ethylene oxide

All the surfactants discussed so far, including soap, are **anionic surfactants**; the working part of the molecules is an anion with a nonpolar part and an ionic end. Some liquid detergents contain **nonionic surfactants**. Examples are the alcohol ethoxylates and alkylphenol ethoxylates. The oxygen atoms, by their attraction (hydrogen bonding) for water molecules, make their end of the molecule water soluble, just like the ionic end of an anionic surfactant. Structures of several of these detergent molecules are given in Figure 17.15. Nonionic surfactants are great for removing oily soil from fabrics. They are not as good as the anionic surfactants at keeping dirt particles in suspension. The alcohol ethoxylates have the unusual property of being more soluble in cold water than in hot. This makes them particularly suitable for cool-water laundering.

Dishwashing Detergents

Liquid detergents for washing dishes by hand generally contain one or more surfactants as the only active ingredients. Perfumes, colors, and additives supposed to soften and smooth hands are frequently added. Their main function, though, seems to be the establishment of a base for exaggerated advertising claims. Surfactants used include LAS as the sodium salt or the triethanolamine salt (or both). Some use nonionic surfactants. Another nonionic type is the amides made from fatty acids and diethanolamine. Cocamido DEA is an example (see Figure 17.15). Few contain phosphates or other builders. Those that do usually have only small amounts.

Most liquid dishwashing detergents differ significantly only in the concentration or effectiveness of the surfactant. In any formulation, the surfactant loosens the greasy food residues so they can be easily removed. It also traps the oily mess so it doesn't redeposit on dishes. The detergent eventually be-

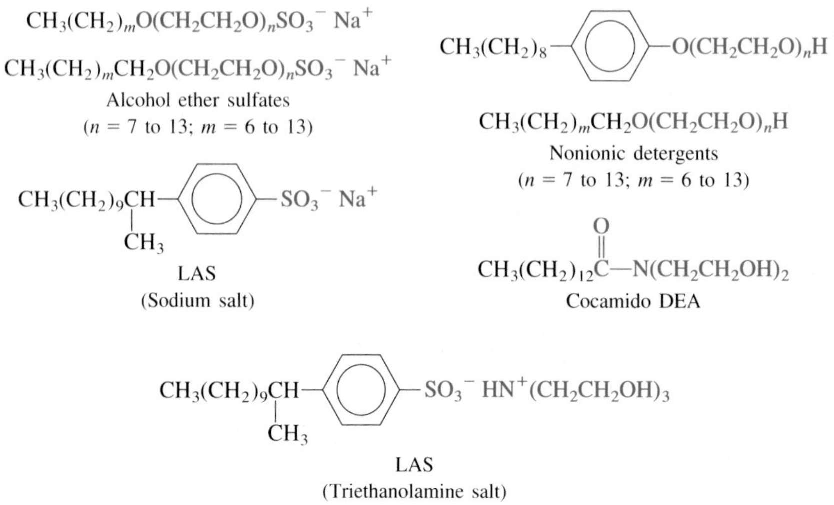

$CH_3(CH_2)_mO(CH_2CH_2O)_nSO_3^-\ Na^+$

$CH_3(CH_2)_mCH_2O(CH_2CH_2O)_nSO_3^-\ Na^+$

Alcohol ether sulfates
($n = 7$ to 13; $m = 6$ to 13)

$CH_3(CH_2)_9CH(CH_3)-C_6H_4-SO_3^-\ Na^+$

LAS
(Sodium salt)

$CH_3(CH_2)_8-C_6H_4-O(CH_2CH_2O)_nH$

$CH_3(CH_2)_mCH_2O(CH_2CH_2O)_nH$

Nonionic detergents
($n = 7$ to 13; $m = 6$ to 13)

$CH_3(CH_2)_{12}C(=O)-N(CH_2CH_2OH)_2$

Cocamido DEA

$CH_3(CH_2)_9CH(CH_3)-C_6H_4-SO_3^-\ HN^+(CH_2CH_2OH)_3$

LAS
(Triethanolamine salt)

Figure 17.15 Some of the kinds of molecules used in various detergent formulations. Nonoxynol-9, the nonylphenol ether in which $n = 9$, is a spermicide. It is used in contraceptives such as the Today Sponge.

comes saturated, though, and breaks down, releasing its greasy load. The weaker the detergent formulation, the sooner it will break down—or the more you will have to use.

Detergents for automatic dishwashers are quite another matter. They often are quite caustic and should never be used for hand dish washing. They contain sodium tripolyphosphate ($Na_5P_3O_{10}$), sodium metasilicate (Na_2SiO_3), sodium sulfate (Na_2SO_4), a chlorine bleach, and only small amounts of surfactant, usually a nonionic type. They depend mainly on their strong alkalis and the vigorous agitation of the machines for the removal of soil.

SECTION 17.7

Quaternary Ammonium Salts: Dead Germs and Soft Fabrics

Earlier in this chapter, we discussed anionic detergents and nonionic surfactants. There is a third type, called **cationic surfactants**, in which the working part of the substance is a positive ion. The most common of these are called quaternary ammonium salts because they have four groups attached to a nitrogen atom that bears a positive charge. An example of such a cationic surfactant is hexadecyltrimethylammonium chloride.

$$CH_3(CH_2)_{14}CH_2-\overset{\displaystyle CH_3}{\underset{\displaystyle CH_3}{\overset{|}{\underset{|}{N^+}}}}-CH_3 \quad Cl^-$$

Hexadecyltrimethylammonium chloride
(a cationic surfactant)

These cationics are not good detergents, but they have a degree of germicidal action. Sometimes they are used along with nonionic surfactants as cleansers and disinfectants in the food and dairy industries. Cationics cannot be used with anionic surfactants: the ions of opposite charge would clump together and precipitate from the solution, and neither surfactant could act as a detergent. Cationic surfactants make up only 3% of the chemicals used in detergent formulations. Anionic surfactants make up 18%, and nonionic surfactants 14% (Figure 17.16).

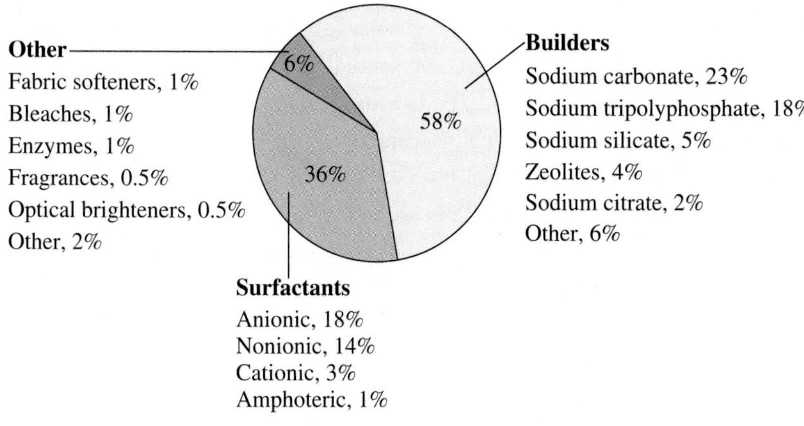

Figure 17.16 Chemicals used in formulating detergents in the United States.

$$\begin{array}{c} \quad CH_3 \quad Cl^- \\ | \\ CH_3(CH_2)_{16}CH_2{-}N^+{-}CH_3 \\ | \\ CH_3(CH_2)_{16}CH_2 \end{array}$$

Dioctadecyldimethylammonium chloride

Another kind of quaternary salt with *two* long carbon chains and two smaller groups on nitrogen is used as a fabric softener. An example is dioctadecyldimethylammonium chloride. These compounds are strongly adsorbed by the fabric, forming a film on the fabric's surface one molecule thick. The long hydrocarbon chains lubricate the fibers, imparting increased flexibility and softness to the fabric.

SECTION 17.8

Bleaches: Whiter Whites

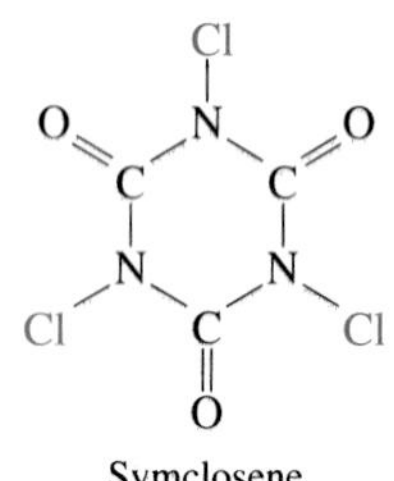

Symclosene

Bleaches are oxidizing agents (Chapter 8). The familiar liquid laundry bleaches are all 5.25% sodium hypochlorite (NaOCl) solutions. They differ only in price. Purex, Clorox, store brands, and generic versions are widely available. Hypochlorite bleaches release chlorine rapidly, and the high concentrations of corrosive chlorine can be quite damaging to fabrics. These bleaches do not work well on polyester fabrics, often causing a yellowing rather than the desired whitening.

Other bleaches are available in solid forms that release chlorine slowly in water. Symclosene is an example of a cyanurate-type bleach. The slow release

Mixing bleach with other household chemicals can be quite dangerous. For example, mixing a hypochlorite bleach with hydrochloric acid produces poisonous chlorine gas.

$$2\,HCl + ClO^- \longrightarrow Cl^- + H_2O + Cl_2$$

Mixing bleach with toilet bowl cleaner (HCl) is especially dangerous. Most bathrooms are small and poorly ventilated, and generating chlorine in such a limited space can be very hazardous. Chlorine can do enormous damage to the throat and the entire respiratory tract. If the concentration of chlorine is high enough, it can kill.

Mixing bleach with ammonia is also extremely hazardous. Two of the gases produced are chloramine (NH_2Cl) and hydrazine (NH_2NH_2), both of which are violently toxic.

We repeat: *never* mix bleach with other chemicals without specific directions to do so. In fact, it is a good rule not to mix any chemicals unless you are well informed about what you are doing.

of chlorine minimizes damage to fabrics because the concentration of chlorine is never very high.

The oxygen-releasing bleaches usually contain sodium perborate ($NaBO_2 \cdot H_2O_2$). As indicated by the formula, this compound is a complex of $NaBO_2$ and hydrogen peroxide (H_2O_2). Above 65 °C, the hydrogen peroxide is liberated and acts as a bleach. It decomposes in turn to liberate oxygen.

$$2\,H_2O_2 \longrightarrow 2\,H_2O + O_2$$

Borates are somewhat toxic. Perborate bleaches are less active than chlorine bleaches and require higher temperatures, higher alkalinity and higher concentrations to do an equivalent job. They are used mainly for bleaching white, resin-treated polyester–cotton fabrics. These fabrics last much longer with oxygen bleaching than with chlorine bleaching. Also, properly used, the oxygen bleaches get fabrics whiter than do chlorine bleaches.

SECTION 17.9

All-Purpose Cleaning Products

A variety of all-purpose cleaning products are available for use on walls, floors, countertops, appliances, and other tough, durable surfaces. Those for use in water may contain surfactants, sodium carbonate, ammonia, solvent-type grease cutters, disinfectants, deodorants, and other ingredients. Be sure you check the labels on such products. Some are great for certain jobs but not very good for others. They damage some surfaces and work especially well on others. Most important, they may be harmful to you when they are used improperly.

Household ammonia solutions, straight from the bottle, are good for loosening baked-on grease or burned-on food. Just soak the object overnight. Diluted with water, household ammonia cleans mirrors, windows, and other glass surfaces. Mixed with detergent, ammonia rapidly removes wax from vinyl floor coverings. Ammonia vapors are highly irritating. This cleanser should never be used in a closed room. Ammonia should not be used on asphalt tile, wood surfaces, or aluminum because it may stain, pit, or erode these materials.

Baking soda (sodium bicarbonate, $NaHCO_3$), straight from the box, is a mild abrasive cleanser. It absorbs food odors readily, making it good for cleaning the inside of a refrigerator. Vinegar (acetic acid) cuts grease film. It should not be used on marble, because it reacts with marble, pitting the surface.

$$\underset{\text{Marble}}{CaCO_3} + \underset{\text{Vinegar}}{2\,CH_3COOH} \longrightarrow Ca^{2+} + 2\,CH_3COO^- + CO_2 + H_2O$$

SECTION 17.10

Special-Purpose Cleaners

There are many highly specialized cleaning products on the market. For metals there are various kinds of chrome cleaners, brass cleaners, copper cleaners, and silver cleaners. There are lime removers, rust removers, and grease removers. There are cleaners specifically for wood surfaces, for vinyl floor coverings, or for ceramic tile. Let us look at just a few special purpose cleaners that you would find in almost any home.

Toilet Bowl Cleaners

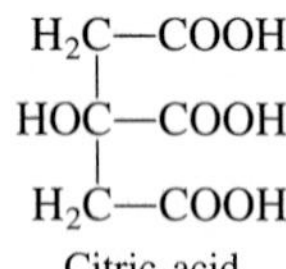

Citric acid

The buildup that forms in toilet bowls is mainly calcium carbonate ($CaCO_3$), deposited from hard water, along with discolorations due to such things as iron compounds and fungal growth. Since calcium carbonate is readily dissolved by acid, toilet bowl cleaners tend to be strongly acidic. The solid crystalline products are usually sodium bisulfate ($NaHSO_4$). The liquid cleaners are hydrochloric acid (HCl), citric acid, or some other acidic material.

Scouring Powder

Most powdered cleansers contain an abrasive such as silica (SiO_2) for rubbing stains from hard surfaces. They also usually contain a surfactant to dissolve grease, and some feature a chlorine-releasing bleach. These cleansers are mainly intended for removing stains from porcelain tubs and sinks. Such abrasive cleansers may scratch the finish on appliances, countertops, and metal utensils. They may even scratch the surfaces of sinks, toilet bowls, and bathtubs. Dirt gets into the scratches and makes cleaning even more difficult. Use these cleansers (with care) only on surfaces that can withstand the abrasion, or on surfaces where scratches won't matter.

Glass Cleaners

Cleaners for window panes and mirrors should be volatile liquids that will evaporate without leaving a residue. The most common glass cleaner is alcohol, usually isopropyl alcohol, diluted with water. Sometimes ammonia is added for greater cleaning power.

Drain Cleaners

When drains become clogged, it is usually because the pipes have gotten caked up with grease, which is usually made up of various kinds of fats (triglycerides). Drain cleaners often contain sodium hydroxide (NaOH), either in the solid form or as a very concentrated liquid. The sodium hydroxide reacts with the water in the pipe to generate heat, which melts much of the grease. The sodium hydroxide then reacts with some of the fat, converting it to soap, which cleans out the rest of the grease in the pipe.

In some products there are also bits of aluminum metal that react with the sodium hydroxide solution to form hydrogen gas, which bubbles out of the clogged area of the drain, creating a stirring action.

Oven Cleaners

Most oven cleaners also contain sodium hydroxide. Several popular products dispense it as an aerosol foam. The greasy deposits on the oven walls are converted to soaps when they react with the sodium hydroxide. The resulting mixture can then be washed off with a wet sponge. (You must wear rubber gloves, of course. Sodium hydroxide is extremely caustic and very hard on the skin!)

SECTION 17.11

Organic Solvents in the Home

A variety of solvents are used in the home. They may be used to remove paint, varnish, adhesives, waxes, and other materials. Some cleansers also contain organic solvents. Perhaps the best known are those containing pine oil. This oil consists mainly of terpenes, compounds with 10 carbon atoms that occur widely in nature. The pine-oil terpenes usually have a ring structure and one or more alcohol functions. An example is terpineol. Pine oil acts as a mild disinfectant, and it helps to dissolve grease. In moderate concentrations, it has a pleasant odor.

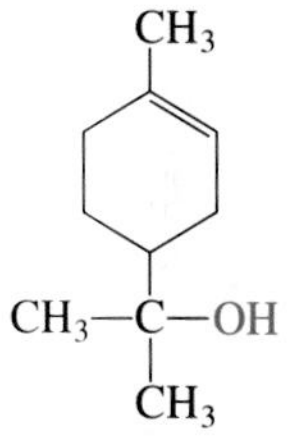

CH_3

$CH_3—C—OH$

CH_3

Terpineol

Petroleum distillates are added to some all-purpose cleansers as grease cutters. These are hydrocarbons—much like gasoline—derived from petroleum. They dissolve grease readily, but like gasoline, are highly flammable. They are also deadly when swallowed. The lungs become saturated with hydrocarbon vapors, fill with fluid, and fail to function.

A troubling problem connected with household solvents is the practice of inhaling fumes. The popularity of solvent sniffing seems to be related mainly to peer pressure, especially among young teenagers. They tell each other that sniffing a certain glue or solvent is a good way to get high, but sometimes the results are deadly. Long-term sniffing can cause permanent damage to vital organs, the lungs in particular. Often there is irreversible brain damage as well.

Most of the organic solvents used around the home are volatile and flammable. Many have toxic fumes; nearly all are narcotic at high concentrations. Some people, trying to get their kicks from sniffing glue or other solvents, have died of heart failure. Such solvents should be used only with adequate ventilation. They should never be used around a flame. Be sure to read—and heed—all precautions before you use any solvent. And never use gasoline for cleaning; it is too hazardous in too many ways.

SECTION 17.12

Paints

Paint is a broad term that covers a wide variety of products—lacquers, enamels, varnishes, oil-base coatings, and a number of different water-base finishes. Any or all of these materials might be found in any home.

White lead is basic lead carbonate, $2PbCO_3 \cdot Pb(OH)_2$.

A paint contains three basic ingredients: a pigment, a binder, and a solvent. The universal pigment today is titanium dioxide, TiO_2 (which has taken the place of "white lead," the poisonous white pigment that was finally banned in 1977). Titanium dioxide is a brilliant white pigment with great stability and excellent hiding power. All ordinary paints are pigmented with titanium dioxide. For colored paints small amounts of colored pigments or dyes are added to the white base mixture. The binder, or film former, is the substance that binds the pigment particles together and holds them on the painted surface. In oil paints the binder is usually tung oil or linseed oil. In water-base paints it is a polymer of some kind. Most interior paints have polyvinyl acetate as the binder. Exterior water-base paints use acrylic resins as binders. Acrylic latex paints are much more resistant to rain and sunlight. The solvent is added in order to keep the paint fluid until it is applied to a surface. The solvent might be an alcohol, a hydrocarbon, an ester (or some mixture thereof), or water.

The paint may also contain additives: a drier (or activator) to make the paint dry faster, a fungicide to act as a preservative, a thickener to increase the paint's viscosity, an antiskinning agent to keep the paint from forming a skin inside the can, and perhaps a surfactant to stabilize the mixture.

SECTION 17.13

Waxes

Waxes are esters of long-chain organic acids (fatty acids) with long-chain alcohols. They are produced by plants and animals mainly as protective coatings. (These are not to be confused with paraffin wax, which is made up of hydrocarbons.)

Beeswax is the material from which bees build their honeycombs. The following ester is a typical molecule in beeswax.

$$CH_3(CH_2)_{24}\overset{\overset{\displaystyle O}{\|}}{C}{-}OCH_2(CH_2)_{32}CH_3$$

Beeswax is used in such household products as candles and shoe polish.

Carnauba wax is a coating that forms on the leaves of certain palm trees in Brazil. It is a mixture of esters similar to those in beeswax. It is used in making automobile wax, floor wax, and furniture polish.

Spermaceti wax, which is extracted from the head of the sperm whale, is largely cetyl palmitate.

$$CH_3(CH_2)_{14}\overset{\overset{\displaystyle O}{\|}}{C}{-}OCH_2(CH_2)_{14}CH_3$$

Once widely used in making cosmetics and other products, spermaceti wax is now in short supply because the sperm whale has been hunted almost to extinction.

Lanolin, the grease in sheep's wool, is also a wax. Since it forms stable emulsions with water, it is very useful for making various skin creams and lotions.

SECTION 17.14

Cosmetics: Personal Care Chemicals

Ages ago, primitive people used materials from nature for cleansing, beautifying, and otherwise altering their appearance. Evidence indicates that Egyptians, 7000 years ago, used powdered antimony (Sb) and the green copper ore malachite as eye shadow. Egyptian pharaohs used perfumed hair oils as far back as 3500 B.C. Claudius Galen, a Greek physician of the second century A.D., is said to have invented cold cream. Dandy gentlemen of seventeenth-century Europe used cosmetics lavishly, often to cover the fact that they seldom bathed. Ladies of eighteenth-century Europe whitened their faces with lead carbonate ($PbCO_3$), and many died from lead poisoning.

The use of cosmetics has a long and interesting history, but nothing in the past comes close to the amounts and varieties of cosmetics used by people in the modern industrial world. Each year we spend billions of dollars on everything from hair sprays to toenail polishes, from mouth washes to foot powders.

What is a cosmetic? The United States Food, Drug, and Cosmetic Act of 1938 defines **cosmetics** as "articles intended to be rubbed, poured, sprinkled or sprayed on, introduced into, or otherwise applied to the human body or any part thereof, for cleansing, beautifying, promoting attractiveness or altering the appearance. . . ." Soap, although obviously used for cleansing, is specifically excluded from coverage by the law. Also excluded are substances that affect the body's structure or functions. Antiperspirants, products that reduce perspiration, are legally classified as drugs. So are antidandruff shampoos. The main difference between drugs and cosmetics is that drugs must be proven "safe and effective" before they are marketed; cosmetics generally do not have to be tested before they're marketed. Most brands of a given type of cosmetic contain the same (or quite similar) active ingredients. Thus, advertising is usually geared to selling a name, a container, or a fragrance, rather than the actual product itself.

About 8000 different chemicals are used in cosmetics. They are sold in 20,000 to 40,000 different combinations.

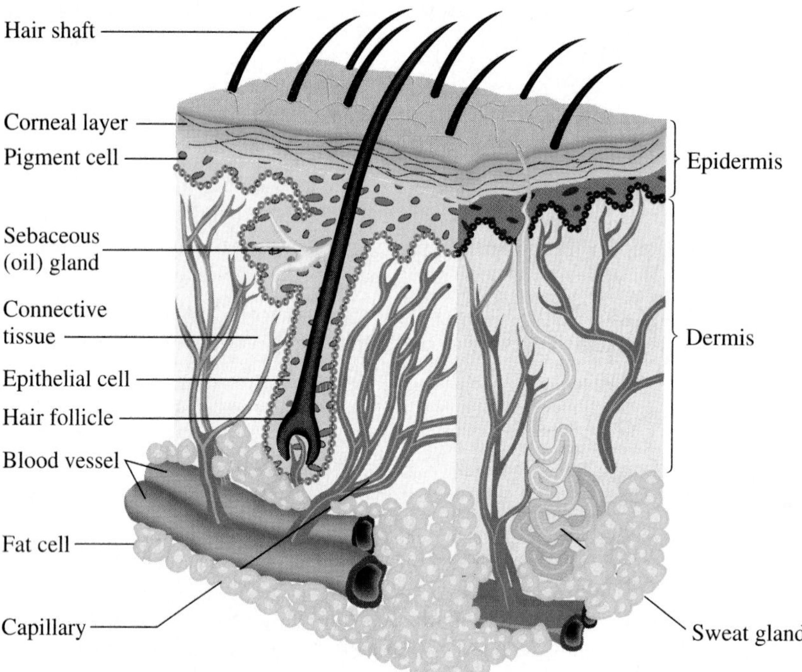

Figure 17.17 Cross section of an area of skin. Cosmetics affect only the outer corneal layer of dead cells.

Skin Creams and Lotions

Skin is a complex organ that encloses our bodies (Figure 17.17). The outer layer of skin is called the **epidermis**. The epidermis, in turn, is divided into two parts: dead cells on the outside (the corneal layer), and living cells beneath the corneal layer that continually replace corneal cells, which are sloughed off. Cosmetics are applied to the dead cells of the corneal layer. About $20 billion a year is spent in the United States on various preparations applied to the skin.

The corneal layer is composed mainly of a tough, fibrous protein called **keratin**. Keratin has a moisture content of about 10%. Below 10% moisture, the skin is dry and flaky. Above 10%, conditions are ideal for the growth of harmful microorganisms. Skin is protected from loss of moisture by **sebum**, an oily secretion of the sebaceous glands. One interesting property of skin is that it is insoluble in water (otherwise we would dissolve in the shower). It is, however, slightly permeable to water and light. Exposure to sun and wind may leave the skin dry and scaly, and too frequent washing also removes natural skin oils.

A variety of lotions and creams are available to treat the skin. A **lotion** is an emulsion of tiny oil droplets dispersed in water. A **cream** is the opposite; tiny water droplets are dispersed in oil (Figure 17.18). The essential ingredient

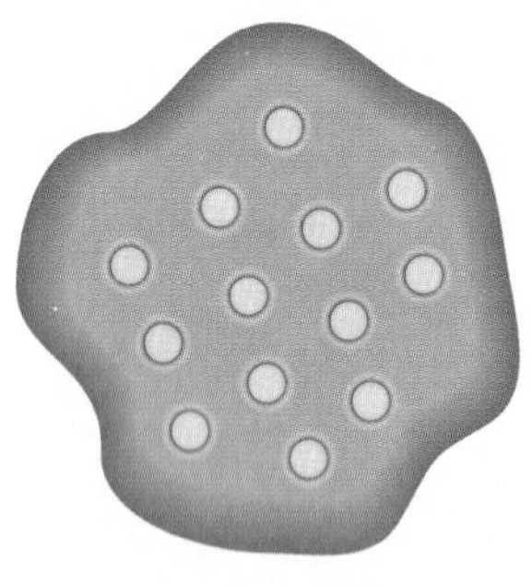
(a)

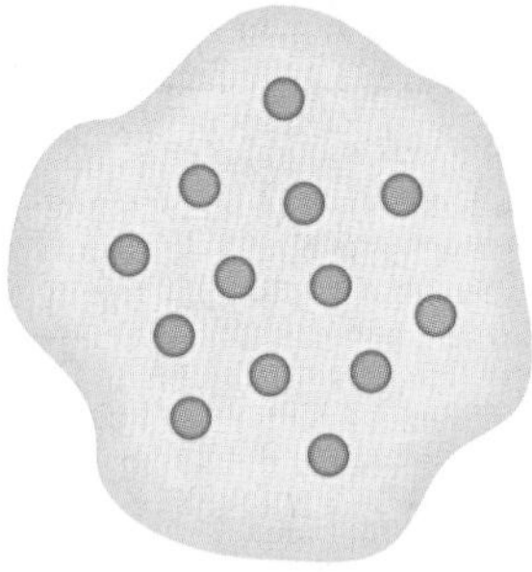
(b)

Figure 17.18 A lotion is an emulsion of an oil in water (a). A cream is an emulsion of water in an oil (b). A lotion feels cool because evaporating water removes heat from the skin. A cream feels greasy.

of each is a fatty or oily substance that forms a protective film over the skin. Typical ingredients are mineral oil and petroleum jelly; sometimes both are used in the same preparation. These are mixtures of alkanes obtained from petroleum. Other ingredients include natural fats and oils, perfumes, waxes, water, and emulsifiers (compounds that keep the oily portions from separating from the water). Natural materials used on the skin include lanolin, a fat obtained from sheep's wool, and olive oil. Often, beeswax is added to harden the product.

Some creams have been formulated with hormones, queen bee jelly, and other strange ingredients. None of these has been found to confer any particular benefit. Creams and lotions protect the skin by providing a protective coating and by softening it. Such skin softeners are called **emollients**. You could just as well use petroleum jelly (one trade name is Vaseline) or a good grade of white mineral oil (sometimes called baby oil) as the fancy creams.

Some cosmetics contain *humectants* (Chapter 16) such as glycerol. Glycerol, with its three hydroxyl groups, holds water molecules by hydrogen bonding. Emollients such as petroleum jelly keep the skin moist by forming a physical barrier that hinders evaporation of water from the skin.

It may seem strange that gasoline, a mixture of alkanes, dries out the skin, and that the higher alkanes in mineral oil and petroleum jelly soften it. Keep in mind, though, that gasoline is a thin, free-flowing liquid. It dissolves the natural skin oils and carries them away. The higher alkanes are viscous. They stay right on the skin and serve as emollients.

Moisturizers, which hold moisture to the skin, are usually substances such as lanolin. Collagen (the protein in connective tissue) has also been found to be an effective moisturizer in skin lotions.

Skin is damaged by the ultraviolet radiation from the sun. The shorter wavelengths are more energetic and are especially harmful. **Sunscreen lotions** are used to block this radiation while letting through the less energetic long-wave ultraviolet rays that promote tanning. The active ingredient in many of these preparations is *para*-aminobenzoic acid. Various concentrations of this compound in the lotion provide the **skin protection factor** (SPF) ratings. SPF values vary from 2 to 35 or more. An SPF of 10, for example, means that you can stay in the sun without burning 10 times as long as you could with unprotected skin.

$$H_2N-C_6H_4-\overset{O}{\overset{\|}{C}}-OH$$

para-Aminobenzoic acid

Ultraviolet rays in sunlight turn light skin darker by triggering the production of the pigment **melanin**. The dark melanin then protects the deeper layers of the skin from damage. Excessive exposure to ultraviolet radiation causes premature aging of the skin and leads to skin cancer. Our quest for tanned skin has us now in the midst of an epidemic of skin cancer.

Cigarette smoking also leads to premature aging of the skin. Nicotine causes the constriction of the tiny blood vessels that feed the skin. Repeated constrictions several times a day over the years cause the skin to lose its elasticity and become wrinkled. The best treatment for wrinkling of the skin is prevention. Stay out of the sun and don't smoke cigarettes.

Lipstick

Lipstick is quite similar to skin creams in composition. It is made of an oil and a wax. To keep the lipstick firm, a higher proportion of wax is used than in creams. Dyes and pigments provide color. The oil is frequently castor oil. Waxes often employed are beeswax, carnauba, and candelilla. Perfumes are added to cover up the unpleasant fatty odor of the oil. Antioxidants are also employed to retard rancidity. Bromo acid dyes such as tetrabromofluorescein, a bluish red compound, are responsible for the color of most modern lipsticks.

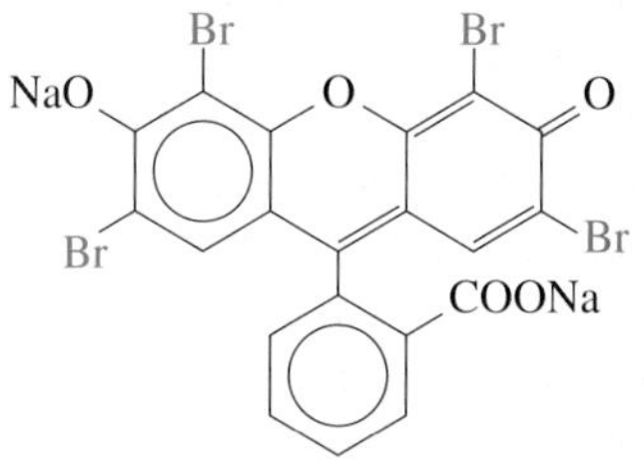

Tetrabromofluorescein

Because it has little in the way of protective oils, the skin of the lips is easily dried out, leading to chapped lips. With or without coloring, lipsticks do protect the lips from drying.

Eye Makeup

A variety of chemicals are used to decorate the eyes. **Mascara** is used to darken eyelashes. It is composed of a base of soap, oils, fats, and waxes. Mascara is colored brown by iron oxide pigments, black by carbon (lampblack), green by chromium(III) oxide (Cr_2O_3), or blue by ultramarine (a silicate that contains some sulfide ions). A typical composition is 40% wax, 50% soap, 5% lanolin, and 5% coloring matter. Eyebrow pencils, which have about the same base, are colored by lampblack or iron oxides.

Eye shadow has a base of petroleum jelly with the usual fats, oils, and waxes. It is colored by dyes or made white by zinc oxide (ZnO) or titanium dioxide (TiO_2) pigments. A typical composition is 60% petroleum jelly, 10% fats and waxes, 6% lanolin, with the remainder dyes or pigments or both.

Some people have allergic reactions to ingredients in eye makeup. A more serious problem is eye infections from bacterial contamination. It is recommended that eye makeup be discarded after three months. Mascara is of special concern because of the wand applicator. A slip of the hand and it can scratch the cornea.

Deodorants and Antiperspirants

Deodorants are products formulated so as to kill odor-causing bacteria. Bacteria act on perspiration residue and sebum (natural body oil) to produce various malodorous compounds, such as short-chain fatty acids and amines.

The active ingredient in a deodorant is usually a long-chain quarternary ammonium salt (Section 17.7).

Antiperspirants are usually deodorants as well, but they also stop or retard perspiration. The active antiperspirant ingredient is a variable complex of chlorides and hydroxides of aluminum and zirconium. (Aluminum chlorohydrate is a typical compound found in many antiperspirants.) The zirconium and aluminum chlorides and hydroxides function as an **astringent**, which constricts the openings of the sweat glands, thus restricting the amount of perspiration that can escape. Antiperspirants can be formulated into creams or lotions, or they can be dissolved in alcohol and applied as sprays.

$Al_2(OH)_5Cl \cdot 2H_2O$
Aluminum chlorohydrate

Sweating is a natural, healthy body process, and to stop it on a routine basis is probably not very wise. Regular bathing and changing of clothes should make antiperspirants unnecessary. But if you still feel the need for underarm protection, try the simpler deodorant product.

SECTION 17.15

Toothpaste: Soap with Grit and Flavor

After soap (which doesn't count, because the law says it isn't a cosmetic), toothpaste is the most important cosmetic product. The only essential components of toothpaste are a detergent and an abrasive (Figure 17.19). Soap and sodium bicarbonate would do the job quite well but would be rather unpalat-

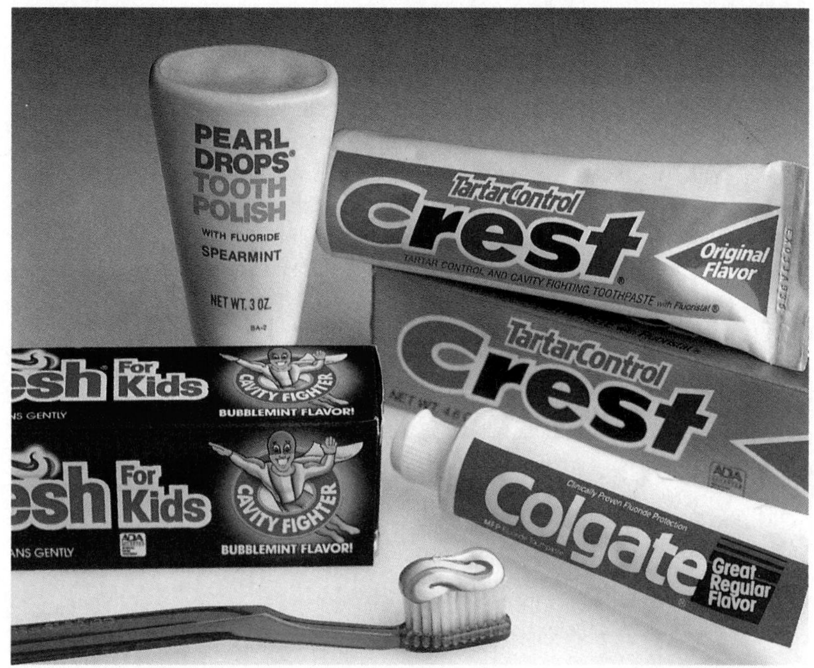

Figure 17.19 Toothpastes are available under many brand names. The only essential ingredients in any toothpaste are a detergent and an abrasive. [*Richard Megna/Fundamental Photographs.*]

Table 17.1 Abrasives Commonly Used in Toothpaste

Name	Chemical Formula
Precipitated calcium carbonate	$CaCO_3$
Insoluble sodium metaphosphate	$(NaPO_3)_n$
Calcium hydrogen phosphate	$CaHPO_4$
Titanium dioxide	TiO_2
Tricalcium phosphate	$Ca_3(PO_4)_2$
Calcium pyrophosphate	$Ca_2P_2O_7$
Hydrated alumina	$Al_2O_3{\cdot}nH_2O$
Hydrated silica	$SiO_2{\cdot}nH_2O$

able. The ideal abrasive should be hard enough to clean the teeth but not hard enough to damage the tooth enamel. Abrasives frequently used in toothpaste are listed in Table 17.1. Some have been criticized as being too harsh.

A typical detergent is sodium dodecyl sulfate (sodium lauryl sulfate).

$$CH_3CH_2CH_2CH_2CH_2CH_2CH_2CH_2CH_2CH_2CH_2CH_2OSO_3^- \ Na^+$$

Any pharmaceutical grade of soap or detergent probably would work satisfactorily. Most toothpastes today are full of minty flavors, colors, aromas, and sweet tastes. Ingredients include sweeteners such as sorbitol, glycerol (glycerin), and saccharin (Chapter 16); flavors such as wintergreen and peppermint; thickeners such as cellulose gum and polyethylene glycols; and preservatives such as sodium benzoate. Table 17.2 gives a typical recipe for toothpaste.

Tooth decay is caused primarily by bacteria that convert sugars to sticky dextrans or plaque and to acids such as lactic acid ($CH_3CHOHCOOH$). Acids dissolve tooth enamel. Brushing and flossing remove plaque and thus prevent decay. Decay can be minimized by eating sugars only at meals rather than in snacks, and by brushing immediately after eating. Acids such as the phosphoric acid (H_3PO_4) in beer and pop may also dissolve the tooth enamel of people who consume these beverages in large amounts.

Many modern toothpastes contain fluorine compounds, such as stannous

Table 17.2 A Typical Recipe for Toothpaste

Ingredient	Function	Amount
Precipitated calcium carbonate	Abrasive	46 g
Castile soap or sodium dodecyl sulfate	Detergent	4 g
Glycerol (glycerin)	Sweetener	20 g
Gum tragacanth or gum cellulose	Thickener	1 g
Oil of peppermint (or peppermint extract)	Flavoring	1 mL
Water	—	28 mL

fluoride (SnF_2), shown to be effective in reducing the incidence of tooth decay. The enamel of teeth is similar to hydroxyapatite in composition. Fluoride from drinking water and toothpaste converts a part of the enamel to fluorapatite. Fluorapatite is a stronger material than hydroxyapatite and is more resistant to decay.

$Ca_5(PO_4)_3OH$
Hydroxyapatite

$Ca_5(PO_4)_3F$
Fluorapatite

A major cause of tooth loss in adults is gum disease. Toothpaste formulations that include ingredients such as baking soda ($NaHCO_3$) and hydrogen peroxide (H_2O_2) are purported to prevent gum disease.

SECTION 17.16

Perfumes, Colognes, and Aftershaves

No one wants to smell bad. Many people like to give off the pleasant aroma of a fruit or a flower, perhaps moderated a bit to avoid overwhelming a neighbor's nose. **Perfumes** are among the most ancient and the most widely used of the cosmetics. Their chemistry, however, is exceedingly complex.

Originally, perfumes were extracted from natural sources. Nowadays, chemists have identified many of the components and synthesized them in the laboratory. The best perfumes, perhaps, are still made from natural materials because chemists have so far been unable to identify all the many important, but minor, ingredients.

A good perfume may have a hundred or more constituents. Often the components are divided into three categories, called *notes,* based on differences in volatility. The most volatile fraction (that which vaporizes most readily) is called the **top note**. This fraction, made up of relatively small molecules, is responsible for the odor when a perfume is first applied. The **middle note** is intermediate in volatility. It is responsible for the lingering aroma after most of the top-note compounds have vaporized. The **end-note** fraction has low volatility and is made up of compounds with large molecules.

Note that the fragrance molecules (Table 17.3) are fat-soluble. They dissolve in skin oils and are retained for hours.

Several compounds with flowery or fruity odors are synthesized in large quantities for use in perfumes. Some of these, with their approximate odors, are listed in Table 17.3. Odors vary with dilution. A concentrated solution (lots of compound in a small amount of water or other solvent) may be unpleasant, yet a dilute solution (a small amount in lots of solvent) of the same compound may have a pleasant aroma. Most of the compounds exist in several isomeric forms.

Many flowery or fruity odors are sickeningly sweet, even in dilute solutions. Compounds such as the musks often are added to moderate the odor. Musks and similar compounds have extremely disagreeable odors when concentrated, but often are pleasant at extreme dilution. Several of these compounds are described in Table 17.4.

The civet cat, from which the compound civetone is obtained, is a skunk-like animal of eastern Africa. Its secretion, like that of the skunk, is a defen-

Table 17.3 Compounds with Flowery and Fruity Odors Used in Perfumes

Name	Structure*	Odor
Citral	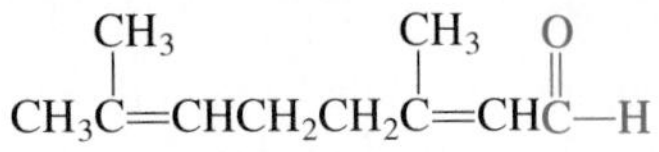 $CH_3\overset{CH_3}{\overset{\vert}{C}}{=}CHCH_2CH_2\overset{CH_3}{\overset{\vert}{C}}{=}CH\overset{O}{\overset{\Vert}{C}}{-}H$	Lemon
Irone	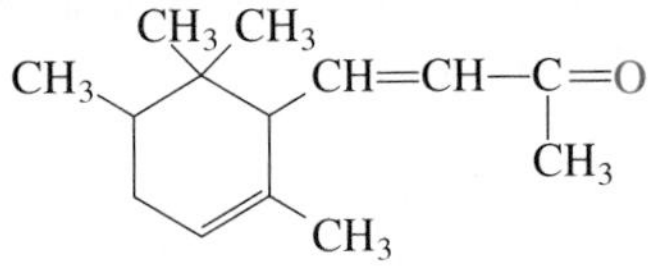 ring structure with CH_3, CH_3, CH_3, CH_3 substituents and $CH{=}CH{-}\underset{CH_3}{\underset{\vert}{C}}{=}O$	Violet
Jasmone	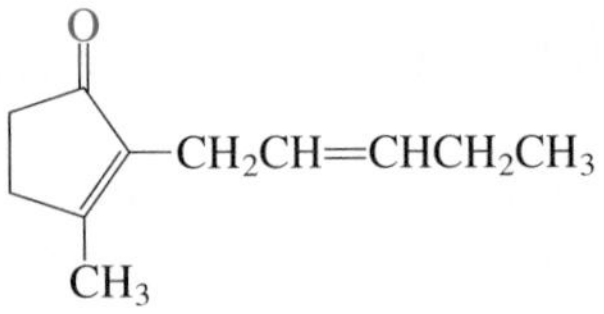 cyclopentenone ring (O) with $-CH_2CH{=}CHCH_2CH_3$ and CH_3	Jasmine
Geraniol	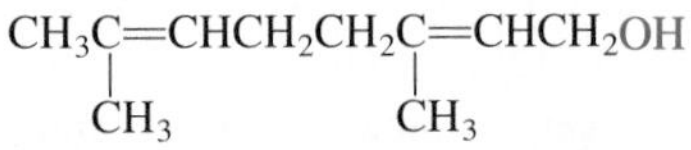 $CH_3\underset{CH_3}{\underset{\vert}{C}}{=}CHCH_2CH_2\underset{CH_3}{\underset{\vert}{C}}{=}CHCH_2OH$	Rose

*Note that each compound has only one oxygen atom for ten or more carbon atoms.

Table 17.4 Unpleasant Compounds Used to Fix Delicate Odors in Perfumes

Compound	Structure	Natural Source
Civetone	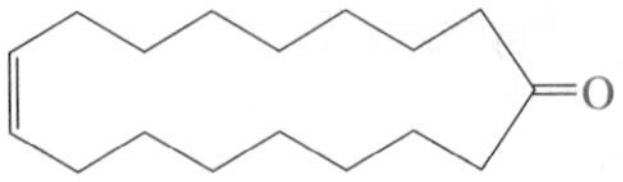 large ring with C=C and =O	Civet cat
Muscone	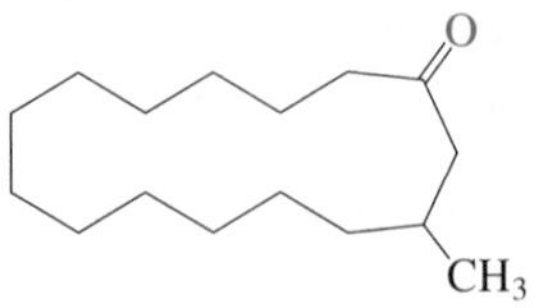 large ring with O and CH_3	Musk deer
Indole	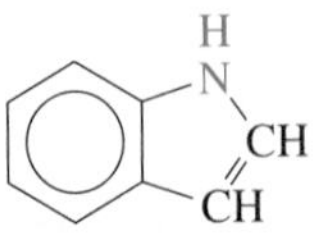 benzene ring fused to ring with N–H, CH, CH	Feces

sive weapon. The secretion from musk deer is thought to be a pheromone (Chapter 16) used as a sex attractant.

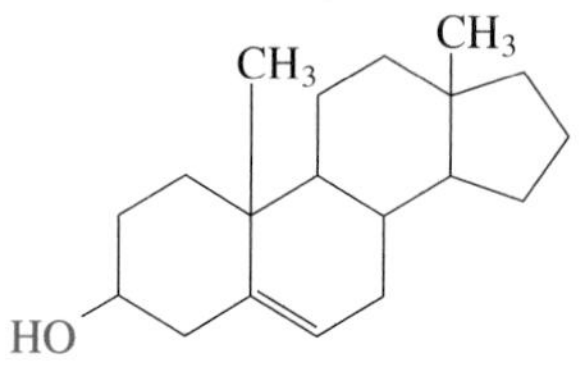

Perhaps the ultimate in perfumes is Andron by Jovan. Studies hint that α-androstenol may act as a sex attractant between male and female humans. α-Androstenol is a steroid that occurs naturally in human hair and urine. The evidence is far from conclusive, but Jovan puts a minute amount of the compound in Andron. Jovan claims that α-androstenol is active at concentrations as low as 6 ppm. Competitors call the whole thing an advertising ploy.

CH_3 CH_3 HO

α-Androstenol

It is of some interest that sex attractants have been identified in primates. Indeed, the same set of pheromones (a mixture of organic acids) that operate as sex attractants in rhesus monkeys have been isolated from the vaginal secretions of human females. It is doubtful, however, that sex attractants have much influence on human behavior. We probably have overriding cultural constraints. Anyway, we seem to prefer the sex attractant of the musk deer.

A perfume usually consists of 10 to 25% odorous compounds and fixatives. The remainder is ethyl alcohol, which serves as a solvent. **Colognes** are diluted perfumes. They often contain only 1 or 2% perfume essence. Thus, colognes are about 10% as strong as perfumes. Dilution can be made with ethyl alcohol alone or with alcohol–water mixtures.

Aftershave lotions are similar to colognes. Most are about 50 to 70% ethanol with the remainder water, a perfume, and food coloring. Some have menthol added for a cooling effect on the skin; others add an emollient of some sort to soothe chapped skin.

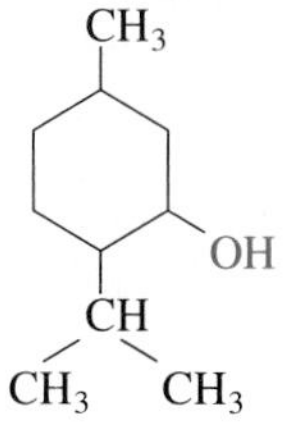

CH_3 OH CH CH_3 CH_3

Menthol

Perfumes are the source of many of the allergic reactions associated with cosmetics and other consumer products. **Hypoallergenic cosmetics** are those that purport to cause fewer allergic reactions than regular products. The term has no legal meaning, but most "hypoallergenic" cosmetics leave out the perfume.

SECTION 17.17

Some Hairy Chemistry

Like skin, hair is composed primarily of the fibrous protein keratin. Recall (Chapter 15) that protein molecules are made up of chains of amino acids.

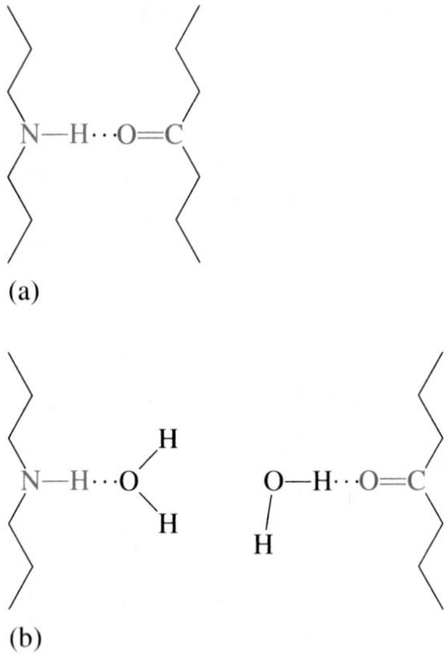

N—H···O=C

(a)

N—H···O(H)(H)   O(H)—H···O=C

(b)

Figure 17.20 In hair adjacent protein chains are held together by hydrogen bonds that link the carbonyl group of one chain to the amide group of another (a). When wet, these groups can hydrogen-bond with water rather than with each other (b), disrupting the hydrogen bonds between chains.

These chains are held to one another by four types of forces: hydrogen bonds, salt bridges, disulfide linkages, and hydrophobic interactions (see Figure 15.18). Of these, hydrogen bonds and salt bridges are important to our understanding of the actions of shampoos and conditioners. Hydrogen bonds between chains are disrupted by water (Figure 17.20). Salt bridges are destroyed by changes in pH (Figure 17.21). Disulfide linkages are broken and restored when you get a permanent wave or have your hair straightened.

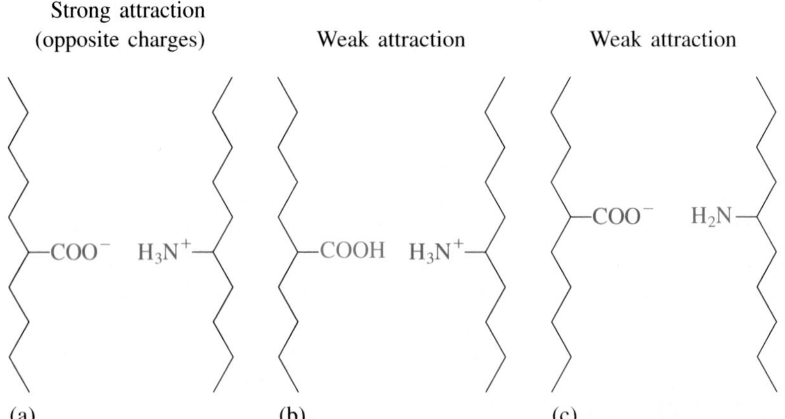

Figure 17.21 Strong electrostatic forces hold a protein chain with an ionized carboxyl group to another chain with an ionized amino function (a). A change in pH can disrupt this salt bridge. If the pH drops, the carboxyl group takes on a proton and loses its charge (b). If the pH rises, a proton is removed from the amino group and it is left uncharged (c). In either case, the forces between the protein chains are weakened considerably.

Shampoo

The keratin of hair has five or six times as many disulfide linkages as the keratin of skin. When hair is washed, the keratin absorbs water and is softened and made more stretchable. The water disrupts hydrogen bonds and some of the salt bridges. Acids and bases are particularly disruptive to salt bridges, making control of pH important in hair care. The maximum number of salt bridges exist at a pH of 4.1.

The visible portion of hair is dead; only the root is alive. The hair shaft is lubricated by sebum. Washing the hair removes this oil and any dirt adhering to it.

Before World War II, the cleansing agent in shampoos was soap. Soap-based shampoos worked well in soft water but left a dulling film on the hair in hard water. Often, people removed the film by using a rinse containing vinegar or lemon juice. Such a rinse usually is not needed with today's products.

Modern shampoos use a synthetic detergent as a cleansing agent. In shampoos for adults, the detergent is often an anionic type, such as sodium dodecyl sulfate. (Yes, that is the same detergent used in many toothpastes.) For shampoos used on babies and children, the detergent is often an amphoteric type that is less irritating to the eyes. Amphoteric detergents react with both acids and bases. This molecule is an example.

$$\underbrace{CH_3CH_2CH_2CH_2CH_2CH_2CH_2CH_2CH_2CH_2CH_2CH_2CH_2CH_2CH_2CH_2}_{\text{Hydrocarbon chain}}\underbrace{\overset{H}{\underset{H}{\overset{|}{\underset{|}{N^+}}}}{-}CH_2\overset{O}{\overset{\|}{C}}{-}O^-}_{\text{Ionic end}}$$

In acidic solution, it can accept a proton on the negatively charged oxygen. In basic solution, it can give up one of the protons from nitrogen.

The only essential ingredient in shampoo is a detergent of some sort (Figure 17.22). What, then, is all the advertising about? You can buy shampoos that are fruit and herb flavored, protein enriched, pH balanced, and made for oily and dry hair. Let's have a look at some of the gimmicks.

Hair is protein with acidic and basic groups on the protein chain. It stands to reason that the acidity or basicity of a shampoo would affect hair. Hair and skin are slightly acidic. Highly basic (high-pH) or strongly acidic (low-pH) shampoos would damage the hair. More important, such products would irritate the skin and eyes. Most shampoos, however, have pH values between 5 and 8, close to neutral or very slightly acidic.

Because hair is protein, protein-enriched shampoo does give the hair more body. The protein (usually keratin or collagen) coats the hair and literally glues split ends together.

Protein is often added to conditioners, too. Conditioners are mainly long-

Figure 17.22 Shampoos are available in many forms and colors and under many brand names. The only essential ingredient in any shampoo is a detergent. [*Richard Megna/Fundamental Photographs.*]

chain alcohols or long-chain quaternary ammonium salts. The "quats" are similar to the compounds used in fabric softeners (Section 17.7), and they work in much the same way to coat the hair fibers. Although sometimes combined with shampoo, conditioners are normally used after the shampoo.

Shampoos for oily, normal, and dry hair seem to differ primarily in the concentration of the detergent. Shampoo for oily hair is more concentrated; dry hair shampoo is more dilute.

How about all those flavors and fragrances? Ample evidence indicates that such "natural" ingredients as milk, honey, strawberries, herbs, cucumbers, and lemons add nothing to the usefulness of shampoos or other cosmetics. Why are they there? Smells sell, and there is an appeal to those taken by the back-to-nature movement. There is one hazard to the use of such fragrances: bees, mosquitoes, and other insects like fruit and flower odors, too. Using such products before going on a picnic or a hike could lead to a bee in your bonnet.

Hair Coloring

The color of hair and skin is determined by the relative amounts of two pigments: **melanin**, a brownish black pigment, and **phaeomelanin**, a red-brown pigment that colors the hair and skin of redheads. Brunettes have lots of melanin. Blondes have little of either pigment. Brunettes who would like to become blondes can do so by oxidizing the colored pigments in their hair to colorless products. Hydrogen peroxide (Chapter 8) is the oxidizing agent most frequently employed to bleach hair.

Hair dyeing is a good deal more complicated than bleaching. The color may be temporary; water-soluble dyes that can be washed out the next time the hair is washed are sometimes used. More permanent dyes penetrate the hair and remain there. These dyes often are used in the form of a water-soluble, often colorless, precursor that soaks into the hair. The chemical then is oxidized by hydrogen peroxide to a colored compound.

$H_2N-C_6H_4-NH_2$

para-Phenylenediamine

Permanent dyes often are derivatives of an aromatic amine called *para*-phenylenediamine. Variations in color can be obtained by placing a variety of substituents on this molecule. *para*-Phenylenediamine itself produces a black color. The derivative *para*-aminodiphenylaminesulfonic acid is used in blonde formulations. Intermediate colors can be obtained by the use of other derivatives. One such derivative, *para*-methoxy-*meta*-phenylenediamine (MMPD), and its sulfate salt have been shown to be carcinogenic when fed to rats and mice. The hazard to those who use these dyes to color their hair is not yet known. It is interesting to note that one substitute for MMPD was its homolog, *para*-ethoxy-*meta*-phenylenediamine (EMPD). Screening revealed that

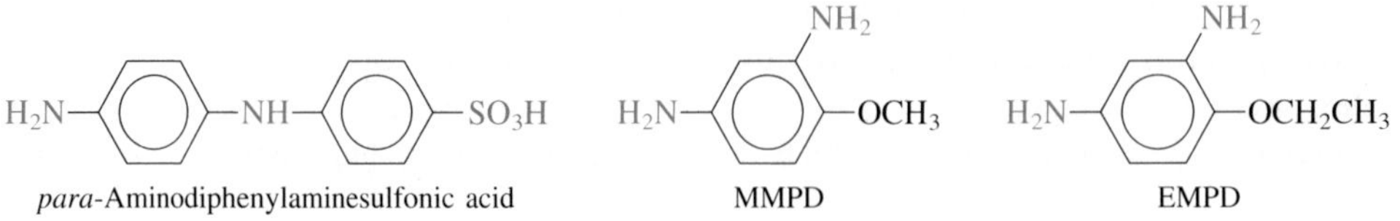

$H_2N-C_6H_4-NH-C_6H_4-SO_3H$

para-Aminodiphenylaminesulfonic acid

$H_2N-C_6H_3(NH_2)-OCH_3$

MMPD

$H_2N-C_6H_3(NH_2)-OCH_2CH_3$

EMPD

EMPD causes mutations in bacteria. Such mutagens often are also carcinogens. The FDA cannot ban a cosmetic without proof of harm, but testing for carcinogenicity would cost at least \$500,000 and take two years. We consider tests of this sort in Chapter 20.

The chemistry of colored oxidation products is quite complex. The products probably include quinones and nitro compounds, among others. These are well-known products of the oxidation of aromatic amines. It should be noted that even permanent dyes affect only the dead outer portion of the hair shaft. New hair, as it grows from the scalp, has its natural color.

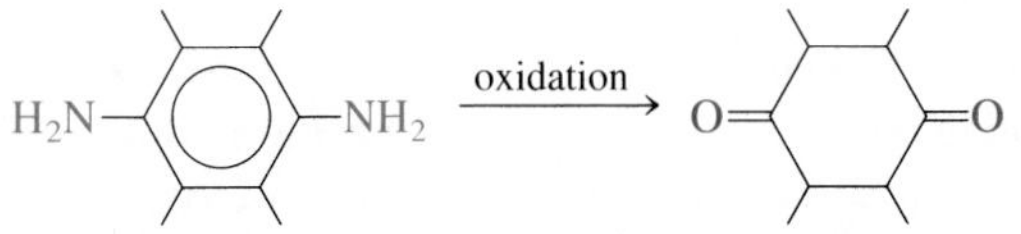

H_2N—(ring)—NH_2 $\xrightarrow{\text{oxidation}}$ O=(ring)=O

A diamine A quinone

(ring)—NH_2 $\xrightarrow{\text{oxidation}}$ (ring)—NO_2

An amine A nitro compound

Hair treatments, such as Grecian Formula, that develop color gradually use a rather simple chemistry. A solution containing colorless lead acetate $[Pb(CH_3COO)_2]$ is rubbed on the hair. As it penetrates the hair shaft, the Pb^{2+} ions react with sulfur atoms in the hair to form black lead sulfide (PbS). Repeated applications lead to darker colors as more lead sulfide is formed.

Permanent Waving: Chemistry to Curl Your Hair

The chemistry of curly hair is interesting. Hair is protein, and adjacent protein chains are held together by disulfide linkages. To put a permanent wave in the hair, you use a lotion containing a reducing agent such as thioglycolic acid. This wave lotion ruptures the disulfide linkages (Figure 17.23), allowing the protein chains to be pulled apart as the hair is held in a curled position on rollers. The hair is then neutralized with a mild oxidizing agent such as hydrogen peroxide. Disulfide linkages are formed in new positions to give shape to the hair.

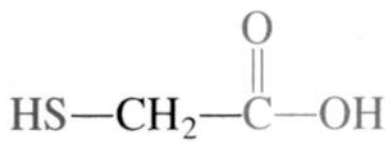

$$HS{-}CH_2{-}\overset{\overset{\large O}{\|}}{C}{-}OH$$

Thioglycolic acid

The same chemical process can be used to straighten naturally curly hair. The change in curliness depends only on how you arrange the hair after the disulfide bonds have been reduced and before the linkages have been restored. As with permanent dyes, permanent curls grow out as new hair is formed.

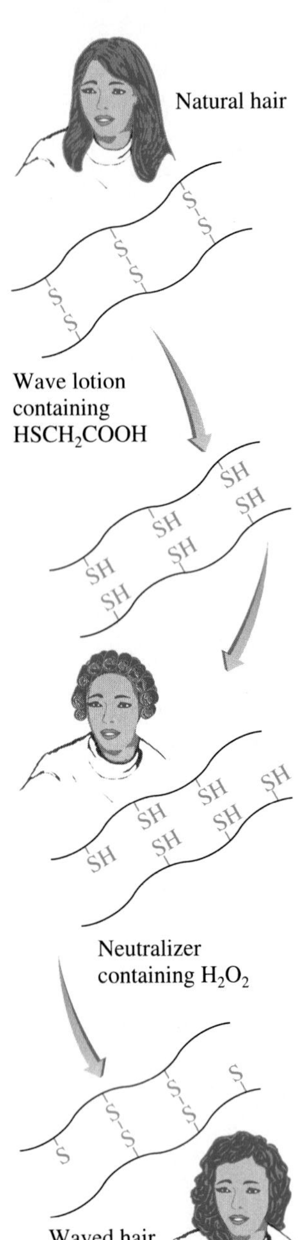

Figure 17.23 Permanent waving of hair is accomplished by breaking disulfide bonds and then reforming them in new positions.

Hair Sprays

Hair can be held in place by using **resins**, solid or semisolid organic materials that form a sticky film on the hair. A common resin used on hair is polyvinylpyrrolidone (PVP).

$$\cdots\underset{\substack{|\\ \text{N-pyrrolidone (C=O)}}}{CH}-CH_2-\underset{\substack{|\\ \text{N-pyrrolidone (C=O)}}}{CH}-CH_2-\underset{\substack{|\\ \text{N-pyrrolidone (C=O)}}}{CH}-CH_2-\underset{\substack{|\\ \text{N-pyrrolidone (C=O)}}}{CH}-CH_2-\underset{\substack{|\\ \text{N-pyrrolidone (C=O)}}}{CH}-CH_2\cdots$$

PVP is dissolved in solvent and sprayed on the hair, where the solvent evaporates. The propellants usually used are flammable hydrocarbons.

Holding resins are also available as mousses. A **mousse** is simply a foam or froth. The active ingredients, like those of the hair sprays, are resins such as PVP. Coloring agents and conditioners are also available as mousses.

Hair Removers

Chemicals that remove unwanted hair are called **depilatories**. Most of them contain a soluble sulfur compound, such as sodium sulfide or calcium thioglycolate $[Ca(CH_2SHCOO)_2]$, formulated into a cream or lotion. They are strongly basic mixtures that destroy some of the peptide bonds in the hair, so that it can be washed off. But remember that skin is made of protein, too, so any chemical that attacks the hair can also damage the skin.

Hair Restorers

For years men have been searching for some way to make hair grow on bald spots. Women suffer from baldness, too, but wigs are much more acceptable to women than artificial hairpieces are to men.

Minoxidil was first introduced as a drug for treating high blood pressure. It acts by dilating the blood vessels. When people who were taking the drug started growing hair on various parts of their bodies, it was applied to the scalps of people who were becoming bald. Minoxidil can produce a growth of fine hair anyplace on the skin where there are hair follicles.

In Canada and Europe minoxidil is sold over-the-counter under the trade name Rogaine. In the United States it is a prescription drug. It needs to be used on a continuous basis, and the cost approaches $1000 per year.

SECTION 17.18

The Well-Informed Consumer

No attempt has been made here to tell you everything about the many chemical products you have in your home. Books have been written about some of these products, and there are many volumes in the library that can help you if you want to know more about any of these familiar chemicals. We hope, however, that the knowledge you have gained here, coupled with what you can read on product labels, will make you a better informed consumer.

When you look over the many different brands of a product at the supermarket, remember that the most expensive one is not necessarily better than the others. Consider the field of soaps and detergents. That is an industry that does a lot of advertising. Each laundry powder and shampoo claims to be superior to all its competitors, even though in most cases all the brands are rather similar. Advertising is expensive, and so the brands that are most widely advertised probably cost the most, too. Of course, it sometimes happens that one brand really is better than another, but the better product might actually be the less expensive one.

If you should be very unhappy with a product you have bought, don't hesitate to tell the company that made it. The cosmetic industry gets around 5000 complaints from customers every year. In addition, people sometimes complain to the Food and Drug Administration (FDA). In 1990 there were about 100 reports to the FDA regarding adverse effects of cosmetic products. About 25% were about makeup. The others had to do mainly with skin care products, fragrances, and various kinds of hair care products.

Most cosmetics are made from inexpensive ingredients, and yet many highly advertised cosmetics have extremely high price tags. Are they worth it? That is a judgment that lies beyond the realm of chemistry. The product inside that fancy little $100 bottle with the famous label may have cost the manufacturer less than five dollars. On the other hand, if the product makes you look better or feel better, perhaps the price is not important. Only you can decide.

Summary

1. Although *soap* has been known for many centuries, it is only in the past two centuries that it has come into common use.
2. *Soap* is made from fat by reaction with lye (sodium hydroxide).
3. A soap molecule has a hydrocarbon tail (soluble in oil or grease) and a polar head (soluble in water), so it can disperse oil or grease in water.
4. In hard water, which contains calcium and magnesium ions, soap forms an insoluble scum.
5. Water softeners, such as sodium carbonate, remove the ions that cause water to be hard as insoluble salts.
6. *Synthetic detergents* are salts of long-chain acids that are not natural fatty acids. Usually they are sulfuric acid derivatives.
7. When the hydrocarbon portion of a detergent molecule has a branched chain structure, it is not easily biodegraded. Biodegradable detergents need linear hydrocarbon chains.
8. In addition to the surface active agent, detergent formulations also include *builders* and *optical brighteners.*
9. Whereas most solid detergents are *anionic surfactants,* liquid laundry detergents are usually *nonionic surfactants.*
10. *Cationic surfactants* (alkylammonium salts) are useful as disinfectants. The ones with two hydrocarbon chains are used as fabric softeners.
11. *Bleaches* are oxidizing agents. Hypochlorite bleaches release chlorine; perborate bleaches release oxygen.
12. Any organic solvents around the house should be

carefully stored because most are flammable and most are toxic.

13. A *paint* contains three basic ingredients: a pigment, a binder, and a solvent.

14. *Waxes* are esters of fatty acids with long chain alcohols.

15. The skin is made of a tough fibrous protein called *keratin.*

16. Skin lotions are emulsions of oil in water; skin creams are emulsions of water in oil.

17. Eye makeup and lipstick are mainly blends of oil, waxes, and pigments.

18. Toothpaste contains detergent, abrasive, thickener, flavoring, and often an added fluoride compound.

19. *Perfumes* can usually be split into three basic fractions: the *top note* (which is most volatile), the *middle note* (of intermediate volatility), and the *end note* (of lowest volatility). *Colognes* are diluted perfumes.

20. Shampoos are synthetic detergents blended with perfume ingredients and sometimes mixed with protein or other conditioners to give the hair more body.

21. Dark hair contains the pigment *melanin;* red hair contains *phaeomelanin;* and blonde hair contains very little of either pigment.

22. Permanent waving of the hair uses a reducing agent to break the disulfide linkages and then an oxidizing neutralizer to reform the disulfide bonds.

23. The most expensive brand of a given product is not necessarily the best.

Key Terms

anionic surfactants 17.6
antiperspirants 17.14
astringent 17.14
bleaches 17.8
builder 17.5
cationic surfactants 17.7
colognes 17.16
cosmetics 17.14
cream 17.14
deodorants 17.14
depilatories 17.17
emollients 17.14
end note 17.16
epidermis 17.14
eye shadow 17.14
hypoallergenic cosmetics 17.16
keratin 17.14
lanolin 17.13
lotion 17.14
mascara 17.14
melanin 17.14
middle note 17.16
moisturizers 17.14
mousse 17.17
nonionic surfactants 17.6
optical brighteners 17.5
paint 17.12
perfumes 17.16
phaeomelanin 17.17
resins 17.17
saponins 17.1
sebum 17.14
skin protection factor 17.14
soap 17.2
sunscreen lotion 17.14
surface-active agent 17.5
synthetic detergents 17.4
top note 17.16
waxes 17.13

Review Questions

1. What are saponins?
2. How do carbonate ions make water basic?
3. What is a soap?
4. What products are formed when tripalmitin reacts with lye (sodium hydroxide)?
5. List some ingredients found in toilet soaps.
6. What kind of ingredients are used in scouring soaps?
7. How does a potassium soap differ from a sodium soap?
8. Give the structural formula for each of the following.
   a. potassium stearate b. sodium palmitate
9. How does soap (or detergent) clean a dirty surface?
10. Why does soap fail to work in acidic water?
11. What is hard water?
12. How does hard water affect the action of soap?
13. How does trisodium phosphate soften water? (List two ways.)
14. How does a water-softening tank work?

15. What are some advantages of soap?
16. What are the advantages of synthetic detergents (such as sodium lauryl sulfate)?
17. What is an LAS detergent?
18. What is a surface-active agent (surfactant)?
19. What is a (detergent) builder?
20. How does sodium tripolyphosphate aid the cleaning action of a soap or detergent?
21. How do zeolites aid the cleaning action of a soap or detergent?
22. List two advantages of sodium carbonate as a replacement for phosphates as a builder in detergent formulations.
23. What advantages do zeolites have over carbonates as a builder?
24. What is an optical brightener? How do these compounds work?
25. What is an anionic surfactant?
26. What is an nonionic detergent?
27. What ingredients are used in liquid dishwashing detergents (for hand dishwashing)?
28. How do detergents for automatic dishwashers differ from those used for hand dishwashing?
29. How do cyanurate bleaches (such as Symclosene) work?
30. How does sodium perborate work as a bleach?
31. List some uses of diluted household ammonia.
32. What safety precautions should be followed when using ammonia?
33. What sort of surfaces should not be cleaned with ammonia? Why?
34. List two properties of baking soda that make it useful as a cleaning agent.
35. List a use of vinegar in cleaning.
36. Why should marble surfaces not be cleaned with vinegar?
37. List some uses of organic solvents in the home.
38. List two useful properties of pine oil cleansers.
39. What are the main hazards of using solvents in the home?
40. Should gasoline be used as a cleaning solvent? Why or why not?
41. What are cationic surfactants?
42. Cationic surfactants are not particularly good detergents, yet they are widely used, especially in the food industry. Why?
43. Cationic surfactants often are used along with nonionic surfactants, but seldom are used with anionic surfactants. Why not?
44. Give an example of a compound that acts as a fabric softener.
45. How do fabric softeners work?
46. List the essential ingredients in each of the following household products.
    a. scouring cleaner
    b. window cleaner
    c. oven cleaner
    d. toilet bowl cleaner
    e. liquid drain cleaner
    f. solid drain cleaner
    g. paint remover
47. What is the legal definition of a cosmetic?
48. Is soap a cosmetic?
49. What are the essential ingredients in a toothpaste?
50. How do fluorides strengthen tooth enamel?
51. What is the principal cause of tooth decay?
52. What are the best ways to prevent tooth decay?
53. What is the principal material of the corneal layer of skin?
54. What is sebum?
55. What is an emollient?
56. What is the ideal moisture content of skin?
57. What are the principal components of creams and lotions?
58. What is a skin moisturizer?
59. How does lipstick differ from a skin cream?
60. What materials are used to color lipsticks?
61. What is a perfume?
62. What are the three fractions of a perfume? How do these fractions differ at the molecular level?
63. What is musk? Why are musks added to perfumes?
64. What are colognes?
65. What are the principal ingredients of an aftershave lotion?
66. Why is menthol added to some aftershave lotions?
67. What is a deodorant?

68. What is an antiperspirant?

69. What is the only essential ingredient of shampoos?

70. What type of detergent is used in baby shampoos? Why?

71. What chemical substances determine the color of hair and skin?

72. What chemical compound is used most often to bleach hair?

73. What is the difference between a temporary and a permanent hair dye? Why is a permanent dye not really permanent?

74. What kind of chemical reagent is used to break disulfide linkages in hair?

75. What kind of chemical reagent is used to restore disulfide linkages in hair?

76. What are resins?

Problems

Detergents: Structures and Properties

Problems 77 through 81 refer to the following structures.

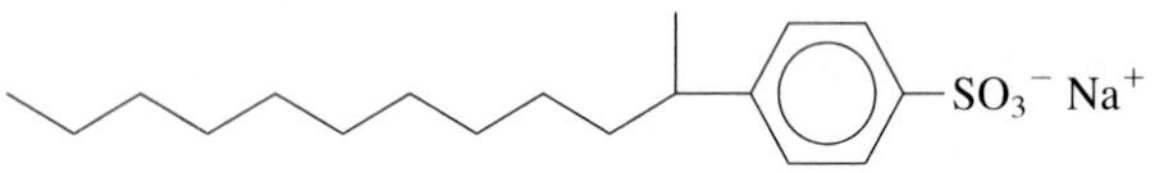

$—SO_3^-\ Na^+$

I

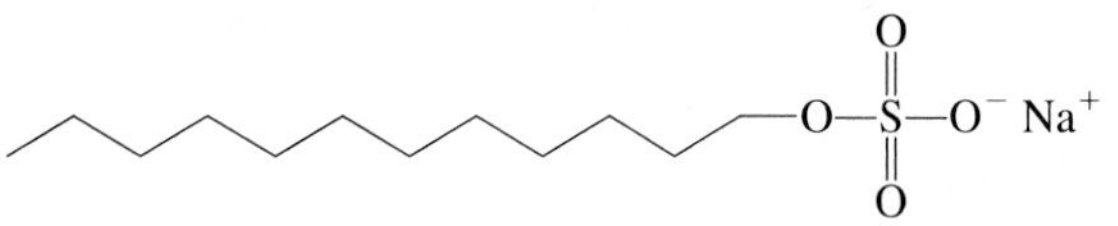

$—O—\overset{O}{\underset{O}{\overset{\|}{\underset{\|}{S}}}}—O^-\ Na^+$

II

$—\overset{CH_3}{\underset{CH_3}{\overset{|}{\underset{|}{N^+}}}}—CH_3\ Cl^-$

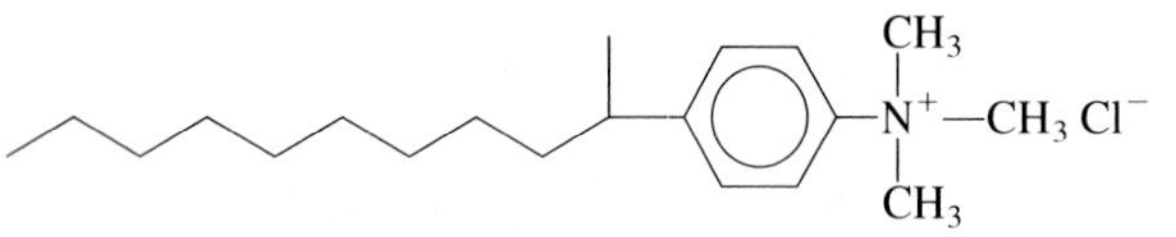

III

77. Which are biodegradable?

78. Which is an LAS detergent?

79. Which is sodium lauryl sulfate?

80. Which are anionic detergents?

81. Which is a cationic detergent?

Chemical Equations

82. Write the equation that shows how carbonate ion reacts with water to form a basic solution.

83. Write the equation that shows how phosphate ion reacts with water to form a basic solution.

84. Give the equation that shows how carbonate ions precipitate hard water ions. You may use M^{2+} to represent the hard water ion.

85. Give the equation that shows how phosphate ions precipitate hard water ions. You may use M^{2+} to represent the hard water ion.

86. Give the structures of the products formed in the following reaction.

$$\begin{array}{l} CH_2O—\overset{O}{\overset{\|}{C}}(CH_2)_{10}CH_3 \\ | \\ CH—O—\overset{O}{\overset{\|}{C}}(CH_2)_{10}CH_3 \ + \ 3\ NaOH \longrightarrow \\ | \\ CH_2O—\overset{O}{\overset{\|}{C}}(CH_2)_{10}CH_3 \end{array}$$

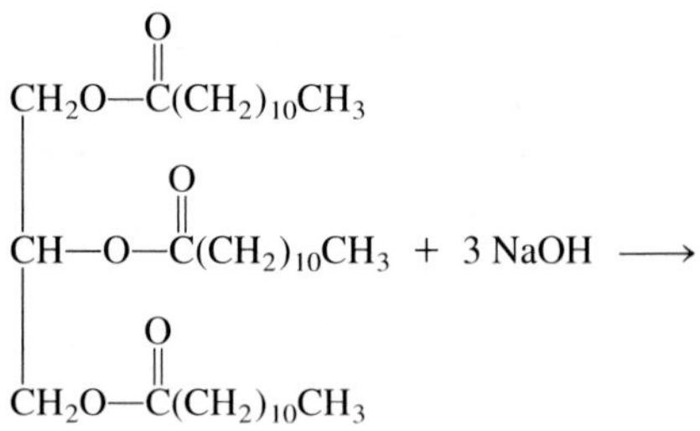

87. Give the structures of the products formed in the following reaction.

$$\begin{array}{l} CH_2O—\overset{O}{\overset{\|}{C}}(CH_2)_6CH_3 \\ | \\ CH—O—\overset{O}{\overset{\|}{C}}(CH_2)_4CH_3 \ + \ 3\ NaOH \longrightarrow \\ | \\ CH_2O—\overset{O}{\overset{\|}{C}}(CH_2)_8CH_3 \end{array}$$

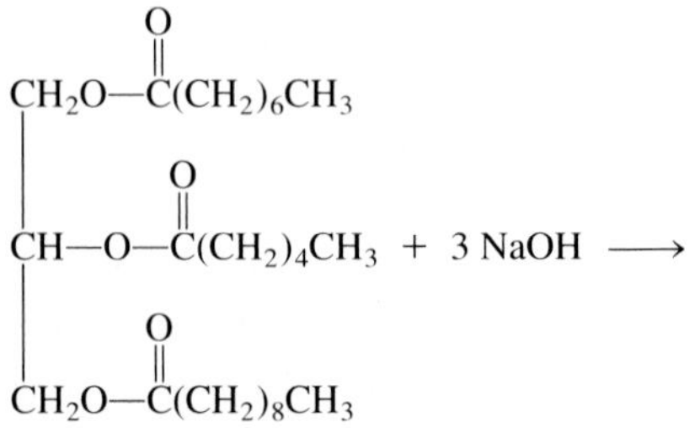

Types of Chemical Reactions

88. What kind of chemical reaction is involved when disulfide linkages in hair are broken?

89. What kind of chemical reaction is involved when new disulfide linkages are formed in hair?

90. What kind of chemical reaction is involved when a colorless hair dye compound is converted to a colored dye?

91. What kind of chemical reaction is involved in bleaching hair?

Chemical Structures and Properties

Problems 92 through 95 refer to structures I through IV below.

92. Which are synthetic detergents?

93. Which are soaps?

94. Which is an amphoteric detergent?

95. Which is a principal ingredient in baby shampoos?

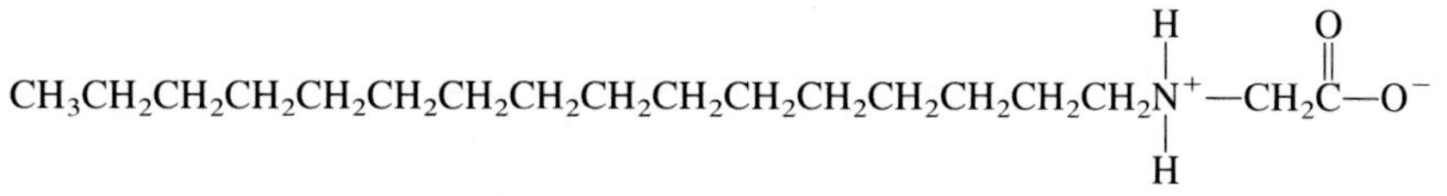

$$CH_3CH_2CH_2CH_2CH_2CH_2CH_2CH_2CH_2CH_2CH_2CH_2CH_2CH_2CH_2CH_2\overset{H}{\underset{H}{|}}N^+{-}CH_2\overset{O}{\overset{\|}{C}}{-}O^-$$

I

$$CH_3CH_2CH_2CH_2CH_2CH_2CH_2CH_2CH_2CH_2CH_2COO^-\ Na^+$$

II

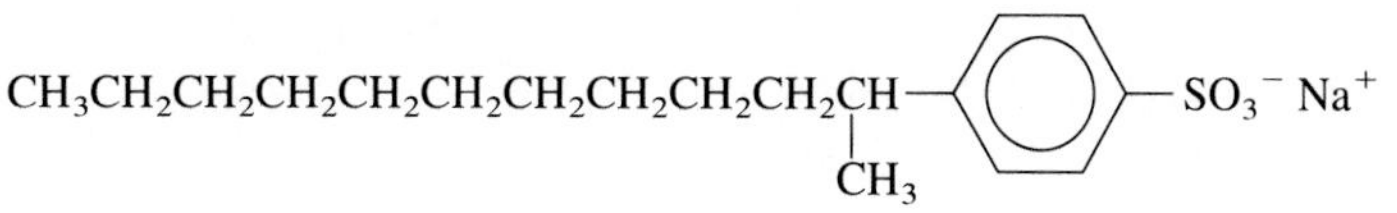

$$CH_3CH_2CH_2CH_2CH_2CH_2CH_2CH_2CH_2CH_2\underset{CH_3}{\underset{|}{CH}}{-}C_6H_4{-}SO_3^-\ Na^+$$

III

$$CH_3CH_2CH_2CH_2CH_2CH_2CH_2CH_2CH_2CH_2CH_2CH_2{-}C_6H_4{-}\overset{CH_3}{\underset{CH_3}{N^+}}{-}CH_3\ Cl^-$$

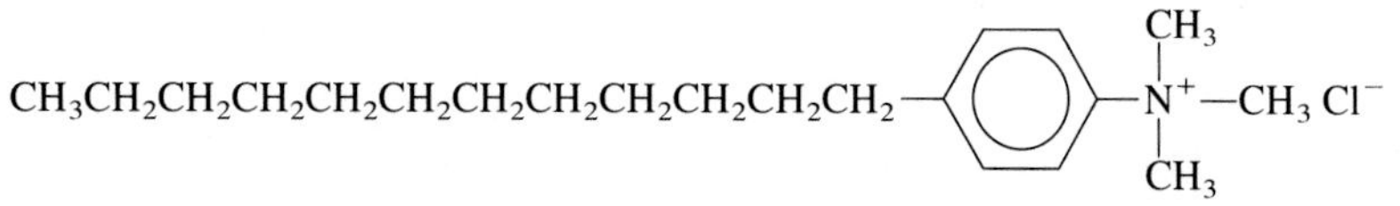

IV

Projects

96. The Clearwater Chemical Company has announced the development of a biodegradable detergent. What tests, if any, should be made before the detergent is marketed? Justify your recommendations.

97. Maxisuds, Inc., has announced that it has found a replacement for phosphate builders in detergents. What tests, if any, should be made before the builder is used in detergent formulations? Should Maxisuds be allowed to market the builder until it is proven harmful? Can a product ever be proven safe? Justify your position.

98. Read the labels of five brands of toothpaste. List the ingredients of each. If you can, classify each ingredient, according to its function—that is, a detergent, an abrasive, and so on. You may find a reference book such as *The Merck Index* (reference 3) to be helpful.

99. Read the labels on five brands of lotions or creams (those designed for use on the skin, whether face, hands, or body). List the ingredients in each brand. If you can, tell the function of each ingredient.

100. Read the labels on five brands of shampoos. List the ingredients in each brand. Which ingredients are detergents? If you can, tell the function of each nondetergent ingredient.

References and Readings

1. "All-Purpose Cleaners." *Consumer Reports,* August 1988, pp. 519–521. See also "Cleansers: One Stood Out," *Consumer Reports,* January 1990, pp. 61–63.
2. "Baldness: Is There Hope?" *Consumer Reports,* September 1988, pp. 543–547.
3. Budavari, Susan (Ed.). *The Merck Index,* 11th ed. Rahway, NJ: Merck and Co., 1989.
4. "Chemistry of Cosmetics." *Journal of Chemical Education,* December 1978, pp. 802–803.
5. Commoner, Barry. *The Closing Circle.* New York: Bantam Books, 1972. See pp. 151–156 for comments on LAS degradation.
6. "Dealing With Your Drains." *Consumer Reports,* January 1994, pp. 44–48.
7. "Dishwashing Liquids." *Consumer Reports,* March 1988, pp. 176–178.
8. Drozd, Joseph C. "An Introduction to Light Duty (Dishwashing) Liquids." *Chemical Times and Trends,* July 1984, pp. 29–32.
9. "Facial Cleansers." *Consumer Reports,* June 1989, pp. 408–409. They take off makeup and do little else.
10. Fowkes, Frederick M. "Principles of Detergency." *Chemical Times and Trends,* January 1983, pp. 31–33, 53.
11. "Furniture Cleaners and Polishes." *Consumer Reports,* June 1993, pp. 358–360.
12. Goldschmiedt, Henry. *Practical Formulas for Hobby and Profit.* New York: Chemical Publishing Company, 1973. Chapter 3 gives recipes for cosmetics.
13. Greek, Bruce F. "Detergent Industry Ponders Products for New Decade." *Chemical and Engineering News,* 29 January 1990, pp. 37–56.
14. Griffin, John J., Robert F. Corcoran, and Kenn K. Akana. "The pH of Hair Shampoos." *Journal of Chemical Education,* September 1977, pp. 553–554.
15. "Hair Conditioners." *Consumer Reports,* January 1986, pp. 56–59.
16. "Hair Mousse." *Consumer Reports,* October 1988, pp. 638–641.
17. "Hand and Bath Soaps." *Consumer Reports,* January 1985, pp. 52–55.
18. "Hand Lotions and Creams." *Consumer Reports,* August 1977, pp. 448–451.
19. Hart, J. Roger. "EDTA-Type Chelating Agents in Everyday Consumer Products: Some Medicinal and Personal Care Products." *Journal of Chemical Education,* December 1984, pp. 1060–1061.
20. "Hazardous Waste at Home: Handle with Care." *Consumer Reports,* February 1994, p. 101.
21. "Interior Latex Paints." *Consumer Reports,* February 1994, pp. 127–131.
22. Layman, Patricia L. "Cosmetics." *Chemical and Engineering News,* 29 April 1985, pp. 19–46.
23. Layman, Patricia L. "Detergents Shift Focus of Zeolites Market." *Chemical and Engineering News,* 27 September 1982, pp. 10–15.
24. "Lipsticks." *Consumer Reports,* February 1988, pp. 75–78.
25. Marmion, D. *Handbook of Colorants for Food, Drugs, and Cosmetics,* 3rd ed. New York: John Wiley, 1991.
26. Morris, E. T. *Fragrance: The Story of Perfume from Cleopatra to Chanel.* New York: Charles Scribner's Sons, 1990.
27. Muller, P. M., and D. Lamparsky (Eds.). *Perfumes: Art, Science and Technology.* Essex: Elsevier, 1991.
28. Murphy, Lillian S. "Don't Buy from the Shelf When You Can Make It Yourself." *Modern Maturity,* October–November 1982, pp. 17–18.
29. "New Ways to Save Your Teeth?" *Consumer Reports,* August 1989, pp. 504–509. Evaluates toothpastes, mouthwashes, and other products that claim to help save teeth.
30. Niemark, Jill. "Dew Gooders." *American Health,* October 1984, pp. 22–23. Discusses moisturizers.
31. "Painting the House." *Consumer Reports,* September 1993, pp. 610–615.
32. Rinzler, Carol Ann. "The New and Improved Chemistry of Cosmetics." *Science 82,* April 1982, pp. 54–61.
33. Schamper, Tom. "Chemical Aspects of Antiperspi-

rants and Deodorants.'' *Journal of Chemical Education,* March 1993, pp. 242–244.

34. ''Shampoos and Conditioners: Heading Off the Hype.'' *Consumer Reports,* June 1992, pp. 395–403.

35. Stehlin, Dori. ''Cosmetic Safety: More Complex Than At First Blush.'' *FDA Consumer,* November 1991, pp. 16–23.

36. ''Sunscreens.'' *Consumer Reports,* June 1988, pp. 370–374.

37. ''Toothpastes.'' *Consumer Reports,* March 1986, pp. 144–147.

38. Traven, Beatrice. *Here's Egg on Your Face.* New York: Pocket Books, 1971. Tells you how to make cosmetics with ingredients from the grocery store and drugstore.

39. Williams, D. F., and W. H. Schmitt (Eds.). *Chemistry and Technology of the Cosmetics and Toiletries Industry.* New York: Routledge, Chapman & Hall, 1992.

18 Fitness and Health

The Chemical Connection

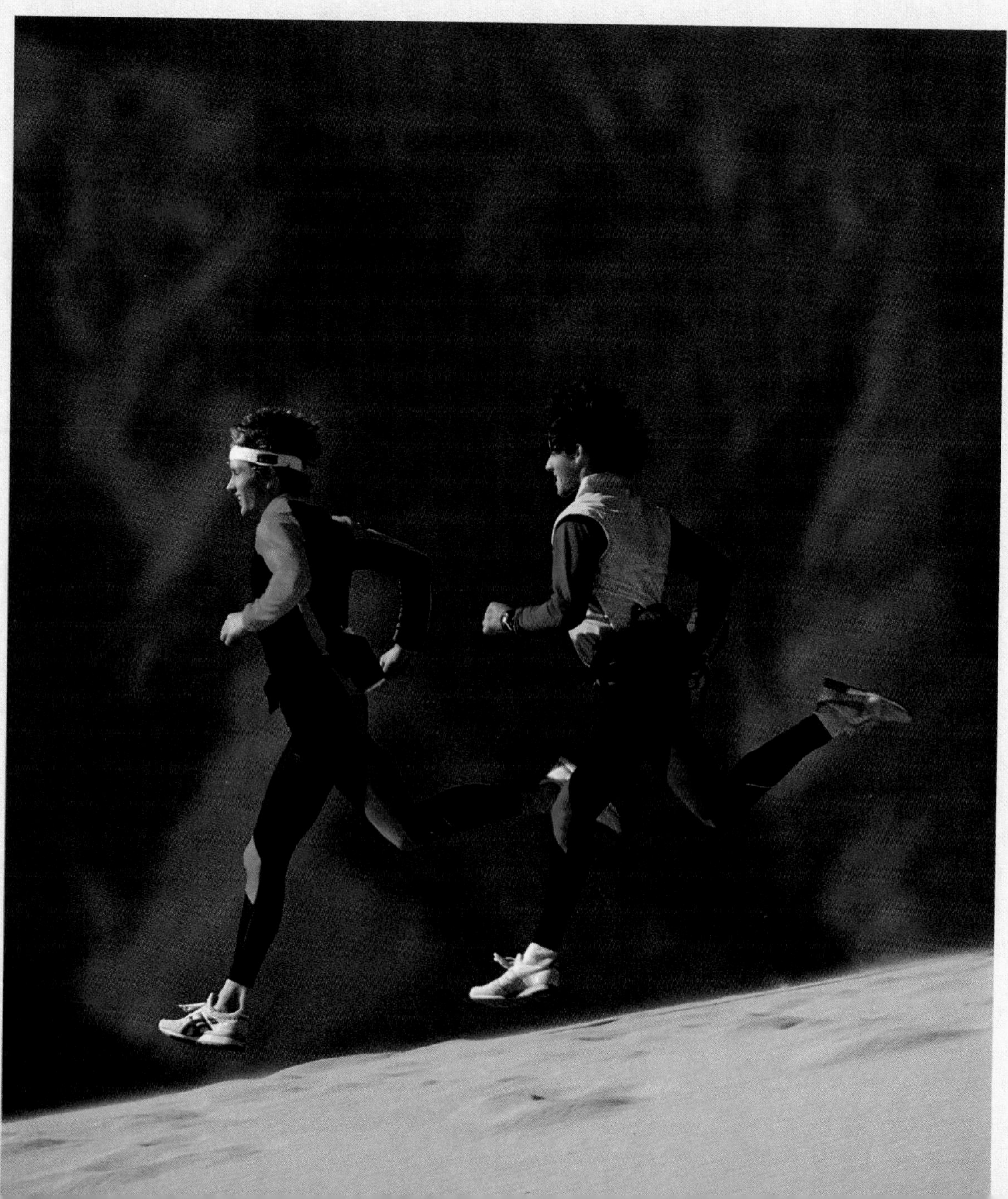

The key to health and fitness
Is just to realize
You need a healthy diet
And lots of exercise.

We in the Western World have spent centuries developing labor-saving devices. Now many of us earn our daily bread with little physical exertion. That presents us with a different kind of problem: how to stay healthy and physically fit. We now put great effort into keeping our bodies fit through exercise and proper diet.

In our culture today, thin is in. Many former sedentary people have joined the ranks of joggers, walkers, tennis players, body-builders, and bikers in pursuit of the bulgeless body. About a million people per week are treated in weight-loss clubs and clinics. Countless others treat themselves with crash diets, dietary supplements, diet pills, and other programs. Diet books populate the bestseller lists. Many of these diet plans are simply fads. They are more likely to increase their creator's wealth than to decrease your weight. Many are grossly unbalanced. An unbalanced diet, especially over an extended period of time, causes a variety of nutritional deficiencies, a decrease in resistance to disease, and a decline in general health.

Through modern science we have a much better understanding of what it means to be physically fit and what it takes to achieve fitness (Figure 18.1). Athletes are larger, stronger, and faster than ever before. Quackery abounds, however. We read of diets, drugs, and exercise plans that promise health and longer life. Often these schemes require little effort on our part. It is difficult sometimes to sort science from nonsense. In this chapter we examine some of the ways chemistry contributes to athletic achievement and to our understanding of health and fitness. We examine muscle action, diet, electrolyte balance, drugs, and other items of concern to those interested in their own well being.

SECTION 18.1

Nutrition

It hardly need be said that good health requires a nutritious diet. After all, "you are what you eat." Good nutrition means a sufficient intake of carbohy-

Exercise is an important component of a lifestyle that fosters good health and fitness. [*John Kelly/The Image Bank.*]

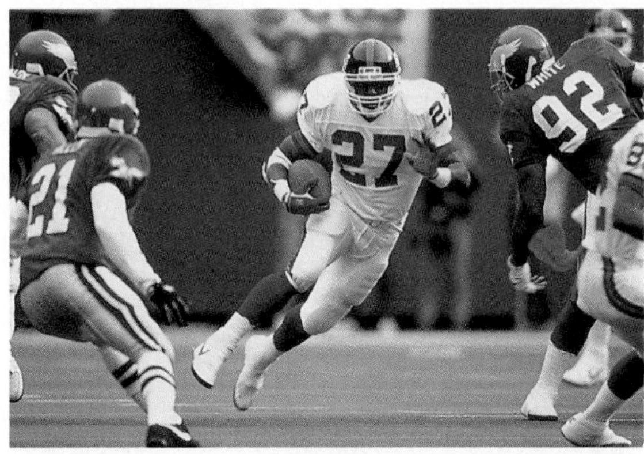

Figure 18.1 Athletic performance has been transformed in recent years. Biochemistry has brought a better understanding of the athlete's body and the way it works. Chemistry has provided drugs that keep athletes healthier (but increase the potential for abuse) and materials that make sports safer and increase proficiency. [*Top to bottom: Al Tielemans, Steven E. Sutton, and Paul J. Sutton/Duomo.*]

drates, fats, and proteins, including all the essential minerals and vitamins, along with a measure of dietary fiber and plenty of water.

Fewer Calories

For people in many parts of the world, food is in such short supply that obtaining anything at all to eat is a daily struggle. Meanwhile, in the United States one of our most serious health problems is overeating.

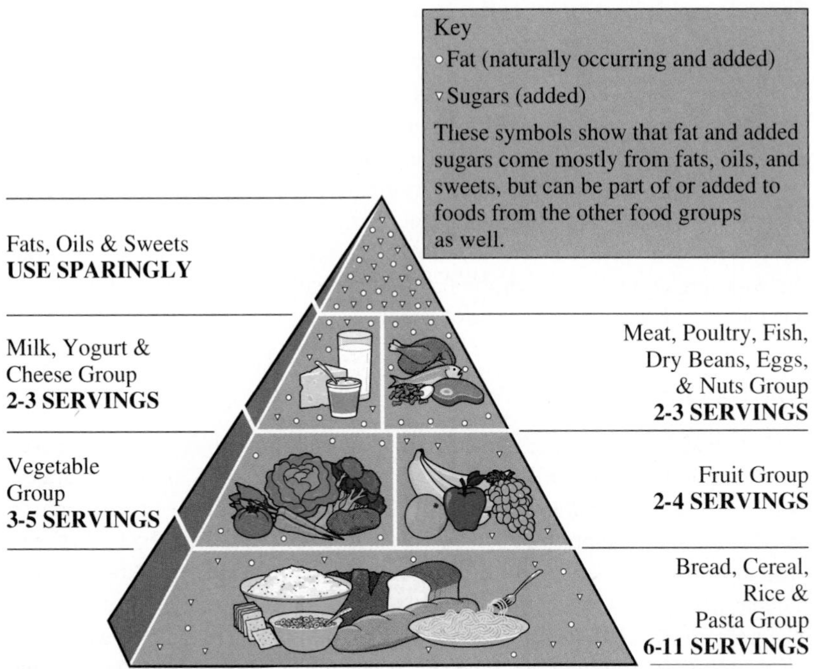

Figure 18.2 The daily food guide pyramid. Start with bread, cereal, rice, and pasta, along with fruits and vegetables; then add some servings from the milk and meat groups; and go easy on fats, oils, and sweets. [*U.S. Department of Agriculture/U.S. Department of Health and Human Services.*]

USDA serving sizes *Milk, yogurt, and cheese:* 1 cup of milk or yogurt, 1½ oz of natural cheese, 2 oz of process cheese. *Meat:* 2 to 3 oz of cooked lean meat, fish, or poultry; 1 to 1½ cups of cooked dry beans; 2 to 3 eggs; 4 to 6 tbs of peanut butter. *Vegetables:* 1 cup of raw, leafy vegetables, ½ cup of other vegetables (cooked or chopped raw), ¾ cup of vegetable juice. *Fruit:* 1 medium apple, banana, or orange; ½ cup of chopped, cooked, or canned fruit; ¾ cup of fruit juice. *Bread, cereal, rice, and pasta:* 1 slice of bread; 1 oz of ready-to-eat cereal; ½ cup of cooked cereal, rice, or pasta.

Research has shown repeatedly that undereating is much healthier than overeating. Massive studies with mice have shown that when mice are given limited food (40% less than the control group chose) they live longer and have fewer tumors. Humans also appear to stay healthier when they eat a bit too little than when they eat too much.

Total calories consumed is important, but even more important is the distribution of calories. Modern advice is that good nutrition should include generous portions of complex carbohydrates (starches), along with fruits and vegetables, modest portions of protein, and very small amounts of fats, oils, and simple carbohydrates (sugars). See the food guide pyramid in Figure 18.2.

Less Fat But More Starch

The recommended diet contains no more than 25% of calories from fat. Unfortunately, the typical American diet is 34% fat. (For some it runs as high as 45%.) In China people normally consume about 20% more calories than Americans do, but they are 25% thinner. The big difference is that the Chinese eat twice as much starch as Americans do, but only one-third as much fat.

Less Protein—Especially Red Meat

Americans also consume more meat than they need. It appears that for people with a high protein diet, weight gain is faster and weight loss is more difficult. (Studies with rats indicate that when protein is strictly limited, the rats can eat as much as they want with very little change in weight.)

It does appear that Americans are trying to reduce their consumption of red meat. In 1976 the average consumption of beef per person was 95 lb per year. By 1994 it had been reduced to 70 lb per person. This probably results from the widespread publication of the fact that a steady diet of red meat increases the rate of colon cancer.

Nutrition and the Athlete

Athletes need more calories because they expend more than the average sedentary individual. The athlete should get those extra calories as carbohydrates. Foods rich in carbohydrates, especially the starches, are the cheapest source of calories, and are the preferred source of energy for the healthy body. Fat- and protein-rich foods also supply calories, but they are more expensive; protein metabolism also produces more toxic wastes that tax the liver and kidneys. The pregame steak dinner consumed by football players in the past was based on a myth that protein builds muscles; if athletes want more muscle, says the myth, they should eat more protein. This is just not true. Although athletes do need the RDA for protein (based on grams per kilogram of body weight), they do not need an excess. Protein consumed in amounts greater than needed for synthesis and repair of tissue will only make the athlete fatter (due to excessive calorie intake) and not more muscular.

Some extreme endurance athletes such as triathletes and ultramarathoners may need a bit more than the RDA for protein. Since nearly all Americans eat 50% more protein than they need, even these special athletes seldom need protein supplements.

Muscles are built through exercise, not through eating excess protein. When a muscle contracts against a resistance, an amino acid called creatine is released. Creatine stimulates the production of the protein myosin, thus building more muscle tissue. If the exercise stops, the muscle begins to shrink after about 2 days. After about 2 months without exercise, muscle built through the exercise program is almost completely gone. (The muscle does *not* turn to fat, as some athletes believe. Former athletes often get fat, though, because they continue to take in the same number of calories and expend fewer.)

$$HN{=}C(NH_2){-}N(CH_3){-}CH_2C({=}O){-}OH$$

Creatine

SECTION 18.2

Crash Diets: Quick = Quack

Most quick weight-loss diets depend on factors other than fat metabolism to hook a prospective customer. Many contain a **diuretic**, such as caffeine, to increase the output of urine. Weight loss is water loss; weight is regained when the body is rehydrated.

Other such diets depend on depleting the body's stores of glycogen. When carbohydrates are eliminated from the diet, the body draws on its glycogen reserves, depleting them in about 24 hr. Recall (Chapter 15) that glycogen is a polymer of glucose. Glycogen molecules have lots of hydroxyl (—OH) groups that can form hydrogen bonds to water molecules. Each pound of glycogen carries with it about 3 lb of water held to it by these hydrogen bonds. Depleting the pound or so of glycogen results in a weight loss of about 4 lb (1 lb glycogen + 3 lb water). No fat is lost, and the weight is quickly regained when the dieter resumes eating carbohydrates.

If your normal energy expenditure is 2400 kcal/day, the most fat you could lose by total fasting for a day would be 0.69 lb (2400 kcal/day divided by 3500 kcal/lb adipose tissue). That assumes your body would burn nothing but fat. It won't. Recall (Chapter 15) that the brain runs on glucose, and if that glucose isn't supplied in the diet, it is obtained from protein. On any diet that restricts carbohydrate intake, you lose muscle mass as well as fat. When you gain the weight back (as 90% of all dieters do), you gain only fat. People who diet and then gain back the lost weight are replacing metabolically active tissue (muscle) with inactive fat. Weight loss becomes harder with each subsequent attempted diet.

On a weight-loss diet (without exercise), about 65% of the loss is fat. About 11% is protein (muscle tissue). The rest is water and a little glycogen.

▶ EXAMPLE 18.1

If you ordinarily expend 2200 kcal/day, and you go on a diet of 1500 kcal/day, how long would it take to diet off 1 lb of fat?

SOLUTION

You would use 700 kcal/day more than you consume. There are 3500 kcal in 1 lb of fat, so it would take

$$3500\ \cancel{\text{kcal}} \times \frac{1\ \text{day}}{700\ \cancel{\text{kcal}}} = 5\ \text{days}$$

(Keep in mind, though, that your weight loss would not be all fat. You would probably lose more than 1 lb, but it would be mostly water with some protein and a little glycogen.)

Exercise 18.1

A person who expends 1800 kcal/day goes on a diet of 1200 kcal/day without a change in activities. How much fat will she lose if she stays on a diet for 3 weeks?

Any weight-loss program that promises a loss of more than a pound or two a week is likely to be dangerous quackery.

It is possible to lose weight through dieting. If you reduce your intake by 100 kcal/day and keep your activity constant, you will burn off a pound of adipose tissue (3500 kcal) in 35 days. Unfortunately, people are seldom so patient, and they resort to more stringent diets. To achieve their goals more rapidly, they exclude certain foods and reduce the amounts of others. Such diets are harmful. Diets with fewer than 1200 kcal/day are likely to be deficient in necessary nutrients, particularly in B vitamins and iron. Further, dieting slows down metabolism. Weight loss through dieting is quickly regained when the dieter resumes old eating habits.

According to one theory, hunger is regulated by the hypothalamus. When the hypothalamus senses that the level of fatty acids circulating in the bloodstream is low, it triggers the hunger mechanism. According to the **set-point theory**, each of us has a unique level at which the hypothalamus acts. Some of us might have a higher level of body fat than others to avoid constant hunger. This is consistent with the fact that obesity seems to be inherited. It is also rather grim news for those of us who would like to lose weight; we can do so only by being constantly hungry. There is some good news, however. It appears that our set point can be lowered through exercise.

SECTION 18.3

Diet and Exercise

People who do not increase their food intake when they begin an exercise program will lose weight. And contrary to myth, exercise (up to 1 hr a day) will not cause an increase in appetite. Most of the weight loss from exercise is due to the increase in metabolic rate during the activity, but the increased metabolic rate continues for several hours after completion of the exercise. Exercise helps us maintain both fitness and thinness.

▶ EXAMPLE 18.2

A fast game of tennis burns off about 10 kcal/min. How long would you have to play to burn off 1 lb of adipose tissue (fat).

SOLUTION

One pound of adipose tissue stores 3500 kcal of energy. To burn it off at 10 kcal/min requires

$$3500\,\cancel{\text{kcal}} \times \frac{1\text{ min}}{10\,\cancel{\text{kcal}}} = 350\text{ min}$$

It takes 350 min (4 hr 50 min) to burn off 1 lb of fat, even with a fast-paced game of tennis. (You don't have to do it all in one day.)

Exercise 18.2

Walking a mile burns off about 100 kcal. How far do you have to walk to burn off 1 lb of fat?

The most sensible approach to weight loss is to adhere to a balanced low-calorie diet that meets the RDA for essential nutrients, and to engage in a reasonable, consistent, individualized exercise program. Hence, the principles of weight loss are met by decreasing intake and increasing output. All people, athletes included, should be aware of the possible risks in dieting. Vitamin deficiency often develops slowly over months or years. This problem can be corrected or avoided by careful nutritional planning or by taking vitamin supplements.

Minerals such as iron, calcium, and potassium also are deficient in many crash diets. A deficiency of these minerals can interrupt the smooth function of nerve impulse transmission to muscle. This impairs athletic performance. Impulse transmission to vital organs also may be impaired in cases of severe restriction. Several deaths from cardiac arrest have resulted from variations of the ''liquid protein diet.''

Weight loss or gain is based on the law of conservation of energy (Chapter 14). If we take in more calories than we use up, the excess calories are stored as fat. If we take in fewer calories than we need for our activities, our bodies burn some of the stored fat to make up for the deficit. One pound of adipose (fatty) tissue contains about 3500 kcal. This tissue requires 200 miles of blood capillaries to serve its cells. Excess fat therefore puts extra strain on the heart; it has to work harder to supply blood to the extra tissue.

A moderately active person can calculate the calories needed each day to maintain proper weight by multiplying the desired weight (in pounds) by 15 kcal/lb. If you want to maintain a weight of 120 lb, you need (120 lb × 15 kcal/lb) = 1800 kcal per day.

How much fat is enough? The male body requires about 3% essential body fat; the average female 10 to 12%. It is difficult to measure percent body fat accurately. Skinfold calipers are quite inaccurate; they measure water retention as well as fat. Dunk tanks for measuring body density (Figure 18.3) are better, but results vary with the amount of air in the lungs. Fat is less dense (0.903 g/mL) than the water (1.000 g/mL) that makes up most of the mass of

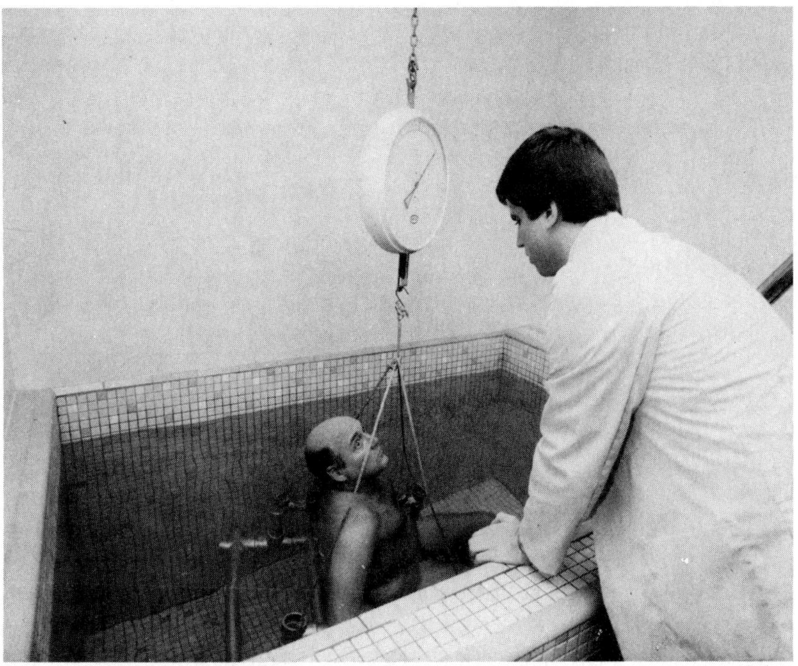

Figure 18.3 Percent of body fat can be determined by weighing a person completely submerged in water. The calculation must include a correction for the volume of air in the lungs. [*Will & Deni McIntyre/Photo Researchers, Inc.*]

our bodies. The higher the proportion of body fat in a person, the more buoyant that person is in water.

Body Mass Index

The body mass index (BMI) is a commonly used measure of fatness. It is defined as follows:

$$\text{BMI} = \frac{700 \times \text{Body weight (lb)}}{[\text{Height (in.)}]^2}$$

In other words, for a person who is 5 ft 10 in. tall (70 in.) and weighs 180 lb, the body mass index is

$$\text{BMI} = \frac{700 \times 180}{70 \times 70} = 26$$

Average BMI values for adults are between 20 and 25. Values greater than 27 may be associated with obesity as well as other related problems, such as high blood pressure or diabetes.

EXAMPLE 18.3

What is the body mass index for a person who is 6 ft 2 in. tall and weighs 230 lb?

SOLUTION

The person's height is $6 \times 12 + 2 = 72 + 2 = 74$ in.

$$\text{BMI} = \frac{700 \times 230}{74 \times 74} = \frac{161{,}000}{5500} = 29$$

The body mass index is 29.

Exercise 18.3

What is the body mass index for a person who is 5 ft tall and weighs 150 lb?

SECTION 18.4

Vitamins and Minerals

Medical spokesmen often insist that vitamin and mineral supplements are unnecessary when the diet is well balanced and includes the **recommended daily allowance** (RDA) for each nutrient. However, for those who choose to take such supplements, reasonable doses of vitamin and mineral tablets and capsules certainly appear to do no harm. In fact, they can sometimes be quite beneficial. There are times when the body has an increased demand for vitamins and minerals and supplements are advisable. During periods of rapid growth, during pregnancy and lactation, and during periods of trauma and recovery from disease, for example, there is increased need for many essential nutrients.

A study of retired people 65 and older has shown that a daily dose of 18 vitamins and minerals seems to strengthen the immune system. The number of infections in the group being studied was less than 50% of the control group, which did not receive vitamin and mineral supplements.

Vitamins and minerals are both discussed in Chapter 16. See Table 16.5 for a list of the essential minerals and trace minerals. Table 16.6 lists most of the vitamins along with their molecular structures, food sources, and deficiency symptoms.

The RDA values for the vitamins and minerals are set by the Food and Nutrition Board of the U.S. National Academy of Sciences and the National Research Council. The values are quite modest, and any well-balanced diet should have no trouble supplying the RDA for each of the vitamins and minerals.

The RDA values for the vitamins are amounts that will prevent deficiency diseases such as beriberi, pellagra, and scurvy. Some scientists believe that *optimum* intakes of the vitamins are much higher than their RDA values. In Table 18.1 the daily intakes recommended by four different scientists are compared with the RDA values. Most notable among these scientists is Linus Pauling.

Table 18.1 Recommended Daily Adult Intakes of Vitamins

	RDA	Williams[a]	Allen[b]	Leibovitz[c]	Pauling[d]
Vitamin C	60 mg	2500 mg	1500 mg	2500 mg	1000–18,000 mg
Vitamin E	10 IU	400 IU	600 IU	300 IU	800 IU
Vitamin A	5000 IU	15,000 IU	15,000 IU	20,000 IU	20,000–40,000 IU
Vitamin K	None	100 mg	None	None	None
Vitamin D	400 IU	400 IU	300 IU	800 IU	800 IU
Thiamine, B_1	1.5 mg	20 mg	300 mg	100 mg	50–100 mg
Riboflavin, B_2	1.7 mg	20 mg	200 mg	100 mg	50–100 mg
Niacinamide, B_3	18 mg	200 mg	750 mg	300 mg	300–600 mg
Pyridoxine, B_6	2.2 mg	30 mg	350 mg	100 mg	50–100 mg
Cobalamin, B_{12}	3 mg	90 mg	1000 mg	100 mg	100–200 mg
Folacin	400 mg	400 mg	400 mg	400 mg	400–800 mg
Pantothenic acid	None	150 mg	500 mg	200 mg	100–200 mg

[a]Roger J. Williams, 1975
[b]Mary B. Allen, 1981
[c]Brian Leibovitz, 1984
[d]Linus Pauling, 1986

SECTION 18.5

Linus Pauling, Vitamin C, and the Common Cold

In 1970 Linus Pauling, winner of two Nobel prizes (for chemistry in 1954 and for peace in 1962) suggested that ascorbic acid, vitamin C, could be used as a weapon against the common cold. He recommended daily doses of 250 to 15,000 mg, depending on the person and the circumstances. Vitamin C, he said, could prevent colds, or at least lessen their severity. Pauling's claims were greeted with skepticism—and even ridicule—by many in the scientific and medical communities. However, in recent years he has been getting increasing medical support. Public support has always been strong, as indicated by the zooming sales of vitamin C ever since he made his first announcement.

According to Pauling, the RDA of 60 mg of vitamin C is enough to prevent scurvy, but vitamin C does much more than just prevent scurvy. He believes that the optimum intake of vitamin C is much higher than the RDA, and probably varies considerably from one person to another.

He points out that most species manufacture their own ascorbic acid. Pauling believes that primates once had that ability but lost it somewhere along the evolutionary trail, so that now ascorbic acid is a vitamin (a molecule the human body needs but cannot make).

It is Pauling's contention that vitamin C not only helps prevent (or lessen the effects of) the common cold, but it also has value in preventing and treating influenza, as well as a number of other viral diseases. Vitamin C, he believes, has many functions, including the strengthening of the immune system. It does promote the healing of wounds and burns, and it has demonstrated an ability to heal gastric ulcers. It also seems to play an important role in maintaining the body's collagen supply. Like vitamin E, it is an antioxidant, and these two vitamins, along with beta-carotene, have been making headlines because of their antioxidant activity. Antioxidants often act as anticarcinogens.

Pauling has long insisted that the power of vitamin C extends to anticancer activity, and many in the medical community are beginning to agree. Indeed, vitamin C does inhibit the formation of nitrosamines (which are known carcinogens) under conditions such as those in the stomach.

Linus Pauling, winner of two Nobel prizes, advocates massive doses of vitamin C to prevent the common cold. [*AP/Wide World Photos.*]

The common cold still mystifies us.
When cure for it finally arises,
If it's vitamin C,
Linus Pauling may be
The first to win three Nobel prizes.

There is mounting evidence that vitamin C is essential for the efficient functioning of the immune system. The interferons have been recognized as agents in the immune system. (Interferons are large molecules formed by the action of viruses on their host cells. By producing an interferon, a virus can interfere with the growth of another virus.) It has been demonstrated that increased vitamin C increases the body's production of interferons.

Low levels of vitamin C have been linked to the formation of cataracts, and also to the development of glaucoma.

For whatever reasons, whether it is to prevent colds, to protect against cancer, or just to maintain an optimum state of health, there are many people who are now taking vitamin C supplements on a regular basis. Rarely do they take the larger megadoses that Pauling has recommended, but they do continue to take the supplements year after year. If vitamin C is not helping all these people to stay healthier, at least they seem to believe that it is.

Low intake of vitamin C can also lead to gingivitis and periodontal disease.

In a national survey, people were asked to rate their own state of health on a scale from excellent to poor. Those who had high blood levels of vitamin C were most likely to rate their health as ''excellent'' or ''very good.'' Those with low vitamin C levels tended to rate their health as ''fair'' or ''poor.''

SECTION 18.6

Other Vitamins

If it is true that the optimum daily amount of vitamin C is greater than the RDA, then what about the other vitamins? Do you need more than the RDA of all the vitamins? Should you take vitamin supplements? Are supplementary vitamins ever harmful? Are they expensive?

If you have a well-balanced diet and you are in good health, you probably do not need vitamin supplements. But if you did take them, they probably would do no harm. In small doses vitamins are not toxic, but in extremely large doses they might be, especially if they are fat soluble. Vitamin tablets and capsules vary widely in price, and some are fairly expensive. When you take daily doses of vitamins that you don't need, you are probably paying more than you should for vitamins, no matter what the price.

We have already considered vitamin C. Now let's look at a few of the other vitamins.

Vitamin A

Vitamin A is essential for good vision, bone development, and skin maintenance. There is also evidence that vitamin A may confer resistance against certain kinds of cancer. This may help to explain the anticancer activity that has been noted for cruciferous vegetables, such as broccoli, cauliflower, brussels sprouts, and cabbage. Although they do not contain vitamin A as such, all these vegetables are rich in beta-carotene, which is a precursor to vitamin A.

Of course, it should be remembered that vitamin A is a fat soluble vitamin that is stored in the fatty tissues of the body, especially in the liver. Very large doses can be toxic.

Dorothy Hodgkin, an X-ray crystallographer at Oxford University, received the 1964 Nobel prize in chemistry for her work on the structure of vitamin B_{12}. [*AP/Wide World Photos.*]

Vitamin B Complex

There are at least eight members of the vitamin B family (see Table 16.6). Since they are all water soluble, any excess intake is excreted in the urine. There appears to be no toxicity connected with the B vitamins, with the possible exception of vitamin B_6, which apparently can cause neurological damage in some people if taken in extremely large daily doses. As a group the B vitamins are important for maintaining the skin and the nervous system.

Niacin (vitamin B_3), in addition to preventing the skin lesions of pellagra, offers some relief from arthritis and rheumatism. It also helps in lowering the blood cholesterol level. Very large daily doses (5 to 30 g) have been taken for years by schizophrenic patients without any toxic effects.

Vitamin B_6, pyridoxine, has been found to be helpful for people with arthritis by shrinking the synovial membranes that line the joints. It also reduces the swelling and pain of the wrists in carpal tunnel syndrome. Vitamin B_6 is known to be a coenzyme for more than 100 different enzymes.

Vitamin B_{12} is not found in plants, and vegetarians are apt to be deficient in this vitamin. A deficiency of vitamin B_{12} can lead to pernicious anemia. Vitamin B_{12}, also called cyanocobalamine, is a very complicated molecule. (See the molecular structure given in Table 16.6). The structure was finally determined through X-ray crystallography. Dorothy Hodgkin at Oxford University received the 1964 Nobel prize in chemistry for this work. (Only three Nobel prizes in chemistry have ever gone to women. The other two went to Marie Curie in 1911 and to her daughter, Irene, in 1935.)

Vitamin D

Vitamin D is a steroid-type vitamin that protects children against rickets. It promotes the absorption of calcium and phosphorus from foods to produce and maintain healthy bones.

There is less need for vitamin D in adults. In fact, too much vitamin D can lead to excessive calcium and phosphorus absorption, with subsequent formation of calcium deposits in various soft body tissues, including those of the heart. The RDA for vitamin D is 400 IU, and it is wise not to exceed that amount of this fat-soluble vitamin.

Vitamin E

Recent evidence shows that it is an *oxidized* form of LDL cholesterol that is deposited in arteries. Vitamin E acts to prevent the oxidation of cholesterol.

Vitamin E is a mixture of tocopherols (see Table 16.6). It seems to have considerable value in maintaining the cardiovascular system. Vitamin E has been used to treat coronary heart disease, angina, rheumatic heart disease, high blood pressure, arteriosclerosis, varicose veins, and a number of other cardiovascular problems. It is also an anticoagulant that has been useful in preventing blood clots after surgery.

Rats deprived of vitamin E become sterile. Because of this, vitamin E is sometimes called the antisterility vitamin. Low intake of vitamin E can also lead to muscular dystrophy, a disease of the skeletal muscles.

Vitamin E is a potent antioxidant that is able to inactivate free radicals. It is generally believed that much of the physiological damage from aging is due to the production of free radicals. Indeed, vitamin E has also been referred to as the antiaging vitamin.

A vitamin E deficiency can lead to a deficiency in vitamin A. Vitamin A can be oxidized to an inactive form when vitamin E is no longer present to act as an antioxidant. Vitamin E also collaborates with vitamin C in protecting blood vessels and other tissues against oxidation. Vitamin E is the fat-soluble antioxidant vitamin, and vitamin C is the water soluble antioxidant vitamin. Oxidation of unsaturated fatty acids in the cell membrane can be prevented or reversed by vitamin E, which is itself oxidized in the process. Vitamin C can then restore vitamin E to its unoxidized form.

Figure 18.4 The best replacement for fluids lost during exercise is plain water. Electrolyte replacement fluids help little if any, and salt tablets will likely do more harm than good. [*Dave Caulkin/AP/Wide World Photos.*]

SECTION 18.7

Body Fluids: Electrolytes

There is yet another aspect of chemistry and nutrition that you should be aware of: the balance between fluid and electrolyte intake. **Electrolytes** are substances that conduct electricity when dissolved in water. Sodium ions (Na^+), potassium ions (K^+), and chloride ions (Cl^-) are the major electrolytes essential for proper cellular function. Water is also an essential nutrient, a fact obvious to anyone who has been deprived of it. Modern Western people seem to prefer to meet the body's need for water by consuming soda pop, coffee, beer, fruit juice, and other beverages. Many of these—such as beer, caffeinated soda pop, and coffee—actually contain drugs that impair the body's use of the water in the beverages. The alcohol in beer and other alcoholic drinks promotes water loss by blocking the action of the *antidiuretic hormone* (ADH). The caffeine in colas and other soda pops and in coffee and tea has a diuretic effect on the kidneys; it promotes urine formation and consequent water loss.

Based on available information, the best way to replace water loss as it occurs through sweat, tears, respiration, and urination is to drink water (Figure 18.4). Unfortunately, thirst is often a *delayed* response to water loss, and it may be masked by such symptoms as exhaustion, confusion, headache, and nausea (the symptoms are a result of dehydration). Body sweat contains 99% water, some sodium ions (Na^+) and chloride ions (Cl^-), minute amounts of calcium (Ca^{2+}) ions and potassium (K^+) ions, urea, lactic acid, and body oils. The normal American diet probably contains too much sodium chloride ($NaCl$). Thus, it makes the most sense to replace the water component of lost sweat with pure water itself.

Carbohydrate replacement drinks may be of some slight benefit to runners and cyclists who exercise vigorously for more than 2 hr at a time. These drinks delay the onset of exhaustion by a few minutes.

Commercial "thirst quenchers" are quite popular with both serious and weekend athletes. These drinks, designed to replace the salts and water your body loses during long, sweat-provoking periods of exercise, may be too concentrated—a hazard that could lead to diarrhea. At best, the thirst quenchers are of marginal value.

How do you know if you are drinking enough water and are in a proper state of hydration? Just monitor a source of water loss. The urine is a good indicator of hydration. When you are dehydrated, your urine is cloudy and yellow because your kidneys are trying to conserve water. The body water shifts to conserve the shrinking blood volume and to prevent shock. As dehydration worsens, muscles tire and cramp. Dizziness and fainting may follow, and brain cells may shrink, resulting in mental disturbances. The heat regulatory system may also fail, causing **heat stroke**. Without prompt medical attention, a victim of heat stroke may die.

SECTION 18.8

No Smoking

One thing you should *not* do if you want to stay fit and healthy is to smoke cigarettes. Smoking is associated with many serious health problems. It causes emphysema, chronic bronchitis, and lung cancer. And lung cancer is not the only malignancy caused by smoking. It can also cause cancers of the mouth, esophagus, and throat. Smoking is also strongly correlated with cancers of the pancreas, bladder, kidney, and cervix.

But the major killers of smokers are heart attacks and strokes. Cigarette smokers have 70% higher risk of heart attacks than nonsmokers. When they are also overweight, with high blood pressure and high cholesterol, their risk is 200% higher. At every age the death rate is higher among smokers than nonsmokers.

Women who smoke during pregnancy have more still-born babies, more premature babies, and more babies who die before they are a month old. Women who take birth control pills and also smoke have an abnormally high risk of strokes. Children of parents who smoke miss twice as much school time from respiratory infections as do children of nonsmokers.

An interesting study at Harvard University involving military veterans concluded that smokers were 4.3 times as likely as nonsmokers to develop Alzheimer's disease.

One of the gases in cigarette smoke is carbon monoxide, which ties up about 8% of the hemoglobin in the blood. Therefore, smokers' breathing is less efficient. They work harder to breathe, but they get less oxygen to their cells. It is no wonder that fatigue is a common problem for smokers.

In the lining of the lungs there are cilia (hairs) that help to sweep out particulate matter breathed into the lungs. But in smokers the tar from their cigarette smoke forms a sticky coating that keeps the cilia from moving freely. The reduced action of the cilia may explain the smoker's cough. Pathologists will tell you that during an autopsy it is usually easy to tell if the body belonged to a smoker. Lungs are normally pink; a smoker's lungs are black.

Unfortunately, the smoking of cigarettes does not harm only the smoker. Nonsmokers who are regularly exposed to the "second hand" smoke of other people can suffer some of the same health problems as the smokers themselves.

In recent years there have been some efforts to lessen the indoor pollution caused by smoking. Many restaurants and government buildings have been declared nonsmoking, and Congress could outlaw smoking in *all* public places. Smoking has never been a smart or a healthy thing to do, but now it isn't even fashionable.

SECTION 18.9

Some Chemistry of Muscles

Studies have shown that frequent exercise prolongs life and lowers the incidence of disease. Humans have about 600 muscles each. Exercise can make muscles stronger, more flexible, and more efficient in their use of oxygen. Strong muscles can do more work than weak ones. That is good, because the heart is an organ comprised mainly of muscle. A strong heart is a healthy heart. With regular exercise, resting pulse and blood pressure usually decline. After several months of an effective exercise program, pulse and blood pressure remain lower even *during* exercise. The net result, called the **training effect**, is that a person who exercises regularly is able to do more physical work with less strain.

People who expand their capacity to do more physical work under less strain often begin to think of doing more—faster and with more agility and accuracy—of whatever they do. These people become athletes. Exercise is an art, but it is increasingly also a science—a science in which chemistry plays a vital role.

Energy for Muscle Contraction: ATP

When cells metabolize glucose or fatty acids, only a part of the chemical energy in those substances is converted to heat. Some of it is stored in another chemical compound called adenosine triphosphate (ATP).

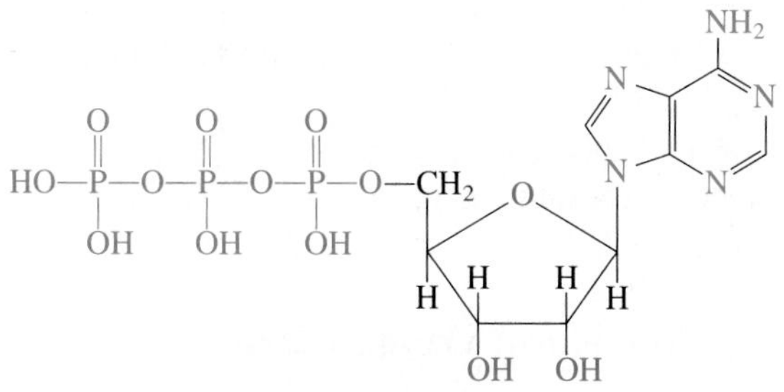

Adenosine triphosphate

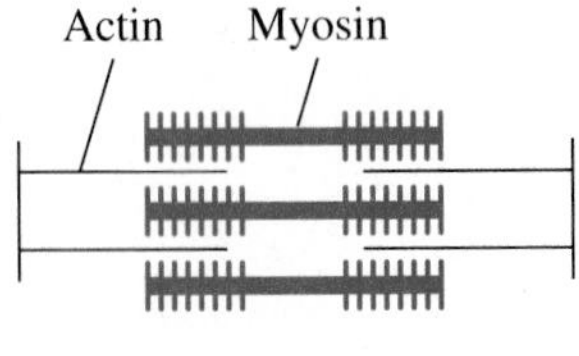

(a)

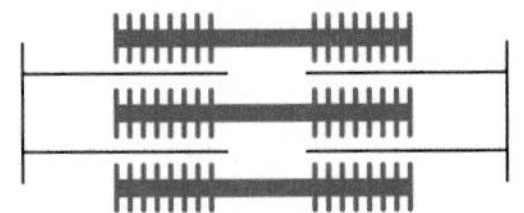

(b)

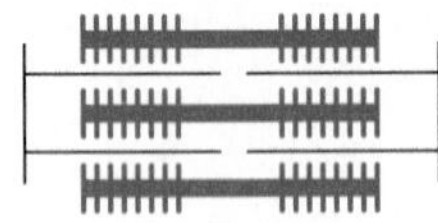

(c)

Figure 18.5 Diagram of actomyosin complex in muscle. (a) Extended muscle. (b) Resting muscle. (c) Partially contracted muscle.

The stimulation of muscle causes it to contract; that contraction is work and requires energy. The immediate source of energy for muscle contraction is ATP. The energy stored in this molecule powers the physical movement of muscle tissue. Two proteins, actin and myosin, play important roles in this process. Together actin and myosin form a loose complex called **actomyosin**, the contractile protein of which muscles are made (Figure 18.5). When ATP is added to isolated actomyosin, the protein fibers contract. It seems likely that the same process occurs *in vivo,* that is, in muscle in living animals. Not only does myosin, serve as part of the structural complex in muscles, it also acts as an enzyme for the removal of a phosphate group from ATP. Thus, it is directly involved in liberating the energy required for the contraction.

In the resting person, muscle activity (including that of the heart muscle) accounts for only about 15 to 30% of the energy requirements of the body. Other activities, such as cell repair or the transmission of nerve impulses or even the maintenance of body temperature, account for the remaining energy needs. During intense physical activity, the energy requirements of muscle may be more than 200 times the resting level.

Aerobic Exercise: Plenty of Oxygen

The ATP in muscle tissue is sufficient for activities lasting at most a few seconds. Fortunately muscles store a more extensive energy supply in the form of **glycogen**. This starch is a storage form of dietary carbohydrate that has been ingested, digested to glucose, and absorbed. This blood glucose is stored as glycogen when not needed immediately for energy.

When muscle contraction begins, the glycogen is converted by muscle cells in a series of steps to pyruvic acid.

$$\underset{\text{Glycogen}}{(C_6H_{11}O_5)_n} \longrightarrow \underset{\text{Pyruvic acid}}{2n\ CH_3\overset{\overset{\displaystyle O}{\|}}{C}-\overset{\overset{\displaystyle O}{\|}}{C}OH}$$

Then, if sufficient oxygen and other factors are readily available, the pyruvic acid is oxidized in a series of steps to carbon dioxide and water.

$$2\ CH_3\overset{\overset{\displaystyle O}{\|}}{C}-\overset{\overset{\displaystyle O}{\|}}{C}OH + 5\,O_2 \longrightarrow 6\,CO_2 + 4\,H_2O$$

Muscle contractions that occur under these circumstances—that is, in the presence of oxygen—constitute **aerobic exercise** (Figure 18.6).

Anaerobic Exercise and Oxygen Debt

If sufficient oxygen is not available, pyruvic acid is reduced to form lactic acid.

$$CH_3COCOOH + [2\,H] \longrightarrow CH_3CHOHCOOH$$

This is **anaerobic exercise**. If it persists, an excess of lactic acid results. The buildup of lactic acid causes a weaker response of muscle cells to the stimuli that originally caused them to contract; hydronium ions (H_3O^+) from lactic acid are embedded in the muscle cell tissue. Athletes recognize this weakness as fatigue and sometimes as pain. At this point they are tempted to quit, and if they do, the **oxygen debt** is repaid (Figure 18.7). After exercise ends, the cells' demand for oxygen decreases, making more oxygen available to oxidize the lactic acid that results from anaerobic metabolism back to pyruvic acid. This acid is then converted to carbon dioxide, water, and energy.

Athletes usually emphasize one type of training (anaerobic or aerobic) over the other. For example, an athlete training for a 60-m dash will do mainly anaerobic work, but one planning to run a 10-km race will do mainly aerobic training. Sprinting and weight lifting are largely anaerobic activities; a marathon run is largely aerobic. During a marathon, athletes must set a pace to run for more than 2 hours. Their muscle cells depend on slow, steady aerobic conversion of carbohydrates to energy. During anaerobic activities, however, the muscle cells use almost no oxygen. Rather, they need the quick energy provided by anaerobic metabolism.

After glycogen stores are depleted, muscle cells can switch over to fat metabolism. Fats are the main source of energy for sustained activity of low or moderate intensity, such as the last part of a marathon run (42 km).

Figure 18.6 Aerobic exercise is carried out at a pace that allows us to get sufficient oxygen to our muscle cells so they can oxidize pyruvic acid to carbon dioxide and water. Aerobic dance is a popular form of such exercise. [*David Madison/Duomo.*]

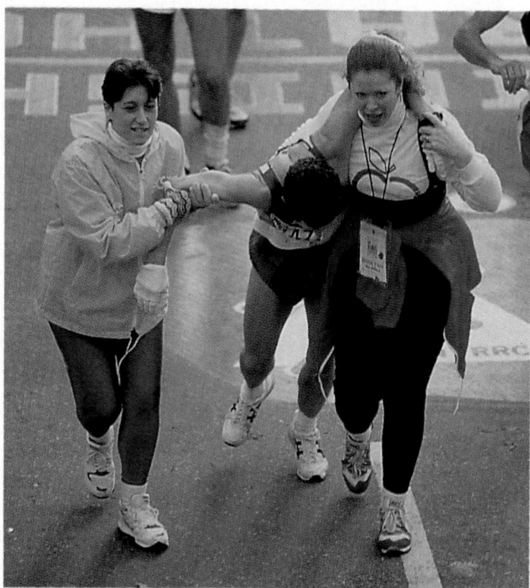

Figure 18.7 The race is over, but the metabolism that fueled the effort continues. The athlete gasps air to repay her oxygen debt. [*Paul J. Sutton/Sutton/Duomo.*]

Table 18.2 A Comparison of Types of Muscle Fiber

	Type I	Type IIB*
Category	Slow twitch	Fast twitch
Color	Red	White
Respiratory capacity	High	Low
Myoglobin level	High	Low
Catalytic activity of actomyosin	Low	High
Capacity for glycogen use	Low	High

*A type IIA fiber exists that resembles Type I in some respects and Type IIB in others. We will discuss only the two types described in the table.

SECTION 18.10

Muscle Fibers: Twitch Kind Do You Have?

The quality and type of the muscle fiber that an athlete has affects his or her athletic performance. The quality of machinery plays a role in how work is done. For example, we can get snow off a driveway in more than one way. We can remove it in 10 min with a snowblower, or we can spend 50 min shoveling. Muscles are tools with which we do work. They are classified according to the speed and effort required to get work done.

Slow-Twitch Fibers: Endurance Activities

The two classes of muscle fibers are **fast-twitch fibers** (those stronger and larger and most suited for anaerobic activity) and **slow-twitch fibers** (those best for aerobic work). Table 18.2 lists some characteristics of these two types of muscle fibers. The Type I (slow-twitch) fibers described in Table 18.2 are called on during activity of light or moderate intensity. The respiratory capacity of these fibers is high, which means they can provide much energy via aerobic pathways. Notice that for Type I fibers myoglobin levels are also high. Aerobic oxidation requires oxygen, and muscle tissue rich in slow-twitch fibers is geared to supply high levels of oxygen.

Myoglobin is the heme-containing protein in muscle that stores oxygen (similar to the way hemoglobin transports oxygen in the blood).

The capacity of Type I muscle fibers for use of glycogen is low. This tissue is not geared to anaerobic generation of energy and does not require the hydrolysis of glycogen. The catalytic activity of the actomyosin complex is low. Remember that actomyosin is not only the structural unit in muscle that actually undergoes contraction; it is also responsible for catalyzing the hydrolysis of ATP to provide energy for the contraction. Low catalytic activity means that the energy is parceled out more slowly, which is not good if you want to lift 200 kg but is perfect for a jog of 15 km.

Fast-Twitch Fibers: Bursts of Power

The Type IIB (fast-twitch) fibers described in Table 18.2 have characteristics just the opposite of Type I fibers. Low respiratory capacity and low myoglobin levels argue against aerobic oxidation. A high capacity for glycogen use and high catalytic activity of actomyosin allow tissue rich in fast-twitch fibers to generate ATP rapidly and also to hydrolyze that ATP rapidly in intense muscle activity. Thus, this type of muscle tissue gives you the capacity to do short bursts of vigorous work. We say *bursts* because this type of muscle fatigues relatively quickly. A period of recovery in which lactic acid is cleared from the muscle is required between brief periods of activity.

Figure 18.8 The New York City Marathon, a test of the respiratory capacity of muscle. [*David Madison/Duomo.*]

Building Muscles

Endurance exercise increases myoglobin levels in skeletal muscles. This provides for faster oxygen transport and increased respiratory capacity (Figure 18.8). These changes are apparent within 2 weeks. Endurance training does not necessarily increase the size of muscles. If you want larger muscles, try weight lifting. Weight training (Figure 18.9) develops fast-twitch muscle fibers. These fibers increase in size and strength with repeated anaerobic exercise. Weight training does *not* increase respiratory capacity, however.

Muscle-fiber type seems to be inherited. Research shows that world-class marathon runners may possess up to 80 to 90% slow-twitch fibers, compared with championship sprinters who may have up to 70% fast-twitch muscle fiber. Some exceptions have been noted, however.

Figure 18.9 Strength exercises build muscle mass, but do not increase respiratory capacity of muscles. [*David Madison/Duomo.*]

Thus, factors such as training and body composition are important in athletics. Other factors are equally important, including nutrition, fluid and electrolyte balance, and drug use or misuse.

SECTION 18.11

Carbohydrate Loading and Blood Doping

Athletes, it seems, will try almost anything to gain a competitive edge. Those engaged in endurance activities, such as long distance running, often try a technique called **carbohydrate loading**. Muscles run on glycogen as the preferred fuel, but the human body can store only a limited amount, perhaps about 500 g in a 70-kg person.

In carbohydrate loading, the athlete cuts back on training a few days before competition and eats a diet high in carbohydrates. Under this regimen, the body will store more glycogen than usual. The benefits of carbohydrate loading, even for top athletes, are marginal. For casual athletes, the technique is probably ridiculous.

A more radical attempt to improve athletic performance is called **blood doping**. A quantity of blood is withdrawn from the athlete, then stored for 6 weeks while the person's body manufactures replacement blood. Then, a few days before competition, the red blood cells from the stored blood are returned to the athlete. This increased supply of red cells enables the athlete's cardiovascular system to transport oxygen more efficiently. In controlled studies, blood-doped runners were able to improve their times by 9 seconds in an 8-km treadmill race, compared with a group that took a saline solution. These runners were already well trained. Blood doping will not convert a mediocre runner into an outstanding one. Several sports groups have banned blood doping.

SECTION 18.12

Drugs and the Athlete

Muscle chemistry and metabolism, aerobic and anaerobic exercise, and nutrition and fluid balance all relate to normal internal processes. It seems unlikely that properly fed, well-trained athletes can safely improve their performance with magic, superstition, or drugs. Many athletes, however, are not satisfied with the idea that hard work and proper nutrition are the best answers to improved performance. They turn instead to dangerous drugs such as the stimulants amphetamine and cocaine, and to anabolic steroids to build muscles.

Restorative Drugs

Athletes also use drugs to alleviate pain or soreness that result from overtraining and to treat injury. These **restorative drugs** include analgesics (painkillers), such as aspirin and acetaminophen (Chapter 19), and anti-inflammatory drugs, such as aspirin, ibuprofen, and cortisone. Cortisone derivatives are often injected to reduce swelling in damaged joints and tissues. Relief is often transitory, however, and side effects are often severe. Prolonged use of these drugs can cause fluid retention, hypertension, ulcers, disturbance of the sex hormone balance, and other problems. Another substance, one that can be bought over the counter and applied externally, is methyl salicylate, or oil of wintergreen. This substance causes a mild burning sensation when applied to the skin, thus serving as a counterirritant for sore muscles. An aspirin derivative, it also acts as an analgesic.

Stimulant Drugs

Some athletes use **stimulants** (Chapter 19) in the hope that these drugs will improve performance. The caffeine in a strong cup of coffee may be of some benefit. Caffeine triggers the release of fatty acids. Metabolism of these compounds could conserve glycogen, but any such effect is small. The caffeine also increases the heart rate, speeds metabolism, and increases urine output. The latter could lead to dehydration.

Other athletes resort to stronger stimulants, including amphetamines and cocaine. A multipurpose drug, cocaine was first used in the late 1800s as a local anesthetic to deaden pain during dental work or minor surgery. It is also a powerful stimulant, the reason for its current popularity among athletes. Like amphetamines, cocaine stimulates the central nervous system, increasing alertness, respiration, blood pressure, muscle tension, and heart rate. Both stimulants mask symptoms of fatigue and give the athlete a sense of increased stamina. The death from cocaine intoxication of several thousand people, including prominent professional athletes, brought the problem to public attention. Other athletes have been arrested, suspended, and banned for cocaine use.

Anabolic Steroids

Many strenuous athletic performances depend on well-developed muscles. Muscle mass depends on levels of the male hormone **testosterone** (Chapter 19). When boys reach puberty, testosterone levels rise and the boys begin to become more muscular if they exercise. Men generally have larger muscles than women because they have more testosterone.

Testosterone and some of its semisynthetic derivatives are taken by athletes in an attempt to build muscle mass more rapidly (Figure 18.10). Steroid hormones used to increase muscle mass are called **anabolic steroids**. These chemicals aid in the building (anabolism) of body proteins and thus of muscle tissue.

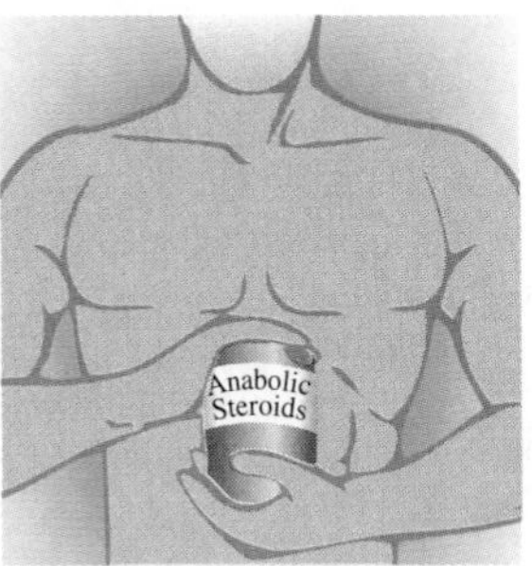

Figure 18.10 Anabolic steroids help speed the development of muscle mass. These drugs are illegal and dangerous, however, and they have devastating side effects.

Not all athletes are interested in growing larger. Female gymnasts sometimes use androgen antagonists such as cyproterone along with the progestin medroxyprogesterone to delay puberty and retard growth.

There are no good controlled studies that demonstrate the effectiveness of anabolic steroids. They do seem to work—at least for some people—but the side effects are many. In males, side effects include testicular atrophy and loss of function, impotence, acne, liver damage, edema (swelling), elevated cholesterol levels, and growth of breasts. Liver cancer is now showing up at an alarming rate in athletes who began using steroids in the 1960s.

Anabolic steroids act as male hormones (androgens). They make women more masculine. They help women build larger muscles, but they also induce balding, cause them to develop extra body hair and deep voices, and cause menstrual irregularities. What price will an athlete pay for improved performance?

Drugs, Athletic Performance, and Drug Screening

The use of drugs to increase athletic performance is not only illegal and physically dangerous, but it can also be psychologically damaging. Stimulant drugs give the user a false sense of confidence by affecting the person's ego, giving the delusion of invincibility. Such "greatness" often ends as the competition begins, however, because the stimulant effect is short-lived (it lasts for only half an hour to an hour for cocaine). What usually follows is exhaustion and extreme depression (Figure 18.11). More stimulants are needed to combat these "down" feelings, and the user then runs the risk of addiction or overdose.

Because drug use for the enhancement or improvement of performance is illegal, chemists have another role to play in sports: screening blood and urine samples for illegal drugs. Using sophisticated instruments, chemists can detect minute amounts of illegal drugs. Drug testing has been used at the Olympic games since 1968. It is rapidly becoming standard practice for athletes in college and professional sports.

Drugs are used at great risk to the athlete and with little evidence of benefit in most cases. Use of illegal drugs violates the spirit of fair athletic competition. Let us work toward a world in which events and medals are won by drug-free athletes who are highly motivated, well nourished, and well trained.

Figure 18.11 Cocaine is a powerful stimulant that gives athletes a sense of increased stamina. The effect is short-lived, however, and users soon experience extreme fatigue.

SECTION 18.13

Exercise and the Brain

Hard training improves athletic performance. Few chemical substances help very much. Our own bodies, however, manufacture a variety of substances that do help. Examples are the pain-relieving **endorphins** (Section 19.17).

After vigorous activity, athletes have an increased level of endorphins in their blood. Stimulated by exercise, deep sensory nerves release endorphins to block the pain message. Exercise extended over a long period of time, such as distance running, causes in the athlete many of the same symptoms experienced by opiate users. Runners get a euphoric high during or after a hard run. Unfortunately, both the natural endorphins and their analogs, like morphine, seem to be addictive. Athletes, especially runners, tend to suffer from withdrawal: they feel bad when they don't get the vigorous exercise of a long, hard run. This can be positive because exercise is important to the maintenance of good health. It can be negative if the person exercises when injured or to the exclusion of family or work obligations. Athletes also seem to develop a tolerance to their own endorphins: they have to run farther and farther to get that euphoric feeling.

Good physical condition, daily vigorous exercise, and experiencing a natural, cheap, legal high seem safer than resorting to illegal drugs. You may get hooked on exercise, but that seems far better than getting hooked on narcotics.

Naloxone, a morphine antagonist (Section 19.16), blocks the effect of endorphins also.

The good feeling that you get from an altruistic act may be due to the release of endorphins.

SECTION 18.14

Chemistry of Sports Materials

The athlete's world is shaped by chemicals. The ability to perform is a reflection of physiology and biochemistry within the athlete. Chemicals taken internally may hinder or enhance performance to some degree. Deep sensory nerves, stimulated by rigorous exercise, trigger the release of the brain's own painkillers, the endorphins, which may block pain and provide a euphoric high—a reward for strenuous endeavor and reinforcement for future challenge.

There is yet another chemical dimension to the sports world: the chemicals involved in providing athletic clothing and equipment (Table 18.3). Athletic clothing gets its shape and protective qualities from chemicals. Swimsuits, ski pants, and elastic supports stretch because of synthetic fibers. Joggers run in thunderstorms and windstorms in Gore-Tex suits without getting cold and wet. Gore-Tex is a thin, membranous material made by stretching fibers of polytetrafluoroethylene (Teflon). The material has billions of tiny holes that are too small to pass drops of water but readily pass water vapor. Raindrops falling on the outside are held together by surface tension and run off rather than penetrating the fabric and wetting the wearer. Water vapor from the

Table 18.3 Some Sports and Recreational Items Made from Petroleum

Golf balls	Whistles	Tote bags	Hockey pucks
Wet suits	Motorcycle helmets	Darts	Ice buckets
Parachutes	Dune-buggy bodies	Tents	Fishing nets
Card tables	Checkers	Stadium cushions	Hiking boots
Golf-cart bodies	Chess boards	Finger paints	Frisbees
Warm-up suits	Shorts	Foul-weather gear	Fishing boots
Ping-Pong paddles	Tennis shoes	Foot pads	Diving masks
Rafts	Paddles	Visors	Guitar picks
Uniforms	Decoys	Swimming-pool liners	Beach balls
Phonographs	Volley balls	Vinyl tops for cars	Sunglasses
Racks	Sleeping bags	Ice chests	Dog leashes
Track shoes	Sports-car bodies	Life jackets	Dice
Dominoes	Tennis balls	Audio tape	Pole-vaulting poles
Windbreakers	Reclining chairs	Model planes	Aquariums
Sails	Insect repellent		

sweating skin can pass from the warm area inside the suit where vapor concentration is high to the cooler area outside the suit where the vapor concentration is lower. The runner stays relatively dry inside the suit (Figure 18.12).

The jogger is not the only athlete pampered by equipment made of materials from the chemical industry. Football, hockey, and baseball players are protected by plastic helmets. Protective pads of synthetic foamed rubbers help these players avoid injuries to the torso and legs. Brightly dyed nylon uniforms with synthetic colors add to the glamour of amateur and professional teams. Sports events are played on artificial turf, a carpet of nylon or polypropylene with an under-pad of synthetic foamed polymers. Stadiums are protected from the weather by Teflon covers reinforced with glass fibers and held

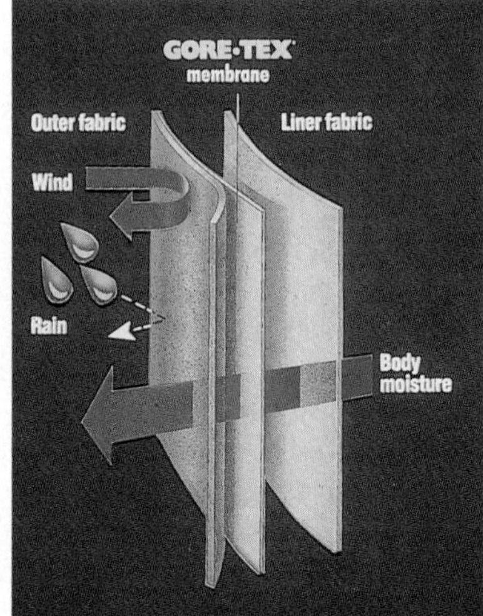

Figure 18.12 Imaginative use of synthetic fibers and other materials has made it possible for athletes to perform in relative comfort even in inclement weather. Gore-Tex fabrics repel wind and raindrops, but allow body moisture to escape. [*Courtesy of Gore Association.*]

Figure 18.13 Baseball in a domed stadium on an artificial surface is quite different from a game played outside on grass.

aloft by air pressure. Baseball or football on artificial turf in air-conditioned, enclosed stadiums is quite different from games played outside on grass (Figure 18.13). Balls bounce differently and players wear shoes especially designed to cope with the artificial surface.

The dramatic effect of new materials is perhaps best illustrated in pole vaulting. Between 1940 and 1960, the world-record height for this event increased only 23.5 cm. After the development of plastic poles reinforced with glass fibers, however, the record rapidly rose another 100 cm. Other sports have been affected similarly, if less dramatically (see references at the end of the chapter for examples).

From the soles of sports shoes, to the wax used on cross-country skis, to tennis sweaters that stretch yet retain their shape, athletes are immersed in a world of chemicals. Even the gum they chew contains polyvinyl acetate resin and synthetic flavors and colors. Athletes (and nonathletes) who chew tobacco chew on leaves modified by natural and synthetic flavors, moisturizers, and other chemical additives. From within, athletes can strive to achieve goals and records by monitoring their body chemistry. They will be comfortable and protected by materials designed exclusively for them. With almost half the population exercising on a regular basis, and many of the rest involved at least in watching athletic events, chemistry will continue to play an active role in our activity-oriented culture.

Summary

1. Good health and fitness require a nutritious diet and exercise.
2. A good diet consists largely of complex carbohydrates (starches) with plenty of fruits and vegetables, modest amounts of dairy and meat products, and very little fat and sugar.
3. Most Americans should eat less fat and less protein but more starch.
4. Athletes need more calories than sedentary people, and they should get them in the form of carbohydrates.

5. Quick weight-loss diets do not usually lead to permanent weight loss.
6. If you begin an exercise program and do not increase your food intake, you will lose weight.
7. Each pound of excess adipose tissue requires 200 extra miles of blood capillaries.
8. The body mass index (a measure of fatness) can be calculated from a person's height and weight.
9. Linus Pauling and others suggest daily vitamin intakes much higher than the *recommended daily allowance* (RDA) values.
10. Vitamin C, vitamin E, and beta-carotene are antioxidants, which help to prevent damage by free radicals.
11. If you wish to maintain a state of optimum health, then do not smoke.
12. Exercise strengthens muscles, prolongs life, and lowers the incidence of disease.
13. *Aerobic exercise* involves muscle contraction in the presence of oxygen. A long, steady run is aerobic exercise.
14. *Anaerobic exercise* involves quick bursts of energy. A 60-m dash is anaerobic exercise.
15. There are two classes of muscle fibers, *fast-twitch* (which are larger and stronger and good for rapid bursts of power) and *slow-twitch* (which are geared for steady exercise of long endurance).
16. Weight training develops fast-twitch muscles, but does not increase respiratory capacity.
17. Long distance running develops slow-twitch muscles and increases respiratory capacity.
18. *Carbohydrate loading* is done before endurance activities in order to maximize glycogen storage, but the benefits are marginal.
19. *Blood doping* has sometimes enabled runners to improve their times, but it does not improve the quality of the runner.
20. Water is the best drink to prevent dehydration.
21. Athletes may use *restorative drugs* such as aspirin, and stimulant drugs such as caffeine, but drugs such as cocaine are illegal and can be deadly.
22. *Anabolic steroids* may increase muscle mass, but they have many bad side effects, including impotence and liver cancer.
23. It appears that athletes can become addicted to their own *endorphins.*
24. From modern sports clothing and equipment to artificial turf, chemistry has contributed heavily to the world of sports.

Key Terms

actomyosin 18.9
aerobic exercise 18.9
anabolic steroids 18.12
anaerobic exercise 18.9
antioxidants 18.6
blood doping 18.11
carbohydrate loading 18.11
diuretic 18.2
electrolytes 18.7
endorphins 18.13
fast-twitch fibers 18.10
glycogen 18.9
heat stroke 18.7
oxygen debt 18.8
recommended daily allowance 18.4
restorative drugs 18.12
set-point theory 18.2
slow-twitch fibers 18.10
stimulants 18.12
testosterone 18.12
training effect 18.9

Review Questions

1. According to the food guide pyramid, what foodstuff should make up the largest part of your diet?
2. What foodstuffs should be eaten only sparingly?
3. The recommended diet contains what percent of its calories as fat?
4. The typical American diet contains what percent fat?
5. Mention two good reasons for not going on a high-protein diet.
6. Why have many people cut back on their consumption of red meat?
7. Do athletes need more protein than other people?

8. Muscles are made of protein. Does eating more protein result in larger muscles?
9. Do you think it is likely that obesity is inherited?
10. How might you lose 5 lb in just one week?
11. What are recommended daily allowance (RDA) values?
12. Do we ever need more of a given vitamin than its RDA?
13. What are some of the benefits of vitamin A?
14. In addition to preventing pellagra, what does vitamin B_3 (niacin) do?
15. What kinds of problems are helped by vitamin B_6 (pyridoxine)?
16. What benefit is derived from vitamin B_{12}?
17. Is taking a daily megadose of vitamin D a good idea?
18. What are the benefits of vitamin D?
19. Vitamins C and E are antioxidants. What does that mean?
20. Mention several health problems that vitamin E can be used to treat.
21. What are some of the health problems related to smoking?
22. Why is it important for pregnant women not to smoke?
23. List three ways that chemistry has had an impact upon sports.
24. How many muscles do humans have?
25. Describe the training effect.
26. What is the *immediate* source of energy for muscle contraction?
27. What two proteins make up the actomyosin complex?
28. What are the two functions of the actomyosin protein complex?
29. What is aerobic exercise?
30. What is anaerobic exercise?
31. Which type of metabolism (aerobic or anaerobic) is primarily responsible for providing energy for intense bursts of vigorous activity?
32. Which type of metabolism (aerobic or anaerobic) is primarily responsible for providing energy for prolonged low levels of activity?
33. What is meant by *oxygen debt?*
34. Identify Type I and Type IIB muscle fibers as
    a. fast twitch or slow twitch
    b. suited to aerobic oxidation or to anaerobic use of glycogen
35. Explain why high levels of myoglobin are appropriate for muscle tissue that is geared to aerobic oxidation.
36. Why does the high catalytic activity of actomyosin in Type IIB fibers suggest that these are the muscle fibers engaged in brief, intense physical activity?
37. Why can the muscle tissue that uses anaerobic glycogen metabolism for its primary source of energy be called on only for *brief* periods of intense activity?
38. Which type of muscle fiber is most affected by endurance training exercises? What changes occur in the muscle tissue?
39. Birds use large, well-developed breast muscles for flying. Pheasants can fly 80 km/hr, but only for short distances. Great blue herons can fly only about 35 km/hr but can cruise great distances. What kind of fibers would each have in its breast muscles?
40. How do the nutritional needs of an athlete differ from those of a sedentary individual? How is that extra need best met?
41. Muscle is protein. Does an athlete need extra protein (above the RDA) to build muscles? What is the only way to build muscles?
42. Describe the biochemical process by which muscles are built.
43. List two ways to determine percent body fat. Describe a limitation of each method.
44. List some problems that result from low-calorie diets.
45. How many calories of energy are stored in 1 lb of adipose tissue?
46. Why does excess body fat put a strain on the heart?
47. Describe the set-point theory of body weight.
48. Why does a diet that restricts carbohydrate intake lead to loss of muscle mass as well as fat?
49. List two ways in which fad diets lead to a ''quick weight loss.'' Why is this weight rapidly regained?
50. How much glycogen can the average human body store?
51. What is carbohydrate loading? How does it aid an athlete in an endurance event?
52. What is blood doping? How does it work?

53. List the three major electrolytes essential for proper cellular function?
54. What is a diuretic?
55. Describe the function of antidiuretic hormone (ADH).
56. What fluid is best for replacing water lost during exercise?
57. Why are beer and cola drinks not recommended for fluid replacement?
58. Is thirst a good indicator of dehydration? Are you always thirsty when dehydrated?
59. How is the appearance of urine related to dehydration?
60. What is heat stroke?
61. What are restorative drugs?
62. How is cortisone used in sports medicine? What are some of the side effects of its use?
63. What are anabolic steroids?
64. List some side effects of the use of anabolic steroids in males.
65. What are the effects of anabolic steroids on females?
66. Does cocaine enhance athletic performance? Explain fully.
67. What happens when the effect of cocaine wears off?
68. What is the role of chemists in control of drug use by athletes?
69. Give a biochemical explanation for the ''runners' high.''
70. List the evidence that indicates that long-distance running is addictive.
71. Give a biochemical explanation of addiction to long-distance running.
72. How does Gore-Tex clothing work? What is it made of?

Problems

73. What is the body mass index for a 6-ft baseball player who weighs 175 lb?
74. If you are moderately active and want to maintain a weight of 150 lb, about how many calories do you need each day?

Diet and Exercise

75. A giant double-decker hamburger provides 600 kcal of energy. How long would you have to walk to burn off that energy if 1 hr of walking uses about 300 kcal?
76. One kilogram of fat tissue stores about 7700 kcal of energy. An average person burns about 40 kcal walking 1 km. If that person walks 5 km a day, how much fat will be burned in 1 year?
77. How long would you have to run to burn off the 110 kcal in one glass of beer if 1 hr of running burns off 1100 kcal?
78. How far would you have to run to burn off 5 kg of fat if your running burns off 100 kcal/km? (Assume that there are 7700 kcal in 1 kg of fat.)
79. The RDA for protein is about 0.8 g per kilogram of body weight. How much protein is required each day by a 125-kg football player?
80. How much protein is required each day by a 50-kg gymnast (See Problem 79)?
81. A 70-kg man can store about 2000 kcal as glycogen. How far could the man run on this stored starch if he expends 100 kcal/km while running and the glycogen was his only source of energy?
82. A 70-kg man can store about 100,000 kcal of energy as fat. How far could the man run on this stored fat if he expends 100 kcal/km while running and the fat was his only source of energy?

Additional Problems

83. An athlete can run a 400-m race in 45 seconds. Her maximum oxygen intake is 4 L per minute, yet working muscles at maximum exertion require about 0.2 L of O_2 per minute for each kilogram of body weight. If an athlete weighs 50 kg, what oxygen debt will she incur?

84. Fat tissue has a density of about 0.9 g/mL, lean tissue a density of about 1.1 g/mL. Calculate the density of a person with a body volume of 80 L who weighs 85 kg. Is the person fat or lean?

Project

85. List as many synthetic materials as you can that are used in each of the following sports.

a. baseball
b. basketball
c. football
d. ice hockey
e. golf
f. tennis
g. swimming
h. sailing
i. skiing
j. track and field
k. jogging

References and Readings

1. Aruoma, Okezie I., and Barry Halliwell (Eds.). *Free Radicals and Food Additives.* London: Taylor and Francis, 1991.
2. Bent, Harry A. ''Energy and Exercise.'' *Journal of Chemical Education,* July 1978, pp. 456–458; August 1978, pp. 526–528; September 1978, pp. 586–587; October 1978, pp. 659–660; November 1978, pp. 726–727; December 1978, pp. 796–797.
3. Bishop, Marvin. ''Food In/Energy Out.'' *ChemTech,* August 1983, pp. 494–496.
4. ''Boats to Badminton: Chemicals Build Athletic Products.'' *ChemEcology,* April 1982, pp. 1–13.
5. Duthie, Garry G. ''Vitamin E and Antioxidants.'' *Chemistry and Industry,* 17 August 1992, pp. 598–601.
6. Goode, Erica E. ''Getting Slim.'' *U.S. News and World Report,* 14 May 1990, pp. 56–65.
7. Gurin, Joel. ''What's Your Natural Weight?'' *American Health,* May 1984, pp. 43–47.
8. Hill, John W. ''Weight-Loss Diets and the Law of Conservation of Energy.'' *Journal of Chemical Education,* December 1981, p. 996.
9. Jentz, Kathy, ''Exercise for Weight Control.'' *Running and Fitnews,* April 1990, p. 4.
10. Layman, Donald K. (Ed.). *Nutrition and Aerobic Exercise.* Washington, DC: American Chemical Society, 1986.
11. Lineback, David R. ''Nutrition (Diet) and Exercise.'' *Journal of Chemical Education,* June 1984, pp. 536–539.
12. Looney, Douglas S. ''A Test with Nothing But Tough Questions.'' *Sports Illustrated,* 9 August 1982, pp. 24–29. About testing athletes for drug use.
13. Pauling, Linus. *How to Live Longer and Feel Better.* New York: Freeman, 1986.
14. Rapaport, Roger. ''The Blade Runners.'' *Science 82,* November 1982, pp. 96–97. Talks about artificial turf.
15. Smart, Joanne McAllister. ''Antioxidants for the Uninitiated.'' *Vegetarian Times,* April 1994, p. 26.
16. Smith, Trevor. ''Chemistry, Exercise, and Weight Control.'' *Today's Chemist,* February 1990, pp. 10–11, 21.
17. Smith, Trevor. ''Chemistry in Sports Medicine.'' *Today's Chemist,* October 1990, pp. 14–17. Discusses caffeine, ozone, fats, and more.
18. Sprague, Ken. *The Athlete's Body.* Los Angeles: J. P. Tarcher, 1981.
19. Stifler, John. ''Drink to Your Health.'' *Runner's World,* July 1989, pp. 72–78. Discusses beverages for runners.
20. Upton, Arthur C., and Eden Graber. *Staying Healthy in a Risky Environment.* New York: Simon and Schuster, 1993.
21. Zuckerman, Sam. ''Food for Sport.'' *Nutrition Action,* July–August 1984, pp. 6–11.
22. Zurer, Pamela S. ''Drugs in Sports.'' *Chemical and Engineering News,* 30 April 1984, pp. 69–78.

19 Drugs

Chemical Cures, Comforts, and Cautions

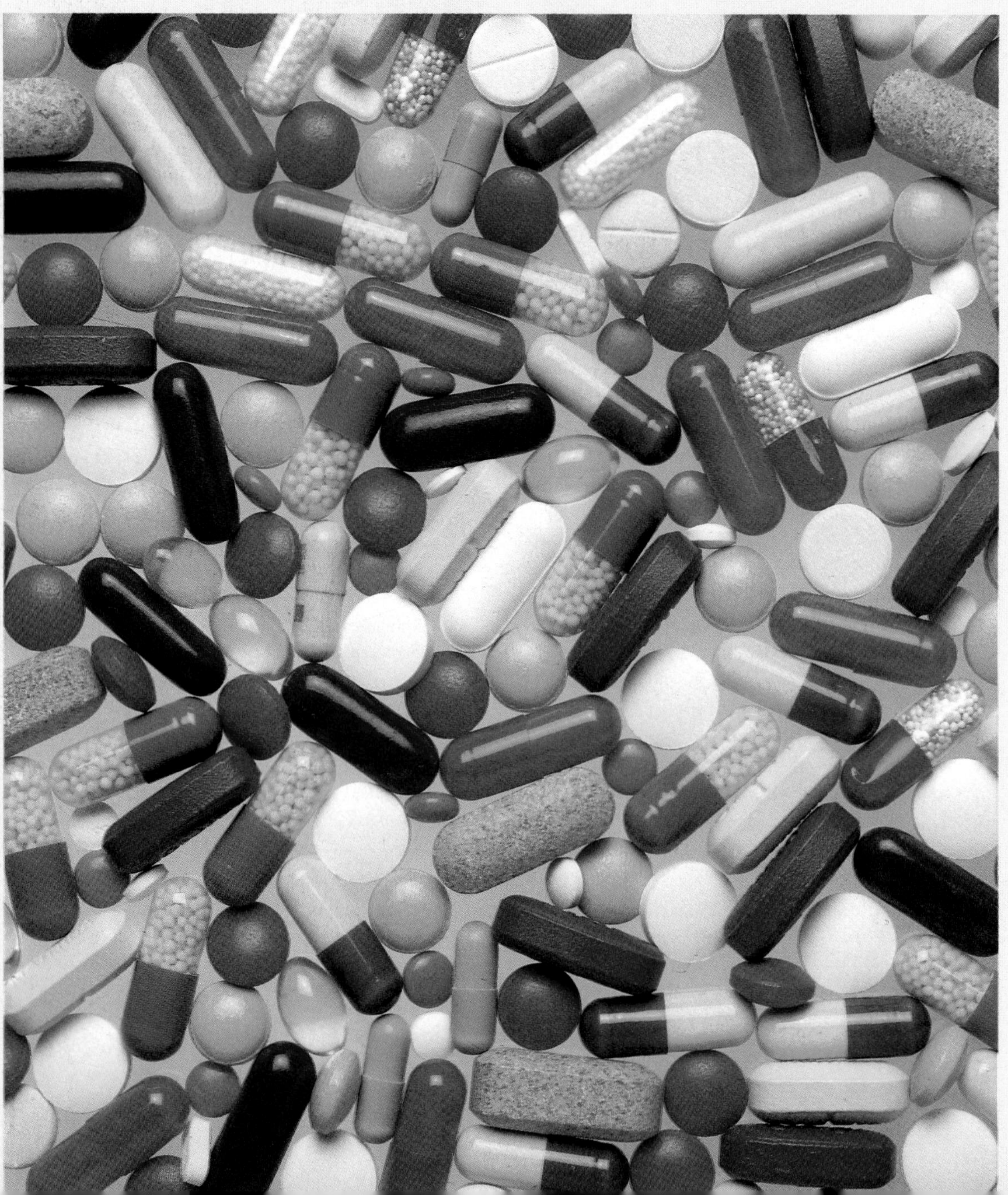

Some wake you up; some make you sleep;

Some cure you when you're ill.

If you're in need of pain relief,

Just choose the proper pill.

Ours is a society that uses drugs—many different kinds of drugs. We use drugs to fight infections, to relieve pain, to stay awake, to get to sleep, to prevent conception, to heal ulcers, to lower blood pressure, to treat arthritis, and to do any number of other things to improve our state of health or our sense of well being.

Ours is not the first society to use drugs. The use of chemicals in attempts to relieve pain and cure illnesses dates to prehistoric times. In some societies, the use of drugs was excessive, but the variety of drugs was limited. Most societies used ethyl alcohol, a depressant. The use of marijuana (*Cannabis sativa*) goes back at least to 3000 B.C. The narcotic effect of the opium poppy was known to the Greeks in the third century B.C. (Figure 19.1). The Indians of the Andes Mountains have long chewed the leaves of the coca plant (*Erythroxylon coca*) for the stimulating effect of the cocaine and the other alkaloids in its leaves.

A scientific rationale for the use of chemical substances to treat diseases was developed early in this century. In 1904, Paul Ehrlich, a German chemist, realized that certain chemicals were more toxic to disease organisms than to human cells. These chemicals could be used to control or cure infectious diseases. Ehrlich coined the term **chemotherapy**, a shorter version of the term "chemical therapy." Ehrlich found that certain dyes used to stain bacteria to make them more visible under a microscope also could be used to kill the bacteria. He used dyes against the organism that causes African sleeping sickness and synthesized an arsenic compound effective against the organism that causes syphilis. Ehrlich was awarded the Nobel prize in 1908.

Figure 19.1 This ancient Greek coin shows an opium poppy capsule. Some cults in ancient Greece worshipped opium. [*Courtesy of the Bureau of Narcotics and Dangerous Drugs, Washington DC.*]

The use of drugs is not new. What is new is the vast array of drugs available to people in modern society. Many of these drugs are available without a prescription from the shelves of supermarkets and drugstores. They are "pushed" on television with the latest Madison Avenue techniques. Illegal drugs in great variety are readily available in most places. Even with prescription drugs, prepared and tested under the supervision of the FDA, there are some serious problems. Physicians prescribe drugs that had not yet been dis-

Modern chemistry provides medicines for many different kinds of ailments, from headaches and infectious diseases to mental illness and cancer. [*Roy Schneider/The Stock Market.*]

covered when they were in medical school. New drugs are being discovered, tested, and placed on the market each year.

Sales of over-the-counter drugs total more than $10 billion each year in the United States.

There are more than 25,000 prescription medicines available in the United States, and about 300,000 over-the-counter drugs. Some of these drugs mean the difference between life and death for the people who use them. On the other hand, there are drugs that are purely recreational. Some of these drugs can be quite destructive. Some are mind-bending, and many are addictive. Most of these drugs are also illegal.

In this chapter we take a look at some of the vast array of drugs that are available today.

SECTION 19.1

Pain Relievers: Aspirin

Freedom from pain has long been a human goal. Alcohol, opium, cocaine, and Indian hemp (marijuana) were used as medicines for pain relief in some early societies. The first successful synthetic pain relievers were derivatives of salicylic acid (Figure 19.2). Salicylic acid was first isolated from willow bark in 1860, although an English clergyman, Edward Stone, had reported to the Royal Society as early as 1763 that an extract of willow bark was useful in reducing fever. Salicylic acid is itself a good **analgesic** (pain reliever) and **antipyretic** (fever reducer), but it is sour and irritating when taken orally. Chemists sought to modify the structure of the molecule to remove this undesirable property while retaining (or even improving) the desirable properties.

The first modification was simple neutralization of the acid (Chapter 7). The salt sodium salicylate was first used in 1875. It was less unpleasant to swallow, but it proved to be highly irritating to the lining of the stomach. Phenyl salicylate (salol) was introduced in 1886. It passed unchanged through the stomach. In the small intestine, it was hydrolyzed to the desired salicylic acid, but phenol, which is rather toxic, also was formed.

Acetylsalicylic acid, to which the German Bayer Company assigned the trade name Aspirin, was first introduced in 1899. It soon became the best-selling drug in the world. Over 55 billion aspirin tablets (under many trade names) are now consumed annually in North America. Another derivative of

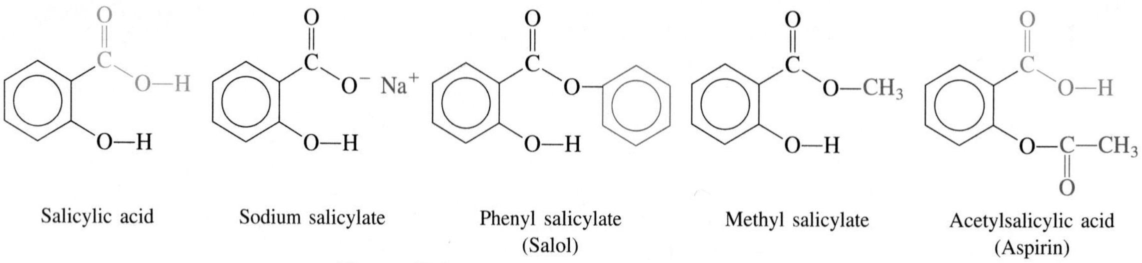

Salicylic acid | Sodium salicylate | Phenyl salicylate (Salol) | Methyl salicylate | Acetylsalicylic acid (Aspirin)

Figure 19.2 Salicylic acid and some of its derivatives.

salicylic acid, methyl salicylate (oil of wintergreen), is used as a flavoring agent. It is also used in rubbing compounds. It causes a mild burning sensation when applied to the skin, thus serving as a counterirritant for sore muscles.

Aspirin is the most widely used drug in the world. In the United States 14 million kg is produced annually. Each aspirin tablet typically contains 325 mg of acetylsalicylic acid, held together with an inert binder (usually starch). "Extra strength" formulations just have 500 mg of aspirin instead of the usual 325 mg.

"Buffered" aspirin simply contains added antacids. The FDA experts who evaluated Bufferin, the most highly advertised brand of aspirin, found that there was no basis for the claim that Bufferin is "twice as fast as aspirin" or that it "helps prevent the stomach upset often caused by aspirin." Some people experience mild stomach irritation when they take aspirin on an empty stomach. Eating a little food first, or drinking a glass of water with the aspirin is just as effective as taking "buffered" tablets.

Aspirin: Pain Relief and Anti-inflammatory Action

Aspirin relieves minor aches and pains and suppresses inflammation. It works by inhibiting the production of *prostaglandins,* compounds that are involved in sending pain messages to the brain. Aspirin doesn't cure whatever is causing the pain; it merely kills the messenger.

Inflammation results from an overproduction of prostaglandin derivatives. The **anti-inflammatory** action of aspirin results from its inhibition of prostaglandin synthesis. Aspirin often is the initial drug of choice for treatment of arthritis, a disease characterized by the inflammation of joints and connective tissues.

Prostaglandins are produced in all body tissues and help to regulate many biochemical processes. These compounds are discussed in some detail in Section 19.8.

Aspirin as an Anticoagulant: Heart Attack and Stroke Prevention

Aspirin also acts as an **anticoagulant**; it inhibits the clotting of blood. Studies indicate that there is some intestinal bleeding every time aspirin is ingested. The blood loss is usually minor (0.5 to 2.0 mL per two-tablet dose), but it may be substantial in some cases. Aspirin should not be used by people facing surgery, childbirth, or other hazards involving the possible loss of blood (nonuse should start a week before the hazard). On the other hand, small daily doses seem to lower the risk of coronary heart attack and stroke, presumably by the same anticoagulant action that causes bleeding in the stomach.

Aspirin and Fever Reduction

Fevers are induced by substances called *pyrogens.* These compounds are produced by and released from leukocytes and other circulating cells. Pyrogens usually use prostaglandins as secondary mediators. Fevers therefore can be reduced by aspirin and other prostaglandin inhibitors.

Some pyrogens do not work through prostaglandins. Aspirin doesn't affect those pyrogens.

Table 19.1 Acute Toxicities of Chemicals Presently or Formerly Used in Over-the-Counter Drugs

Chemical Compound	LD_{50}*
Acetaminophen	338†
Acetanilide	800
Aspirin	1500
Caffeine	355 (246)
Diphenhydramine	500
Ibuprofen	(1050)
Methyl salicylate	887
Naproxen	534
Phenacetin	1650
Piroxicam	360†

*LD_{50} values are for oral administration of the drug to rats in milligrams per kilogram of body weight (unless otherwise noted). Values in parentheses are for male rats.
†Orally in mice.
Source: Susan Budavari (Ed.), *The Merck Index,* 11th ed. Rahway, NJ: Merck and Co., 1989.

Our bodies try to fight off infections by elevating the temperature. Mild fevers (those below 39 °C or 102 °F) usually are therefore best left untreated. High fevers can cause brain damage and require immediate treatment.

Limitations of Aspirin

Use of aspirin to treat children feverish with flu or chickenpox is associated with *Reye's syndrome.* This syndrome is characterized by vomiting, lethargy, confusion, and irritability. Fatty degeneration of the liver and other organs can lead to death unless treatment is begun promptly. Just how aspirin use enhances the onset of Reye's syndrome is not known, but the correlation is quite strong. Aspirin products bear a warning not to use aspirin to treat children with fevers.

Aspirin is not effective for severe pain—for example, pain from a migraine headache. Prolonged use of aspirin, as for arthritic pain, can lead to gastrointestinal disorders. Like all drugs, aspirin is somewhat toxic. It is the drug most often involved in the accidental poisoning of children. The toxicities of aspirin and other drugs are listed in Table 19.1 (Also see Section 20.13.)

Still another hazard associated with aspirin use is allergic reaction. In some people, an allergy to aspirin can cause skin rashes, asthmatic attacks, and even loss of consciousness. Some doctors claim that the allergic reaction may be delayed 3 to 5 hours, so the victim may not associate the reaction with aspirin. Susceptible individuals must be careful to avoid aspirin by itself or in any combination with other drugs.

Aspirin Substitutes: Acetaminophen and Ibuprofen

Some people who are allergic to aspirin may safely take substitute medicines (Figure 19.3). The most common is acetaminophen. This compound gives relief of pain and reduction of fever comparable to the action of aspirin. Unlike aspirin, however, it is not effective against inflammation, and it

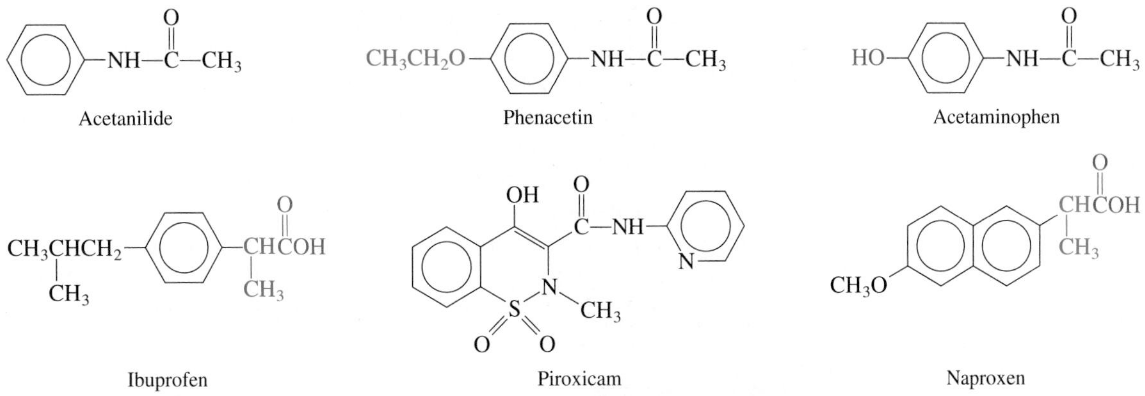

Figure 19.3 Some aspirin substitutes. Phenacetin and acetaminophen are derivatives of acetanilide. Ibuprofen and naproxen are derivatives of proprionic acid. Acetanilide and phenacetin are no longer used.

doesn't induce bleeding. Thus, it is of limited use to people with arthritis, but it is often used to relieve the pain that follows minor surgery. Acetaminophen usually costs more than aspirin, especially in the form of highly advertised brands such as Tylenol, Panadol, and Anacin-3. Regular acetaminophen tablets are 325 mg; "extra-strength" forms are 500 mg. Overuse of acetaminophen has been linked to liver and kidney damage.

Ibuprofen, an anti-inflammatory drug available for decades by prescription under the trade name Motrin, has been available for over-the-counter use since 1984. Brand names are Advil, Nuprin, and Motrin IB. Ibuprofen also is available in generic form. The usual nonprescription dosage is 200 mg. Ibuprofen is perhaps superior to aspirin in its action against inflammation. It also relieves mild pain and reduces fevers but is unlikely to be any better than aspirin for those purposes. It usually is much more expensive than aspirin, and many people who are sensitive to aspirin are also sensitive to ibuprofen.

Two more recently developed anti-inflammatory drugs are piroxicam, which has a much more complicated chemical structure, and naproxen, which is a derivative of naphthalene. Naprosyn (naproxen), was formerly a prescription drug, but has been approved for over-the-counter marketing.

Combination Pain Relievers

Much of what we spend on analgesics is spent on combinations of aspirin and other drugs. Is our money well spent? What are those other drugs?

For many years, the most familiar combination was aspirin, phenacetin, and caffeine (APC), available under a variety of trade names. Phenacetin has about the same effectiveness as aspirin in reducing fever and relieving minor aches and pains. It has been implicated in damage to the kidneys, in blood abnormalities, and as a likely carcinogen, however. The U.S. FDA banned further use of phenacetin in 1983.

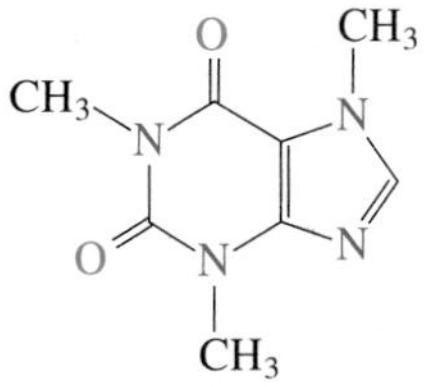

Caffeine

Anacin, which, along with Empirin and Excedrin, was once an APC formulation, now contains only aspirin and caffeine. Caffeine is a mild stimulant found in coffee, tea, and cola syrup. There is no reliable evidence that caffeine significantly enhances the effect of aspirin. In fact, evidence indicates that for fever reduction, caffeine *counteracts* the action of aspirin. Combinations containing caffeine are therefore *less effective* than plain aspirin for this use. Excedrin also contains caffeine, along with aspirin and acetaminophen.

Many other combination products are available. Brands and formulations change frequently. Extensive studies show repeatedly that, for most people, plain aspirin is the cheapest, safest, and most effective product.

SECTION 19.2

Chemistry, Chicken Soup, and the Common Cold

We spend a lot of money for cold remedies, but as yet there is no real cure for the common cold. Colds are caused by as many as 100 related viruses. No antibiotic or other drug is effective against any of them. If we can't cure colds, why do we use so much medicine for them?

Most cold remedies treat the symptoms. Some, perhaps, give worthwhile relief. None prevent or cure the cold or shorten its duration. Included in the arsenal of weapons used against the cold are cough suppressants, expectorants, bronchodilators, anticholinergics, nasal decongestants, and antihistamines. (Figure 19.4 gives chemical structures for several compounds used in cold medicines.) An advisory panel to the FDA reviews these compounds periodically for safety and effectiveness.

An **anticholinergic** is a drug that acts on the nerves that use acetylcholine as a neurotransmitter. (Neurotransmitters are discussed in Section 19.18.)

Antihistamines and Allergies

Antihistamines (such as diphenhydramine, chlorpheniramine, and promethazine) relieve the symptoms of allergies: sneezing, itchy eyes, and runny nose. Antihistamines are not effective against colds, although they may temporarily relieve some cold symptoms.

Diphenhydramine (Benadryl) is also used as a topical drug for treating itchy skin.

Antihistamines do relieve the symptoms of allergies. When an **allergen** (a substance that triggers an allergic reaction) binds to the surface of certain cells, it triggers the release of histamine, which causes the redness, swelling, and itching associated with allergies. Antihistamines block the release of histamine, but most also enter the brain and act upon the cells controlling sleep.

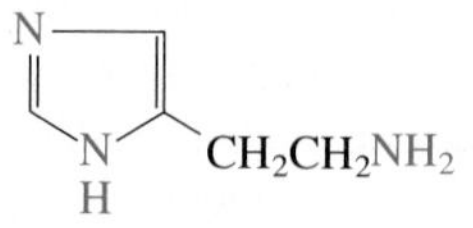

N, N, H, $CH_2CH_2NH_2$

Histamine

Chlorpheniramine Diphenhydramine Promethazine

(a) **Antihistamines**

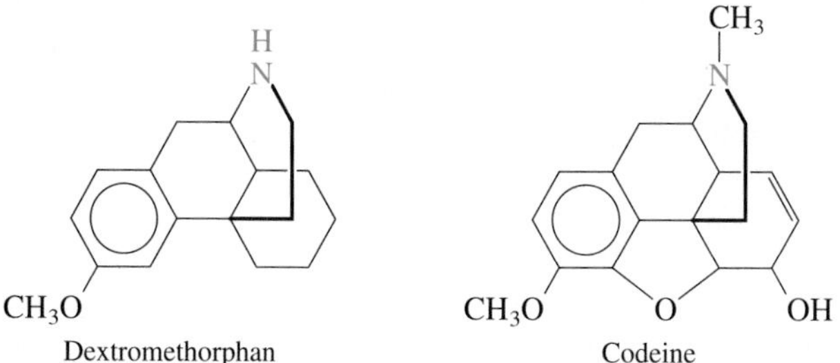

Dextromethorphan Codeine

(b) **Cough suppressants**

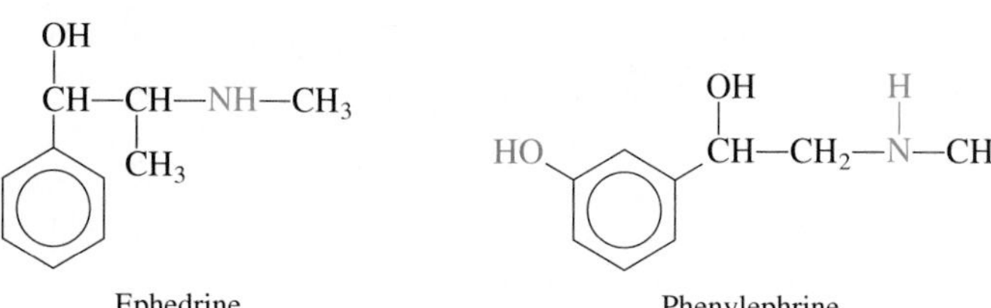

Ephedrine Phenylephrine

(c) **Nasal decongestants**

Figure 19.4 Some chemicals used to treat colds: antihistamines (a), cough suppressants (b), and some nasal decongestants (c). Diphenhydramine, an antihistamine, also is used as a cough suppressant. All of these compounds are amines, but they are often used as their hydrochloride salts, since the salts are more soluble than the free bases.

There are two prescription drugs, terfenadine (Seldane) and astimizole (Hismanal), that inhibit release of histamine but cannot enter the brain, so they do not cause drowsiness.

Cough Suppressants (Antitussives)

The FDA panel has found only three compounds effective for suppressing coughs. Two of these, codeine and dextromethorphan, are narcotics (Section 19.15). The other, diphenhydramine, is an antihistamine.

Should a cough be suppressed? Usually, no. The respiratory tract is using the cough mechanism to rid itself of congestion. When the cough is dry or interferes with needed rest, temporary suppression may be advisable.

Expectorants and Nasal Decongestants

OCH_3 / OCH_2CHCH_2OH / OH

Guaifenesin

Expectorants help bring up mucus from the bronchial passages. The FDA panel has found only guaifenesin to be safe and effective, and its effectiveness is only marginal.

On the other hand, the panel found a variety of **nasal decongestants** to be safe and effective for occasional use. Repeated use leads to a rebound effect in which the nasal passages swell and make congestion seem worse. Examples of nasal decongestants are phenylephrine and ephedrine and their salts.

What should you take for the common cold? Chicken soup is probably as good as anything. You should drink plenty of liquids and get lots of rest. If you are in good physical condition and have a strong immune system, colds seem to strike less often and the symptoms can be less severe. Many people claim to get some relief from cold symptoms with over-the-counter drugs.

SECTION 19.3

Antibacterial Drugs

There is no cure for colds, but colds seldom kill. Less than a century ago, however, infectious diseases were the principal cause of death (Figure 19.5); today only one of those infectious disease categories (pneumonia and influenza) remains among the ten leading causes of death. [However, the list now includes a terrible new kind of viral infection that causes AIDS (acquired immune deficiency syndrome)]. Many diseases have been brought under control by the use of *antibacterial drugs.* The first of these were sulfa drugs, the prototype of which was discovered in 1935 by the German chemist Gerhard Domagk. Sulfa drugs were used extensively during World War II to prevent infection in wounds. Many soldiers lived who, with similar wounds, would have died in earlier wars.

The action of sulfanilamide, the simplest of the sulfa drugs, was one of the first to be understood at the molecular level. Its effectiveness is based on a case of mistaken identity. Bacteria need *para*-aminobenzoic acid (PABA) to make folic acid. Folic acid is essential for the formation of certain compounds the bacteria require for proper growth. But the bacterial enzymes can't tell the difference between sulfanilamide and PABA, because the substances are similar in structure (Figure 19.6). When sulfanilamide is applied to an infection in large amounts, the bacteria incorporate it into a pseudofolic acid. These false molecules cannot perform the growth-enhancing function of folic acid; hence, the bacteria cease to grow.

Sulfanilamide analogs have been developed and tested by the thousands. Only a few are used today. The structures of two common ones are given in Figure 19.6. Introduction of a heterocyclic ring generally increases the activity of a sulfa drug. Some of the sulfa drugs tend to cause damage to the kidneys by crystallizing there. Some present other toxicity problems. For many purposes, sulfa drugs have been replaced by newer antibacterial drugs.

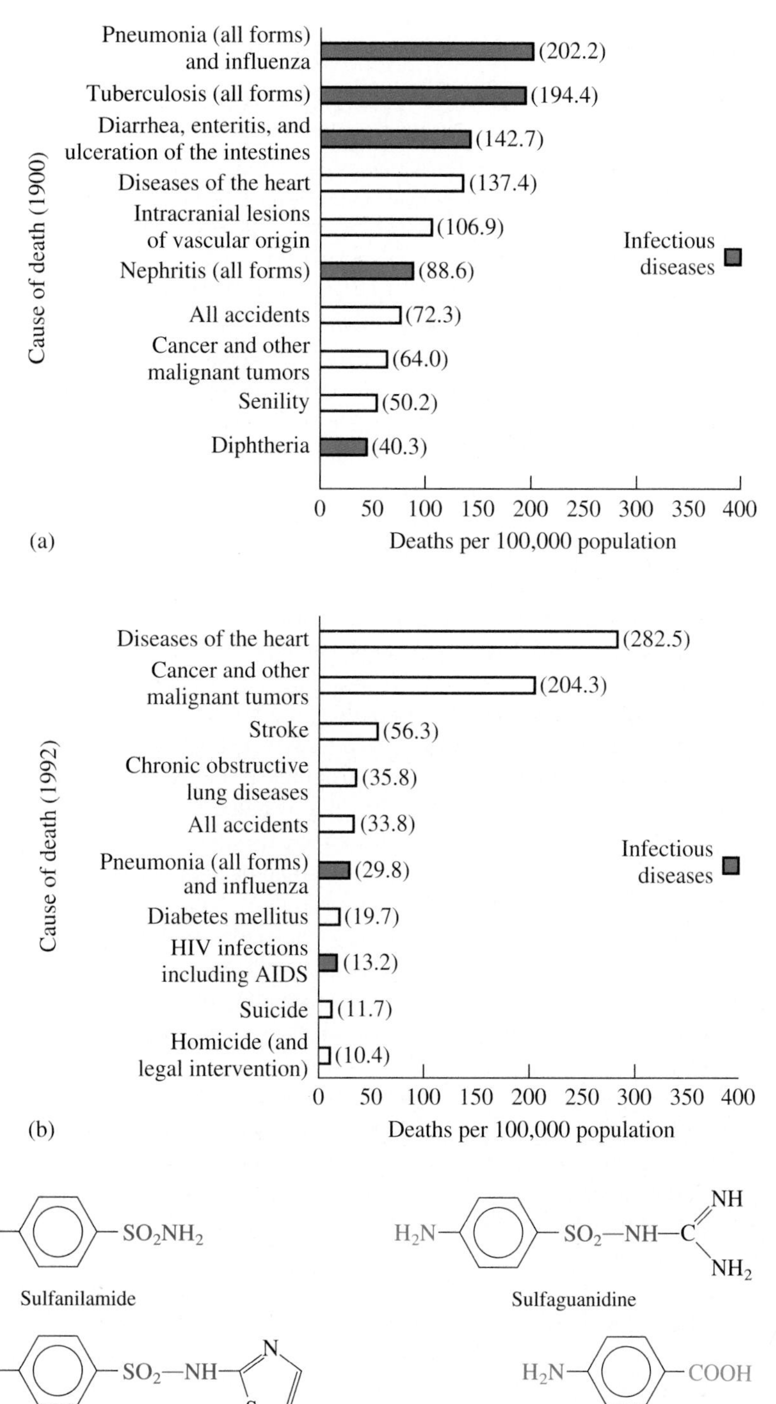

Figure 19.5 (a) In 1900 half of the 10 leading causes of death were infectious diseases. (b) By 1992 only one of those infectious disease categories was still listed as a leading cause of death. However, a new viral infection that causes AIDS has now been added to the list.

Figure 19.6 Sulfanilamide and two other sulfa drugs. *para*-Aminobenzoic acid has a similar structure.

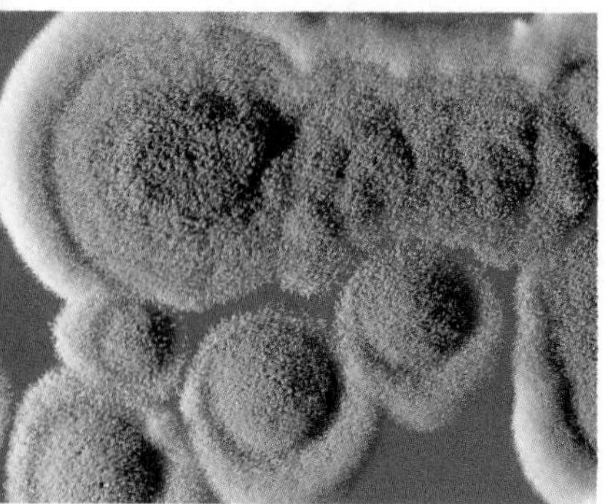

Figure 19.7 Penicillin molds. These symmetrical colonies of mold are *Penicillium chrysogenum,* a mutant form of which now produces almost all of the world's commercial penicillin. [*Dr. Jeremy Burgess/ Science Photo Library/Photo Researchers, Inc.*]

SECTION 19.4

Penicillins and Cephalosporins

The next important discovery was that of penicillin, an antibiotic. **Antibiotics** are soluble substances (derived from molds or bacteria) that inhibit the growth of other microorganisms (Figure 19.7). Penicillin was first discovered in 1928, but it was not tried on humans until 1941. Alexander Fleming, a Scottish microbiologist then working at the University of London, first observed the antibacterial action of a mold, *Penicillium notatum.* Fleming was studying an infectious bacterium, *Staphylococcus aureus.* One of his cultures became contaminated with blue mold. Contaminated cultures generally are useless. Most investigators probably would have destroyed the culture and started over, but Fleming noted that the bacterial colonies had been destroyed in the vicinity of the mold.

Fleming was able to make crude extracts of the active substance. This materia!, later called penicillin, was further purified and improved by Howard Florey (an Australian) and Ernst Boris Chain (a refugee from Nazi Germany). Florey and Chain were working at Oxford University. Fleming, Florey, and Chain shared the 1945 Nobel prize in physiology and medicine for their work on penicillin.

It soon became apparent that penicillin was not a single compound but a group of compounds with related structures. Chemists recognized that by deliberately designing molecules with different structures they could vary the properties of the drugs. The penicillins that resulted from these experiments vary in effectiveness. Some must be injected; others can be taken orally. Bacteria resistant to one penicillin may be killed by another. The structures of several common penicillins are shown in Figure 19.8.

A knowledge of the structure of drug molecules enables chemists to be more efficient when they design drugs. A knowledge of the structure of the drug molecule is not sufficient for an understanding of the molecular basis of the drug's action, however. It is also necessary for chemists to understand the structure of the molecules of the human body, of bacteria, and of viruses.

Sir Alexander Fleming, discoverer of penicillin, shown here on a 1981 Hungarian stamp. [*Professor Marvin Lang and Mr. Gary Shulfer, University of Wisconsin–Stevens Point (Scott Cat. Hungary #2699).*]

The mode of action of penicillin has been unraveled. In certain bacteria, the cell walls are made up of mucoproteins, polymers in which amino sugars are combined with protein molecules. Penicillin prevents cross-linking between these large molecules and thereby prevents the synthesis of cell walls in bacteria. Cells of higher animals do not have mucoprotein walls. Instead, they have external membranes that differ in composition from the cell walls of bacteria. These membranes are not affected by penicillin. Thus, penicillin can destroy bacteria without harming human cells. Many people—perhaps as many as 5% of the population—are allergic to penicillin, however.

During their early history, antibiotics came to be known as miracle drugs. The number of deaths from blood poisoning, pneumonia, and other infectious diseases was reduced substantially by the use of antibiotics. Only six decades ago, a person with a major infection almost always died. Today, such deaths

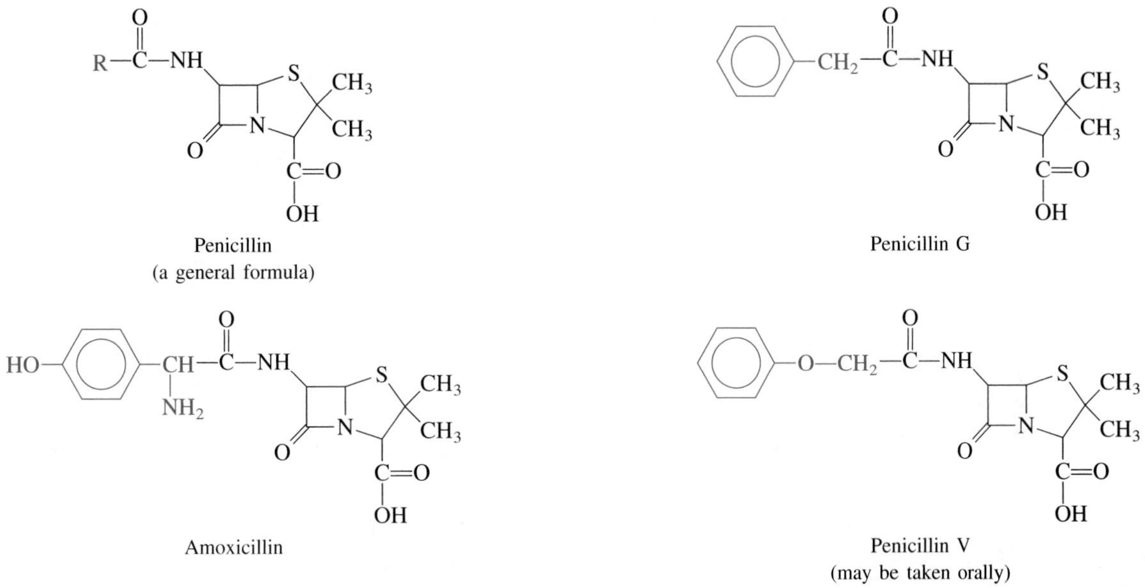

Penicillin
(a general formula)

Penicillin G

Amoxicillin

Penicillin V
(may be taken orally)

Figure 19.8 Some common penicillins. Amoxicillin was the most widely prescribed drug in the United States in 1988. However, by 1992, it had dropped to 30th place.

are rare. Six decades ago, pneumonia was a dread killer of people of all ages. Today, it kills only the very old or those ill from other causes. The antibiotics have indeed worked miracles in our time, but even miracle drugs are not without problems. It wasn't long after the drugs were first used that disease organisms began to develop strains resistant to the drugs. Before erythromycin (an antibiotic obtained from *Streptomyces erythreus*) was used very much, all strains of staphylococci could be handled readily by the drug. After it had been put into extensive use, resistant strains began to appear. Staph infections became a serious problem in hospitals. People who had gone to the hospital to be cured got serious bacterial infections instead. Some even died from staph infections picked up at hospitals.

Drug-resistant tuberculosis now accounts for 1 out of every 7 new cases. (More than 5% of these resistant cases have already died.)

In a race to stay ahead of resistant strains of bacteria, scientists continue to seek new antibiotics. The penicillins now have partially been displaced by related compounds called cephalosporins. Perhaps the most notable of these is cephalexin (Keflex). Unfortunately, some strains of bacteria already show resistance to cephalexin.

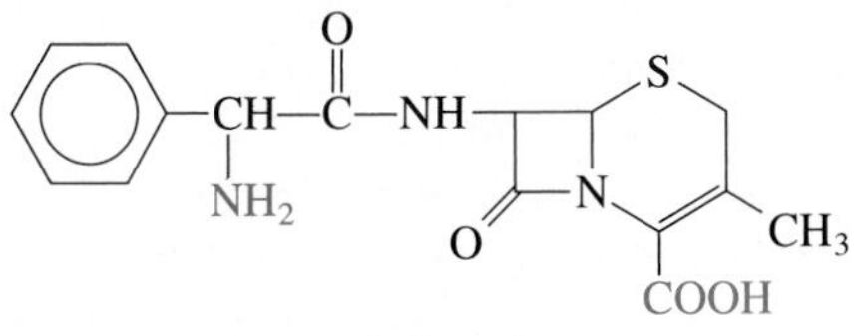

Cephalexin

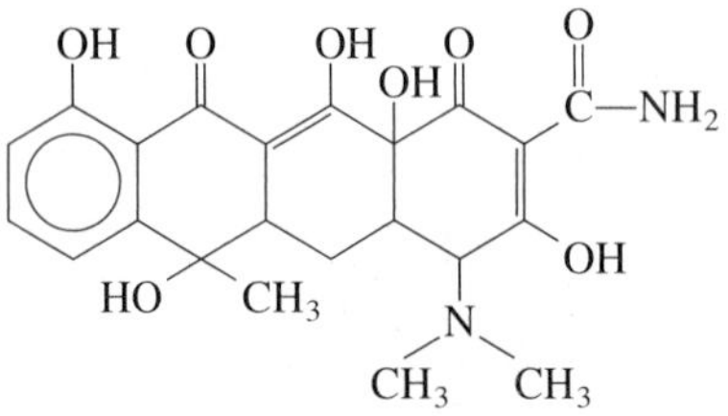

Tetracycline

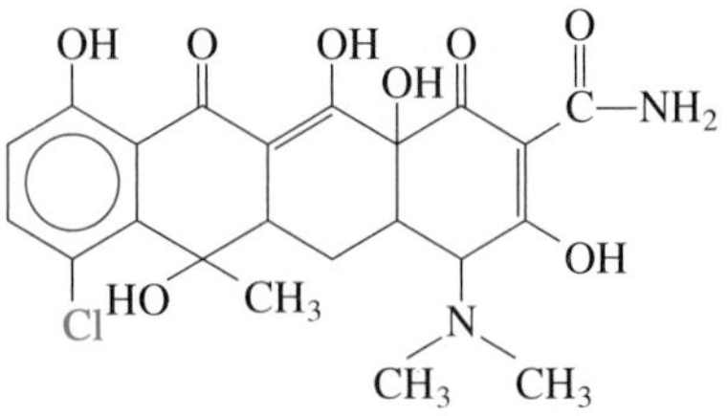

Aureomycin
(Chlorotetracycline)

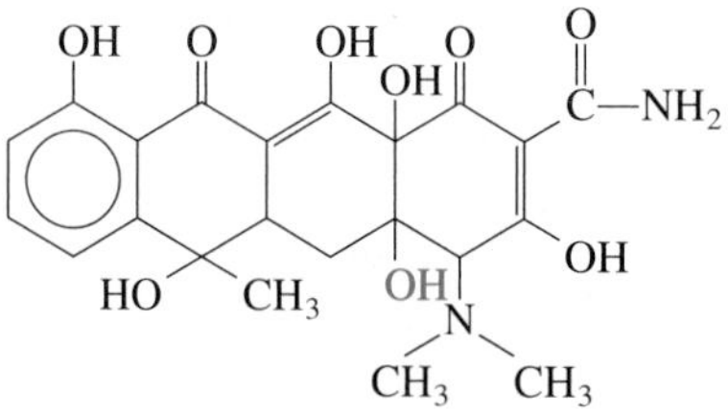

Terramycin
(Oxytetracycline)

Figure 19.9 Three tetracycline antibiotics.

SECTION 19.5

Tetracyclines: Four Rings and Other Things

Another important development in the field of antibiotics was the discovery of a group of related compounds called **tetracyclines**. The first of these, called aureomycin, was isolated in 1948 by Benjamin Duggor from *Streptomyces aureofaciens.* A group of scientists at Pfizer Laboratories isolated terramycin from *Steptomyces rimosus* in 1950 after testing 116,000 different soil samples. Both drugs later were found to be derivatives of tetracycline, a compound now obtained from *Streptomyces viridifaciens.* These three compounds (Figure 19.9) are called **broad-spectrum antibiotics** because they are effective against a wide variety of microorganisms.

Tetracyclines bind to bacterial ribosomes. This inhibits bacterial protein synthesis and thus blocks bacterial growth. The tetracyclines do not bind to mammalian ribosomes and thus do not affect protein synthesis in host cells. Several microorganisms have developed strains resistant to tetracyclines.

When given to young children, the tetracyclines can cause a discoloration of the permanent teeth, even though the teeth may not appear until several years later. This probably results from the interaction of tetracyclines with calcium during the period of tooth development. Calcium ions in milk and other foods combine with hydroxyl groups on the tetracycline molecules.

SECTION 19.6

Viruses and Antiviral Drugs

Antibiotics have taken much of the terror out of bacterial infections such as pneumonia and diphtheria. We still worry about resistant strains of bacteria, but we hope those problems are not insurmountable. Viral diseases cannot be cured by antibiotics, however, and viral infections including colds, genital herpes, and **acquired immune deficiency syndrome (AIDS)** still plague us.

Some viral infections—such as poliomyelitis, mumps, measles, and smallpox—can be prevented by vaccination. Influenza vaccines are available, but because there are so many strains of flu viruses (with new ones appearing periodically), the vaccines are limited to use against a few recurrent strains.

DNA Viruses and RNA Viruses

Viruses are composed of nucleic acids and proteins. They have an external coat with a repetitive pattern of protein molecules. Some have coats that also include a lipid membrane. Others have sugar-protein combinations called glycoproteins.

The genetic material of a virus may be either DNA or RNA. A *DNA virus* enters a host cell where the DNA is replicated and where it directs the host cell to produce viral proteins. The viral proteins and viral DNA assemble into new viruses that are released by the host cell. These new viruses can then invade other cells and continue the process.

Most *RNA viruses* use their nucleic acids in much the same way as the DNA viruses. The virus penetrates a host cell where the RNA strands are replicated and where they induce the synthesis of viral proteins. The new RNA strands and viral proteins are then assembled into new viruses. Some RNA viruses, called **retroviruses**, synthesize DNA in the host cell. This process is the opposite of the transcription of a DNA code into RNA (Chapter 15) that normally occurs in cells. The synthesis of DNA from an RNA template is catalyzed by an enzyme called *reverse transcriptase.* The human immunodeficiency virus (HIV) that causes AIDS is perhaps the best known of the retroviruses. The HIV invades and eventually destroys T cells, a group of white blood cells that normally help protect the body from infections. With the T cells destroyed, the AIDS victim often succumbs to pneumonia or other infection.

AZT apparently acts by substituting for thymine. The reverse transcriptase enzyme incorporates AZT into the DNA chain, blocking the synthesis of the DNA. Unfortunately, the toxicity of AZT prevents its use in quantities sufficient to stop HIV replication completely. Other drugs similar to AZT are undergoing testing. We can only hope these new medicines are less toxic and even more effective than AZT.

Antiviral Drugs

In recent years scientists have developed drugs that are effective against some viruses. Structures of some familiar ones are given in Figure 19.10. Amantadine helps prevent some influenza A infections. Acyclovir (Zovirax) controls flareups of the herpes viruses that cause genital sores, chickenpox, shingles, mononucleosis, and cold sores. Azidothymidine (AZT) slows the onslaught of

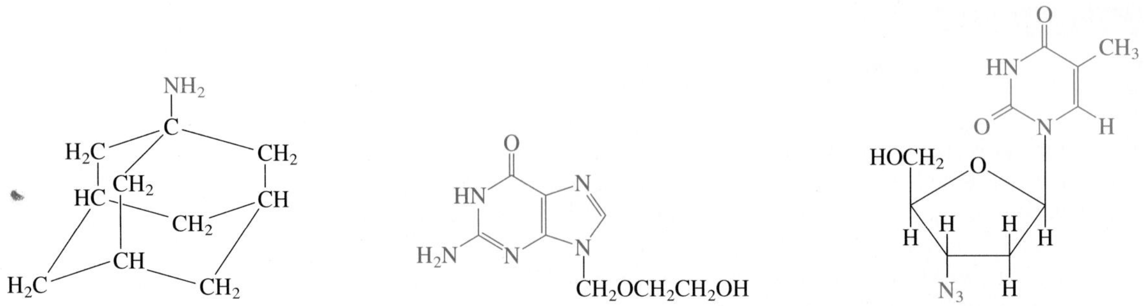

Figure 19.10 Some antiviral drugs. Amantadine is totally synthetic. Acyclovir is derived from the purine base guanine, and azidothymidine is derived from a thymine base attached to a deoxyribose sugar.

the dread AIDS virus, but the disease is still relentlessly fatal to those infected by the virus.

Basic Research and Drug Development

Elion, Hitchings, and Black shared the 1988 Nobel prize for physiology and medicine.

Scientists had to learn the normal biochemistry of cells before they could develop drugs that treat abnormal conditions caused by the invasion of viruses. Gertrude Elion and George Hitchings of Burroughs Wellcome Research Laboratories in North Carolina and James Black of Kings College in London did the basic biochemistry that led to the development of antiviral drugs (and many of the anticancer drugs, Section 19.7). They determined the shapes of cell membrane receptors, and they learned how normal cells work. Then they and other scientists were able to design drugs to block receptors in infected cells. Today scientists use powerful computers to design molecules to fit receptors. Drug design often was hit or miss in its early decades. It is now becoming a more precise science.

SECTION 19.7

Chemicals Against Cancer

Chemotherapy also affects body cells that undergo rapid replacement. These include those cells that line the digestive tract and those that produce hair. Side effects of the therapy include nausea and loss of hair. Eventually the normal cells are affected to such a degree that treatment must be discontinued.

Chemists have designed molecules to relieve headache, cure infectious diseases, and prevent conception. Why can't they do something about cancer? They have done a lot, but much remains to be done. Treatment with drugs, radiation, and surgery has led to high rates of cures for a few forms of cancer (for example, one form of skin cancer). For other forms, such as lung cancer, the rate of cure is still quite low. Over 30 different chemical substances are used widely in the treatment of cancer. That number will no doubt increase rapidly as our understanding of basic cell chemistry increases. We examine a few representative anticancer drugs here.

Antimetabolites: Inhibition of Nucleic Acid Synthesis

In cancer chemotherapy, **antimetabolites** usually are compounds that inhibit the synthesis of nucleic acids. Rapidly dividing cells, characteristic of cancer, require large quantities of DNA. The anticancer metabolites block DNA synthesis and therefore block the increase of the number of cancer cells. Because cancer cells are undergoing rapid growth and cell division, they generally are affected to a greater extent than normal cells.

The most widely used anticancer drug is Cisplatin, a platinum-containing compound. Cisplatin binds to DNA and blocks its replication.

Two other prominent antimetabolites are 5-fluorouracil and its deoxyribose derivative, 5-fluorodeoxyuridine. In the body, both of these can be incorporated into a phosphate–sugar–base unit called a *nucleotide* (Chapter 15). The fluorine-containing nucleotide inhibits the formation of thymine-containing nucleotides required for DNA synthesis. Thus, both compounds slow the division of cancer cells. These compounds have been employed against a variety of cancers, especially those of the breast and the digestive tract.

Another common antimetabolite is 6-mercaptopurine. This compound can substitute for adenine in a nucleotide. The pseudonucleotide then inhibits the synthesis of nucleotides incorporating adenine and guanine. Hence, DNA synthesis and cell division are slowed. 6-Mercaptopurine has been used in the treatment of leukemia.

Another antimetabolite, methotrexate, acts in a somewhat different manner. Note the similarity between its structure and that of folic acid. Like the pseudofolic acid formed from sulfanilamide, methotrexate competes successfully with folic acid for an enzyme but cannot perform the growth-enhancing function of folic acid. Again, cell division is slowed and cancer growth retarded. Methotrexate is used frequently against leukemia.

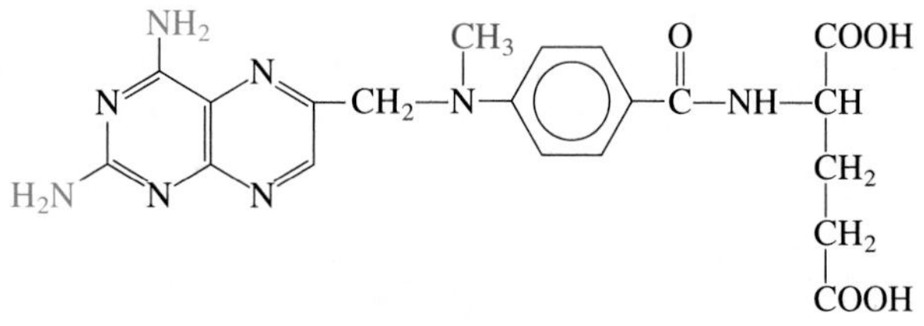

Methotrexate

Folic acid

Transplatin, an isomer of Cisplatin, is ineffective. The *shape* of the molecule is all important.

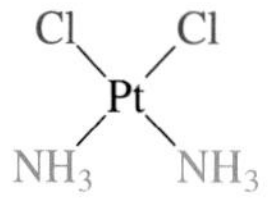

Cisplatin

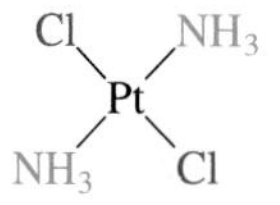

Transplatin

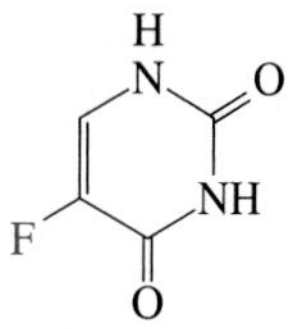

5-Fluorouracil

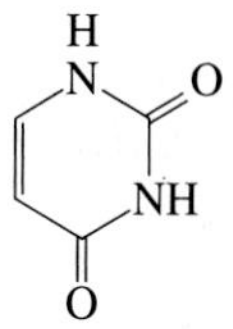

Uracil

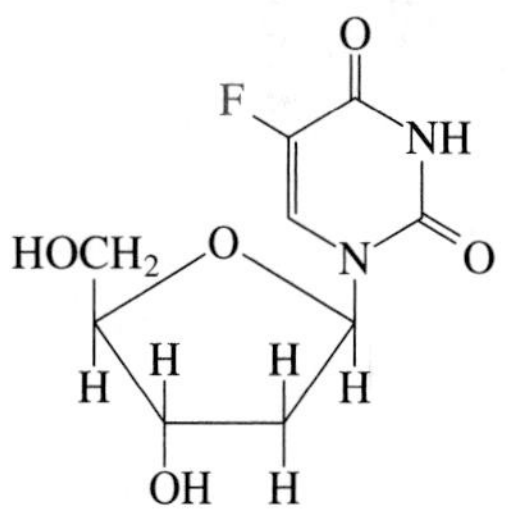

5-Fluorodeoxyuridine

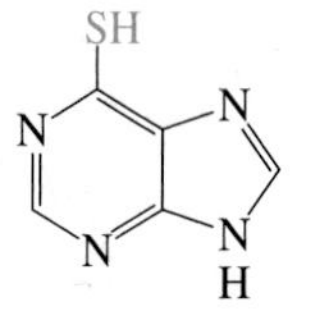

6-Mercaptopurine

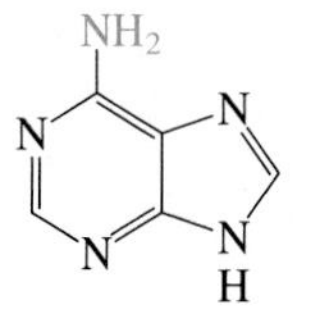

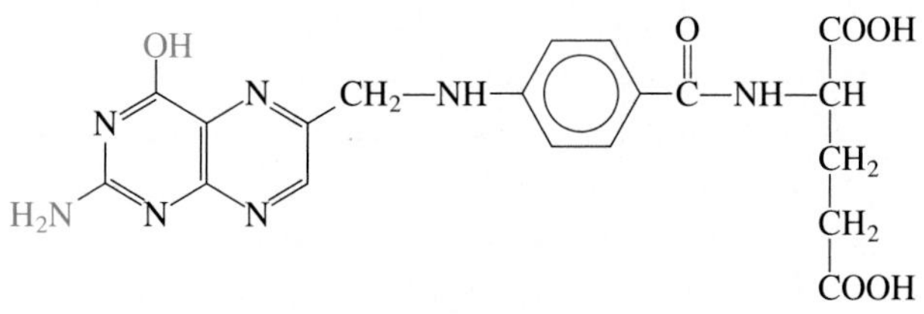

Adenine

Alkylating Agents: Turning War Gases on Cancer

Alkylating agents are highly reactive compounds that can transfer alkyl groups to compounds of biological importance. These foreign alkyl groups then block the usual action of the biological molecules. A variety of alkylating agents are used against cancer. Typical among these are the nitrogen mustards, compounds that arose out of chemical warfare research.

The original mustard "gas" was a sulfur-containing blister agent used in chemical warfare during World War I. Contact with either the liquid or the vapor causes blisters that are painful and slow to heal. It is easily detected, though, by its garlic or horseradish odor. Mustard gas is denoted by the military symbol H.

$$Cl{-}CH_2CH_2{-}S{-}CH_2CH_2{-}Cl$$

Mustard gas
H

The nitrogen mustards (symbol HN) were developed about 1935. Though not quite as effective overall as mustard gas, the nitrogen mustards produce greater eye damage and don't have an obvious odor. Structurally, the nitrogen mustards are chlorinated amines.

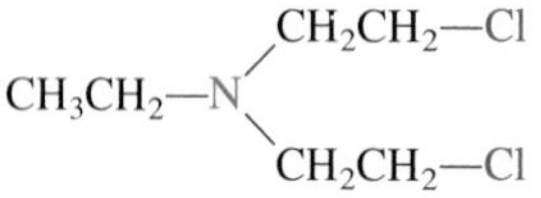

$$CH_3CH_2{-}N(CH_2CH_2{-}Cl)_2$$

HN_1

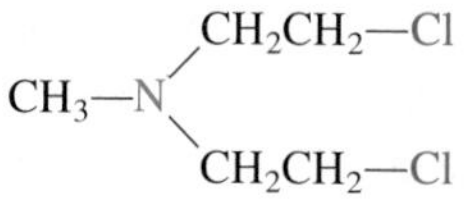

$$CH_3{-}N(CH_2CH_2{-}Cl)_2$$

HN_2

$$Cl{-}CH_2CH_2{-}N(CH_2CH_2{-}Cl)_2$$

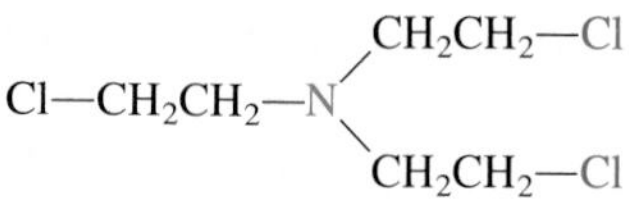

HN_3

To our pleasant surprise, the nitrogen mustards have been found to be effective in the treatment of certain types of cancer. We have seen in earlier chapters how, on occasion, a chemical designed for one (beneficial) purpose was later found to have undesirable side effects. For example, we saw how DDT, once thought to be the perfect insecticide, came under attack for threatening entire species of higher animals. With the nitrogen mustards, we see just the opposite. These molecules, designed for war, have been used most successfully against a formerly fatal type of skin cancer. Complete remission is often obtained by bathing patients in a solution of nitrogen mustards.

Through this example, we see that knowledge gained through science is neither good nor evil. The knowledge can be used either for our benefit or for our destruction.

Chemical agents were used extensively during World War I. Over 30 such substances were employed, killing 91,000 and wounding 1.2 million (many of them for life). Fritz Haber supervised the release of chlorine gas in the first attack by the Germans. Adolf Hitler was among those wounded when the British retaliated with phosgene a few days later. Haber considered gas warfare to be "a higher form of killing." Fortunately, his views have not become widely accepted. The use of chemical warfare agents was largely avoided during World War II. They have been used in smaller wars, most recently by Iraq in its war with Iran.

The nitrogen mustard of choice for cancer therapy nowadays is a compound called cyclophosphamide (Cytoxan). It is of interest to note that cyclophosphamide can defleece sheep chemically, saving the high cost of shearing.

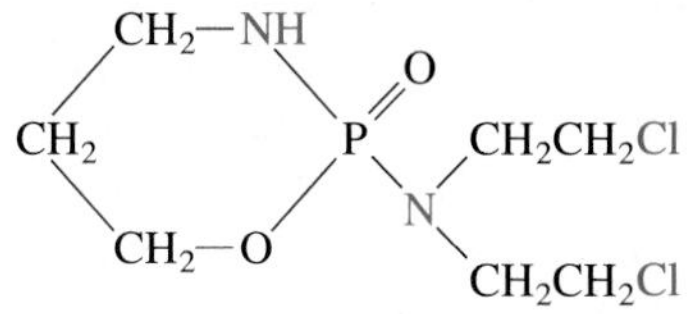

```
      CH2—NH
     /      \    O
    /        \  //
CH2           P      CH2CH2Cl
    \        /  \   /
     \      /     N
      CH2—O         \
                     CH2CH2Cl
```

Cyclophosphamide

After treatment with the drug, the wool can be removed almost as easily as taking off an overcoat.

It should be noted that alkylating agents can cause cancer as well as cure it. For example, the nitrogen mustard HN_2 causes lung, mammary, and liver tumors when injected into mice; yet, it can be used with some success in the management of certain human tumors. There is still a lot of mystery—and seeming contradiction—about the causes and cures of cancer.

Miscellaneous Anticancer Agents

There is a bewildering variety of anticancer agents that defy ready classification. Alkaloids from vinca plants have been shown effective against leukemia and Hodgkin's disease. Actinomycin, a mixture of complex compounds obtained from the molds *Streptomyces antibioticus* and *Streptomyces parvus,* is used against Hodgkin's disease and other types of cancer. It is quite effective but extremely toxic. Actinomycin acts by binding to the double helix of DNA, thus blocking the replication of RNA on the DNA template. Protein synthesis is inhibited.

Sex hormones can be used against cancers of the reproductive system. For example, the female hormones estradiol (a natural hormone) and DES (a synthetic hormone) can be used against cancer of the prostate gland. Conversely, male hormones such as testosterone can be used against breast cancer. Such treatment often brings about a temporary cessation—or even a regression—in the growth of cancer cells.

The food additive butylated hydroxy toluene (BHT) has been shown to be anticarcinogenic in tests involving laboratory animals. Some people involved in cancer research speculate that the use of this additive as a preservative in foods may account for the declining rate of stomach cancer in the United States.

Similarly, there is some evidence that vitamin A may confer a resistance to some cancers. For example, persons suffering from vitamin A deficiencies exhibit a higher incidence of lung cancer. Vitamin C also may have an anticancer function. It has been shown to inhibit the formation of nitrosamines under conditions similar to those in the human stomach.

Chemotherapy is only a part of the treatment of cancer. Surgical removal of tumors and radiation treatment remain major weapons in the war on cancer. It is unlikely that a single agent will be found to cure all cancers. Steady prog-

Table 19.2 Some Human Hormones and Their Physiological Effects			
Name	Gland and Tissue	Chemical Nature	Effect
Various releasing and inhibitory factors	Hypothalamus	Peptide	Triggers or inhibits release of pituitary hormones
Human growth hormone (HGH)	Pituitary, anterior lobe	Protein	Controls the general body growth; controls bone growth
Thyroid-stimulating hormone (TSH)	Pituitary, anterior lobe	Protein	Stimulates growth of the thyroid gland and production of thyroxin
Adrenocorticotrophic hormone (ACTH)	Pituitary, anterior lobe	Protein	Stimulates growth of the adrenal cortex and production of cortical hormones
Follicle-stimulating hormone (FSH)	Pituitary, anterior lobe	Protein	Stimulates growth of follicles in ovaries of females, sperm cells in testes of males
Luteinizing hormone (LH)	Pituitary, anterior lobe	Protein	Controls production and release of estrogens and progesterone from ovaries, testosterone from testes
Prolactin	Pituitary, anterior lobe	Protein	Maintains the production of estrogens and progesterone; stimulates the formation of milk
Vasopressin	Pituitary, posterior lobe	Peptide	Stimulates contractions of smooth muscle; regulates water uptake by the kidneys
Oxytocin	Pituitary, posterior lobe	Peptide	Stimulates contraction of the smooth muscle of the uterus; stimulates secretion of milk
Parathyroid	Parathyroid	Protein	Controls the metabolism of phosphorus and calcium
Thyroxine	Thyroid	Amino acid derivative	Increases rate of cellular metabolism
Insulin	Pancreas, beta cells	Protein	Increases cell usage of glucose; increases glycogen storage
Glucagon	Pancreas, alpha cells	Protein	Stimulates conversion of liver glycogen to glucose
Cortisol	Adrenal gland, cortex	Steroid	Stimulates conversion of proteins to carbohydrates
Aldosterone	Adrenal gland, cortex	Steroid	Regulates salt metabolism; stimulates kidneys to retain Na^+ and excrete K^+
Epinephrine (adrenaline)	Adrenal gland, medulla	Amino acid derivative	Stimulates a variety of mechanisms to prepare the body for emergency action including the conversion of glycogen to glucose
Norepinephrine (noradrenaline)	Adrenal gland, medulla	Amino acid derivative	Stimulates sympathetic nervous system; constricts blood vessels, stimulates other glands
Estradiol	Ovary, follicle	Steroid	Stimulates female sex characteristics; regulates changes during menstrual cycle
Progesterone	Ovary, corpus luteum	Steroid	Regulates menstrual cycle; maintains pregnancy
Testosterone	Testis	Steroid	Stimulates and maintains male sex characteristics

ress is being made, however. Rates of cure should improve as research progresses. Perhaps a greater hope lies in the prevention of cancer. Much active research is underway on the mechanisms of carcinogenesis. The more we learn about what causes cancer, the better equipped we will be in trying to prevent it.

SECTION 19.8

Hormones: The Regulators

Before we can discuss the next group of drugs, we must take a brief look at the human endocrine system and some of the chemical compounds, called **hormones**, that the system produces. Hormones are chemical messengers produced in the endocrine glands (Figure 19.11). Those released in one part of the body signal profound physiological changes in other parts of the body. By causing reactions to speed up or slow down, they control growth, metabolism, reproduction, and many other functions of body and mind.

Some of the more important human hormones and their physiological effects are listed in Table 19.2. Closely related to the hormones is another group of powerful chemical substances called **prostaglandins**, which are hormone mediators.

Prostaglandins: Hormone Mediators

People engaged in medical research are excited about the possibilities of drugs related to the prostaglandins. Thousands of papers are published each year on these compounds. Our bodies synthesize prostaglandins from arachidonic acid, a fatty acid with 20 carbon atoms (Figure 19.12). There are six primary prostaglandins, and many others have been identified. These compounds are widely distributed throughout the body at extremely low concentrations. They

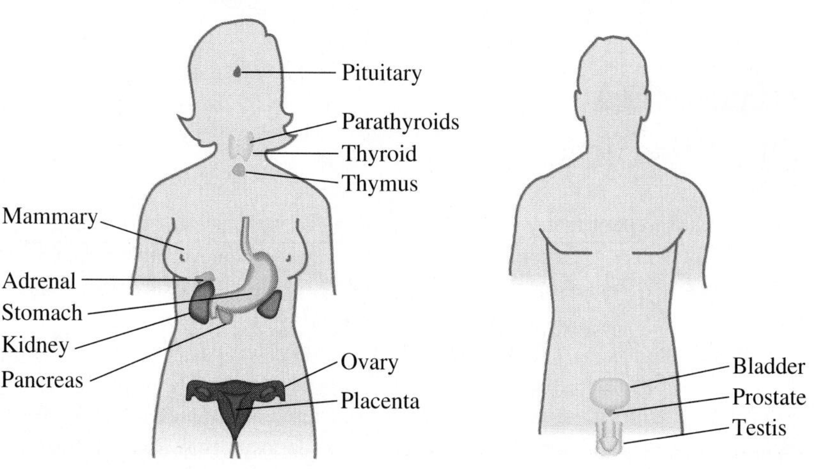

Figure 19.11 The approximate locations of the endocrine glands in the human body.

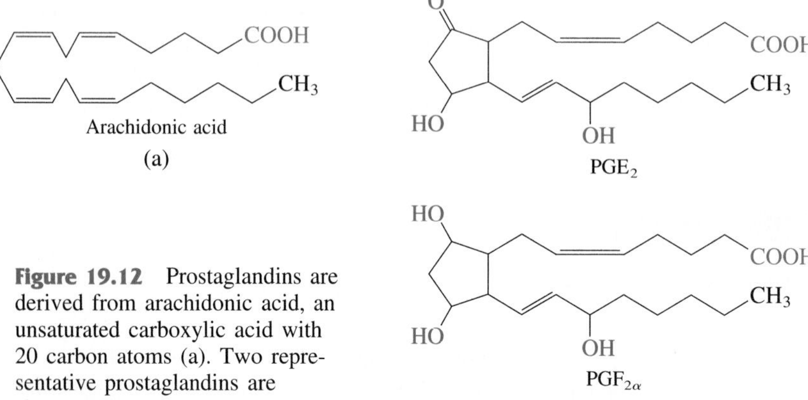

Figure 19.12 Prostaglandins are derived from arachidonic acid, an unsaturated carboxylic acid with 20 carbon atoms (a). Two representative prostaglandins are shown (b).

are among the most potent biological chemicals, and extremely small doses can elicit marked changes.

The prostaglandins act as mediators of hormone action. They regulate such things as smooth-muscle activity, blood flow, and secretion of various substances. This range of physiological activity led to the synthesis of hundreds of prostaglandins and their analogs. Two such derivatives are now in use in the United States to induce labor. Others have been employed clinically to lower or increase blood pressure, to inhibit stomach secretions, to relieve nasal congestion, to provide relief from asthma, and to prevent the formation of the blood clots associated with heart attacks and strokes.

$PGF_{2\alpha}$ is used in cattle breeding. A prize cow is treated with a hormone and with $PGF_{2\alpha}$ to induce the release of many ova. The ova are then fertilized with sperm from a champion bull. The developing embryos are implanted in less valuable cows. This enables a farmer to get several calves a year from one outstanding cow.

Over 30 prostaglandins and related compounds are now under investigation as possible drugs. They hold great promise for a variety of purposes.

SECTION 19.9

The Steroids

Chemists have spent many years determining the structures of hormones. As indicated in Table 19.2, some hormones have complicated protein structures. Others are, by comparison, rather simple. The **steroids**, for example, all have the same skeletal, four-ring structure (Figure 19.13).

Not all steroids have hormonal activity. Cholesterol is a common component of all animal tissues. The brain is about 10% cholesterol, but cholesterol's function there is not known. Cholesterol is a major component of certain types of gallstones. It also is found in deposits in hardened arteries.

Steroid skeletal structure
(a)

Cholesterol
(b)

Cortisone

Prednisone

(c)

Figure 19.13 Steroids all have the same basic four-ring structure (a). Cholesterol (b) is a component of all animal cells. Cortisone and prednisone (c) are anti-inflammatory substances.

Another steroid of some interest is the adrenal hormone cortisone. Applied topically or injected, cortisone acts in the body to reduce inflammation. It was once widely used in the treatment of arthritis, but relief was transitory and repeated application caused serious side effects. Cortisone has been largely replaced by the related compound prednisone. Because prednisone is effective in much smaller doses, its side effects are greatly reduced.

Steroid Drugs

There are various kinds of drugs, both natural and synthetic, that are based on steroids. Some of these are the anabolic steroids (Section 18.12) that are taken by many athletes to improve their performance. Some are anti-inflammatory drugs used to treat such conditions as arthritis, bronchial asthma, dermatitis, and eye infections. And some are progestins used in formulating birth control pills (Section 19.10).

Serendipity often plays a role in the discovery of new drugs. Percy Julian was a chemist at the Glidden Paint Company doing research on soybeans when his work led to the development of some new steroid-based drugs. It often happens that new drugs are discovered by chemists who are not really looking for them.

In 1992 the U.S. Postal Service issued this stamp honoring Percy Lavon Julian (1899–1975). Born in Alabama, he earned his Ph.D. in Vienna and later won acclaim for his synthesis of physostigmine, a drug used to treat glaucoma.

The Sex Hormones

Closely related in structure to cholesterol, cortisone, and prednisone are the sex hormones (Figure 19.14). It is interesting to note that male sex hormones

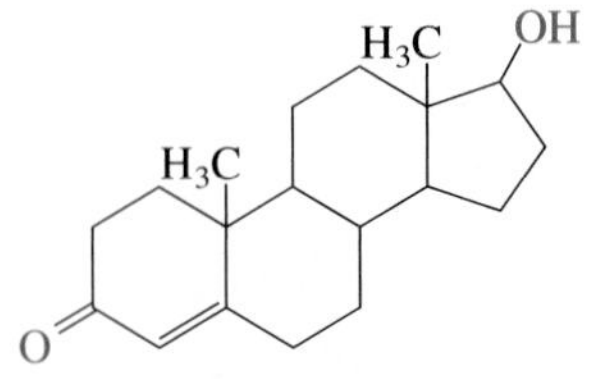

Testosterone

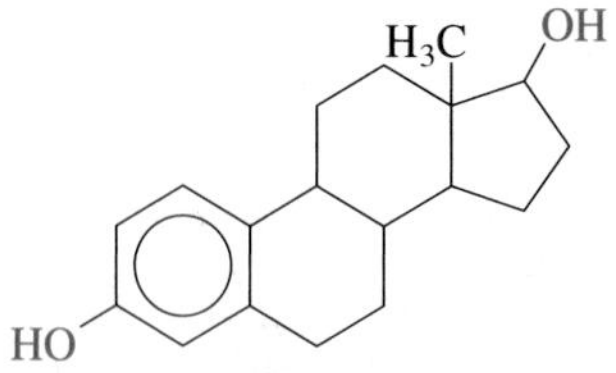

Estradiol

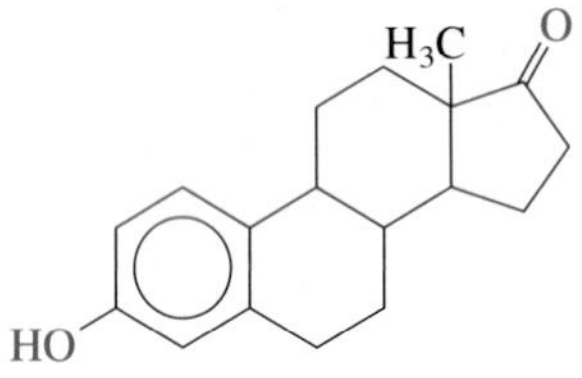

Estrone

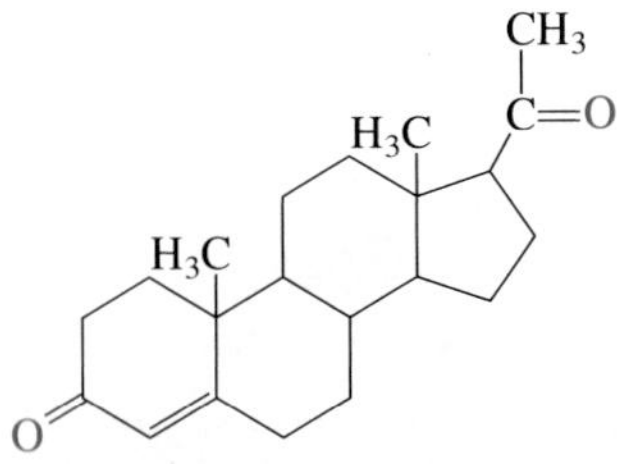

Progesterone

Figure 19.14 The principal sex hormones. Testosterone is the main male sex hormone. Estradiol and estrone are female sex hormones. Progesterone, also found in females, is essential to the maintenance of pregnancy.

differ only slightly in structure from female sex hormones. In fact, the female hormone progesterone can be converted by a simple biochemical reaction into the male hormone testosterone. The physiological actions of these structurally similar compounds are markedly different.

Male sex hormones, called **androgens**, are secreted by the testes. These hormones are responsible for the development of the sex organs and for secondary sexual characteristics such as voice and hair distribution. The most important male hormone is testosterone.

There are two important groups of female sex hormones. The **estrogens** are produced mainly in the ovaries. They control female sexual functions such as the menstrual cycle, the development of breasts, and other secondary sexual characteristics. Two important estrogens are estradiol and estrone. Another female sex hormone is progesterone, which prepares the uterus for pregnancy and prevents the further release of eggs from the ovaries during pregnancy.

Sex hormones—both natural and synthetic—sometimes are used therapeutically. For example, a woman who has had her ovaries removed may be given female hormones to compensate for those no longer being produced by her ovaries. Some of the earliest compounds employed in cancer chemotherapy were sex hormones. The male hormone testosterone was used to treat carcinoma of the breast in females; estrogens, female sex hormones, were given to males to treat carcinoma of the prostate.

SECTION 19.10

Chemistry and Social Revolution: The Pill

When administered by injection, progesterone serves as an effective birth-control drug. This knowledge led to attempts by chemists to design a contraceptive that would be effective when taken orally.

The structure of progesterone was determined in 1934 by Adolf Butenandt. Just four years later, Hans Inhoffen synthesized the first oral contraceptive, ethisterone. Ethisterone had to be taken in large doses to be effective, and it was never widely used as an oral contraceptive.

The next breakthrough came in 1951, when Carl Djerassi synthesized 19-norprogesterone, progesterone with one of its methyl groups missing. This compound was four to eight times as effective as progesterone as a birth-control agent. Like progesterone itself, it had to be given by injection, an undesirable property. Djerassi then put it all together. Removal of a methyl group made the drug more effective. The ethynyl group (—C≡CH) allowed oral administration. He then synthesized 17α-ethynyl-19-nortestosterone, mercifully known by the trade name Norlutin. This drug proved effective when it was taken in small doses. Djerassi's work was published in 1954, and a patent was issued to him in 1956. At about the same time, Frank Colton synthesized norethynodrel. Note that the two substances differ only in the

Much of our synthetic progesterone has been made starting with a naturally occurring steroid (diosgenin) found in the roots of some wild yams that grow in the tropical forests of southern Mexico. This is just one of many examples of plant materials that have been converted to useful pharmaceutical products.

Tropical rainforests are tree-filled areas that are warm and moist and very fertile. More plants and animals live in tropical rainforests than in any other places on Earth. These forests are living chemical factories. They are invaluable and irreplaceable natural resources. Unfortunately, the world's rainforests are rapidly disappearing. In the decade between 1980 and 1990 the rate at which tropical rainforests were being destroyed increased by 50%.

Carl Djerassi, professor of chemistry at Stanford University and president of Zoecon Corporation, Palo Alto, California.

position of a double bond (Figure 19.15). Patents were issued to Colton's employer, G. D. Searle, in 1954 and 1955. Searle produced the first approved contraceptive, Enovid, in 1960. The pill contained 9.85 mg of norethynodrel

19-Norprogesterone

Ethisterone

Norethindrone

Norethynodrel

Mestranol

Figure 19.15 Some synthetic steroids.

Frank Colton, a chemist at G. D. Searle Co., synthesized norethynodrel, a progestin used in Enovid, the first birth-control pill.

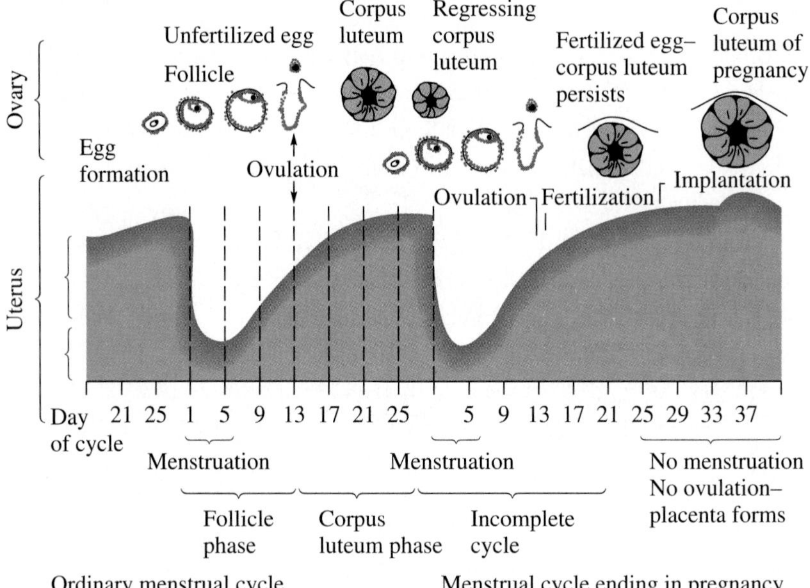

Figure 19.16 Changes in the ovary and the uterus during the menstrual cycle. Pregnancy—or the pseudo-pregnancy caused by the birth-control pill—prevents ovulation.

and 150 μg of mestranol. Norethynodrel, norethindrone, and related compounds are called **progestins** because they mimic the action of progesterone. Mestranol is a synthetic estrogen, added to regulate the menstrual cycle. The progestin acts by establishing a state of false pregnancy (Figure 19.16). A woman does not ovulate when she is pregnant (or in the state of false pregnancy established by the progestin). Since the woman does not ovulate, she cannot conceive.

DES: A Missed-Period Pill

Diethylstilbestrol (DES) has been used for two seemingly opposite purposes. It was used at low doses in an attempt to help women with a history of miscarriages to maintain pregnancy, and at high doses it is used to induce abortion. We know now that it was not effective in maintaining pregnancy, but many doctors prescribed it for that purpose during the 1950s and 1960s. It caused increased vaginal cancer in daughters and sterility in sons of women who took DES to prevent miscarriage. DES is effective at high doses in inducing abortions, but unpleasant side effects frequently are noted.

DES also has been used as a growth promoter in cattle and poultry (Chapter

CH_3 OH

HO H_3C

DES

16). It was banned from that use by the FDA when residues of DES were found (as incidental additives) in the meat of those animals.

How can the FDA approve the use of a substance as a drug and ban it as an additive in the feed of animals intended as human food? The answer lies in the way the law is written and in the way the chemical is used. The Delaney clause *requires* the banning of any food additive that causes cancer in humans or laboratory animals. DES had been shown to cause vaginal cancer. Therefore, it had to be banned as a food additive. There is no such requirement for drugs; DES could be approved for use in inducing abortions. It also was pointed out that, if the drug were effective, there would be no offspring to have cancer.

DES is not a steroid, but the shape of the molecule shows some similarity to the structure of estradiol (see Figure 19.15). Thus, DES is a synthetic female hormone.

RU-486: Convenience and Controversy

In the third world countries, 200,000 women die each year from botched attempts at self-induced abortion. Availability of mifepristone presumably could reduce the toll substantially.

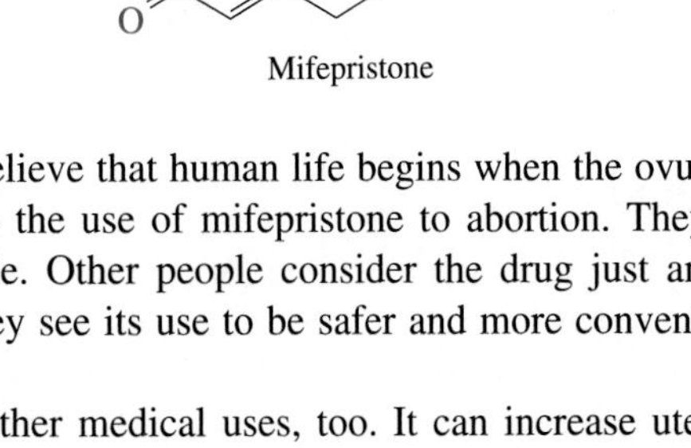

Progesterone is essential for the maintenance of pregnancy. If its action is blocked, pregnancy could not be established or maintained. Rousel-Uclaf, a French subsidiary of Hoescht, has developed a drug that blocks the action of progesterone. Mifepristone (RU-486) is available in France and China, where it has replaced a substantial number of surgical abortions. A woman who wishes to abort a pregnancy takes three mifepristone tablets, followed in a few days by an injection of a prostaglandin (Section 19.8). The lining of the uterine wall and the implanted fertilized egg are sloughed off and pregnancy is terminated.

CH_3
CH_3—N
OH
H_3C
C≡CCH_3
O

Mifepristone

People who believe that human life begins when the ovum is fertilized by the sperm equate the use of mifepristone to abortion. They oppose making the drug available. Other people consider the drug just another method of birth control. They see its use to be safer and more convenient than surgical abortion.

RU-486 has other medical uses, too. It can increase uterine contractions when labor has stalled during childbirth. It appears to trigger lactation in mothers and increase milk production. It seems to slow the growth of certain types of cancer. And it is also being studied as a treatment for Cushing's syndrome, which results from too much production of cortisone.

Risks of Taking Birth Control Pills

Oral contraceptives have been used in the United States since 1960 by millions of women. They appear to be safe in most cases, but some women experience hypertension, acne, or abnormal bleeding. The pills increase the risk of blood clotting in some women, but so does pregnancy. Blood clots can clog arteries and cause death by stroke or heart attack. The death rate associated with birth control pills is about 3 in 100,000, only one-tenth of that associated with childbirth (which is about 30 per 100,000). The FDA advises women over 40, and any women who smoke, to use some other method of contraception. For these women, the risk of using birth control pills is greater than the risk of childbirth.

The Minipill

Since most of the side effects of oral contraceptives are associated with the estrogen component, the amount of estrogen in these pills has been greatly reduced over the years. Today's pills contain only a fraction of a milligram of estrogen. In fact, minipills are now available that contain only small amounts of progestin and no estrogen at all. Minipills are not quite as effective as the combination pills, but they do have fewer side effects.

Birth Control by Implant

In 1990 the FDA approved the use of an under-the-skin implant for birth control. Levonorgestrel (Norplant), a synthetic analog of progesterone, is incorporated into six plastic capsules, each about 2.5 cm long and the size of a match stick. These are then implanted under the skin of the upper arm. Release of the progestin over time prevents pregnancy for five years with a failure rate of only 0.2%. It has about the same side effects as the progestin-only minipill. (Levonorgestrel is the progestin used in some ordinary birth control pills.)

Some form of chemical contraceptive seems to be with us to stay. The pill has already brought revolutionary social changes. The impact of science on society is perhaps nowhere more evident.

A Pill for Males?

Why do females have to bear responsibility for contraception? Why not a pill for males? Recent studies have shown that many men—including a majority of the younger ones—are willing to share the risks and the responsibility of contraception. Nevertheless, there are biological reasons for females to bear the burden. Women get pregnant when contraception fails, and in females, contraception has only to interfere with one monthly event—ovulation. On the other hand, males produce sperm continuously.

A good deal of research has gone into the development of contraceptives for men, but results so far have not been very successful. One product was an effective contraceptive, but an alcoholic drink would bring on a state of extreme nausea. Estrogens would work as male contraceptives, but they would also cause development of female characteristics in the men, and a loss of interest in sexual relations with women. One of the prostaglandins, PGE_2, has shown some promise as a means of birth control for men.

Probably the most widely used male contraceptive so far has been gossypol, a natural pigment found in cottonseed.

Gossypol

This has been used extensively in China, where overpopulation has reached the point that the government now sanctions only one child per married couple and penalizes couples who have more.

The availability of vasectomies—simple surgical procedures that block the emission of sperm—has lessened the demand for a male contraceptive. A major drawback to vasectomy, however, is that it is often irreversible. A safe, cheap, effective male contraceptive is still a goal of biochemical research.

SECTION 19.11

Drugs and the Human Mind

Psychotropic drugs are those that affect the human mind. Probably the first drugs to be used by primitive peoples influenced the mind. Alcohol, marijuana, opium, cocaine, peyote, and other plant materials have been used for thousands of years. Generally, only one or two of these were used in any one society. Today, however, thousands of drugs that affect the mind are readily available to us (Figure 19.17). Some still come from plants, but the great majority are synthetic.

There is no clear distinction between drugs that affect the mind and those that affect the body. Most of the drugs we take probably affect our minds as well as our bodies. It can still be useful, however, to distinguish between those drugs that act primarily on the body and those that affect primarily the mind.

Figure 19.17 We can change our moods with chemicals. A few drugs—such as caffeine and nicotine—are available in vending machines, and many others are easily available (from legal and illegal sources).

Drugs that affect the mind can be divided into three classes. The **stimulants** increase alertness, speed up mental processes, and generally elevate the mood. This group, which includes the amphetamines and related compounds, sometimes is referred to as "uppers." Another category—which includes alcohol, barbiturates, and minor tranquilizers—is the **depressants**. These often are called "downers." They reduce the level of consciousness and the intensity of reactions to environmental stimuli. In general, depressants dull emotional responses. The third category—variously called **hallucinogenics**, **psychotomimetics**, and **psychedelics** (because they induce hallucinations, psychoses, and colorful visions)—is popularly called "mindbenders." These drugs don't so much depress or stimulate the mind as they alter qualitatively the way we perceive things. Notable among the mindbenders are LSD, mescaline, and marijuana.

Let us examine some of the major drugs in each category. But first, let's take another look at the most widely used, and most heavily abused, drug of them all—alcohol.

SECTION 19.12

Alcohol

We have already discussed ethyl alcohol in some detail (Section 9.8), but we discuss it again in this chapter on drugs, since alcohol is by far the most used and abused drug in the world.

Ethyl alcohol was probably the first synthetic chemical to be made by man. We know that the Egyptians as early as 3700 B.C. were fermenting fruit to make wine, and the Babylonians were making beer from barley. People have been fermenting fruits and grains ever since.

Many people imagine that they are being stimulated when they take a drink of alcohol, but alcohol is actually a depressant. It slows down both physical and mental activity. There does seem to be a positive side to drinking alcohol. Longevity studies indicate that those who use alcohol only moderately (no more than a drink or two a day) live longer than nondrinkers. This is probably due to the relaxing effect of the alcohol. Heavier drinking, however, can cause many health problems and will shorten the life span.

About 100 million people in the United States drink alcohol, roughly two-thirds of the population. Some 10 million of them are alcoholics. About 80% of alcoholics are men; 20% are women. Among adolescents about 70% already drink alcohol (62% of seventh graders; 80% of twelfth graders).

Driving after drinking alcohol is the fifth leading cause of death in the United States. For the under-35 age group, it is the leading cause of death. Alcoholism is the third leading health problem in the United States, right below cancer and heart disease. The suicide rate for alcoholics is 58 times greater than for other people.

Alcohol plays a major role in more than half of all highway fatalities, and

Table 19.3 Drugs Mentioned Most Frequently by Emergency Rooms in 1990

Rank	Drug Name	Number of Mentions	Percent of Total Episodes
1	Alcohol-in-combination	115,162	31.02
2	Cocaine	80,355	21.65
3	Heroin/morphine	33,884	9.13
4	Acetaminophen	25,422	6.85
5	Aspirin	19,188	5.17
6	Ibuprofen	16,299	4.39
7	Alprazolam	15,846	4.27
8	Marijuana/hashish	15,706	4.23
9	Diazepam	14,836	4.00
10	Amitriptyline	8,642	2.33
11	Acetaminophen with codeine	8,222	2.21
12	OTC sleep aids	7,984	2.15
13	Lorazepam	7,625	2.05
14	D-Propoxyphene	7,417	2.00
15	Fluoxetine	6,917	1.86
16	Diphenhydramine	6,483	1.75
17	Methamphetamine/speed	5,236	1.41
18	Oxycodone	4,526	1.22
19	PCP/PCP combinations	4,408	1.19
20	Lithium carbonate	4,402	1.19
21	Clonazepam	4,335	1.17
22	Hydantoin	4,026	1.08
23	Hydrocodone	3,921	1.06
24	LSD	3,869	1.04
25	Triazolam	3,801	1.02

Source: National Institute on Drug Abuse, Drug Abuse Warning Network (DAWN), May 1991.

more than half of all arrests are related to alcohol abuse. Alcohol is also a significant factor in cases involving battered children.

Alcohol is now recognized as a potent teratogen. A baby born to an alcoholic mother might have an abnormally small head and defects of the arms, legs, face, or other body parts. The greater the mother's use of alcohol, the more serious the birth defects of her child are apt to be. In the United States 4000 to 5000 babies are born each year with birth defects due to alcohol.

One way to put drug abuse into perspective is to look at emergency room records. When drug abuse reaches dangerous levels, people often end up in hospital emergency rooms. Table 19.3 lists the top 25 drugs mentioned most frequently by emergency rooms in reporting drug abuse incidents in 1990. Notice that alcohol, often in combination with other drugs, is at the top of the list. Of the estimated 371,208 drug abuse incidents reported by emergency rooms in 1990, about 31% involved alcohol abuse.

SECTION 19.13

Anesthetics: Under and Out

Anesthetics are the ultimate depressants. A **general anesthetic** acts on the brain to produce unconsciousness, along with a general insensitivity to feeling or pain.

History of Anesthesia

$CH_3CH_2—O—CH_2CH_3$
Diethyl ether

Diethyl ether was the first general anesthetic. It was introduced into surgical practice in 1846 by a Boston dentist, William Morton. Inhalation of ether vapor produces unconsciousness by depressing the activity of the central nervous system. Ether is relatively safe. There is a fairly wide gap between the effective level of anesthesia and the lethal dose. The disadvantages are its high flammability and its side effect, nausea.

N=N—O
Nitrous oxide
(Dinitrogen monoxide)

Nitrous oxide, or laughing gas (N_2O), was tried by Morton without success before he tried ether. Nitrous oxide was discovered by Joseph Priestley in 1772. Its narcotic effect was noted, and it soon came to be used widely at laughing gas parties among the nobility (Figure 19.18). Nitrous oxide, mixed with oxygen, finds some use in modern anesthesia. It is quick acting but not very potent. Concentrations of 50% or greater must be used to be effective. When nitrous oxide is mixed with ordinary air instead of oxygen, not enough oxygen gets into the patient's blood and permanent brain damage can result.

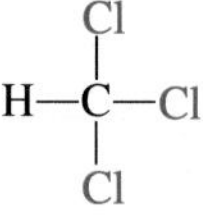

Chloroform
(Trichloromethane)

Chloroform ($CHCl_3$) was introduced as a general anesthetic in 1847. Its use quickly became popular after Queen Victoria gave birth to her eighth child while anesthetized by chloroform in 1853. Chloroform was used widely for

Figure 19.18 An 1802 caricature by James Gillray of a lecture demonstrating the effect of ether vapor and nitrous oxide. [*Courtesy of the Bureau of Narcotics and Dangerous Drugs, Washington DC.*]

years. It is nonflammable and produces effective anesthesia, but it has a number of serious drawbacks. For one, it has a narrow safety margin; the effective dose is close to the lethal dose. It also causes liver damage, and it must be protected from oxygen during storage to prevent the formation of deadly phosgene gas.

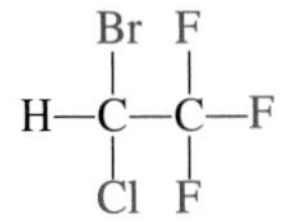

Modern Anesthesia

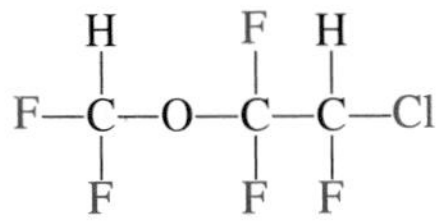

Modern anesthetics include fluorine-containing compounds such as halothane, enflurane, and isoflurane (Figure 19.19). These compounds are nonflammable and relatively safe for the patient. Their safety, particularly that of halothane, for operating-room personnel, however, has been questioned. For example, female operating room workers suffer a higher rate of miscarriages than the general population.

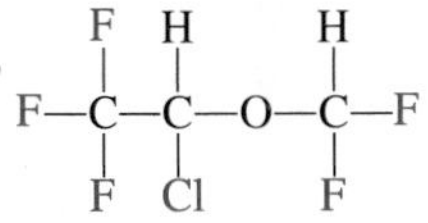

Modern surgical practice has moved away from the use of a single anesthetic. Generally, a patient is given an intravenous anesthetic such as thiopental (Section 19.14) to produce unconsciousness. The gaseous anesthetic then is administered to provide insensitivity to pain and to keep the patient unconscious. A relaxant, such as curare, also may be employed. Curare and related compounds produce profound relaxation; thus, only light anesthesia is required. This practice avoids the hazards of deep anesthesia.

The potency of an anesthetic is related to its solubility in fat. General anesthetics seem to work by dissolving in the fatlike membranes of nerve cells. This changes the permeability of the membranes, and the conductivity of the neurons is depressed.

$$\mathrm{H{-}\overset{\overset{\large Br}{|}}{\underset{\underset{\large Cl}{|}}{C}}{-}\overset{\overset{\large F}{|}}{\underset{\underset{\large F}{|}}{C}}{-}F}$$

Halothane

$$\mathrm{F{-}\overset{\overset{\large H}{|}}{\underset{\underset{\large F}{|}}{C}}{-}O{-}\overset{\overset{\large F}{|}}{\underset{\underset{\large F}{|}}{C}}{-}\overset{\overset{\large H}{|}}{\underset{\underset{\large F}{|}}{C}}{-}Cl}$$

Enflurane

$$\mathrm{F{-}\overset{\overset{\large F}{|}}{\underset{\underset{\large F}{|}}{C}}{-}\overset{\overset{\large H}{|}}{\underset{\underset{\large Cl}{|}}{C}}{-}O{-}\overset{\overset{\large H}{|}}{\underset{\underset{\large F}{|}}{C}}{-}F}$$

Isoflurane

Figure 19.19 Three modern general anesthetics.

Curare is the arrow poison used by South American Indian tribes. Large doses of curare kill by causing a complete relaxation of all muscles. Death occurs because of respiratory failure.

Solvent Sniffing: Self-administered Anesthesia

Nearly all gaseous and volatile-liquid organic compounds exhibit anesthetic action. (Methane is an exception; it appears to be physiologically inert. Nitrous oxide is the only common inorganic anesthetic.) It is the anesthetic action of organic solvents that leads to their abuse. Sniffing glue solvents, gasoline, aerosol propellants, and other inhalants is perhaps the deadliest form of drug abuse. The dose required for intoxication is not far from that which will stop the heart. And it is difficult to measure a dose from a plastic or paper bag—the usual method of ''glue sniffing.'' Also, as with nitrous oxide, sublethal doses can cause permanent brain damage by cutting down the oxygen supply to the brain.

Figure 19.20 Coca leaves and illicit forms of cocaine. [*Courtesy of U.S. Department of Justice Drug Enforcement Administration.*]

Local Anesthetics

A **local anesthetic** renders one part of the body insensitive to pain, but the patient remains conscious. For dental work and minor surgery, it is usually desirable to deaden the pain in one part of the body only. The first local anesthetic to be used successfully was cocaine. This drug was first isolated in 1860 from the leaves of the coca plant (Figure 19.20). Its structure was determined by Richard Willstätter in 1898. Even before Willstätter's work, there

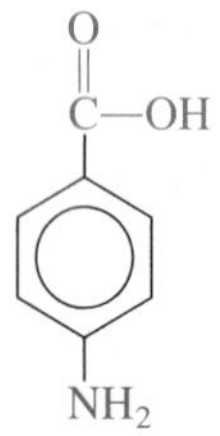

para-Aminobenzoic acid
(a)

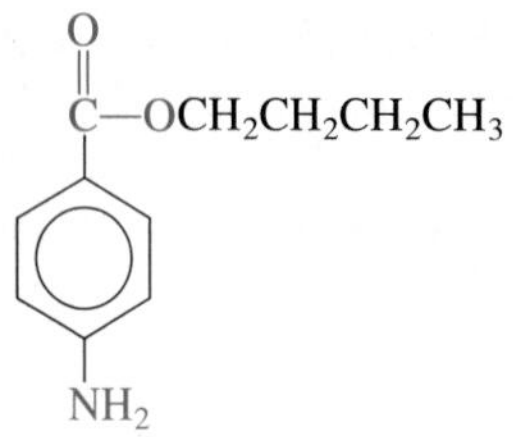

Butyl *para*-aminobenzoate
(Butesin)
(b)

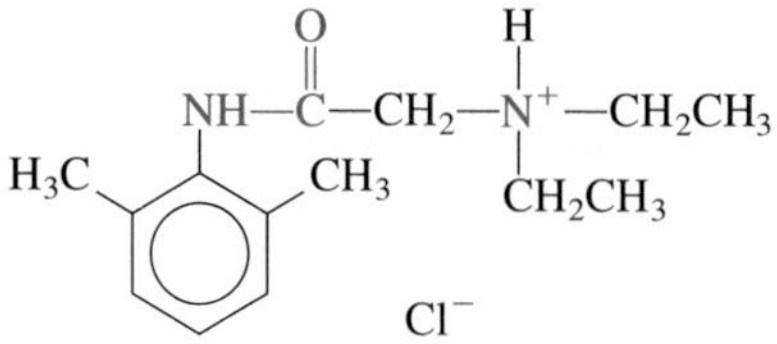

Lidocaine
(Xylocaine)
(c)

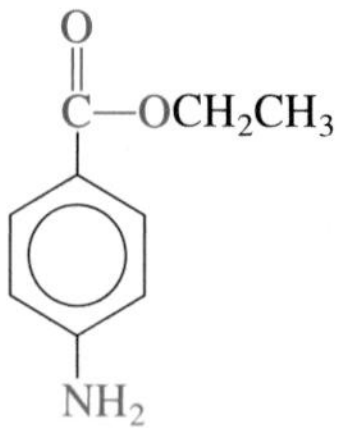

Ethyl *para*-aminobenzoate
(Benzocaine)
(b)

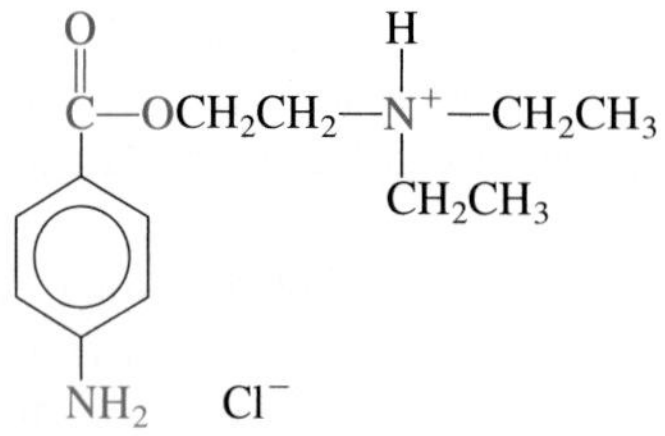

Procaine
(Novocaine)
(b)

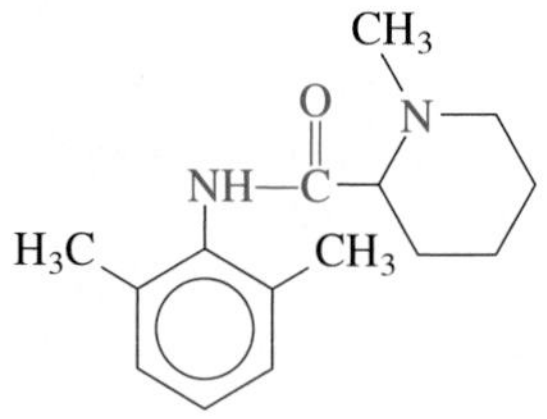

Mepivicaine
(c)

Figure 19.21 Some local anesthetics. Three (b) are derived from *para*-aminobenzoic acid (a). The other two (c) have amide functions. All are often used in the form of the hydrochloride salt, which is more soluble in water than the free base.

were attempts to develop synthetic compounds with similar properties. Cocaine is a powerful stimulant. Its abuse is discussed later in this chapter.

Certain esters of *para*-aminobenzoic acid act as local anesthetics (Figure 19.21). The ethyl and butyl esters are used to relieve the pain of burns and open wounds. These are applied as ointments, usually in the form of picrate salts.

More powerful in their anesthetic action are a series of derivatives with a second nitrogen atom in the alkyl group of the ester. Perhaps the best known of these is procaine (Novocaine), first synthesized by Alfred Einhorn in 1905. Procaine can be injected as a local anesthetic, or it can be injected into the spinal column to deaden the entire lower portion of the body. Local anesthetics work by blocking nerve impulses to the brain. When the block involves the spinal cord, messages of pain from the lower parts of the body are prevented from reaching the brain.

The local anesthetic of choice nowadays is often lidocaine or mepivicaine. Each compound is highly effective and yet has a fairly low toxicity (Table 19.4). Note that lidocaine and mepivicaine are not derivatives of *para*-aminobenzoic acid, but they do share some structural features with the compounds that are.

The toxicities in Table 19.4 are given as LD_{50} values. These are the doses in milligrams (per kilogram of body weight of the test animal) that will kill

Table 19.4 Toxicities of Various Drugs*

	Drug	LD_{50} (mg/kg)	Method of Administration	Experimental Animal
Local anesthetics	Lidocaine	292	Oral	Mice
	Procaine	45	Intravenous	Mice
	Cocaine	17.5	Intravenous	Rats
Barbiturates	Barbital	600	Oral	Mice
	Pentobarbital	118	Oral	Rats
	Phenobarbital	162	Oral	Rats
	Amobarbital	212	Subcutaneous	Mice
	Thiopental	149	Intraperitoneal	Mice
Narcotics	Morphine	500	Subcutaneous	Mice
	Heroin	21.8	Intravenous	Mice
	Meperidine	170	Oral	Rats
Stimulants	Caffeine	355	Oral	Rats
	Nicotine	230	Oral	Mice
		0.3	Intravenous	Mice
	Amphetamine	180	Subcutaneous	Rats
	Methamphetamine	70	Intraperitoneal	Mice
	Mescaline	370	Intraperitoneal	Rats

*Comparisons of toxicities in different animals—and extrapolation to humans—are at best crude approximations. The method of administration can have a profound effect on the observed toxicity.

Source: Susan Budavari (Ed.), *The Merck Index*, 11th ed. Rahway, NJ: Merck and Co., 1989.

50% of a population of test animals. LD_{50} values are discussed later in Section 20.13.

Dissociative Anesthetics: Ketamine and PCP

Ketamine, an intravenous anesthetic, is called a **dissociative anesthetic**: it induces hallucinations similar to those reported by people who have had near-death experiences. They seem to remember observing their rescuers from a vantage point above it all, or moving through a dark tunnel toward a bright light. Unlike thiopental, ketamine seems to affect associative pathways before it hits the brain stem.

O NHCH$_3$ Cl

Ketamine

Little is known of the action of ketamine at the molecular level. If it acts by fitting receptors in the body, we can assume that our bodies produce their own chemicals that fit those receptors. These compounds may be synthesized or released only in extreme circumstances—such as in near-death experiences. Is it possible that we are on the threshold of the discovery of the chemistry of "life after death"?

N

Phencyclidine (PCP)

Closely related to ketamine is phencyclidine (PCP), known on the street as "angel dust." PCP is soluble in fat and has no appreciable water solubility. It is stored in fatty tissue and released when the fat is metabolized; this accounts for the ''flashbacks'' commonly experienced by users.

PCP is an important part of the illegal drug scene. It is cheap and easily prepared. It was tested and found to be too dangerous for human use, but found use as an animal tranquilizer. Many users experience bad ''trips'' with PCP. About 1 in 1000 develops a severe form of schizophrenia. Laboratory tests show that PCP depresses the immune system. This could lead to increased risk of infection. Despite these well-known problems, every few years a new crop of young people appears on the scene to be victimized by this hog tranquilizer.

SECTION 19.14

The Barbiturates: Sedation, Sleep, and Synergism

As a family of related compounds, the barbiturates display a wide variety of properties. They can be employed to produce mild sedation, deep sleep, and even death.

Barbituric acid was first synthesized in 1864 by Adolph von Baeyer, a young student of August Kekulé (Chapter 9). He made it from urea, which occurs in urine, and malonic acid, which occurs in apples. The term **barbiturates**, according to Willstätter, came about because, at the time of the discovery, von Baeyer was infatuated with a girl named Barbara. The word comes from *Barbara* and *urea.*

$$O{=}C(NH_2)_2 + (HO{-}CO)_2CH_2 \longrightarrow O{=}C(NH{-}CO)_2CH_2 + 2\,H_2O$$

Urea — Malonic acid — Barbituric acid

Synergism: Barbiturates and Alcohol

The barbiturates are especially dangerous when ingested along with ethyl alcohol. This combination produces an effect much more drastic than just the sum of the effects of two depressants. The effect of the barbiturate is enhanced by factors of up to 200 when taken with alcoholic beverages. This effect of one drug in enhancing the action of another is called a **synergistic effect.**

Synergism probably has led to many deaths. According to news reports, an autopsy showed that a well-known Hollywood gossip columnist ingested only about 200 mg of barbiturate and 2 oz of alcohol. Either alone would have produced only mild sedation; the combination killed her. Synergistic effects are not limited to alcohol–barbiturate combinations. You should never take two drugs at the same time without competent medical supervision.

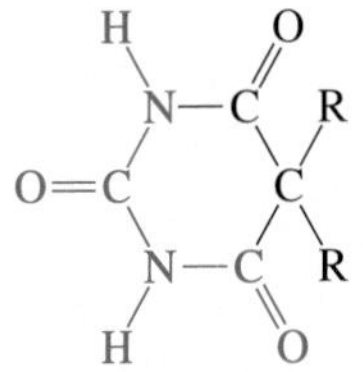

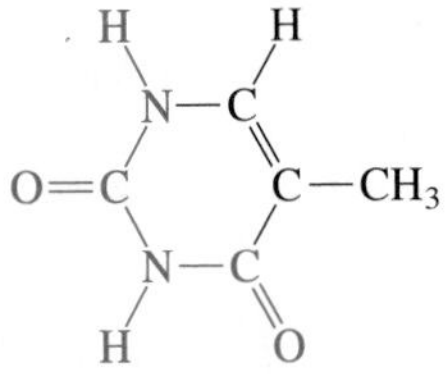

The barbiturates are strongly addictive. Habitual use leads to the development of a tolerance to the drugs and ever-larger doses are required to get the same degree of intoxication. Barbiturates are legally available by prescription only, but they are a part of the illegal drug scene also. They are known as ''downers'' because of their depressant, sleep-inducing effects.

The side effects of barbiturates are similar to those of alcohol. Abuse leads to hangovers, drowsiness, dizziness, and headaches. Withdrawal symptoms are often severe, accompanied by convulsions and delirium. In fact, some medical authorities now say that withdrawal from barbiturates is more dangerous—that is, more likely to cause death—than withdrawal from heroin.

Barbiturates are cyclic amides. Notice, however, that the barbiturate ring resembles that of thymine, one of the bases found in nucleic acids. Evidence indicates that barbiturates may act by substituting for thymine (or cytosine or uracil) in nucleic acids, thus interfering with protein synthesis.

A barbiturate

Thymine

Barbiturates as Medicines

The medicinal value of the barbiturates was discovered in 1903 by Joseph von Mering. A derivative called barbital (Figure 19.22) was found to be useful in putting dogs to sleep. Several thousand barbiturates have been synthesized through the years, but only a few have found widespread use in medicine. Pentobarbital (Nembutal) is employed as a short-acting hypnotic drug. Before the discovery of the modern tranquilizers, pentobarbital was used widely to calm anxiety and other disorders of psychic origin.

Phenobarbital (Luminal) is a long-acting drug. It, too, is a hypnotic and can

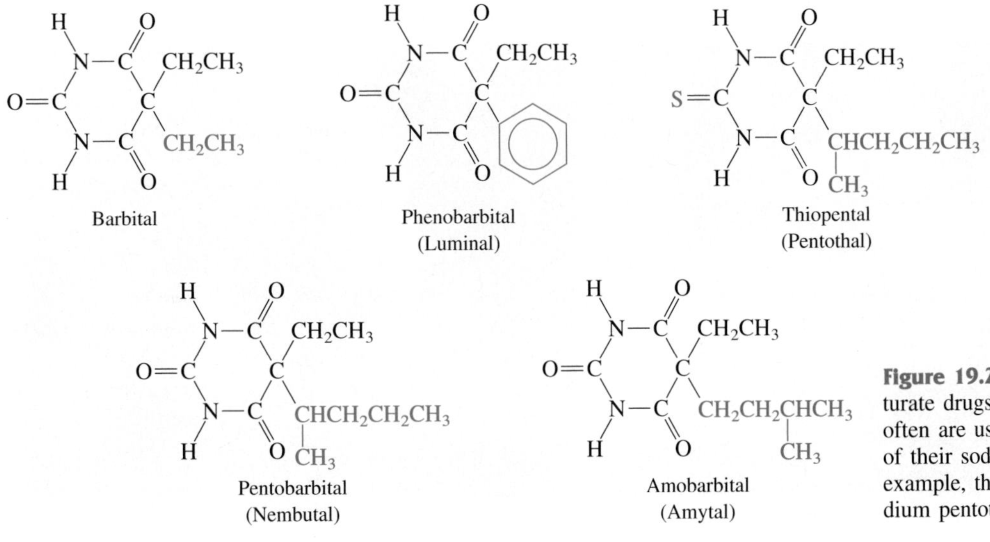

Barbital

Phenobarbital (Luminal)

Thiopental (Pentothal)

Pentobarbital (Nembutal)

Amobarbital (Amytal)

Figure 19.22 Some barbiturate drugs. These drugs often are used in the form of their sodium salts, for example, thiopental as ''sodium pentothal.''

Thiopental has been investigated as a possible truth drug. It does seem to help psychiatric patients recall traumatic experiences. It also helps uncommunicative individuals talk more freely. It does not, however, prevent one from withholding the truth or even from lying. No true truth drug exists.

be used as a sedative. Phenobarbital is employed widely as an anticonvulsant for epileptics and brain-damaged people. The action of amobarbital (Amytal) is intermediate in duration. Thiopental (Pentothal), a compound that differs from pentobarbital only in that an oxygen atom on the ring has been replaced by a sulfur atom, is used widely in anesthesia (Figure 19.22).

The barbiturates were once used in small doses as sedatives. The dosage for sedation was generally a few milligrams. In larger dosages (about 100 mg), the barbiturates induce sleep. They were once the sleeping pills so widely used—and abused—by middle-class, often middle-aged, people. The lethal dose is in the vicinity of 1500 mg (1.5 g). Barbiturates are the drug of choice for many suicides. News reports list the cause of death as ''an overdose of sleeping pills.'' There is also potential for accidental overdose. After a couple of tablets, the person becomes groggy. If the person is unable to remember whether he or she took the sleeping pills or not, he or she may take more pills.

SECTION 19.15

The Opium Alkaloids: Narcotics

Narcotics are drugs that produce narcosis (stupor or general anesthesia) and relief of pain (analgesia). Many drugs produce these effects, but in the United States only those that are also *addictive* are legally classified as narcotics. Their use is regulated by federal law.

Opium and Morphine

Opium is the dried, resinous juice of the unripe seeds of the oriental poppy (*Papaver somniferum*) (Figure 19.23). It is a complex mixture of some 20

(a)

(b)

Figure 19.23 Opium poppy flower (a) and seed pod (b). [*Courtesy of U.S. Department of Justice Drug Enforcement Administration.*]

Figure 19.24 Trade card advertising Mrs. Winslow's soothing syrup. [*Courtesy of U.S. Department of Justice Drug Enforcement Administration.*]

nitrogen-containing organic bases (alkaloids), sugars, resins, waxes, and water. The principal alkaloid, morphine, makes up about 10% of the weight of raw opium. Raw opium was used in many patent medicines of the nineteenth century. Mrs. Winslow's Soothing Syrup (Figure 19.24) was just one of many such products.

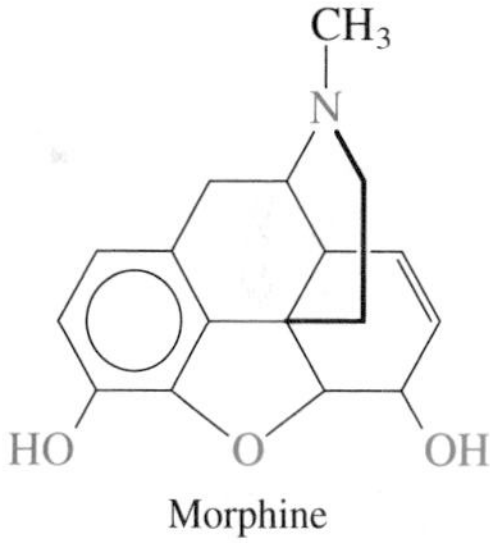

Morphine

Morphine was first isolated in 1805 by Friedrich Sertürner, a German pharmacist. With the invention of the hypodermic syringe (Figure 19.25) in the 1850s, a new method of administration became available. Injection of morphine directly into the bloodstream was more effective for the relief of pain, but it also seriously escalated the problem of addiction. Morphine was used widely during the American Civil War (1861–1865). It was effective for relief of pain caused by battle wounds. One side effect of morphine use is constipation. Noting this, soldiers came to use morphine as a treatment for that other common malady of men on the battlefront—dysentery. During their wartime service, over a hundred thousand soldiers became addicted to morphine. The affliction was so common among veterans that it came to be known as "soldier's disease."

Figure 19.25 An old-fashioned hypodermic syringe. [*Courtesy of the Bureau of Narcotics and Dangerous Drugs, Washington DC.*]

Morphine and other narcotics were placed under the federal government's control by the Harrison Act of 1914. Morphine is still used by prescription for relief of severe pain. It also induces lethargy, drowsiness, confusion, euphoria, chronic constipation, and depression of the respiratory system. It is strongly addictive if administered in amounts greater than the prescribed doses or for a period longer than the prescribed time.

Codeine and Heroin

Slight changes in the molecular architecture of morphine produce altered physiological properties. Replacement of one of the —OH groups by an

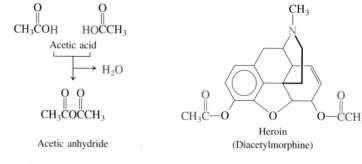

CH_3 — N — CH_3O — O — OH

Codeine
(Methylmorphine)

—OCH_3 group produces codeine. Actually, codeine occurs in opium to an extent of about 0.5%. It is usually synthesized, however, by methylating the more abundant morphine molecules. About 55,000 kg of codeine is produced each year in the United States, enough for 16 doses of 15 mg for every person in the country.

Codeine is similar to morphine in its physiological action, except that it is less potent and has less tendency to induce sleep. It is also thought to be less addictive. In amounts of less than 2.2 mg/mL, codeine is exempt from the stringent narcotics regulations and is used in a few ''controlled substance'' cough syrups.

In the laboratory, reaction of morphine with acetic anhydride (acetic anhydride is derived from acetic acid by removal of water) produces heroin. This morphine derivative was first prepared by chemists at the Bayer Company of Germany in 1874. It received little attention until 1890, when it was proposed as an antidote for morphine addiction. Shortly thereafter, Bayer was widely advertising heroin as a sedative for coughs, often in the same ads as aspirin (Figure 19.26)! It soon was found, however, that heroin induced addiction more quickly than morphine and that heroin addiction was harder to cure.

The physiological action of heroin is similar to that of morphine, except that heroin seems to produce a stronger feeling of euphoria for a longer period of time. Heroin is not legal in the United States, even by prescription. It has, however, been advocated for use in relief of pain in terminal cancer patients and has been so used in Britain.

$$CH_3\overset{O}{\overset{\|}{C}}OH \quad HO\overset{O}{\overset{\|}{C}}CH_3$$

Acetic acid

$$\xrightarrow{-H_2O}$$

$$CH_3\overset{O}{\overset{\|}{C}}O\overset{O}{\overset{\|}{C}}CH_3$$

Acetic anhydride

CH_3 — N — $CH_3\overset{O}{\overset{\|}{C}}—O$ — O — $O—\overset{O}{\overset{\|}{C}}CH_3$

Heroin
(Diacetylmorphine)

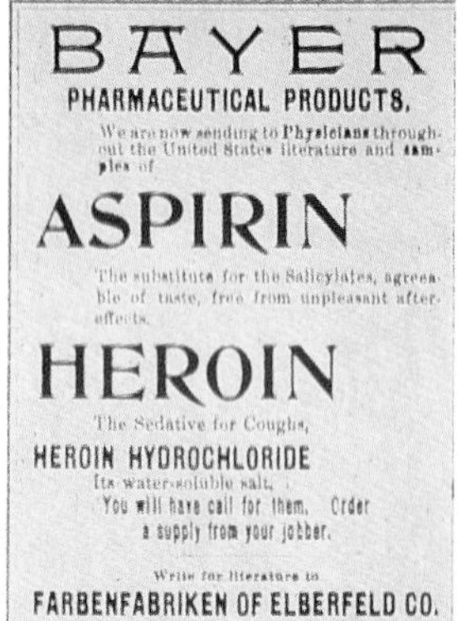

Figure 19.26 Heroin was advocated as a safe medicine in 1900. It was widely marketed as a sedative for coughs. [*Courtesy of U.S. Department of Justice Drug Enforcement Administration.*]

Addiction probably has three components: emotional dependence, physical dependence, and tolerance. Psychological dependence is evident in the uncontrollable desire for the drug. Physical dependence is shown by acute withdrawal symptoms such as convulsions. The body cells are conditioned to the drug and cannot function normally without it. Tolerance for the drug is evidenced by the increasing dosages required to produce in the addict the same degree of narcosis and analgesia.

Deaths from heroin usually are attributed to overdoses, but the situation is not altogether clear. The problem seems to be a matter of quality control. As an illustration, the office of the chief medical examiner for New York City analyzed 132 samples of drugs, supposedly heroin, that had been confiscated on the streets (Figure 19.27). Twelve contained no heroin at all. The remaining 120 varied from 1 to 77% heroin. A user could think he or she was getting a dose of 1 unit, when it was actually 77 times as much—a catastrophic overdose.

Figure 19.27 Forms of heroin found on the street. [*Courtesy of U.S. Department of Justice Drug Enforcement Administration.*]

SECTION 19.16

Synthetic Narcotics: Analgesia and Addiction

Much research has gone into developing a drug that would be as effective as morphine for the relief of pain but that would not be addictive. Perhaps the best known of the synthetic narcotics is meperidine (Demerol) (Figure 19.28). Meperidine is somewhat less effective than morphine, but it has the advantage that it does not cause nausea. Repeated use, unfortunately, does lead to addiction.

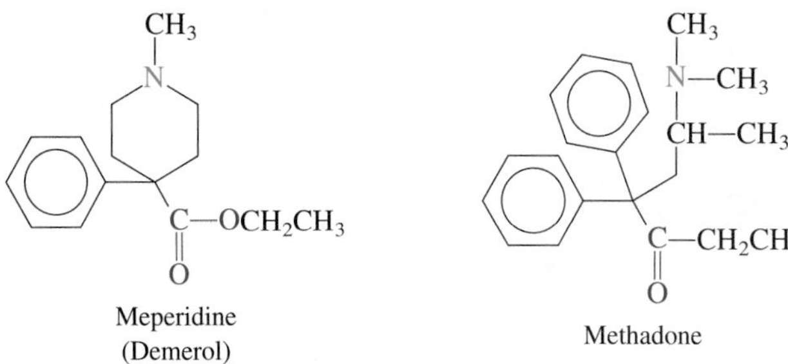
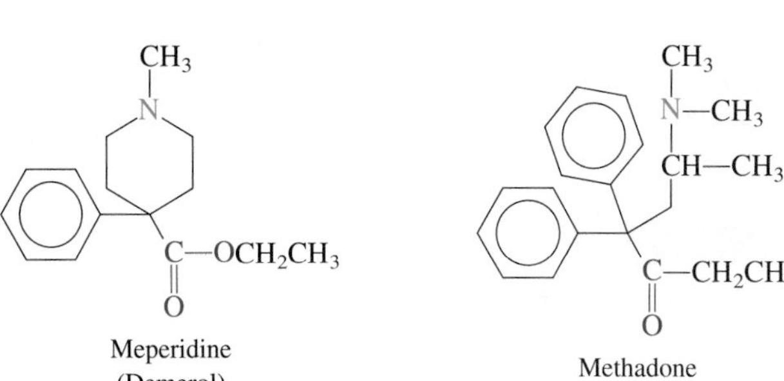

Another synthetic narcotic is methadone. This drug has been widely used to treat heroin addiction. Like heroin, methadone is highly addictive. However, when taken orally, it does not induce the sleepy stupor characteristic of heroin intoxication. Unlike a heroin addict, a person on methadone maintenance usually is able to hold a productive job. Methadone is available free in clinics. If an addict who has been taking methadone reverts to heroin, the methadone in his or her system effectively blocks the euphoric rush normally given by heroin, and so reduces the addict's temptation to use heroin.

Methadone maintenance is not a perfect answer. Perhaps it is not even a good one. When injected into the body, methadone gives an effect similar to that of heroin, and methadone has been diverted for illegal use in this manner. An addict on methadone is still an addict. All the problems of tolerance (and cross-tolerance with heroin and morphine) still exist.

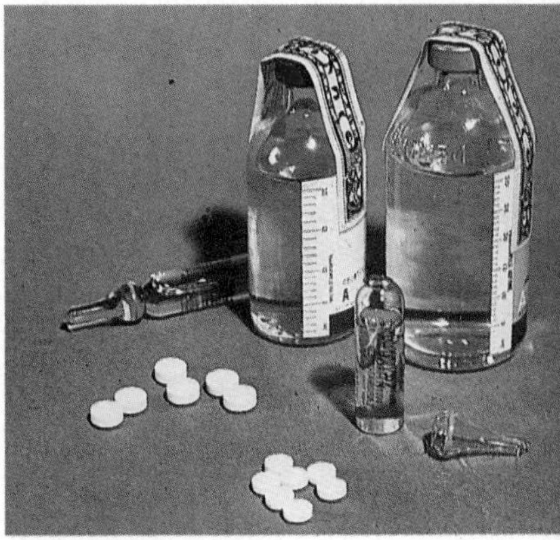

Figure 19.28 Dosage forms of meperidine (Demerol), a synthetic narcotic. [*Courtesy of U.S. Department of Justice Drug Enforcement Administration.*]

More Morphine Analogs: Agonists and Antagonists

Chemists have synthesized thousands of morphine analogs. Only a few have shown significant analgesic activity. Most are addictive. Morphine acts by

An *agonist* is a substance that mimics the action of a drug. An *antagonist* blocks the action of a drug.

binding to receptors in the brain. Those molecules that have morphinelike action are called **agonists**. Morphine **antagonists** are drugs that block the action of morphine, most likely by blocking the receptors. Some molecules have both agonist and antagonist activity. These show great promise as analgesics. An example is pentazocine (Talwin). It is less addictive than morphine and yet it is effective for relief of pain.

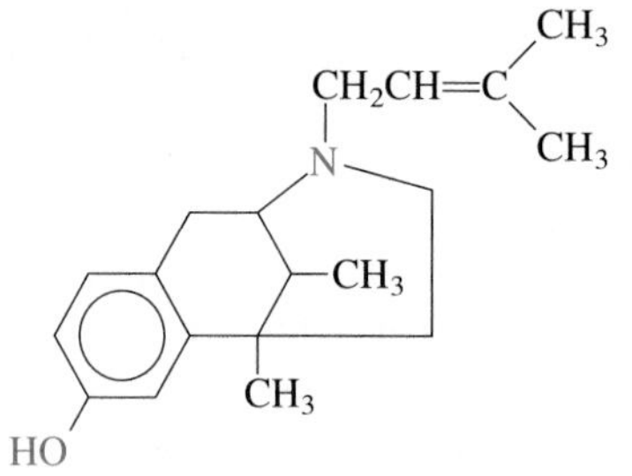

Pentazocine
(Talwin)

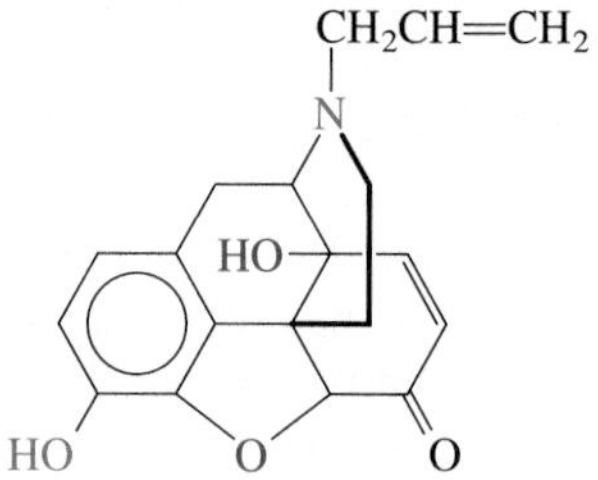

Naloxone

Pure antagonists such as naloxone are of value in treating opiate addicts. Overdosed addicts can be brought back from death's door by an injection with naloxone.

SECTION 19.17

A Natural High: The Brain's Own Opiates

Morphine acts by fitting specific receptor sites in the brain. These morphine receptors were first demonstrated in 1973 by Solomon Synder and Candace Pert at Johns Hopkins University School of Medicine.

Why should the human brain have receptors for a plant-derived drug like morphine? There seemed to be no good reason, so several investigators started a search for morphinelike substances produced by the human body. Not one, but several such substances, called **endorphins** (''endogenous morphines''), soon were found. Each was a short peptide chain composed of amino acid units. Those with five amino acid units are called **enkephalins**. There are two enkephalins, and they differ only in the amino acid at the end of the chain. *Leu*-enkephalin has the sequence Tyr-Gly-Gly-Phe-Leu, and *Met*-enkephalin is Tyr-Gly-Gly-Phe-Met. Other substances with chains of 30 amino acids also were found.

Some of the enkephalins have been synthesized and shown to be potent pain relievers. Their use in medicine is quite limited, however, because, after being injected, they are rapidly broken down by the enzymes that hydrolyze proteins. It is hoped, though, that analogs more resistant to hydrolysis can be employed as morphine substitutes for the relief of pain. Unfortunately, both the natural enkephalins and the analogs, like morphine, seem to be addictive.

It appears that endorphins are released as a response to pain deep in the body. Bruce Pomeranz of the University of Toronto has collected evidence that indicates that acupuncture anesthetizes by stimulating the release of the brain ''opiates.'' The long needles stimulate deep sensory nerves that cause the release of the peptides that then block the pain signals.

Endorphin release also has been used to explain other phenomena once thought to be largely psychological. A soldier, wounded in battle, feels no pain until the skirmish is over. His body has secreted its own painkiller. The production of these compounds during strenuous athletic activity is discussed in Section 18.13.

We shall return shortly to the chemistry of drugs. Before we do so, however, let's examine some chemistry of the nervous system and explore how nerve cells work.

SECTION 19.18

Some Chemistry of the Nervous System

The nervous system is made up of billions of **neurons** (nerve cells) with 10^{15} connections between them. The brain operates with a power output of about 25 W and has capacity for about 10 trillion bits of information. Nerve cells vary a great deal in shape and size. One type is shown in Figure 19.29. The essential parts of each cell are the cell body, the axon, and the dendrites. We discuss here only those nerves that make up the involuntary (autonomic) nervous system. These nerves carry messages between the organs and glands that act involuntarily (such as the heart, the digestive organs, and the lungs) and the brain and spinal column.

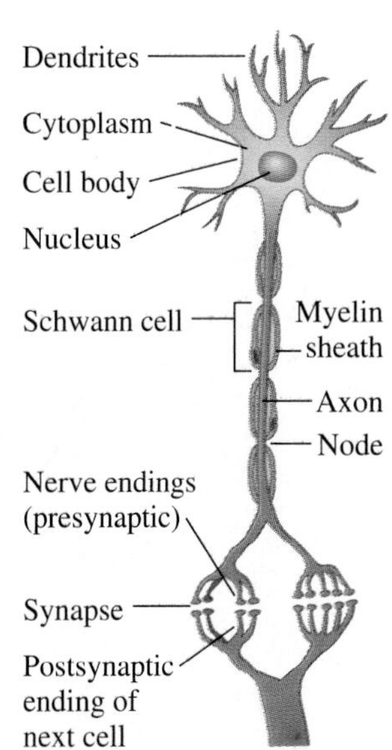

Figure 19.29 Diagram of a human nerve cell.

Although the axons on a given nerve cell may be up to 60 cm long, there is no continuous pathway from an organ to the central nervous system. Messages must be transmitted across tiny, fluid-filled gaps, or **synapses** (Figure 19.30). When an electrical signal from the brain reaches the end of an axon, specific chemicals (called **neurotransmitters**) that carry the impulse across the synapse to the next cell are liberated. There are perhaps a few dozen neurotransmitters. Each has a specific function. Messages are carried to other nerve cells, to muscles, and to the endocrine glands (such as the adrenal glands). Each neurotransmitter fits a specific receptor site on the receptor cell (Figure 19.31). Many drugs (and some poisons) act by mimicking the action of the neurotransmitter. Others act by blocking the receptor and preventing the neurotransmitter from acting on it. Several of the neurotransmitters are amines (Chapter 9), as are some of the drugs that affect the chemistry of our brains.

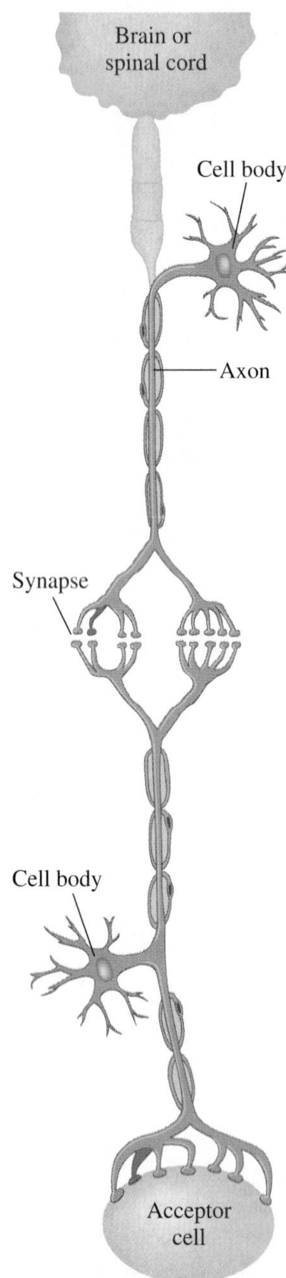

Figure 19.30 Schematic diagram of the pathway by which messages are transmitted to (and from) an acceptor cell in a gland or organ from (and to) the central nervous system.

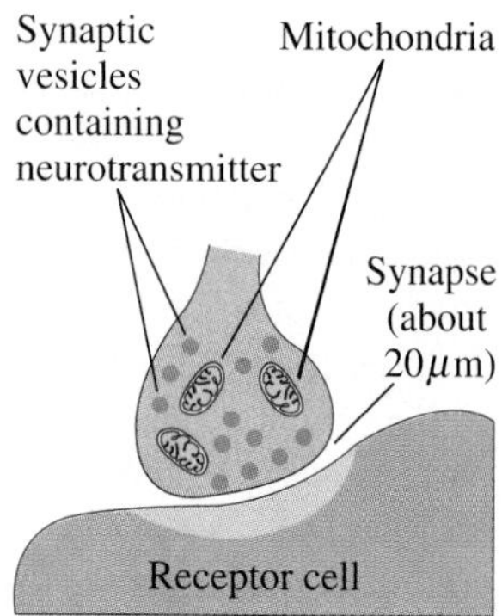

Figure 19.31 Schematic diagram of a synapse. When an electrical signal reaches the nerve endings, neurotransmitter molecules are released from the vesicles. They then migrate across the narrow gap (synapse) and move to the receptor cell, where they fit specific sites.

SECTION 19.19

Brain Amines: Depression and Mania

We all have our ups and downs in life. These moods probably result from multiple causes, but it appears likely that a variety of chemical compounds formed in the brain are involved. Before we consider these ups and downs, though, let's take a look at epinephrine, an amine formed in the adrenal glands.

$$\text{HO}(\text{HO})\text{C}_6\text{H}_3\text{—}\underset{\text{OH}}{\text{CHCH}_2}\text{—NH—CH}_3$$

Epinephrine
(Adrenaline)

$$\text{HO}(\text{HO})\text{C}_6\text{H}_3\text{—}\underset{\text{OH}}{\text{CHCH}_2}\text{—NH}_2$$

Norepinephrine

Commonly called adrenaline, epinephrine is secreted by the adrenal glands. A tiny amount of epinephrine causes a great increase in blood pressure. When a person is under stress or is frightened, the flow of adrenaline prepares the body for fight or flight. Because culturally imposed inhibitions prevent fighting or fleeing in most modern situations, the adrenaline-induced supercharge is not used. This sort of frustration has been implicated in some forms of mental illness.

A Biochemical Theory of Mental Illness

One simplified biochemical theory of mental illness involves two brain amines. One is norepinephrine (NE), a relative of epinephrine. NE is a neurotransmitter formed in the brain. When formed in excess, NE causes the person to be elated—perhaps even hyperactive. In large excess, NE induces a manic state. A deficiency of NE, on the other hand, could cause depression.

The other brain amine is serotonin, also a neurotransmitter. Serotonin is involved in sleep, sensory perception, and the regulation of body temperature. Its exact role in mental illness is not clear. A metabolite of serotonin, 5-hydroxyindoleacetic acid (5-HIAA), is found in unusually *low* levels in the spinal fluid of violent suicide victims. This indicates that abnormal serotonin metabolism may play a role in depression.

Levels of 5-HIAA also are low in murderers and other violent offenders. Levels of this serotonin metabolite are higher than normal in persons with obsessive–compulsive disorders, sociopaths, and those people who have guilt complexes.

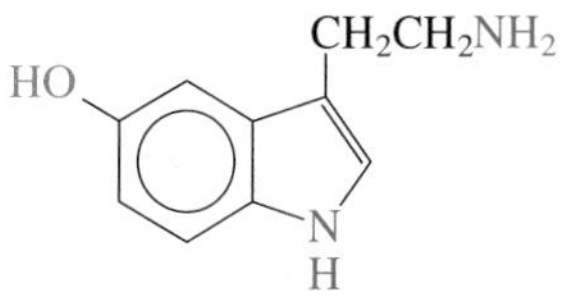

$CH_2CH_2NH_2$, HO, N, H

Serotonin

O, CH_2C—OH, HO, N, H

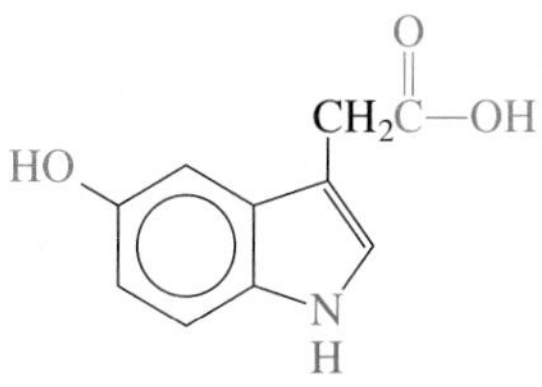

5-HIAA

Brain Amine Agonists and Antagonists in Medicine

Our cells have at least six different receptors that are activated by NE and related compounds. NE agonists (drugs that enhance or mimic its action) are stimulants. NE antagonists (drugs that block the action of NE) slow down various processes. Drugs called beta blockers reduce the stimulant action of epinephrine and NE on various kinds of cells. Propranolol (Inderal) is used to treat cardiac arrhythmias, angina, and hypertension by lessening slightly the force of the heart beat. Unfortunately, it also causes lethargy and depression.

Serotonin agonists are used to treat depression and anxiety. Agonist drugs are being used experimentally to treat obsessive–compulsive disorder. Serotonin antagonists are used to treat migraine headaches and to relieve the nausea caused by cancer chemotherapy.

Brain Amines and Diet: You Feel What You Eat

Richard Wurtman of the Massachusetts Institute of Technology has found a relationship between diet and serotonin levels in the brain. Serotonin is produced in the body from the amino acid tryptophan (Figure 19.32). Wurtman found diets high in carbohydrates lead to high levels of serotonin. High protein lowers the serotonin concentration.

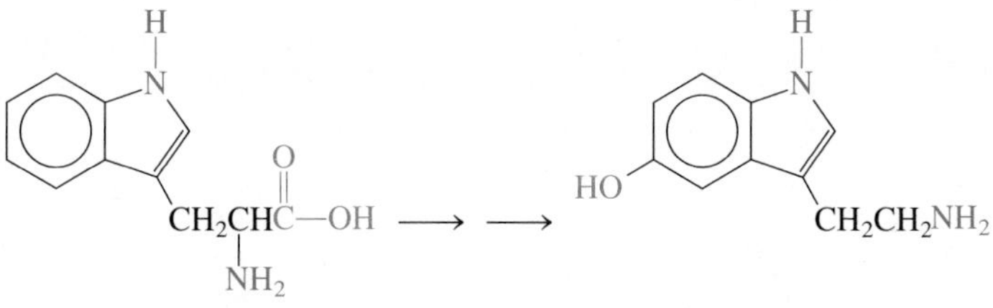

H, N, O, CH_2CHC—OH, NH_2 ⟶ ⟶ H, N, HO, $CH_2CH_2NH_2$

Tryptophan Serotonin

Figure 19.32 Serotonin is produced in the brain from tryptophan.

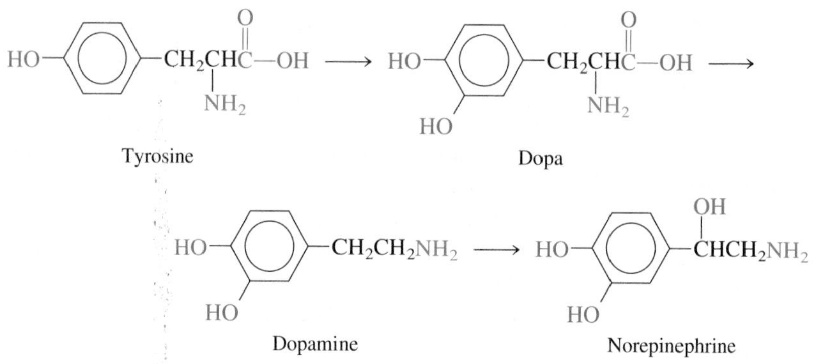

HO–C6H4–$CH_2CH(NH_2)COOH$ ⟶ (HO)2C6H3–$CH_2CH(NH_2)COOH$ ⟶

Tyrosine Dopa

(HO)2C6H3–$CH_2CH_2NH_2$ ⟶ (HO)2C6H3–$CH(OH)CH_2NH_2$

Dopamine Norepinephrine

Figure 19.33 The biosynthesis of norepinephrine from tyrosine.

Norepinephrine also is synthesized in the body from an amino acid. It is derived from tyrosine. The synthesis is complex and proceeds through several intermediates (Figure 19.33). The intermediate compounds also have physiological activity: dopa has been used successfully in the treatment of Parkinson's disease, and dopamine has been employed to treat low blood pressure. Since tyrosine is also a component of our diets, it may well be that our mental state depends to a fair degree on our diet.

Nearly one out of every ten people in the United States suffers from mental illness. Over half the patients in hospitals are there because of mental problems. When the biochemistry of the brain is more fully understood, mental illness may be cured (or at least alleviated) by administration of drugs.

Love: A Chemical Connection

The notion that love might be chemical in origin is an unsettling one. However, it is possible that the emotions that trigger romantic relationships are governed in part by a chemical called phenylethylamine (PEA). PEA functions as a neurotransmitter in the human brain. It appears to create excited, alert feelings and moods. Increased levels of PEA produce a "high" feeling identical to that people describe as "being in love." Not surprisingly, the chemical structure of PEA resembles that of norepinephrine.

How much PEA does it take to get back that old feeling? Levels of PEA in the brain can be estimated by measuring levels of its metabolite, phenylacetic acid, in the urine. Low levels of urinary phenylacetic acid correlate with depression. This has prompted researchers to investigate factors that increase PEA levels in the brain. There are no food sources of PEA, but protein-rich foods contain phenylalanine (Chapter 16), an amino acid precursor of PEA. Perhaps a steak dinner is a way to your true love's heart after all.

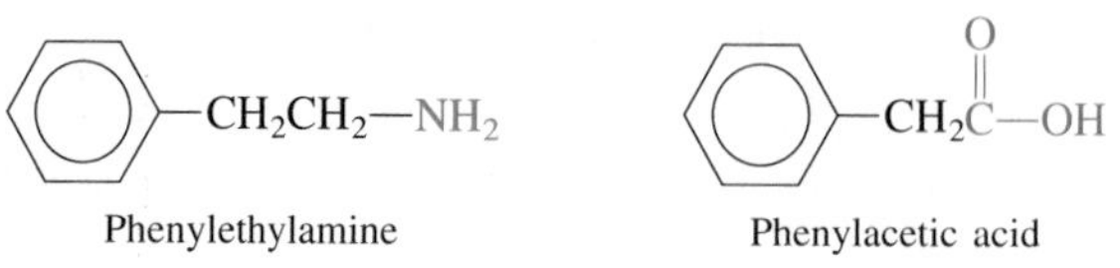

C_6H_5–CH_2CH_2–NH_2 C_6H_5–CH_2COOH

Phenylethylamine Phenylacetic acid

SECTION 19.20

Antianxiety Agents

Antianxiety agents are sometimes called minor tranquilizers.

The hectic pace of life in the modern world has driven people to seek rest and relaxation in chemicals. Ethyl alcohol is undoubtedly the most widely used tranquilizer. The drink before dinner—to ''unwind'' from the tensions of the day—is very much a part of the American way of life. Many people, however, seek their relief in other chemical forms.

Several over-the-counter drugs—Cope, Vanquish, and Compoz among others—claim to be able to help us cope with or vanquish our problems, or at least to compose ourselves in the face of minor adversity. Such products usually contain a little aspirin plus an antihistamine. The latter has a side effect of making one drowsy. These products have come under attack by consumer groups for being worthless at best—and perhaps even dangerous.

Another class of widely used antianxiety drugs are the benzodiazepines, compounds that feature seven-membered heterocyclic rings. Of these, perhaps the best known are diazepam (Valium), and chlorodiazepoxide (Librium). Antianxiety agents make people feel better simply by making them feel dull and insensitive. They do not solve any of the underlying problems that cause anxiety.

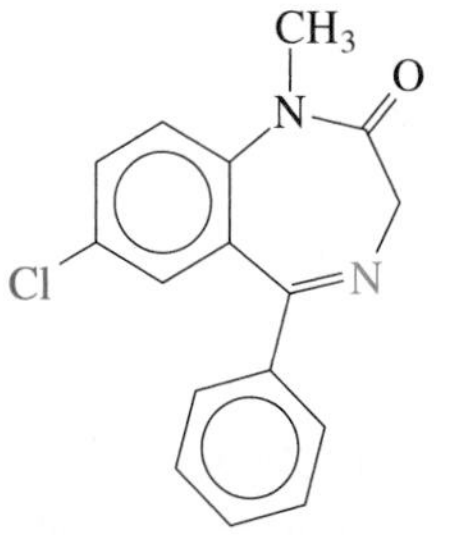

Diazepam (Valium)

Chlordiazepoxide (Librium)

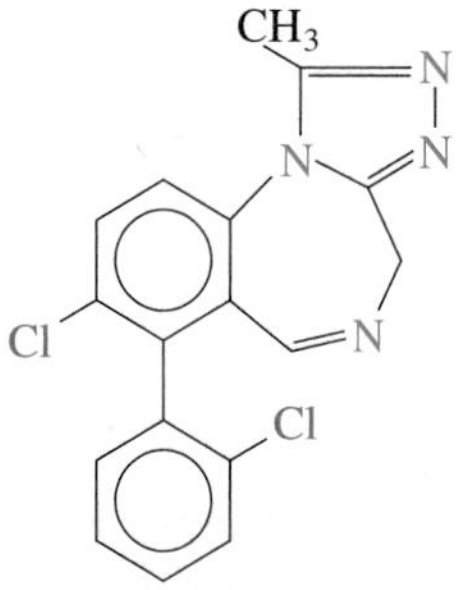

Triazolam (Halcion)

Certain benzodiazepines are also used to treat insomnia, the best known being triazolam (Halcion). It does help a person get to sleep, but the sleep is not restful. The drug does not really cure insomnia. Furthermore, Halcion and similar drugs have been banned by the British because of their dangerous side effects.

Like most other mind-altering drugs, the benzodiazepines act by fitting specific receptors. Presumably our bodies produce compounds that fit these receptors. To date, no such compound has been found. Rather, scientists have found compounds, called *beta*-carbolines, that act on the brain's anxiety receptors to produce *terror*. There is yet so much to learn about the chemistry of the brain.

Anyway, what price tranquility? After 20 years of use, benzodiazepines were found to be addictive. People trying to go off the drugs after prolonged use go into painful withdrawal.

Antipsychotic Agents

Antipsychotic agents are often referred to as major tranquilizers.

For centuries, the people of India used the snakeroot plant, *Rauwolfia serpentina,* to treat a variety of ailments including fever, snakebite, and other poisonings, and—most importantly—to treat maniacal forms of mental illness. Western scientists became interested in the plant near the middle of the twentieth century—after disdaining such remedies as quackery for many generations.

In 1952, rauwolfia was introduced into American medical practice as a hypertensive (blood-pressure-reducing) agent by Robert Wilkins of Massachusetts General Hospital. In the same year, Emil Schlittler of Switzerland isolated an active alkaloid, which he named reserpine, that has an impressive (intimidating?) structure.

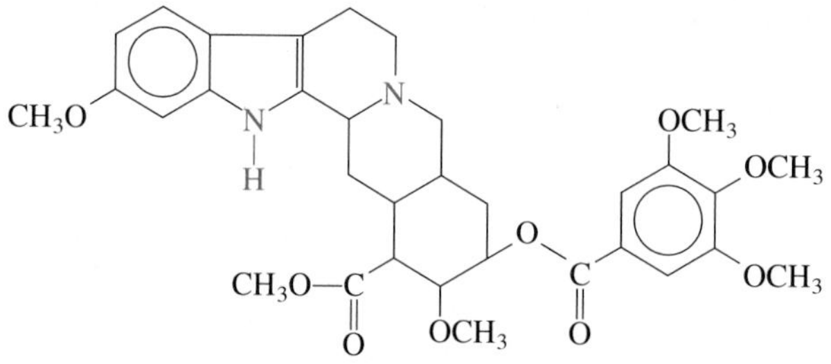

Reserpine

Reserpine was found not only to reduce blood pressure but also to bring about sedation. The latter finding attracted the interest of psychiatrists, who found reserpine so effective that by 1953 it had replaced electroshock therapy for 90% of psychotic patients.

Phenothiazines

Also in 1952 chlorpromazine (Thorazine) was tried as a tranquilizer on psychotic patients in the United States. The drug had been tested in France as an antihistamine. Medical workers there noted it calmed mentally ill people who were being treated for allergies. Chlorpromazine was found to be quite effective in controlling the symptoms of schizophrenia. It truly revolutionized mental illness therapy.

Chlorpromazine is one of a group of related compounds called phenothiazines. Several of these compounds are used in medicine. Promazine (chlorpromazine without the chlorine atom) is also a tranquilizer, but it is much less potent than chlorpromazine.

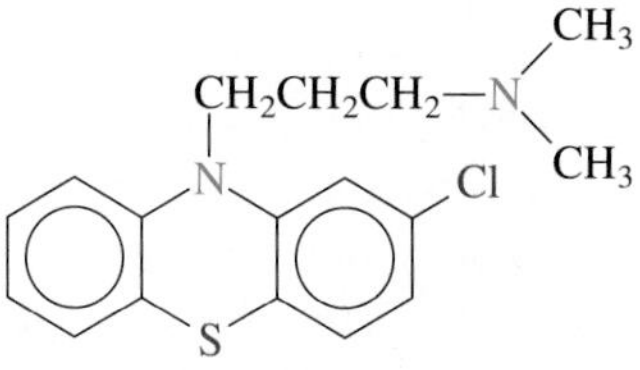

Chlorpromazine
(Thorazine)

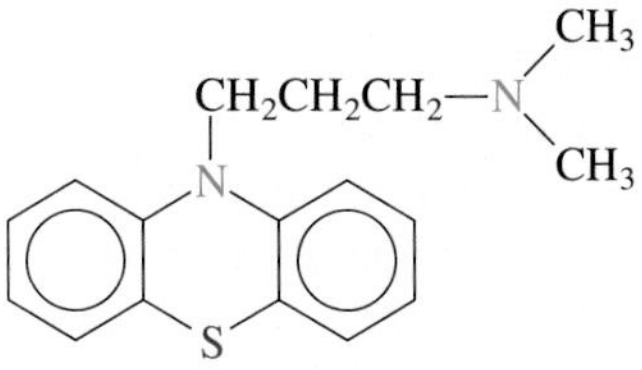

Promazine

The phenothiazines are dopamine antagonists. Dopamine (see Figure 19.33) is important in the control of detailed motion (such as grasping small objects), in memory and emotions, and in exciting the cells of the brain. Some researchers think schizophrenic patients produce too much dopamine, others that they have too many dopamine receptors. In either case, blocking the action of dopamine relieves the symptoms of schizophrenia.

The antipsychotic drugs have been one of the real triumphs of chemical research. They have served to greatly reduce the number of patients confined to mental hospitals by controlling the symptoms of schizophrenia to the extent that 95% of all schizophrenics no longer need hospitalization.

Antidepressants

It is interesting to note that slight changes in structure can result in profound changes in properties. Replacing the sulfur atom of promazine by a $—CH_2CH_2—$ group produces imipramine (Tofranil), a compound that is not a tranquilizer at all. Rather, it is an antidepressant. Another common tricyclic (three-ring) antidepressant drug is amitriptylene (Elavil), in which the ring nitrogen atom also is replaced by a carbon atom.

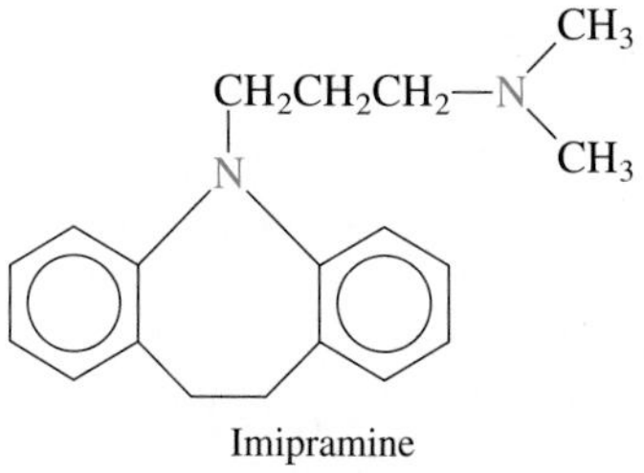

Imipramine
(Tofranil)

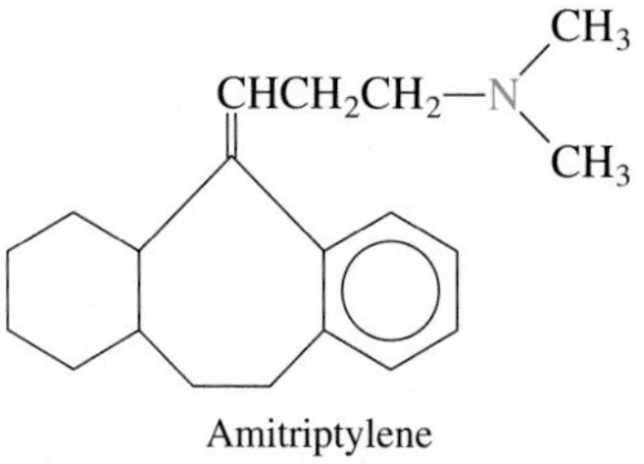

Amitriptylene
(Elavil)

Although the tricyclic antidepressants have been around since the 1950s, they have been only mildly successful. In the first place, there is a very narrow range in which the dose is both safe and effective. Low doses have very little effect, and higher doses quickly become toxic. There are also undesirable side effects, such as nausea, headache, dizziness, loss of appetite, or grogginess, as well as more serious problems such as jaundice or high blood pressure.

Since 1988 several new antidepressants have been introduced, the most popular by far being fluoxetine (Prozac). In 1993 antidepressant sales exceeded \$3 billion, and more than 40% of the sales were for Prozac. Doctors write prescriptions for Prozac at the rate of about a million per month. They prescribe it to help people cope with gambling problems, obesity, fear of public speaking, or premenstrual syndrome (PMS). Sometimes they prescribe it for healthy people just to help them loosen up a little or to have a more cheerful disposition. The drug works by enhancing the effect of serotonin, blocking its reabsorption by the cells. It is safer than the older antidepressants and more easily tolerated.

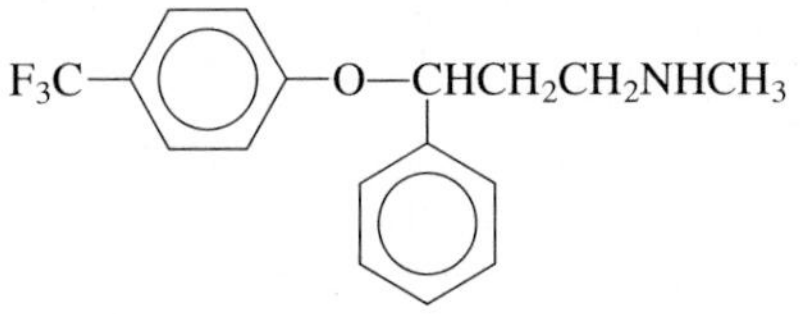

$F_3C-C_6H_4-O-CH(C_6H_5)CH_2CH_2NHCH_3$

Fluoxetine
(Prozac)

SECTION 19.21

Stimulant Drugs: Amphetamines

Among the more widely known stimulant drugs are a variety of synthetic amines related to phenylethylamine (Figure 19.34). Note the similarity of these molecules to those of epinephrine and norepinephrine; all are derived

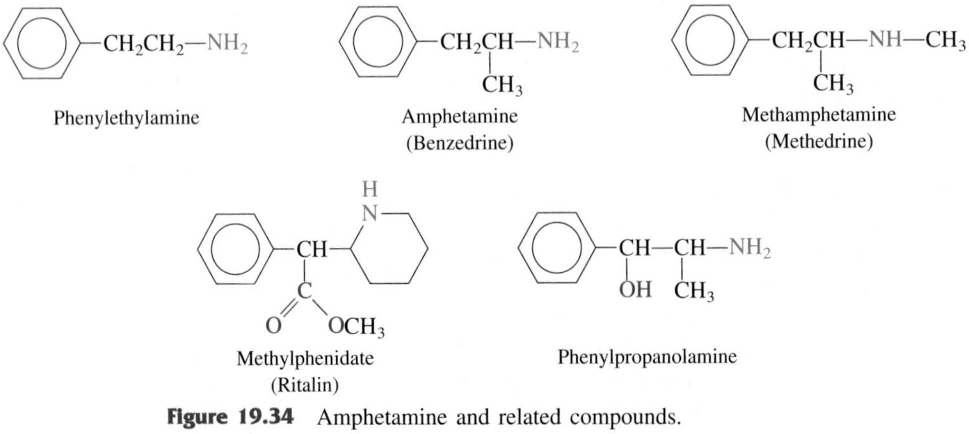

$C_6H_5-CH_2CH_2-NH_2$

Phenylethylamine

$C_6H_5-CH_2CH(CH_3)-NH_2$

Amphetamine
(Benzedrine)

$C_6H_5-CH_2CH(CH_3)-NH-CH_3$

Methamphetamine
(Methedrine)

$C_6H_5-CH(COOCH_3)-C_5H_9NH$ (piperidin-2-yl)

Methylphenidate
(Ritalin)

$C_6H_5-CH(OH)-CH(CH_3)-NH_2$

Phenylpropanolamine

Figure 19.34 Amphetamine and related compounds.

from the basic phenylethylamine structure. The **amphetamines** probably act as stimulants by mimicking the natural brain amines.

Amphetamine and methamphetamine have been widely abused. Amphetamine has been extensively used for weight reduction. It has also been employed for treating mild depression and narcolepsy, a rare form of sleeping sickness. Amphetamine induces excitability, restlessness, tremors, insomnia, dilated pupils, increased pulse rate and blood pressure, hallucinations, and psychoses. It is no longer recommended for weight reduction. It was found that, generally, any weight loss was only temporary. The greatest problem, however, was the diversion of vast quantities of amphetamines into the illegal drug market. Amphetamines are inexpensive. Armed forces personnel, truck drivers, and college students have been among the heavy users.

Like other amine drugs, the amphetamines usually are used in the form of salts called hydrochlorides. A free base form of methamphetamine is used like crack cocaine for smoking. This form is called ''ice'' because it exists in a clear crystalline form that resembles the solid form of water.

Dextroamphetamine (Dexedrine) is another stimulant drug that has been abused widely. It is related to amphetamine in a very subtle way. Actually, amphetamine is a mixture of two isomers. These isomers have the same atoms and groups of atoms, but the relative spatial orientations of these atoms and groups are different in the two isomers. One isomer is the mirror image of the other.

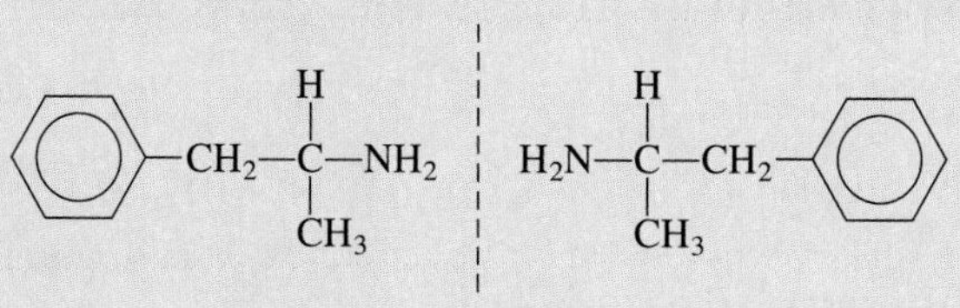

$$C_6H_5{-}CH_2{-}\underset{CH_3}{\overset{H}{C}}{-}NH_2 \quad \Big| \quad H_2N{-}\underset{CH_3}{\overset{H}{C}}{-}CH_2{-}C_6H_5$$

These isomers are not superimposable but are related to one another in much the same way that your right hand is related to your left. You can't fit a right-hand glove on your left hand or vice versa. These mirror-image isomers fit enzymes differently; thus they have different effects. The **dextro** (right-handed) isomer is a stronger stimulant than the **levo** (left-handed). Dexedrine is the trade name for the pure dextro isomer. Benzedrine is the trade name for a mixture of the two isomers in equal amounts. Dexedrine is two to four times as active as Benzedrine.

Mirror-image isomerism is quite common in organic chemicals of biological importance. Any molecule that has a carbon atom with four different kinds of atoms or groups attached to it can exist as mirror-image isomers. More than 40% of current drugs have dextro and levo isomers, and most of them are sold as mixtures of the isomers.

Methamphetamine has a more pronounced psychological effect than amphetamine. Generally, the "speed" that abusers inject into their veins is methamphetamine. Such injections, at least initially, are said to give the abuser a euphoric rush. Shooting methamphetamine is quite dangerous, though, because the drug is rather toxic (see Table 19.4).

Another amphetamine derivative, phenylpropanolamine, is widely used as an over-the-counter appetite suppressant. Like its relatives, this compound is a stimulant. Studies show that it is at best marginally effective as a diet aid, and it poses a threat to people with hypertension. Nevertheless, sales of phenylpropanolamine are 1 billion tablets at $150 million dollars each year.

One controversial use of amphetamines is their employment in the treatment of attention deficit disorder (ADD) in children. The drug of choice is often methylphenidate (Ritalin). Although it is a stimulant, the drug seems to calm kids who otherwise can't sit still. This use has been criticized as "leading to drug abuse" and as "solving the teacher's problem, not the kid's."

Cocaine: The Snow Sniffers

Cocaine, first used as a local anesthetic (Section 19.13), also serves as a powerful stimulant. The drug is obtained from the leaves of a shrub that grows almost exclusively on the eastern slopes of the Andes Mountains. Many of the Indians living in and around the area of cultivation chew coca leaves—mixed with lime and ashes—for their stimulant effect. Cocaine used to arrive in the United States as the salt, cocaine hydrochloride. Now much of it comes in the form of broken lumps of the free base. This form is called *crack cocaine.*

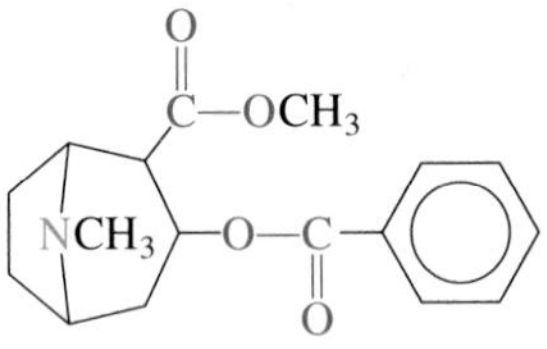

Cocaine

Crack is the same chemical as the free base that occasionally made the news several years ago. When cocaine hydrochloride was the main street form of the drug, some people made free base by reacting the salt with a strong base. The person then used a solvent to extract the free base (pure cocaine) from solution. Some free basers were badly burned when the solvent (such as diethyl ether) was accidentally ignited.

Cocaine hydrochloride is readily absorbed through the watery mucous membrane of the nose. It is the form used by those who snort cocaine. Those who smoke cocaine use the free base (crack), which readily vaporizes at the temperature of a burning cigarette. When smoked, cocaine reaches the brain in 15 seconds. Cocaine acts by preventing the neurotransmitter dopamine from being taken back up after it is released by nerve cells. High levels of dopamine are therefore available to stimulate the pleasure centers of the brain. After the binge, dopamine is depleted in less than an hour. This leaves the user in a pleasureless state and (often) craving more cocaine.

Use of cocaine increases stamina and reduces fatigue, but the effect is short-lived. Stimulation is followed by depression. Once quite expensive and limited to use mainly by the wealthy, cocaine is now available in cheap and potent forms. Hundreds, including several well-known athletes, have died from cocaine overdose.

Caffeine: Coffee, Tea, or Cola

The beverages coffee and tea and some soft drinks contain the mild stimulant caffeine. The effective dose of caffeine is about 200 mg, corresponding to about two cups of strong coffee or tea. Caffeine is also available in tablet form as a stay-awake or keep-alert type of drug. The best known brands are probably No-Doz and Vivarin. No-Doz contains about 100 mg of caffeine per tablet; each Vivarin tablet has 200 mg.

Each year a million kilograms of caffeine are added to food in the United States. Most of it goes into soft drinks (soda pop).

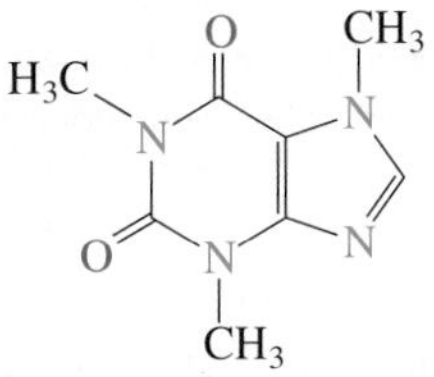

H_3C O CH_3 N N N N O CH_3

Caffeine

Is caffeine addictive? The "morning grouch" syndrome indicates that it may be mildly so. There is also evidence that caffeine may be involved in chromosome damage. To be safe, people in their childbearing years should avoid large quantities of caffeine. Overall, the hazards of caffeine ingestion seem to be slight.

Nicotine: Going Up in Smoke

Another common stimulant is nicotine. This drug is taken by smoking or chewing tobacco. Nicotine is highly toxic to animals (see Table 19.4). It is especially deadly when injected; the lethal dose for a human is estimated to be about 50 mg. Nicotine has been used in agriculture as a contact insecticide. Nicotine seems to have a rather transient effect as a stimulant. This initial response is followed by depression.

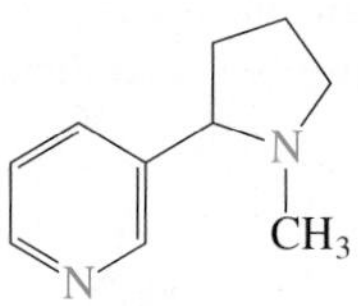

N CH_3 N

Nicotine

Clonidine, a drug used to treat high blood pressure, reduces nicotine withdrawal symptoms.

Is nicotine addictive? Casual observation of a person trying to quit smoking seems to indicate that it is. Consider the 1972 memorandum by a Philip Morris scientist who noted that "no one has ever become a cigarette smoker by

Table 19.5 Use of Drugs Among Smokers and Nonsmokers Aged 12 to 17

Drug	Use Among People Who Never Smoked (%)	Use Among All Smokers (%)	Use Among Heavy Smokers* (%)
Heroin	0.1	1.2	5.1
Cocaine	0.3	15.4	31.7
Crack	0.1	5.7	11.1
Marijuana	2.5	57.5	68.2
Any illicit substance[†]	7.2	67.7	78.9

*Smoke more than one pack a day.

[†]Includes marijuana, cocaine, crack, heroin, PCP, hallucinogens, and, except for medical use, inhalants, sedatives, tranquilizers, stimulants, analgesics, and steroids.

Source: Center on Addiction and Substance Abuse at Columbia University. Data derived from the national household survey on drug abuse by the National Institute of Drug Abuse, 1991.

smoking cigarettes without nicotine.'' He suggested that the company ''think of the cigarette as a dispenser for a dose unit of nicotine.''

It is interesting that teenagers who smoke are much more likely to use hard drugs than those who are nonsmokers. Among heavy smokers from 12 to 17 years old, 4 out of 5 are also drug users (Table 19.5.)

SECTION 19.22

The ''Mindbenders'': LSD

The third major class of drugs is popularly called the ''mindbenders'' because they qualitatively change the way we perceive things. Probably the most powerful of these drugs is LSD (from the German *lysergsäure diethylamid*). The physiological properties of this compound were discovered quite accidentally by Albert Hofmann in 1943. Hofmann, a chemist at the Sandoz Laboratories in Switzerland, unintentionally ingested some LSD. He later took 250 μg, which he considered a small dose, to verify that LSD had caused the symptoms he had experienced. Hofmann had a very rough time for the next few hours, exhibiting such symptoms as visual disturbance and schizophrenic behavior.

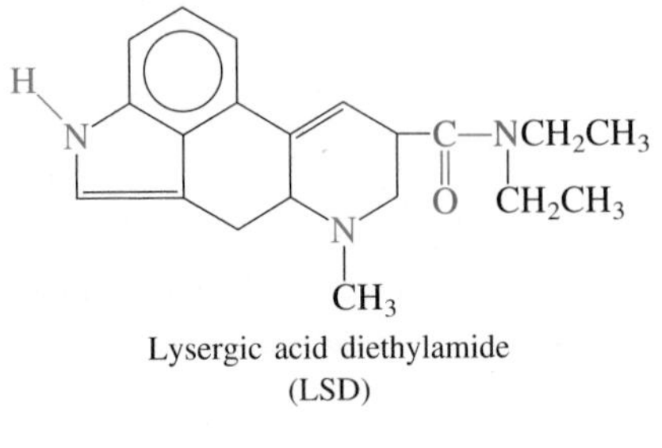

H N, C(=O)—NCH_2CH_3, CH_2CH_3, N, CH_3

Lysergic acid diethylamide (LSD)

Lysergic acid is obtained from ergot, a fungus that grows on rye. It is converted to the diethylamide by treatment with thionyl chloride ($SOCl_2$), followed by diethylamine. Note that a part of the LSD structure (color) resembles that of serotonin. LSD seems to act as a serotonin agonist.

LSD is a potent drug, as indicated by the small amount required for a person to experience its fantastic effects. The usual dosage is probably about 10 to 100 μg. No wonder Hofmann had a bad time with 250 μg! To give you an idea of how small 10 μg is, let's compare that amount of LSD to the amount of aspirin in one tablet—one aspirin tablet contains 325,000 μg of aspirin.

Is LSD a dangerous drug? A few facts are known, but most are disputed. In 1967, Maimon Cohen of the State University of New York at Buffalo reported that LSD damages chromosomes, especially those of the leucocytes (white blood cells). Additional studies produced mixed results. The question still has not been resolved.

Several useful drugs are obtained from the ergot fungus. Ergotamine shrinks blood vessels in the brain; it is used to treat migraine headaches. Ergonovine causes small blood vessels to contract. It is used to induce uterine contractions and thus reduce bleeding after childbirth. Both compounds, like LSD, are amides of lysergic acid.

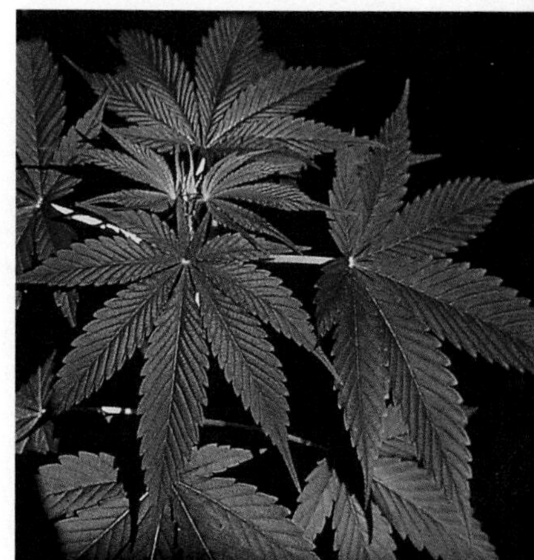

Figure 19.35 The marijuana plant. [*Courtesy of U.S. Department of Justice Drug Enforcement Administration.*]

SECTION 19.23

Marijuana: Some Chemistry of Cannabis

Many complete books have been written about marijuana, yet all we know for certain about the drug would fill only a few pages. Let's look at some chemistry of the weed and at some of the ways chemists are involved with the marijuana problem.

The weed *Cannabis sativa* (Figure 19.35) has long been useful. The stems yield tough fibers for making ropes. *Cannabis* has been used as a drug in tribal religious rituals. Marijuana also has a long history as a medicine, particularly in India. In the United States, marijuana is second only to alcohol in popularity as an intoxicant.

Marijuana: Forms and Potency

The term **marijuana** refers to a preparation made by gathering the leaves, flowers, seeds, and small stems of the plant (Figure 19.36). These are generally dried and smoked. They contain a variety of chemical substances, many of them still unidentified. The principal active ingredient, however, is tetrahydrocannabinol (THC). Actually, there are several active cannabinoids in marijuana; only one is shown here.

Marijuana plants vary considerably in THC content. Most of the marijuana sold in North America has a THC content of about 1%. Plants native to the United States have a low THC content, usually about 0.1%. Potency depends on the genetic variety of plant, not to any significant extent on the climate or the soil where it is grown.

CH_3 OH CH_3 CH_3 O C_5H_{11}

Tetrahydrocannabinol (THC)

Figure 19.36 Retail forms of marijuana. [*Courtesy of U.S. Department of Justice Drug Enforcement Administration.*]

More potent preparations sometimes are made from marijuana. By selecting only the flowering tops and tender top leaves, you get a stronger product called *ganja.* (The ordinary marijuana is called *bhang* in India.) Jamaican *ganja* has a THC content of between 4 and 8%. Indian *ganja* is generally somewhat less potent. By collecting only the resinous secretions of the flowering parts, you get a product called hashish, or "hash" (known as *charas* in India). Hash has a THC content of between 5 and 12%. Liquid hash and hash oil are probably solvent extracts of marijuana.

Effects of Marijuana

The effects of marijuana are difficult to measure, partly because of the variable THC content of different preparations. A variety of standard potency is now grown and supplied for controlled clinical studies. With this standard product, some of the effects of marijuana can be measured in reproducible experiments. Smoking *Cannabis* increases the pulse rate, distorts the sense of time, and impairs some complex motor functions. Marijuana smoking sometimes induces a euphoric sense of lightness—a floating sensation. Sometimes it causes a feeling of anxiety. Often, the user has an impression of brilliance, although studies have shown no mind-expanding effects. Users sometimes experience hallucinations, although these are much less frequent than with LSD. Marijuana seems to heighten one's enjoyment of food, with users relishing beans as much as they normally would enjoy steak.

The gene for a THC receptor was cloned in 1990. The receptors are found in movement control centers, thus explaining the loss of coordination in those intoxicated with the drug. The memory and cognition areas of the brain also are rich in THC receptors. This explains why marijuana users often do poorly on tests. There are few receptors in the brainstem where breathing and heartbeat are controlled. This correlates with the fact that it is difficult to get a lethal dose of marijuana.

The long-term effects of marijuana use are more difficult to evaluate. Some people claim that smoking marijuana leads to the use of harder drugs. There is little objective evidence of this. More heroin addicts start drug use with alcohol than with marijuana.

There is some evidence that marijuana causes brain damage. But if so, it is less extensive than that caused by alcohol.

Some people also have claimed that excessive use of marijuana leads to psychoses. It has long been known to induce short-term psychotic episodes in those already predisposed and in others who take excessive amounts. Long-term psychoses, however, occur at the same rate among regular marijuana users as among the general population.

One of the more interesting reports has come from two surgeons at the Harvard Medical School. Menelaos Aliapoulios and John Harman claim to have treated 13 young men for gynecomastia (enlarged breasts). All were heavy marijuana users. Their breasts also discharged a white milky liquid. The painful swelling receded in three of the men after they stopped smoking *Cannabis,* but three others needed surgery. The two doctors are convinced that marijuana contains a "feminizing ingredient." There is a slight structural similarity of THC to the female hormones. Some studies indicate that THC binds weakly to estrogen receptors.

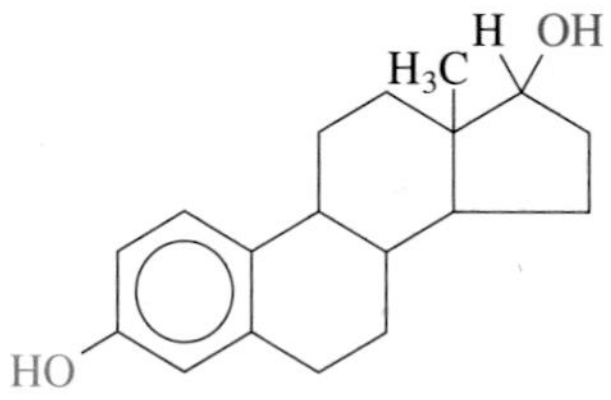

Estradiol

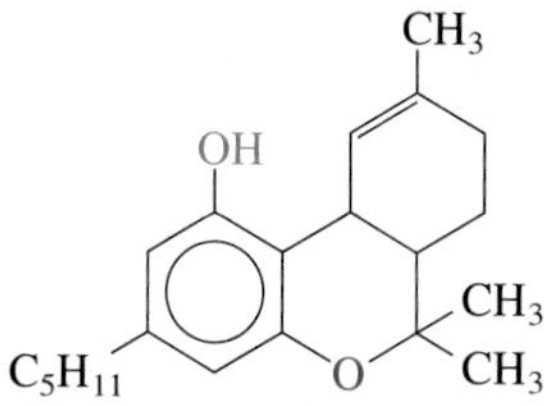

Tetrahydrocannabinol
(THC)

Despite its apparent feminizing properties, THC in both high and low doses causes an initial rise in testosterone levels in men. With high doses, however, the slight increase is followed by a rapid fall to below-normal testosterone levels.

Chemists and Marijuana

What do chemists have to do with all this? Well, they have isolated the active components and synthesized them. They can monitor the THC content of marijuana. They can monitor THC in the bloodstream and identify the products of its breakdown. They cannot, however, tell how it changes the body chemistry or what its long-term effects are.

Perhaps the most significant harm from marijuana comes through its impairment of complex motor functions—such as those used in driving an automobile. In addition to dodging drunks, we have to look out for potheads. Incidentally, chemists have developed a THC detector similar to the one used to test blood alcohol.

Unlike alcohol, THC persists in the bloodstream for several days because it is soluble in fats. The products of its breakdown remain in the blood for as long as 8 days. The persistence of these chemicals inside the body indicates that some of a given dose may still be active in the body at the time another dose is taken.

Marijuana has some legitimate medical uses. It reduces pressure in the eyes of people who have glaucoma. If not treated, the increasing pressure eventually causes blindness. Also, marijuana relieves the nausea that afflicts cancer patients undergoing radiation treatment and chemotherapy.

SECTION 19.24

Drugs and Deception: Chemistry and Quality Control

One recurring problem on the illegal drug scene is that drugs are not always what they are supposed to be. The buyers simply have to trust the sellers—and their own experience—about the identity and quality of the products they buy. Even with marijuana, a readily identified weed, there are problems of quality control. The nonpotent variety often is used to dilute the potent product to increase profits. The nonpotent variety also has been found to be laced with other drugs to produce some kind of physiological effect. There have been numerous reports of synthetic THC on the illegal-drug scene. Little if any has been confirmed by crime laboratories. Most "THC" has been shown to be either LSD or PCP.

Many fraudulent products frequent the illegal drug scene. The scare about chromosomal damage from LSD led to increased interest in mescaline, a drug derived from the peyote cactus. (Note that mescaline is another phenylethylamine.) Mescaline is scarce and expensive; yet a product called mescaline became readily available on the street in some areas. Lab analyses proved that the product was really PCP. In fact, crime labs generally have found nearly two-thirds of all drugs (other than marijuana) brought in for analysis to be something other than what the dealers said they were.

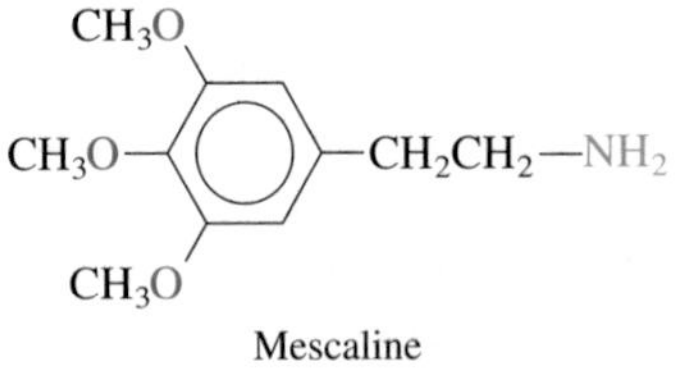

CH_3O

CH_3O—(benzene ring)—CH_2CH_2—NH_2

CH_3O

Mescaline

Use of illegal drugs is costly to society. From 50 to 80% of all personal injuries and accidents in the workplace are drug related. Drug users are absent from work 2.5 times as often as nonusers and 5 times as likely to file claims for compensation. In short, drug users cause a 25% loss of income to their employers.

Drug misuse (for example, using penicillin, which has no effect on viruses, to treat a viral infection) is all too common. **Drug abuse** (using a drug for its intoxicating effect) is even more widespread. Worldwide, people spend more money for illegal drugs than for food. Annual revenues for drug dealers amount to $500 billion or more.

Chemistry can provide drugs of enormous benefit to society. It can also provide drugs that present society with serious problems. But it can't solve the drug problem. However, chemistry can provide information on which intelligent choices can be based. The choices, though, are up to you—as an individual and as a member of society. No recreational drug has ever made anyone a better person or any society a better society.

SECTION 19.25

The Placebo Effect

An interesting phenomenon associated with the testing of drugs is the placebo effect. A **placebo** is an inactive substance given in the form of medication to a patient who thinks it is the real thing.

It is common practice when evaluating a new drug to give the drug to one group of patients and to give placebos to a similar "control" group. The patients have no idea whether they are receiving the real drug or the placebo. Normally they believe that they are really trying out the new drug. Doctors treating the patients usually do not know which ones are receiving the drug and which are getting the placebo until the study has been completed.

It sometimes happens that because a patient thinks he is receiving a certain drug and is expecting positive results, he actually experiences the results he is expecting, even though he has not received the real drug at all. In one study of a new tranquilizer, 40% of the patients who had been given placebos reported that they felt much better and rated the drug as highly effective.

The placebo effect is good evidence for the fact that there is a strong connection between the mind and the body. Medical treatment of the human body should not ignore the enormous potential of the human mind.

Summary

1. Drugs are substances used to relieve pain, to treat an illness, or to improve the state of health or well being.
2. *Analgesics* are pain relievers, aspirin being the foremost example.
3. Drugs used to treat colds include *antihistamines,* cough suppressants (antitussives), and nasal decongestants.
4. The first *antibacterial* drug was sulfanilamide.
5. *Antibiotics* (derived from mold or bacteria) include the penicillins, the cephalosporins, and the *tetracyclines.*
6. Viral diseases, such as measles, mumps, and smallpox, are usually prevented by vaccination, but not all viral diseases have vaccines.
7. Herpes and AIDS are both viral infections that have no cure, but several antiviral drugs have been developed to treat them.
8. Anticancer drugs are mainly *antimetabolites,* which interfere with DNA synthesis, thereby slowing down the growth of cancers.
9. "Birth control" pills are mixtures of an *estrogen* and a *progestin,* which act by creating a state of pseudo pregnancy, during which a woman does not ovulate and therefore cannot conceive.
10. *Psychotropic drugs* are those that affect the mind.

They can be divided into three classes: *depressants, stimulants,* and *psychedelics.*

11. *Anesthetics* are depressants that are used during surgery to produce unconsciousness or local insensitivity to pain.
12. *Narcotics* are drugs that relieve pain and can produce anesthesia, but they are also addictive. Morphine is an example.
13. Natural painkillers called *endorphins* are produced within the body.
14. Tranquilizers are depressants that many people use to relax. Some are over-the-counter drugs.
15. Stimulant drugs include the *amphetamines.*
16. Many drugs are optically active, with *dextro* and *levo* forms. Usually one form is more active than the other, and sometimes the other form is actually harmful.
17. Caffeine in coffee, tea, and soft drinks and nicotine in tobacco are common stimulants.
18. Although cocaine was first used as an anesthetic, it is also a powerful stimulant.
19. The best known among the psychedelic or "mind-bending" drugs are LSD and *marijuana.*
20. Not only do street drugs have a negative effect on one's health, but they are also likely not to be what they are supposed to be. Quality control does not exist in that market.
21. The *placebo effect* occurs when people experience the effects of having taken a drug when they believe that they have taken it, even though they actually have not.

Key Terms

acquired immune deficiency syndrome (AIDS) 19.6
agonist 19.16
allergen 19.2
amphetamines 19.21
analgesic 19.1
androgens 19.9
anesthetic 19.13
antagonist 19.16
antibiotics 19.4
anticholinergic 19.2
anticoagulant 19.1
antihistamines 19.2
anti-inflammatory 19.1
antimetabolites 19.7
antipyretic 19.1
barbiturates 19.14
broad spectrum antibiotics 19.5
buffer 19.1
chemotherapy 19.0
depressants 19.11
dextro 19.21
dissociative anesthetic 19.13
drug abuse 19.24
drug misuse 19.24
endorphins 19.17
enkephalins 19.17
estrogens 19.9
expectorants 19.2
general anesthetic 19.13
hallucinogenic 19.11
hormones 19.8
levo 19.21
local anesthetic 19.13
marijuana 19.23
narcotics 19.15
nasal decongestants 19.2
neurons 19.18
neurotransmitters 19.18
placebo 19.25
progestins 19.10
prostaglandins 19.8
psychedelics 19.11
psychotomimetics 19.11
psychotropic drugs 19.11
retroviruses 19.6
steroids 19.9
stimulants 19.11
synapses 19.18
synergistic effect 19.14
tetracyclines 19.5

Review Questions

1. What is an analgesic?
2. What is an antipyretic?
3. What is the chemical name for aspirin?
4. How does aspirin suppress inflammation?
5. In what ways do aspirin and acetaminophen act similarly?
6. Why is aspirin chosen over acetaminophen for treatment of arthritis?
7. Why is acetaminophen chosen over aspirin for the relief of pain associated with surgical procedures?
8. What is ibuprofen?
9. What is a placebo?

10. What are sulfa drugs? How do they work?

11. What are antibiotics?

12. How can a drug such as penicillin kill bacterial cells without killing human cells?

13. Tetracyclines are broad-spectrum antibiotics. What does that mean?

14. What is a hormone?

15. What is cortisone?

16. What is an androgen?

17. What is an estrogen?

18. How do birth-control pills work?

19. What is DES? What problems have been associated with its use?

20. How does mifepristone (RU-486) work?

21. What are prostaglandins?

22. List two major classes of anticancer drugs.

23. What are psychotropic drugs?

24. What is the difference in drug misuse and drug abuse? Give an example of each.

25. List the three classes of drugs that affect the mind.

26. What are the effects of a stimulant drug?

27. What are the effects of a depressant drug?

28. What are the effects of a hallucinogenic drug?

29. What is a general anesthetic?

30. What is a local anesthetic?

31. Which of these anesthetics are dangerous because of flammability?
   a. diethyl ether b. halothane
   c. chloroform d. cyclopropane

32. For each of the following anesthetics, describe a disadvantage associated with its use. Do not include flammability.
   a. nitrous oxide b. halothane
   c. diethyl ether

33. What application does curare have in modern anesthesiology?

34. Describe two medical uses of barbiturates and the relative dosage level of each.

35. What are the hazards of long-term barbiturate use?

36. Describe a synergistic reaction involving drugs.

37. Name two dissociative anesthetics.

38. What are narcotics?

39. What is opium?

40. How does codeine differ from morphine in its physiological effects? In what ways are the two drugs similar?

41. How does methadone maintenance work? What is your position on the use of methadone maintenance to treat heroin addiction?

42. How are endorphins related to each of the following?
   a. the anesthetic effect of acupuncture
   b. the absence of pain in a badly wounded soldier.

43. What is an agonist?

44. What is an antagonist?

45. Define or identify each of the following.
   a. neuron
   b. synapse
   c. neurotransmitter

46. How may our mental state be related to our diet?

47. Are tranquilizers a cure for schizophrenia?

48. What are some of the problems involved in the clinical evaluation of LSD?

49. What are some of the problems in the clinical evaluation of marijuana?

Problems

Chemical Structures and Properties

50. What two functional groups are present in a salicylic acid molecule?

51. What two functional groups are present in an aspirin (acetylsalicylic acid) molecule?

52. Terfenadine (Seldane) has the chemical formula $C_{32}H_{41}NO_2$. Is the compound water soluble or fat soluble? Explain your answer.

53. Ephedrine hydrochloride has the chemical formula $C_{10}H_{16}NO^+Cl^-$. Is the compound water soluble or fat soluble? Explain your answer.

54. From what carboxylic acid are many local anesthetics derived?

55. What structural feature characterizes the more powerful local anesthetics?

56. What is the basic structure common to all barbiturate molecules? How is the basic structure modified to change the properties of individual barbiturate drugs?

Drug Toxicities

57. If the minimum lethal dose (MLD) of amphetamine is 5 mg/kg, what would be the MLD for a 70-kg person? Can toxicity studies on animals always be extrapolated to humans?

Additional Problems

58. What structural feature makes a birth-control steroid effective orally?

59. Which of the following compounds are classified as steroids?
    a. tristearin b. cholesterol
    c. testosterone d. prostaglandins

60. Give the structure for *para*-aminobenzoic acid.

61. Give the structure for sulfanilamide.

62. What structural feature is shared by tetracyclines?

63. What structural feature is shared by all steroids?

64. When administered intravenously to rats, the LD_{50} of procaine is 50 mg/kg and that of nicotine is 1.0 mg/kg of body weight. Which drug is more toxic? What would be the approximate lethal dose of each for a 50-kg human?

Projects

65. Following are the generic names of several of the most widely prescribed drugs. Use *The Merck Index* or a similar reference to determine (if possible) the chemical structure, medical use, toxicity, and side effects of each.
    a. furosemide b. methyldopa
    c. hydrochlorothiazide d. digoxin
    e. triazolam f. amoxicillin
    g. propranolol h. metoclopramide

66. Examine the labels of three over-the-counter sleeping pills (such as Nytol, Sominex, and Sleep-eze). Make a list of the ingredients in each. Look up the properties (medical uses, dosages, side effects, toxicities) of each in a reference work such as *The Merck Index.*

67. Examine the labels of at least three over-the-counter antitension formulations (such as Cope and Compoz). Make a list of the ingredients in each. Look up the properties (medical uses, dosages, side effects, toxicities) of the ingredients in a reference work such as *The Merck Index.*

68. Do a cost analysis of five brands of plain aspirin. Calculate the cost per gram of each.

69. Compare the cost per gram of an "extra strength" aspirin formulation to that of plain aspirin.

70. Discuss some of the problems involved in proving a drug safe or proving it harmful. Which is easier? Why?

References and Readings

1. Abraham, E. P. "The Beta-Lactam Antibiotics." *Scientific American,* June 1981, pp. 76–86. Discusses penicillins and cephalosporins.

2. Alper, Joseph. "The Microchip Microbe Hunters." *Science,* 16 February 1990, pp. 804–806. Designing drugs by computer to bind to specific sites on viruses.

3. Baum, Rudy M. "New Variety of Street Drugs

Poses Growing Problems.'' *Chemical and Engineering News,* 9 September 1985, pp. 7–16. Discusses designer drugs.

4. Cowan, Ron. ''Receptor Encounters.'' *Science News,* 14 October 1989, pp. 248–252.
5. Davis, Audrey B. ''The Development of Anesthesia.'' *American Scientist,* September–October 1982, pp. 522–528.
6. DiSpezio, Michael A. ''Retroviruses.'' *The Science Teacher,* October 1990, pp. 40–47.
7. Douglas, Kenneth T. ''Anticancer drugs: DNA as a Target.'' *Chemistry and Industry,* 15 October 1984, pp. 738–742.
8. Gilman, A. G., et al. *Goodman and Gilman's The Pharmacological Basis of Therapeutics,* 7th ed. New York: Macmillan, 1985.
9. Harborne, Jeffrey B., and Herbert Baxter, eds. *Phytochemical Dictionary: A Handbook of Bioactive Compounds from Plants.* London: Taylor and Francis, 1992.
10. Hecht, Annabel. ''Sulfa: Yesterday's Hero Is Still Taking Bows.'' *FDA Consumer,* October 1984, pp. 8–11.
11. Hill, John W., and Susan M. Jones. ''Consumer Applications of Chemical Principles: Drugs.'' *Journal of Chemical Education,* April 1985, pp. 328–331.
12. Jackson, Ian. ''Pharmaceutical Revolution: The Pill Generation.'' *Today's Chemist,* October 1990, pp. 17–20.
13. Julien, Robert M. *A Primer of Drug Action,* 5th ed. San Francisco: W. H. Freeman, 1989.
14. Liska, Ken. *Drugs and the Human Body,* 4th ed. New York: Macmillan, 1994.
15. Marx, Jean. ''Marijuana Receptor Gene Cloned.'' *Science,* 10 August 1990, pp. 624–626.
16. Modeland, Vern. ''Modern Anesthesia: Going Under Safely.'' *FDA Consumer,* December 1989–January 1990, pp. 13–17.
17. Nelson, Norman A., Robert C. Kelly, and Roy A. Johnson. ''Prostaglandins and the Arachidonic Acid Cascade.'' *Chemical and Engineering News,* 16 August 1982, pp. 30–44.
18. Palca, Joseph. ''The Pill of Choice?'' *Science,* 22 September 1989, pp. 1319–1323. Discusses RU 486.
19. *Physicians' Desk Reference,* 48th ed. Oradell, NJ: Medical Economics Company, 1994.
20. Revkin, Andrew C. ''Crack in the Cradle.'' *Discover,* September 1989, pp. 62–69. Effects of cocaine on the fetus.
21. Podolsky, Doug. ''Birth Control.'' *U.S. News and World Report,* 24 December 1990, pp. 58–65.
22. Prendergast, Alan. ''Beyond the Pill.'' *American Health,* October 1990, pp. 37–44. Discusses contraceptives for males.
23. Stinson, Stephen C. ''Better Understanding of Arthritis Leading to New Drugs to Treat It.'' *Chemical and Engineering News,* 16 October 1989, pp. 37–70.
24. Ulmann, Andre, Georges Teutsch, and Daniel Philibert. ''RU 486.'' *Scientific American,* June 1990, pp. 42–48.
25. Van Dyck, Craig, and Robert Byck. ''Cocaine.'' *Scientific American,* March 1982, pp. 128–141.
26. Weissmann, Gerald. ''Aspirin.'' *Scientific American,* January 1991, pp. 84–90. Its action is even more complicated than we thought.
27. Yam, Philip. ''Cannabis Comprehended.'' *Scientific American,* October 1990, p. 38. Describes possible new drugs to come from the discovery of the THC receptor.

20 Poisons

Chemical Toxicology

Toxic chemicals can be
quite ordinary stuff.
Anything's a poison if
the dose is big enough.

When the Greek philosopher Socrates was accused of corrupting the youth of Athens in 399 B.C., he was given the choice of exile or death. He chose death and implemented his decision by drinking a cup of hemlock (Figure 20.1).

Poisons from plant and animal sources were well known in the ancient world. Snake and insect venoms and plant alkaloids were used. Today, many primitive tribes still use a variety of poisons in hunting and warfare. Curare, used by certain South American tribes, is one of the more notorious examples.

Although poisonous substances have been known and used for centuries, it is only within the last 150 years that scientists have learned the nature of the chemicals that are the active components of the various poisons. Socrates'

Figure 20.1 Jacques Louis David's painting *The Death of Socrates* (1787) shows Socrates about to drink the cup of hemlock to carry out the death sentence decreed by the rulers of Athens. [*Courtesy of the Metropolitan Museum of Art, Wolfe Fund, 1931. Catherine Lorillard Wolfe Collection (31.45).*]

Abandoned barrels of toxic waste can cause serious environmental problems. [*The Stock Market.*]

Figure 20.2 *Conium maculatum* (poison hemlock).

N—H $CH_2CH_2CH_3$

Coniine

hemlock probably was prepared from the fully grown but unripened fruit of *Conium maculatum* (poison hemlock) (Figure 20.2). Usually, the fruit is dried carefully and then brewed into a "tea." Hemlock contains several alkaloids, but the principal one is coniine. This drug causes nausea, weakness, paralysis, and—as in the case of Socrates—death.

Poisons have always been with us, but our knowledge of them is greater than ever before. Industrial accidents, such as that at Bhopal, India, have made the public acutely aware of problems with toxic substances. At Bhopal, the accidental release of methyl isocyanate ($CH_3—N=C=O$), an intermediate in the synthesis of carbamate insecticides (Chapter 16), killed more than 2000 people and injured countless others. People are also concerned about long-term exposure to toxic substances in the air, in their drinking water, and in their food. Chemists can detect exceedingly tiny quantities of such substances. It is still quite difficult, though, to determine the effect of these trace amounts of toxic materials on human health.

A study of the response of living organisms to drugs is called **pharmacology**. Organisms respond in a variety of ways to the many chemical substances we call drugs. In this chapter, we deal with those substances that are poisonous or otherwise injurious. **Toxicology** is the branch of pharmacology that deals with the effects of poisons, their identification or detection, and the development and use of antidotes.

SECTION 20.1

All Things Are Poisons

What is a poison? Perhaps a better question would be, How much is a poison? Substances may be harmless—or even necessary nutrients—in one amount, and injurious—or even deadly—in another. Even common substances such as salt and sugar can be poisonous when eaten in abnormally large amounts. Too much sugar—candy or sweets—can give a child a stomachache. Too much salt—sodium chloride—can induce vomiting. There have even been cases of fatal poisoning when salt was accidentally substituted for lactose (milk sugar) in formulas for infants. Some substances are obviously more toxic than others, however. It would take a massive dose of salt to kill the average healthy adult, whereas only a few micrograms of some of the nerve poisons can be fatal. Toxicity depends on the chemical nature of the substance.

People also respond differently to the same chemical. To cite an extreme case, a few grams of sugar would cause no acute symptoms in a normal person but would be dangerous to a diabetic. Excessive amounts of salt would be especially serious to a person with edema (swelling due to excessive amounts of fluid in the tissues).

Still another complicating factor is that chemicals behave differently when administered in different ways. Nicotine is more than 50 times as toxic when applied intravenously as when taken orally. (See Table 19.4.) Good, fresh water is delightful when taken orally, but even water can be deadly when inhaled in sufficient quantity. Further complications arise from the fact that even closely related animal species react differently to a given chemical. Even individuals within a species may react to different degrees.

Poisons Around the House

We have already noted that many household chemicals are poisonous (Chapter 17). Drain cleaners, oven cleaners, and toilet bowl cleaners are highly corrosive. Insecticides and rodenticides are extremely toxic. Laundry bleach and ammonia are both very toxic, and when they are mixed together they can be deadly.

Even seemingly harmless products around the house can be dangerous if a young child happens to get into them. A bottle of cough syrup can trigger an emergency visit to the hospital.

Poisons in the Garden

Some of the toxic products used on the farm (Chapter 16) are also used in home gardens. But herbicides and insecticides are not the only poisons you are likely to find in a garden. Sometimes the plants themselves are toxic.

Poisons sometimes come in very pretty packages. Iris are beautiful and so are azaleas and hydrangeas, but all of these popular perennials are poisonous. Holly berries, wisteria seeds, and the leaves and berries of privet hedges are also among the more poisonous products of the home garden.

Even those healthy-looking green plants that beautify the inside of the house can be toxic. One poisonous houseplant, philodendron, is probably the most widely cultivated of all indoor plants.

In our discussion here, we cover only a few of the many toxic substances. These are organized into groups with similar activities. Those more likely to be encountered in everyday life are given priority.

SECTION 20.2

Corrosive Poisons: A Closer Look

In Chapter 7, we examined the corrosive effect of strong acids and bases on human tissue. These chemicals indiscriminately destroy living cells. Corrosive chemicals, in lesser concentrations, also exert a more subtle effect.

Strong Acids and Bases

Both acids and bases, even in dilute solutions, catalyze the hydrolysis of protein molecules in living cells. These reactions involve the breaking of the amide (peptide) linkages in the molecules.

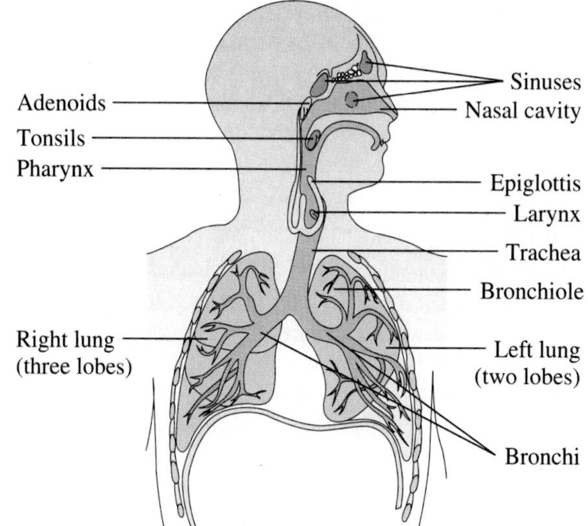

Figure 20.3 The respiratory system, showing the route of air through the nose, pharynx (throat), larynx (voice box), and trachea (windpipe), into the bronchi and bronchioles (bronchial tubes), and ending in the alveoli (air sacs).

$$\text{Protein molecule: } \sim\!\sim\!\text{C(=O)}\!-\!\text{N(H)}\!\sim\!\sim + H_2O \xrightarrow{H^+ \text{ or } OH^-} \sim\!\sim\!\text{C(=O)}\!-\!OH + H\!-\!\text{N(H)}\!\sim\!\sim \text{ (Fragments)}$$

Generally, the fragments are not able to carry out the functions of the original protein. In cases of severe exposure, the fragmentation continues until the tissue has been completely destroyed.

Acids in the lungs are particularly destructive. In Chapter 12, we saw how sulfuric acid is formed when sulfur-containing coal is burned. Acids are also formed when plastics and other wastes are burned. The damage these pollutants do can be explained as the breakdown of lung tissue by the acids (Figures 20.3 and 20.4).

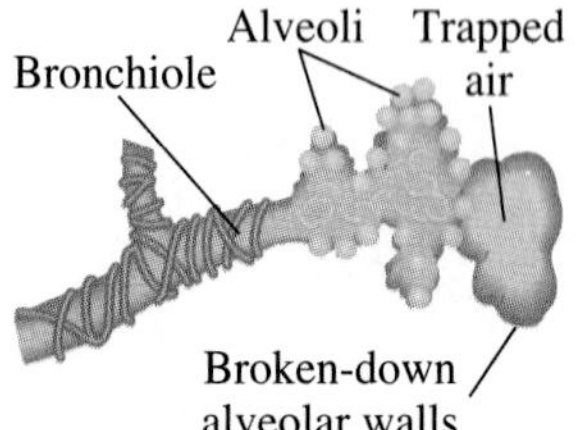

Figure 20.4 Pulmonary emphysema. The loss of elasticity and the deterioration of the alveolar walls deter the exhalation of carbon dioxide.

Oxidizing Agents

Other air pollutants also damage living cells. Ozone, peroxyacetyl nitrates (PAN), and the other oxidizing components of photochemical smog probably do their main damage through the deactivation of enzymes. The active sites of enzymes often incorporate the sulfur-containing amino acids cysteine and methionine. Cysteine is readily oxidized by ozone to cysteic acid.

$$\underset{\text{Cysteine}}{HS\text{—}CH_2\text{—}CH(NH_2)\text{—}COOH} + O_3 \longrightarrow \underset{\text{Cysteic acid}}{HO\text{—}S(=O)_2\text{—}CH_2\text{—}CH(NH_2)\text{—}COOH}$$

Methionine is oxidized to methionine sulfoxide.

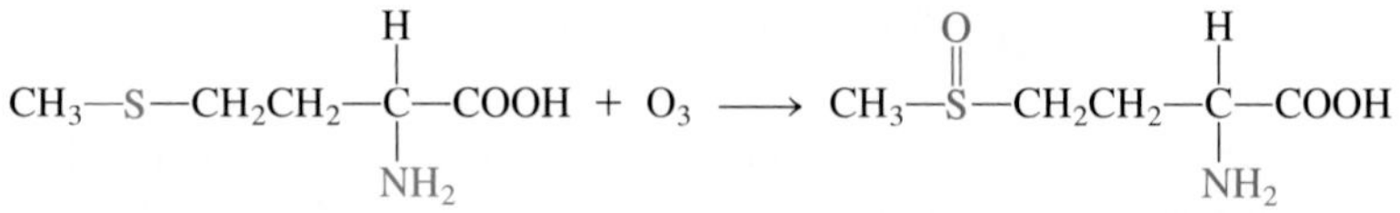

$$CH_3{-}S{-}CH_2CH_2{-}\underset{NH_2}{\overset{H}{C}}{-}COOH + O_3 \longrightarrow CH_3{-}\overset{O}{\overset{\|}{S}}{-}CH_2CH_2{-}\underset{NH_2}{\overset{H}{C}}{-}COOH$$

Methionine — Methionine sulfoxide

Still another amino acid, tryptophan, is known to react with ozone. Tryptophan, which does not contain sulfur, undergoes a ring-opening oxidation.

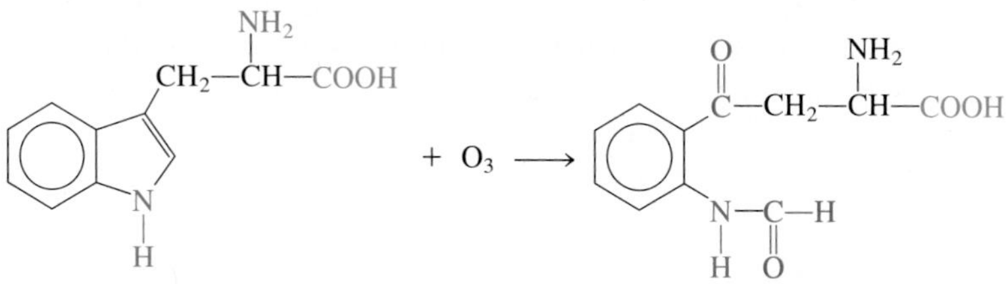

Tryptophan + O_3 ⟶ Oxidation product

No doubt oxidizing agents can break bonds in many of the chemical substances in a cell. Such powerful agents as ozone are more likely to make an indiscriminate attack than to react in a highly specific way.

SECTION 20.3

Blood Agents

Certain chemical substances block the transport of oxygen in the bloodstream and prevent the oxidation of metabolites by oxygen in the cells. All act on the iron atoms in complex protein molecules. Probably the best known of these blood agents is carbon monoxide. Recall that carbon monoxide binds tightly to the iron atom in hemoglobin, blocking the transport of oxygen (Chapter 12).

Nitrates, which occur in dangerous amounts in the groundwater in some agricultural areas (Chapter 13), also serve to diminish the ability of hemoglobin to carry oxygen. Nitrates are reduced by microorganisms in the digestive tract to nitrites.

$$\underset{\text{Nitrate ion}}{NO_3^-} \longrightarrow \underset{\text{Nitrite ion}}{NO_2^-}$$

The nitrite ions oxidize the iron atoms in hemoglobin from Fe^{2+} to Fe^{3+}. The resulting compound, called **methemoglobin**, is incapable of carrying oxygen. The resulting oxygen deficiency disease is called methemoglobinemia. In infants, this disease is called blue baby syndrome.

Hemoglobin is bright red and is responsible for the red color of the blood. Methemoglobin is brown. During cooking, red meat turns brown because of the oxidation of hemoglobin to methemoglobin. Dried bloodstains turn brown for the same reason.

SECTION 20.4

Cyanide: Agent of Death

Cyanide is one of the most notorious poisons in both fact and fiction. It acts almost instantaneously and only a minute amount constitutes a lethal dose. The average fatal dose is only 50 or 60 mg. Cyanide is used as gaseous hydrogen cyanide ($H—C{\equiv}N$) and as solid salts that contain the cyanide ion ($C{\equiv}N^-$). Hydrogen cyanide is used (with great care by specially trained experts) to exterminate insects and rodents in the holds of ships, in warehouses, in railway cars, and on citrus and other fruit trees. Sodium cyanide (NaCN) is used to extract gold and silver from ores. It also is used in electroplating baths. Hydrogen cyanide is generated easily enough from the sodium salt by treatment with an acid.

$$NaCN + H_2SO_4 \longrightarrow HCN + NaHSO_4$$

Some people speculate that life on Earth may have developed from molecules formed by the polymerization of hydrogen cyanide. This gas is readily formed when an electric discharge is passed through a mixture of gases designed to simulate the Earth's early atmosphere. Many of the amino acids and the organic bases necessary to form DNA and RNA can be derived by the polymerization of HCN with the rearrangement of only a few atoms. Glycine, the simplest amino acid, can be derived from aminomalononitrile (a trimer of HCN) by hydrolysis, decarboxylation, and a second hydrolysis.

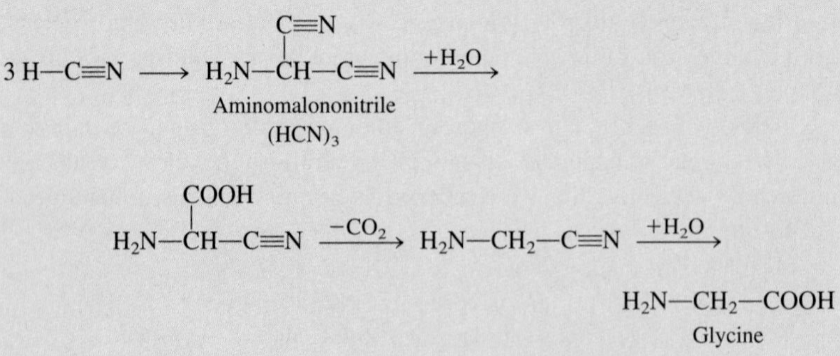

$$3\,H—C{\equiv}N \longrightarrow H_2N—\underset{\text{Aminomalononitrile } (HCN)_3}{CH(C{\equiv}N)—C{\equiv}N} \xrightarrow{+H_2O}$$

$$H_2N—CH(COOH)—C{\equiv}N \xrightarrow{-CO_2} H_2N—CH_2—C{\equiv}N \xrightarrow{+H_2O} \underset{\text{Glycine}}{H_2N—CH_2—COOH}$$

Adenine is a pentamer of HCN. Would it not indeed be ironic if life arose from compounds made from such a deadly poison?

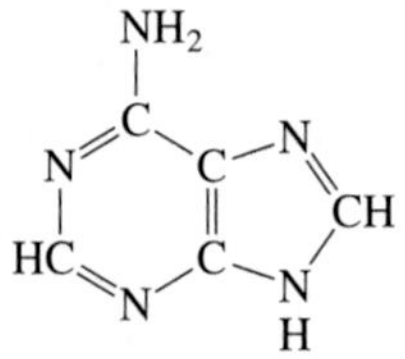

Adenine
$(HCN)_5$

Unlike carbon monoxide, cyanide does not react with hemoglobin. Instead, it blocks the oxidation of glucose inside the cell by forming a stable complex with the oxidation enzymes. The enzymes, called cytochrome oxidases, contain iron and copper atoms. They normally act by providing electrons for the reduction of oxygen in the cell. Cyanide ties up those mobile electrons, rendering them unavailable for the reduction process. Thus, cyanide brings an abrupt end to cellular respiration, causing death in a matter of minutes.

Any antidote for cyanide poisoning must be administered quickly. Sodium thiosulfate (the ''hypo'' used in developing photographic film) is the usual treatment. A sulfur atom is transferred from the thiosulfate ion to the cyanide ion, converting cyanide to relatively innocuous thiocyanate.

$$\underset{\text{Cyanide}}{CN^-} + \underset{\text{Thiosulfate}}{S_2O_3^{2-}} \longrightarrow \underset{\text{Thiocyanate}}{SCN^-} + \underset{\text{Sulfite}}{SO_3^{2-}}$$

Unfortunately, few victims of cyanide poisoning survive long enough to be treated.

Deaths from house fires often are caused by toxic gases. Carbon monoxide is formed in smoldering fires, but hydrogen cyanide (formed by burning plastics, Chapter 10) is perhaps equally important. A medical journal, *The Lancet* (20 August 1988, p. 457), recommends that fire department rescue crews carry spring loaded hypodermic syringes filled with sodium thiosulfate solution for emergency treatment of victims of ''smoke inhalation.''

SECTION 20.5

Make Your Own Poison: Fluoroacetic Acid

Although the body generally acts to detoxify poisons, there are notable exceptions in which the body converts an essentially harmless chemical into a deadly poison. Fluoroacetic acid is one such compound.

The body cells use acetic acid to produce citric acid. The citric acid is then broken down in a series of steps, most of which release energy. When fluoroacetic acid is ingested, it is incorporated into fluorocitric acid. The latter effectively blocks the citric acid cycle by tying up the enzyme that acts on citric acid. Thus, the energy-producing mechanism of the cell is shut off and death rapidly ensues.

Sodium fluoroacetate (Compound 1080) is used to poison rats and predatory animals. It is not selective, and thus is dangerous to humans, pets, and other animals. Sodium fluoroacetate is used by ranchers to poison coyotes, eagles, and other animals suspected of preying on sheep and cattle. Such use drove the eagles nearly to extinction. As a result, the poisoning of predators with sodium fluoroacetate and other deadly chemicals was banned on federal land.

Fluoroacetic acid occurs in nature in a highly poisonous plant called *gifblaar* that grows in South Africa. For years natives have used the plant to poison the tips of their arrows.

CH_3COOH

Acetic acid

$F{-}CH_2COOH$

Fluoroacetic acid

$$\begin{array}{c} CH_2{-}COOH \\ | \\ HO{-}C{-}COOH \\ | \\ CH_2{-}COOH \end{array}$$

Citric acid

$$\begin{array}{c} F{-}CH{-}COOH \\ | \\ HO{-}C{-}COOH \\ | \\ CH_2{-}COOH \end{array}$$

Fluorocitric acid

$F{-}CH_2COO^-Na^+$

Sodium fluoroacetate

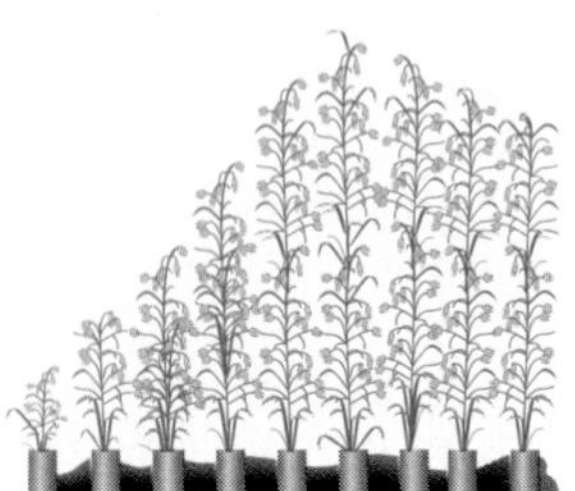

Figure 20.5 The effect of copper on the height of oat seedlings. From left to right, the quantities of copper present are 0, 3, 6, 10, 20, 100, 500, 2000, and 3000 μg/L. Plants on the left show the effect of a deficiency; those on the right exhibit copper toxicity. Plants in the middle have an optimum amount of copper. [*After C. S. Piper,* Journal of Agricultural Science, *32 (1942), 143.*]

SECTION 20.6

Heavy Metal Poisons

People have long used a variety of metals in industry and agriculture and around the home. Most metals and their compounds show some toxicity when ingested in large amounts. Even the essential mineral nutrients can be toxic when taken in excessive amounts (Figure 20.5). Quite often, too little of a metal (deficiency) can be as dangerous as too much (toxicity). For example, the average adult requires 10 to 18 mg of iron every day. If less is taken in, the person suffers from anemia. Yet an overdose can cause vomiting, diarrhea, shock, coma, and even death. As few as 10 to 15 tablets containing 5 grains (324 mg) each of iron (as $FeSO_4$) have been fatal to children.

It is not known exactly how iron poisoning works. The heavy metals—those near the bottom of the periodic table—exert their action primarily by inactivating enzymes. It is well known in inorganic chemistry that heavy metal ions react with hydrogen sulfide to form insoluble sulfides.

$$\underset{\text{Lead ion (in solution)}}{Pb^{2+}} + H_2S \longrightarrow \underset{\text{Lead sulfide (insoluble)}}{PbS} + 2\,H^+$$

$$\underset{\text{Mercury ion (in solution)}}{Hg^{2+}} + H_2S \longrightarrow \underset{\text{Mercury sulfide (insoluble)}}{HgS} + 2\,H^+$$

Most enzymes have amino acids with sulfhydryl (—SH) groups at or near the active sites. Heavy metal ions tie up these groups, rendering the enzymes inactive.

$$\underset{\text{Enzymes with sulfhydryl groups}}{\begin{matrix}\text{SH}\\ \\ \text{SH}\end{matrix}} + Hg^{2+} \longrightarrow \underset{\text{Enzyme (inactive)}}{\begin{matrix}\text{S}\\ |\\ \text{Hg}\\ |\\ \text{S}\end{matrix}} + 2\,H^+$$

SECTION 20.7

Quicksilver → Slow Death

Mercury (Hg) is a most unusual metal. It is the only common metal that occurs as a liquid at room temperature. This bright, silvery, dense liquid—formerly known as quicksilver—has long held the fascination of people. Children

Arsenic is not a metal, but it has some metallic properties. In commercial poisons, arsenic usually is found as arsenate (AsO_4^{3-}) or arsenite (AsO_3^{3-}) ions. These also render enzymes inactive by tying up sulfhydryl groups.

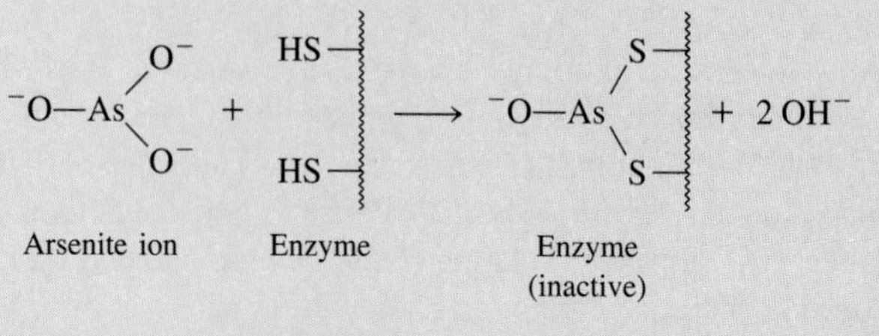

$$^{-}O{-}As(O^{-})_2 + (HS)_2\text{-Enzyme} \longrightarrow {}^{-}O{-}As(S)_2\text{-Enzyme} + 2\,OH^{-}$$

Arsenite ion   Enzyme   Enzyme (inactive)

Organic compounds containing arsenic are well known. One such compound, arsphenamine, was the first antibacterial. It once was used widely in the treatment of syphilis.

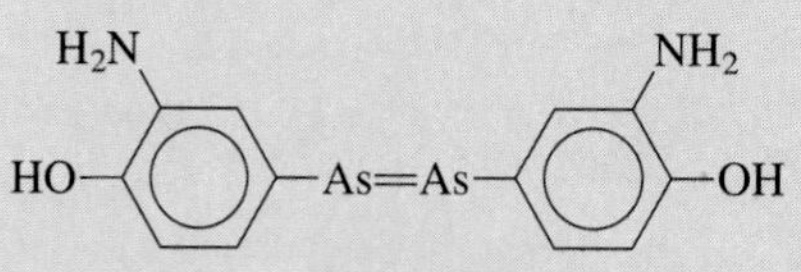

$$HO{-}C_6H_3(NH_2){-}As{=}As{-}C_6H_3(NH_2){-}OH$$

Arsphenamine

Another arsenic compound was developed as a blister agent for use in chemical warfare. This agent was first synthesized by (and named for) W. Lee Lewis. The United States started large-scale production of Lewisite in 1918. Fortunately, World War I came to an end before this gas could be employed.

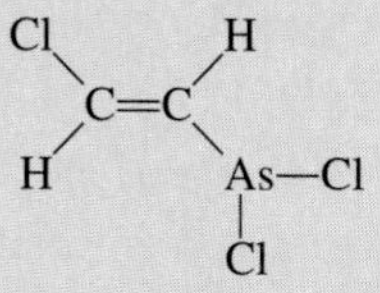

$$ClHC{=}CH{-}AsCl_2$$

Lewisite

sometimes play with the mercury from a broken thermometer. Mercury vapor is hazardous. An open container of mercury or a spill on the floor can put enough mercury vapor into the air to exceed the established maximum safe level by a factor of 200.

Mercury presents a hazard to those who work with it (Figure 20.6). Dentists use it to make amalgams for filling teeth. Laboratory workers use mercury and its compounds in a variety of ways. Farmers use seeds treated with compounds of mercury. Since mercury is a cumulative poison (it takes the

Figure 20.6 Painters once used a variety of toxic materials. Popular yellow paints had cadmium sulfide pigments. Red pigments included mercury(II) sulfide and cadmium selenide. These children are painting with relatively safe paints colored with organic dyes. [*Robert Semeniuk/The Stock Market.*]

body about 70 days to rid itself of *half* a given dose), chronic poisoning is a real threat to those continually exposed.

CH_2—CH—CH_2
| | |
OH SH SH

BAL

Fortunately, there are antidotes available for mercury poisoning. British scientists, searching for an antidote for the arsenic-containing war gas Lewisite, came up with a compound effective for heavy metal poisoning as well. The compound, a derivative of glycerol, came to be known as BAL (British Anti-Lewisite). BAL acts by chelating (Greek *chela:* ''claw'') the metal ion. Thus tied up, the mercury cannot attack vital enzymes.

[CH_2—CH—CH_2
| | |
S S OH
\ /
Hg
/ \
OH S S
| | |
CH_2—CH—CH_2] $^{2-}$

Mercury atom chelated by two BAL molecules

Now for the bad news. The symptoms of mercury poisoning may not show up for several weeks. By the time the symptoms—loss of equilibrium, sight,

▼

Metallic mercury seems not to be very toxic when ingested (swallowed). Most of it passes through the system unchanged. Indeed, there are numerous reports of mercury being given orally in the eighteenth and nineteenth centuries as a remedy for obstruction of the bowels. Doses varied from a few ounces to a pound or more. Reports from the poison control center of the New York City Department of Health confirm the low toxicity of metallic mercury taken orally. Eighteen incidents without serious effects were noted over a 2-year period.

When inhaled, however, mercury vapor is quite hazardous, particularly when exposure occurs over a long period of time. Such chronic exposure usually involves mining, extraction, or regular occupational use of the metal. The body seems able to convert the inhaled mercury, by some as yet unknown mechanism, to Hg^{2+} ions. All the compounds of mercury, except those that are essentially insoluble in water, are poisonous no matter how they are administered.

feeling, hearing—are recognizable, extensive damage has already been done to the brain and the nervous system. Such damage is largely irreversible. The BAL antidote is effective only when a person knows that he or she has been poisoned and seeks treatment right away.

SECTION 20.8

Lead in the Environment

Compounds of the element lead (Pb) are widespread in the environment. This reflects the many uses we have for this soft, dense, corrosion-resistant metal and its compounds. Lead (as Pb^{2+}) is present in many foods, generally in concentrations of less than 0.3 ppm. Lead (again as Pb^{2+}) also gets into our drinking water (up to 0.1 ppm) from lead-sealed pipes. Lead compounds, mainly from automobiles that used to burn leaded gasoline, even permeate the air we breathe. (This exposure has decreased dramatically as we have switched to unleaded gasoline.)

Lead compounds are quite toxic. Metallic lead is generally converted to Pb^{2+} in the body. So, with all that lead, why aren't we dead? We can excrete about 2 mg of lead per day. Our intake from air, food, and water is generally less than that, so generally we do not accumulate toxic levels. If intake exceeds excretion, however, lead builds up in the body and chronic irreversible lead poisoning results.

The U.S. Environmental Protection Agency estimates that lead poisoning contributes to 123,000 cases of hypertension and to 680,000 miscarriages, and it retards the growth of 7000 children each year. These problems add \$635 million to health care costs annually.

Lead poisoning is a major problem with children, particularly those in areas with old, run-down buildings. Some children develop a craving that causes them to eat unusual things. Children with this syndrome (called **pica**) eat chips of peeling, lead-based paints. These children probably also pick up lead compounds from the streets, where they have been deposited by automobile exhausts. They also may get lead from canned milk and other sources. In all, thousands of children suffer from lead poisoning each year. Such poisoning often leads to mental retardation and neurological disorders through damage to the brain and nervous system.

Lead poisoning usually is treated with a combination of BAL and another chelating agent called EDTA (ethylenediaminetetraacetic acid).

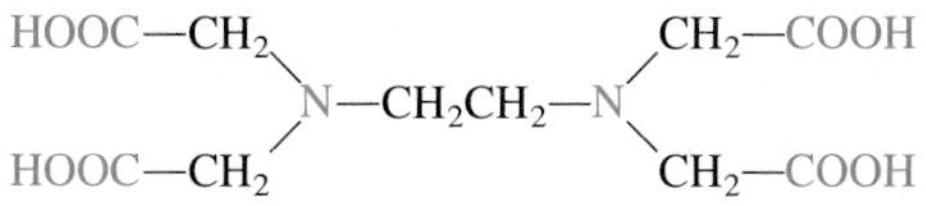

$$(HOOC{-}CH_2)_2N{-}CH_2CH_2{-}N(CH_2{-}COOH)_2$$

EDTA

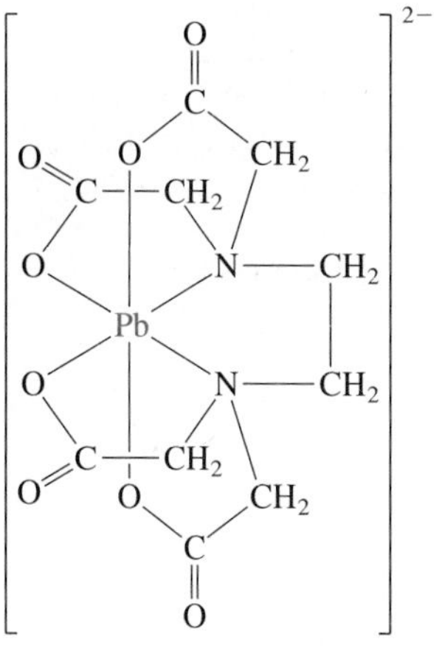

The calcium salt of EDTA is administered intravenously. In the body, calcium ions are displaced by lead ions, which the chelate binds more tightly.

$$CaEDTA^{2-} + Pb^{2+} \longrightarrow PbEDTA^{2-} + Ca^{2+}$$

The lead–EDTA complex is excreted.

Lead–EDTA complex

As with mercury poisoning, the neurological damage done by lead compounds is essentially irreversible. Treatment must be begun early to be effective.

SECTION 20.9

Cadmium: The "Ouch-Ouch" Disease

Cadmium (Cd) is used less extensively than lead or mercury, but it too has caused major catastrophes. Cadmium is used widely in alloys, in the electronics industry, in nickel–cadmium rechargeable batteries, and many other applications. Cadmium poisoning leads to loss of calcium ions (Ca^{2+}) from the bones, leaving them brittle and easily broken. It also causes severe abdominal pain, vomiting, diarrhea, and a choking sensation.

The most notable cases of cadmium poisoning occurred along the upper Zintsu River in Japan. The metal (as Cd^{2+}) entered the water in milling wastes from a mine. Downstream the water was used by farm families for drinking, cooking, and other household uses. It was also used to irrigate the rice fields. Soon the farm folk began to suffer from a strange, painful malady that became known as *itai-itai,* the "ouch-ouch" disease. Over 200 people died, and thousands were disabled.

SECTION 20.10

More Chemistry of the Nervous System

It is interesting to note that manic-depressive people are overly sensitive to acetylcholine; they seem to have too many receptors. This may account for their wild swings in mood.

Some poisons—among them the most toxic substances known—act upon the nervous system. Signals are shuttled across synapses between cells by chemical substances called neurotransmitters. One such chemical messenger is acetylcholine. It is thought to activate the next cell by fitting a specific receptor and thus changing the permeability of the cell membrane to certain ions.

Once acetylcholine has carried the impulse across the synapse, it is rapidly hydrolyzed to acetic acid and choline. This reaction is catalyzed by an enzyme, cholinesterase.

$$\underset{\text{Acetylcholine}}{CH_3\overset{O}{\overset{\|}{C}}OCH_2CH_2{-}\overset{CH_3}{\overset{|}{\underset{CH_3}{\underset{|}{N^+}}}}{-}CH_3} + H_2O \xrightarrow{\text{cholinesterase}} \underset{\text{Acetic acid}}{CH_3\overset{O}{\overset{\|}{C}}{-}OH} + \underset{\text{Choline}}{HOCH_2CH_2{-}\overset{CH_3}{\overset{|}{\underset{CH_3}{\underset{|}{N^+}}}}{-}CH_3}$$

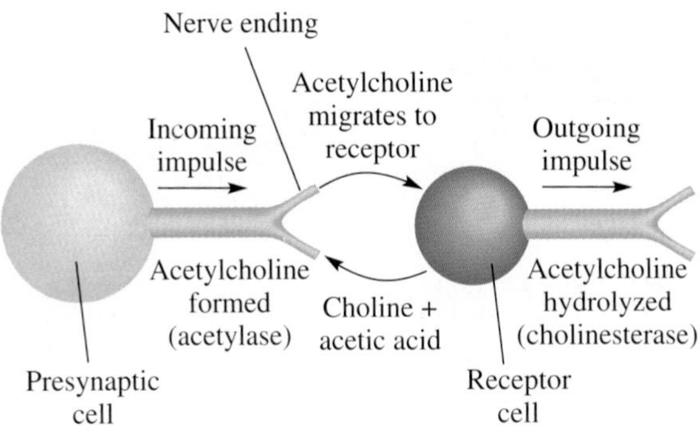

Figure 20.7 The acetylcholine cycle.

Choline is relatively inactive. After this breakdown, the receptor cell releases the hydrolysis products and is then ready to receive further impulses. Other enzymes, such as acetylase, convert the acetic acid and choline back to acetylcholine, completing the cycle (Figure 20.7).

People with Alzheimer's disease are deficient in the enzyme acetylase. They produce too little acetylcholine for proper brain function.

SECTION 20.11

Nerve Poisons: Stopping the Acetylcholine Cycle

Various chemical substances can disrupt the acetylcholine cycle at three different points. Botulin, the deadly toxin given off by *Clostridium botulinum* (an anaerobic bacterium) in improperly processed canned food (Chapter 16), blocks the synthesis of acetylcholine. With no messenger formed, no messages are carried. Paralysis sets in and death occurs, usually by respiratory failure.

Curare, atropine, and some of the local anesthetics (Chapter 19) act by blocking the receptor sites. In this case, the message is sent but not received. In the case of local anesthetics, this can be good for pain relief in a limited area. These drugs, too, can be fatal in sufficient quantity.

Anticholinesterase Poisons

The third category of poisons, called anticholinesterase poisons, inhibit the enzyme cholinesterase (Figure 20.8). The organic phosphorus insecticides (Chapter 16) are well-known nerve poisons. The phosphorus–oxygen linkage is thought to bond tightly to cholinesterase. This blocks the breakdown of acetylcholine. The acetylcholine therefore builds up, causing the receptor nerves to fire repeatedly. This overstimulates the muscles, glands, and organs. The heart beats wildly and irregularly. The victim goes into convulsions and dies quickly.

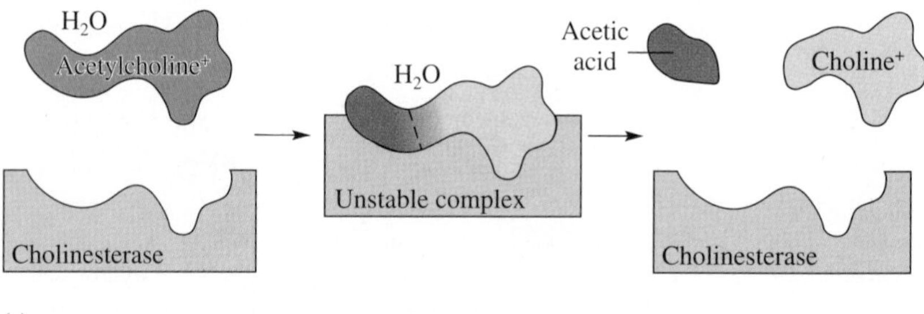

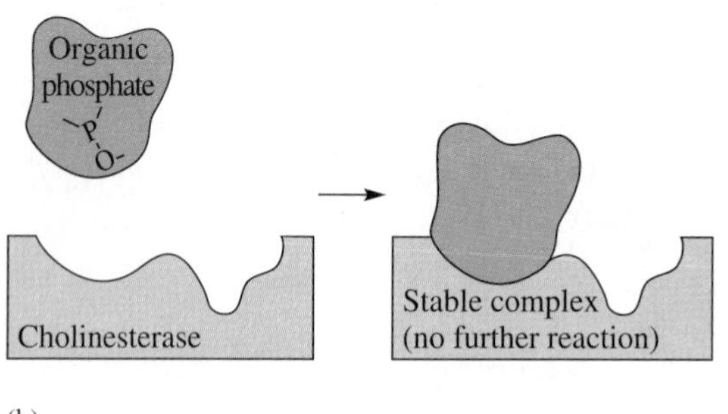

Figure 20.8 (a) Cholinesterase catalyzes the hydrolysis of acetylcholine to acetic acid and choline. (b) An organic phosphate ties up cholinesterase, preventing it from breaking down acetylcholine.

Organic Phosphates: Insecticides and Weapons of War

While doing research on organic phosphates as possible insecticides during World War II, German scientists discovered some extremely toxic compounds with a frightening potential for use in warfare. The Russians captured a German plant that manufactured a compound called Tabun (designated as agent GA by the U.S. Army). The Soviets dismantled the factory and moved it to Russia. The U.S. Army captured some of the stock, however. Presumably, the gas has been produced and stocked by both Germany and Russia.

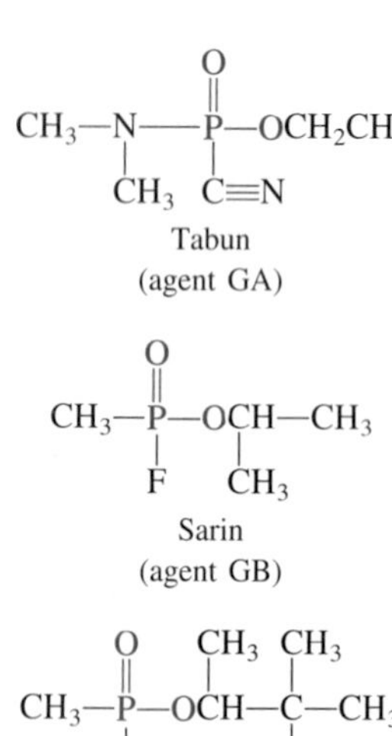

The United States has developed other nerve poisons. One is called Sarin (agent GB). Sarin is four times as toxic as Tabun. It also has the ''advantage'' of being odorless. Tabun has a fruity odor. Another organophosphate nerve poison is Soman (agent GD). It is moderately persistent, whereas Tabun and Sarin are generally nonpersistent.

Note again the variation in structure from one compound to the next. The approach for developing chemical warfare agents is much the same as the approach for developing drugs—find one that works and then synthesize and test structural variations.

The nerve gases are among the most toxic synthetic chemicals known. They kill by inhalation or by absorption through the skin. They result in the complete loss of muscular coordination and subsequent death by cessation of breathing. The usual antidote is atropine injection and artificial respiration. Without the antidote, death may occur in 2 to 10 minutes.

The similarity of insecticides such as malathion and parathion (Chapter 16) to these nerve gases should be recognized. Though somewhat less toxic than the nerve gases, these phosphorus-based insecticides and others like them should be used with extreme caution. Even the relatively safe (to mammals) chlorinated hydrocarbon pesticides act as nerve poisons. Acute DDT poisoning causes tremors, convulsions, and cardiac or respiratory failure. Chronic exposure to DDT leads to the degeneration of the central nervous system. Other chlorinated compounds, such as the PCBs (Chapter 10), act in a similar manner.

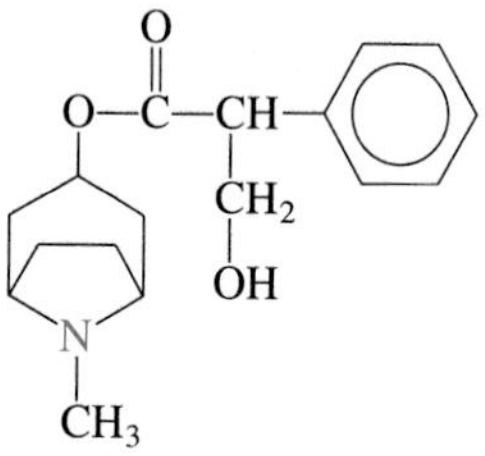

Nerve poisons aren't all bad. Even the nerve gases, with their tremendous potential for death and destruction, are not nearly as toxic as the natural toxin botulin. More importantly, the nerve poisons have helped us gain an understanding of the chemistry of the nervous system. It is that knowledge that enables scientists to design antidotes for the nerve poisons. In addition, our increased understanding should contribute to progress along more positive lines—in the control of pain, for example. Even deadly botulin has found use in medicine. To correct crossed eyes, tiny amounts of botulin are injected into the optic muscles, temporarily blocking muscles that are pulling too hard and causing the eyes to cross.

Atropine

SECTION 20.12

Your Liver: A Detox Tank

The human body can handle a moderate amount of some poisons. The liver is able to detoxify some compounds by oxidation, reduction, or coupling with amino acids or other normal body chemicals.

Perhaps the most common route is oxidation. Ethanol (Chapter 9) is detoxified by oxidation to acetaldehyde which in turn is oxidized to acetic acid, a normal constituent of cells. The acetic acid is then oxidized to carbon dioxide and water.

$$CH_3CH_2OH \longrightarrow CH_3\overset{\overset{\displaystyle O}{\|}}{C}{-}H \longrightarrow CH_3\overset{\overset{\displaystyle O}{\|}}{C}{-}OH \longrightarrow CO_2 + H_2O$$

Highly toxic nicotine from tobacco is detoxified by oxidation to cotinine.

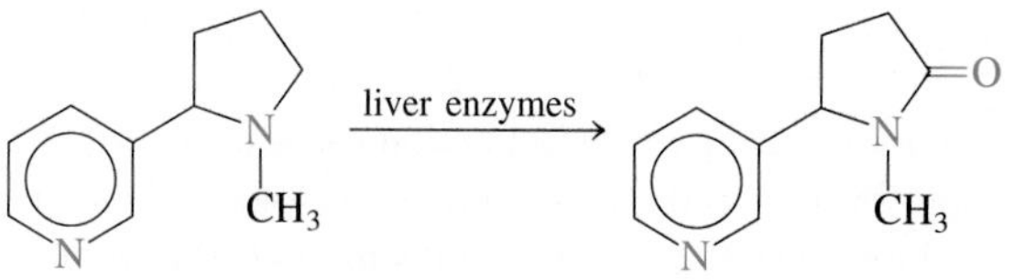

liver enzymes

Nicotine

Cotinine

Cotinine is less toxic than nicotine. The added oxygen atom also makes cotinine more water soluble, and thus more readily excreted in the urine, than nicotine.

The liver is equipped with a system of enzymes, called P-450, that oxidize fat-insoluble substances that are likely to be retained in the body, into water soluble ones that are readily excreted. It can also conjugate compounds with amino acids. For example toluene is essentially insoluble in water. The P-450 enzymes oxidize toluene to more soluble benzoic acid. The latter is then coupled with the amino acid glycine to form hippuric acid, which is still more soluble and is readily excreted.

$$\underset{\text{Toluene}}{C_6H_5\text{—}CH_3} \xrightarrow{\text{oxidation}} \underset{\text{Benzoic acid}}{C_6H_5\text{—}\overset{O}{\overset{\|}{C}}\text{—}OH} \xrightarrow[\text{(Glycine)}]{H_2N\text{—}CH_2\text{—}\overset{O}{\overset{\|}{C}}\text{—}OH} \underset{\text{Hippuric acid}}{C_6H_5\text{—}\overset{O}{\overset{\|}{C}}\text{—}NH\text{—}CH_2\text{—}\overset{O}{\overset{\|}{C}}\text{—}OH}$$

An antidote for methyl alcohol poisoning is ethyl alcohol. Ethyl alcohol is administered intravenously in an attempt to "load up" the liver enzymes with ethyl alcohol and thus to block the oxidation of methyl alcohol until the compound can be excreted.

It should be pointed out that the liver enzymes just oxidize, reduce, or conjugate. The end product is not always less toxic. For example, methanol is oxidized to a more toxic form, formaldehyde. It is probably the formaldehyde that reacts with the protein in the cells to cause blindness, convulsions, respiratory failure, and death.

$$CH_3OH \xrightarrow{\text{liver enzymes}} H\text{—}\overset{O}{\overset{\|}{C}}\text{—}H$$

Note that the same enzymes that oxidize the alcohols deactivate the male hormone testosterone. Buildup of these enzymes in a chronic alcoholic leads to a more rapid destruction of testosterone. Thus, we have the mechanism for alcoholic impotence, one of the well-known characteristics of the disease.

Benzene, because of its general inertness in the body, is not acted upon until it gets to the liver. There it is slowly oxidized to an epoxide.

$$C_6H_6 \xrightarrow[\text{the liver}]{\text{oxidation in}} C_6H_6O \text{ (epoxide)}$$

The epoxide is a highly reactive molecule that can attack certain key proteins. The damage done by this epoxide sometimes results in leukemia.

Carbon tetrachloride, CCl_4, is also quite inert in the body. But when it reaches the liver, it is converted to the very reactive trichloromethyl free radical $Cl_3C\cdot$, which in turn attacks the unsaturated fatty acids in the body. This action can trigger cancer.

Table 20.1 LD_{50} Values for Some Common Substances

Substance	Test Animal	LD_{50} (g/kg)
Ethyl alcohol	Rats	10.3
Vitamin B_1	Mice	8.2
Sodium chloride	Rats	3.75
Aspirin	Mice	1.5
Acetaminophen	Mice	0.34
Nicotine	Mice	0.23
Caffeine	Mice	0.13

Table 20.2 Estimated LD_{50} Values for Some Highly Lethal Poisons

Substance	LD_{50} (mg/kg)
Sodium cyanide (NaCN)	15
Arsenic trioxide (As_2O_3)	15
Aflatoxin B	10
Rotenone	3
Strychnine	0.5
Muscarine	0.2
Tetanus toxin	0.000005
Botulin	0.00000003

SECTION 20.13

The Lethal Dose

There are many substances on Earth that are poisonous, but some are much more poisonous than others. In order to quantify toxicity, scientists use the term LD_{50} to indicate a dosage that will kill 50% of a population of test animals. The reason for choosing the LD_{50} is that some animals are unusually strong and can survive fairly large doses of a given poison, while others are more susceptible and will die from much smaller doses. The dose that will kill 50% is the average lethal dose.

Usually LD_{50} values are given in terms of weight of poison per unit body weight of the test animal. It is assumed that the material is given orally unless specified otherwise. (Intravenous values can be quite different from those measured with oral ingestion.) Although the LD_{50} values are useful in comparing the relative toxicities of various substances, they are not necessarily the same values that would be measured for humans.

The substances listed in Table 20.1 are not usually thought of as poisons. Yet there is a dose for each one that will kill half of a given population of test animals. The larger the LD_{50} value, the less toxic the substance. Nicotine and caffeine are the most toxic substances listed in Table 20.1 because they have the smallest LD_{50} values.

The substances in Table 20.2 are highly lethal poisons. Notice that the doses are given in milligrams per kilogram of body weight, whereas the values in Table 20.1 are in grams per kilogram. The most highly poisonous substance of all is the botulinum toxin.

SECTION 20.14

Chemical Carcinogens: The Slow Poisons

Carcinogens cause the growth of tumors. A tumor is an abnormal growth of new tissue. Tumors may be either benign or malignant. **Benign tumors** are characterized by slow growth; they often regress spontaneously, and they do not invade neighboring tissues. **Malignant tumors** may grow slowly or rapidly, but their growth is generally irreversible. They often are called **cancers**. Malignant growths invade and destroy neighboring tissues. Actually, cancer is not a single disease. It is a catchall term for over a hundred different afflictions. Many are not even closely related to each other.

What Causes Cancer?

There are 400,000 cancer deaths each year in the United States. Of these 150,000 are related to cigarette smoking. Another 150,000 are related to our diet.

The World Health Organization estimates that 80 to 90% of the cases are caused by environmental factors, 10 to 20% by genetic factors and (perhaps) viruses. Included most prominently among those "environmental" causes are cigarette smoking (40%), dietary factors (25 to 30%), and occupational exposure (10%). That leaves 10 to 15% that may be caused by environmental *pollutants*. If you read the newspapers or watch television, you may get the idea that "everything" causes cancer. Even among those chemicals that have been suspect, however, many cannot be shown to be carcinogenic. Only about 30 chemical compounds have been identified as human carcinogens. Another 300 or so have been shown to cause cancer in laboratory animals. Some of the 300 are widely used, though.

Not all carcinogens are synthetic chemicals. Some, such as safrole in sassafras and the aflatoxins produced by molds on foods, occur naturally. Some researchers estimate that 99.99% of all carcinogens that we ingest are natural ones. Plants produce compounds to protect themselves from fungi, insects, and higher animals, including humans. Some of the compounds are carcinogens that are found in mushrooms, basil, celery, figs, mustard, pepper, fennel, parsnips, and citrus oils—almost every place that a curious chemist looks. Carcinogens are also produced during cooking and as products of normal metabolism. We must have some way of protecting ourselves from carcinogens.

How Cancers Develop

How do chemicals cause cancer? Their mechanisms of action are probably as varied as their chemical structure. Some carcinogens chemically modify DNA, thus scrambling the code for replication and for the synthesis of proteins. For example, aflatoxin B is known to bind to guanine residues in DNA. Just how this initiates cancer, however, is not known for sure.

There is a genetic component to the development of many forms of cancer. Certain genes, called oncogenes, seem to trigger or sustain the processes that convert normal cells to cancerous ones. Oncogenes arise from ordinary genes that regulate cell growth and cell division. These oncogenes can be activated by chemical carcinogens, radiation, or perhaps some viruses. It seems that more than one oncogene must be turned on, perhaps at different stages of the process, before a cancer develops. We also have suppressor genes that ordinarily prevent the development of cancers. These genes must be inactivated before a cancer develops. Suppressor gene inactivation can occur through mutation, alteration, or loss. In all, 10 or 15 mutations may be required in a cell before it turns cancerous.

There is hope that someday suppressor genes can be produced through genetic engineering and used in therapy.

Chemical Carcinogens

A variety of widely different chemical compounds are carcinogenic. We could not attempt to cover all the types of carcinogens here, but we do concentrate on a few major classes.

Some of the more notorious carcinogens are the polycyclic aromatic hydrocarbons (Chapter 9), of which 3,4-benzpyrene is perhaps the best known. Carcinogenic hydrocarbons are formed during the incomplete burning of nearly any organic material. They have been found in charcoal-grilled meats, cigarette smoke, automobile exhausts, coffee, burnt sugar, and many other materials. Not all polycyclic aromatic hydrocarbons are carcinogenic. There are strong correlations between carcinogenicity and certain molecular sizes and shapes. The mechanism of their action is currently under intense investigation. It already appears rather certain that the actual carcinogens are not the hydrocarbons themselves but the oxidation products formed in the liver.

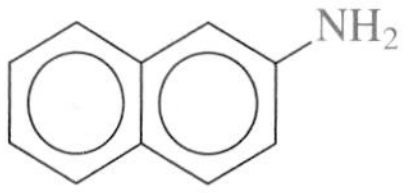

Another important class of carcinogens are the aromatic amines. Two prominent ones are β-naphthylamine and benzidine. These compounds once were used widely in the dye industry. They were responsible for a high incidence of bladder cancer among the workers whose jobs brought them into prolonged contact with the compounds.

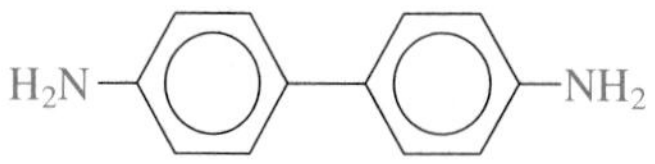

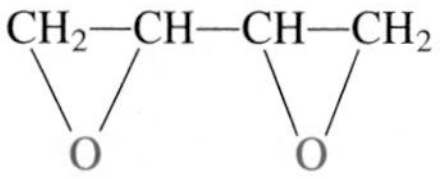

NH_2

β-Naphthylamine

H_2N — — NH_2

Benzidine

$CH_2—CH—CH—CH_2$
O O

Bis(epoxy)butane

H_2C
N—$(CH_2)_{11}CH_3$
H_2C

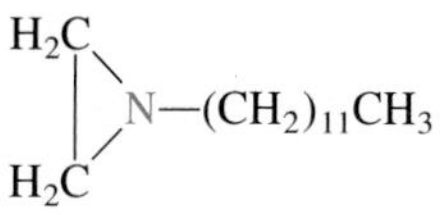

N-Laurylethyleneimine

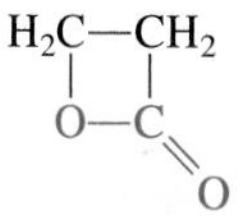

$H_2C—CH_2$
O—C
O

β-Propiolactone

Several aminoazo dyes have been shown to be carcinogenic. An interesting example is 4-dimethylaminoazobenzene. This compound is also known as "butter yellow." It was used widely as a coloring for butter and oleomargarine before its carcinogenicity became known.

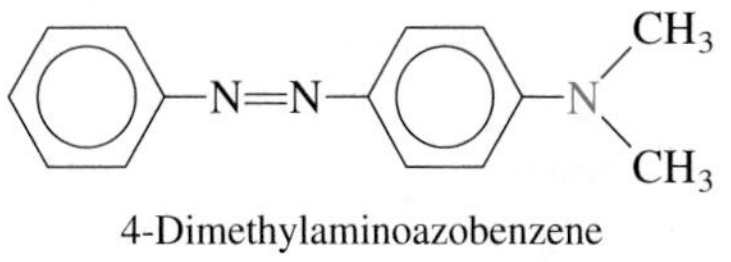

—N=N— —N(CH_3)(CH_3)

4-Dimethylaminoazobenzene

Not all carcinogens are aromatic. Prominent among the aliphatic (nonaromatic) ones are dimethylnitrosamine (Chapter 16) and vinyl chloride (Chapter 10). Others include three- and four-membered heterocyclic rings containing nitrogen or oxygen. The epoxides and derivatives of ethyleneimine are examples. Others are cyclic esters called lactones.

Keep in mind that this list is not all-inclusive. Rather, its purpose is to give you an idea of the kinds of compounds that have tumor-inducing properties—and the list grows almost daily as the results of research are released.

Anticarcinogens

There are many natural carcinogens in our food. So why don't we all get cancer? Perhaps there are **anticarcinogens** in our food as well. Fiber is believed to protect against colon cancer. The food additive BHT (butylated hydroxytoluene) may give protection against stomach cancer. And certain vitamins have been shown to have anticarcinogenic activity.

The vitamins that are antioxidants (vitamin C, vitamin E, and beta-carotene, a precursor to vitamin A) seem to exhibit the strongest anticancer properties. A diet rich in cruciferous vegetables (cabbage, broccoli, brussels sprouts, kale, and cauliflower) has been shown to reduce the incidence of cancer both in animals and in human population groups.

A number of studies with vitamins A, C, and E, used separately or in combination, have confirmed the fact that each of these vitamins has some ability to lower the incidence of cancer. There may be still other anticarcinogens in our food that have not yet been identified.

SECTION 20.15

Testing for Carcinogens: Three Ways

How do we know that a chemical causes cancer? Obviously, we can't experiment on humans to see what happens. That leaves us with no way to prove beyond doubt that a chemical does or does not cause cancer in humans. There are three ways, however, to gain evidence against a compound: bacterial screening for mutagenesis, animal tests, and epidemiological studies.

The Ames Test: Bacterial Screening

The quickest and cheapest way to find out whether or not a substance may be carcinogenic is to use the screening test developed by Bruce N. Ames of the University of California at Berkeley. The Ames test is a simple laboratory procedure that can be carried out in a petri dish. It assumes that most carcinogens are also mutagens, altering the genes in some way. (This usually seems to be the case. About 90% of the chemicals that appear on a list of either mutagens or carcinogens are found on the other list as well.)

The Ames test uses a special strain of Salmonella bacteria that have been modified so that they require histidine as an essential amino acid. The bacteria are placed in an agar medium containing all nutrients except histidine. Incubating the mixture in the presence of a mutagenic chemical causes the bacteria to mutate, so that they no longer require histidine and can grow like normal bacteria. Growth of bacterial colonies in the petri dish means that the chemical added was a mutagen, and probably also a carcinogen.

Animal Testing

Chemicals suspected of being carcinogens can be tested on animals. Tests using low dosages on millions of rats would cost too much, so tests usually are done by using large doses on 30 or so rats. An equal number of rats serve as controls. The control group is exposed to the same diet and environment as the experimental group, except that the control group does not get the suspected carcinogen. A higher incidence of cancer in the experimental animals than in the controls indicates that the compound is carcinogenic.

Animal studies cost about $1 million each and take 2 years.

Animal tests are not conclusive. Humans usually are not exposed to comparable doses; there may be a threshold below which a compound is not carcinogenic. Further, human metabolism is different from the metabolisms of the test animals. The carcinogen might be active in the rat but not in humans (or vice versa!). There is, however, fairly good correlation between animal tests and occurrence of human cancers.

There is only a 70% correlation between the carcinogenesis of a chemical in rats and that in mice. The correlation between carcinogenesis in either rodent and that in humans probably is less.

Epidemiological Studies

The best evidence that a substance causes cancer in humans comes from epidemiological studies. A population that has a higher than normal rate for a particular kind of cancer is studied for common factors in their background. It was this sort of study, for example, that showed that cigarette smoking causes lung cancer, that vinyl chloride causes a rare form of liver cancer, and that asbestos causes cancer of the lining of the pleural cavity (the body cavity that contains the lungs). These studies require sophisticated mathematical analyses. There is always the chance that some other (unknown) factor is involved in the carcinogenesis.

In the early 1990s a series of epidemiological studies was carried out to determine whether or not the electric field surrounding high voltage power lines can trigger leukemia. The results of the study indicate that if there is an effect, it is quite small.

SECTION 20.16

Birth Defects: Teratogens

Another group of toxic chemicals are those that cause birth defects. These substances are called **teratogens**. Perhaps the most notable teratogen is the tranquilizer thalidomide. This drug was linked to a great tragedy during the late 1950s and early 1960s. Thalidomide was considered so safe, based on laboratory studies, that it often was prescribed for pregnant women. In Germany it was available without a prescription. It took several years for the human population to provide evidence that laboratory animals had not. The

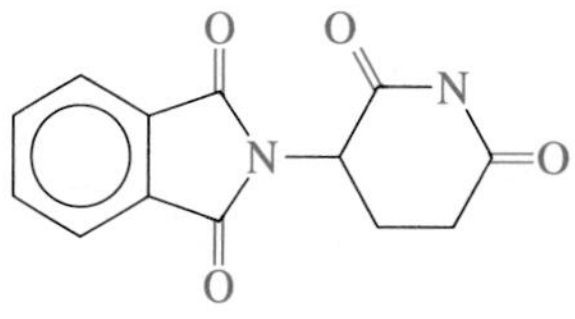

Thalidomide

Frances O. Kelsey, M.D., of the U.S. Food and Drug Administration, refused to approve the tranquilizer drug thalidomide for marketing in the United States. Her action prevented many mothers from facing the tragedy of giving birth to a ''thalidomide baby.'' She is shown here receiving the Distinguished Federal Civilian Service Award from President John F. Kennedy on 7 August 1962. [*Courtesy of* FDA Consumer, *Rockville, Md.*]

drug had a disastrous effect on developing human embryos. Women who had taken the drug during the first 12 weeks of pregnancy had babies that suffered from phocomelia, a condition characterized by shortened or absent arms and legs and other physical defects. The drug was used widely in Germany and Great Britain, and these two countries bore the brunt of the tragedy. The United States escaped relatively unscathed because an official of the FDA had believed there was evidence to doubt the drug's safety and had not, therefore, approved it for use in the United States.

There are other chemicals that act as teratogens. More careful testing, however, has helped us to avoid another thalidomide tragedy. Indeed, some now say that our testing procedures are too strict, expensive, and time consuming, and that they keep needed drugs off the market for years.

SECTION 20.17 Hazardous Wastes

Toxic substances, carcinogens, teratogens—the public has become increasingly concerned in recent years about the problems involving hazardous wastes in the environment. Problems with chemical dumps have made household words out of ''Love Canal'' in New York and ''Valley of the Drums'' in Kentucky. Although often overblown in the news media, serious problems do exist. Hazardous wastes can cause fires or explosions. They can pollute the air. They can contaminate our food and water. Occasionally they poison by direct contact. As long as we want the products our industries produce, though, we will have to deal with the problems of hazardous wastes (Table 20.3).

Table 20.3 Industrial Products and Hazardous-Waste By-products

Product	Associated Waste
Plastics	Organic chlorine compounds
Pesticides	Organic chlorine compounds, organic phosphate compounds
Medicines	Organic solvents and residues, heavy metals (e.g., mercury and zinc)
Paints	Heavy metals, pigments, solvents, organic residues
Oil, gasoline	Oil, phenols and other organic compounds, heavy metals, ammonium salts, acids, caustics
Metals	Heavy metals, fluorides, cyanides, acid and alkaline cleaners, solvents, pigments, abrasives, plating salts, oils, phenols
Leather	Heavy metals, organic solvents
Textiles	Heavy metals, dyes, organic chlorine compounds, solvents

Source: Adapted from ''Everybody's Problem: Hazardous Waste.'' Washington, DC: U.S. Environmental Protection Agency, 1980.

The first step in dealing with any problem is to understand what the problem is. A **hazardous waste** is one that can cause or contribute to death or illness or that threatens human health or the environment when improperly managed. For convenience, hazardous wastes are divided into four types: reactive, flammable, toxic, and corrosive.

Reactive wastes tend to react spontaneously or to react vigorously with air or water. They may generate toxic gases, such as hydrogen cyanide (HCN) or hydrogen sulfide (H_2S), or explode when exposed to shock or heat. Explosives such as trinitrotoluene (TNT) and nitroglycerin obviously are reactive wastes. Another example is sodium metal. Wastes containing sodium caused explosions at Malkins Bank in Great Britain (Figure 20.9). The sodium reacted with water to form hydrogen gas.

Bacteria in the soil break down hazardous wastes, including even TNT and nitroglycerin. Through genetic engineering, scientists are developing new strains of bacteria that can decompose a great variety of wastes.

(a)

(b)

Figure 20.9 (a) A waste dump at Malkins Bank, Cheshire, England, with drums in a lake of chemicals (1970). (b) The same site, cleaned up and restored, is now used as a municipal golf course. [*Harwell/AEA Technology, Oxfordshire, England.*]

$$2\,Na + 2\,H_2O \xrightarrow{\text{fast}} 2\,NaOH + H_2$$

The hydrogen then exploded when ignited in air.

$$2\,H_2 + O_2 \xrightarrow{\text{explosive}} 2\,H_2O$$

Reactive wastes usually can be deactivated before disposal. Sodium can be treated with isopropyl alcohol, with which it reacts slowly, rather than being dumped without treatment.

Flammable wastes are those that burn readily upon ignition, presenting a fire hazard. An example is hexane, a hydrocarbon solvent. In one case, hexane (presumably dumped accidentally by Ralston-Purina) was ignited in the sewers of Louisville, Kentucky.

$$2\,C_6H_{14} + 19\,O_2 \longrightarrow 12\,CO_2 + 14\,H_2O$$

Explosions ripped up several blocks of streets. Flammable wastes usually can be burned safely in incinerators, rather than discarded.

Toxic wastes are those that contain or release toxic substances in quantities sufficient to pose a hazard to human health or the environment. Most of the toxic substances discussed in this chapter would qualify as toxic wastes if they were improperly dumped in the environment. Some toxic wastes can be incinerated safely. PCBs (Chapter 10), for example, are broken down at high temperatures to carbon dioxide, water, and hydrogen chloride. If the hydrogen chloride is removed by scrubbing or is safely diluted and dispersed, this is a satisfactory way to get rid of PCBs. Other toxic wastes cannot be incinerated, however, and must be contained and monitored for years.

Corrosive wastes are those that require special containers because they corrode usual container materials. Acids can't be stored in steel drums because they react with and dissolve the iron.

$$Fe + 2\,H_3O^+ \longrightarrow \underset{\text{(soluble)}}{Fe^{2+}} + H_2 + 2\,H_2O$$

Acid wastes can be neutralized (Chapter 7) before disposal, and lime (CaO) is a cheap base for neutralization.

$$2\,H_3O^+ + CaO \longrightarrow Ca^{2+} + 3\,H_2O$$

The best way to handle hazardous wastes is not to produce them in the first place. Many industries have modified processes to minimize the amount of wastes. Some wastes can be reprocessed to recover energy or materials. Hydrocarbon solvents such as hexane can be purified and reused or burned as a fuel. Sometimes one industry's waste can be a raw material for another industry. For example, waste nitric acid from the metals industry can be converted to fertilizer (Chapter 16). Finally, if a hazardous waste cannot be used or

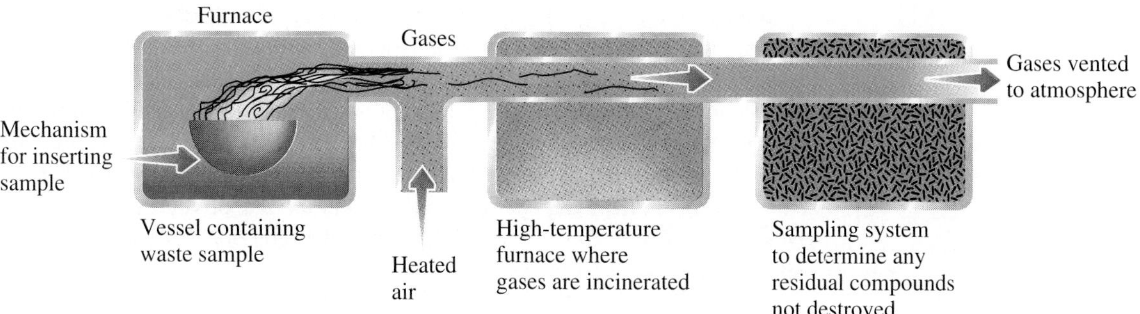

Figure 20.10 A schematic diagram of an incinerator for hazardous wastes. [*After drawing from Midwest Research Institute,* MRI Viewpoint, *Summer 1982.*]

incinerated or treated to render it less hazardous, it must be stored in a secure landfill. Unfortunately, landfills often leak, contaminating the groundwater. We clean up one toxic waste dump and move the materials to another, playing a rather macabre shell game. The best technology at present for treating organic wastes, including chlorinated compounds, is incineration (Figure 20.10). At 1260 °C, 99.9999+% destruction is achieved. Research has identified microorganisms that degrade hydrocarbons such as those in gasoline. Other bacteria, when provided proper nutrients, can degrade chlorinated hydrocarbons. Perhaps biodegradation of the wastes will be the way of the future.

The trouble with incineration is finding a place to build the incinerator. It seems that no one wants to have an incinerator in his own back yard.

SECTION 20.18

What Price Poisons?

We use so many poisons in and around our homes and work places that accidents are bound to happen. Poison-control centers have been established in several cities to help physicians deal with emergency poisonings. Are our insecticides, drugs, cleansers, and other chemicals worth the price we pay in terms of accidental poisonings? That is for you to decide. Generally, it is the *misuse* of these chemicals that leads to tragedy.

Perhaps it is easy to be negative about chemists and chemistry when you think of such horrors as nerve gases, carcinogens, and teratogens. But keep in mind that many toxic chemicals are of enormous benefit to us and that they can be used safely despite their hazardous nature. The plastics industry was able to control vinyl chloride emissions once the hazard was known. We still are able to have valuable vinyl plastics, even though the vinyl chloride from which they are made causes cancer.

Increasingly, we are having to decide whether the benefits we gain from hazardous substances are worth the risks we assume by using them. Many issues involving toxic chemicals are emotional; most of the decisions regarding them are political; but possible solutions to such problems lie mainly in the field of chemistry. We hope that the chemistry you have learned here will help you make intelligent decisions. Most of all, we hope that you will continue to learn more about chemistry throughout the rest of your life, for chemistry affects nearly everything you do. We wish you success and happiness, and may the joy of learning go with you always.

Summary

1. *Pharmacology* is the study of the response of living organisms to drugs. *Toxicology* is the branch of pharmacology that deals with poisons.
2. "The dose makes the poison." Anything can be a poison if the dose is large enough.
3. Strong acids and bases are toxic because they are corrosive.
4. Substances such as ozone are toxic because they are strong oxidizing agents.
5. Carbon monoxide and nitrites are toxic because they interfere with the blood's transport of oxygen.
6. Cyanide is a poison because it shuts down cell respiration.
7. Fluoroacetic acid is a poison because it shuts down the citric acid cycle.
8. Heavy metal poisons, such as lead and mercury, inactivate enzymes by tying up their —SH groups.
9. Nerve poisons, such as organic phosphates, interfere with the acetylcholine cycle.
10. *Carcinogens* are slow poisons that trigger the growth of malignant tumors.
11. *Teratogens* are chemicals of many different types that cause birth defects.
12. *Hazardous waste* includes all kinds of industrial products and by-products that can cause illness or death.

Key Terms

Ames test 20.15
anticarcinogens 20.14
benign tumors 20.14
cancers 20.14
carcinogens 20.14
corrosive wastes 20.17
flammable wastes 20.17
hazardous wastes 20.17
malignant tumors 20.14
methemoglobin 20.3
pharmacology 20.0
pica 20.8
reactive wastes 20.17
teratogens 20.16
toxicology 20.0
toxic wastes 20.17

Review Questions

1. What is pharmacology?
2. What is toxicology?
3. Is sodium chloride (table salt) poisonous? Explain your answer fully.
4. Give an example that shows how the toxicity of a substance depends on the route of administration.
5. Give an example that shows how a substance may be more harmful to one person than another.
6. List three corrosive poisons.
7. How do dilute solutions of acids and bases damage living cells?
8. How does ozone damage living cells?
9. What is a blood agent? List two such poisons.
10. What is methemoglobin?
11. How does cyanide exert its toxic effect?

12. How does sodium thiosulfate act as an antidote for cyanide poisoning?
13. How does fluoroacetic acid exert its toxic effect?
14. Iron (as Fe^{2+}) is a necessary nutrient. What are the effects of too little Fe^{2+}? Of too much?
15. How does mercury (as Hg^{2+}) exert its toxic effect?
16. How does BAL act as an antidote for mercury poisoning?
17. List some sources of mercury poisoning.
18. List some sources of lead poisoning.
19. How does EDTA act as an antidote for lead poisoning?
20. How does cadmium (as Cd^{2+}) exert its toxic effect?
21. What is *itai-itai* disease?
22. What is acetylcholine? Describe its action.
23. How does botulin affect the acetylcholine cycle?
24. How do curare and atropine affect the acetylcholine cycle?
25. What are anticholinesterase poisons? How do they act on the acetylcholine cycle?
26. List three nerve poisons developed for use in warfare.
27. How does atropine act as an antidote for poisoning by organic phosphorus compounds?
28. Describe a use of botulin in medicine.
29. What is the P-450 system? What is its function?
30. How does the liver detoxify ethanol?
31. List two ways that the conversion of nicotine to cotinine in the liver lessens the risk of nicotine poisoning.
32. List two steps in the detoxification of ingested toluene. What is the effect of these steps?
33. Does the P-450 system always detoxify foreign substances?
34. How does ethanol work as an antidote for methanol poisoning?
35. What is a tumor?
36. How are benign and malignant tumors different?
37. What is the single leading cause of cancer?
38. Name several natural carcinogens.
39. What are oncogenes? How are they involved in the development of cancer?
40. What are suppressor genes? How are they involved in the development of cancer?
41. List some conditions under which polycyclic hydrocarbons are formed?
42. Name two aromatic amines that are carcinogens.
43. What is butter yellow? How was it used before it was found to be a carcinogen?
44. Name two aliphatic carcinogens.
45. List some of the limitations involved in testing of compounds for carcinogenicity by using laboratory animals.
46. What is an epidemiological study? Can such a study prove absolutely that a compound causes cancer?
47. What is a mutagen?
48. Describe the Ames test for mutagenicity. What are its limitations as a screening test for carcinogens?
49. What is a teratogen?
50. What is a hazardous waste?
51. Define and give an example of a reactive waste.
52. Define and give an example of a flammable waste.
53. What is a corrosive waste?
54. Why can't waste acids be stored in steel drums?

Projects

55. What is the best method for disposal of toxic organic wastes? Justify fully the method you select.
56. Should carcinogens that occur naturally in foods be subjected to the same tests as are used to evaluate synthetic pesticides? Why or why not?
57. Should substances be tested for toxicity or carcinogenicity on laboratory animals? On people? Explain your answer fully.
58. Are nerve gases less humane than bullets in warfare?
59. In prowling around an old dump, Murgatroyd B. Muckraker finds a metal container filled with 12 oz of liquid. The label indicates the can contains a carcinogen at a concentration of 8.2 mg/fluid oz. On further investigation, he finds that 20 billion such containers are filled and distributed in the United States each year.

a. How many milligrams of the carcinogen are in each can?
b. How many metric tons of the carcinogen are distributed in this manner each year?
c. What should be done about the problem?

60. Choose a disposal method for each of the following wastes. Be as specific as possible, and justify your choice.
a. hydrochloric acid contaminated with iron salts
b. picric acid (an explosive)
c. soybean oil contaminated with PCBs
d. pentane contaminated with residues from penicillin production

References and Readings

1. Abelson, Philip H. ''Treatment of Hazardous Wastes.'' *Science,* 1 August 1986, p. 509.
2. Baker, Scott R., and Chris F. Wilkinson (Eds.). *The Effects of Pesticides on Human Health.* Princeton, NJ: Princeton Scientific, 1990.
3. Bellafante, Ginia. ''Minimizing Household Hazardous Waste.'' *Garbage,* March/April 1990, pp. 44–48.
4. Bernarde, Melvin A. *Our Precarious Habitat.* New York: Wiley, 1989. Chapter 13 discusses hazardous wastes.
5. Bewley, Richard, Barry Ellis, Paul Theile, Ian Viney, and John Rees. ''Microbial Clean-up of Contaminated Soil.'' *Chemistry and Industry,* 4 December 1989, pp. 778–783.
6. Duffus, John H. *Environmental Toxicology.* New York: Halsted Press, 1980.
7. Kamrin, Michael A. *Toxicology.* Chelsea, MI: Lewis Publishers, 1988.
8. Katauskas, Ted. ''PCBs Fall Prey to Ravenous Bugs,'' *R&D Magazine,* January 1990, pp. 19–20. Describes how anaerobic bacteria detoxify PCBs by removing chlorine atoms.
9. Manahan, Stanley E. *Toxicological Chemistry.* Chelsea, MI: Lewis Publishers, 1989.
10. Marshall, Eliot. ''Solving Louisville's Friday-the-Thirteenth Explosion.'' *Science,* 27 March 1981, p. 1405.
11. ''Most Dangerous U.S. Waste Dumps Identified.'' *Chemical and Engineering News,* 3 January 1983, p. 8.
12. Ottoboni, M. Alice. *The Dose Makes the Poison.* Berkeley, CA: Vincente Books, 1984.
13. Palmark, Mogens. ''Future Options for Disposal of Hazardous Wastes.'' *Chemistry and Industry,* 16 June 1985, pp. 416–422.
14. Poole, A., and G. B. Leslie. *A Practical Approach to Toxicological Investigations.* New York: Cambridge University Press, 1990.
15. Reif, Arnold E. ''The Causes of Cancer.'' *American Scientist,* July–August 1981, pp. 437–447.
16. Roberts, Marjorie. ''The Experts and Their X-ray Vision.'' *U.S. News and World Report,* 15 January 1990, p. 66. Radiation from X-rays causes few cases of cancer.
17. Rodricks, Joseph V. *Calculated Risks: Understanding the Toxicity and Human Health Risks of Chemicals in Our Environment.* New York: Cambridge University Press, 1992.
18. Sills, Thomas W. ''Does Beer Drinking Cause Cancer?'' *The Science Teacher,* March 1985, pp. 29–31.
19. Stinson, Stephen C. ''EPA to Evaluate New Technologies for Cleaning Up Hazardous Waste.'' *Chemical and Engineering News,* 25 May 1987, pp. 7–12.
20. Tardiff, Robert G., and Joseph V. Rodricks (Eds.). *Toxic Substances and Human Risk.* New York: Plenum Press, 1987.
21. Wong, John L. ''Cancer and Chemicals . . . and Vegetables.'' *ChemTech,* July 1986, pp. 436–443.

Appendix A
The International System of Measurement

Metric measurement was discussed in some detail in Chapter 1. Further discussion and additional tables are provided here.

The standard unit of length in the International System of Measurement is the **meter**. This distance was once meant to be 0.0000001 of the Earth's quadrant, that is, of the distance from the North Pole to the equator measured along a meridian. The quadrant was difficult to measure accurately. Consequently, for many years the meter was defined as the distance between two etched lines on a metal bar (made of a platinum–iridium alloy) kept in the International Bureau of Weights and Measures at Sevres, France. Today, the meter is defined even more precisely as 1,650,763.73 times the wavelength of the orange-red line in the spectrum of krypton-86.

The primary unit of mass is the **kilogram** (1 kg = 1000 g). It is based on a standard platinum–iridium bar kept at the International Bureau of Weights and Measures. The **gram** is a more convenient unit for many chemical operations.

The basic SI unit of volume is the cubic meter. The unit more frequently employed in chemistry, however, is the **liter** (1 L = 0.001 m^3). All other SI units of length, mass, and volume are derived from these basic units.

Table A.1 Some Metric Units of Length

1 kilometer (km) = 1000 meters (m)
1 meter (m) = 100 centimeters (cm)
1 centimeter (cm) = 10 millimeters (mm)
1 millimeter (mm) = 1000 micrometers (μ)

Table A.2 Some Metric Units of Mass

1 kilogram (kg) = 1000 grams (g)
1 gram (g) = 1000 milligrams (mg)
1 milligram (mg) = 1000 micrograms (μg)

Table A.3	Some Metric Units of Volume
1 liter (L)	= 1000 milliliters (mL)
1 milliliter (mL)	= 1000 microliters (μL)
1 milliliter (mL)	= 1 cubic centimeter (cm^3)

Table A.4	Some Conversions Between Common and Metric Units
Length	
1 mile (mi)	= 1.61 kilometers (km)
1 yard (yd)	= 0.914 meter (m)
1 inch (in.)	= 2.54 centimeters (cm)
Mass	
1 pound (lb)	= 454 grams (g)
1 ounce (oz)	= 28.4 grams (g)
1 pound (lb)	= 0.454 kilograms (kg)
1 grain (gr)	= 0.0648 gram (g)
1 carat (car)	= 200 milligrams (mg)
Volume	
1 U.S. quart (qt)	= 0.946 liter (L)
1 U.S. pint (pt)	= 0.473 liter (L)
1 fluid ounce (fl oz)	= 29.6 milliliters (mL)
1 gallon (gal)	= 3.78 liters (L)

Table A.5	Some Conversion Units for Pressure
1 milliliter of mercury (mm Hg)	= 1 torr
1 atmosphere (atm)	= 760 millimeters of mercury (mm Hg)
	= 760 torr
1 atmosphere (atm)	= 29.9 inches of mercury (in. Hg)
	= 14.7 pounds per square inch (psi)
	= 101 kilopascals (kPa)

Table A.6 Some Temperature Equivalents*

Phenomenon	Fahrenheit	Celsius
Absolute zero	−459.69 °F	−273.16 °C
Nitrogen boils/liquefies	−320.4 °F	−195.8 °C
Carbon dioxide solidifies/sublimes	−109.3 °F	−78.5 °C
Bitter cold night, northern Minnesota	−40 °F	−40 °C
Cold night, Indiana	0 °F	−18 °C
Water freezes/ice melts	32 °F	0 °C
Pleasant room temperature	72 °F	22 °C
Body temperature	98.6 °F	37.0 °C
Very hot day	100 °F	38 °C
Water boils/steam condenses	212 °F	100 °C
Temperature for baking biscuits	450 °F	232 °C

*Temperature conversions can be made using the following equations:

$$°F = \left(\frac{9}{5} \times °C\right) + 32$$

$$°C = \frac{5}{9}(°F - 32)$$

Table A.7 Some Conversion Units for Energy

1 calorie (cal)	= 4.184 joules (J)
1 British thermal unit (Btu)	= 1053 joules (J)
	= 252 calories (cal)
1 food "Calorie"	= 1 kilocalorie (kcal)
	= 1000 calories (cal)
	= 4184 joules (J)

Appendix B
Exponential Notation

Scientists often use numbers that are so large—or so small—that they boggle the mind. For example, light travels at 300,000,000 m/s. There are 602,200,000,000,000,000,000,000 carbon atoms in 12 g of carbon. On the small side, the diameter of an atom is about 0.0000000001 m. The diameter of an atomic nucleus is about 0.000000000000001 m.

It is obviously difficult to keep track of the zeros in such quantities. Scientists find it convenient to express such numbers as *powers of ten*. Tables B.1 and B.2 contain partial lists of such numbers.

The speed of light is usually expressed as 3×10^8 (i.e., $3 \times 10 \times 10 \times 10 \times 10 \times 10 \times 10 \times 10 \times 10$) m/s. The mass of an atom of cesium (Cs) is expressed as 2.21×10^{-22} g, that is, as

$$2.21 \times \frac{1}{10,000,000,000,000,000,000,000} \text{ g}$$

Numbers such as 10^6 are called exponential numbers, where 10 is the *base* and 6 is the *exponent*. Numbers in the form 6.02×10^{23} are said to be written in *scientific notation*.

Exponential numbers are often used in calculations. The most common operations are multiplication and division. Two rules must be followed: (1) to *multiply* exponentials, *add* the exponents, and (2) to *divide* exponentials, *subtract* the exponents. These rules can be stated algebraically as

Table B.1 Positive Powers of Ten
$10^0 = 1$
$10^1 = 10$
$10^2 = 10 \times 10 = 100$
$10^3 = 10 \times 10 \times 10 = 1000$
$10^4 = 10 \times 10 \times 10 \times 10 = 10,000$
$10^5 = 10 \times 10 \times 10 \times 10 \times 10 = 100,000$
$10^6 = 10 \times 10 \times 10 \times 10 \times 10 \times 10 = 1,000,000$
$10^{23} = 100,000,000,000,000,000,000,000$

Table B.2 Negative Powers of Ten
$10^{-1} = 1/10 = 0.1$
$10^{-2} = 1/100 = 0.01$
$10^{-3} = 1/1{,}000 = 0.001$
$10^{-4} = 1/10{,}000 = 0.000{,}1$
$10^{-5} = 1/100{,}000 = 0.000{,}01$
$10^{-6} = 1/1{,}000{,}000 = 0.000{,}001$
$\vdots$
$10^{-13} = 1/10{,}000{,}000{,}000{,}000, = 0.0000000000001$

$$(x^a)(x^b) = x^{a+b}$$

$$\frac{x^a}{x^b} = x^{a-b}$$

Some examples follow.

$$(10^6)(10^4) = 10^{6+4} = 10^{10}$$

$$(10^6)(10^{-4}) = 10^{6+(-4)} = 10^{6-4} = 10^2$$

$$(10^{-5})(10^2) = 10^{(-5)+2} = 10^{-5+2} = 10^{-3}$$

$$(10^{-7})(10^{-3}) = 10^{(-7)+(-3)} = 10^{-7-3} = 10^{-10}$$

$$\frac{10^{14}}{10^6} = 10^{14-6} = 10^8$$

$$\frac{10^6}{10^{23}} = 10^{6-23} = 10^{-17}$$

$$\frac{10^{-10}}{10^{-6}} = 10^{(-10)-(-6)} = 10^{-10+6} = 10^{-4}$$

$$\frac{10^3}{10^{-2}} = 10^{3-(-2)} = 10^{3+2} = 10^5$$

$$\frac{10^{-8}}{10^4} = 10^{(-8)-4} = 10^{-12}$$

$$\frac{10^7}{10^7} = 10^{7-7} = 10^0 = 1$$

Problems involving both a coefficient (a numerical part) and an exponential are solved by multiplying (or dividing) coefficients and exponentials separately.

▶ EXAMPLE B.1

To what is the following expression equivalent?

$$(1.2 \times 10^5)(2.0 \times 10^9)$$

SOLUTION

First, multiply the coefficients.

$$1.2 \times 2.0 = 2.4$$

Then multiply the exponentials.

$$10^5 \times 10^9 = 10^{5+9} = 10^{14}$$

The complete answer is

$$2.4 \times 10^{14}$$

▶ EXAMPLE B.2

To what is the following expression equivalent?

$$\frac{(8.0 \times 10^{11})}{(1.6 \times 10^4)}$$

SOLUTION

First, divide the coefficients.

$$\frac{8.0}{1.6} = 5.0$$

Then divide the exponentials.

$$\frac{10^{11}}{10^4} = 10^{11-4} = 10^7$$

The answer is

$$5.0 \times 10^7$$

EXAMPLE B.3

Give an equivalent for the following expression.

$$\frac{(1.2 \times 10^{14})}{(4.0 \times 10^{6})}$$

SOLUTION

Before carrying out the division, it is convenient to rewrite the dividend (the numerator) so that the coefficient is larger than that of the divisor (the denominator).

$$1.2 \times 10^{14} = 12 \times 10^{13}$$

Note that the coefficient was made larger by a factor of 10 and the exponential was made smaller by a factor of 10. The quantity as a whole is unchanged. Now divide.

$$\frac{12 \times 10^{13}}{4.0 \times 10^{6}} = 3.0 \times 10^{7}$$

EXAMPLE B.4

Give an equivalent for the following expression.

$$\frac{(3 \times 10^{7})(8 \times 10^{-3})}{(6 \times 10^{2})(2 \times 10^{-1})}$$

SOLUTION

In problems such as this, you can carry out the multiplications specified in the numerator and in the denominator separately and then divide the resulting numbers.

$$(3 \times 10^{7})(8 \times 10^{-3}) = 24 \times 10^{4}$$

$$(6 \times 10^{2})(2 \times 10^{-1}) = 12 \times 10^{1}$$

$$\frac{24 \times 10^{4}}{12 \times 10^{1}} = 2 \times 10^{3}$$

The multiplications and divisions in problems like this can be carried out in any convenient order.

Only one other mathematical function involving exponentials is of importance to us. What happens when you raise an exponential to a power? You just multiply the exponent by the power. To illustrate:

$$(10^3)^3 = 10^9$$

$$(10^{-2})^4 = 10^{-8}$$

$$(10^{-5})^{-3} = 10^{15}$$

If the exponential is combined with a coefficient, the two parts of the number are dealt with separately, as in the following example.

$$(2 \times 10^3)^2 = 2^2 \times (10^3)^2 = 4 \times 10^6$$

For a further discussion of—and more practice with—exponential numbers, see the following reference.

Goldish, Dorothy M. *Basic Mathematics for Beginning Chemistry*, 4th ed. New York: Macmillan, 1990. Chapter 3 covers exponential notation.

Problems

1. Express each of the following in scientific notation.

a. 0.000,01
b. 10,000,000
c. 0.0034
d. 0.000,010,7
e. 4,500,000,000
f. 406,000
g. 0.02
h. 124 $\times$ 10^3

2. Carry out the following operations. Express the numbers in scientific notation.

a. $(4.5 \times 10^{13})(1.9 \times 10^{-5})$
b. $(6.2 \times 10^{-5})(4.1 \times 10^{-12})$
c. $(2.1 \times 10^{-6})^2$
d. $\dfrac{(4.6 \times 10^{-12})}{(2.1 \times 10^{3})}$
e. $\dfrac{(9.3 \times 10^{9})}{(3.7 \times 10^{-7})}$
f. $\dfrac{(2.1 \times 10^{5})}{(9.8 \times 10^{7})}$
g. $\dfrac{(4.3 \times 10^{-7})}{(7.6 \times 10^{22})}$

Appendix C
Solving Problems by Unit Conversions

Problems in chemistry, such as those in Chapter 1, are often solved by a method that involves converting from one kind of unit to another. This approach is called the **unit-conversion method** or the **factor-label method**. (It is also often called **dimensional analysis**, although the units employed are not always dimensions.) Whatever we call it, the method employs units—such as L, mi/hr, cm, ft, or g/cm^3—as aids in setting up and solving problems. The general approach is to multiply the known quantity (and its units!) by one or more conversion factors so that the answer is obtained in the desired units.

$$\text{Known quantity and unit} \times \text{Conversion factor} = \text{Answer (in desired unit)}$$

The method is best learned by practice. We urge you to learn it now to save yourself a lot of time and wasted effort later.

Conversions Within a System

Quantities can be expressed in a variety of units. For example, you can buy beverages by the 12-oz can or by the pint, quart, gallon, or liter. If you wish to compare prices, you must be able to convert from one unit to another. Such a conversion changes the numbers and units, but it does not change the quantity. Your actual weight, for example, remains unchanged whether it is expressed in pounds, ounces, or kilograms.

You know that multiplying a number by 1 doesn't change its value. Multiplying by a fraction equal to 1 also leaves the value unchanged. A fraction is equal to 1 when the numerator is equal to the denominator. For example, we know that

$$1 \text{ ft} = 12 \text{ in.}$$

Therefore,

$$\frac{1 \text{ ft}}{12 \text{ in.}} = 1$$

Similarly,

$$\frac{12 \text{ in.}}{1 \text{ ft}} = 1$$

Now, if you want to convert an answer from inches to feet, you can do so by choosing one of the above fractions as a **conversion factor**. Which one do you choose? The one that gives you an answer with the right units! Let's illustrate by an example.

▶ EXAMPLE C.1

My bed is 72 in. long. What is its length in feet?

SOLUTION

You know the answer, of course, but let us proceed to show *how* the answer is obtained by using unit conversions. We need to multiply 72 in. by one of the above fractions. Which one? The known quantity and unit is 72 in.

$$72 \text{ in.} \times \text{conversion factor} = ?\text{ ft}$$

For the conversion factor, choose the fraction that, when inserted in the equation, cancels the unit *in.* and becomes the unit *ft.*

$$72 \cancel{\text{in.}} \times \frac{1 \text{ ft}}{12 \cancel{\text{in.}}} = 6.0 \text{ ft}$$

Just for kicks, let us try the other conversion factor.

$$72 \text{ in.} \times \frac{12 \text{ in.}}{1 \text{ ft}} = \frac{860 \text{ in.}^2}{\text{ft}}$$

Absurd! How can a bed be 860 in.^{2}/ft? You should have no difficulty in choosing between the two possible answers.

One of the advantages of the metric system is the ease of conversion between units. Let us try to demonstrate that with a few examples. Remember that a list of equivalent values is actually a list of conversion factors. Thus, the equality

$$1 \text{ kg} = 1000 \text{ g}$$

can be rearranged into two useful conversion factors.

$$\frac{1\ \text{kg}}{1000\ \text{g}} \quad \text{and} \quad \frac{1000\ \text{g}}{1\ \text{kg}}$$

▶ EXAMPLE C.2

Convert 0.371 kg to grams.

SOLUTION

$$0.371\ \cancel{\text{kg}} \times \frac{1000\ \text{g}}{1\ \cancel{\text{kg}}} = 371\ \text{g}$$

▶ EXAMPLE C.3

Convert 0.371 lb to ounces.

SOLUTION

$$0.371\ \cancel{\text{lb}} \times \frac{16\ \text{oz}}{1\ \cancel{\text{lb}}} = 5.94\ \text{oz}$$

▶ EXAMPLE C.4

Convert 2429 cm to meters.

SOLUTION

$$2429\ \cancel{\text{cm}} \times \frac{1\ \text{m}}{100\ \cancel{\text{cm}}} = 24.29\ \text{m}$$

▶ EXAMPLE C.5

Convert 2429 in. to yards.

SOLUTION

$$2429\ \cancel{\text{in.}} \times \frac{1\ \text{yd}}{36\ \cancel{\text{in.}}} = 67.47\ \text{yd.}$$

In conversions of customary units, you multiply and divide by factors such as 16 or 36. In metric conversions, you multiply and divide by 100 or 1000 and so on. You need only shift the decimal point when doing metric conversions.

Conversion factors are not usually given in a problem. They may be obtained from listings such as those in Appendix A. However, you would be wise to learn to convert within the metric system without the need of tables.

Also, you should remember that this equality

$$1 \text{ cm} = 0.01 \text{ m}$$

is equivalent to

$$100 \text{ cm} = 1 \text{ m}$$

All these fractions

$$\frac{1 \text{ cm}}{0.01 \text{ m}} \qquad \frac{0.01 \text{ m}}{1 \text{ cm}} \qquad \frac{100 \text{ cm}}{1 \text{ m}} \qquad \frac{1 \text{ m}}{100 \text{ cm}}$$

are valid conversion factors.

▶ EXAMPLE C.6

Knowing that 1 mL = 0.001 L, write four conversion factors relating milliliters and liters.

SOLUTION

The first two conversion factors can be formed by arranging the two sides of the equality in the form of a fraction.

$$\frac{1 \text{ mL}}{0.001 \text{ L}} \qquad \frac{0.001 \text{ L}}{1 \text{ mL}}$$

To derive the other two conversion factors, first multiply both sides of the equality by 1000 (in order to obtain the equality in terms of 1 L).

$$1000 \times 1 \text{ mL} = 1000 \times 0.001 \text{ L}$$

$$1000 \text{ mL} = 1 \text{ L}$$

Now just arrange this last equality in fractional form.

$$\frac{1000 \text{ mL}}{1 \text{ L}} \qquad \frac{1 \text{ L}}{1000 \text{ mL}}$$

The conversion factors 1 mL/0.001 L and 1000 mL/1 L would give exactly the same answer if used in a problem. The only difference is convenience. Some people would rather multiply by 1000 than divide by 0.001. In this age of the electronic calculator, perhaps even this difference is no longer significant.

Let us try some more conversions within the metric system.

▶ EXAMPLE C.7

How many milliliters are there in a 2-L bottle of soda pop?

SOLUTION

From memory or from the tables in Appendix A you find

$$1\ \text{L} = 1000\ \text{mL}$$

$$2\ \not{\text{L}} \times \frac{1000\ \text{mL}}{1\ \not{\text{L}}} = 2000\ \text{mL}$$

Notice that we picked the conversion factor that allowed us to cancel liters and obtain an answer in the desired units, milliliters.

Sometimes it is necessary to carry out more than one conversion in a problem.

▶ EXAMPLE C.8

In the United States, the usual soda pop can holds 360 mL. How many such cans could be filled from one 2.0-L bottle?

SOLUTION

The problem tells us that

$$1\ \text{can} = 360\ \text{mL}$$

Using that equivalence, we can calculate the answer.

$$2.0\ \not{\text{L}} \times \frac{1000\ \cancel{\text{mL}}}{1\ \not{\text{L}}} \times \frac{1\ \text{can}}{360\ \cancel{\text{mL}}} = 5.6\ \text{cans}$$

▶ EXAMPLE C.9

How many 325-mg aspirin tablets can be made from 875 g of aspirin?

SOLUTION

The problem asks us to convert the *given* value of 875 g to tablets. The problem also includes a necessary conversion factor.

$$1\ \text{tablet} = 325\ \text{mg}$$

From memory or the tables, we have another required conversion factor.

$$1\ \text{g} = 1000\ \text{mg}$$

By multiplying the given value by the appropriately arranged conversion factors, we arrive at the answer.

$$875\ \cancel{g} \times \frac{1000\ \cancel{mg}}{1\ \cancel{g}} \times \frac{1\ \text{tablet}}{325\ \cancel{mg}} = 2690\ \text{tablets}$$

Conversions Between Systems

To convert from one system of measurement to another, you need a list of equivalents such as that in Appendix A. Let us plunge right in and work some examples.

▶ EXAMPLE C.10

How many kilograms are there in 33 lb?

SOLUTION

$$33\ \cancel{lb} \times \frac{1.0\ \text{kg}}{2.2\ \cancel{lb}} = 15\ \text{kg}$$

▶ EXAMPLE C.11

You know that your weight is 142 lb, but the job application form asks for your weight in kilograms. What is it?

SOLUTION

From the table we find

$$1.00\ \text{lb} = 0.454\ \text{kg}$$

The solution is simple.

$$142\ \cancel{lb} \times \frac{0.454\ \text{kg}}{1.00\ \cancel{lb}} = 64.5\ \text{kg}$$

▶ EXAMPLE C.12

A recipe calls for 750 mL of milk, but your measuring cup is calibrated in fluid ounces. How many ounces of milk will you need?

SOLUTION

$$750\ \cancel{mL} \times \frac{1.00\ \text{fl oz}}{29.6\ \cancel{mL}} = 25.3\ \text{fl oz}$$

EXAMPLE C.13

How many meters are there in 764 ft (1.00 m = 39.4 in.)?

SOLUTION

$$764\,\cancel{\text{ft}} \times \frac{12\,\cancel{\text{in.}}}{1\,\cancel{\text{ft}}} \times \frac{1.00\text{ m}}{39.4\,\cancel{\text{in.}}} = 233\text{ m}$$

EXAMPLE C.14

How would you describe a young man who is 1.6 m tall and weighs 91 kg?

SOLUTION

$$1.6\,\cancel{\text{m}} \times \frac{39\,\cancel{\text{in.}}}{1.0\,\cancel{\text{m}}} \times \frac{1\text{ ft}}{12\,\cancel{\text{in.}}} = 5.2\text{ ft}$$

$$91\,\cancel{\text{kg}} \times \frac{2.2\text{ lb}}{1.0\,\cancel{\text{kg}}} = 200\text{ lb}$$

The young man is 5 ft 2 in. tall and weighs 200 lb. Let us be generous and say that he is well muscled.

It is possible (and frequently necessary) to manipulate units in the denominator as well as the numerator of a problem. Just remember to use conversion factors in such a way that the unwanted units cancel.

EXAMPLE C.15

A sprinter runs the 100-m dash in 11 s. What is her speed in kilometers per hour?

SOLUTION

The given speed is 100 m per 11 s.

$$\frac{100\text{ m}}{11\text{ s}}$$

The conversion factors that we need are found in the tables or recalled from memory.

$$\frac{100\,\cancel{\text{m}}}{11\,\cancel{\text{s}}} \times \frac{1\text{ km}}{1000\,\cancel{\text{m}}} \times \frac{60\,\cancel{\text{s}}}{1\,\cancel{\text{min}}} \times \frac{60\,\cancel{\text{min}}}{1\text{ hr}} = 33\text{ km/hr}$$

Note that the first conversion factor changes *m* to *km*. It takes two factors to change *s* to *hr*. Note also that we could have first applied the factors that convert *s* to *hr* and then converted *m* to *km*. The answer would have been the same.

▶ EXAMPLE C.16

If your heart beats at a rate of 72 times per minute, and your lifetime will be 70 years, how many times will your heart beat during your lifetime?

SOLUTION

Two equivalences are given in the problem.

$$72 \text{ beats} = 1 \text{ min}$$

$$1 \text{ lifetime} = 70 \text{ yr}$$

Three others that you will need you can recall from memory.

$$1 \text{ yr} = 365 \text{ days}$$

$$1 \text{ day} = 24 \text{ hr}$$

$$1 \text{ hr} = 60 \text{ min}$$

Start now with the factor 72 beats/1 min (the known quantities and units) and apply the conversion factors as needed to get an answer in beats/lifetime (the desired units).

$$\frac{72 \text{ beats}}{1 \cancel{\text{min}}} \times \frac{60 \cancel{\text{min}}}{1 \cancel{\text{hr}}} \times \frac{24 \cancel{\text{hr}}}{1 \cancel{\text{day}}} \times \frac{365 \cancel{\text{days}}}{1 \cancel{\text{yr}}} \times \frac{70 \cancel{\text{yr}}}{1 \text{ lifetime}}$$

$$= 2{,}600{,}000{,}000 \text{ beats/lifetime}$$

Appendix D
Significant Figures

Unlike counting, measurement is never exact. You can *count* exactly ten people in a room. If you asked each of those people to *measure* the length of the room to the nearest 0.01 m, however, the values they determine are likely to differ slightly. Table D.1 presents such a set of measurements.

Note that all ten students agree on the first three digits of the measurement; differences occur in the fourth digit. Which values are correct? Actually, all are accurate within the accepted range of uncertainty for this physical measurement. The accuracy of measurement depends on the type of measuring instrument and the skill and care of the person making the measurement. Measured values are usually recorded with the last digit regarded as uncertain. The data in Table D.1 allow us to state that the length of the room is between 14.1 m and 14.2 m, but we are not sure of the fourth digit. The measurements in the table have four *significant figures*, which means that the first three are known with confidence and the fourth conveys an approximate value. **Significant figures** include all digits known with certainty plus one uncertain digit.

In any properly reported measurement, all nonzero digits are significant. The zero presents problems, however, because it can be used in two ways: to position the decimal point or to indicate a measured value. For zeros, follow these rules.

1. A zero between two other digits is always significant.
   Examples: The number 1107 contains four significant figures.
   The number 50.002 contains five significant figures.
2. Zeros to the left of *all* nonzero digits are not significant.
   Examples: The number 0.000**163** has three significant figures.
   The number 0.0**6801** has four significant figures.

Table D.1 A Set of Measurements of the Length of a Room

Student	Length (m)	Student	Length (m)
1	14.14	6	14.14
2	14.15	7	14.17
3	14.17	8	14.17
4	14.14	9	14.16
5	14.16	10	14.17

3. Zeros that are *both* to the right of the decimal point *and* to the right of nonzero digits are significant.
   Examples: The number 0.**2000** has four significant figures.
   The number 0.0**50120** has five significant figures.
   The number **802.760** has six significant figures.

4. Zeros in numbers such as 40,000 (that is, zeros to the right of *all* nonzero digits in a number that is written without a decimal point) may or may not be significant. Without more information, we simply do not know whether 40,000 was measured to the nearest unit or ten or hundred or thousand or ten-thousand. To avoid this confusion, scientists use exponential notation (Appendix B) for writing numbers. In exponential notation, 40,000 would be recorded as 4×10^4 or 4.0×10^4 or 4.0000×10^4 to indicate one, two, and five significant figures, respectively.

Addition or Subtraction

In addition or subtraction, the result should contain no more digits to the right of the decimal point than the quantity that has the least digits to the right of the decimal point. Align the quantities to be added or subtracted on the decimal point, then perform the operation, assuming blank spaces are zeros. Determine the correct number of digits after the decimal point in the answer and round off to this number. In rounding off, you should increase the last significant figure by one if the following digit is 5 or greater.

▶ EXAMPLE D.1

Add the following numbers: 49.146, 72.13, 5.9432.

SOLUTION

Align the numbers on the decimal point and carry out the addition.

$$\begin{array}{r} 49.146 \\ 72.13 \\ 5.9432 \\ \hline 127.2192 \end{array}$$

The quantity with the fewest digits after the decimal point is 72.13. The answer should have only two digits after the decimal point. Since the third digit after the decimal point is 9, the second digit after the decimal point is rounded up to 2. Correct answer: 127.22

▶ EXAMPLE D.2

Add the following numbers: 744, 2.6, 14.812.

SOLUTION

$$\begin{array}{r} 744 \\ 2.6 \\ 14.812 \\ \hline 761.412 \end{array} \longrightarrow 761$$

The first quantity has no digits to the right of the decimal point (which is understood to be at the right of the digits). The answer must therefore be rounded so that it, too, contains no digits to the right of the decimal point.

EXAMPLE D.3

Subtract 9.143 from 71.12496.

SOLUTION

$$\begin{array}{r} 71.12496 \\ -9.143 \\ \hline 61.98196 \end{array} \longrightarrow 61.982$$

Since the second quantity has only three digits to the right of the decimal point, so must the answer.

Multiplication and Division

In multiplication and division, our answer can have no more significant figures than the factor that has the least number of significant figures. In these operations, the *position* of the decimal point makes no difference.

EXAMPLE D.4

Multiply 10.4 by 3.1416.

SOLUTION

$$10.4 \times 3.1416 = 32.672{,}64 \longrightarrow 32.7$$

The answer has only three significant figures because the first term has only three.

EXAMPLE D.5

Divide 5.973 by 3.0.

SOLUTION

$$\frac{5.973}{3.0} = 1.991 \longrightarrow 2.0$$

The answer has only two significant figures because the divisor has only two.

Exact Values

Some quantities are not measured but defined. A kilometer is defined as 1000 meters: 1 km = 1000 m. Similarly, 1 foot can be defined as 12 inches: 1 ft = 12 in. The "1 km" should not be regarded as containing one significant figure; nor should "12 in." be considered to have two significant figures. In fact, these values can be considered to have an infinite number of significant figures (1.000,000,000,000,000,0. . .) or, more correctly, to be *exact* values. Such defined values are frequently used as conversion factors in problems (Chapter 1). When you are determining the number of significant figures for the answer to a problem, you should ignore such exact values. Use only the measured quantities in the problem to determine the number of significant figures in the answer.

Problems

Perform the indicated operations and give answers with the proper number of significant figures.

1. a. 48.2 m + 3.82 m + 48.4394 m
b. 151 g + 2.39 g + 0.0124 g
c. 15.436 mL + 9.1 mL + 105 mL

2. a. 100.53 cm − 46.1 cm
b. 451 g − 15.46 g
c. 19.71 L − 10.4 L

3. a. 73 m × 1.340 m × 0.41 m
b. 0.137 cm × 1.43 cm
c. 3.146 cm × 5.4 cm

4. a. $\frac{5.179 \text{ g}}{4.6 \text{ mL}}$ b. $\frac{4561 \text{ g}}{3.1 \text{ mol}}$ c. $\frac{40.00 \text{ g}}{3.2 \text{ mL}}$

5. $\frac{1.426 \text{ mL} \times 373 \text{ K}}{204 \text{ K}}$

Appendix E Glossary

absolute scale A temperature scale in which the zero point is the coldest temperature possible, or absolute zero.

acid A proton donor.

acid anhydride A substance that reacts with water to produce an acid; a nonmetal oxide.

acidosis Condition that results when the pH of the blood falls below 7.35 and oxygen transport is hindered.

acid rain Rain having a pH less than 5.6.

activated sludge method A combination of primary and secondary sewage treatment methods in which some sludge is recycled.

active site The spot on a molecule at which reaction occurs.

actomysin The contractile protein of which muscles are made.

addition polymer A polymer that contains all the atoms of the starting monomers.

adequate protein A protein that supplies all the essential amino acids in the quantities needed for the growth and repair of body tissues.

adipose tissue Fatty tissue.

advanced sewage treatment Sewage treatment designed to remove phosphates, nitrates, and other soluble impurities.

aeration A process in which water is sprayed into the air to remove odors and improve taste.

aerobic Occurring in the presence of oxygen.

aerobic exercise Exercise in which muscle contractions occur in the presence of oxygen.

aerosol Particles of 1 micrometer diameter, or less, dispersed in air.

aflatoxins Compounds produced by molds growing on stored peanuts and grains.

Agent Orange A combination of 2,4-D and 2,4,5-T used extensively in Vietnam to remove enemy cover and destroy crops that maintained enemy armies.

agonist A molecule that fits and activates a specific receptor.

air The mixture of gases that makes up Earth's atmosphere; mainly nitrogen and oxygen.

alchemy A mystical chemistry that flourished in Europe during the Middle Ages (A.D. 500 to 1500).

alcohol A compound composed of an alkyl group and a hydroxyl group.

aldehyde An organic molecule with a carbonyl at one end.

alkali metal An element in Group IA of the periodic table.

alkaline earth metal An element in Group IIA of the periodic table.

alkaloid A nitrogen-containing organic compound obtained from plants.

alkalosis A physiological condition in which the pH of the blood rises to life-threatening levels.

alkane A hydrocarbon with only single bonds; a saturated hydrocarbon.

alkene A hydrocarbon containing one or more double bonds.

alkyne A hydrocarbon containing one or more triple bonds.

allergen A substance that triggers an allergic reaction.

alloy A mixture of two or more elements, at least one of which is a metal; an alloy has metallic properties.

alpha helix A secondary structure of a protein molecule in which the molecule has a spiral arrangement.

alpha linkage The arrangement of the O atom that bridges two monosaccharide units in a disaccharide (or polysaccharide) in which the O is pointed down relative to the O atom in the ring.

alpha particle A cluster of 2 protons and 2 neutrons; a helium nucleus.

Ames test A laboratory test for mutagens, which are usually also carcinogens.

amide An organic compound having the functional group CON in which the carbon is double-bonded to the oxygen and single-bonded to the nitrogen.

amine A compound that contains the elements carbon, hydrogen, and nitrogen; derived from ammonia by replacing one, two, or three of the hydrogen atoms by alkyl group(s).

amino acid An organic compound that contains both an amino group and a carboxylic acid group; amino acids combine to produce proteins.

amino group An NH_2 unit.

ammonia A volatile polar compound of nitrogen and hydrogen, NH_3, used primarily as fertilizer.

amphetamines Stimulant drugs widely used, and abused.

amylopectin A starch with branched chains of glucose units.

amylose A starch with the glucose units joined in a continuous chain.

anabolic steroid A drug that aids in the building (anabolism) of body proteins and thus of muscle tissue.

anabolism The buildup of body tissues from simpler molecules.

anaerobic Occurring in the absence of oxygen.

anaerobic exercise Exercise involving muscle contractions without sufficient amounts of oxygen.

analgesic A pain reliever.

androgen A male sex hormone.

anesthetic A substance that causes loss of feeling and pain.

anhydride A substance from which water has been removed.

anhydrous Without water.

anion A negatively charged ion.

anionic surfactant A surfactant with a hydrocarbon tail and a water-soluble head that bears a negative charge.

anode An electrode that bears a positive charge.

antacid An alkaline product used to neutralize excess stomach acid.

antagonist A drug that blocks the action of an agonist, by blocking the receptors.

antibiotic A soluble substance, produced by a mold or bacterium, that inhibits growth of other microorganisms.

anticarcinogen A substance that inhibits the formation of cancer.

anticholinergic Drugs that act on nerves using acetylcholine as a neurotransmitter.

anticoagulant A substance that inhibits the clotting of blood.

anticodon The sequence of three adjacent nucleotides in a tRNA molecule that is complement to a codon on mRNA.

antidiuretic A water-conserving substance.

antihistamine A substance that relieves the symptoms of allergies: sneezing, itchy eyes, and runny nose.

anti-inflammatory A substance that inhibits inflammation.

antimetabolite A compound that inhibits the synthesis of nucleic acids.

antioxidants Chemicals so easily oxidized that they protect other substances from oxidation; reducing agents.

antiperspirant A formulation to stop or retard perspiration.

antipyretic A fever-reducing substance.

antiseptic A compound applied to living tissue to kill or prevent the growth of microorganisms.

apoenzyme The pure protein part of an enzyme.

applied research Work oriented toward the solution of a particular problem in industry or the environment.

arithmetic growth A process in which a constant amount is added during each growth period.

aromatic compound Any organic compound that contains a benzene ring.

arteriosclerosis Hardening of the arteries.

artificial transmutation A process by which one element is changed into another by an artificial means.

asbestos A group of related fibrous silicates.

asbestosis A severe respiratory disease caused by inhalation of asbestos fibers 5 to 50 micrometers long over a period of 10 to 20 years.

aspartame A synthetic sweetener comprising a dipeptide of aspartic acid and phenylalanine.

astringent A substance that constricts the opening of the sweat glands, thus reducing the amount of perspiration that escapes.

atmosphere The gaseous mass surrounding the Earth.

atmospheric inversion A warm layer of air above a cool, stagnant lower layer.

atom Smallest characteristic particle of an element.

atomic mass unit The unit of relative atomic weights.

atomic nucleus Concentrated, positively charged matter at the center of an atom; composed of protons and neutrons.

atomic number The number of protons in the nucleus of an atom of an element.

atomic theory A model that offers a logical explanation for the law of multiple proportions and the law of constant composition by stating that all elements are composed of atoms, all atoms of a given element are identical, but the atoms of one element differ from the atoms of any other element; that atoms of different elements can combine to give compounds and a chemical reaction involves a change not in the atoms themselves, but in the way atoms are combined to form compounds.

Avogadro's hypothesis Equal volumes of gases, regardless of their composition, contain equal numbers of molecules under the same conditions of temperature and pressure.

Avogadro's number The number of atoms in exactly 12 g of pure carbon-12; 6.02×10^{23}.

background radiation Ever-present radiation from cosmic rays and from natural radioactive isotopes in air, water, soil, and rocks.

bag filtration A method for removing particulate matter from power plant stack gases by filtering through a bag.

Bakelite A polymer of phenol and formaldehyde.

barbiturate A depressant anticonvulsant drug.

base A proton acceptor.

base triplet The sequence of 3 bases on a tRNA molecule that determine which amino acid it can carry.

basic anhydride A substance that reacts with water to produce a basic solution; a metal oxide.

basic research The search for knowledge for its own sake.

battery A series of electrochemical cells.

benign tumor A tumor characterized by slow growth; often regresses spontaneously; does not invade neighboring tissues.

beta linkage The arrangement of the O atom that bridges two monosaccharide units in a disaccharide (or polysaccharide) in which the O is pointed up relative to the O atom in the ring.

beta particle An electron emitted by a radioactive material.

binary compound A compound consisting of two elements.

binding energy Energy derived from the conversion of mass to energy when neutrons and protons are put together to form nuclei.

biochemical oxygen demand The quantity of oxygen required by microorganisms to remove organic matter from water.

biochemistry A study of the chemistry of living systems.

biocides Substances that kill living organisms.

biological magnification An increase in concentration of a substance as it moves up the food chain.

biological oxidation demand A measure of the amount of oxygen needed for the degradation of organic material.

biomass The total mass of plants and animals; in energy studies, usually means plant material used as a source of fuel.

bitumen A hydrocarbon mixture obtained from tar sands by heating.

bleach A compound used to remove unwanted color from fabrics, hair, or other materials.

blood doping A method used by athletes to improve their performance. A quantity of blood is withdrawn from the athlete and stored for six weeks while the person's body manufactures replacement blood. Then, a few days before competition, the red blood cells from the stored blood are returned to the athlete.

blood sugar Glucose, a simple sugar, circulated in the bloodstream.

bonding pair A pair of electrons that comprise a chemical bond.

breeder reactor A nuclear reactor that is designed to convert nonfissile uranium-238 to fissile plutonium.

broad-spectrum antibiotic An antibiotic effective against a wide variety of microorganisms.

broad-spectrum insecticide An insecticide that kills many kinds of insects.

bronze An alloy of copper and tin.

buffer A compound that reacts with either acid or base to keep the pH of a solution essentially constant.

builder (in detergent formulations) Any substance added to a surfactant to increase its detergency.

Calorie (Cal) 1000 calories (or 1 kcal); used to measure the energy content of foods.

calorie (cal) The amount of heat required to raise the temperature of 1 g of water 1°C.

cancer A malignant tumor; tumor that grows and invades other tissues.

carbohydrate A compound consisting of carbon, hydrogen, and oxygen; a starch or sugar.

carbohydrate loading A method used by athletes to improve the storage of glycogen. Glycogen stores usually are first depleted by limited carbohydrate intake and vigorous training. Then, a few days before competition, the athlete cuts back on training and eats a diet high in carbohydrates.

carbonate A metal combined with carbon and oxygen.

carbon-14 dating A technique for determining the age of artifacts based on the half-life of carbon-14.

carbon monoxide The poisonous product of incomplete combustion of carbon; CO.

carbonyl group A carbon atom double-bonded to an oxygen atom.

carboxyl group —COOH, the functional group of the organic acids.

carboxylic acid An organic compound that contains the —COOH functional group.

carcinogens Substances that produce tumors.

catabolism The metabolic process in which complex compounds are broken down to simpler substances.

catalyst A substance that increases the rate of a chemical reaction without itself being used up.

catalytic converter A device containing a catalyst for oxidizing carbon monoxide and hydrocarbons to carbon dioxide.

catalytic reforming A process that converts low-octane alkanes to high-octane aromatic compounds.

catenation A process in which carbon atoms join together to form long chains of hundreds or even thousands of atoms.

cathode An electrode that bears a negative charge.

cathode ray A stream of high-speed electrons.

cation A positively charged ion.

cationic surfactant Surfactant with a hydrocarbon tail and a water-soluble head that bears a positive charge.

cell The structural unit of plant and animal life.

cell membrane The lipid bilayer that encloses a cell.

cell nucleus The organelle (small structure) within a cell that contains the genetic material of the cell.

celluloid Cellulose nitrate, a synthetic material derived from natural cellulose by reaction with nitric acid.

cellulose A glucose polymer that is beta-linked.

Celsius scale A temperature scale on which water freezes at 0° and boils at 100°.

cement A mixture of lime, clay, and water that hardens like stone when it dries.

ceramic A produce made from clay or similar materials.

chain reaction A self-sustaining change in which one or more products of one event cause one or more new events.

chemical bond The force of attraction that holds atoms together in compounds.

chemical change A change in chemical composition.

chemical equation A before-and-after description in which chemical formulas and coefficients represent a chemical reaction.

chemical properties A description of the way in which a substance reacts with another substance to change its composition.

chemical symbol The abbreviation consisting of one, two, or three letters that stands for an element.

chemistry The study of matter and the changes it undergoes.

chemotherapy The use of chemicals to control or cure diseases.

chlorination Treatment with chlorine, in the case of water to kill bacteria.

chlorofluorocarbon A carbon compound that contains fluorine and chlorine.

chlorophyll The green plant pigment that absorbs light energy to initiate the process of photosynthesis.

chloroplast The structural part (organelle) of a plant cell in which photosynthesis occurs.

cholesterol A steroid made by the body, which uses it to make all the other steroids.

chromosome A structure in the cell nucleus composed of DNA and proteins.

chrysotile A magnesium silicate, a form of asbestos.

clay A complex aluminum silicate that is plastic when wet and can therefore be shaped into pottery, tiles, and so on.

clinkers Mineral matter left behind after the burning of coal.

clone An organism reproduced in identical form.

coal A natural black rock composed mainly of carbon.

coal tar The condensed volatile materials given off in the coking process of coal; a source of organic chemicals, particularly aromatic compounds.

codon A sequence of three adjacent nucleotides in mRNA that specifies one amino acid.

coenzyme An organic molecule (often a vitamin) that combines with an apoenzyme to make a complete, functioning enzyme.

cofactor An inorganic component that combines with an apoenzyme to make a complete, functioning enzyme.

coke A relatively pure form of carbon left behind after coal is heated to drive off volatile matter.

cologne A diluted perfume.

complete fertilizer A fertilizer containing the three main nutrients: nitrogen, phosphorus, and potassium.

compound A pure substance made up of two or more elements combined in fixed proportions.

condensation The reverse of vaporization; the change from the gaseous state to the liquid state.

condensation polymer A polymer that does not contain all the atoms in the starting monomers because water is split out as the polymer is formed.

configuration (electron) The arrangement of an atom's electrons in space.

continuous spectrum A spectrum in which there is a continuous variation from one color to another.

control rod A rod used to absorb neutrons, thus controlling the rate of fission in a nuclear reactor.

copolymer A polymer that is formed by the combination of two or more different monomer units.

core (Earth's) A region at the center of the Earth thought to consist largely of iron and nickel.

corrosive waste A waste that requires a special container because it corrodes usual container materials.

cosmetics Substances defined in the 1938 U.S. Food, Drug, and Cosmetic Act as "articles intended to be rubbed, poured, sprinkled or sprayed on, introduced into, or otherwise applied to the human body or any part thereof, for cleaning, beautifying, promoting attractiveness or altering the appearance. . . ."

cosmic rays Extremely high energy rays from outer space.

covalent bond A bond formed by a shared pair of electrons.

cracking process Splitting molecules into fragments by heat in the absence of air.

cream An emulsion of tiny water droplets in oil.

critical mass The mass of an isotope above which a self-sustaining chain reaction can occur.

crust The outer shell of the Earth composed of the lithosphere, the hydrosphere, and the atmosphere.

crystal A solid with plane surfaces at definite angles.

C terminal The end of a peptide or protein that has a free carboxyl group.

cyclic Ring-containing.

cyclone separator A device that removes particulate matter from smokestack gases by swirling so that centrifugal force throws the heavier particles against the walls, where they settle out.

Dacron A polyester made from terephthalic acid and ethylene glycol.

daughter isotopes Isotopes formed by the radioactive decay of another isotope.

DDT Dichlorodiphenyltrichloroethane, an insecticide.

decongestant A cold remedy that loosens or dries up mucus, thus reducing congestion.

defoliation Premature removal of leaves from plants.

Delaney Amendment A 1958 amendment to the Food and Drug Act that automatically bans any chemical shown to induce cancer in laboratory animals.

denatured alcohol Ethyl alcohol to which some noxious substance has been added to render it unfit for drinking.

density The amount of mass (or weight) per unit volume.

deodorant A product designed to prevent body odor by killing odor-causing bacteria.

deoxyribonucleic acid The type of nucleic acid found primarily in the nuclei of cells.

depilatories Hair removers.

depressant A drug that slows down both physical and mental activity.

destructive distillation The process of decomposition by heat with volatile substances distilled off.

deuterium An isotope of hydrogen with a proton and a neutron in the nucleus (mass of 2 amu).

dextro (isomer) A "right-handed" isomer.

dextrose D-Glucose; blood sugar.

dimensional analysis Problem solving by arranging factors so that units cancel to give an answer in proper units.

dioxins Chlorinated cyclic compounds once found as contaminants in herbicides.

dipeptide A compound composed of two bonded amino acids.

dipole A molecule that has a positive end and a negative end.

dipole forces The attractive forces that exist among polar covalent molecules.

disaccharide A sugar that upon hydrolysis yields two monosaccharide molecules per molecule of disaccharide.

dispersion forces The momentary, usually weak, attractive forces between molecules.

dissociative anesthetic A substance that induces hallucinations similar to those reported by people who have had near-death experiences.

dissolved oxygen Oxygen dissolved in water; a measure of that water's ability to support fish and other aquatic life.

distillation The boiling off of volatile compounds (such as alcohol and water), leaving behind solids and high-boiling compounds.

disulfide linkage A covalent linkage through two sulfur atoms.

diuretic A substance that increases the body's output of urine.

double bond The sharing of two pairs of electrons between two atoms.

doubling time The time it takes a population to double in size.

drug abuse Use of a drug for its intoxicating effect.

drug misuse Use of a drug for a purpose other than its intended use.

ductile Able to be drawn into wire.

elastomer A synthetic polymer with rubberlike properties.

electric current A flow of electrons.

electrochemical cell A device that produces electricity by means of a chemical reaction.

electrodes The carbon rods or metal strips inserted into an electrolytic or electrochemical cell.

electrolysis The process of splitting a compound by means of electricity.

electrolyte A compound that, in water solution, conducts an electric current.

electron The unit of negative charge.

electron configuration The arrangement of an atom's electrons in space.

electron-dot formula The structural formula of a molecule with valence electrons of all the atoms indicated by dots.

electron-dot symbol The symbol of an element surrounded by dots representing the atom's outermost electrons.

electronegativity The ability of an atom to attract electron density toward itself when joined to another atom by a chemical bond.

electrostatic precipitator A device that removes particulate matter from smokestack gases by forming an electric charge on the particles, which are then removed by attraction to a surface of opposite charge.

element A fundamental substance in which all atoms have the same number of protons.

emollient An oil or grease used as a skin softener.

emulsion A suspension of submicroscopic particles of fat or oil in water.

end note The portion of perfume that has low volatility; composed of large molecules.

endorphins The naturally occurring peptides that bond to the same receptor sites as the opiate drugs.

endothermic reaction A chemical reaction to which energy must be supplied as heat.

energy The capacity for doing work.

energy levels The specific, quantized energy levels that an electron may have in an atom.

enkephalins Compounds composed of peptide chains of five amino acid units; morphinelike substances produced by the body.

enrichment (food) Replacement of nutrients lost from a food during processing.

enrichment (isotope) The process by which the proportion of one isotope of an element is increased relative to the others.

entropy A measure of the randomness of a system.

enzyme A biological catalyst.

epidermis The outer layer of skin.

essential amino acid One of eight amino acids not produced in the body that must be included in the human diet.

ester A compound derived from carboxylic acids and alcohols; the —OH of the acid replaced by an —OR group.

estrogen A female sex hormone.

ether A molecule with 2 alkyl groups attached to the same oxygen atom.

eutrophication The excessive growth of plants in a body of water that causes some of the plants to die because of a lack of light. The water becomes choked, depleted of oxygen, and useless for fish or recreational purposes.

excited state That state in which an atom is supplied energy and an electron is moved from a lower to a higher energy level.

exothermic reaction A chemical reaction that releases heat.

expectorant A compound that is supposed to bring mucus up out of the bronchial passages.

experiment (v.) Try out a new idea or activity. (n.) An investigation in which variables are controlled.

eye shadow A product composed of a base of petroleum jelly with fats, oils, and waxes and colored by dyes or by zinc oxide or titanium dioxide pigments; used to color eyelids.

factor-label method Problem solving by arranging factors so that units cancel to give an answer in proper units.

families of elements Vertical columns of the periodic table; also called groups.

fast-twitch fibers The stronger and larger kind of muscle fibers that are suited for anaerobic work.

fat A compound formed by the reaction of glycerol with three fatty acid units; a triglyceride.

fat depots Storage places for fats in the body.

fat-soluble vitamins Nonpolar vitamins with a high proportion of hydrocarbon structural elements; dissolve in the fatty tissue of the body and are stored for future use.

fatty acid A carboxylic acid that contains 4 to 20 or more carbon atoms in a chain.

fermentation The process by which yeast produces alcohol from fruit or grains.

fertilizers Materials put on the soil to improve the quality and quantity of plant growth.

first law of thermodynamics Energy is neither created nor destroyed.

fission reaction The splitting of a large unstable atomic nucleus into smaller fragments.

fixed nitrogen Nitrogen combined with another element.

flammable waste A waste that burns readily upon ignition, presenting a fire hazard.

flotation method A coal-cleaning process that makes use of the different densities of coal and its major impurities.

fluorescence A phenomenon in which, after exposure to sunlight, certain chemicals continue to glow even when taken into a dark room.

fly ash Fine solid particles of soot and dust carried out from burning coal by the draft.

food additive Any substance other than basic foodstuffs that is present in food as a result of some aspect of production, processing, packaging, or storage.

formula Representation of a chemical compound.

formula mass The sum of the atomic masses of all of the atoms represented in the chemical formula.

fossil fuels Coal, petroleum, and natural gas.

free radical A reactive neutral chemical species that contains an unpaired electron.

freezing The reverse of melting; the change from liquid to solid state.

Freon A carbon compound containing fluorine, as well as chlorine; Du Pont trade name for chlorofluorocarbons.

fructose A simple sugar found in fruits and honey or made by isomerization of glucose.

fruit sugar Fructose.

fuels Substances that burn readily with the release of significant amounts of energy.

fuel cell A device in which chemical reactions are used to produce electricity directly from fuels and oxygen.

functional group The atom or group of atoms that conveys characteristic properties upon a molecule.

fundamental particles Basic units from which more complicated structures can be fashioned: protons, neutrons, and electrons.

fusion reaction The combining of small atomic nuclei to produce larger ones.

galactosemia An inherited disorder characterized by the inability to convert galactose (from milk sugar) to glucose. The buildup of galactose causes mental retardation and other problems.

gamma rays Rays similar to X rays that are emitted from radioactive substances; have higher energy and are more penetrating than X rays.

gas The state of matter in which the substance maintains neither shape nor volume.

gasoline The fraction of petroleum containing C_5 to C_{12} alkanes, used as automotive fuel.

gene The segment of a nucleic acid molecule that contains the information necessary to produce a protein; the smallest unit of hereditary information.

general anesthetic A depressant that acts on the brain to produce unconsciousness as well as insensitivity to pain.

geometric growth A doubling in size for each growth period.

geothermal energy Energy derived from the heat of Earth's interior.

glass A noncrystalline material obtained by melting sand with soda, lime, and various other metal oxides.

glass transition temperature An important parameter of polymers; above this temperature the polymer is rubbery and tough, while below it the polymer is like glass—hard, stiff, and brittle.

globular proteins Protein molecules that fold into roughly spherical or ovoid shapes and that can be dispersed in water.

glucose The simple sugar that is circulated in the bloodstream; also called dextrose or blood sugar.

glycogen An animal starch composed of branched chains of glucose units.

gram One one-thousandth of a kilogram; a unit of mass equal to 0.03527 oz.

GRAS list The list, established by the U.S. Congress in 1958, of additives generally recognized as safe.

greenhouse effect The retention of the sun's heat energy by the Earth as a result of excess carbon dioxide or other substances in the atmosphere; causes an increase in the Earth's atmospheric and surface temperature.

ground state The state of an atom in which all electrons are in the lowest possible energy level.

group A vertical column of the periodic table; a family of elements.

groundwater Water located underground in porous rock strata and soil.

half-life The amount of time required for one-half the radioactive nuclei in a sample to decay.

hallucinogen A drug that produces visions and sensations that are not part of reality.

halogen An element in Group VIIA of the periodic table.

hard water Water containing ions of calcium, magnesium, and iron.

hazardous waste A waste that, when improperly managed, can cause or contribute to death or illness or threaten human health or the environment.

heat A measure of a quantity of energy; of how much energy a sample contains.

heat capacity (of a substance) The amount of heat needed to change the temperature of the substance by 1 °C.

heat of vaporization (of a substance) The amount of heat involved in the evaporation or condensation of 1 g of the substance.

heat stroke A failure of the body's heat regulatory system; unless the victim is treated promptly, the rapid rise of body temperature causes brain damage or death.

herbicide A chemical used to kill weeds.

heterocyclic compound A cyclic compound in which one or more atoms in the ring is not carbon.

high-density polyethylene A polyethylene composed of linear molecules; a strong, rigid plastic.

high-fructose corn syrup A sweetener made by the hydrolysis of starch and the isomerization of a part of the resulting glucose to fructose.

homogeneous The same throughout; property of a sample with the same composition in all parts.

homologous series A series of compounds in which adjacent members of the series differ by fixed unit of structure.

hormone A chemical messenger that is secreted into the blood by an endocrine gland.

humectant A moistening agent.

hydrocarbon An organic compound that contains only carbon and hydrogen.

hydronium ion A water molecule to which a hydrogen ion (H^+) has been added; the characteristic ion of an aqueous acid.

hydrophobic interactions The forces (usually dispersion forces) that exist between nonpolar molecules or nonpolar groups within a molecule.

hydrosphere The oceans, seas, rivers, and lakes of the Earth.

hydroxide ion OH^-; responsible for the properties of base in water.

hydroxyl group The —OH group.

hyperacidity An excess of acid in the stomach.

hypoallergenic cosmetics Cosmetics claimed to cause fewer allergic reactions than regular products.

hypotheses Guesses that can be tested by experiment.

hydrogen The lightest substance on Earth with the smallest atoms. The most abundant element in the universe.

hydrogen bomb A bomb based on the nuclear fusion of isotopes of hydrogen.

hydrogen bond The dipole interaction between a hydrogen atom bonded to F, O, or N in a donor molecule and an F, O, or N atom in a receptor molecule.

hydrolysis The reaction of a substance with water; literally, a splitting by water.

incidental additive A food additive accidentally introduced during production, processing or packaging or during storage.

incineration Destruction by burning.

indicator A substance that is one color in acid and another color in base.

inorganic Mineral; composed of compounds other than those of carbon.

inorganic chemistry The study of the compounds of all elements other than carbon.

insecticide A substance that kills insects.

intentional additive A substance purposefully put into a food product to perform some specific function.

interionic forces The electrostatic forces between ions.

iodine number The number of grams of iodine that will be consumed by 100 g of fat or oil; an indication of the degree of unsaturation.

ion A charged atom or group of atoms.

ion–dipole force Electrostatic force between an ion and the end of a dipolar molecule of opposite charge.

ionic bond The chemical bond that results when electrons are transferred from a metal to a nonmetal; the electrostatic attraction between ions of opposite charge.

ionizing radiation Radiation that produces ions as it passes through matter.

isomerization The conversion of a compound into one or more of its isomers.

isomers Compounds that have the same molecular formula but different structural formulas and properties.

isotopes Atoms that have the same number of protons but different numbers of neutrons.

joule (J) The SI unit of heat (1 J = 0.239 cal).

juvenile hormones Hormones that control the rate of development of the young; used to prevent insects from maturing.

kelvin The SI unit of temperature. Zero on the Kelvin scale is absolute zero.

keratin The tough, fibrous protein that comprises most of the outermost layer of the epidermis.

kerogen The complex material found in oil shale; has an approximate composition of $(C_6H_8O)_n$, where n is a large number.

ketone An organic compound with a carbonyl between 2 carbon atoms.

ketone bodies Acetoacetic acid, β-hydroxybutyric acid, and acetone.

ketosis A condition characterized by the presence of excess ketones in the blood and urine.

kilocalorie Measurement of energy content, equal to 1000 calories.

kilogram The SI unit of mass, a quantity slightly greater than two pounds.

kinetic energy The energy of motion.

kinetic-molecular theory A model that uses the motion of molecules to explain the behavior of the three states of matter.

kwashiorkor A disease caused by protein deficiency.

lactose The sugar found in milk; composed of two simpler sugars, glucose and galactose.

lactose intolerance The inability of some individuals to break down the sugar lactose; caused by the absence of the necessary enzyme.

lakes Colored complexes formed by adhering acid dye compounds to metal ions.

landfill (sanitary) A place for disposal of wastes in an excavated area; the wastes are compacted and covered with fill.

lanolin A natural wax obtained from sheep's wool.

law of combining volumes The volumes of gaseous reactants and products are in a small whole number ratio when all measurements are made at the same temperature and pressure.

law of conservation of energy The amount of energy within the universe is constant; energy cannot be created or destroyed, only transformed.

law of conservation of mass Matter is neither created nor destroyed during a chemical change.

law of constant composition A compound always contains elements in certain definite proportions, never in any other combination; also called the law of definite proportions.

law of definite proportions A compound always contains elements in certain definite proportions, never in any other combination; also called the law of constant composition.

law of multiple proportions Elements may combine in more than one proportion to form more than one compound, for example, CO and CO_2.

LD_{50} The dosage that would be lethal to 50% of the population of test animals.

levo (isomer) A "left-handed" isomer.

limiting reagent The reactant that is used up first in a reaction, after which the reaction ceases no matter how much remains of other reactants.

line spectrum The pattern of colored lines emitted by each element.

lipid A substance from animal or plant cells that is soluble in nonpolar solvents and insoluble in water.

lipoprotein A protein combined with a lipid, such as a triglyceride or cholesterol.

liquid A state of matter in which the substance assumes the shape of its container, flows readily, and maintains a fairly constant volume.

liter Measurement of volume, equal to a cubic decimeter.

lithosphere The solid portion of the Earth.

local anesthetic A substance that renders a part of the body insensitive to pain while leaving the patient conscious.

London smog Smog, usually from burning coal, consisting mainly of sulfur dioxide, sulfuric acid, ash, soot, smoke, and fog.

Los Angeles smog Smog created by action of sunlight on unburned hydrocarbons and nitrogen oxides, mainly from automobiles; photochemical smog.

lotion An emulsion of submicroscopic fat or oil droplets dispersed in water.

low-density polyethylene A waxy, semirigid, transparent, or translucent plastic that is resistant to chemicals; composed of branched chains of carbon atoms.

LUST Acronym for leaking underground storage tanks.

macromolecule A molecule with a very high molecular mass; a polymer.

main group elements The elements in the A groups of the periodic table that is customary in the United States and in Groups 1, 2, and 13 to 18 in the periodic table recommended by IUPAC.

malignant tumor Cancer, a tumor characterized by irreversible growth.

malleable Able to be forged and welded.

mantle (Earth's) A region of the Earth between the core and crust; thought to be composed mostly of silicates.

marijuana A preparation made from the leaves, flowers, seeds, and small stems of the *Cannabis* plant.

mascara A product composed of a base of soap, oils, fats, and waxes and colored by iron oxide pigments, carbon, chromium oxide, or ultramarine; used to darken eyelashes.

mass A measure of the quantity of matter.

mass-energy equation Einstein's equation $E = mc^2$ in which E is energy, m is mass, and c is the speed of light.

mass number Nucleon number, the sum of the numbers of protons and of neutrons in the nucleus of an atom.

matter Stuff of which all materials are made; has mass and occupies space.

mechanism A series of individual steps in a chemical reaction that gives the net overall change.

melanin A brownish black pigment that determines the color of the skin and hair.

melting point The temperature at which a substance changes from the solid to the liquid state.

messenger RNA (mRNA) The type of RNA that contains the codons for a protein; mRNA travels from the nucleus of the cell to a ribosome.

metabolism The sum of all the chemical reactions by which the protoplasm of an organism grows, is maintained, obtains energy, and is degraded.

metalloid An element with properties intermediate between those of metals and nonmetals.

metals The group of elements to the left of the heavy, stepped, diagonal line on the periodic table.

methane Simplest compound of C and H (CH_4); primary component of natural gas.

meter The SI unit of length, slightly longer than three feet.

methemoglobin Hemoglobin in which the iron ion has a 3+ charge.

micas Minerals composed of SiO_4 tetrahedra arranged in a two-dimensional, sheetlike array.

micronutrients Substances needed by the body only in tiny amounts.

middle note The portion of perfume intermediate in volatility; responsible for the lingering aroma after most top-note compounds have vaporized.

mineral acids Acids derived from inorganic materials.

minerals (dietary) The inorganic substances required in the diet for good health.

mitochondria Structures within a cell (organelles) in which aerobic metabolism takes place.

mixture Matter with a variable composition.

moderator A substance used to slow down the fission neutrons within a nuclear reactor.

moisturizer (skin) A substance that adds moisture to skin or acts to retain moisture in the skin.

molar mass The formula weight expressed in grams.

molar volume The volume occupied by 1 mol of a substance under specified conditions.

mole The formula weight in grams of an element or compound; or a quantity of chemical substance that contains 6.02×10^{23} units of the substance.

molecule Two or more atoms joined together by covalent bonds; the smallest fundamental unit of an element or compound that retains the characteristic properties of that element or compound.

monomer A molecule of relatively low molecular mass. Monomers are combined to make polymers.

monounsaturated fatty acids Fatty acids that contain one double bond per molecule.

mousse A foam or froth, a hair care product composed of holding resins to hold hair in place.

mutagen Any entity that causes changes in genes without destroying the genetic material.

mutations Changes in the molecules of heredity in reproductive cells.

narcotic A depressant, analgesic drug that induces narcosis (sleep).

narrow-spectrum insecticides Insect-killing chemicals that are directed at one or a few insect species.

nasal decongestant A cold remedy that loosens or dries up mucus in the nostrils, thus relieving congestion.

natural gas A mixture of gases, mainly methane, found in many underground deposits.

natural philosophy Philosophical speculation about nature.

neurons Nerve cells.

neurotransmitters Chemicals that carry the impulse across the synapse from one nerve cell to the next.

neutralization The reaction of an acid and a base to produce a salt and water.

neutron A fundamental particle with a mass of approximately 1 amu and no electrical charge.

noble gases Generally unreactive gases that appear in the far right column of the periodic table.

nonbonding pairs Pairs of electrons not involved in a bond.

nonionic surfactant A surfactant with a hydrocarbon tail and a polar head whose oxygen atoms attract water molecules and make the head water soluble; bears no ionic charge.

nonmetals The group of elements to the right of the heavy, stepped, diagonal line on the periodic table.

nonpolar covalent bond A covalent bond in which there is an equal sharing of electrons.

note A fraction of a perfume based on differences in volatility (*see also* top note, middle note, *and* end note).

N terminal The end of a peptide or protein that has a free amino group.

nuclear fission The splitting of the nucleus into two large fragments.

nuclear fusion Combustion of two small nuclei to produce one larger nucleus.

nuclear winter A period of dark, cold weather that may be caused by the dust and smoke entering the atmosphere from the explosion of nuclear bombs.

nuclear reactor A plant that produces energy by nuclear fission.

nucleic acids Nucleotide polymers, DNA and RNA.

nucleon number The total number of protons and neutrons in an atom; the mass number.

nucleons Protons and neutrons in an atomic nucleus.

nucleotide A combination of a heterocyclic amine, a pentose sugar, and phosphoric acid; the monomer unit of nucleic acids.

nucleus *See* atomic nucleus; cell nucleus.

nylon A synthetic polyamide.

octane rating The antiknock quality of a gasoline compared to mixtures of isooctane (with a rating of 100) and heptane (with a rating of 0).

octet rule Atoms seek an arrangement that will surround them with eight electrons in the outermost energy level.

oils Substances formed from glycerol and three unsaturated fatty acids; liquids at room temperature.

oil shale Fossil rock containing kerogen from which oil can be obtained at high cost by distillation.

open dumps Open solid-waste-disposal areas; often occupied by rats, flies, and other pests.

optical brightener A compound that absorbs the invisible ultraviolet component of sunlight and re-emits it as visible light at the blue end of the spectrum.

orbital A region of space in an atom occupied by one or two electrons.

organic Derived from or pertaining to a living organism; pertaining to the carbon-containing compounds.

organic chemistry The study of the compounds of carbon.

organic farming Farming without synthetic fertilizers or pesticides.

osteoporosis A disease characterized by a reduction in the quantity of bone.

oxidation Combination of elements and compounds with oxygen, loss of hydrogen; loss of electrons.

oxidizing agent A substance that causes oxidation and is itself reduced.

oxygen The most abundant element in Earth's crust.

oxygen debt The demand for oxygen in muscle cells during anaerobic exercise.

ozone A highly reactive allotropic form of oxygen; O_3.

ozone layer The layer of the stratosphere that contains ozone and shields living creatures on Earth from deadly ultraviolet radiation.

paint A surface coating that contains a pigment, a binder, and a solvent.

particulate matter A pollutant composed of solid and liquid particles of greater than molecular size.

pathogenic Disease causing.

peptide bond The amide linkage that bonds amino acids in chains of proteins, polypeptides, and peptides.

perfluorocarbon A compound in which all hydrogen atoms have been replaced by fluorine atoms.

perfumes Fragrant mixtures of plant extracts and other chemicals dissolved in alcohol.

periodic table A systematic arrangement of the elements in columns and rows; elements in a given column have similar properties.

periods The horizontal rows of the periodic table.

persistent pesticide One that does not break down readily in the environment.

pesticide A substance that kills some kind of pest (weeds, insects, rodents, etc.).

petroleum A dark, oily mixture of hydrocarbons, mainly alkanes, occurring in various deposits around the world.

PET scans Images of internal organs obtained through positron emission tomography.

pH The negative logarithm of hydronium ion concentration.

phaeomelanin A red-brown pigment that colors the hair and skin of redheads.

pharmacology The study of the response of living organisms to drugs.

phenol An OH group attached to a benzene ring.

phenyl group C_6H_5—, the benzene ring as a substituent.

pheromones Natural chemicals secreted by organisms to mark a trail, send out an alarm, or attract a mate.

photochemical smog Smog created by the action of sunlight on unburned hydrocarbons and nitrogen oxides, mainly from automobiles; also called Los Angeles smog.

photon A unit particle of energy.

photoscan A permanent visual record showing the differential uptake of a radioisotope by various tissues.

photosynthesis The chemical process used by green plants to convert solar energy into chemical energy by reducing carbon dioxide.

photovoltaic cell A solar cell; a cell that converts sunlight directly to electrical energy.

pH scale An exponential scale of acidity; below 7, acidic; 7, neutral; above 7, basic.

phthalate esters Esters of phthalic acid; used as plasticizers.

physical change A change in physical state or form.

physical properties A description of the qualities of a substance that can be demonstrated without changing the composition of the substance.

pica A disorder in which a person eats dirt, paint chips, and other materials generally regarded as inedible.

placebo A substance that looks and tastes like a real drug but has no active ingredients.

plasma A state of matter similar to a gas but composed of isolated electrons and nuclei rather than discrete whole atoms or molecules.

plasmid A circular DNA molecule that is separate from the chromosomes.

plastic A material that can be shaped when soft and then hardened.

plasticizer A chemical substance added to some plastics, such as vinyl, that makes them more flexible and easier to work with.

pleated sheet A secondary protein structure characterized by antiparallel molecules with zigzag structure.

polar covalent bond A covalent bond in which more than half of the bond's negative charge is concentrated around one of the two atoms.

polar molecule A molecule that has a dipole moment.

pollutant A chemical in the wrong place and/or in the wrong concentration.

pollution Contamination of the environment.

polyatomic ion A charged particle containing two or more covalently bonded atoms.

polychlorinated biphenyls (PCBs) Compounds derived from the hydrocarbon biphenyl by the replacement of from one to ten of the hydrogen atoms with chlorine atoms.

polyester A polymer made from a dicarboxylic acid and a dialcohol.

polyethylene Plastic made up of giant alkane molecules.

polymer A molecule with a large molecular mass; a chain formed of repeating smaller units.

polymerization Making very large molecules from small monomers.

polystyrene A hydrocarbon chain with benzene rings attached.

primary plant nutrients Nitrogen, phosphorus, and potassium.

proton acceptor A base.

polypeptide A polymer of amino acids usually of lower molecular mass than a protein.

polysaccharides Starches and celluloses; carbohydrates one molecule of which yields many molecules of monosaccharide(s) upon hydrolysis.

polyunsaturated Containing two or more double bonds.

positron A positively charged particle with the mass of an electron.

potential energy Energy by virtue of position or composition.

preemergent herbicide A herbicide that is rapidly broken down in the soil and can therefore be used to kill weed plants before the crop seedlings emerge.

primary sewage treatment A type of plant with a holding pond intended to remove some of the sewage solids as sludge by settling.

primary structure The amino acid sequence in a protein or of nucleotides in a nucleic acid.

product A substance produced by a chemical reaction and whose formula follows the arrow in a chemical equation.

progestin A compound that mimics the action of progesterone.

proof Twice the percentage of alcohol by volume.

prostaglandins Hormonelike compounds derived from arachidonic acid that are involved in increased blood pressure, the contractions of smooth muscle, and other physiological processes.

proton The unit of positive charge in the nucleus of an atom; the hydrogen nucleus in acid–base chemistry.

proton donor An acid; a substance that gives up an H^+ (proton).

psychedelics Drugs that induce colorful visions.

psychotomimetics Drugs that induce psychoses.

psychotropic drugs Drugs that affect the mind.

pure substance Matter with a definite, or fixed, composition.

purines Bases with two fused rings found in nucleic acids.

pyrimidines One-ring bases found in nucleic acids.

quantum A packet of specific size; one photon of energy.

quartz A compound composed of silicon and oxygen.

quaternary structure The arrangement of protein subunits in geometric shapes.

radioactive decay Disintegration of an unstable atomic nucleus by spontaneous emission of radiation.

radioactivity Spontaneous emission of alpha, beta, and gamma rays by the disintegration of the nuclei of atoms.

radioisotopes Radioactive isotopes.

radon A radioactive gas; heaviest of the noble gases.

reactant A starting material or original substance in a chemical change and whose formula precedes the arrow in a chemical equation.

reactive wastes Wastes that tend to react spontaneously or to react vigorously with air or water.

recombinant DNA DNA in an organism that contains genetic material from another organism.

recommended daily allowance (RDA) The recommended level of a nutrient necessary for a balanced diet.

recycling Turning trash into useful products.

redox reaction A reaction in which oxidation and reduction occur.

reducing agent A substance that causes reduction and is itself oxidized.

reduction A gain of electrons; a loss of oxygen; a gain of hydrogen.

renewable resource A resource that is replenished by natural events.

replication Copying or duplication; the process by which DNA reproduces itself.

research A study or investigation to find new information.

resin A polymeric material, usually a sticky solid or semisolid organic material.

respiration "Burning" of glucose by living cells; oxygen is absorbed and carbon dioxide is given off.

restorative drugs Drugs used to relieve the pain and reduce inflammation from overuse of muscles.

restriction endonucleases Enzymes that cleave DNA molecules at the location of specific base sequences.

restriction fragment length polymorphisms (RFLPs) A pattern of segments of genetic material produced by restriction endonucleases.

retroviruses RNA viruses that synthesize DNA in the host cells.

reverse osmosis A method of pressure filtration through a semipermeable membrane; water flows from an area of high salt concentration to an area of low salt concentration.

ribonucleic acid (RNA) The form of nucleic acid found mainly in the cytoplasm, but also present in all other parts of the cell.

ribosomes The locations in the cell where protein synthesis occurs.

risk–benefit analysis An approach that involves the calculation of a desirability quotient by dividing the benefits by the risks.

rubber An elastomer; a polymer that stretches.

Rule of 70 A mathematical formula that gives the doubling time for a population growing geometrically; 70 divided by annual rate equals doubling time.

salt An ionic compound produced by the reaction of an acid with a base.

salt bridge An interaction between an acidic side chain on one amino acid residue and a basic side chain on another; the resulting charges serve as an ionic bond between two peptide chains or between two parts of the same chain.

saponins Natural chemical compounds that produce a soapy lather.

saturated fat A fat composed of a large proportion of saturated fatty acids esterified with glycerol.

saturated fatty acid A fatty acid that contains no carbon-to-carbon double bonds.

saturated hydrocarbon A compound of carbon and hydrogen with only single bonds.

science A branch of knowledge based on the laws of nature.

scientific law A summary of experimental data; often expressed in the form of a mathematical equation.

scientific model A representation that serves to explain a scientific phenomenon.

sebum An oily secretion of the body that protects skin from moisture loss.

secondary plant nutrients Magnesium, calcium, and sulfur.

secondary sewage treatment Passing effluent from a primary treatment plant through gravel and sand filters to aerate the water and remove suspended solids.

secondary structure Arrangement of polypeptide chains in a protein, e.g., helix or pleated sheet.

second law of thermodynamics The degree of randomness in the universe increases in any spontaneous process.

segmer The repeating unit within a polymer.

set-point theory The explanation of the difficulty of losing weight by diet that holds that each person has an individual level of circulating fatty acids below which he or she is constantly hungry.

sex attractant A substance or mixture of substances released by an organism to attract members of the opposite sex of the same species for mating.

significant figures Those measured digits that are known with certainty plus one uncertain digit.

silicate minerals Compounds of metals with silicon and oxygen.

silicone A polymer with a base chain of alternating Si and O atoms.

single bond The sharing of one pair of electrons.

skin protection factor (SPF) The rating of a sunscreen's ability to limit the penetration of ultraviolet radiation.

slag Product of the reaction of limestone with silicate impurities in iron ore.

slow-twitch fibers Muscle fibers that are suited for aerobic work.

smog The combination of smoke and fog; polluted air.

soaps Salts (usually sodium salts) of long-chain carboxylic acids.

solar cell A device used for converting sunlight into electricity; a photoelectric cell.

solid A state of matter in which the substance maintains its shape and volume.

solute The substance that is dissolved in another substance (solvent) to form a solution; usually present in a smaller amount than the solvent.

solution A homogeneous mixture of two or more substances.

solvent The substance that dissolves another substance (solute) to form a solution; usually present in a larger amount than the solute.

specific heat (of a substance) The amount of heat required to raise the temperature of 1 g of the substance by 1 °C.

standards of identity The minimum required proportions of ingredients to pass FDA standards for a particular food.

starvation The withholding of nutrition from the body whether voluntary or involuntary.

state of matter A condition or stage in the physical being of matter; solid, liquid, or gas.

steels Alloys formed by adjusting the carbon content of iron.

steroid A molecule that has a 4-ring skeletal structure, with 1 cyclopentane and 3 cyclohexane fused rings.

stimulant A drug that increases alertness, speeds up mental processes, and generally elevates the mood.

stoichiometry Quantity relationships between reactants and products in a chemical reaction.

STP Standard temperature (0 °C) and pressure (1 atm).

straight-run gasoline Gasoline hydrocarbons as they come from the distilling column that separates crude oil into various fractions.

stratosphere The portion of the atmosphere that contains the ozone layer; the next layer above the troposphere.

strong acid An acid that reacts completely; 100% ionized in water; a powerful proton donor.

strong base A base that dissociates 100% in water; powerful proton acceptor.

structural formula A chemical formula that shows how the atoms of a molecule are arranged, to which other atom(s) they are bonded, and the kinds of bonds.

sublevel A subdivision of electron energy levels in an atom.

substrate The substance that bonds to the active site of an enzyme; the substance acted upon.

sucrose Common table sugar derived principally from sugar cane or sugar beets; composed of two simpler sugars, glucose and fructose.

sulfide minerals Compounds of metals with sulfur.

sunscreen lotion A type of lotion that promotes tanning of the skin by blocking out the short-wave ultraviolet radiation while allowing the longer wave ultraviolet radiation to pass through.

surface-active agent Any agent that stabilizes the suspension in water of nonpolar substances such as grease and oil.

synapses The gaps between nerve fibers.

synergistic effect An effect much greater than just the sum of the expected effects.

tailings The waste products of ore processing.

tar sands Sands that contain bitumen, a thick hydrocarbon material.

technology The sum total of processes by which humans modify the materials of nature to better satisfy their needs and wants.

temperature A measure of intensity, or how energetic the particles of a sample are.

teratogens Toxic substances that cause birth defects when introduced into the body of the mother.

tertiary structure The folds, bends, and twists in protein or nucleic acid structure.

tertiary treatment The third stage of sewage treatment; designed to remove impurities not affected by primary or secondary treatment.

testosterone The principal male sex hormone.

tetracyclines Antibacterial drugs with four fused rings.

theories Detailed explanations of the behavior of matter based on experiments; may be revised if new data warrant.

thermal pollution The energy released into the environment that causes undesirable changes in the environment; released during energy conversions for other purposes such as the generation of electricity.

thermonuclear reactions Nuclear fusion reactions that require extremely high temperatures.

thermoplastic A kind of polymer that can be heated and reshaped.

thermoset plastic A kind of polymer that cannot be softened and remolded.

top note The portion of perfume that vaporizes the quickest; composed of relatively small molecules; responsible for odor when perfume is first applied.

toxicology The division of pharmacology that deals with the effects of poisons on the body, their identification and detection, and remedies against them.

toxic waste A waste that contains or releases poisonous substances in large enough amounts to threaten human health or the environment.

tracers Radioactive isotopes used to trace movement or locate the site of radioactivity in physical, chemical, and biological systems.

training effect The net effect, acquired through repeated exercise, of being able to do more physical work with less strain.

transcription The process by which DNA directs the synthesis of an mRNA molecule during protein synthesis.

transfer RNA (tRNA) A small molecule that contains the anticodon nucleotides; the RNA molecule that bonds to an amino acid.

transition elements Metallic elements situated in the center portion of the periodic table in the B groups.

translation The process by which the information contained in the codon of an mRNA molecule is converted to a protein structure.

transmutation Change of one element into another.

tripeptide A compound composed of three bonded amino acids.

triple bond The sharing of three pairs of electrons between two atoms.

triplet Three-base sequence in a nucleic acid; a codon.

tritium A rare radioactive isotope of hydrogen with two neutrons and one proton in the nucleus (a mass of 3 amu).

triglyceride A chemical combination of glycerol with 3 fatty acids.

troposphere The layer of the atmosphere nearest the Earth; harbors almost all living things.

tumor The abnormal growth of new tissues.

unit-conversion method A method of problem solving that considers the units as a part of a quantity and uses conversion factors in such a way that an answer with proper units is obtained.

unsaturated hydrocarbon A hydrocarbon containing a double or a triple bond.

unsaturated molecule A molecule containing a double or triple carbon–carbon double bond.

valence The number of covalent bonds that an atom can form.

valence electrons Electrons in the outermost shell.

valence shell Outer shell of electrons of an atom.

valence-shell electron-pair repulsion theory A theory of chemical bonding; states that valence-shell electrons locate themselves as far apart as possible.

vaporization The process in which a substance changes from the liquid to the gaseous (vapor) state.

vinyl polymer Usually refers to polyvinyl chloride.

vitamins Organic compounds that the body cannot produce in the amounts required for good body health.

volatile organic chemicals (VOC) A class of pollutants consisting of organic compounds that are easily vaporized.

volt The unit of measurement of electrical potential or of the tendency of electrons in a system to flow.

VSEPR theory Valence shell electron pair repulsion theory for determining the shapes of molecules.

vulcanization The process of making naturally soft rubber harder by reaction with sulfur.

water Nature's most abundant liquid, made up of bent, polar H_2O molecules.

water-soluble vitamins Vitamins with a high proportion of oxygen and nitrogen that are able to form hydrogen bonds with water.

waxes Esters of long chain (fatty) acids with long chain alcohols.

weak acid An acid that reacts only slightly; a poor proton donor with a low percentage ionization in solution.

weak base A base that ionizes to a small degree and produces only a few OH^- ions in solution; a poor proton acceptor.

weight A measure of the force of attraction of the Earth for an object.

wet scrubber A pollution control device that uses water or solutions to remove pollutants from smokestack gases.

X ray Radiation similar to visible light but of much higher energy and much more penetrating.

zwitterion A compound that contains both a positive and negative charge; a dipolar ion.

Appendix F
Answers

Answers are provided for all in-chapter exercises, for selected odd-numbered review questions, for the odd-numbered problems in the matched sets, and for selected additional problems.

Note: for numerical problems, your answer may vary slightly from ours because of rounding and the use of significant figures (Appendix D).

Chapter 1

1.1 Society considers the cheap energy from burning coal worth the risks. Environmentalists disagree.

1.2 a. 2.4 kg b. 480

1.3 physical: b, d; chemical: a, c

1.4 elements: He, Hf, No; compounds: CuO, NO, KI

1.5 0.0063

1.6 1.5 lb

1.7 2500

1.8 78 in.

1.9 310 K

1.10 19.3 g/cm^3

1.11 3.7 mL

1. The study of matter and the changes it undergoes.

3. matter: a, b, d, e

19. benefits divided by risk

29. both

33. a. chemical change b. physical change

39. pure substances: a, b; mixture: c

41. pure substance: b; mixture: a

43. elements: a, b, d; compound: c

45. a. C b. Cl c. P d. Ca e. K f. Pu

51. 1000 mm; 100 cm

53. a. L b. same size

55. a. 7500 mm b. 460 mm

57. a. 45 g b. 86 mg

61. a. 298 K b. 546 K

63. a. 820 cal b. 3550 cal

65. a. 25,000 m b. 0.675 m c. 8300 g d. 0.027 L e. 0.289 ms f. 11.8 cm

67. 1.34 g/mL

69. 0.940 g/mL

71. 66.0 g

73. 84.2 mL

Chapter 2

2.1 63 g

2.2 4 atoms hydrogen/1 atom carbon

1. The atomistic theory assumes that matter is made up of small unit particles that cannot be further subdivided and still be the same kind of matter.

3. the discoveries summarized in the laws of definite proportions and multiple proportions

5. atomistic: c, d, e; continuous: a, b, f

7. When a chemical change occurs, matter is neither created nor destroyed.

9. If a substance could be broken down into simpler substances, it was not an element.

11. The copper atoms have not been destroyed; they are in solution.

13. The atoms of the aspirin have not been destroyed; they are in solution.

15. The report is inaccurate; the chlorine atoms have not been accounted for.

19. 11.00 g CO_2; the law of definite proportions

23. the law of conservation of mass
25. the law of multiple proportions
27. No. Dalton assumed that atoms of different elements had *different* masses.
35. Yes; all three samples have the same composition
37. The mass of product plus recovered zinc equals the mass of starting materials; the results obey the law of conservation of mass.
39. 40 g hydrogen
41. 270 g carbon
43. 4 hydrogen atoms per silicon atom

Chapter 3

3.1 32

3.2 (Mg) 2) 8) 2); (P) 2) 8) 5)

3.3 $1s^2 2s^2 2p^5$

3.4 $1s^2 2s^2 2p^6 3s^2 3p^5$

3.5 Rb: $5s^1$; Se: $4s^2 4p^4$

7. A and D are isotopes; B and C are isotopes.
13. neither attract nor repel
15. electrons
17. 10
19. argon
23. 32
27. the atom in which the electron moved from the first to the third level
29. a. 2 b. 11 c. 17 d. 8 e. 12 f. 16
31. A neutral atom with 3 protons would have only 3 electrons
37. metal: chromium; nonmetals: sulfur, iodine
39. a. period 3 b. period 6 c. period 1
41. lack of chemical reactivity
43. alkali metal: K
45. transition metals: Ti, Tc
47. 2
49. 2 electrons; spherical orbital; one orbital
51. $1s^2 2s^2$
53. a. Be b. N c. Al
61. argon (Ar)

Chapter 4

4.1 52

4.2 48

4.3 $^{90}_{37}X$ and $^{88}_{37}X$ are isotopes of element 37; $^{90}_{38}X$ and $^{93}_{38}X$ are isotopes of element 38

4.4 californium-246

4.5 bromine-85

4.6 0.038 mg

4.7 0.5 mg

4.8 a neutron

4.9 $^{188}_{79}Au \rightarrow ^{188}_{78}Pt + ^{0}_{1+}e$

4.10 22,920 yr

5. $^{1}_{1}H$, $^{2}_{1}H$, $^{3}_{1}H$
7. $^{83}_{35}Br$
9. a. $^{69}_{31}Ga$ b. $^{98}_{42}Mo$ c. $^{99}_{42}Mo$ d. $^{98}_{43}Tc$
11. isotopes: b
13. 97
15. 10.8 amu
17. The nucleon number is unchanged; the atomic number increases by 1.
19. Both the nucleon number and the atomic number decrease by 1.
21. Neither the nucleon number nor the atomic number change.
27. alpha particles
29. Alpha particles are more massive and carry a greater charge than beta particles.
31. use shielding; move farther away from the source.
35. medical X rays
39. $^{31}_{16}S \rightarrow ^{31}_{15}P + ^{0}_{1+}e$
41. $^{21}_{12}Mg \rightarrow ^{20}_{11}Na + ^{1}_{1}H$
43. $^{225}_{90}Th \rightarrow ^{221}_{88}Ra + ^{4}_{2}He$
45. a. $^{179}_{79}Au \rightarrow ^{175}_{77}Ir + ^{4}_{2}He$
    b. $^{23}_{10}Ne \rightarrow ^{23}_{11}Na + ^{0}_{1-}e$
47. $^{24}_{11}Na$
49. $^{215}_{85}At$
51. 1500; 750
53. 6.25 mg
55. 11,460 yr
57. 5730 yr; 11,460 yr

59. $^{18}_{8}O$

61. atomic number 105; nucleon number 258

63. 420 cm^3

Chapter 5

5.1 a. :Ar: b. ·Ca· c. :F· d. :N· e. K· f. :S·

5.2 Li· + :F· → Li^+ + :F:$^-$

5.3 2 ·Al· + 3 :O· → 2 Al^{3+} + 3 :O:$^{2-}$

5.4 CaF_2

5.5 a. :Br· + :Br· → :Br:Br:
b. H· + :Br· → H:Br:
c. :I· + :Cl· → :I:Cl:

5.6
```
   H H
H:C:C:Cl:
   H H
```

5.7 :F:O:F:

5.8
```
[   H   ]+
[ H:P:H ]
[   H   ]
```

5.9 :O::N:O:F:

5.10 linear

5.11 pyramidal

1. the noble gases

3. Sodium metal is quite reactive; sodium ions are quite stable.

7. a. ·C· b. K· c. ·Mg·
d. :Cl· e. :N·

15. :I· + ·I: → :I:I:

17. ionic: a, c; polar covalent: b

19. nonpolar covalent: Br_2, F_2; polar covalent: HCl

21. polar: all

25. polar

27. C, N, O

29. molecules with odd number of electrons (NO_2), those with incomplete octets (BF_3), and those with an expanded octets (SF_6)

31. dispersion forces (N_2), dipole interactions (HCl), hydrogen bonds (H_2O), and interionic forces (NaCl)

33. a. melting b. vaporization

43. insoluble in water; soluble in benzene

45. a. ·Al· Al^{3+} b. :Br· :Br:$^-$
c. ·Mg· Mg^{2+} d. :O:$^{2-}$:Ne:

49. a. Mg^{2+} 2 :F:$^-$ b. Ca^{2+} 2 :Cl:$^-$
c. 2 Na^+ :O:$^{2-}$ d. 2 K^+ :S:$^{2-}$

51. a. Mg^{2+} :O:$^{2-}$ b. Al^{3+} :N:$^{3-}$
c. 2 Al^{3+} 3 :S:$^{2-}$

53.
```
·P· + 3 H· → H:P:H
               H
```

55.
```
                 :F:
·C· + 4 ·F: → :F:C:F:
                 :F:
```

59.
```
a.   H           b.
   H:C:O:H         H:N:O:H
     H               H

c.   H           d.
   H:C:N:H         H:N:N:H
     H H             H H
```

61.
```
a.   H           b.
     |
   H—C—O—H         H—N—O—H
     |               |
     H               H

c.   H           d.
     |
   H—C—N—H         H—N—N—H
     | |             | |
     H H             H H
```

63.
```
a.   O:          b.  :Cl:
   :F:C:F:         :Cl:P:Cl:

c. H:O:P:O:H     d. H:C:::N:
     :O:
      H
```

65.
```
a. [:Cl:O:]⁻      b.     :O:
                     [H:O:P:O:]²⁻
                         :O:

c. [:O:Cl:O:]⁻    d. [:O:Br:O:]⁻
                         :O:
```

67. a. :N::O: b. :I:Be:I:
c. :Cl:B:Cl:
:Cl:

69. a. tetrahedral b. bent

71. a. pyramidal b. tetrahedral

73. a. pyramidal b. bent

75. Neon has an octet of valence electrons (a full outer shell)

77. a.
```
    H H             H   H
    ..  ..  ..      ..  ..  ..
 H:C:C:O:H and H:C:O:C:H
    ..  ..  ..      ..  ..  ..
    H H             H   H
```

83. a. $1s^22s^22p^63s^23p^63d^{10}4s^24p^64d^{10}5s^25p^6$ (the xenon configuration)
b. $1s^22s^22p^63s^23p^6$ (the argon configuration)
c. $1s^22s^22p^63s^23p^63d^{10}4s^24p^6$ (the krypton configuration)

Chapter 6

6.1 K_2O

6.2 Ca_3N_2

6.3 CaS

6.4 calcium fluoride

6.5 copper(II) bromide

6.6 K_3PO_4

6.7 $Ca(CH_3CO_2)_2$

6.8 $CaHPO_4$

6.9 calcium carbonate

6.10 potassium dichromate

6.11 iron(III) carbonate

6.12 bromine trifluoride; bromine pentafluoride

6.13 N_2O_5

6.14 P_4Se_3

6.15 $3\ P_4 + 6\ Hz \rightarrow 4\ PH_3$

6.16 $4\ V + 5\ O_2 \rightarrow 2\ V_2O_5$

6.17 $C_3H_8 + 5\ O_2 \rightarrow 3\ CO_2 + 4\ H_2O$

6.18 $2\ H_3PO_4 + 3\ Ca(OH)_2 \rightarrow Ca_3(PO_4)_2 + 6\ H_2O$

6.19 50.0 L

6.20 5.0 L

6.21 1.79 g/L

6.22 74 g/mol

6.23 88 g/mol

6.24 121 g

6.25 a. 0.0640 mol b. 0.776 mol c. 1.04 mol d. 0.0103 mol

6.26 2 molecules of H_2S react with 3 molecules of O_2 to form 2 molecules of SO_2 and 2 molecules of H_2O; 2 mol of H_2S react with 3 mol of O_2 to form 2 mol of SO_2 and 2 mol of H_2O; 68.2 g of H_2S react with 96.0 g of O_2 to form 128 g of SO_2 and 36.0 g of H_2O

6.27 a. 1.59 mol b. 305 mol c. 0.612 mol

6.28 0.333 g

6.29 176 g

6.30 8.02 g

6.31 a. 0.968 g b. 221 g c. 21.4 g

5. 6.02×10^{23} O_2 molecules; 12.04×10^{23} O atoms

7. 44.0 amu; 44.0 g/mol

9. a. 22.4 L b. 22.4 L c. 22.4 L

11. a. 3+ b. 2− c. 1+ d. 1−

13. ClO_2

15. AlP; Mg_3P_2

17. a. 9 b. 9

19. Al: 2; C: 12; H: 18; O: 12

21. balanced: a, d, e; not balanced: b, c

23. balanced: b; not balanced: a

27. a. potassium ion
b. calcium ion
c. zinc ion
d. bromide ion
e. lithium ion
f. sulfide ion

29. a. iron(II) ion
b. copper(I) ion
c. iodide ion

31. a. Na^+ b. Al^{3+}
c. O^{2-} d. Cu^{2+}

33. a. carbonate ion
b. monohydrogen phosphate ion
c. permanganate ion d. hydroxide ion

35. a. NH_4^+ b. HSO_4^- c. CN^- d. NO_2^-

37. a. sodium bromide b. calcium chloride
c. iron(III) chloride d. lithium iodide
e. potassium sulfide f. copper(I) bromide

39. a. $MgSO_4$ b. $NaHCO_3$
c. KNO_3 d. $CaHPO_4$

41. a. $Fe_3(PO_4)_2$ b. $K_2Cr_2O_7$
c. CuI d. NH_4NO_2

43. a. potassium nitrite b. lithium cyanide
c. ammonium iodide d. sodium nitrate
e. potassium permanganate f. calcium sulfate

45. a. sodium monohydrogen phosphate
b. ammonium phosphate
c. aluminum nitrate
d. ammonium nitrate

47. a. N_2O b. P_4S_3 c. PCl_5 d. SF_6

49. a. carbon disulfide
b. dinitrogen tetrasulfide
c. phosphorus pentafluoride
d. disulfur decafluoride

51. a. $Cl_2O_5 + H_2O \rightarrow 2\ HClO_3$
b. $V_2O_5 + 2\ H_2 \rightarrow V_2O_3 + 2\ H_2O$
c. $4\ Al + 3\ O_2 \rightarrow 2\ Al_2O_3$
d. $Sn + 2\ NaOH \rightarrow Na_2SnO_2 + H_2$
e. $PCl_5 + 4\ H_2O \rightarrow H_3PO_4 + 5\ HCl$

53. a. $Na_3P + 3\ H_2O \rightarrow 3\ NaOH + PH_3$
b. $Cl_2O + H_2O \rightarrow 2\ HClO$
c. $2\ CH_3OH + 3\ O_2 \rightarrow 2\ CO_2 + 4\ H_2O$
d. $3\ Zn(OH)_2 + 2\ H_3PO_4 \rightarrow Zn_3(PO_4)_2 + 6\ H_2O$
e. $C_3H_8 + 5\ O_2 \rightarrow 3\ CO_2 + 4\ H_2O$

55. a. 16.0 amu b. 84.0 amu c. 352.0 amu

57. a. 156.9 amu b. 178.0 amu c. 294 amu
d. 342 amu

59. a. 1.59 L b. 80.5 L

61. 1.78 g/L

63. 52.6 g/mol

65. 67.6 g/mol

67. 26.0 g

69. 0.0731 mol

71. a. 0.435 g b. 47.0 g c. 45.0 g d. 615 g

73. a. 16.7 mol b. 55.9 mol

75. 2490 g NH_3

77. 11.3 g O_2

79. 2050 g

81. 3600 g

83. 0.861 g

85. 2.20 g

87. 0.0521 g

Chapter 7

7.1 HNO_3

7.2 KOH

7.3 $Ca(OH)_2 + 2\ HCl \rightarrow CaCl_2 + 2\ H_2O$

7.4 11

7.5 0.01 M

7.6 pH 11

5. hydroxide ion (OH^-)

7. strong bases: NaOH (sodium hydroxide), KOH (potassium hydroxide); weak base: NH_3 (ammonia)

21. acidic

23. sulfuric acid

27. a. base b. acid c. acid

29. a. HCl b. H_2SO_4 c. H_2CO_3 d. HCN

31. a. LiOH b. $Mg(OH)_2$ c. NaOH

33. $HCl + H_2O \rightarrow H_3O^+ + Cl^-$; hydrochloric acid

35. H_2SO_4; an acid

37. KOH; a base

39. strong base

41. weak acid

43. $NaOH - HCl \rightarrow NaCl + H_2O$

45. $Ca(OH)_2 + 2\ HCl \rightarrow CaCl_2 + 2\ H_2O$

47. $H_3PO_4 + 3\ NaOH \rightarrow Na_3PO_4 + 3\ H_2O$

49. 3

51. pH 11

53. 1×10^{-5}

55. 1×10^{-6}

57. 100 g

59. $CaO + H_2SO_4 \rightarrow CaSO_4 + H_2O$; 5.7 t

Chapter 8

8.1 $2\ Zn + O_2 \rightarrow 2\ ZnO$

8.2 $2\ PbS + 3\ O_2 \rightarrow 2\ PbO + 2\ SO_2$

8.3 oxidation: a, b, c, d

8.4 oxidation: b

8.5 oxidation: b, c, d

8.6 a. oxidizing agent: O_2; reducing agent: Se
b. oxidizing agent: $CH_3C{\equiv}N$; reducing agent: H_2
c. oxidizing agent: V_2O_5; reducing agent: H_2
d. oxidizing agent: Br_2; reducing agent: K

13. carbon (C), hydrogen (H_2), hydroquinone [$C_6H_4(OH)_2$]

17. oxidized: a; reduced: b

19. neither

21. a. oxidizing agent: O_2; reducing agent: Al
b. oxidizing agent: O_2; reducing agent: SO_2

23. a. oxidizing agent: HCl (specifically H^+); reducing agent: Fe
b. oxidizing agent: O_2; reducing agent: CS_2

25. a. S is oxidized; N is reduced
b. I is oxidized; Cr is reduced

27. reduced; it gains hydrogen

29. I^- is oxidized; Cl_2 is reduced

31. reduced

33. Zr was oxidized; water was the oxidizing agent

35. oxidized

37. a. SO_2 b. H_2O

39. a. $CO_2 + SO_2$ b. $CO_2 + H_2O$

41. 160 L

43. Water is oxidized; CO_2 is the oxidizing agent. CO_2 is reduced; water is the reducing agent.

Chapter 9

9.1
```
     H  H  H  H  H  H  H  H
     |  |  |  |  |  |  |  |
H—C—C—C—C—C—C—C—C—H
     |  |  |  |  |  |  |  |
     H  H  H  H  H  H  H  H
```
$CH_3CH_2CH_2CH_2CH_2CH_2CH_2CH_3$

9.2
```
      CH2
     /   \
  CH2     CH2
   |       |
  CH2 —— CH2
```

9.3 alcohol: a; ethers: b, c; phenol: d

9.4 $CH_3CH_2OCH_3$

9.5 aldehydes: b, c; ketone: a

9.6
```
      H  H  H  O
      |  |  |  ‖
a. H—C—C—C—C—OH
      |  |  |
      H  H  H

      H  H  H
      |  |  |
b. H—C—C—C—OH
      |  |  |
      H  H  H

      H  O
      |  ‖
c. H—C—C—H
      |
      H

      H  H  O  H  H
      |  |  ‖  |  |
d. H—C—C—C—C—C—H
      |  |     |  |
      H  H     H  H
```

9.7 a. $CH_3CH_2CH_2CH_2NH_2$
b. $CH_3CH_2NHCH_2CH_3$
c. $CH_3NHCH_2CH_2CH_3$
d.
```
CH3CHNHCH3
    |
    CH3
```

9.8 heterocyclic compounds: a, d

11. Alkanes are lighter.

15. anesthetic; solvent

17. Aldehydes have a hydrogen atom attached to the carbonyl carbon atom.

19. Esters are sweeter, often fruity.

21. organic: a, b; inorganic: c, d

23. a. 2 b. 4 c. 7 d. 5

25. a. ethane b. acetylene c. ethylene

27.
```
CH3CH2CH2CH3   CH3CHCH3
                   |
                   CH3
```

29. a. methanol b. isopropyl alcohol

31.
```
a. CH3CH2—   b. CH3CH—
                   |
                   CH3
```

33. (benzene ring)—OH

35.
```
          O          O
          ‖          ‖
a. CH3—C—H   b. H—C—H
```

37. a. propyl alcohol b. propionic acid

39.
```
          O             O
          ‖             ‖
a. CH3—C—OH   b. CH3—C—CH3
```

41.
```
          O
          ‖
a. CH3—C—OCH2CH3

                 O
                 ‖
b. CH3CH2CH2C—OCH3
```

43. $CH_3CH_2NH_2$ b. CH_2NHCH_3

45. a. propylamine b. diethylamine

47. a. same compound
b. same compound

c. isomers

49. a. homologs b. none of these

51. unsaturated: a; saturated: b

53. a. alkene b. alkane

55. $-\overset{\overset{\displaystyle O}{\|}}{C}-$

57. the amino group

59. heterocyclic compounds: a, c

61. a. $CH_3\underset{}{\overset{\overset{\displaystyle OH}{|}}{C}}HCH_3$ b. CH_3OH c. CH_3CH_2OH

63. homology

65. a. homologs b. none of these
c. identical d. isomers

67. Equation is balanced. 601 g ethyl alcohol

Chapter 10

10.1 $\sim CH_2\underset{\underset{\displaystyle CN}{|}}{C}H \sim$

10.2 $\sim CH_2CCl_2CH_2CCl_2CH_2CCl_2CH_2CCl_2 \sim$

11. a double bond

13. polystyrene

15. polyethylene (high density)

19. by crosslinking

27. cannot be melted and remolded

43. $CH_2{=}CHCl$

45. $\sim CH_2\underset{\underset{\displaystyle C_6H_5}{|}}{C}H-CH_2\underset{\underset{\displaystyle C_6H_5}{|}}{C}H-CH_2\underset{\underset{\displaystyle C_6H_5}{|}}{C}H-CH_2\underset{\underset{\displaystyle C_6H_5}{|}}{C}H \sim$ (each C_6H_5 drawn as a benzene ring)

47. $\sim CH_2\underset{\underset{\displaystyle CN}{|}}{C}HCH_2\underset{\underset{\displaystyle CN}{|}}{C}HCH_2\underset{\underset{\displaystyle CN}{|}}{C}HCH_2\underset{\underset{\displaystyle CN}{|}}{C}H \sim$

49. $\sim -CH_2-\underset{\underset{\displaystyle C_4H_9}{|}}{C}H-CH_2-\underset{\underset{\displaystyle C_4H_9}{|}}{C}H-CH_2-\underset{\underset{\displaystyle C_4H_9}{|}}{C}H-CH_2-\underset{\underset{\displaystyle C_4H_9}{|}}{C}H- \sim$

53. $CF_2{=}CF_2$

55. $CH_2{=}\underset{\underset{\displaystyle CH_3}{|}}{C}-CH{=}CH_2$

57. $\sim NH(CH_2)_4NH\overset{\overset{\displaystyle O}{\|}}{C}(CH_2)_4\overset{\overset{\displaystyle O}{\|}}{C}NH(CH_2)_4NH\overset{\overset{\displaystyle O}{\|}}{C}(CH_2)_4\overset{\overset{\displaystyle O}{\|}}{C} \sim$

59. $\sim O-\overset{\overset{\displaystyle O}{\|}}{C}-C_6H_4-\overset{\overset{\displaystyle O}{\|}}{C}-O-CH_2-C_6H_{10}-CH_2 \sim$ (benzene ring and cyclohexane ring drawn)

61. monomer: a; segmer: b; polymer: c

63. condensation polymer

67. $HO-\overset{\overset{\displaystyle CH_3}{|}}{C}H-CH_2-\overset{\overset{\displaystyle O}{\|}}{C}-OH$

Chapter 11

1. crust, mantle, core

3. iron (Fe) and nickel (Ni)

5. compounds of metals with silicon and oxygen

7. the chemistry of elements other than carbon

19. iron ore, coal, and limestone

23. oxidation

27. iron

33. 3.8 million t Zn; 0.3 million t Pb; 0.83 million t Cu

35. a cube 16 m on each side

37. 1.15 billion t

39. 120 million kg Al_2O_3; 260 million kg bauxite

41. $TiCl_4$ is reduced; Na is the reducing agent; Na is oxidized; $TiCl_4$ is the oxidizing agent.

43. ThO_2 is reduced; Ca is the reducing agent; Ca is oxidized; ThO_2 is the oxidizing agent.

45. $AlCl_3$ is reduced; Mn is the reducing agent; Mn is oxidized; $AlCl_3$ is the oxidizing agent.

47. 114 g

49. 0.668 g

Chapter 12

1. the troposphere

3. 78% N_2, 21% O_2, 1% Ar

7. by plants through photosynthesis

11. No. Volcanoes, dust storms, and other natural events contribute to air pollution.

15. cool, damp, cloudy

25. to make cement and mineral wool

27. hydrocarbons, NO_x, O_3, aldehydes, PAN

29. automobiles

35. increase in skin cancer

37. automobiles

41. rain with a pH below 5.6

53. $O_2 + O \rightarrow O_3$

55. $S + O_2 \rightarrow SO_2$

57. $SO_3 + H_2O \rightarrow H_2SO_4$

59. $N_2 + O_2 \rightarrow 2\,NO$

61. $NO_2 \rightarrow NO + O$

63. $Fe + H_2SO_4 \rightarrow H_2 + FeSO_4$

65. 8000 μg (or 8 mg)

67. 0.19 g; 2.6%

Chapter 13

3. The gasoline would not dissolve; it would float.

9. nearly 98%

13. Ca^{2+}, Mg^{2+}, Na^+, K^+

15. water that contains Ca^{2+}, Mg^{2+}, (sometimes Fe^{2+})

17. those that cause disease

25. acid precipitation; drainage from mines

29. neutralize the water with a base

37. to kill pathogenic bacteria

41. hydrocarbons and chlorinated hydrocarbons

49. fertilizer; drainage from feedlots

51. $2\,H_3O^+ + CaCO_3 \rightarrow Ca^{2+} + CO_2 + 3\,H_2O$

53. $PO_4^{3-} + Al^{3+} \rightarrow AlPO_4(s)$

55. Chlorine is reduced; it is the oxidizing agent. Sulfur dioxide is the reducing agent.

Chapter 14

14.1 360,000 J

14.2 21.6 kcal

14.3 0.0815 kcal

1. a material easily burned as a source of energy

3. Reaction goes faster at a higher temperature.

5. a reaction that gives off energy

9. randomness; disorder

11. coal

15. ancient marine animals

17. antiknock performance as compared with isooctane (100) and heptane (0)

19. wood

21. petroleum

23. increased

25. the sun

27. 22%

29. no

31. no

37. a mixture of atomic nuclei and electrons at temperatures such as that on the sun

39. a cell that converts light directly to electricity

47. a and b

49. $2\,C + O_2 \rightarrow 2\,CO$

51. $2\,CH_4 + 3\,O_2 \rightarrow 2\,CO + 4\,H_2O$

53. 3800 kcal

55. 4.6 years

57. 9.3 years

61. $4\,{}^1_1H \rightarrow {}^4_2He + 2\,e$

63. $C + H_2O \rightarrow CO + H_2$

65. $2\,H_2 + O_2 \rightarrow 2\,H_2O$

67. 19.4 W

69. 499 g

Chapter 15

15.1 sugar: ribose; base: uracil

3. They make food by photosynthesis.

9. The acetal linkages between glucose units are alpha in starch and beta in cellulose.

13. long-chain carboxylic acids

19. the carbon-to-carbon double bond

21. in every cell

23. Proteins are polyamides.

25. amino groups and carboxylic acid groups

27. inner salts or dipolar ions

29. an amide function that joins two amino acid units

33. the sequence of amino acids

39. a protein with molecules that fold into a spheroid or ovoid shape

43. DNA (found mainly in the nucleus) and RNA

45. in DNA: guanine, cytosine, adenine, and thymine; in RNA: guanine, cytosine, adenine, and uracil

47. hydrogen bonding

53. DNA and mRNA

55. mRNA

63. monosaccharides: c, d

65. disaccharides: 64c, 64d

67. a. glucose b. galactose and glucose c. glucose

73. saturated: d, e

75. monounsaturated: c

77. a. 18 b. 16 c. 18

79. corn oil

85. a. $H_3N^+CH_2\overset{\overset{O}{\|}}{C}-NH-\underset{\underset{CH_3}{|}}{CH}-COO^-$

b. $H_3N^+\underset{\underset{CH_3}{|}}{CH}\overset{\overset{O}{\|}}{C}-NH-\underset{\underset{CH_2OH}{|}}{CH}-COO^-$

87. nucleotide: c

89. ribose: a, b; deoxyribose: c

91. ribose; uracil

93. a. guanine b. thymine c. cystosine d. adenine

95. TTAAGC

97. AGGCTA

99. a. AAC b. CUU

101. pyrimidine: a; purine: b

103. a. a purine b. in RNA

105. lipid: b

Chapter 16

16.1 47.5%

16.2 750 kcal; 250 kcal

16.3 10%

16.4 0.91 g

16.5 by 2011

1. carbohydrates, fats, proteins

3. a. glucose b. sucrose c. fructose

7. maltose, then glucose

9. about 4 kcal

11. energy, thermal insulation, protection of vital organs

13. more solid; more saturated

15. adipose tissue

17. 10%

23. yes

27. a. thyroid gland b. hemoglobin c. bones, teeth, blood clotting d. nucleic acids, bones, teeth

29. vitamin A: retinol; vitamin B_{12}: cyanocobalamine; vitamin C: ascorbic acid; vitamin D: calciferol; vitamin E: tocopherol

31. a. C b. D c. A

33. water soluble: b, c, d; fat soluble: a, e

35. fat soluble

37. glycogen

39. intentional and incidental

43. a. sweetener b. prevents rickets c. preservative

45. monosodium glutamate, a flavor enhancer

47. reducing agents

49. vitamin E

51. It is a dipeptide.

53. potent carcinogens found in stored peanuts and grain

59. water and carbon dioxide

61. as NO_3^- or NH_4^+

65. NH_3 without water; fertilizer

71. It interferes with calcium metabolism and formation of egg shells.

75. chemicals secreted to mark a trail, send an alarm, or attract a mate

79. causes leaves to fall off plants

83. that population growth would exceed food production

87. $6\,CO_2 + 6\,H_2O \rightarrow C_6H_{12}O_6 + 6\,O_2$

91. 40 g

93. 32.4 g

95. 11.3%

97. 300 mg

99. 8192; geometric

103. alkene

105. ether, alkene, ester

Chapter 17

3. a salt of a long-chain carboxylic acid

11. water containing Ca^{2+}, Mg^{2+}, or Fe^{2+} ions

25. a surfactant molecule with a negative charge on its water-soluble "head"

41. a surfactant molecule with a positive charge on its water-soluble "head"

51. bacteria

53. keratin

55. skin softener

57. an oil and water

59. harder; more wax

63. sex attractant of musk deer

65. alcohol, perfume, coloring

67. product that kills bacteria that cause odor

69. detergent

71. melanin and phaeomelanin

73. Temporary dyes are water soluble; can be washed out.

75. oxidizing agent

77. all three

79. II

81. III

83. $PO_4^{3-} + H_2O \rightarrow HPO_4^{2-} + OH^-$

85. $2\ PO_4^{3-} + 3\ M^{2+} \rightarrow M_3(PO_4)_2$

87. $$\begin{matrix} CH_2OH \\ | \\ CH{-}OH \\ | \\ CH_2OH \end{matrix} + CH_3(CH_2)_6\overset{\overset{O}{\|}}{C}O^-Na^+ + CH_3(CH_2)_4\overset{\overset{O}{\|}}{C}O^-Na^+ + CH_3(CH_2)_8\overset{\overset{O}{\|}}{C}O^-Na^+$$

Chapter 18

18.1 35 mi

18.2 3.6 lb

1. starch

3. 25%

11. recommended daily allowance

17. No

19. They are reducing agents.

27. actin and myosin

31. anaerobic

39. pheasants: Type IIB; herons: Type I

41. no; exercise

43. skin calipers; dunk tank (body density)

45. about 3500 kcal

49. glycogen depletion; dehydration

53. Na^+, K^+, and Ca^{2+}

75. 2 hr

77. 0.1 hr (or 6 min)

79. 100 g

81. 20 km

83. 4.5 L

Chapter 19

1. a pain reliever

3. acetyl salicylic acid

5. Both are analgesics and antipyretics

7. Aspirin enhances bleeding; acetaminophen does not.

9. preparation given to a patient who thinks it contains medication, although it actually does not

11. antibacterial substance produced, for example, by a mold

13. They are effective against a wide variety of bacteria.

15. a steroid produced in the adrenal cortex

17. a female sex hormone

19. diethylstilbestrol

21. hormone mediators derived from arachidonic acid

23. drugs that affect the mind

25. stimulants, depressants, and hallucinogens

29. a drug that acts on the brain to produce unconsciousness

31. diethyl ether and cyclopropane

37. ketamine and PCP

39. dried juice of unripe seeds of the oriental poppy

43. An agonist mimics the action of a drug.

47. No

51. phenolic and carboxylic acid groups

53. water soluble

57. 350 mg; no

59. b, c

61. 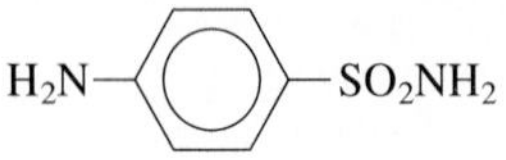 H_2N—◯—SO_2NH_2

63. the 4-ring structure

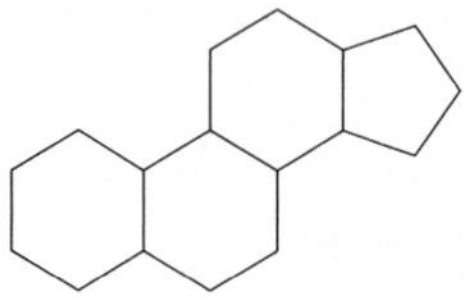

Chapter 20

1. a study of the response of living organisms to drugs

3. yes; in extremely large amounts

5. Sugar is more harmful to a diabetic than to a nondiabetic.

7. by breaking down proteins

11. by blocking the oxidation of glucose, thus stopping cellular respiration

13. by blocking the citric acid cycle

15. by tying up sulfhydryl groups, thus deactivating enzymes

19. by chelating lead ions, thus enhancing their excretion

21. cadmium poisoning

23. Botulin blocks the synthesis of acetylcholine.

25. nerve poisons; by inhibiting cholinesterase

27. by blocking acetylcholine receptor sites

29. a group of enzymes that oxidizes various substances

31. Cotinine is less toxic and more water soluble than nicotine.

33. No. Some are made more toxic.

35. an abnormal growth of new tissue

37. cigarette smoking

39. genes that regulate cell growth; they sustain the abnormal growth characteristic of cancer

43. an azo dye, once used to color butter

47. a chemical substance that causes mutations

49. a chemical substance that causes birth defects

Appendix B

1. a. 1×10^{-5} b. 1×10^{7} c. 3.4×10^{-3} d. 1.07×10^{-5} e. 4.5×10^{9} f. 4.06×10^{5} g. 2×10^{-2} h. 1.24×10^{5}

2. a. 8.6×10^{8} b. 2.5×10^{-16} c. 4.4×10^{-12} d. 2.2×10^{-15} e. 2.5×10^{16} f. 2.1×10^{-3} g. 5.7×10^{-30}

Appendix D

1. a. 100.5 m b. 153 g c. 130 mL

2. a. 54.4 cm b. 436 g c. 9.3 L

3. a. 40 m^3 b. 0.196 cm^2 c. 17 cm^2

4. a. 1.1 g/mL b. 1500 g/mol c. 13 g/mL

5. 2.61 mL

Index

D

I

N

O

Symbols and Names for Some Simple Ions

Group	Element	Name of Ion	Symbol for Ion
IA	Hydrogen	Hydrogen ion*	H^+
	Lithium	Lithium ion	Li^+
	Sodium	Sodium ion	Na^+
	Potassium	Potassium ion	K^+
IIA	Magnesium	Magnesium ion	Mg^{2+}
	Calcium	Calcium ion	Ca^{2+}
IIIA	Aluminum	Aluminum ion	Al^{3+}
VA	Nitrogen	Nitride ion	N^{3-}
VIA	Oxygen	Oxide ion	O^{2-}
	Sulfur	Sulfide ion	S^{2-}
VIIA	Fluorine	Fluoride ion	F^-
	Chlorine	Chloride ion	Cl^-
	Bromine	Bromide ion	Br^-
	Iodine	Iodide ion	I^-
IB	Copper	Copper(I) ion (cuprous ion)	Cu^+
		Copper(II) ion (cupric ion)	Cu^{2+}
	Silver	Silver ion	Ag^+
IIB	Zinc	Zinc ion	Zn^{2+}
VIIIB	Iron	Iron(II) ion (ferrous ion)	Fe^{2+}
		Iron(III) ion (ferric ion)	Fe^{3+}

*Does not exist independently in aqueous solution.

Some Common Polyatomic Ions

Charge	Name	Formula
1+	Ammonium ion	NH_4^+
	Hydronium ion	H_3O^+
1−	Hydrogen carbonate (bicarbonate) ion	HCO_3^-
	Hydrogen sulfate (bisulfate) ion	HSO_4^-
	Acetate ion	$CH_3CO_2^-$ (or $C_2H_3O_2^-$)
	Nitrite ion	NO_2^-
	Nitrate ion	NO_3^-
	Cyanide ion	CN^-
	Hydroxide ion	OH^-
	Dihydrogen phosphate ion	$H_2PO_4^-$
	Permanganate ion	MnO_4^-
2−	Carbonate ion	CO_3^{2-}
	Sulfate ion	SO_4^{2-}
	Monohydrogen phosphate ion	HPO_4^{2-}
	Oxalate ion	$C_2O_4^{2-}$
	Dichromate ion	$Cr_2O_7^{2-}$
3−	Phosphate ion	PO_4^{3-}